Robotics

Robotics

Science and Systems VI

edited by Yoky Matsuoka, Hugh Durrant-Whyte, and José Neira

The MIT Press
Cambridge, Massachusetts
London, England

For information about special quantity discounts, please email special_sales@mitpress.mit.edu or write to Special Sales Department, The MIT Press, 55 Hayward Street, Cambridge, MA 02142.

Printed and bound in the United States of America.

Library of Congress Cataloging-in-Publication Data

Robotics: Science and Systems Conference (6th : 2010 : Zaragoza, Spain)
Robotics : science and systems VI / edited by Yoky Matsuoka, Hugh Durrant-Whyte, and José Neira.
p. cm.
"This volume contains the 40 papers presented at Robotics: Science and Systems (RSS) 2010, held at the University of Zaragoza in Spain, from June 27 to June 30, 2010"—Pref.
Includes bibliographical references.
ISBN 978-0-262-51681-5 (pbk. : alk. paper)
1. Robotics—Congresses. I. Matsuoka, Yoky. II. Durrant-Whyte, Hugh F., 1961- III. Neira, José. VI. Title.
TJ210.3.R6435 2011
629.8'92—dc22 2011007422

10 9 8 7 6 5 4 3 2 1

Contents

Preface

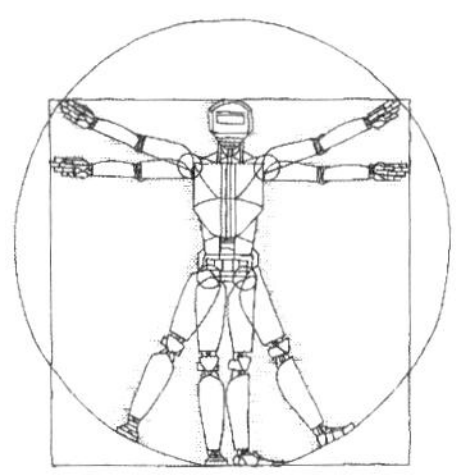

This volume contains the 40 papers presented at *Robotics: Science and Systems (RSS) 2010*, held at the University of Zaragoza in Spain, from June 27 to June 30, 2010. A record of 239 paper were submitted to RSS 2010. RSS takes pride in having a rigorous reviewing process: the 218 members of the program committee wrote more than 1,000 high-quality reviews so that each paper received at least four. The authors were invited to rebut the reviews, and after further discussion between the program committee members and the 16 area chairs, the reviews were finalized. Final acceptance and presentation category (20 poster presentations, 20 podium presentations) was decided during the full-day area chair meeting in Los Angeles, making the acceptance rate of RSS be under 17%.

The selected papers cover a wide range of topics in robotics spanning mechanisms, kinematics, dynamics and control, human-robot interaction and human-centered systems, distributed systems, mobile systems and mobility, manipulation, field robotics, medical robotics, biological robotics, robot perception, and estimation and learning in robotic systems.

The conference spanned three and a half days. There were four invited talks by leaders in fields that inspire robotics:

- Prof. Christopher M. Bishop from Microsoft Research Cambridge and the University of Edinburgh gave a talk titled "*Third Generation Machine Intelligence*".
- Prof. Deborah M. Gordon from Stanford University gave a talk titled "*Interaction Networks as Distributed Algorithms in Ants*".
- Prof. Larry Matthies from the Jet Propulsion Laboratory and the University of Southern California gave a talk titled "*Robotic Solar System Exploration: Progress and Challenges*".
- Prof. Philip H. S. Torr from Oxford Brookes University gave a talk titled "*Towards Global Energy Models for Scene Understanding*".

There were two early carrer spotlight talks this year, given by rising stars in the robotics community:

- Russ Tedrake from Massachusetts Institute of Technology gave a talk titled "*Dynamic walking on rough terrain and flying like a bird: a computational approach to exploiting nonlinear dynamics*".
- Pieter Abbeel from University of California at Berkeley gave a talk titled "*Apprenticeship Learning for High-Performance Robot Control*".

The Workshop Chair, Kevin Lynch (Northwestern University), together with a group of very competent reviewers, selected twelve workshops that were extremely well attended: more than 260 people attended the eight Sunday workshops, and more than 110 attended the four Monday morning workshops. The workshops were the following:

- *OMNIVIS Omnidirectional Vision, Camera Networks and Non-classical Cameras* organized by Ryad Benosman, Christopher Geyer, and Olivier Koch;
- *Towards Closing the Loop: Active Learning for Robotics* organized by Rubén Martínez-Cantín, Jan Peters and Andreas Krause;
- *RGB-D: Advanced Reasoning with Depth Cameras* organized by Xiaofeng Ren, Dieter Fox, Jana Kosecka and Kurt Konolige;
- *Predictive Models in Humanoid Gaze Control and Locomotion* organized by Paolo Dario, Alain Berthoz, Jose Santos-Victor and Atsuo Takanishi;
- *Strategies and Evaluation for Mobile Manipulation in Household Environments* organized by Antonio Morales, Mario Prats, Siddhartha Srinivasa and Radu Bogdan Rusu;
- *Motion Planning: from Theory to Practice* organized by Kris Hauser, Ron Alterovitz, Kostas Bekris and Juan Cortes;
- *Grasp Acquisition: How to Realize Good Grasps* organized by Jeff Trinkle, Patrick van der Smagt and Thomas Wimboeck;
- *Learning for Human-Robot Interaction Modeling* organized by Mohamed Chetouani and Adriana Tapus;
- *Enabling Technologies for Image-Guided Robotic Interventional Procedures* organized by Gregory Fischer and Robert Webster;
- *Representations for Object Grasping and Manipulation in Single and Dual Arm Tasks* organized by Danica Kragic, Bruno Siciliano and Vronique Perdereau;
- *Good Experimental Methodology in Robotics and Replicable Robotics Research* organized by Fabio Bonsignorio, John Hallam and Angel del Pobil;
- *Non-Smooth Contact Modeling in Robotic Simulation and Control* organized by Katsu Yamane and Abderrahmane Kheddar.

RSS 2010 was a big success thanks to the efforts of many people. We gratefully acknowledge the enormous effort and time spent by the program committee and the 16 area chairs, whose joint expertise covered an extraordinary broad swath of the robotics landscape. The area chairs were: Alin Albu-Schffer (DLR, Germany), José Castellanos (Universidad de Zaragoza), Peter Corke (CSIRO, Australia), Gregory Dudek (McGill University), Vincent Hayward (Université Pierre et Marie Curie), David Hsu (National University of Singapore), Dan Koditschek (University of Pennsylvania), Jana Kosecka (George Mason University), Danica Kragic (Royal Institute of Technology, Sweden), James Kuffner (Carnegie Mellon University), Jan Peters (Max Planck Institute for Biological Cybernetics), Sanjiv Singh (Carnegie Mellon University), Cyrill Stachniss (University of Freiburg), Russ Tedrake (Massachusetts Institute of Technology), Stefan Williams (University of Sydney) and Katsu Yamane (Disney Research).

Our warmest thanks also to the publicity chair, Katherine Kuchenbecker (University of Pennsylvania), who did a great job in attracting the largest yet number of authors and attendees to an RSS conference. The local arrangements chairs, José Neira, Juan D. Tardós and Luis Montano (University of Zaragoza), chose an amazing venue and organized everything beautifully, making sure that attendees got the best value for their money. Enormous thanks also to our Webmaster, David Ribas (University of Girona), who designed and maintained an elegant and ergonomic webpage. We also want to thank the conference support staff the Instituto de Investigación en Ingeniería de Aragón for handling local details, even on a Sunday. Last but not least, we would like to thank the team of twenty three volunteers that were always enthusiastically at hand when something needed to be done.

RSS 2010 was possible because of industrial and institutional sponsoring; thanks to Willow Garage, Google, the Spanish Ministry of Science and Innovation, the Regional Government of Aragón, Heartland Robotics, Aldebaran Robotics, Barrett Technology and Robotnik for providing funds for the general conference. Thanks to Springer for funding the best student paper award, and to Willow Garage for funding the best open source code award. We would also like to thank our technical sponsors: IEEE Robotics and Automation Society, the Association for the Advancement of Artificial Intelligence, the International Foundation of Robotics Research and the Robotics Society of Japan.

Finally, we would like to thank again the robotics community for adopting RSS and its philosophy. RSS 2010 had a record attendance of 315 researchers from 25 countries in North America, Europe, Asia and Australia. The attendance rate of 8 persons per presented paper confirms RSS as one of the highest quality single-track mainstream robotics conferences. We enthusiastically look forward to yet more exciting meetings in the years to come.

The online version of these proceedings (including color and links) can be found at: `http://www.roboticsproceedings.org/rss06/index.html`

Yoky Matsuoka, University of Washington

Hugh Durrant-Whyte, University of Sydney

José Neira, University of Zaragoza

July 2010

Organizing Committee

General Chair	Yoky Matsuoka, University of Washington
Program Chair	Hugh Durrant-Whyte, University of Sydney
Local Arrangement Co-Chairs	José Neira, Universidad de Zaragoza Juan Tardós, Universidad de Zaragoza Luis Montano, Universidad de Zaragoza
Publicity Chair	Katherine Kuchenbecker, University of Pennsylvania
Publications Chair	José Neira, Universidad de Zaragoza
Workshop Chair	Kevin Lynch, Northwestern University
Web Master	David Ribas, Universitat de Girona
Area Chairs	Alin Albu-Schffer, DLR, Germany José Castellanos, Universidad de Zaragoza Peter Corke, CSIRO, Australia Gregory Dudek, McGill University Vincent Hayward, Université Pierre et Marie Curie David Hsu, National University of Singapore Dan Koditschek, University of Pennsylvania Jana Kosecka, George Mason University Danica Kragic, Royal Institute of Technology, Sweden James Kuffner, Carnegie Mellon University Jan Peters, Max Planck Institute for Biological Cybernetics Sanjiv Singh, Carnegie Mellon University Cyrill Stachniss, University of Freiburg Russ Tedrake, Massachusetts Institute of Technology Stefan Williams, University of Sydney Katsu Yamane, Disney Research
RSS Foundation Board	
President	Sebastian Thrun, Stanford University
Directors	Oliver Brock, Technische Universitt Berlin Dieter Fox, University of Washington Lydia Kavraki, Rice University Sven Koenig, University of Southern California John Leonard, Massachusetts Institute of Technology Daniela Rus, Massachusetts Institute of Technology Stefan Schaal, University of Southern California Gaurav Sukhatme, University of Southern California Jeff Trinkle, Rensselaer Polytechnic Institute
Treasurer	Wolfram Burgard, University of Freiburg
Secretary	Nick Roy, Massachusetts Institute of Technology

Program Committee

Abbeel, Pieter
Adams, Martin
Akella, Srinivas
Andrade-Cetto, Juan
Antonelli, Gianluca
Argyros, Antonis
Arras, Kai Oliver
Asano, Fumihiko
Asfour, Tamim
Atkeson, Chris
Bagnell, James
Bailey, Tim
Barkby, Stephen
Bennewitz, Maren
Bergbreiter, Sarah
Bingham, Brian
Birchfield, Stan
Blaschko, Matthew
Borst, Christoph
Bosse, Michael
Brennan, Sean
Brock, Oliver
Brooks, Alex
Bryson, Mitch
Buehler, Martin
Burschka, Darius
Byl, Katie
Carloni, Raffaella
Censi, Andrea
Chestnutt, Joel
Christensen , Henrik
Civera, Javier
Clark, Jonathan
Cowan, Noah
Cutkosky, Mark
Davison, Andrew
Deisenroth, Marc
Dellaert, Frank
Dissanayake, Gamini
Douillard, Bertrand
Etienne , Burdet
Eustice, Ryan
Everett, Hazel
Fairfield, Nathanial
Fermuller, Cornelia
Ferrie, Frank
Fitch, Robert
Fox, Dieter
Frisoli, Antonio
Geyer, Chris
Ghrist, Robert
Gillespie, Brent
Goldman, Daniel
Gomez, Manuel
Gould, Steve
Grizzle, Jessy
Grocholsky, Ben
Grollman, Dan
Grosse-Wentrup, Moritz
Gruppen, Rod
Guerrero Campo, José
Hafner, Roland
Harders, Matthias
Hauser, Kris
Hirai, Shinichi
Hirche, Sandra
Hoburg, Warren
Hollinger, Geoff
Hosoda, Koh
Howard, Andrew
Huebner, Kai
Hyon, sang-ho
Iida, Fumiya
Inamura, Tetsunari
Jadbabaie, Ali
Jakuba, Michael
Jenkin, Michael
Jenkins, Chad
Jensfelt, Patric
Johnson-Roberson, Matthew
Kagami, Satoshi
Kajita, Shuuji
Kersting, Kristian
Kikuuwe, Ryo
Kim, Sangbae
Kober, Jens
Kolter, Zico
Kootstra, Gert
Kroemer, Oliver
Kuipers, Benjamin
Kulic, Dana
Kurniawati, Hanna
Kyrki, Ville
Lacroix, Simon
Lampert, Christoph
Lane, David
LaValle, Steven
Lee, Dongheui
Leonard, John
Leonardis, Alex
Lepetit, Vincent
Li, Tsai-Yen
Lien, Jyh-Ming
Likachev, Maxim
Lilienthal, Achim
Lopes, Manuel
López-Nicolás, Gonzalo
MacDonald, Bruce
MacLean, Karon
Macnab, Chris
Mahon, Ian
Mansard, Nicolas
Martínez-Cantín, Rubén
Matsuoka, Yoky
Merino, Luis
Metta, Giorgio
Michael, Nathan
Mochiyama, Hiromi
Moll, Mark
Montesano, Luis
Montiel, José M. M.
Morales, Antonio
Morimoto, Jun
Morse, Bryan
Muelling, Katharine

Murillo, Ana
Murrieta, Rafael
Negahdaripour, Shahriar
Neira, José
Nenchev, Dragomir
Neumann, Gerhard
Newman, Paul
Nguyen-Tuong, Duy
Niemeyer, Gunter
Nishiwaki, Koichi
Nuske, Steve
O'Kane, Jason
Ogata, Tetsuya
Okada, Masafumi
Olson, Edwin
Ott, Christian
Oztop, Erhan
Palli, Gianluca
Papanikolopoulos, Nikolaos
Paz, Lina
Petillot, Yvan
Pfaff, Patrick
Piater, Justus
Pinies, Pedro
Pizarro, Oscar
Plagemann, Christian
Plaku, Erion
Platt, Robert
Posner, Ingmar
Pradalier, Cédric
Pratichizzo, Domenico
Pratt, Jerry
Rajan, Kanna
Ramos, Fabio
Redon, Stephane
Rekleitis, Ioannis
Riedmiller, Martin
Roberts, Jonathan
Robuffo Giordano, Paolo
Rock, Stephen
Rodríguez-Losada, Diego
Roman, Chris
Roumeliotis, Stergios
Roy, Nicholas
Sagüés, Carlos
Sanz, Pedro
Saranli, Uluc
Saripalli, Srikanth
Saxena, Ashutosh
Schaal, Stefan
Schiele, Bernt
Sentis, Luis
Shibata, Tomohiro
Shimoda, Shingo
Shiriaev, Anton
Sim, Robert
Simeon, Thierry
Singh, Surya
Singh, Hanumant
Smart, Bill
Solà, Joan
Spenko, Matt
Spletzer, John
Srinivasa, Siddhartha
Stasse, Olivier
Stilman, Mike
Sugihara, Tomomichi
Sukhatme, Gaurav
Sukkarieh, Salah
Tardós, Juan
Taylor, CJ
Ting, Jo-Anne
Tipaldi, Diego
Torres-Mendez, Luz
Ude, Ales
Ueda, Jun
van den Berg, Jur
van der Stappen, Frank
Vanderborght, Bram
Velagapuddi, Pras
Venture, Gentiane
Villani, Luigi
Vona, Marsette
Wang, Zhikun
Whitcomb, Louis
Wingate, David
Wyeth, Gordon
Yim, Mark
Yokokohji, Yasuyoshi
Zha, Hong-Bin
Zhang, Hong
Zillich, Michael
Zlot, Robert

Sponsors

The organizers of Robotics Science and Systems 2010 gratefully acknowledge the following conference sponsors:

- Gold Sponsors:

  Willow Garage

  Google

- Silver Sponsors:

  HEARTLAND ROBOTICS

  ALDEBARAN Robotics

- Bronze Sponsors:

  Barrett Technology® Inc.

  Robotnik

- Awards Sponsors:

  Springer

  Willow Garage

- Institutional Sponsors:

  GOBIERNO DE ESPAÑA MINISTERIO DE CIENCIA E INNOVACIÓN

  GOBIERNO DE ARAGON
  Departamento de Ciencia, Tecnología y Universidad

- Technical Sponsors:

- Organized by:

Biophysically Inspired Development of a Sand-Swimming Robot

Ryan D. Maladen*, Yang Ding†, Paul B. Umbanhowar‡, Adam Kamor† and Daniel I. Goldman*†
*Bioengineering Program, †School of Physics, Georgia Institute of Technology, Atlanta, Georgia 30332–0250
‡Department of Mechanical Engineering, Northwestern University, Evanston, IL 60208
email: daniel.goldman@physics.gatech.edu

Abstract— **Previous study of a sand-swimming lizard, the sandfish, *Scincus scincus*, revealed that the animal swims within granular media at speeds up to 0.4 body-lengths/cycle using body undulation (approximately a single period sinusoidal traveling wave) without limb use [1]. Inspired by this biological experiment and challenged by the absence of robotic devices with comparable subterranean locomotor abilities, we developed a numerical simulation of a robot swimming in a granular medium (modeled using a multi-particle discrete element method simulation) to guide the design of a physical sand-swimming device built with off-the-shelf servo motors. Both in simulation and experiment the robot swims limblessly subsurface and, like the animal, increases its speed by increasing its oscillation frequency. It was able to achieve speeds of up to 0.3 body-lengths/cycle. The performance of the robot measured in terms of its wave efficiency, the ratio of its forward speed to wave speed, was 0.34 ± 0.02, within $8\ \%$ of the simulation prediction. Our work provides a validated simulation tool and a functional initial design for the development of robots that can move within yielding terrestrial substrates.**

I. Introduction

There is a need for robots that can move within complex material like sand, rubble, and loose debris. For example such robots could help locate hazardous chemical leaks [2], function as self propelled inspection devices [3], and search for victims in disaster sites [4, 5, 6]. Limbless robots that use their bodies to move appear better suited to navigate complex terrains than traditional wheeled [7, 8, 9, 10] and legged [11, 12, 13, 14, 15, 16, 17] robots which are often impeded by the size or shape of their appendages which can result in entrapment or failure. Previous terrestrial limbless robots utilized serpentine locomotion to move on the surface of media. Of these, most were tested on rigid surfaces [18, 19, 20, 21, 22] with only a few developed for and tested in unstructured environments [23, 24, 25].

Advances in creating high performing flying and swimming devices [22, 26, 27] in aerial and aquatic domains and wheeled/tracked vehicles on relatively structured terrestrial terrain have occurred mainly because the respective fields of aerodynamics, fluid dynamics, and terramechanics [28, 26] provide accurate models of locomotor-media interaction which are used in turn to design improved wings, fins, wheels, and legs. However, a major hurdle arises when one attempts to design robots to move on and within complex flowing particulate environments (e.g. sand, soil, and leaf-litter) that can display both solid and fluid-like behavior in response to stress. In such materials, comparable and comprehensive validated analytic continuum theories at the level of the Navier-Stokes equations [29] for fluids do not exist. However, it is possible to understand the interaction between the locomotor and the media by using numerical and physical modeling approaches [30, 31, 32].

In the absence of theory, the biological world is a fruitful source of principles of movement that can be incorporated into the design of robots that navigate within complex substrates. Many desert organisms like scorpions, snakes, and lizards burrow and swim effectively in sand [33, 34, 35, 36, 37] to escape heat and predators, and hunt for prey [38, 39]. It has been hypothesized that many of these animals have evolved morphological adaptations like marked body elongation and limb reduction to deal with deformable terrain [40, 41]. Our recent high speed x-ray imaging study investigating the subsurface locomotion of the sandfish *Scincus scincus*, a small ($\sim$ 10 cm long snout - tail tip) lizard that inhabits the Saharan desert [1] (Fig. 1), reveals that once within the media the animal no longer uses limbs for propulsion but "swims" forward by propagating a sinusoidal traveling wave posteriorly from head to tail.

Motivated by the subsurface locomotion of the sandfish, the present work utilizes a numerical simulation of a sandfish inspired undulator as a design tool to build an appendage-less sand-swimming device. The robot is driven by a simple open loop controller which, like the animal kinematics, varies the joint position trajectories to create a sinusoidal wave that travels posteriorly along the device. The robot swims within a model laboratory granular medium of plastic particles and displays locomotion features similar to the organism and predicted by the numerical robot simulation.

II. Previous work

A. Biological Experiment

The biological experiments presented in [1] model the subsurface undulatory motion of the sandfish with a posteriorly traveling single-period sinusoidal wave

$$y = A \sin \frac{2\pi}{\lambda}(x + v_w t) \qquad (1)$$

with x the position along the sandfish, y the body displacement from the midline of the animal, A the amplitude, λ the wavelength and $v_w = f\lambda$ the wave speed where f is the wave frequency. The spatial characteristics, A and λ, did

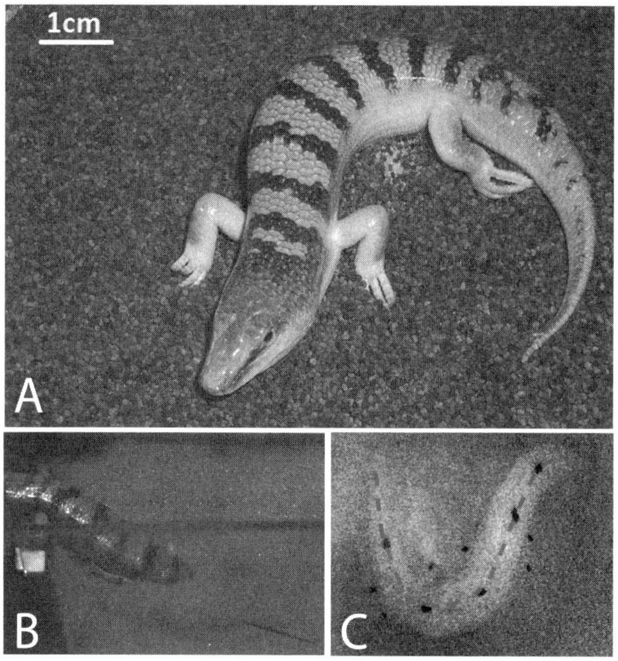

Fig. 1. (A) The sandfish *Scincus scincus*, a sand-swimming lizard that inhabits the Saharan desert, (B) burying into granular media (0.3 mm spherical glass beads), and (C) swimming subsurface where the x-ray image shows the body (light area) and opaque markers fixed to limbs and midline. Red dashed line indicates tracked midline.

not vary significantly with media preparation and their ratio was approximately 0.2 implying that the animal increased its forward velocity by increasing its oscillation frequency.

A measure of undulatory performance is the wave efficiency, η, the ratio between the forward speed of the swimmer, v_x, and the velocity of the wave traveling down its body, v_w, or equivalently the slope of the velocity-frequency relationship for velocity measured in wavelengths per second. Typical wave efficiencies of undulatory organisms moving in fluids at low Reynolds number (such as nematodes in water) are 0.25 [42, 43, 44], whereas $\eta \approx 0.8 - 0.9$ for organisms undulating (creeping) along solid – air interfaces [45, 46, 47]. Locomotion with $\eta = 1$ is equivalent to movement within a rigid tube. For the sandfish swimming in glass beads, $\eta \approx 0.5$ independent of particle size and media preparation (i.e. packing density).

B. *Resistive Force Theory for Granular Media*

An empirical resistive force theory (RFT) was developed to predict wave efficiency η for undulatory subsurface granular locomotion [1]. The RFT, inspired by theory used to predict swimming speeds of microorganisms in fluids [42], partitions the body of the organism into infinitesimal segments each of which generates thrust and experiences drag when moving through the medium. These segmental forces are integrated over the entire body, and, by setting the net forward force to zero (assuming a constant average velocity), η is solved for numerically.

Unlike fluids, in granular media no validated theory exists in the regime relevant to sand-swimming to estimate the force on individual segments moving through the medium. Previously, Maladen et al. [1] obtained these forces empirically by dragging a rod (representative segment) through the media the animal was tested in. With these forces as input and by propagating a sinusoidal traveling wave along the body, the RFT shows that translational motion within granular media without limb use is possible. Also, the RFT accurately predicts that the sandfish swims with $\eta \approx 0.5$ within a granular media of 0.3 mm glass particles (comparable in size and density to desert sand [48]).

While the RFT qualitatively describes some features of sand-swimming, it is based on several assumptions: e.g. the measured drag force on a rod is representative of the average force on a segment of the sandfish, the forces generated by a segment are localized, and the center of mass of the animal does not oscillate laterally. Since the assumptions of the RFT have not been rigourously tested and applying the RFT to different treatments (particle friction, particle size, body design, etc.) require force laws to be measured for each condition, we instead use numerical simulation techniques as a general robotic design tool.

A numerical simulation approach, once validated against experiment, can provide an understanding of body generated drag and thrust forces from the particle perspective and can be used to generate empirical drag laws for input into the RFT. Our numerical simulation is a flexible design tool that accurately predicts robot performance and allows easy variation of physical and design parameters such as particle-particle friction and number of segments.

III. Numerical Simulation of Sand-Swimming Robot

A. *Development and Validation*

To design a sand-swimming robot, we developed a numerical simulation of a laboratory scale device with a finite number of discrete, rigid segments to test if it could swim within granular medium. The simulation couples a numerical model of the robot to a model of the granular medium. We model the granular material using a multi-particle discrete element method (DEM) simulation [30]. To compute the robot-particle and particle-particle interaction forces we calculate the normal force [49], F_n, and the Coulombic tangential force, F_s, acting at each contact with

$$\begin{aligned} F_n &= k\delta^{3/2} - G_n v_n \delta^{1/2} \\ F_s &= \mu F_n, \end{aligned} \tag{2}$$

where δ is the virtual overlap between particles or between particle and robot segment, v_n is the normal component of relative velocity, and $k = 2 \times 10^5$ kg s^{-2}m$^{-1/2}$ and $G_n = 5$ kg s^{-1}m$^{-1/2}$ are the hardness and viscoelastic constant. μ quantifies the particle-particle ($\mu_{pp} = 0.08$) or body-particle ($\mu_{bp} = 0.27$) friction coefficient depending on which elements are in contact. μ_{bp} was measured between the robot skin and plastic particles used in the physical experiments. To reduce the required torque in the physical experiments and to decrease the computational time, we used a granular medium composed of 4.7×10^5 spherical plastic particles with diameter 6 mm and density 1.03 ± 0.04 g/cm^3 in experiment, and 3.2×10^5 bi-disperse (50:50) sphere mixture (5.81, 5.93 mm) and density 1.06 g/cm^3 in simulation. The 35 particle deep bed of particles in experiment and 24 particle deep bed of particles in simulation were held in a container with horizontal cross section of 188×62 particle diameters. To validate the simulated medium and obtain the values of μ_{pp}, k, and G_n given above, we dropped an aluminum ball (diameter 6.35 cm and mass 385 g) into the plastic particles with varying impact velocity ($0.5 - 3$ m/s) in both experiment and simulation and set grain interaction parameters to best match the measured and simulated penetration force during the impact collision as a function of time (Fig. 2). With parameters determined from impact at $v = 1.4$ m/s, the force profile fit well at other impact velocities. In additional experiments, we directly measured μ_{pp} and the coefficient of restitution (determined by G_n with fixed k) for the plastic particles and found them to be within 5% and 10% of the fitted values respectively. For simplicity we used the same normal force parameters for both particle-particle and body-particle interactions.

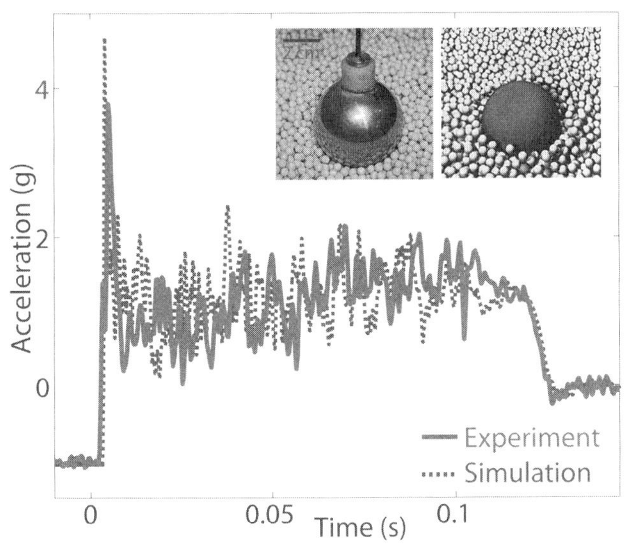

Fig. 2. Validation of the multi-particle discrete element method (DEM) simulation of the granular medium using measured acceleration of a sphere during vertical impact after free-fall. Acceleration vs. time in simulation (blue dashed trace) and experiment (red solid trace) agree well. The impact velocity for this representative run is 1.4 m/s. Acceleration is in units of g, the acceleration due to gravity. (Left inset) Aluminium ball instrumented with accelerometer resting on 6 mm plastic particles. (Right inset) Ball and particles in simulation.

To model the sand-swimming device we used the commercial multi-body simulator software package Working Model (WM) 2D (Design Simulation Technologies). Modeling the device in a 2D simulation environment is sufficient to capture the dynamics since the sand-swimming robot moves roughly in a horizontal plane at fixed depth in this study, and there was excellent agreement between experiment and simulation in preliminary studies. The simulated robot was sized for easy testing of the corresponding physical device in the laboratory. Since the sandfish does not use its limbs to move subsurface and RFT had shown that body undulation was sufficient for propulsion [1], the simulated robot did not include limbs. No tapering along the device was considered. The simulated robot consisted of 49 cuboidal segments interconnected and actuated by virtual motors (vertical cylinders) of the same height (Fig. 3). Depending on the number of segments (N) employed, every $48/N$ motor was driven with an open loop signal to generate a sinusoidal wave traveling posteriorly from head to tail while the remaining motors were immobilized to form a straight segment of length $48\ /\ N$ cm. To approximate a sinusoidal traveling wave, the angle between adjacent segments is modulated using

$$\beta(i,t) = \beta_0 \xi \sin(2\pi\xi i/N - 2\pi f t), \tag{3}$$

with $\beta(i,t)$ the motor angle of the ith motor at time t, β_0 the angular amplitude which determines A/λ, ξ the number of wavelengths along the body (period), and N the number of motors.

WM integrates the equations of motion of the coupled links and the DEM calculates the resultant forces from both the particle-particle and body-particle interactions. For each time step, the net force from particles on each segment is passed to WM, and velocity and position information transferred back to DEM. Roll and pitch were not modeled.

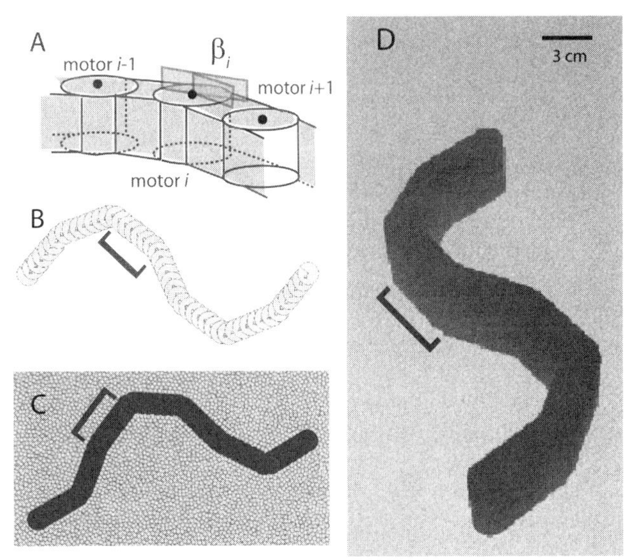

Fig. 3. Simulation of a sand-swimming robot. (A,B) Side and top view of the robot modeled with 49 inter-connected motor segments and one head segment. The angle between adjacent motors (β_i) is modulated using Eqn. 3 to reproduce the sandfish's sinusoidal traveling wave kinematics. (C) Top view of the device submerged in 6 mm particles with particles above the robot rendered transparent. (D) Rendering of the simulated robot for the same parameters used in robot experiment (see Fig. 4). The brackets ([ ) indicate a single robot segment.

Using Eqn. 3 the simulated robot with 7 total segments moved forward within 6 mm plastic particles and increased its forward speed linearly with oscillation frequency (Fig. 7). The wave efficiency was $\eta = 0.36 \pm 0.02$, less than that of the sandfish lizard. Motivated by these results we built a physical instantiation of the scaled model.

IV. Sand Swimming Robot

A. *Design and Control*

The basic mechanical design of our device was adapted from previously developed snake robots [25] which consisted of repeated modules (motors) each with a single joint that permit angular excursions in a plane and connected via identical links. In our design, each module consists of a servomotor attached to an aluminium bracket and is connected to adjacent motors via the brackets. The wire bundle that routes power and control signals to each motor was run atop each module over the length of the device and strain relieved at the last (tail) segment. For convenience and to maintain a reasonable size, our device employed 6 standard size servomotors and a dummy segment (the head) with the same weight and form factor as the motor segments for a total of 7 segments (Fig. 4A).

The simulation found that the peak torque required to swim subsurface at a depth of 4 cm was 0.7 N m. To verify this finding we dragged an object with the same form factor as a motor through the 6 mm plastic medium at 0.25 m/s. The measured force at a depth of 4 cm was 3.2 N. Since the maximum torque occurs at the middle motor (0.23 m to either end) we estimated the maximum possible total force along an effective segment extending from the middle servo to either the tail or the head (length 0.23 m) to be 18 N with a corresponding maximum torque of 2.0 Nm. We selected a servomotor that exceeded both torque estimates, see Table I.

Servomotors are powered in parallel from a 7.4 V, 30 A supply. The pulse width based control signal for each motor is generated in LabVIEW using Eqn. 3 as a multiplexed signal, output from a PCI-card (NI-6230), and connected to the clock input of a decade counter (CD4017BC) which functions as a demultiplexer and distributes a control pulse to each motor every 20 ms.

Since the robot operates in a granular medium it is critical to encase it in a material that prevents particles from getting between the motor segments but allows the device to easily undulate. After testing several materials we found that a 2-layer encasement consisting of an outer Lycra spandex sleeve with a single seam (located at the top of the device) enclosing an inner thin latex sleeve that fit tautly around the motors was satisfactory (Fig. 4).

B. *Experimental Methods*

We tested the robot in a container of the same dimensions as used in the robot simulations and filled with 6 mm plastic particles prepared in an as-poured state [50]. Overhead video (100 fps) was collected for each condition tested. To facilitate subsurface tracking the first and last module were fixed with a mast with a visible marker. The wire bundle was run up the mast on the last segment and tethered above the container. The kinematics of the subsurface motion of the robot were also obtained using x-ray imaging for a representative condition ($f = 0.25$ Hz, $A/\lambda = 0.2$), see Fig. 5A-C. For each test the top of the robot was submerged 4 cm below the surface and the surface leveled. Due to the servomotor angular velocity limits the maximum oscillation frequency was 1 Hz. For each frequency, $1 - 2$ cycles of motion were collected.

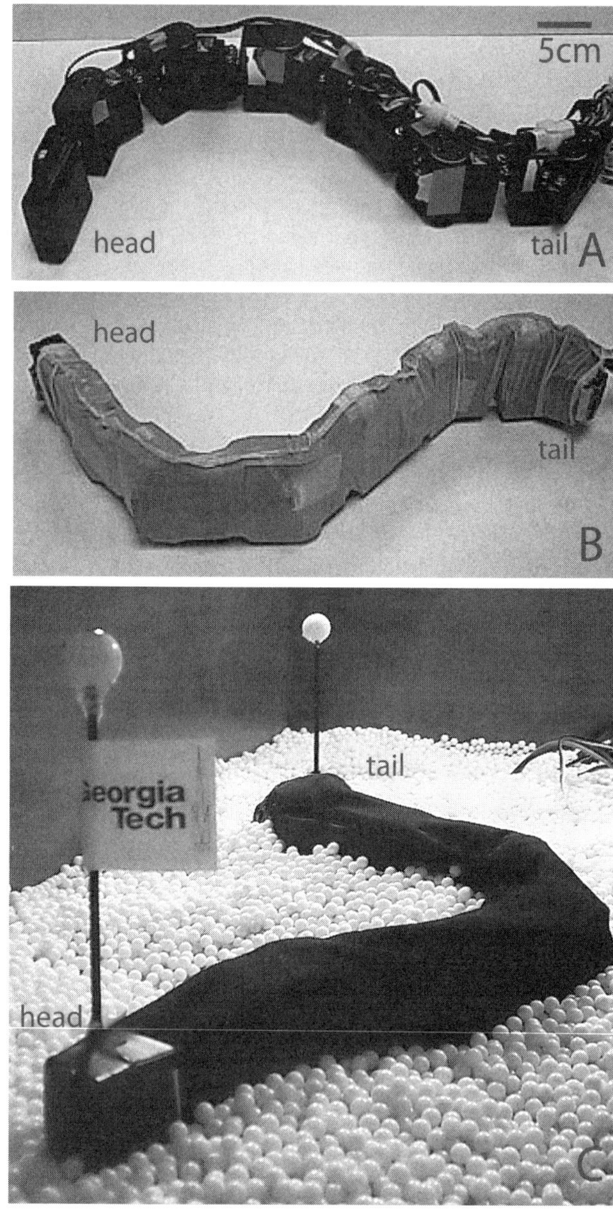

Fig. 4. Prototype of the sand-swimming robot. (A) basic construction (servomotors and aluminium brackets with power wires running along the top of the device). The robot has a double layer skin: (B) tight fitting thin latex inner layer, and (C) Lycra spandex outer layer. Balls on narrow masts on the head and tail segments allow subsurface motion tracking.

Dimensions	$0.48 \times 0.028 \times 0.054$ m^3
Mass	0.83 kg
Motor	HSR-5980SG
Motor Torque	2.94 N m
Number of Motors	6
Total Segments	7

TABLE I
PHYSICAL ROBOT CHARACTERISTICS

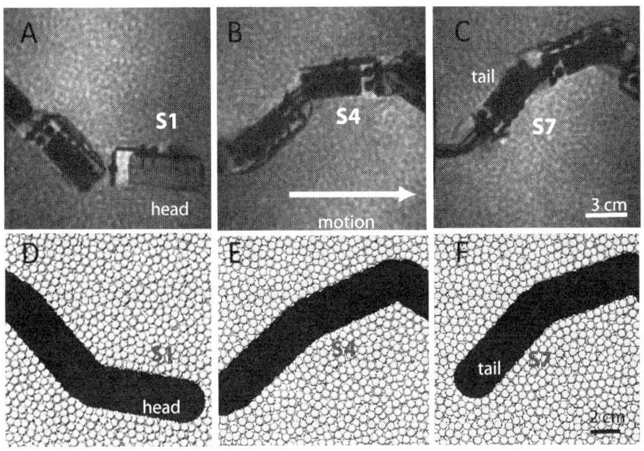

Fig. 5. Subsurface swimming in experiment and simulation. (A-C) Sequential x-ray images of the robot swimming in 6 mm particles, and (D-F) robot swimming in simulation. Segments from head to tail are denoted as S1 to S7.

C. Robot Performance

To calibrate the device we placed it on a rigid surface and used video to track the position of the segments from which we determined the mapping between the maximum relative segment angle β_0 (Eqn. 3) and A/λ. Within the granular material, the forward velocity of the device monotonically increased with increasing oscillation frequency (Fig. 7) for $A/\lambda = 0.2$ and a single period wave. The slope of this relationship (η) was 0.34 ± 0.02. For the same parameters the simulation predicted $\eta = 0.36 \pm 0.02$.

V. Discussion

Like the sandfish, the robot swims within granular media by propagating a traveling sinusoidal wave posteriorly from head to tail without limb use. The physical device demonstrates that subsurface locomotion in granular media using a relatively low degree of freedom device and a open loop control scheme is possible. However, the robot does not move forward as fast (normalized by body-length) or with the same wave efficiency as the animal.

In the biological experiments, η for a range of granular material preparations and bead size was approximately 0.5. The robot in both experiment and simulation performed below this value. We hypothesized that the number of segments (for a fixed length device) affected both η and the forward speed of the device. Increasing the number of segments in the robot simulation caused the device to move forward faster and with greater wave efficiency until $N \sim 15$ where η plateaued (Fig. 8). Interestingly, the maximum $\eta \approx 0.5$ is the same as measured in the animal experiment. We utilized our previously developed RFT to predict the performance of the sand-swimming device with parameters set to match those for the plastic particles used in the robot experiment. We estimated $\eta = 0.56$ for a smooth profiled undulator which corresponds to the numerical robot simulation prediction for $N > 15$ (gray band, Fig. 8).

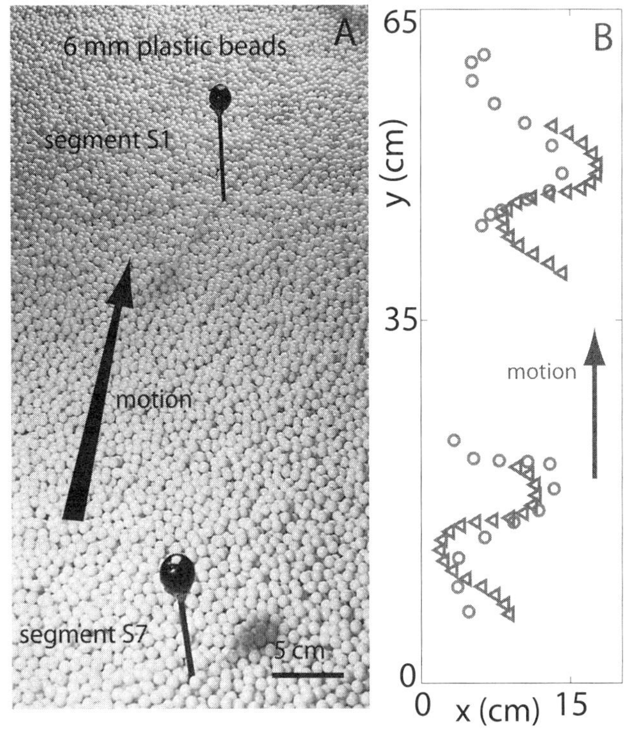

Fig. 6. Subsurface swimming in experiment and simulation. (A) Robot submerged in a container filled with 6 mm plastic particles. Masts with spherical markers are attached to the first and last module. (B) Kinematics of the first and last segment of the robot in experiment (green circles) and simulation (blue triangles).

Increasing N allowed the device to better match a sinusoidal wave and increased η This suggests that deviation from the smooth form of a traveling sinusoidal wave reduces performance. A seven segment robot operates below the minimum N required to achieve maximum η. As a design criterion, N is important when the length of the device is fixed as increasing the number of motors beyond the critical N requires motors with smaller dimensions but capable of producing the same torque.

We used the numeric robot simulation to measure the time varying torque required to move within the medium. As expected, the torque was approximately sinusoidal for all motors and the torque amplitude generated by the central motors (3 and 4) was larger than the torque from the motors nearest the ends, see Fig. 9. As noted earlier, the maximum torque in the simulation of 0.7 N m was well below the maximum of the motors used in experiment (see Table 1).

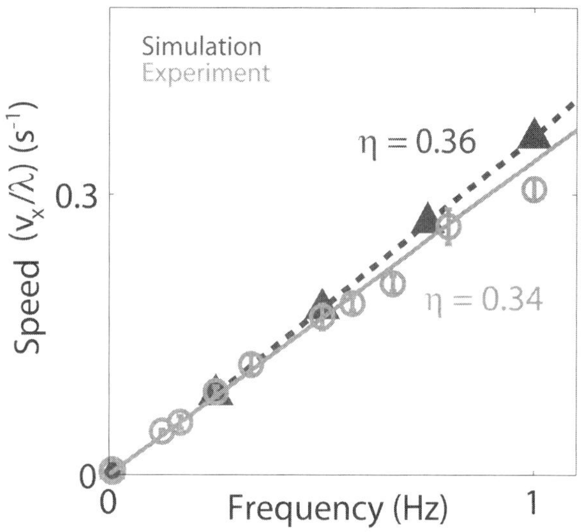

Fig. 7. Forward velocity vs. oscillation frequency for the robot in experiment (green circles) and simulation (blue triangles) ($A/\lambda = 0.2$). The slope of the dashed (simulation) and solid (experiment) fit lines gives the wave efficiency η.

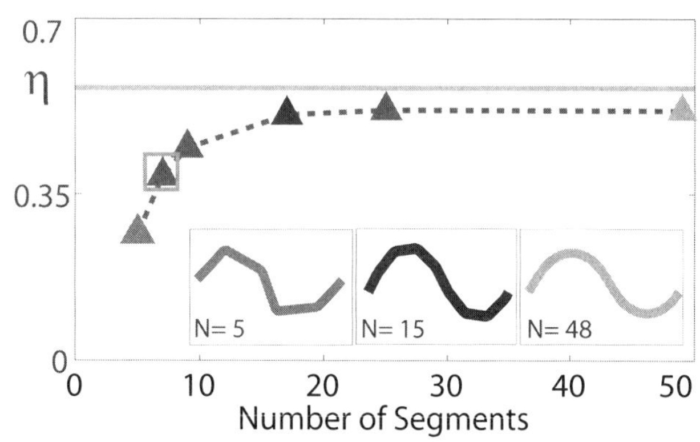

Fig. 8. Wave efficiency increases with number of segments for a fixed length robot in simulation (blue dashed curve)($f = 1$ Hz and $A/\lambda = 0.2$). The red, black, and cyan triangles correspond to 5, 15, and 48 segment robots respectively. The green square corresponds to the seven segment physical robot, and the grey line indicates η predicted by the RFT solved for a continuous body profile (see text for details).

Also, the fluctuations in torque at frequencies higher than the oscillation frequency of the robot were small in comparison to the torque amplitude.

VI. Future Work

A sand swimming robot combined with a proven simulation tool opens many avenues for further research. Of immediate interest is testing the RFT prediction that an optimal spatial form (ratio of amplitude to wavelength) maximizes forward speed of an undulatory sand swimmer [1]. The effect of the predicted optimal kinematics can also be evaluated by measuring the mechanical cost of transport. In conjunction with the numerical simulation the robot can test the effect of motion profiles (wave shapes) on performance. Since the

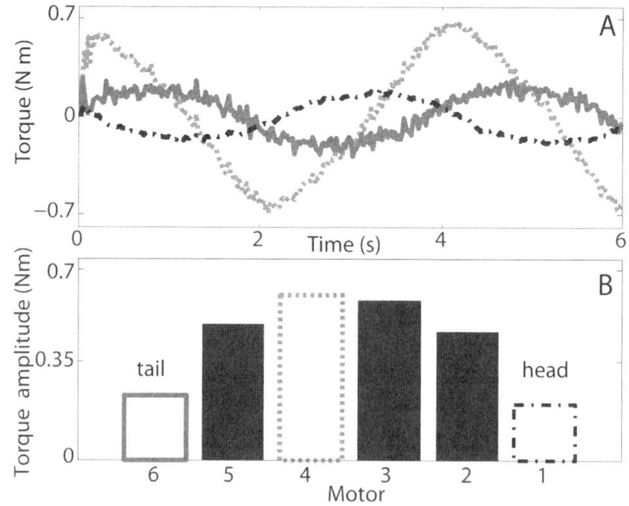

Fig. 9. Motor torque for the simulated 7 segment, 6 motor robot ($f = 1$ Hz) (A) varying with time. (B) Torque amplitude vs. motor position; orange (solid curve), green (dotted curve) , and black (dash-dot curve) correspond to motor 6 (tail), 4, and 1 (head) with motor position 1 denoting segment number 2 in Fig. 5 and 6.

sandfish uses the same kinematics to move in a variety of media, duplicating the animals control methods and sensing modalities in a robot could lead to more effective locomotion.

The sandfish has a non-trivial shape which suggests changing the morphological characteristics of the robotic device. For example, the cross sectional shape of the sandfish (flat belly and rounded top) have been hypothesized to aid rapid burial into granular media [51]. Our simulated and physical robot can be used to explore the influence of this morphology along with body taper on performance. The robotic simulation can also tune parameters like skin friction and body compliance to identify optimal values which could then be tested with our robot.

VII. Conclusion

Motivated by biological experiments revealing rapid subsurface sand-swimming in the sandfish lizard, we have used numerical simulation as a design tool to build an undulatory sand-swimming device. We used our robot simulation to test whether a device with a finite number of segments (7) could advance using a simple open loop (traveling wave sinusoid) control scheme and calculated the motor torque requirements for the robot. We then built and tested a prototype of the device to validate the biological observations and predictions from the RFT [1] and simulations that limbless body undulations were sufficient to propel the robot forward. Our findings show that the device can swim, and that it translates faster by increasing its oscillation frequency just as the sandfish does. The design tools (numerical model and robot) we developed can generate testable hypotheses of neuromechanical control [52] and improve our understanding of how organisms exploit the solid and fluid-like properties of granular media, enabling the construction of robots that can locomote effectively within complex environments.

Acknowledgments

We thank Nick Gravish and Chen Li for help with the physics experiments, and we thank Daniel Cohen and Andrew Slatton for help with development of the numerical simulation. We also acknowledge our funding from The Burroughs Wellcome Fund Career Award at the Scientific Interface, NSF Physics of Living Systems grant PHY-0749991, and the Army Research Laboratory (ARL) Micro Autonomous Systems and Technology (MAST) Collaborative Technology Alliance (CTA) under cooperative agreement number W911NF-08 − 2 − 0004.

References

[1] R. Maladen, Y. Ding, C. Li, and D. Goldman, "Undulatory Swimming in Sand: Subsurface Locomotion of the Sandfish Lizard," *Science*, vol. 325, no. 5938, p. 314, 2009.

[2] C. Humphrey and J. Adams, "Robotic Tasks for Chemical, Biological, Radiological, Nuclear and Explosive Incident Response," *Advanced Robotics*, vol. 23, no. 9, pp. 1217–1232, 2009.

[3] J. McKean, S. Buechel, and L. Gaydos, "Remote sensing and landslide hazard assessment," *Photogrammetric engineering and remote sensing*, vol. 57, no. 9, pp. 1185–1193, 1991.

[4] G. Metternicht, L. Hurni, and R. Gogu, "Remote sensing of landslides: An analysis of the potential contribution to geo-spatial systems for hazard assessment in mountainous environments," *Remote sensing of Environment*, vol. 98, no. 2-3, pp. 284–303, 2005.

[5] A. Ashcheulov, I. Gutsul, and V. Maevski, "Device for monitoring the radiation temperature in coal mines," *Journal of Optical Technology*, vol. 67, no. 3, p. 281, 2000.

[6] W. Marcus, C. Legleiter, R. Aspinall, J. Boardman, and R. Crabtree, "High spatial resolution hyperspectral mapping of in-stream habitats, depths, and woody debris in mountain streams," *Geomorphology*, vol. 55, no. 1-4, pp. 363–380, 2003.

[7] R. Ritzmann, R. Quinn, and M. Fischer, "Convergent evolution and locomotion through complex terrain by insects, vertebrates and robots," *Arthropod structure and development*, vol. 33, no. 3, pp. 361–379, 2004.

[8] R. Siegwart, P. Lamon, T. Estier, M. Lauria, and R. Piguet, "Innovative design for wheeled locomotion in rough terrain," *Robotics and Autonomous systems*, vol. 40, no. 2-3, pp. 151–162, 2002.

[9] J. Wong, "On the study of wheel-soil interaction," *Journal of Terramechanics*, vol. 21, no. 2, pp. 117–131, 1984.

[10] P. Arena, P. Di Giamberardino, L. Fortuna, F. La Gala, S. Monaco, G. Muscato, A. Rizzo, and R. Ronchini, "Toward a mobile autonomous robotic system for Mars exploration," *Planetary and Space Science*, vol. 52, no. 1-3, pp. 23–30, 2004.

[11] C. Li, P. B. Umbanhowar, H. Komsuoglu, D. E. Koditschek, and D. I. Goldman, "Sensitive dependence of the motion of a legged robot on granular media," *Proceedings of the National Academy of Science*, vol. 106, no. 9, pp. 3029–3034, 2009.

[12] C. Li, P. Umbanhowar, H. Komsuoglu, and D. Goldman, "The effect of limb kinematics on the speed of a legged robot on granular media," *Experimental Mechanics*, pp. 1–11, 2010.

[13] U. Saranli, M. Buehler, and D. Koditschek, "RHex: A simple and highly mobile hexapod robot," *The International Journal of Robotics Research*, vol. 20, no. 7, p. 616, 2001.

[14] R. Playter, M. Buehler, and M. Raibert, "BigDog," in *Unmanned Ground Vehicle Technology VIII*, ser. Proceedings of SPIE, D. W. G. Grant R. Gerhart, Charles M. Shoemaker, Ed., vol. 6230, 2006, pp. 62 302O1–62 302O6.

[15] A. Hoover, E. Steltz, and R. Fearing, "RoACH: An autonomous 2.4 g crawling hexapod robot," in *IEEE Intelligent Robots and Systems Conference Proceeding*, 2008, pp. 26–33.

[16] A. Saunders, D. I. Goldman, R. J. Full, and M. Buehler, "The rise climbing robot: body and leg design," in *Unmanned Systems Technology VIII*, G. R. Gerhart, C. M. Shoemaker, and D. W. Gage, Eds., vol. 6230. SPIE, 2006, p. 623017.

[17] S. Kim, J. E. Clark, and M. R. Cutkosky, "iSprawl: Design and tuning for high-speed autonomous open-loop running," *International Journal of Robotics Research*, vol. 25, no. 9, 2006.

[18] S. Hirose and A. Morishima, "Design and control of a mobile robot with an articulated body," *The International Journal of Robotics Research*, vol. 9, no. 2, p. 99, 1990.

[19] Y. Shan and Y. Koren, "Obstacle accommodation motion planning," in *Intelligent autonomous systems, IAS–3: proceedings of the international conference, Pittsburgh, Pennsylvania, February 15-18, 1993.* Ios Pr Inc, 1993, p. 94.

[20] R. Desai, C. Rosenberg, J. Jones, and I. Inc, "Kaa: an autonomous serpentine robot utilizes behavior control," in *Proceedings of the 1995 International Conference on Intelligent Robots and Systems, Pittsburgh.* Citeseer, 1995.

[21] B. Klaassen and K. Paap, "GMD-SNAKE 2: a snake-like robot driven by wheels and a method for motion control," in *Proceedings from the IEEE International Conference On Robotics And Automation*, vol. 4, 1999, pp. 3014–3019.

[22] A. Crespi, A. Badertscher, A. Guignard, and A. Ijspeert, "Swimming and crawling with an amphibious snake robot," in *IEEE International Conference On Robotics And Automation*, vol. 3. Citeseer, 2005, p. 3024.

[23] G. Chirikjian and J. Burdick, "The kinematics of hyper-redundant robot locomotion," *IEEE transactions on Robotics and Automation Proceedings*, vol. 11, no. 6, pp. 781–793, 1995.

[24] H. Choset, J. Luntz, E. Shammas, T. Rached, D. Hull, and C. Dent, "Design and motion planning for serpentine robots," in *Proceedings of SPIE*, vol. 3990, 2000, p. 148.

[25] K. Dowling, "Limbless locomotion: learning to crawl," in *IEEE International Conference on Robotics and Automation Proceedings*, vol. 4, 1999.

[26] H. Kim, D. Shim, and S. Sastry, "Flying robots: modeling, control and decision making," in *IEEE International Conference on Robotics and Automation Proceedings.*, vol. 1, 2002.

[27] J. Zufferey, A. Klaptocz, A. Beyeler, J. Nicoud, and D. Floreano, "A 10-gram vision-based flying robot," *Advanced Robotics*, vol. 21, no. 14, pp. 1671–1684, 2007.

[28] M. Bekker, *Theory of Land Locomotion*. The University of Michigan Press, 1956.

[29] D. Tritton, *Physical Fluid Dynamics*. Oxford University Press, 1989.

[30] D. C. Rapaport, *The Art of Molecular Dynamics Simulation*, 2nd ed. Cambridge University Press, 2004.

[31] K. Gerritsen, A. van den Bogert, and B. Nigg, "Direct dynamics simulation of the impact phase in heel-toe running," *Journal of Biomechanics*, vol. 28, no. 6, pp. 661–668, 1995.

[32] A. Mouazen and M. Nemenyi, "Tillage tool design by the finite element method: Part 1. finite element modelling of soil plastic behaviour," *Journal of Agricultural Engineering Research*, vol. 72, no. 1, pp. 37–51, 1999.

[33] F. Meysman, J. Middelburg, and C. Heip, "Bioturbation: a fresh look at Darwin's last idea," *Trends in Ecology & Evolution*, vol. 21, no. 12, pp. 688–695, 2006.

[34] R. Seymour, P. Withers, and W. Weathers, "Energetics of burrowing, running, and free-living in the Namib Desert golden mole (Eremitalpa namibensis)," *Journal of Zoology*, vol. 244, no. 01, pp. 107–117, 2001.

[35] K. Dorgan, S. Arwade, and P. Jumars, "Burrowing in marine muds by crack propagation: kinematics and forces," *Journal of Experimental Biology*, vol. 210, no. 23, p. 4198, 2007.

[36] L. Fielden, "Home range and movement of the Namib Desert golden mole, *Eremitalpa granti namibensis* (Chrysochloridae)." *Journal of Zoology*, vol. 223, no. 4, pp. 675–686, 1991.

[37] S. Hawke and R. Farley, "Ecology and behavior of the desert burrowing cockroach, *Arenivaga sp.* (Dictyoptera, Polyphagidae)," *Oecologia*, vol. 11, no. 3, pp. 263–279, 1973.

[38] E. Arnold, "Identifying the effects of history on adaptation: origins of different sand-diving techniques in lizards," *J. Zool., Lond*, vol. 235, pp. 351–388, 1995.

[39] C. White, "The influence of foraging mode and arid adaptation on the basal metabolic rates of burrowing mammals," *Physiological and Biochemical Zoology*, vol. 76, no. 1, pp. 122–134, 2003.

[40] C. Gans, "Tetrapod limblessness: evolution and functional corollaries," *Integrative and Comparative Biology*, vol. 15, no. 2, p. 455, 1975.

[41] S. Carranza, E. Arnold, P. Geniez, J. Roca, and J. Mateo, "Radiation, multiple dispersal and parallelism in the skinks, *Chalcides* and *Sphenops*

(Squamata: Scincidae), with comments on Scincus and Scincopus and the age of the Sahara Desert," *Molecular phylogenetics and evolution*, vol. 46, no. 3, pp. 1071–1094, 2008.

[42] J. Gray and G. Hancock, "The propulsion of sea-urchin spermatozoa," *Journal of Experimental Biology*, vol. 32, no. 4, p. 802, 1955.

[43] N. Cohen and J. Boyle, "Swimming at low Reynolds number: a beginners guide to undulatory locomotion," *Contemporary Physics*, vol. 51, no. 2, pp. 103–123, 2010.

[44] S. Berri, J. Boyle, M. Tassieri, I. Hope, and N. Cohen, "C. elegans locomotion: a unified multidisciplinary perspective," *BMC Neuroscience*, vol. 10, no. Suppl 1, p. P16, 2009.

[45] J. Gray and H. Lissmann, "The locomotion of nematodes," *Journal of Experimental Biology*, vol. 41, no. 1, p. 135, 1964.

[46] B. Jayne, "Kinematics of terrestrial snake locomotion," *Copeia*, pp. 915–927, 1986.

[47] J. Gray, "The mechanism of locomotion in snakes," *Journal of Experimental Biology*, vol. 23, no. 2, p. 101, 1946.

[48] R. A. Bagnold, *The Physics of Blown Sand and Desert Dunes*. Methuen and Co. Ltd., 1954.

[49] J. Lee and H. Herrmann, "Angle of repose and angle of marginal stability: molecular dynamics of granular particles," *Journal of Physics A: Mathematical and General*, vol. 26, pp. 373–383, 1993.

[50] R. Nedderman, *Statics and kinematics of granular materials*. Cambridge Univ. Pr., 1992.

[51] W. Mosauer, "Adaptive convergence in the sand reptiles of the Sahara and of California: a study in structure and behavior," *Copeia*, pp. 72–78, 1932.

[52] K. Nishikawa, A. A. Biewener, P. Aerts, A. N. Ahn, H. J. Chiel, M. A. Daley, T. L. Daniel, R. J. Full, M. E. Hale, T. L. Hedrick, A. K. Lappin, T. R. Nichols, R. D. Quinn, R. A. Satterlie, and B. Szymik, "Neuromechanics: an integrative approach for understanding motor control," *Integrative and Comparative Biology*, vol. 47, no. 1, pp. 16–54, 2007.

Passive Torque Regulation in an Underactuated Flapping Wing Robotic Insect

P. S. Sreetharan and R. J. Wood
School of Engineering and Applied Science
Harvard University
Cambridge, Massachusetts 02138
pratheev@post.harvard.edu

Abstract—**Recent developments in millimeter-scale fabrication processes have led to rapid progress towards creating airborne flapping wing robots based on Dipteran (two-winged) insects. Previous work to regulate reaction forces and torques generated by two flapping wings has largely focused on wing trajectory control. An alternative approach introduces additional degrees of freedom to the wing flapping mechanism to passively regulate these forces and torques. The resulting 'mechanically intelligent' devices can execute wing trajectory corrections to realize desired body forces and torques without the intervention of an active controller.**

This paper describes an insect-scale flapping wing aeromechanical structure consisting of a piezoelectric bimorph power actuator, an underactuated transmission mechanism, and passively rotating wings. The transmission is designed to passively modulate wing stroke velocity to eliminate the net roll torque imparted to the airframe.

The system is modeled as having four degrees of freedom driven open-loop by a single power actuator. The theoretical model predicts lift-generating wing trajectories as well as a passive reduction in roll torque experienced by the airframe. An at-scale structure constructed using Smart Composite Microstructure (SCM) fabrication techniques provides experimental support for the theoretical model.

I. Introduction

Advances in millimeter scale fabrication processes have enabled rapid progress towards the development of flapping wing micro air vehicles (FWMAVs) with system mass on the order of 100mg [1]. However, flight stability and control mechanisms for these mass and power limited systems remain active areas of research.

Investigation into the aerodynamics of biological insect flight has produced approximate aerodynamic models allowing computationally inexpensive prediction of aerodynamic forces and torques from wing trajectories [2], [3]. Accordingly, research into transmission and control mechanisms of flapping wing robotic insects has focused on control of wing trajectory. For example, the Berkeley Micromechanical Flying Insect (MFI) is a FWMAV platform with the ability to execute a range of predetermined wing trajectories using a fully actuated wing drive mechanism, neglecting elastic deformation of the transmission and wings [4]. In one notable exception, the Harvard Microrobotic Fly (HMF) has proven capable of realizing qualitatively biomimetic wing trajectories using passive compliance to allow variation of wing angles of attack [1]. The associated reduction in complexity has allowed this aeromechanical platform to achieve a lift to weight ratio greater than unity.

However, the benefits of underactuation and passive compliance can extend beyond simple reduction of mechanical complexity, in particular for devices in which the distribution of forces and torques is of fundamental importance. A ubiquitous example is the automobile differential, an underactuated mechanism commonly used to distribute engine power to two wheels. The differential incorporates an additional degree of freedom q_2 to balance the torque delivered to each wheel (see Fig. 1). The differential fundamentally operates on wheel torques instead of rotations; aided by passive mechanisms, the wheels can rotate along complex relative trajectories, maintaining traction on the ground without closed loop active control.

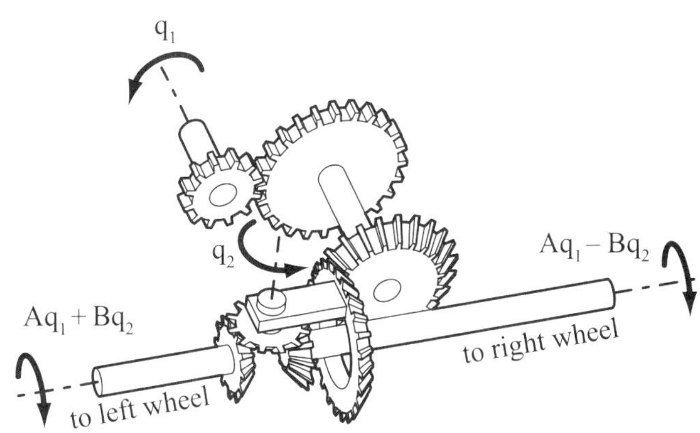

Fig. 1: A car differential balances output torques using an underactuated mechanism. Degree of freedom q_1 receives engine torque while q_2 is unactuated.

Previous work has introduced the concept of Passive Aeromechanical Regulation of Imbalanced Torques (PARITy) in the context of insect-scale FWMAV design [5]. Embodying the PARITy concept, the 'Drag PARITy' is an underactuated two degree of freedom FWMAV transmission that, analogous to an automobile differential, passively distributes power from a single actuator to balance torques delivered to two wings. Previous work has demonstrated its effectiveness within a planar two degree of freedom system with fixed 90° wing angles of attacks. Though the transmission was shown to passively balance drag induced roll torques, a fixed 90° angle of attack prevents the wings from generating lift.

This paper describes a lift-generating FWMAV design integrating passively rotating wings with the Drag PARITy transmission. Variation of wing angle of attack is achieved by incorporating a compliant 'wing hinge' (Fig. 3c) into the wing, similar to the approach taken by [1]. The complete design described in this paper has four degrees of freedom, a significant increase in dynamic complexity over the two degree of freedom system previously demonstrated in [5]. A single power actuator applies an oscillatory force, exciting motion in all four degrees of freedom through a variety of aerodynamic and inertial effects.

The underactuated flapping wing system is shown to execute stable, qualitatively biomimetic, lift-generating wing trajectories, indicating that the Drag PARITy is a viable transmission design for insect-scale FWMAVs. A theoretical model of the system is developed to investigate torque balancing characteristics in simulation. A control ('Uncut') trial with a nominally symmetric system demonstrates passive balancing of roll torques imparted from each wing, compensating for fabrication variation. In '1-Cut' and '2-Cut' trials, the system is simulated with successive removal of planform area from one wing (to provide an asymmetric disturbance) and is shown to continue successfully balancing roll torques, compensating for large inertial and aerodynamic wing asymmetries. By passively diverting more power to an underperforming wing, the design is also shown to indirectly compensate for imbalanced lift generation. Finally, an at-scale test device is constructed and observed to execute wing trajectories supporting theoretical predictions. Prior to describing the experiment, however, the PARITy methodology for FWMAV control which motivates this investigation will first be outlined.

II. The PARITy Methodology

Though it would allow for a highly capable FWMAV, fully-actuated high-bandwidth control of wing trajectories has not been achieved on a 100mg platform. Millimeter-scale fabrication techniques have not yet demonstrated the requisite complexity within mass constraints. Furthermore, power and mass constraints are likely to limit the bandwidth of electronic sensing and control systems on these lightweight platforms.

Acknowledging these limitations, research has been conducted into 'time-averaged' wing control, seeking to control average forces and torques by applying kinematic wing trajectory corrections on a long timescale (longer than a wing flapping period) [6]. Assuming that active control will not be attempted at short (sub-wingbeat) timescales, the question of the ideal short timescale behavior of a wing flapping mechanism is raised. Traditional kinematic control approaches tacitly assume that rigid specification of wing trajectory is a preferred short timescale behavior.

However, the specific wing trajectory executed is not fundamentally important to an FWMAV. Rather, an active flight control system for a robotic flier uses the wings as a tool to generate desired reaction forces and torques on its airframe.

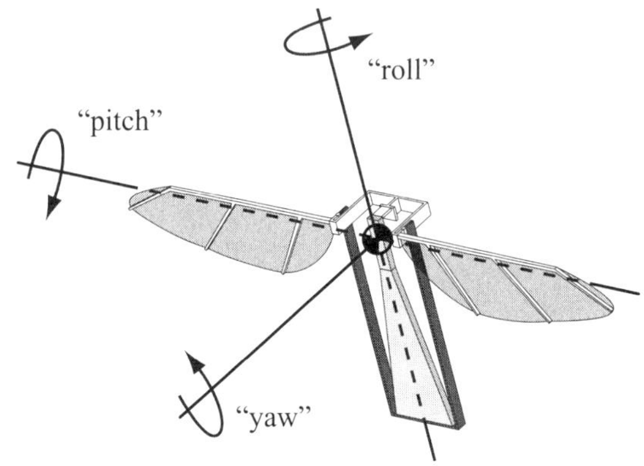

Fig. 2: Definition of roll, pitch, and yaw in the body frame.

Ideally, the wings should execute whatever trajectories are necessary to realize these desired forces and torques.[1] A drivetrain that passively regulates these forces and torques at a short timescale may simplify a longer timescale flight controller.

This alternative short term behaviour is conjectured to produce systems that reject short timescale disturbances passively, alleviating requirements on active control systems. It is also expected to compensate for some fabrication asymmetries, passively realizing the necessary adjustments to wing trajectory. This feature is extremely attractive, since fabrication variation is a major concern for devices manufactured at the millimeter scale.

Under the PARITy methodology, long timescale control is achieved not by altering the wing trajectories directly, but by modulating the dynamics of the short timescale passive system. In the context of PARITy based FWMAV designs, control inputs would perturb the setpoint of short timescale system dynamics. For example, the 'Drag PARITy' drivetrain analyzed in this paper passively balances body roll torques imparted by each wing. However, actuation of an active control input could bias system dynamics such that the roll torque imparted by one wing is passively regulated to be 10% higher than that from the other wing. This local passive regulation may enable direct active force and torque control at long timescales, simplifying the control problem for mass-limited flapping wing aeromechanical platforms.

Such active control mechanisms are the subject of future work and will not be discussed in depth in this paper, but their brief mention serves to motivate the detailed analysis of simpler PARITy drivetrains without control capability. The following sections analyze a specific FWMAV system introducing passively rotating wings to an actuated Drag PARITy transmission.

[1]The specific trajectory may be important for efficiency concerns, but is irrelevant for the purposes of stabilizing and controlling the airframe.

III. The Mechanism

A. Actuation

Piezoelectric actuation has been chosen due to its high bandwidth and high power density [7]. The actuator is a bimorph PZT cantilever, with a peak-to-peak actuation stroke of approximately 500μm. The base of the cantilever is grounded to the FWMAV airframe, while the output is affixed to the transmission input (Fig. 3).

B. The Drag PARITy transmission

The Drag PARITy transmission is a millimeter scale planar linkage constructed using Smart Composite Microstructure (SCM) fabrication techniques [8]. Unidirectional carbon fiber beams form rigid links, while revolute joints are realized by polymer flexure interconnects. The transmission has a single actuated input q_1 and dual outputs driving the stroke angles of each wing. The right wing stroke angle ϕ^R is illustrated in Fig. 3c, while the left wing stroke angle ϕ^L (not shown) is the analogous angle on the opposing wing.

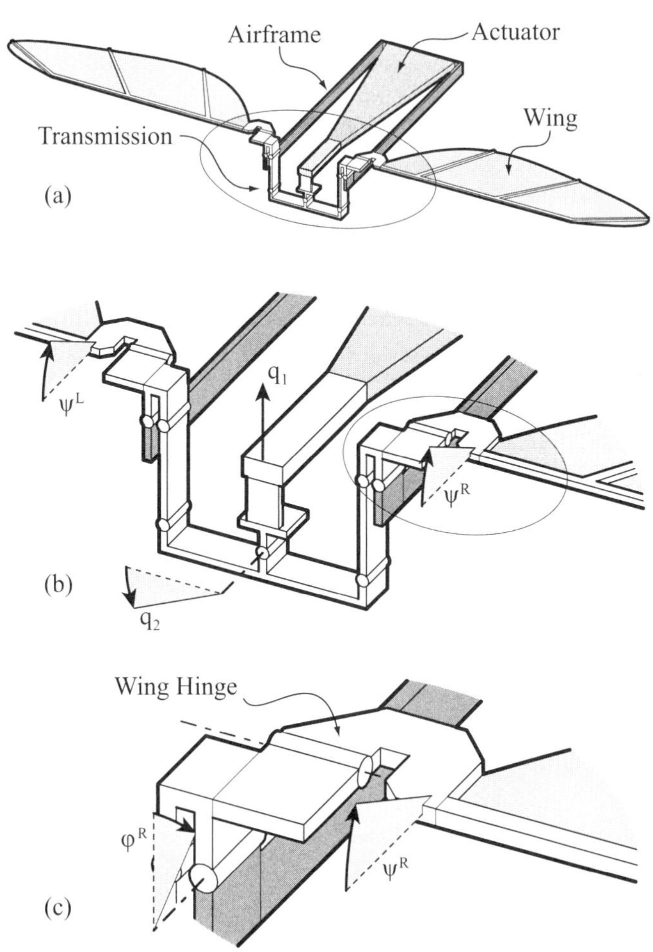

Fig. 3: (a) Diagram of the FWMAV design. (b) The four degrees of freedom q_1, q_2, ψ^L, and ψ^R with respect to airframe ground. (c) A view of the shoulder clarifying rotation angle ψ^R. Right wing stroke angle ϕ^R can be determined from q_1 and q_2, as can ϕ^L of the left wing (not shown).

The transmission mechanism has two degrees of freedom; referring to Fig. 3b, q_1 is actuated and allows power to be injected into the system, while q_2 is passively determined. The degree of freedom q_2 couples the upstroke of one wing to the downstroke of the other, allowing the mechanism to passively modulate wing stroke velocities to balance the roll torques imparted by the wings on the airframe. An invertible kinematic mapping relates q_1 and q_2 to ϕ^R and ϕ^L; either pair of coordinates can be used to describe the configuration of the transmission. A more detailed description of this mechanism is available in [5].

C. Wings

Wings consist of a 1.5μm polyester membrane supported by carbon fiber venation, shown in Fig. 4. Fabricated wings have masses under 1mg and are effectively rigid plates, exhibiting limited deformation while flapping. In an approach pioneered by [1], each wing is attached to a transmission output in series with a polymer flexure 'wing hinge' that allows the rigid wing to passively rotate around its longitudinal axis (see Fig. 3c). Compliance around the wing hinge axis allows the angle of attack of each wing to vary passively while flapping.

IV. Theoretical Simulation

A. Actuation

Actuator drive voltage is the single input to the simulation model. Using results from a laminate plate theory analysis, the first bending mode of the cantilever power actuator has been modeled as a grounded spring in parallel with a voltage-proportional force [7]. The cantilever beam has a linear spring constant of 467mN/mm, and under a 100V amplitude sinusoidal drive signal, the actuator exerts a 120mN amplitude sinusoidal force. The drive signal is applied at 110Hz, near mechanical resonance to increase stroke amplitude and limit reactive power.

B. Mechanical model

The transmission mechanism along with the wing hinge has been treated using a pseudo rigid body model [9]. All carbon fiber links are assumed to be infinitely stiff, while polymer flexure interconnects have been modeled as perfect revolute joints in parallel with linear torsion springs. Spring constants for the transmission joints and wing hinges have been calculated using classical beam theory, and no damping or other internal loss mechanisms have been modeled.

The wings themselves are the only significant inertias within the system and are the only inertias considered in the model. The mass of the SCM linkage mechanism is neglected. Though the piezoelectric actuator mass is significant, due to the large transmission ratio the effective inertia of the actuator is negligible and has been omitted from the model.

The final theoretical system has four degrees of freedom: two are contained within the Drag PARITy transmission, while the two wings each add a degree of freedom from their respective wing hinges. The orientation of each wing can be fully described by the angle of the corresponding transmission

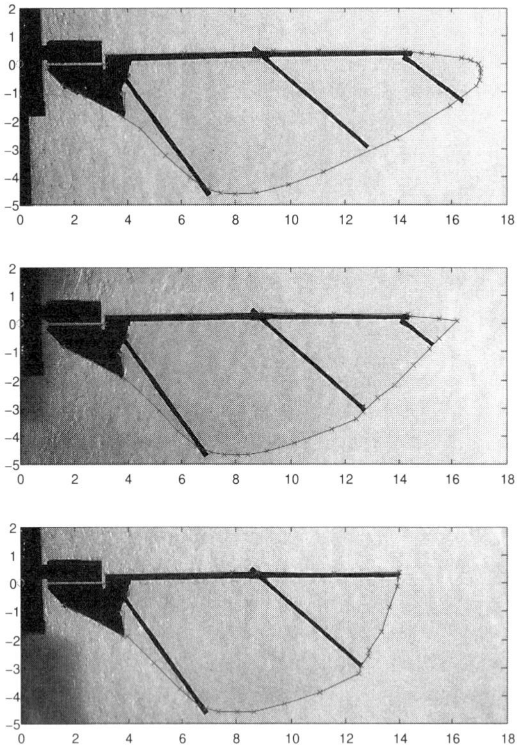

Fig. 4: The wing with membrane outline indicated for the Uncut, 1-Cut, and 2-Cut trials, from top to bottom. Axis units are in mm. For inertial components in Table I, The z and x coordinate axes correspond to horizontal and vertical image axes, respectively.

Wing Trial	Left *All*	Right *Uncut*	Right *1-Cut*	Right *2-Cut*
I_{xx}	49.0	47.1	40.6	32.8
I_{yy}	50.5	48.6	42.0	34.1
I_{zz}	1.56	1.49	1.43	1.29
I_{xz}	4.20	3.87	3.38	2.97
Ω_1	46.8	38.2	28.7	21.2
Ω_2	0.587	0.438	0.419	0.387
Ω_3	17.5	13.5	10.6	9.10
Ω_4	0.712	0.952	0.787	0.691

TABLE I: Inertial and aerodynamic parameters used for the left and right wings for the Uncut, 1-Cut, and 2-Cut trials. All values have units of mg·mm^2. The coordinate frame for inertial components is described in Figure 4.

output (the 'stroke angle' ϕ) and the deflection angle of the wing hinge (the 'rotation angle' ψ), illustrated for the right wing in Fig. 3c.

C. Aerodynamic model

Aerodynamic effects have been simulated using a model derived from the blade element method, assuming a perfectly rigid wing planform. As modeled, lift and drag torques are proportional to $\dot{\phi}^2$, the square of stroke velocity. Averaged lift and drag coefficients, strong functions of the rotation angle ψ, were taken from experimental data collected from dynamically scaled models of a fruit fly (*Drosophila melanogaster*) wing flapping in mineral oil [2]. Calculation of wing rotational moments, important for realizing passive wing rotation, relies on additional experimental work quantifying a non-dimensional center of pressure location $\hat{d}_{cp}$ of fruit fly wings [3], [10]. Rotational damping, proportional to $\dot{\psi}^2$, the square of wing rotational velocity, has been modeled in accordance with experimental and theoretical work on tumbling plates [11].

The complete aerodynamic model can be distilled into the following four aerodynamic moments applied to each wing:

$$M_N = -\Omega_1 \text{sgn}(\dot{\phi})\dot{\phi}^2 C_N(\psi) \quad (1)$$
$$M_T = -\Omega_1 \text{sgn}(\dot{\phi})\dot{\phi}^2 C_T(\psi) \quad (2)$$
$$M_{rd} = -\Omega_2 \text{sgn}(\dot{\psi})\dot{\psi}^2 C_{rd} \quad (3)$$
$$M_r = -\left(\Omega_3 \hat{d}_{cp}(\psi) - \Omega_4\right) \text{sgn}(\dot{\phi})\dot{\phi}^2 C_N(\psi) \quad (4)$$

In the previous set of equations, M_T acts about an axis perpendicular to the wing plane and is the result of aerodynamic forces acting in the wing plane. M_r and M_{rd} are the rotational and rotational damping moments, respectively, both acting on the wing around the wing hinge axis. M_N acts about an axis perpendicular to both the wing plane normal and the hinge axis, and results from aerodynamic forces normal to the wing. The three aerodynamic coefficients, related to tangential (C_T), normal (C_N), and rotational damping (C_{rd}) aerodynamic forces, are described in [10].

The parameters Ω_1, Ω_2, Ω_3, and Ω_4 have units of mg·mm^2 and can be calculated from the air density ρ and the specific wing morphology (see Table I for calculated values). Related work has produced extensive experimental data verifying that this aerodynamic model adequately describes passive rotation of a single wing executing a predetermined stroke angle trajectory, along with generated lift forces [10]. This referenced work contains a detailed description of the aerodynamic model briefly summarized here.

D. Mathematical formulation

For the theoretical model, the four coordinates specifying the device configuration were taken to be the left and right wing stroke angles (ϕ^L and ϕ^R, respectively) along with the left and right wing rotation angles (ψ^L and ψ^R, respectively). These four quantities and their time derivatives $\dot{\phi}^L$, $\dot{\phi}^R$, $\dot{\psi}^L$, and $\dot{\psi}^R$ form the full eight element state vector of the dynamic system.

The body inertia of a robotic 100mg FWMAV is assumed to be orders of magnitude larger than the wing inertias. This assumption is representative of many biological insects, though some exceptions exist (e.g. butterflies). Accordingly, the body frame has been treated as an inertial reference frame for the purpose of predicting wing dynamics. This assumption accurately represents the grounded-airframe experiment undertaken in Section VI. Furthermore, theoretical wing dynamic predictions are not expected to be impacted significantly by the non-inertial nature of the body frame of a free flying FWMAV.

The equations of motion for the wings were derived from an Euler-Lagrange formulation assuming a fixed body frame.

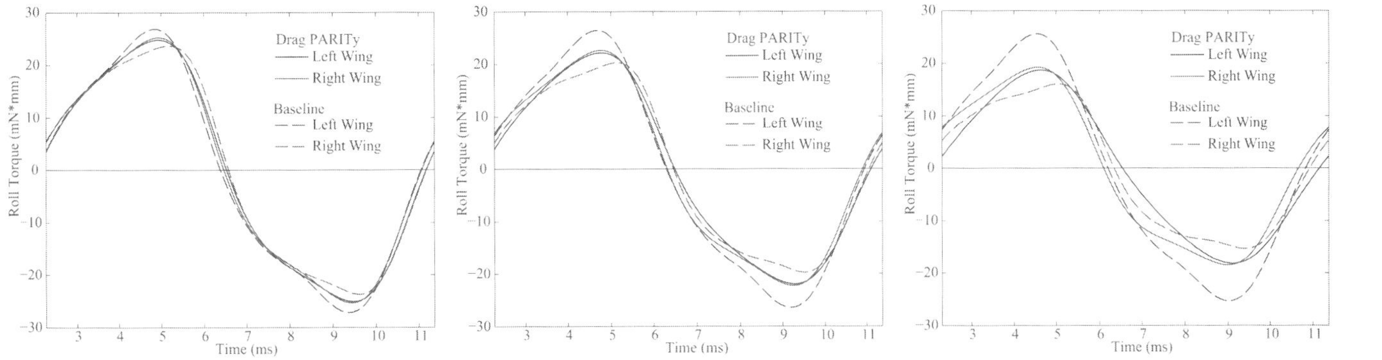

(a) (b) (c)

Fig. 5: Theoretical roll torque experienced by the airframe in the (a) Uncut, (b) 1-Cut, and (c) 2-Cut trials.

Since the only modeled inertias in the system are those of the two wings, the form of kinetic energy T is straightforward:

$$T = \frac{1}{2}\left(\vec{\omega}^L\right)^T \boldsymbol{I}^L \vec{\omega}^L + \frac{1}{2}\left(\vec{\omega}^R\right)^T \boldsymbol{I}^R \vec{\omega}^R \quad (5)$$

In the preceding equation, $\boldsymbol{I}^L$ and $\boldsymbol{I}^R$ are the inertial tensors of each wing, constant in the wing frame and calculated about an origin defined by the closest point on the wing hinge axis to the shoulder axis (see Table I for calculated values). The small shoulder offset of the Drag PARITy transmission has been neglected, thus no translational kinetic energy terms are present. The term $\vec{\omega}^L$ is the angular velocity of the left wing, a function of ϕ^L, ψ^L, $\dot{\phi}^L$, and $\dot{\psi}^L$. An analogous statement applies to the right wing angular velocity $\vec{\omega}^R$.

The potential energy V has the following form:

$$V = \frac{1}{2}\sum_{i=1}^{9} k_i \gamma_i^2 + \frac{1}{2} k_a q_1^2 \quad (6)$$

The full device contains nine polymer flexure joints: two wing hinges along with seven internal to the Drag PARITy transmission. The quantities γ_i represent the angular deflection of each flexure joint, functions of ϕ^L, ψ^L,ϕ^R, and ψ^R. The constants k_i represent the linearized torsional spring constants for each polymer flexure joint. The constant k_a is a linear spring constant describing the restoring force of the actuator in response to its linear deflection q_1, itself a function of ϕ^L and ϕ^R.

The Lagrangian L is defined as $L = T - V$, and the equations of motion are derived from the Euler-Lagrange equations for each of the four coordinates $p_i \in \{\phi^L, \psi^L, \phi^R, \psi^R\}$:

$$\frac{\mathrm{d}}{\mathrm{d}t}\frac{\partial L}{\partial \dot{p}_i} - \frac{\partial L}{\partial p_i} = \tau_i \quad (7)$$

Actuation force as well as aerodynamic torques appear in the model as generalized forces τ_i. Actuation occurs along q_1 (see Fig. 3) and aerodynamic torques are more naturally calculated in the wing frame, so the appropriate Jacobians have been used to map these forces onto the configuration variables.

All necessary Jacobian matrices and partial derivatives have been derived in closed analytical form for use within the model, but the details have been omitted for brevity. The four 2nd order differential equations produced from (7) were expressed as a first order system of eight coupled nonlinear differential equations. All theoretical results for the Drag PARITy design are the result of numerically integrating these differential equations using a Runge-Kutta based method as implemented by the MATLAB function ode45.

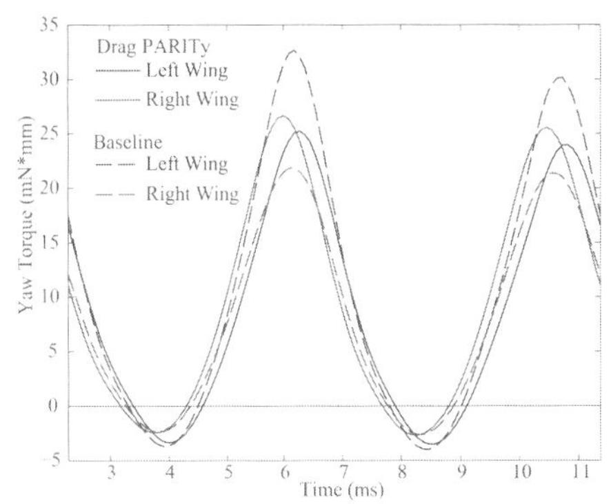

Fig. 6: Yaw torque in the 1-Cut Trial.

The theoretical dynamics model has been used to investigate the reaction torque regulating properties of the system in response to wing asymmetry. The system was compared with a baseline design in which the Drag PARITy transmission has been replaced with a conventional transmission characterized by fully actuated wing stroke angles. This baseline design is realized by freezing the degree of freedom q_2 of the Drag PARITy to $q_2 = 0$ (equivalent to the constraint $\phi^L = \phi^R$). This constraint is accommodated by introducing a time-dependent Lagrange multiplier to the Lagrangian:

$$L = T - V + \lambda(\phi^L - \phi^R) \quad (8)$$

In the modified equations of motion given by (7), $\lambda(t)$ represents an internal constraint force and is calculated algebraically at each timestep to satisfy the kinematic constraint. It is to be noted that this baseline three degree of freedom design is identical to that of the HMF [1].

A control trial, which will be called the 'Uncut' trial, was simulated using a structure mechanically and aerodynamically symmetric to the tolerances achievable with the SCM manufacturing process. Two additional trials were undertaken with intentionally asymmetric wing parameters, realized by removing successive amounts of planform area from the distal extent of the right wing. These trials will be called the 1-Cut and 2-Cut trials, respectively. Images of the right wing planform for all three trials are shown in Fig. 4. The left wing is nominally identical to the Uncut right wing for all three trials.

All wing parameters were measured directly from wings fabricated for the experimental test structure. Inertia tensors for the wing were constructed using a baseline mass measurement coupled with a photogrammetric process to determine the spatial distribution of wing mass. Aerodynamic parameters for the wings were calculated using the photogrammetrically determined wing planform areas shown in Fig. 4. Since both inertial and aerodynamic properties are highly sensitive to wing mounting accuracy, the photogrammetric techniques were conducted in situ to avoid disturbing the device. Fabrication variation has resulted in measurable asymmetry even in the Uncut case, apparent in theoretical and experimental results. See Table I for all calculated inertial and aerodynamic parameters.

In all trials, the Drag PARITy design is observed to execute stable wing trajectories qualitatively similar to those executed by biological insects. Wing stroke angles ϕ^L and ϕ^R oscillate over approximately $100°$ with a rotation angles ψ^L and ψ^R oscillating between $\pm 60°$, approximately $90°$ out of phase. These rotation angles correspond to an angle of attack $\alpha = 90°$ at stroke extents and $\alpha = 30°$ midstroke. Theoretical wing trajectories are plotted in Figs. 9a and 9c.

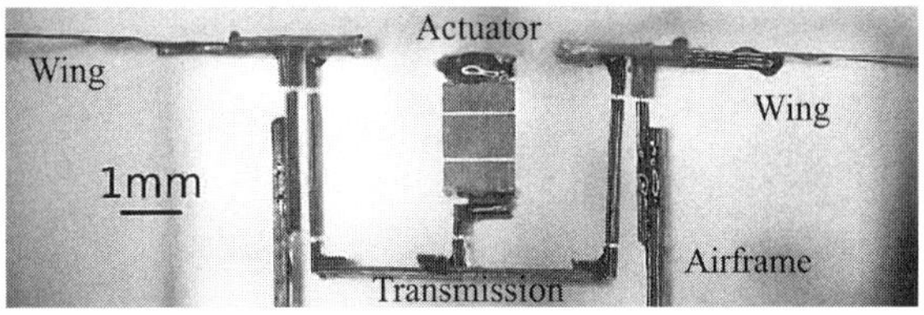

Fig. 7: Front view of the experimental device.

V. Passive Body Torque Regulation

The Drag PARITy design distinguishes itself from the baseline design in the theoretical reaction torques imparted by the flapping wings on the airframe. The Drag PARITy is designed to balance the roll reaction torques imparted by the flapping wings. Fig. 5 plots the theoretical roll torques experienced by the body of a FWMAV using a Drag PARITy transmission compared to that experienced by a FWMAV using a conventional transmission. In all three trials, it is apparent that the Drag PARITy transmission has succeeded in balancing the roll torques experienced by the body due to each wing. The results are especially striking in the Uncut trial (Fig. 5a), where the transmission has passively compensated for fabrication and assembly error present in the nominally symmetric structure.

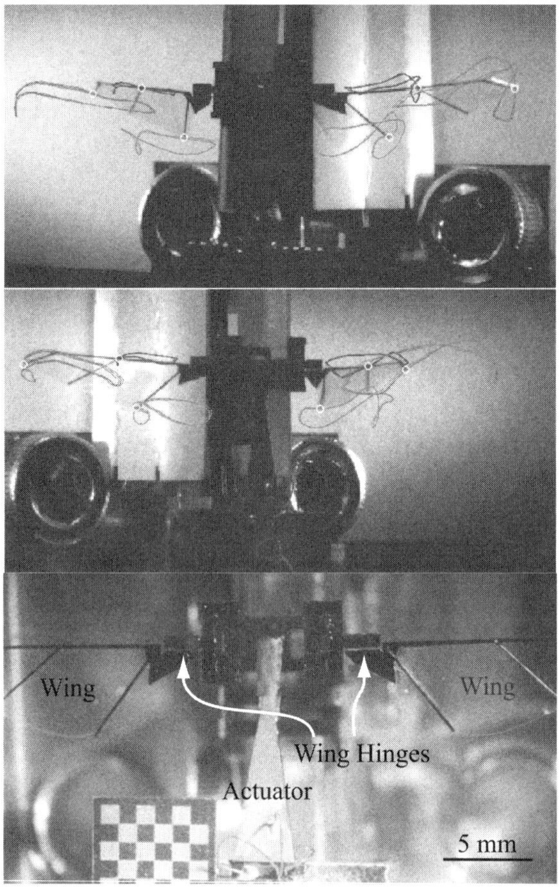

Fig. 8: Upper images are synchronized frames from the two cameras during the Uncut trial. Tracked points are indicated along with their trajectories over the course of the video. Lower image illustrates test structure.

Another interesting result of this theoretical study is the indirect balancing of lift dominated yaw torques. The results for the 1-Cut trial are presented in Fig. 6. In the drastically asymmetric 1-Cut and 2-Cut trials, use of the Drag PARITy design reduces the large average yaw torque imparted on the airframe by 71% and 72% respectively. However, in the nominally symmetric Uncut case, the Drag PARITy did not reduce the small average yaw error torque.

VI. Experimental Verification

After acquiring data for the Uncut trial, the right wing was cut in situ to conduct the 1-Cut and 2-Cut trials without perturbing the alignment of the wing on the transmission output. As previously mentioned, all dynamic wing parameters were measured without disturbing the device and are presented in Table I.

Two high speed video cameras were positioned such that each obtained a clear view of both wings over the entire flapping motion. Prior to acquiring video, the cameras were

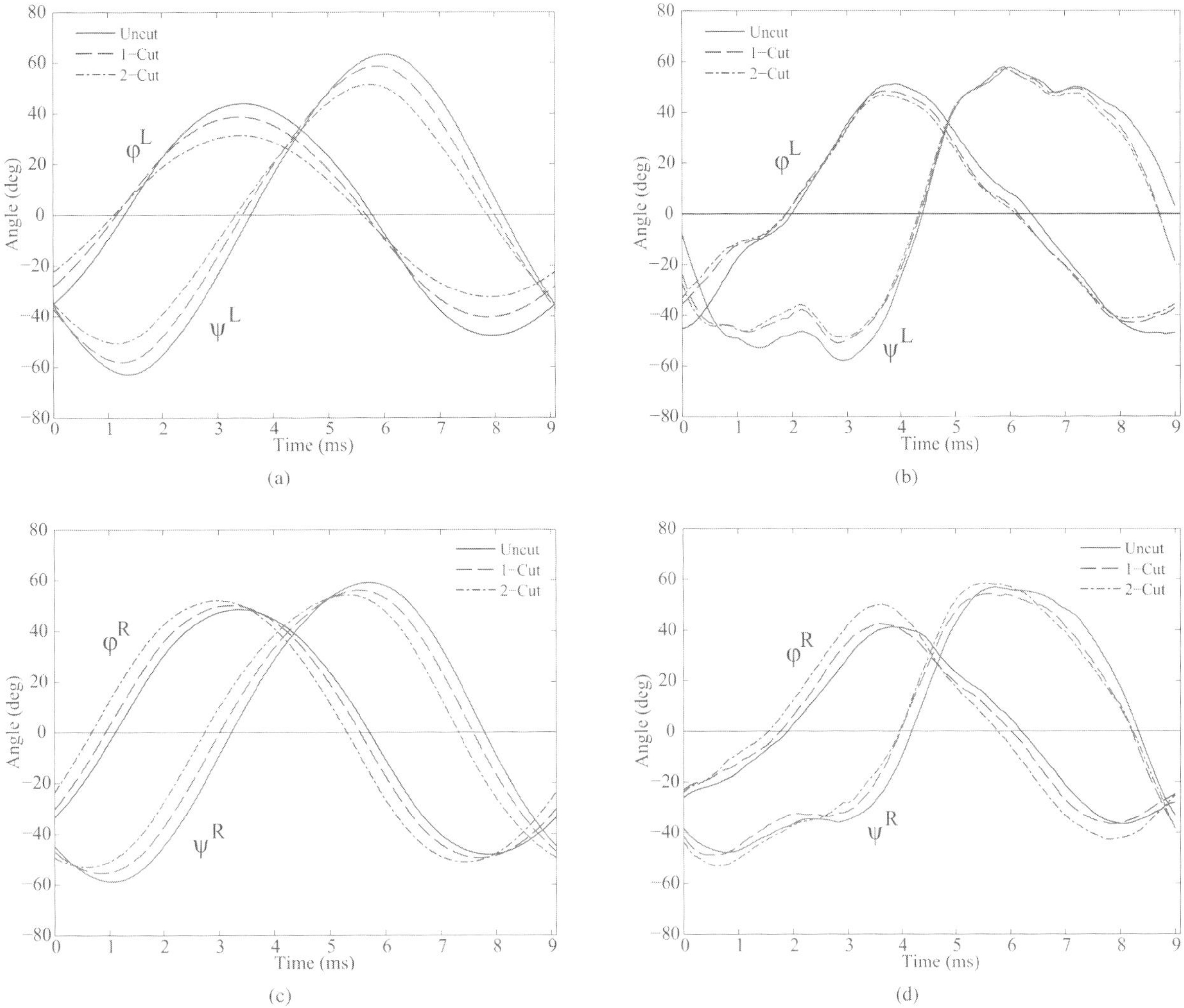

Fig. 9: Left wing trajectories (a) predicted by theory and (b) observed experimentally, along with (c) theoretical and (d) experimental right wing trajectories. Each plot shows results from the Uncut, 1-Cut, and 2-Cut trials.

calibrated using routines from the CalTech Camera Calibration Toolbox for MATLAB [12]. Once calibrated, the toolbox allows reconstruction of three dimensional coordinates of points identified in both camera views.

A 110Hz 200V (peak to peak) sinusoidal voltage was applied to the power actuator and synchronized high speed video was acquired from both video cameras at 10,000fps, or 91 frames per wingstroke period. Sample still frames are shown in Fig. 8.

Three easily distinguished features of the wing venation pattern were manually tracked across 300 frames for each trial. Identification of all three points in two camera views allows stereophotogrammetric reconstruction of the full wing orientation. The sinusoidal drive voltage applied to the actuator has been recorded and digitized at 5kHz, synchronized with the high speed video stream.

The observed stroke and rotation angles are plotted as a function of time in Fig. 9, along with predictions produced by the theoretical model. Time synchronization has been achieved by aligning the theoretical and experimental drive signals, omitted from the plots for clarity. The functional form of the applied voltage signal as a function of time (in seconds) is:

$$V(t) = 100V + 100V \cdot \sin(110 \cdot 2\pi t) \qquad (9)$$

From Fig. 9, it is immediately apparent that the theoretical model accurately captures qualitative characteristics of the experimental model, with rotation angle ψ exhibiting an approximately 90° phase lag behind the stroke angle ϕ. Furthermore, the theory also accurately predicts oscillation amplitudes from applied drive voltage, an achievement considering the complexity of this nonlinear dynamic system.

Theoretically predicted trends in wing trajectories as the right wing planform is altered are apparent in experimental data. The model predicts a monotonic increase in $\phi^R(t)$ amplitude as planform area is successively removed from the right wing, coupled with an associated decrease in the amplitude of $\phi^L(t)$. This trend is reflected in the experimental data as the Drag PARITy transmission passively diverts additional power to the underperforming right wing. The model also predicts a successive decrease in the amplitudes of both wing rotations $\psi^L(t)$ and $\psi^R(t)$ as wing membrane is removed. This trend is apparent in the observed trajectory of $\psi^L(t)$, though somewhat ambiguous in the observed trajectory of $\psi^R(t)$.

Among features not predicted by this simulation model are the square-wave appearance of observed wing rotations and the complex non-sinusoidal details of stroke angle trajectories. In future work, it is hoped that these discrepancies will be reduced by a more detailed theoretical model including, for example, mechanical loss mechanisms and nonlinear descriptions of polymer flexures to better predict dynamic characteristics at large joint angles. The transmission design itself will be refined to limit unintended and difficult to model behavior. For example, one source of error in this experimental trial was off-axis transmission compliance, resulting in measurable deviation of the wings from their mean stroke planes.

VII. Conclusion and Future Work

This paper has presented further evidence supporting the utility of passive underactuated mechanisms in FWMAVs. Significantly extending previous work, the load balancing Drag PARITy transmission has been shown to be compatible with longitudinally compliant wing hinges allowing passive variation of wing angle of attack. The resulting singly actuated four degree of freedom system has been shown to execute stable qualitatively biomimetic flapping wing trajectories well described by the associated theoretical model. Furthermore, the Drag PARITy transmission is shown to maintain its load balancing capabilities, passively altering wing trajectories so as to balance roll torques experienced by the FWMAV airframe.

Future work exploring the PARITy methodology will proceed along two parallel tracks:

1) Demonstrating long timescale control mechanisms
2) Expanding passive regulation to larger subsets of body forces and torques

As previously mentioned, long timescale control in PARITy enabled FWMAVs will be achieved not by direct modulation of wing kinematic trajectories, but by active modification of system dynamics. For example, the Drag PARITy transmission described in this paper exhibits short timescale dynamics that balance roll torques from each wing. An active control input could be introduced to bias these dynamics such that they passively regulate the ratio of roll torques τ_{roll}^L and τ_{roll}^R from the left and right wings, respectively, to a specified setpoint q_3:

$$\tau_{\text{roll}}^L / \tau_{\text{roll}}^R = q_3 \tag{10}$$

Note that q_3 is fixed at unity for the simple Drag PARITy transmission. A variety of dynamic parameters within the transmission, such as spring constants and link lengths, can be actively modulated at long timescales to realize this biased short timescale behavior. The potential for simple control relationships such as (10), bypassing wing kinematics to directly concern airframe forces and torques, is an exciting result of the PARITy methodology. Demonstration of such control features will motivate one track of future work.

A second research track involves introducing alternative or additional passive degrees of freedom to an FWMAV drivetrain to regulate different or expanded subsets of the body forces and torques produced by the wings. The Drag PARITy drivetrain is a mechanically intelligent device that has demonstrated regulation of body roll torques, arising in part from aerodynamic drag. In one nascent concept, careful introduction of passive features may enable an FWMAV drivetrain that directly regulates yaw torques arising from aerodynamic lift in addition to roll torques. The design space of such mechanically intelligent structures is vast, and future work will attempt to produce a variety of force and torque regulating FWMAV structures.

The PARITy methodology has the potential to simplify flight control of insect-scale robotic FWMAVs. It is hoped that future research into this novel methodology will provide tools to increase aerodynamic performance and reduce requisite system complexity, hastening the arrival of an autonomous 100mg-scale robotic FWMAV.

Acknowledgement

The authors gratefully acknowledge support from the National Science Foundation (Award No. CMMI-07466 38). Any opinions, findings and conclusions or recommendations expressed in this material are those of the authors and do not necessarily reflect those of the National Science Foundation.

References

[1] R. J. Wood, "The first takeoff of a biologically inspired at-scale robotic insect," *IEEE Trans. Rob.*, vol. 24, pp. 341–347, 2008.
[2] M. H. Dickinson, F. O. Lehmann, and S. P. Sane, "Wing rotation and the aerodynamic basis of insect flight," *Science*, vol. 284, pp. 1954–1960, 1999.
[3] W. B. Dickson, A. D. Straw, C. Poelma, and M. H. Dickinson, "An integrative model of insect flight control," in *Proc. AIAA Aerospace Sciences Meeting and Exhibit*, Reno, NV, January 2006.
[4] R. S. Fearing, K. H. Chiang, M. H. Dickinson, D. L. Pick, M. Sitti, and J. Yan, "Wing transmission for a micromechanical flying insect," *J. Micromechatronics*, vol. 1, no. 3, pp. 221–237, 2001.
[5] P. S. Sreetharan and R. J. Wood, "Passive aerodynamic drag balancing in a flapping wing robotic insect," *J. Mech. Des.*, vol. 132, 2010.
[6] X. Deng, L. Schenato, and S. Sastry, "Flapping flight for biomimetic robotic insects: Part ii-flight control design," *IEEE Trans. Rob.*, vol. 22, no. 4, pp. 789–803, 2006.
[7] R. J. Wood, E. Steltz, and R. S. Fearing, "Optimal energy density piezoelectric bending actuators," *Sensors & Actuators: A. Physical*, vol. 119, no. 2, pp. 476–488, 2005.
[8] R. J. Wood, S. Avadhanula, R. Sahai, E. Steltz, and R. S. Fearing, "Microrobot design using fiber reinforced composites," *J. Mech. Des.*, vol. 130, 2008.
[9] L. L. Howell, *Compliant Mechanisms*. John Wiley and Sons, Inc., 2001.
[10] J. P. Whitney and R. J. Wood, "Aeromechanics of passive rotation in flapping flight," *J. Fluid Mech.*, vol. In Press, 2010.
[11] A. Andersen, U. Pesavento, and Z. Wang, "Unsteady aerodynamics of fluttering and tumbling plates," *J. Fluid Mech.*, vol. 541, pp. 65–90, 2005.
[12] J. Y. Bouguet, "Camera calibration toolbox for matlab," 2008, http://www.vision.caltech.edu/bouguetj/calib_doc/index.html.

Color-Accurate Underwater Imaging Using Perceptual Adaptive Illumination

Iuliu Vasilescu, Carrick Detweiler, and Daniela Rus
Distributed Robotics Laboratory, CSAIL
Massachusetts Institute of Technology
Cambridge, MA 02139
Email: iuliuv@tiaresearch.com, {carrick, rus}@csail.mit.edu

Abstract— **Capturing color in water is challenging due to the heavy non-uniform attenuation of light in water across the visible spectrum, which results in dramatic hue shifts toward blue. Yet observing color in water is important for monitoring and surveillance as well as marine biology studies related to species identification, individual and group behavior, and ecosystem health and activity monitoring. Underwater robots are equipped with motor control for large scale transects but they lack sensors that enable capturing color-accurate underwater images. We present a method for color-accurate imaging in water called perceptual adaptive illumination. This method dynamically mixes the illumination of an object in a distance-dependent way using a controllable multi-color light source. The color mix compensates correctly for color loss and results in an image whose color composition is equivalent to rendering the object in air. Experiments were conducted with a color palette in the pool and at three different coral reefs sites, and with an underwater robot collecting image data with the new sensor.**

Fig. 1. Comparison of underwater imaging using white flash and adaptive illumination in a coral reef setting and a wreck setting. The color row in the middle shows color patches extracted from these images for visualization without the scene context. The left image shows a clown-fish and sea anemone in Moorea with white flash (Top) and adaptive illumination (Bottom). The pictures were taken at 20m depth by Melissa Holbrook Schmitt. The scene was at a distance of 3m. The right image shows a coral formation on a ship wreck in the Grand Cayman with existing white flash (Top) and adaptive illumination (Bottom). The pictures were taken at 15m depth by Cathy Church. The scene was at a distance of 3m.

I. Introduction

Underwater monitoring of natural as well as human-engineered environments can be greatly enhanced by automated capture of underwater color images using underwater robots [8, 13]. Our goal is to develop color-accurate automated imaging of underwater environments using robots capable of capturing, mosaicing, mapping, and analyzing color images of the space. In our previous work [12] we have described a robot capable of such a task. Figure 7 shows this underwater robot at the end of an underwater color imaging mission. In this paper we discuss an algorithm and instrument that enables robots to image objects such as coral heads and wrecks by automatically collecting a succession of underwater color-accurate images. Color encodes important information and plays a key role in underwater object identification, monitoring and surveillance, and marine biology—for example a healthy sea anemone is pigmented by zooxenthella, while an unhealthy sea anemone is bleached.

Current cameras are not able to capture color in water at distances greater than 1m, which makes color-dependent underwater studies challenging [6]. Part of the challenge for underwater imaging is that color is unavailable as an accurate data parameter for distances greater than 1m due to the uneven attenuation of the color spectrum as light travels through water. Longer wavelengths (e.g., red light) attenuate more rapidly than shorter wavelengths (e.g., blue light), resulting in dramatic hue shifts toward blue. Absorption, the primary cause for color loss in water, is exponential with respect to the propagation distance. The exponential varies according to the wavelength of the light [9]. Other factors such as scattering contribute to image loss but are substantially independent of the wavelength of the visible light and do not affect color. Current approaches to underwater imaging rely on flooding objects with white light from a very close distance (e.g. less than 0.5m), using color filters, and doing manual post-processing. These techniques are cumbersome, do not render accurate colors, and work only for fixed setups, as color loss is distance-dependent. For example, Figure 1(Top) shows underwater images taken with current techniques at 3m distance.

In this paper we describe a new method for color-accurate imaging in water called *perceptual adaptive illumination*. Our method uses knowledge about the physical processes that lead to the color shift. It computes light loss given distance to the imaged object and compensates for light loss using an illuminant whose radiation spectrum is controlled to be approximately the inverse of the water transfer function. The human color perception models are used for computing the optimal radiation spectrum. The full dynamic range of the camera is used and post-processing is not necessary. The end result is the color-accurate presentation of the object's image to the camera's CCD. Color-accurate imaging renders the object as if the image was taken in air, without the color loss side effects of water.

Figure 1 shows the comparison between images taken in water, using white flash and using perceptual adaptive illumination. The top pictures were taken with a white Xenon flash. The bottom figures were taken using the perceptual adaptive illumination method described in this paper. The color row in between the pictures shows color patches extracted manually from the images, presented without the scene context.

Underwater imaging with perceptual adaptive illumination requires a controllable illumination source capable of estimating distances. We have developed an instrument, algorithm, and software for capturing color-accurate images underwater by adaptive illumination which includes a spectrum-controllable light source. The light composition is calculated so that this composition is transformed into white light by the water between the camera and the subject. The energy content of the light source is calculated using the optical properties of water and the distance to the subject. The light is generated by a source composed of several filtered Xenon light bulbs. Varying the relative power of the Xenon flashes effectively results in a light source with variable spectrum. Intuitively, the device senses the distance of the object and mixes the light of a multi-color flash so as to compensate for each component of the color spectrum according to known physics of how that particular light frequency dissipates in water. Distance sensing is accomplished with an acoustic sonar, by using the distance information from the camera's auto-focus system or can be manually entered by the operator.

We present data from a suite of pool experiments and ocean experiments using our system. We use a color palette and the L*a*b* color metric for evaluating color accuracy. Our experiments demonstrate color-accurate imaging in one image plane at distances ranging from 1m to 5m and at depths ranging from 5m to 30m. Distance to the imaged object was measured in three different ways: using a measuring tape, using an external distance sensor, and using auto-focus information from the camera. We have attached the imaging apparatus to our underwater robot (see Figure 7(Right)) and used it to collect underwater images. We do not report in detail on the underwater robot missions because the focus of the paper is the underwater imaging system.

A. Related Work

Color plays a very important role in underwater monitoring and surveillance, as well as in marine biology studies related to species and behavior identification [2]. Many recognized species of coral reef fishes exhibit two or more color variants [7]. A recent study [1] found that fish use color to communicate.

Current approaches to color imaging underwater rely on flooding objects with white light from close distances (e.g. less than 0.5m [3]), possibly followed by post-processing [15, 11]. The post-processing step approximates the color of the image by manually setting a white point and correcting the image uniformly so that the selected point appears white. Since the color attenuation is not uniform this technique will not render accurate colors. These methods address the blue color shift by making assumptions about the properties of the picture (e.g. tagging known white patches in the picture). They do not use any knowledge of the physical process that led to color shift.

When objects of known color are present in the picture, various statistical or learning approaches have been applied [11]. Another option is to add a color filter to the camera to block much of the blue light, correcting for the loss of red [3]. Since absorption is distance-dependent, a different filter is needed for each distance. All these current techniques are cumbersome, do not render accurate colors, and produce color approximations for fixed distances and configurations only. What is needed is a method for adaptively illuminating an underwater scene with the correct light mix for each distance and depth. This will achieve color-accurate images without post processing. Accurate coloring in water refers to the equivalence between the color spectrum that reaches the imaging device in water to the color spectrum for the same object in air.

A recent paper by Yamashita et al. [15] considers the problem from a fundamental perspective and models the water effects on the color using the wavelength-dependent absorption coefficients and distance between camera and subject. The computed inverse function is applied to the captured image. The function is only a very coarse approximation that takes into account only the effect on three particular light wavelengths instead of the continuous spectrum. This simplification limits the performance and is not scalable to significant distances.

B. Outline

This paper is organized as follows. Section II introduces the physics of color perception in water. Section III discusses color perception by the human visual system and its implications to computation. Section IV presents the adaptive illumination algorithm. Section V describes the adaptive illumination instrument. Finally, Section VI presents our experimental data and evaluation.

II. Color Perception in Water

In this section we discuss the physics of color perception in water and the intuition behind how knowledge about color-absorption in water can be used to do color-accurate imaging in water.

There are two important phenomena that affect imaging in water: scattering and absorption. Scattering is the physical process whereby light is forced to deviate from straight trajectory

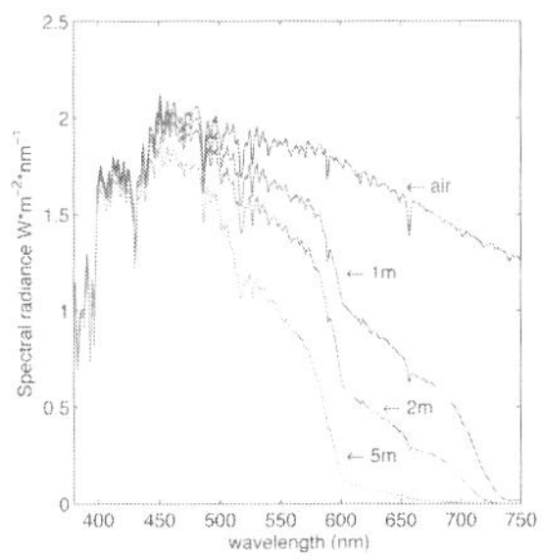

Fig. 2. The simulated spectrum of sunlight after it travels through 1m, 2m, and 5m of water. The simulation uses published sun spectrum and published light attenuation coefficients of water [9].

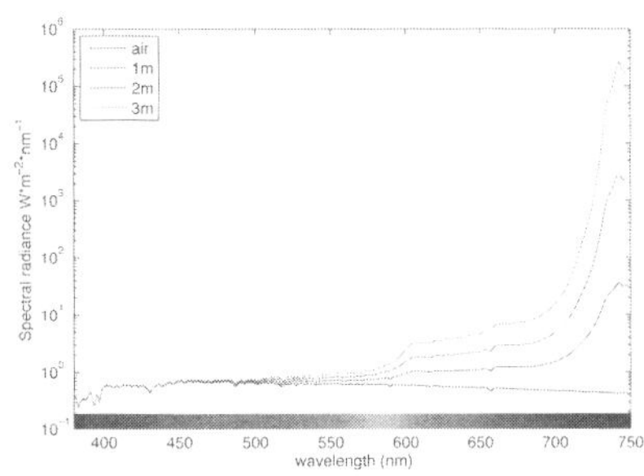

Fig. 3. The light required to compensate for color absorption in water. This is the inverse of water light transfer function plotted in Figure 2. The data is presented on a logarithmic scale due to its high dynamic range.

as it travels through water. The effect is caused predominantly by solid particles suspended in water, but also by the water molecules and substances dissolved in water. Scattering is wavelength independent and does not affect the color balance of the underwater image. This paper is not concerned with scattering.

Absorption is caused by water molecules and dissolved substances which convert light energy into heat. The water absorption coefficient, a_w, is wavelength dependent. Within the visible spectrum, longer wavelengths are attenuated stronger than shorter wavelengths (see Figure 2). Absorption is responsible for the color shift in underwater imaging. The absorption law describing the light energy transmitted by water is exponential.

$$I(\lambda, d) = I(\lambda, 0)e^{-a_w(\lambda)d}$$

where:

λ	light wavelength
$I(\lambda, 0)$	spectral power distribution at source
$a_w(\lambda)$	water absorption coefficient
$I(\lambda, d)$	spectral power distribution at distance d from the source, in water

The distance light travels through water (and is attenuated) is double the distance between subject and camera d_{sc}.

One solution is to compensate exactly for light absorption (see Figure 3). For example, given the camera to subject distance, the $650nm$ red light is attenuated 5 times more than the $530nm$ green light; thus, the light source should output 5 times more power at $650nm$ than at $530nm$. In general the spectral output power of the light source should be:

$$I_{ec}(\lambda, d_{sc}) = \frac{I_{D65}(\lambda)}{e^{-2a_w(\lambda)d_{sc}}} = I_{D65}(\lambda)e^{2a_w(\lambda)d_{sc}}$$

Such a light source compensates exactly for the light loss at the specified camera to subject distance d_{sc}. The subject appears as though illuminated by mid-day Sun light ($D65$ illuminant in air).

The usefulness of the brute-force method that relies on $I_{ec}(\lambda, d_{sc})$ is very limited in practice. Fabricating a light source with the required spectral power distribution is challenging as the amount of optical power needed rises very sharply with distance (e.g. at $3m$ the power required to take the picture in water is over 10^5 times the power required to take the picture in air). In addition, existing light sources generally have fixed output spectra, which implies the use of filters to generate the required spectral distribution. This increases the power requirements by another order of magnitude.

Alternatively, we can exploit the scope of human color vision to save power.

III. Color Perception by the Human Visual System

A light source that can compensate exactly for light loss in water yields color-accurate pictures (at least for relatively planar scenes, where the distance between the subject and the camera or light source is relatively constant). However, this method is both wasteful in terms of power and resources, and not necessary as the human vision system is not a "perfect" light sampler. This section discusses the physics of the color vision system of humans. We use the properties of this system to create a low complexity and power efficient light source for color-accurate underwater imaging.

Creating the exact inverse of water attenuation will concentrate the majority of the energy in the deep red part of the spectrum where the water attenuation is the highest. However, it turns out that the human eye sensitivity is very low in this part of the spectrum. Very little extra accuracy would be achieved by exactly compensating for the deep red attenuation.

Exact compensation is not necessary since the eyes (and cameras) sample the spectrum with only three discrete color sensors, collapsing the infinite-dimensional color space into a three dimensional space. To preserve the human-perceived color accuracy, it is only necessary to preserve the color coordinates for naturally occurring colors in the three dimensional color space of humans. This allows for some tolerance in the light source spectrum.

Under typical conditions the objects around us are illuminated by a white light source $I_0(\lambda)$, which contains energy at all visible spectrum wavelengths. The color of an object is given by its wavelength-dependent reflection coefficient $R(\lambda)$ which represent the fraction of incident power that is reflected. Our eyes receive the reflected light as shown in Figure 4. The spectral distribution of the reflected light is given by the equation: $E(\lambda) = R(\lambda)I_0(\lambda)$.

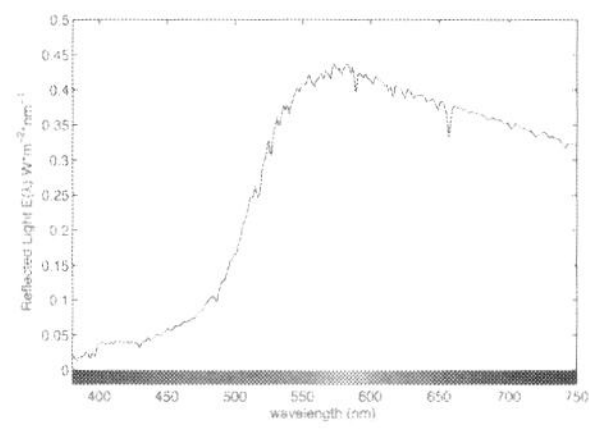

Fig. 4. Reflected spectral power of red color sample when lighted by a standard light (CIE illuminant D65).

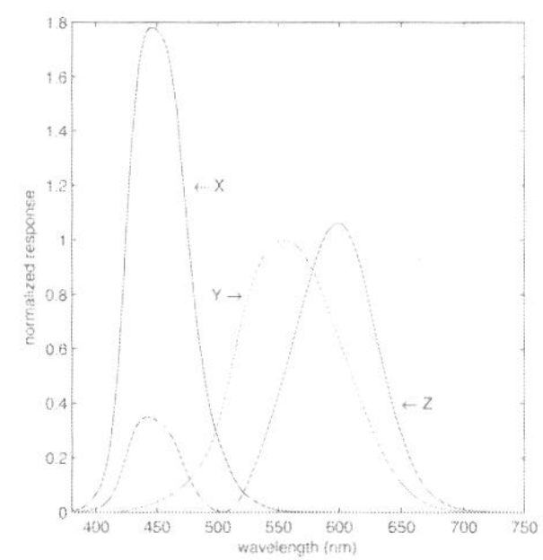

Fig. 5. The human visual system color response as modeled by the CIE 1931 2° Standard Observer [4]. The 3 curves model the response of the three types of cone cell in human retina.

The human vision system perceives colors by sampling the light spectrum with three different types of cone cells (ρ, γ and β). Each type of cell is sensitive to a different region of the visible spectrum — long, medium and short wavelengths. The three normalized sensitivity curves $S_\rho(\lambda)$, $S_\gamma(\lambda)$ and $S_\beta(\lambda)$ are plotted in Figure 6. The responses of these three types of cells T_ρ, T_γ and T_β are used by the central nervous system to associate colors to objects. For example, in the case of the Munsell $5Y8/10$ color sample, illuminated with white light the responses of the three types of cone cells can be calculated as:

$$T_\rho^{5Y8/10} = \int_{380nm}^{750nm} S_\rho(\lambda)E(\lambda)d\lambda =$$

$$\int_{380nm}^{750nm} S_\rho(\lambda)R_{5Y8/10}(\lambda)I_0(\lambda)d\lambda$$

$$T_\gamma^{5Y8/10} = \int_{380nm}^{750nm} S_\gamma(\lambda)E(\lambda)d\lambda =$$

$$\int_{380nm}^{750nm} S_\gamma(\lambda)R_{5Y8/10}(\lambda)I_0(\lambda)d\lambda$$

$$T_\beta^{5Y8/10} = \int_{380nm}^{750nm} S_\beta(\lambda)E(\lambda)d\lambda =$$

$$\int_{380nm}^{750nm} S_\beta(\lambda)R_{5Y8/10}(\lambda)I_0(\lambda)d\lambda$$

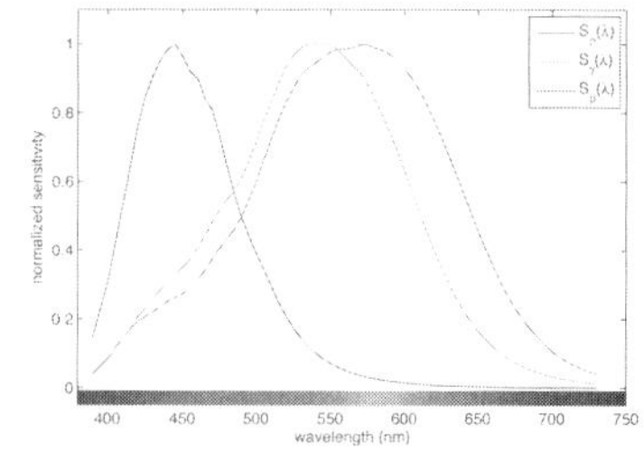

Fig. 6. Normalized sensitivities of the three types of cone cells present in human retina (ρ, γ and β) [10].

The brain receives a tuple of three numbers $(T_\rho, T_\gamma, T_\beta)$ and associates colors to objects based on this tuple. This results in an apparent three dimensional color space for humans. Any two objects that generate the same respective response in the three types of cone cells will appear to human observers as having the same color. This property can be exploited to reduce the complexity and power requirements of the underwater light source.

Using this observation, we develop a light source that preserves the coordinates of most naturally occurring colors in the human three dimensional color space, when used underwater. With this approach, the coordinates are also preserved in the color space of current cameras, since they closely model the human vision system color space.

Artificial light sources do not have the same spectral composition as the sun light. Therefore, they do not render the color of objects in the same way sunlight does. The Color Rendering Index (CRI) was developed by Commission Internationale de l'Éclairage (CIE) as a metric for comparing ability of light sources to render color accurately. The metric is based on comparing the color coordinates of 15 sample colors illuminated with the reference light and the tested light. The color samples have been chosen as representative for naturally occurring colors. The Euclidean distance between the color coordinates of two colors represented in CIE 1964 $U^*V^*W^*$ color space [14] is used as a quantitative value for color difference. $U^*V^*W^*$ is a uniform color space, meaning the Euclidean distance is a consistent measure of perceptual color difference, across the entire color space.

Our algorithm selects the power levels of a multi-color flash to maximize the CRI for objects illuminated in water. Formally, CRI is defined as:

$$\mathrm{CRI} = 100 - 4.6\Delta\bar{E}_{U*V*W*}$$

$$\Delta\bar{E}_{U*V*W*} = \frac{1}{15}\sum_{i=1}^{15}\Delta E_i$$

$$\Delta E_i = \sqrt{\Delta U_i^2 + \Delta V_i^2 + \Delta W_i^2}$$

CRI has values between 0% and 100%. A CRI of 100% corresponds to a perfect light source which renders colors identically to the reference light source.

IV. Perceptual Adaptive Illumination

Perceptual adaptive illumination is a power-efficient method for color-accurate imaging underwater. Given the distance to the object, the wavelength-dependent attenuation is computed using the known optical properties of water [9]. The light source spectrum is adjusted to compensate for this attenuation. The required light spectra is generated by a variable spectrum light source composed of several basic light sources, with fixed but distinct spectral power distributions. The variable spectrum is achieved by varying the relative power of the composing light sources. The spectral power distribution is optimized such that when filtered by the water between the camera and the subject, it will render the subject's human-perceived colors in the same way that natural light would render it in air. The optimization is performed by maximizing the apparent CRI of the light source when filtered by the water.

Given the camera-subject distance d_{sc}, adaptive illumination computes the optimal spectrum that can be generated by a source composed of n light sources with fixed output spectra $I_i(\lambda)$ and independently adjustable output power p_i. The total output spectrum is a linear combination of the composing light sources:

$$I(\lambda) = \sum_{i=1}^{n} p_i I_i(\lambda)$$

The spectra of the component light sources can be viewed as the basis functions that are used to generate the required spectral power distribution. The scaling factors, p_i, correspond to the output power setting of the component light sources.

The generated light is filtered by the water between the light source and the subject, and also by water as it travels back from the subject to the camera. Thus, the total travel distance is $2d_{sc}$, where d_{sc} is the distance between subject and camera. Water attenuation makes the illumination equivalent to a light source in air:

$$\hat{I}(\lambda) = e^{-2a_w(\lambda)d_{sc}} I(\lambda) = e^{-2a_w(\lambda)d_{sc}} \sum_{i=1}^{n} p_i I_i(\lambda)$$

From the camera (and observer) perspective, the subject looks as though it is illuminated by $\hat{I}(\lambda)$ in air. The problem can be posed as adjusting $\hat{I}(\lambda)$ so that the colors are rendered accurately in the observer's color space.

The power settings are found by solving the following optimization problem for the subject to camera distance parameter d_{sc}:

$$p_1..p_n = \underset{p_1..p_n \in [0..1]}{\arg\max}\ \mathrm{CRI}\left(\hat{I}(\lambda)\right) =$$

$$\underset{p_1..p_n \in [0..1]}{\arg\max}\ \mathrm{CRI}\left(e^{-2a_w(\lambda)d_{sc}} \sum_{i=1}^{n} p_i I_i(\lambda)\right)$$

This optimization problem can be solved using any existing numerical optimization software. Algorithm 1 summarizes the computation for the optimal light.

Algorithm 1 Light optimization

Require: $I_i(\lambda, t_i)$ Output characteristics of the component light sources
Require: d_{max} Maximum distance for which the light will be used
for $d_{sc} = 0m$ to d_{max} *in* $0.1m$ *steps* **do**
  compute $t_1(d_{sc})$ for optimal illumination at distance d_{sc} in air
  compute $t_2(d_{sc})..t_n(d_{sc}) = \arg\max_{t_2..t_n \in [0..1]} \mathrm{CRI}\left(e^{-2a_w(\lambda)d_{sc}} \sum_{i=1}^{6} I_i(\lambda, t_i)\right)$
end for

Fig. 7. (Left) The adaptive illumination device consists of six colored flashes with adjustable power. The device is attached to a standard digital camera in an underwater housing. (Right) Imaging apparatus attached to an underwater robot at the end of a mission.

The range of obtainable Color Rendering Index for $\hat{I}(\lambda)$ depends on the choice of the source light components $I_1(\lambda)..I_n(\lambda)$. This is restricted by the available light source technologies and filters, and is discussed in Section V. The aim is to obtain a CRI of at least 90%, which is equivalent to the best artificial light sources in air.

V. Imaging Apparatus

We developed and built a light source prototype capable of adaptive illumination called *AquaLight*. The light source can be used as an external flash for a digital SLR camera (as seen in Figure 7(Left)), as a free-standing illumination unit, or attached to an underwater robot or sensor node. Figure 7(Right) shows the apparatus attached to an underwater robot.

The light source is composed of 6 independently controlled Xenon strobes. Each strobe is capable of energy discharge up to 50J. The strobes have different but fixed spectra. One is unfiltered and capable of full light spectrum illumination. The other five strobes are filtered with increasingly longer cut-off long pass optical filters. The choice of filters was based on availability and simulation. We optimized for maximum corrected light output for objects at practical underwater photographic range (e.g distances d_{sc} between 0m and 5m). The output spectrum of each flash was measured with a calibrated radiometric spectrometer. Adaptive illumination is achieved by controlling the strobes' output power, or, equivalently, the energy discharged by each individual flash. Energy discharge is controlled by timing the flash so as to achieve the desired CRI.

More specifically, a Xenon discharge lamp is used as the light source. We chose for a Xenon lamp for its uniform output power across the visible spectrum, which is very similar to sun light. The Xenon lamp is capable of the highest instantaneous output power for its size (thus its widespread use in general photography). One of the flashes is unfiltered and provides the power required for the short wavelengths of the spectrum. Each of the other 5 component light sources has a Xenon discharge lamp and a long-pass filter. Filters are needed to generate different spectra for the 6 light sources. Long-pass filters are used in order to improve the overall efficiency of the device. Since the attenuation in water increases with the wavelength, more power is needed at longer wavelengths than the shorter wavelengths for any distance. The filters were chosen using simulation and maximization of the CRI and illumination power. The cut-off wavelengths for the 5 filters are 475nm, 575nm, 600nm, 600nm, and 630nm.

For the duration of their discharge, the Xenon lamps are powered by a capacitor bank. Each lamp has an associated 1080μF capacitor, for a maximum discharge energy of 33J. The capacitors are charged to 350V by an inverter. The entire unit is powered by a 11.7V, 2.1Ah Lithium-Polymer battery which provides enough power for approximatively 300 full power flashes.

An NXP LPC2148 CPU together with Xilinx XC2C256 CPLD control the timing of the 6 flashes. The timing is based on the distance to the subject and a precomputed lookup table of distance vs power requirements. In our system, the distance to subject can be input to the unit through a user interface (e.g. distance can be set manually using magnet to set a counter implemented with magnetic switches), camera auto-focus information, or it can be automatically determined using a distance sensor such as an ultrasonic pinger. The data we report in this paper was collected using manually set distance and auto-focus information using an Olympus camera. We found that the ultrasonic pinger (Teledyne/Benthos PSA-916) performed poorly for our application (swimming pool conditions as well typical reef environment).

The flash electronics are housed in a water and pressure resistant enclosure made out of clear acrylic. Two underwater connectors are placed on the back of the enclosure. One is used for the electrical connection between the camera and the flash (for synchronization). The other is used for interface with the ultrasonic pinger, programming and battery re-charging. The flash is attached to the camera through a rigid fixture, which ensures they point in the same direction (see Figure 7).

VI. Evaluation

Adaptive illumination has been evaluated in a series of the pool and ocean experiments. The goal was to measure the color accuracy and to compare the results with existing methods.

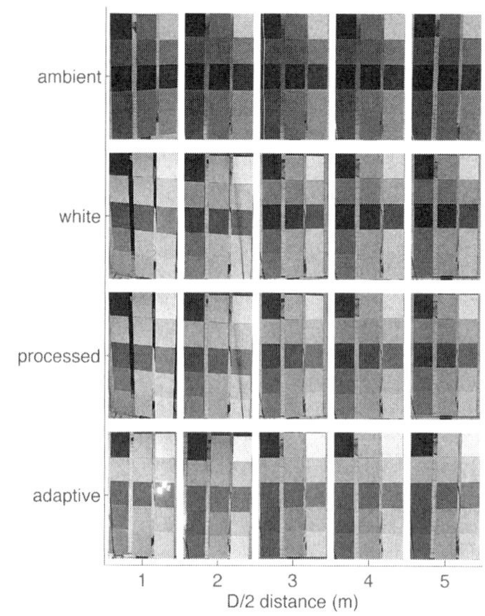

Fig. 8. Results of imaging the color palette in water. The columns correspond to the distance between the light source and the object: 1m, 2m, 3m, 4m, and 5m (note this corresponds to d_{sc} in the main document. The rows show the image obtained using ambient light (first row), white strobe (second row), post-processed white strobe image (third row), and adaptive illumination (fourth row).

We ran the adaptive illumination optimization algorithm for distances between 0.1m and 5 m (in 0.1m increments) and determined the optimal power ratios between the flashes. Table I shows the relative power of six flashes for illuminating an object at distance d_{sc} away in water and the resulting CRI. We note that the predicted CRI is above 95% in all cases. The best available artificial light sources are halogen incandescent bulbs which have a CRI of 98% to 100%, while fluorescent illumination has a CRI of 75%. In Table I F1 is the white flash which was kept at a constant power. Very little power is required from F1 relative to the other flashes. F2 is the yellow flash which contributes to the overall high CRI, but for longer distances a light red flash would have been more effective. The choice of filters was done empirically.

The flash power settings for each distance are stored in the microcontroller's memory of the adaptive illumination device. When taking a picture, given the distance to the desired object, the corresponding power settings are selected from this table.

The experimental methodology consisted of imaging the test object at distance d_{sc} varying between 1m and 5m (in 1m increments) using (1) ambient light, (2) white light strobe, (3) white light strobe and post-processing, and (4) adaptive illumination. The test object was also imaged in air for ground-truth. The test object was a custom made waterproof color palette with 15 colors distributed evenly in the color space (see Figure 8). Distance was determined with a measuring tape and was set manually using the system's magnetic switches. We also used the camera's auto-focus information to automatically

TABLE I

THE RELATIVE POWER OF THE 6 FLASHES AT DIFFERENT DISTANCES AND THE EXPECTED CRI. NOT THAT THE OBJECT IS AT DISTANCE $\frac{D}{2}$ FROM THE LIGHT SOURCE.

$\frac{D}{2}$	F_1	F_2	F_3	F_4	F_5	F_6	CRI
1m	0.0652	0.0227	0.0405	0.0285	0.0272	0.0188	99.26
2m	0.0652	0.0317	0.0789	0.0710	0.0517	0.0302	98.56
3m	0.0652	0.0421	0.1798	0.1518	0.1425	0.1090	97.97
4m	0.0652	0.0475	1.0000	1.0000	0.9802	0.6584	96.16

determine the distance to the imaged object. The standard white flash pictures were post processed with the current state of the art method: equalization using a manually marked white sample patch as reference. This method adjusts the R, G, B channels proportionally to yield the white color on the selected sample.

Color accuracy was measured by computing the color distance between manually selected patches of the color palette and the corresponding patches of the image captured in air. For each pair of color patches we converted the colors to the L*a*b* color space [5] and computed the resulting Euclidean distance. Since the L*a*b* color space was specifically designed to preserve the perceptual color distance, the Euclidean distance is an accurate representation of the perceptual color difference. The smaller the distance the better the color accuracy. We used L*a*b* color space due to the difficulty of measuring CRI experimentally (i.e. procuring the precise color samples used in CRI's definition).

Figure 10 shows the average L*a*b* error for all 15 colors in the palette for each distance. The numerical values for these averages is shown in Table II. The ambient light performs the worst as expected. White flash performs well at 1m (and below), but the performance decreases significantly for greater distances. The post processed white flash under-performs adaptive illumination. Its performance decreases steadily with a lower slope than the white flash. Adaptive illumination has no significant decrease in quality up to the measured distance.

Figures 9 show experimental data for four colors from the palette: yellow, red, green and blue. We note that the images taken with the adaptive illumination method have low constant L*a*b* distances from ground-truth for all distances (accurate colors), while the images taken with the other methods have L*a*b* distances that increase with distance (that is, the rendering is increasingly inaccurate in color).

As expected, not all the colors are affected in the same way. (see Figure 9). Red is most affected by water and it is hardly recoverable by post-processing. Blue is the least affected color. All methods render blue well. Green and the brown - the most commonly occurring natural colors are significantly distorted. This figure also shows distance-dependent L*a*b* values and the perceived colors the L*a*b* values map to, as computed by ambient light (top row), white flash (second row), post-processed white flash (third row), and adaptive illumination (fourth row).

An additional suite of pool experiments was done using the camera auto-focus information for automatically setting

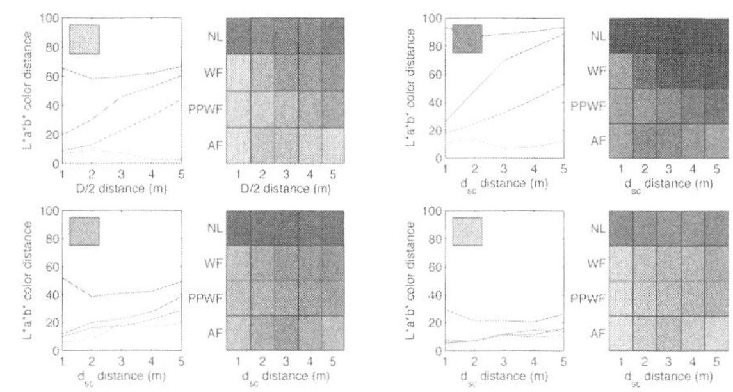

Fig. 9. The L*a*b* color error (left graphs) and visual appearance (right graphs) for four colors extracted from the color palette (yellow, red, green and blue), as captured by the four imaging methods: ambient light (blue curve), white strobe (green curve), white strobe followed by post-processing (red curve), and adaptive illumination (cyan curve). In the left graphs, the smaller the error, the more accurate the color. The ambient light method performs poorly, the white flash and postprocessed white flash methods degrade with distance, and the adaptive illumination curve remains constant for all distances. The right graphs present the colors extracted from the color palette, as captured by the camera. The columns of the matrix correspond to distance. The rows correspond to the method used (NL for ambient light, WF is white flash, PPWF is postprocessed white flash and AF is adaptive illumination).

the distance to the imaged object. The camera used in these experiments was an Olympus E520 SLR camera. Data from these tests is very similar to the data from the tests where the distance was measured externally.

Underwater imaging experiments have also been conducted in the field at four sites: Fiji, Tahiti, Hawaii, and Grand Cayman. In each experiment images were taken at measured distances using the white flash and the adaptive flash. The images were compared visually. Figure 1 shows typical images using white strobe and adaptive illumination.

We installed the adaptive illumination device on our underwater robot and commanded the robot to swim around a coral head taking a dense sequence of color-accurate photographs. These photographs can be mosaiced to reconstruct the coral head. The robot experiment demonstrated that color-accurate underwater images can be gathered automatically.

VII. CONCLUSIONS

Adaptive illumination computes the color dissipation at a given distance and compensates for color loss by introducing the correct color mix for that distance into the scene. Perceptual adaptive illumination was tested in a variety of environments and was shown to significantly outperform existing methods for underwater photography with respect to image

$Dist/2$	$1m$	$2m$	$3m$	$4m$	$5m$
ambient	59.4586	52.2793	53.8530	54.2241	58.6122
white	14.9747	26.4679	37.4749	43.9231	51.0177
white post-processed	9.2605	14.3705	18.9246	27.9815	34.8542
adaptive	8.3305	11.0851	11.3560	8.7389	9.9966

TABLE II

THE L*A*B* VALUE AVERAGED OVER 15 COLORS.

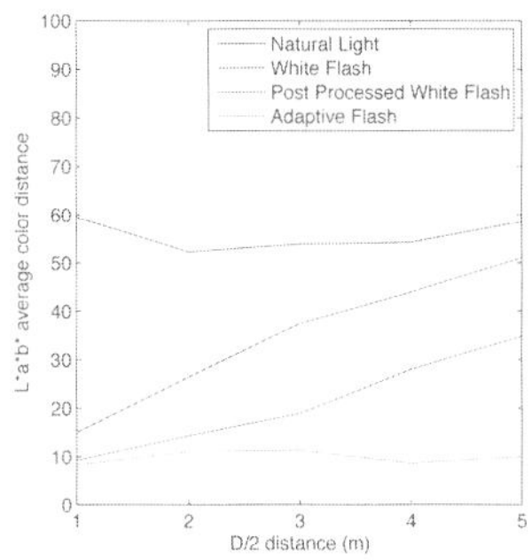

Fig. 10. L*a*b* average values for 15 colors at 1m, 2m, 3m, 4m, and 5m, as computed by 4 different imaging methods: ambient light (blue curve), white strobe (green curve), white strobe with post-processing (red curve), and adaptive illumination (cyan curve).

color accuracy using the L*a*b* metric space for measuring colors. Color encodes important information about underwater environments and habitats. This work enables the capture of color-accurate underwater images and thus opens the door to automating underwater monitoring and surveillance operations based on color.

Our work demonstrates that color-adaptive illumination is effective for color-accurate imaging in water for objects at a specified or computed distance. Our experimental data provides support for this claim for imaging objects that are up to 5m away from the camera. Current methods are cumbersome and do not produce accurate colors at distances greater than 2m. Our imaging instrument was prototyped in our lab using inexpensive components to demonstrate the concept. The instrument was not optimized for distance. Imaging at distances greater than 5m is achievable if the instrument uses more powerful light sources.

Perceptual adaptive illumination renders color accurately for all the objects at the given distance d in the image. Objects further away or closer to the camera will not be rendered correctly because adaptive illumination is distance-dependent. We are currently extending perceptual adaptive illumination from color-accuracy in one image plane to color-accuracy in multiple image planes. We are also working on extensions from still images to video.

Our current work includes field experiments to mosaic underwater scenes using robots equipped with the perceptual adaptive illumination device described in this paper.

REFERENCES

[1] D. Bellwood. The 'lost' language of fishes. http://www.coralcoe.org.au/news_stories/fishcolour.html.

[2] A.C. Crook. Colour patterns in a coral reef fish: Is background complexity important? *J. of eperimental marine biology and ecology*, 217(2), 1997.

[3] Martin Edge. The underwater photographer, third edition: Digital and traditional techniques (paperback), 2006.

[4] Commission internationale de l'éclairage. Cie 1931 2° standard observer, 1931.

[5] Commission internationale de l'éclairage. Supplement no. 2 to cie publication no. 15, colorimetry, 1976.

[6] Jules Jaffe, Kad Moore, John McLean, and Michael Strand. Underwater optical imaging: Status and prospects. *Oceanography*, 14, 2007.

[7] V. Messmer, G.P. Jones, L. van Herwerden, and P.L. Munday. Genetic and ecological characterisation of colour dimorphism in a coral reef fish. *Evironmental Biology of Fishes*, 74, 2005.

[8] Hanumant Singh, Roy Armstrong, Fernando Gilbes, Ryan Eustice, Chris Roman, Oscar Pizarro, and Juan Torres. Imaging coral i: Imaging coral habitats with the seabed auv. *Subsurface Sensing Technologies and Applications*, 5, 2004.

[9] R. C. Smith and K. S. Baker. Optical properties of the clearest natural waters (200–800 nm). *Appl. Opt.*, 20:177–184, 1981.

[10] A. Stockman, D.I.A. MacLeod, and N.E. Johnson. Spectral sensitivities of human cones. *Journal of the Optical Society of America*, 10:2491–2521, 1993.

[11] L.A. Torres-Mndez and G. Dudek. A statistical learning-based method for color correction of underwater images. *Research on Computer Science*, 17, 2005.

[12] Iuliu Vasilescu, Carrick Detweiler, Marek Doniec, Daniel Gurdan, Steffan Sosnowski, Jan Stumpf, , and Daniela Rus. Amour v: a hovering energy efficient underwater robot capable of dynamic payloads. *The International Journal of Robotics Research*, 29(5):547–570, 2010.

[13] Stefan B. Williams, Oscar Pizarro, Martin How, Duncan Mercer, George Powell, Justin Marshall, and Roger Hanlon. Surveying noctural cuttlefish camouflage behaviour using an auv. In *ICRA*, pages 214–219, 2009.

[14] G. Wyszecki. Proposal for a new color-difference formula. *JOSA*, 53:1318–1319, 1963.

[15] A. Yamashita, M. Fujii, and T. Kaneko. Color registration of underwater images for underwater sensing with consideration of light attenuation. *Robotics and Automation, 2007 IEEE International Conference on*, pages 4570–4575, April 2007.

Probabilistic Lane Estimation Using Basis Curves

Albert S. Huang Seth Teller
MIT Computer Science and Artificial Intelligence Laboratory
Cambridge, MA 02139
albert@csail.mit.edu teller@mit.edu

Abstract— **Lane estimation for autonomous driving can be formulated as a curve estimation problem, where local sensor data provides partial and noisy observations of spatial curves. The number of curves to estimate may be initially unknown and many of the observations may be outliers or false detections (due e.g. to to tree shadows or lens flare). The challenges lie in detecting lanes when and where they exist, and updating lane estimates as new observations are made.**

This paper describes an efficient probabilistic lane estimation algorithm based on a novel curve representation. The key advance is a principled mechanism to describe many similar curves as variations of a single *basis curve*. Locally observed road paint and curb features are then fused to detect and estimate all nearby travel lanes. The system handles roads with complex geometries and makes no assumptions about the position and orientation of the vehicle with respect to the roadway.

We evaluate our algorithm with a ground truth dataset containing manually-labeled, fine-grained lane geometries for vehicle travel in two large and diverse datasets that include more than 300,000 images and 44 km of roadway. The results illustrate the capabilities of our algorithm for robust lane estimation in the face of challenging conditions and unknown roadways.

I. Introduction

A system able to automatically and reliably estimate the roadway and its lanes from a moving vehicle using on-board sensors such as cameras or laser range scanners (LIDAR) would have enormous benefits for land-based travel. It could be used for tasks ranging from wide-scale road and lane quality assessment, to providing inputs to a driver assistance safety or navigation system, to serving as a situational awareness component in a fully autonomous vehicle.

Aspects of the lane estimation problem have been studied for decades in the context of autonomous land vehicle development [6, 13] and driver-assistance technologies [3, 4, 1, 9]. The majority of lane estimation systems have been developed for lane keeping and lane departure warning (LDW) in highway environments [11]. The goal of these systems is to estimate the vehicle's current travel lane and, in the case of lane keeping systems, apply minor corrective steering when the vehicle detects an unsignaled lane departure.

LDW systems have achieved success by restricting the problem domain in numerous ways, such as assuming the vehicle begins in a lane and is oriented accordingly, estimating only the vehicle's current travel lane, and optimizing for well-marked highway environments with low road curvature and few, if any, inflection points. High failure rates are acceptable, as a human driver can choose to ignore the LDW output.

However, these simplifications also limit applicability of LDW systems to fully autonomous driving, which has a significantly lower margin for error, and requires more situational awareness than an estimate of the current travel lane. Autonomous vehicle operation on arbitrary road networks will require estimates of all nearby lane geometries, even under challenging environmental conditions.

II. Recent Work

Recent work on lane estimation has focused on operation in more varied and complex environments. Wang et al. used B-Splines to model roads with significant curvature [15], while still retaining the single lane and other LDW assumptions. Matsushita and Miura use a particle filter to detect certain types of intersections [10], in a step towards estimating road network topology. Our work in the DARPA Urban Challenge used a RANSAC-based method to estimate multiple travel lanes from sensor data [7]. The primary challenge to all of these approaches lies in determining which (noisy) sensor observations correspond to lane markings.

In previous work [8], we described the lateral uncertainty algorithm for estimating potential lane boundaries. The algorithm uses probabilistic methods to estimate individual curves from noisy observations, but does not address how to group curves to form lane estimates or track whole lanes over time. Neither does it distinguish true lane boundaries from long tree shadows or other non-boundary painted lines such as stop lines and pedestrian crossings.

This paper generalizes the lateral uncertainty method to perform joint inference of multiple curves (i.e., the left and right boundaries of a lane). It provides a principled framework for ignoring sensor observations that are similar to, but do not correspond to, lane boundaries, and for using observations of one curve to update estimates of another (e.g., when one lane boundary is faded or occluded by traffic). We formulate lane estimation as a curve estimation problem, describe a novel representation for open 2D curves, and present a Bayesian lane estimation algorithm that is robust to the noise and outliers typical of image and LIDAR data.

III. Method Overview

The geometry of a travel lane can be described by a single centerline curve $\mathbf{f}(s) : [s_1, s_n] \rightarrow \mathbb{R}^3$ whose width varies along the length of the curve:

$$\mathbf{f}(s) = (f_x(s), f_y(s), f_h(s))^\top \qquad (1)$$

where $f_x(s)$ and $f_y(s)$ describe the lane centerline, and $f_h(s)$ describes the lane half-width, all of which are defined on the domain $s \in [s_1, s_n]$. The interpretation of s depends on the curve model chosen, and may not have any physical meaning. We describe the curve in a coordinate system fixed with respect to the local environment, such that lane geometry is nearly invariant to time and the vehicle's own state.

Initially, the number of lanes and their geometries are unknown. The goal of our system is to detect all nearby lanes, estimate their geometries, and update these estimates as the vehicle travels. The system takes as input curve fragments corresponding to potential lane boundary observations, such as those detected by a vision-based road-paint detection algorithm, or a LIDAR-based curb-detection algorithm. While many of these curve fragments are true observations of the road and lanes, some may arise from tree shadows or lens flare, and others may simply be due to painted markings that do not correspond to lane boundaries (e.g., crosswalk stripes, parking spot delimiters, or stop lines).

We approximate $\mathbf{f}(s)$ as piecewise-linear, such that it can be described by a series of control points connected by line segments, and note that approximation error decreases with control point density. To reason about lanes, their boundaries, and observations, we introduce the notion of *basis curves*.

The insight behind this representation is that nearby curves often share common structure. Factoring out this structure leaves a set of residuals which are computationally convenient to manipulate. Within the space of these residuals, which we call the basis curve space, lane boundary observations can be expressed as linear transformations of the true lane geometry, and the Gaussian distribution can be used to model uncertainty over curve geometry. Standard estimation algorithms such as the Kalman filter can then be used to estimate complex lane geometries [2].

IV. Basis Curves

Consider a piecewise-linear curve $\mathbf{b}$, represented by the matrix of control points $\mathrm{B} = (\mathbf{b}_1, \mathbf{b}_2, \ldots, \mathbf{b}_n)^\top$. Denote the unit normal vectors to the curve at each of the control points[1] by the matrix $\bar{\mathrm{B}} = (\bar{\mathbf{b}}_1, \bar{\mathbf{b}}_2, \ldots, \bar{\mathbf{b}}_n)^\top$. We consider a curve $\mathbf{g}$, represented as $\mathrm{G} = (\mathbf{g}_1, \mathbf{g}_2, \ldots, \mathbf{g}_n)^\top$, to be a *variation* of $\mathbf{b}$ if each $\mathbf{g}_i, i \in \{1 \ldots n\}$, can be expressed as:

$$\mathbf{g}_i = \mathbf{b}_i + g_i^{\mathrm{B}} \bar{\mathbf{b}}_i \tag{2}$$

where each g_i^{B} is a signed scalar that indicates how much $\mathbf{g}$ varies from $\mathbf{b}$ along the normal vector at a particular control point. In this context, $\mathbf{b}$ is a *basis curve*, and the vector $\mathbf{g}^{\mathrm{B}} = (g_1^{\mathrm{B}}, g_2^{\mathrm{B}}, \ldots, g_n^{\mathrm{B}})^\top$ is the *offset vector* of $\mathbf{g}$ from $\mathbf{b}$.

Here we have defined $\mathbf{g}$ as a variation of the basis curve $\mathbf{b}$. Suppose $\mathbf{g}$ is instead expressed using a different curve representation, possibly a spline or even a polyline with a different number of control points. If $\mathbf{g}$ is geometrically similar to $\mathbf{b}$, then we can choose a polyline approximation of $\mathbf{g}$ as a variation of $\mathbf{b}$. When the matrix of control points G is defined in this manner, we call the resulting offset vector $\mathbf{g}^{\mathrm{B}}$ the *projection of* $\mathbf{g}$ *onto* $\mathbf{b}$ (Fig. 1).

[1] We use $\bar{\mathbf{b}}_i = \begin{bmatrix} 0 & -1 \\ 1 & 0 \end{bmatrix} \frac{\mathbf{b}_{i+1} - \mathbf{b}_i}{||\mathbf{b}_{i+1} - \mathbf{b}_i||}, i \in \{1 \ldots n-1\}$ and $\bar{\mathbf{b}}_n = \bar{\mathbf{b}}_{n-1}$.

(a) A basis curve (blue) represented by polyline B, and a uniform quadratic B-Spline $\mathbf{g}$.

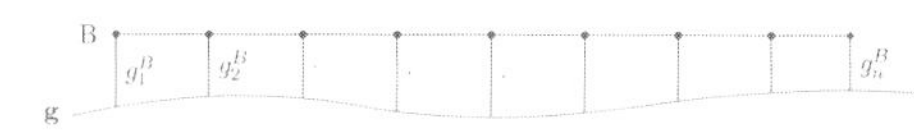

(b) The projection of $\mathbf{g}$ onto $\mathbf{b}$, producing the g_i^{B}.

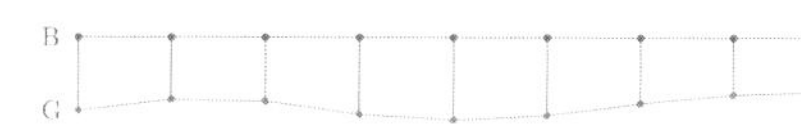

(c) Approximation of curve $\mathbf{g}$ as a variation G of the basis curve $\mathbf{b}$.

Fig. 1. A uniform quadratic B-Spline (magenta) defined by the green control points can be projected onto a basis curve (blue). Curves using other representations can be projected using analogous intersection procedures.

The error of this approximation (e.g., as measured by the area between the polyline approximation and the original curve) is small if $\mathbf{g}$ and $\mathbf{b}$ have similar geometry, and if the control point spacing of $\mathbf{b}$ is small relative to the curvatures of the two curves. If the curves are very different (e.g., nearly orthogonal or without overlap), it may not be meaningful to project one onto the other.

To review, thus far we have defined:

$\mathbf{b}$	A basis curve.
B	The $n \times 2$ matrix of control points representing $\mathbf{b}$.
$\mathbf{b}_i$	The i-th control point of B.
$\bar{\mathrm{B}}$	The $n \times 2$ matrix of unit normal vectors of $\mathbf{b}$ at each control point.
$\bar{\mathbf{b}}_i$	The i-th unit normal vector of $\bar{\mathrm{B}}$.
$\mathbf{g}$	A curve geometrically similar to $\mathbf{b}$.
G	The $n \times 2$ matrix of control points approximating or representing $\mathbf{g}$.
$\mathbf{g}_i$	The i-th control point of G.
$\mathbf{g}^{\mathrm{B}}$	The projection of $\mathbf{g}$ onto $\mathbf{b}$, also referred to as the offset of $\mathbf{g}$ from $\mathbf{b}$.

We can now use $\mathbf{b}$ to represent common structure shared by many curves, and offset vectors to reason about the differences among curves. One interpretation of this model is that $\mathbf{b}$ induces an n-dimensional vector space, into which we can project any curve. We call this the *basis curve space* of $\mathbf{b}$. When projected into a basis curve space, a curve can be represented by a single n-dimensional point – its offset vector.

In general, any curve that spans the length of $\mathbf{b}$ can be projected onto $\mathbf{b}$. It is also useful to reason about curves that do not run the length of $\mathbf{b}$, but do have some "longitudinal overlap" in that they intersect with one or more support lines of normal vectors of $\mathbf{b}$. In this case, each of the $\frac{n(n-1)}{2}$ consecutive subsets of the control points of $\mathbf{b}$ can be treated as an individual basis curve with its own basis curve space. These smaller basis curve spaces are subspaces of the basis curve space of $\mathbf{b}$ (Fig. 2).

(a) Variation of the first four control points of a basis curve.

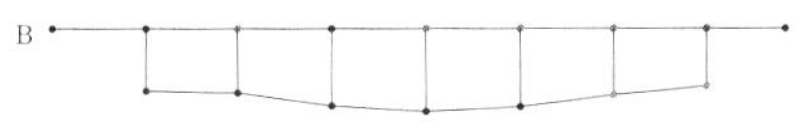

(b) Variation of a mid-section of a basis curve.

Fig. 2. Variations (magenta) of a subset of the basis curve (blue) can be described using subspaces of the vector space of the full basis curve.

A. *Basis curve normal distributions*

Next, we consider probability distributions on curves. Specifically, consider a random variation of a basis curve $\mathbf{b}$, where the offset vector $\mathbf{g}^{\mathrm{B}}$ is normally distributed according to $\mathbf{g}^{\mathrm{B}} \sim \mathcal{N}(\boldsymbol{\mu}, \Sigma)$. Fig. 3 shows a basis curve and several hundred variations, where the offset vectors of each variation are drawn from a multivariate Gaussian.

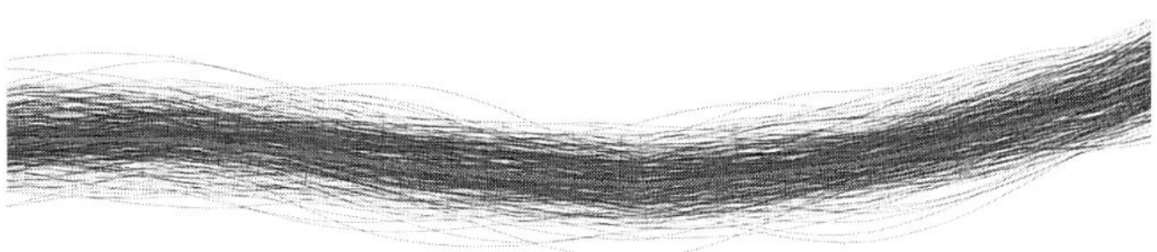

Fig. 3. Three hundred normally distributed variations of a basis curve (red).

Together, $\mathbf{b}$, $\boldsymbol{\mu}$, and Σ define a distribution over curves. We refer to a distribution of this form as a *basis curve normal distribution*, and represent it with the term $\tilde{\mathcal{N}}(\mathbf{b}, \boldsymbol{\mu}, \Sigma)$. The probability density of a curve $\mathbf{g}$ is then defined as:

$$P_{\mathbf{g}}(\mathbf{g}; \mathbf{b}, \boldsymbol{\mu}, \Sigma) = \mathcal{N}(\mathbf{g}^{\mathrm{B}}; \boldsymbol{\mu}, \Sigma) \tag{3}$$

where $\mathcal{N}(\mathbf{x}; \boldsymbol{\mu}, \Sigma)$ refers to the probability density function of the normal distribution with mean $\boldsymbol{\mu}$ and covariance Σ.

The intuition behind this formulation is that control point uncertainty is expressed only along the curve normals of the basis curve. Thus, each control point has one degree of freedom instead of two; even though G has $2n$ distinct components, it has only n degrees of freedom. The covariance matrix Σ represents the degree and structure of variation between curves drawn from the distribution.

If a distribution is instead defined on the entire control point space such that each control point has 2 degrees of freedom [5], it is possible to sample two curves that have the same shape, but different probability densities. Thus, distributions over the entire space of control points are not necessarily good distributions over curve shape.

Restricting each control point to one degree of freedom has useful properties. In our case, two random curves drawn from a basis curve normal distribution with different probability densities will also exhibit different shapes. Evaluating the likelihood that an arbitrary curve is drawn from a distribution becomes a simple process of computing curve and ray intersections. There are some exceptions to this that arise when the covariance at a control point is large relative to the curvature, but this has not had a noticeable impact in our usage for lane estimation.

B. *Changing basis curves*

It is sometimes desirable to switch the basis curve upon which a curve distribution is defined, but without changing the underlying distribution. In general, it is not possible to match the original distribution exactly, and the approximation error introduced by the reparameterization is directly related to the amount by which the basis curve normal vectors change: approximation error is smallest when the new basis curve is locally parallel to the original basis curve. However, if the new basis curve $\mathbf{b}'$ is similar to the original, then reasonable choices of a new mean and covariance, $\boldsymbol{\mu}'$ and Σ', are possible.

Variation: If $\mathbf{b}'$ is a variation of the original basis curve $\mathbf{b}$, then B' can be described by:

$$\mathrm{B}' = \mathrm{B} + \mathrm{diag}(\mathbf{b}'^{\mathrm{B}})\bar{\mathrm{B}} \tag{4}$$

where $\mathbf{b}'^{\mathrm{B}}$ is the projection of $\mathbf{b}'$ onto $\mathbf{b}$. In this case, a new mean can be defined by subtracting the projection of the new basis curve, and the covariance can remain unchanged:

$$\begin{aligned} \boldsymbol{\mu}' &= \boldsymbol{\mu} - \mathbf{b}'^{\mathrm{B}} \\ \Sigma' &= \Sigma \end{aligned} \tag{5}$$

Re-sampling: If $\mathbf{b}'$ is defined by a re-sampling of the control points of $\mathbf{b}$, then each of the m control points of $\mathbf{b}'$ is a convex linear combination of two neighboring control points on the original curve $\mathbf{b}$. Thus, B' can be related to B by:

$$\mathrm{B}' = \mathrm{HB} \tag{6}$$

where H is a $m \times n$ matrix whose rows have at most two positive entries that are adjacent and sum to unity (representing the convex linear combination of neighboring control points). Additionally, $\mathrm{H}_{i,j}$ may be non-zero only if $\mathrm{H}_{i-1,k}$ is zero for all $k > j$ (control points may not be re-sampled out of order).

Since every point on a polyline is either a control point or an interpolation of neighboring control points, we can define a new mean and covariance by applying the re-sampling transformation:

$$\begin{aligned} \boldsymbol{\mu}' &= \mathrm{H}\boldsymbol{\mu} \\ \Sigma' &= \mathrm{H}\Sigma\mathrm{H}^{\top} \end{aligned} \tag{7}$$

V. Lane Boundary Estimation

To detect and estimate lanes, our method first estimates potential lane boundaries, then groups boundary curves together to form lane estimates. Here, we develop a recursive algorithm that estimates a single unknown curve $\mathbf{f}$, which we use to model a potential lane boundary. The algorithm takes as input an initial curve estimate and a noisy observation of the curve, and produces an updated estimate that reflects the novel information provided by the observation.

We refer to our initial belief about $\mathbf{f}$ using the basis curve normal distribution $\tilde{\mathcal{N}}(\mathbf{b}, \boldsymbol{\mu}, \Sigma)$. The matrix of control points $\hat{\mathrm{F}} = (\hat{\mathbf{f}}_1, \hat{\mathbf{f}}_2, \ldots, \hat{\mathbf{f}}_n)^{\top}$ describes our mean estimate $\hat{\mathbf{f}}$, and is obtained from $\mathbf{b}$ and $\boldsymbol{\mu}$ as:

$$\hat{\mathbf{f}}_i = \mathbf{b}_i + \mu_i \bar{\mathbf{b}}_i \tag{8}$$

We use $\mathbf{f}^{\mathrm{B}} = (f_1^{\mathrm{B}}, f_2^{\mathrm{B}}, \ldots, f_n^{\mathrm{B}})^\top$ to denote the projection of $\mathbf{f}$ onto $\mathbf{b}$. Given a fixed basis curve $\mathbf{b}$, the vector $\boldsymbol{\mu}$ is effectively an estimate of $\mathbf{f}^{\mathrm{B}}$. In this sense, there are two estimates: the estimate $\boldsymbol{\mu}$ of $\mathbf{f}^{\mathrm{B}}$, and the estimate $\hat{\mathbf{f}}$ of $\mathbf{f}$, where $\hat{\mathbf{f}}$ is defined in terms of $\boldsymbol{\mu}$ and $\mathbf{b}$ as in Eq. (8).

A. Observations

We model an observation $\mathbf{z}$ of $\mathbf{f}$ as a subset of $\mathbf{f}$, corrupted by zero-mean i.i.d. noise. Note that $\mathbf{z}$ is typically a partial observation of $\mathbf{f}$, since only a portion of a painted road boundary may be observed at any given moment. For simplicity, we first consider the case where $\mathbf{z}$ is a full observation of $\mathbf{f}$, and later consider partial observations.

If we project $\mathbf{z}$ onto $\mathbf{b}$ and refer to the piecewise linear approximation of $\mathbf{z}$ as $\mathrm{Z} = (\mathbf{z}_1, \mathbf{z}_2, \ldots, \mathbf{z}_n)^\top$, then each $\mathbf{z}_i$ can be expressed as:

$$\begin{aligned} \mathbf{z}_i &= \mathbf{b}_i + (f_i^{\mathrm{B}} + v_i)\bar{\mathbf{b}}_i \\ &= \mathbf{b}_i + z_i^{\mathrm{B}}\bar{\mathbf{b}}_i \end{aligned} \tag{9}$$

where we model the noise terms $\mathbf{v} = (v_1, v_2, \ldots, v_n)^\top$ as jointly Gaussian with zero mean and covariance R, such that $\mathbf{v} \sim \mathcal{N}(\mathbf{0}, \mathrm{R})$. The projection $\mathbf{z}^{\mathrm{B}} = (z_1^{\mathrm{B}}, z_2^{\mathrm{B}}, \ldots, z_n^{\mathrm{B}})^\top$ of $\mathbf{z}$ onto $\mathbf{b}$ can then be described as:

$$z_i^{\mathrm{B}} = f_i^{\mathrm{B}} + v_i \tag{10}$$

and:

$$\mathbf{z}^{\mathrm{B}} = \mathrm{A}\mathbf{f}^{\mathrm{B}} + \mathbf{v} \tag{11}$$

where the observation matrix $\mathrm{A} = \mathrm{I}_{n \times n}$ is a trivial linear transformation. If $\mathbf{z}$ is a partial observation of $\mathbf{f}$, then it can still be related to $\mathbf{f}$ by choosing an observation matrix A of size $m \times n$, where $m \leq n$, and A has an $m \times m$ identity sub-matrix and is zero elsewhere. For example, if the basis curve has three control points and the observation curve spans only the first 2 of these control points, then the corresponding observation matrix $\mathrm{A}_{2\times 3}$ is:

$$\mathrm{A}_{2\times 3} = \begin{bmatrix} 1 & 0 & 0 \\ 0 & 1 & 0 \end{bmatrix} \tag{12}$$

B. Data association

Not all observations correspond to the desired curve $\mathbf{f}$. To avoid corrupting the curve estimate, only true observations should be incorporated, and non-observations ignored or used to update a different curve estimate. Determining if an observation corresponds to a tracked curve is the well-known data association problem [14].

We use a standard χ^2 outlier rejection approach. Define the scalar random variable y as the following Mahalanobis distance [2]:

$$y = (\mathbf{z}^{\mathrm{B}} - \mathrm{A}\boldsymbol{\mu})^\top (\mathrm{R} + \mathrm{A}\Sigma\mathrm{A}^\top)^{-1} (\mathbf{z}^{\mathrm{B}} - \mathrm{A}\boldsymbol{\mu}) \tag{13}$$

If $\mathbf{z}$ is an observation of the curve $\mathbf{f}$, then y obeys a χ^2 distribution with m degrees of freedom. Observations that fall in the extremities of this distribution are rejected as outliers.

When simultaneously estimating and tracking multiple curves, we apply a gated greedy matching procedure, in which each observation is associated with the curve that best "explains" that observation. Other techniques, such as joint-compatibility branch and bound [12], may also be used. If no tracked curve is likely to have generated the observation according to the current estimates, then the observation is used to initialize a new curve for tracking.

C. Curvature prediction

As the vehicle travels, it may receive observations that extend beyond the end of its current curve estimate. To incorporate these observations, the observation and curve estimate are first extended (Fig. 4b). We use a method described in previous work [8], briefly summarized here.

Once we have observed and estimated one portion of a curve, we can reliably predict the shape of the nearby unobserved parts of the curve. To do so, we fit a first-order Markov model of road curvature to a dataset containing the manually annotated coarse geometry of existing public roadways. Given the signed curvature at one point in a road, the model predicts a distribution over the curvature at a further point along the road. This model can then be used to extend both the curve estimates and observations as the vehicle travels.

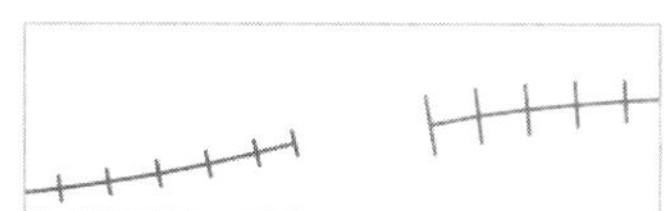

(a) Curve estimate (left) and a non-overlapping observation (right)

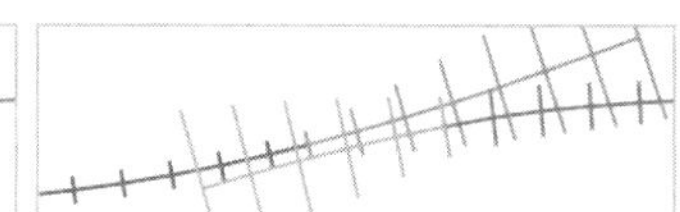

(b) Predicted extensions of both curves

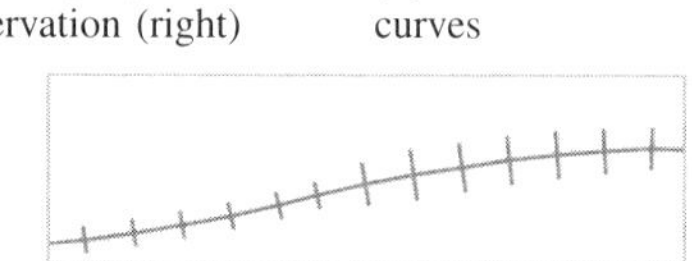

(c) Final curve estimate

Fig. 4. A curvature prediction model allows us to associate observations with existing curve estimates when there is little or no overlap. The short line segments perpendicular to each curve indicate 1-σ uncertainty.

D. Update

Once an observation $\mathbf{z}$ has been associated with an existing curve estimate, the algorithm revises that estimate using the update step of a standard Kalman filter [2].

As mentioned earlier, approximation errors resulting from projection onto a basis curve are minimized when the basis curve geometry matches the true curve geometry. Therefore, we reparameterize the curve estimate, so that the basis curve coincides with the newly updated maximum likelihood curve estimate. Since this estimate is a variation of the current basis curve, reparameterization consists of offsetting the basis curve by the mean vector, then setting the mean vector to zero (Sec. IV-B). We also re-sample the basis curve control points to maintain nearly uniform control point spacing.

Finally, we note that propagating the curve estimate forward through time consists of a simple identity transform, a consequence of expressing the curve geometry in a coordinate system fixed to the local environment. Algorithm `update_curve_estimate` shows a full update cycle. It takes as input the initial belief $\tilde{\mathcal{N}}(\mathbf{b}, \boldsymbol{\mu}, \Sigma)$ over a single curve, and an observation. If the observation passes the data association test, then it is used to generate the updated curve estimate $\tilde{\mathcal{N}}(\mathbf{b}^{+}, \boldsymbol{\mu}^{+}, \Sigma^{+})$. Fig. 4 illustrates an update step for a single curve estimate.

Algorithm `update_curve_estimate`

Input: an initial belief $\tilde{\mathcal{N}}(\mathbf{b}, \boldsymbol{\mu}, \Sigma)$
Input: an observation $\tilde{\mathcal{N}}(\mathbf{b}, \mathbf{z}^{\mathrm{B}}, \mathrm{R})$
Input: an observation matrix A
Output: the new belief $\tilde{\mathcal{N}}(\mathbf{b}^{+}, \boldsymbol{\mu}^{+}, \Sigma^{+})$

Augment the initial estimate and observation as necessary via curve prediction
Let m be the dimensionality of $\mathbf{z}^{\mathrm{B}}$
Let k be the 0.95 percentile of the χ^2_m distribution
if $(\mathbf{z}^{\mathrm{B}} - \mathrm{A}\boldsymbol{\mu})^{\top}(\mathrm{R} + \mathrm{A}\Sigma\mathrm{A}^{\top})^{-1}(\mathbf{z}^{\mathrm{B}} - \mathrm{A}\boldsymbol{\mu}) > k$ **then**
  Reject observation as an outlier
  return $\tilde{\mathcal{N}}(\mathbf{b}, \boldsymbol{\mu}, \Sigma)$
$\mathrm{K} \leftarrow \Sigma\mathrm{A}^{\top}(\mathrm{A}\Sigma\mathrm{A}^{\top} + \mathrm{R})^{-1}$
$\tilde{\boldsymbol{\mu}} \leftarrow \boldsymbol{\mu} + \mathrm{K}(\mathbf{z}^{\mathrm{B}} - \mathrm{A}\boldsymbol{\mu})$
$\tilde{\Sigma} \leftarrow (\mathrm{I} - \mathrm{KA})\Sigma$
$\mathbf{b}^{+} \leftarrow \mathrm{B} + \mathrm{diag}(\tilde{\boldsymbol{\mu}})\bar{\mathrm{B}}$
$\boldsymbol{\mu}^{+} \leftarrow \mathbf{0}$
$\Sigma^{+} \leftarrow \tilde{\Sigma}$
return $\tilde{\mathcal{N}}(\mathbf{b}^{+}, \boldsymbol{\mu}^{+}, \Sigma^{+})$

VI. Lane Estimation

The boundary curves of a single lane are highly correlated, and information about one boundary gives valuable information about the other. We represent a lane as a piecewise linear centerline curve whose width varies along the curve, and describe its control points with the matrix $\mathrm{F} = (\mathbf{f}_1, \mathbf{f}_2, \ldots, \mathbf{f}_n)^{\top}$, where each $\mathbf{f}_i$ is defined as $\mathbf{f}_i = (f_{xi}, f_{yi}, f_{hi})^{\top}$.

Using the convention that the normal vectors of a curve point "left", two points $\mathbf{f}_{li}$ and $\mathbf{f}_{ri}$ on the left and right boundaries, respectively, can be described as:

$$\mathbf{f}_{li} = \begin{bmatrix} f_{xi} + f_{hi}\bar{f}_{xi} \\ f_{yi} + f_{hi}\bar{f}_{yi} \end{bmatrix} \qquad \mathbf{f}_{ri} = \begin{bmatrix} f_{xi} - f_{hi}\bar{f}_{xi} \\ f_{yi} - f_{hi}\bar{f}_{yi} \end{bmatrix} \tag{14}$$

where $\bar{\mathbf{f}}_i = (\bar{f}_{xi}, \bar{f}_{yi})^{\top}$ is the normal vector to the centerline curve at point i.

A. Lane distributions

As with zero-width curves, a basis curve can be used to represent and approximate lanes. We describe the projection $\mathbf{f}^{\mathrm{B}}$ of $\mathbf{f}$ onto $\mathbf{b}$ as:

$$\mathbf{f}^{\mathrm{B}} = (f^{\mathrm{B}}_{c1}, f_{h1}, f^{\mathrm{B}}_{c2}, f_{h2}, \ldots, f^{\mathrm{B}}_{cn}, f_{hn})^{\top} \tag{15}$$

where $f^{\mathrm{B}}_{c1}, \ldots, f^{\mathrm{B}}_{cn}$ is the projection of the centerline of $\mathbf{f}$ onto $\mathbf{b}$. Thus, the projection of a lane onto $\mathbf{b}$ is simply the projection of its centerline augmented by its half-width terms.

A normal distribution over the projections of $\mathbf{f}$ onto $\mathbf{b}$ defines a distribution over lanes. We use such a distribution, parameterized by $\boldsymbol{\mu}$ and Σ, to represent a belief over the true geometry of $\mathbf{f}$. The mean estimate can be represented by a matrix of control points $\hat{\mathrm{F}} = (\hat{\mathbf{f}}_1, \hat{\mathbf{f}}_2, \ldots, \hat{\mathbf{f}}_n)^{\top}$, where each control point $\hat{\mathbf{f}}_i = (\hat{f}_{xi}, \hat{f}_{yi}, \hat{f}_{hi})^{\top}$ can be expressed as:

$$\hat{\mathbf{f}}_i = \begin{bmatrix} \hat{f}_{xi} \\ \hat{f}_{yi} \\ \hat{f}_{hi} \end{bmatrix} = \begin{bmatrix} b_{xi} + \mu_{ci}\bar{b}_{xi} \\ b_{yi} + \mu_{ci}\bar{b}_{yi} \\ \mu_{hi} \end{bmatrix} \tag{16}$$

where each μ_{ci} describes the mean offset of a centerline control point, and each μ_{hi} describes the mean half-width of the lane estimate at the control point.

It is sometimes useful to change the basis curve upon which a lane distribution has been defined, while incurring minimal changes to the actual distribution. Choosing a new mean and covariance in the cases of re-sampled and offset basis curves follows the same procedure as in Sec. IV-B, with minor modifications. When the new basis curve is a variation of the original basis curve, the width components of the mean lane do not change. When the new basis curve is a re-sampling of the original basis curve, the re-sampling matrix H must account for re-sampling the width components in addition to the centerline offset values.

B. Observation model, data association, and update

A full boundary observation of $\mathbf{f}$ is a curve, which we describe with the matrix of control points $\mathrm{Z} = (\mathbf{z}_1, \mathbf{z}_2, \ldots, \mathbf{z}_n)^{\top}$, where each $\mathbf{z}_i$ can be written:

$$\begin{aligned} \mathbf{z}_i &= \mathbf{b}_i + (f^{\mathrm{B}}_{ci} + af_{hi} + v_i)\bar{\mathbf{b}}_i \\ &= \mathbf{b}_i + z^{\mathrm{B}}_i\bar{\mathbf{b}}_i \end{aligned} \tag{17}$$

where a has value $+1$ or -1 for an observation of the left or right boundary, respectively, and we model the noise terms $\mathbf{v} = (v_1, v_2, \ldots, v_n)^{\top}$ jointly as $\mathbf{v} \sim \mathcal{N}(\mathbf{0}, \mathrm{R})$.

Collectively, the offset vector $\mathbf{z}^{\mathrm{B}} = (z^{\mathrm{B}}_1, z^{\mathrm{B}}_2, \ldots, z^{\mathrm{B}}_n)^{\top}$ can then be expressed as:

$$\mathbf{z}^{\mathrm{B}} = \mathrm{A}\mathbf{f}^{\mathrm{B}} + \mathbf{v} \tag{18}$$

where the elements of the observation matrix A are chosen to satisfy Eq. (17). If $\mathbf{z}$ is a partial observation of the boundary, such that it projects onto only m control points of $\mathbf{b}$, then A has size $2m \times 2n$, similar to the case for zero-width curves.

Data association and update steps can be approached in the same way as for zero-width curves. Given a lane distribution and observation as expressed above, we can apply a χ^2 test to determine if $\mathbf{z}$ is an observation of $\mathbf{f}$. When estimating multiple lanes, we use a gated greedy assignment procedure to assign observations to lanes. Once an observation has been associated with a lane estimate, the standard Kalman update steps are used to update the mean and covariance.

After the updated estimates have been computed, we once again reparameterize the distribution such that the basis curve coincides with the updated maximum likelihood estimate, to minimize approximation error in future update steps. Fig. 5 shows a full update cycle, where an observation of a lane boundary is used to both update and extend the lane.

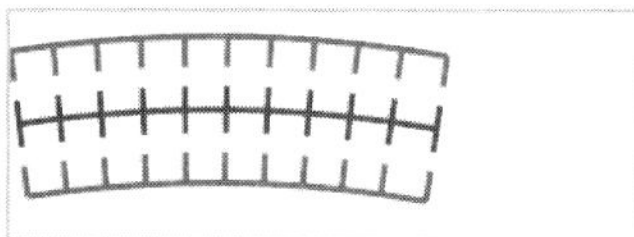

(a) A lane estimate

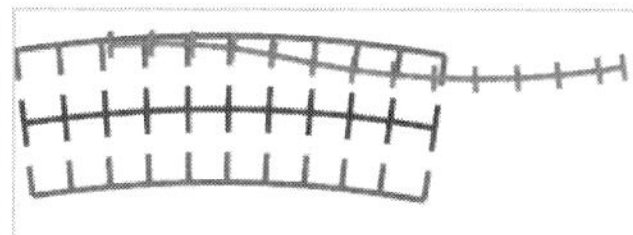

(b) A boundary observation

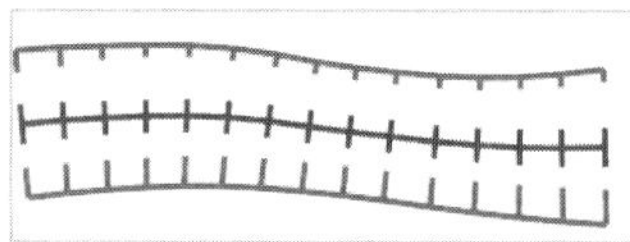

(c) The updated estimate

Fig. 5. A boundary observation is used to update the lane estimate. The middle (black) curve marks the lane centerline, and the outer (blue) curves mark the left and right boundary marginal distributions. Short line segments along the curves mark control points, and the length of these segments indicate 1-σ uncertainty. Note that both boundaries are updated even though only one boundary is observed.

C. Initial estimate

We initialize a lane estimate by independently estimating many zero-width curves as described in Sec. V, while periodically searching for curve pairs that are sufficiently long, parallel, and separated by an appropriate distance. Once a suitable pair of boundary curves is identified, they are used to initialize a lane estimate. The initial lane basis curve $\mathbf{b}$ is chosen by projecting one boundary curve onto the other and scaling the offset vector by 0.5 (an approximation of the medial axis), and both curve estimates are then reparameterized with $\mathbf{b}$. Referring to the left and right curve estimates as $\tilde{\mathcal{N}}(\mathbf{b}, \boldsymbol{\mu}_l, \Sigma_l)$ and $\tilde{\mathcal{N}}(\mathbf{b}, \boldsymbol{\mu}_r, \Sigma_r)$, we treat them as independent observations of the same lane, and express them jointly as:

$$\mathbf{z} = \begin{bmatrix} \boldsymbol{\mu}_l \\ \boldsymbol{\mu}_r \end{bmatrix} = \begin{bmatrix} \mathrm{A}_l \\ \mathrm{A}_r \end{bmatrix} \mathbf{f}^{\mathrm{B}} + \mathbf{v} = \mathrm{A}_z \mathbf{f}^{\mathrm{B}} + \mathbf{v} \quad (19)$$

where A_l and A_r are the transformation matrices relating a lane to its left and right boundary observations (Sec. VI-B), $\mathbf{f}^{\mathrm{B}}$ is the projection of the unobserved true lane onto $\mathbf{b}$, and $\mathbf{v} \sim \mathcal{N}(\mathbf{0}, \Sigma_z)$ is a noise term described by:

$$\Sigma_z = \begin{bmatrix} \Sigma_l & 0 \\ 0 & \Sigma_r \end{bmatrix} \quad (20)$$

Using the information filter [14], we can see that the initial distribution parameters best representing the information provided by the boundary curves can be expressed as:

$$\begin{aligned} \Sigma_0 &= (\mathrm{A}_z^\top \Sigma_z^{-1} \mathrm{A}_z)^{-1} \\ \boldsymbol{\mu}_0 &= \Sigma \mathrm{A}_z^\top \Sigma_z^{-1} \mathbf{z} \end{aligned} \quad (21)$$

VII. Experiments

To quantitatively assess the performance of our system, we evaluated it against ground truth across two datasets containing data from a forward-facing wide-angle camera (Point Grey Firefly MV, 752x480 @22.8 Hz), and a Velodyne HDL-64E laser range scanner. As input to our system, we used vision- and LIDAR-based road paint and curb detection algorithms described in previous work [7].

The first dataset consists of 30.2 km of travel in 182 minutes, and can be characterized by wide suburban lanes, no pedestrians, generally well-marked roads, few vehicles, and a bright early morning sun. The vehicle also traverses a short 0.4 km dirt road and a 1.7 km stretch of highway. The second dataset consists of 13.6 km of travel in 58 minutes through a densely populated city during afternoon rush hour. This dataset can be characterized by roads of varying quality, large numbers of parked and moving vehicles, and many pedestrians.

To produce ground truth, we annotated high-resolution geo-registered ortho-rectified aerial imagery with lane geometry. The vehicle's GPS estimates during a data collection provide an initial guess as to the vehicle's pose; these were corrected by manually aligning sensor data (i.e., camera and LIDAR data) with the aerial imagery at various points in the data collection. The result is a dataset containing ground truth lane geometry relative to the vehicle at every moment of travel. We emphasize that our algorithm uses only local sensor data – GPS and the ground truth map were used only for evaluation purposes.

We compare the results of our algorithm, which we refer to as the basis curve (BasCurv) algorithm, with our previous work in the DARPA Urban Challenge [7], which we refer to as the evidence image (EvImg) algorithm. The evidence image algorithm can be used as a standalone lane estimation system by using the output of the first of its two stages, which performs lane detection from sensor data only. Both algorithms use the same features as input.

For computational speed, our implementation of the basis curve algorithm used diagonal covariance matrices when estimating lane boundaries, and block-diagonal covariance matrices (2×2 blocks) for lane estimation. This introduces additional approximation errors, but yielded good performance in our experiments. After each observation update, basis curves are re-sampled to maintain a uniform (1 m) control point spacing. Parameters such as covariances and data association thresholds were determined experimentally. The basis curve algorithm was implemented in Java and runs at real-time speeds.

A. Centerline error

The centerline error of a lane estimate at a given point on the estimate is defined as the shortest distance from the estimated lane centerline point to the true centerline of the nearest lane. Fig. 6 shows the 50th and 90th percentile values for the centerline error of the two algorithms as a function of distance from the vehicle. The basis curve algorithm has significantly lower error at all distances.

Fig. 7 shows the centerline error as a function of true lane width. The evidence image algorithm assumes a fixed lane width of 3.66 m, and its performance degrades as the true lane width departs from this assumed value. Since the basis curve algorithm jointly estimates lane width and centerline geometry, it remains relatively invariant to changes in lane width.

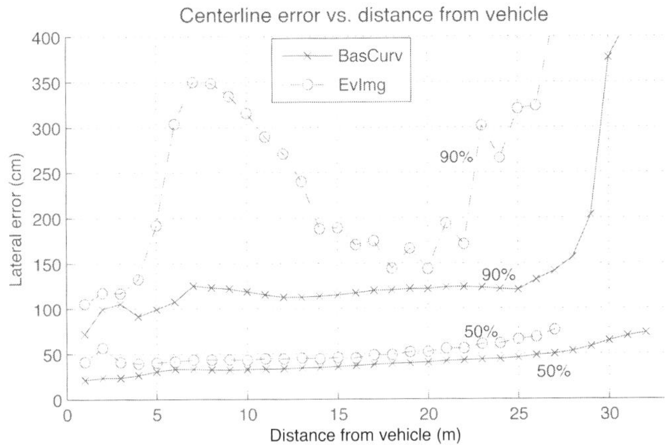

Fig. 6. The 50 and 90 percentile values for centerline lateral error, as a function of increasing distance from the vehicle.

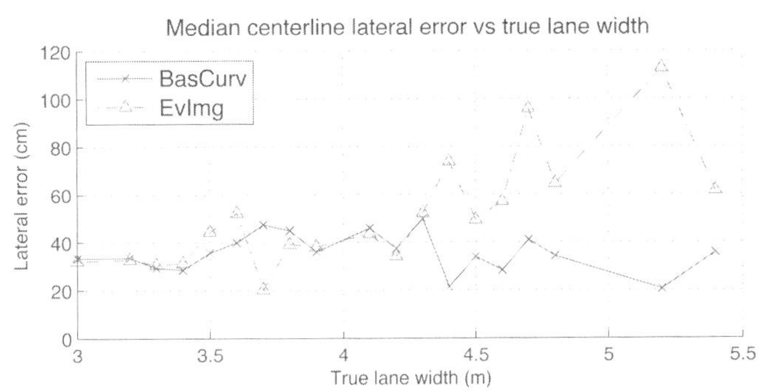

Fig. 7. Median centerline lateral error as a function of true lane width.

B. Lookahead distance and time

The *lookahead distance* and *lookahead time* metrics measure how much farther the vehicle can travel before reaching the end of its current lane estimate, assuming constant speed. Lookahead distance is computed by measuring the distance from the vehicle to the farthest point ahead on the current lane estimate, and lookahead time is computed by dividing the lookahead distance by the vehicle's instantaneous speed.

Fig. 8 aggregates lookahead statistics over both datasets, and shows the lookahead distance and lookahead time cumulative distributions for the two algorithms. In all cases, the basis curve algorithm outperforms the evidence image algorithm. For example, the basis curve algorithm provided some lane estimate forward of the vehicle for 71% of distance traveled, compared to 36% for the evidence image algorithm.

C. Qualitative results

Fig. 9 shows the output of the basis curve lane estimation algorithm in a variety of challenging scenarios. In (a) and (b), tree shadows and an erroneous curb detection are detected and rejected as outliers, leaving the lane estimates intact. In (b), the median strip separating opposite lanes of traffic is correctly excluded from the lane estimates. In (c) and (d), correctly detected road paint is successfully excluded from

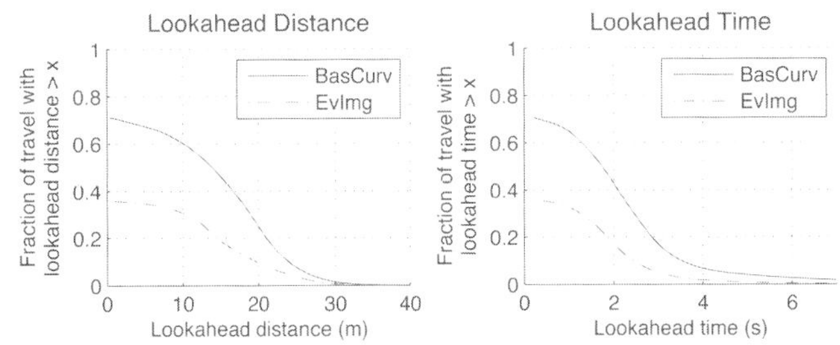

Fig. 8. Lane estimate lookahead distance (left) and time (right) CDFs.

the lane estimates. In (e-g), our method performs well despite adverse lighting scenarios and a snow-covered environment.

VIII. Discussion

Overall, the basis curve algorithm provides lane estimates of equal or better accuracy to those produced by the evidence image approach, and does so more often and with a greater lookahead. We attribute this to the data association and outlier rejection properties of the basis curve algorithm, and to the joint estimation of lane width and centerline geometry. The evidence image algorithm performs no outlier rejection, attempting to fit lanes to both true lane boundary detections and false detections such as tree shadows and non-boundary road paint; nor does it estimate lane width.

We have formulated the lane estimation problem in such a way that standard estimation and tracking algorithms based on the Kalman filter can be used for complex lane geometries. In doing so, we gain the advantages of the Kalman filter, and also invite all of its shortcomings. Cases where outliers appear very similar to inliers, such as long shadows nearly parallel to the road, can cause the lane estimates to diverge. Another failure mode arises when one road marking appears initially to be the correct boundary, but the true lane boundary then comes into view. In this case, our method will converge upon the first marking as the lane boundary, since the unimodal nature of the Kalman filter will prevent it from assigning substantial weight to the true boundary.

These difficulties are similar to those studied in other estimation domains, and it should also be possible to apply lessons learned in those domains to lane estimation with basis curves. One approach is particle filtering, which has been successfully applied in many estimation and tracking problems to model complex distributions and enable multi-hypothesis tracking. However, the high dimensionality of the lane estimates would require careful treatment.

Finally, in using a Gaussian noise model, we are simplifying and approximating the true system errors. This has proved successful in practice, although more careful study is required to understand the extent to which our simplifications and approximations are valid, and when other models may be more appropriate.

IX. Conclusion

This paper introduced the notion of basis curves for curve estimation, and described an application to the lane estimation

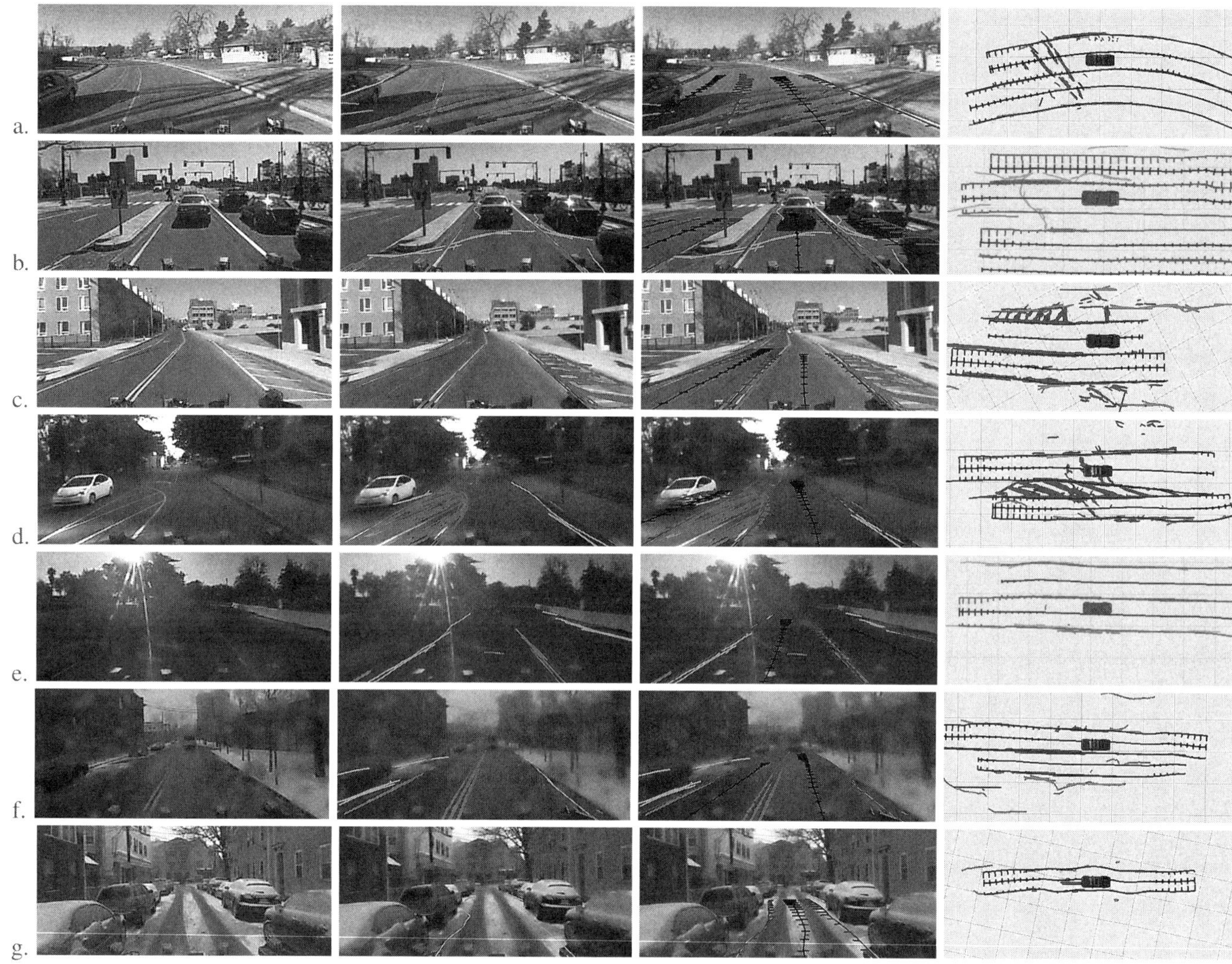

Fig. 9. Lane estimation in a variety of environments. *Column 1:* Camera images. *Column 2:* Detections of road paint (magenta) and curbs (green). *Column 3:* Lane centerline estimates (black) and boundary curve estimates (blue) projected into the image. *Column 4:* Synthesized overhead view of lane estimates.

problem. A detailed evaluation of our method's performance on a real-world dataset, and a quantitative comparison against ground truth and a previous approach, shows distinct advantages of the basis curve algorithm, particularly for estimating lanes using partial observations, for handling noisy data with high false-positive rates, and for jointly estimating centerline geometry and lane width.

REFERENCES

[1] N. Apostoloff and A. Zelinsky. Vision in and out of vehicles: Integrated driver and road scene monitoring. *Int. Journal of Robotics Research*, 23(4-5):513–538, Apr. 2004.

[2] Y. Bar-Shalom and X.-R. Li. *Estimation with Applications to Tracking and Navigation*. John Wiley & Sons, Inc., 2001.

[3] M. Bertozzi and A. Broggi. GOLD: a parallel real-time stereo vision system for generic obstacle and lane detection. *IEEE Transactions on Image Processing*, 7(1):62–80, Jan. 1998.

[4] M. Bertozzi, A. Broggi, and A. Fascioli. Vision-based intelligent vehicles: State of the art and perspectives. *Robotics and Autonomous Systems*, 1:1–16, 2000.

[5] A. Blake and M. Isard. *Active Contours*. Springer-Verlag, 1998.

[6] E. Dickmanns and B. Mysliwetz. Recursive 3-D road and ego-state recognition. *IEEE Trans. Pattern Analysis and Machine Intelligence*, 14(2):199–213, Feb. 1992.

[7] A. S. Huang, D. Moore, M. Antone, E. Olson, and S. Teller. Finding multiple lanes in urban road networks with vision and lidar. *Autonomous Robots*, 26(2-3):103–122, Apr. 2009.

[8] A. S. Huang and S. Teller. Lane boundary and curb estimation with lateral uncertainties. In *Proc. IEEE Int. Conf. on Intelligent Robots and Systems*, St. Louis, Missouri, Oct. 2009.

[9] Z. Kim. Robust lane detection and tracking in challenging scenarios. *IEEE Trans. Intelligent Transportation Systems*, 9(1):16–26, Mar. 2008.

[10] Y. Matsushita and J. Miura. On-line road boundary modeling with multiple sensory features, flexible road model, and particle filter. In *Proc. European Conference on Mobile Robots*, Sep. 2009.

[11] J. C. McCall and M. M. Trivedi. Video-based lane estimation and tracking for driver assistance: Survey, system, and evaluation. *IEEE Transactions on Intelligent Transport Systems*, 7(1):20– 37, Mar. 2006.

[12] J. Neira and J. D. Tardos. Data association in stochastic mapping using the joint compatibility test. *IEEE Trans. Robotics and Automation*, 17(6):890–897, Dec 2001.

[13] C. Thorpe, M. Hebert, T. Kanade, and S. Shafer. Vision and navigation for the Carnegie-Mellon Navlab. *IEEE Transactions on Pattern Analysis and Machine Intelligence*, 10(3):362–373, May 1988.

[14] S. Thrun, W. Burgard, and D. Fox. *Probabilistic Robotics*. MIT Press, 2005.

[15] Y. Wang, E. K. Teoh, and D. Shen. Lane detection and tracking using B-Snake. *Image and Vision Computing*, 22(4):269 – 280, 2004.

Reinforcement Learning to Adjust Robot Movements to New Situations

Jens Kober
MPI for Biol. Cybernetics, Germany
Email: jens.kober@tuebingen.mpg.de

Erhan Oztop
ATR Comput. Neuroscience Labs, Japan
Email: erhan@atr.jp

Jan Peters
MPI for Biol. Cybernetics, Germany
Email: jan.peters@tuebingen.mpg.de

Abstract—**Many complex robot motor skills can be represented using elementary movements, and there exist efficient techniques for learning parametrized motor plans using demonstrations and self-improvement. However, in many cases, the robot currently needs to learn a new elementary movement even if a parametrized motor plan exists that covers a similar, related situation. Clearly, a method is needed that modulates the elementary movement through the meta-parameters of its representation. In this paper, we show how to learn such mappings from circumstances to meta-parameters using reinforcement learning. We introduce an appropriate reinforcement learning algorithm based on a kernelized version of the reward-weighted regression. We compare this algorithm to several previous methods on a toy example and show that it performs well in comparison to standard algorithms. Subsequently, we show two robot applications of the presented setup; i.e., the generalization of throwing movements in darts, and of hitting movements in table tennis. We show that both tasks can be learned successfully using simulated and real robots.**

I. Introduction

In robot learning, motor primitives based on dynamical systems [1], [2] allow acquiring new behaviors quickly and reliably both by imitation and reinforcement learning. Resulting successes have shown that it is possible to rapidly learn motor primitives for complex behaviors such as tennis-like swings [1], T-ball batting [3], drumming [4], biped locomotion [5], ball-in-a-cup [6], and even in tasks with potential industrial applications [7]. The dynamical system motor primitives [1] can be adapted both spatially and temporally without changing the overall shape of the motion. While the examples are impressive, they do not address how a motor primitive can be generalized to a different behavior by trial and error without re-learning the task. For example, if the string length has been changed in a ball-in-a-cup [6] movement[1], the behavior has to be re-learned by modifying the movements parameters. Given that the behavior will not drastically change due to a string length variation of a few centimeters, it would be better to generalize that learned behavior to the modified task. Such generalization of behaviors can be achieved by adapting the meta-parameters of the movement representation[2].

In machine learning, there have been many attempts to use meta-parameters in order to generalize between tasks [8].

[1]In this movement, the system has to jerk a ball into a cup where the ball is connected to the bottom of the cup with a string.

[2]Note that the tennis-like swings [1] could only hit a static ball at the end of their trajectory, and T-ball batting [3] was accomplished by changing the policy's parameters.

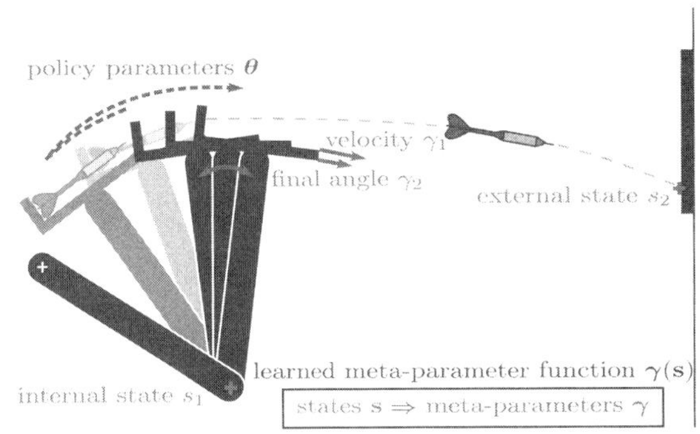

Figure 1: This figure illustrates a 2D dart throwing task. The situation, described by the state s corresponds to the relative height. The meta-parameters γ are the velocity and the angle at which the dart leaves the launcher. The policy parameters represent the backward motion and the movement on the arc. The meta-parameter function $\gamma(\mathrm{s})$, which maps the state to the meta-parameters, is learned.

Particularly, in grid-world domains, significant speed-up could be achieved by adjusting policies by modifying their meta-parameters (e.g., re-using options with different subgoals) [9]. In robotics, such meta-parameter learning could be particularly helpful due to the complexity of reinforcement learning for complex motor skills with high dimensional states and actions. The cost of experience is high as sample generation is time consuming and often requires human interaction (e.g., in cart-pole, for placing the pole back on the robots hand) or supervision (e.g., for safety during the execution of the trial). Generalizing a teacher's demonstration or a previously learned policy to new situations may reduce both the complexity of the task and the number of required samples. For example, the overall shape of table tennis forehands are very similar when the swing is adapted to varied trajectories of the incoming ball and a different targets on the opponent's court. Here, the human player has learned by trial and error how he has to adapt the global parameters of a generic strike to various situations [10]. Hence, a reinforcement learning method for acquiring and refining meta-parameters of pre-structured primitive movements becomes an essential next step, which we will address in this paper.

We present current work on automatic meta-parameter acquisition for motor primitives by reinforcement learning. We focus on learning the mapping from situations to meta-parameters and how to employ these in dynamical systems motor primitives. We extend the motor primitives of [1] with a learned meta-parameter function and re-frame the problem as an episodic reinforcement learning scenario. In order to obtain an algorithm for fast reinforcement learning of meta-parameters, we view reinforcement learning as a reward-weighted self-imitation [11], [6].

As it may be hard to realize a parametrized representation for meta-parameter determination, we reformulate the reward-weighted regression [11] in order to obtain a Cost-regularized Kernel Regression (CrKR) that is related to Gaussian process regression [12]. We compare the Cost-regularized Kernel Regression with a traditional policy gradient algorithm [3] and the reward-weighted regression [11] on a toy problem in order to show that it outperforms available previously developed approaches. As complex motor control scenarios, we evaluate the algorithm in the acquisition of flexible motor primitives for dart games such as *Around the Clock* [13] and for *table tennis*.

II. Meta-Parameter Learning for Motor Primitives

The goal of this paper is to show that elementary movements can be generalized by modifying only the meta-parameters of the primitives using learned mappings. In Section II-A, we first review how a single primitive movement can be represented and learned. We discuss how such meta-parameters may be able to adapt the motor primitive spatially and temporally to the new situation. In order to develop algorithms that learn to automatically adjust such motor primitives, we model meta-parameter self-improvement as an episodic reinforcement learning problem in Section II-B. While this problem could in theory be treated with arbitrary reinforcement learning methods, the availability of few samples suggests that more efficient, task appropriate reinforcement learning approaches are needed. To avoid the limitations of parametric function approximation, we aim for a kernel-based approach. When a movement is generalized, new parameter settings need to be explored. Hence, a predictive distribution over the meta-parameters is required to serve as an exploratory policy. These requirements lead to the method which we derive in Section II-C and employ for meta-parameter learning in Section II-D.

A. Motor Primitives with Meta-Parameters

In this section, we review how the dynamical systems motor primitives [1], [2] can be used for meta-parameter learning. The dynamical system motor primitives [1] are a powerful movement representation that allows ensuring the stability of the movement, choosing between a rhythmic and a discrete movement and is invariant under rescaling of both duration and movement amplitude. These modification parameters can become part of the meta-parameters of the movement.

In this paper, we focus on single stroke movements which appear frequently in human motor control [14], [2]. Therefore, we will always focus on the discrete version of the dynamical system motor primitives in this paper (however, the results may generalize well to rhythmic motor primitives and hybrid settings). We use the most recent formulation of the discrete dynamical systems motor primitives [2] where the phase z of the movement is represented by a single first order system

$$\dot{z} = -\tau \alpha_z z. \quad (1)$$

This canonical system has the time constant $\tau = 1/T$ where T is the duration of the motor primitive and a parameter α_z, which is chosen such that $z \approx 0$ at T. Subsequently, the internal state $\mathbf{x}$ of a second system is chosen such that positions $\mathbf{q}$ of all degrees of freedom are given by $\mathbf{q} = \mathbf{x}_1$, the velocities by $\dot{\mathbf{q}} = \tau \mathbf{x}_2 = \dot{\mathbf{x}}_1$ and the accelerations by $\ddot{\mathbf{q}} = \tau \dot{\mathbf{x}}_2$. The learned dynamics of Ijspeert motor primitives can be expressed in the following form

$$\begin{aligned} \dot{\mathbf{x}}_2 &= \tau \alpha_x \left(\beta_x \left(\mathbf{g} - \mathbf{x}_1\right) - \mathbf{x}_2\right) + \tau \mathbf{A} \mathbf{f}\left(z\right), \quad (2) \\ \dot{\mathbf{x}}_1 &= \tau \mathbf{x}_2. \end{aligned}$$

This set of differential equations has the same time constant τ as the canonical system and parameters α_x, β_x are set such that the system is critically damped. The goal parameter $\mathbf{g}$, a transformation function $\mathbf{f}$ and an amplitude matrix $\mathbf{A} = \mathrm{diag}\left(a_1, a_2, \ldots, a_I\right)$, with the amplitude modifier $\mathbf{a} = \left[a_1, a_2, \ldots, a_I\right]$ allow representing complex movements. In [2], the authors use $\mathbf{a} = \mathbf{g} - \mathbf{x}_1^0$, with the initial position $\mathbf{x}_1^0$, which ensures linear scaling. Other choices are possibly better suited for specific tasks, see for example [15]. The transformation function $\mathbf{f}\left(z\right)$ alters the output of the first system, in Equation (1), so that the second system in Equation (2), can represent complex nonlinear patterns and is given by

$$\mathbf{f}\left(z\right) = \textstyle\sum_{n=1}^{N} \psi_n\left(z\right) \boldsymbol{\theta}_n z. \quad (3)$$

Here, $\boldsymbol{\theta}_n$ contains the nth adjustable parameter of all degrees of freedom, N is the number of parameters per degree of freedom, and $\psi_n(z)$ are the corresponding weighting functions [2]. Normalized Gaussian kernels are used as weighting functions given by

$$\psi_n = \frac{\exp\left(-h_n\left(z - c_n\right)^2\right)}{\sum_{m=1}^{N} \exp\left(-h_m\left(z - c_m\right)^2\right)}. \quad (4)$$

These weighting functions localize the interaction in phase space using the centers c_n and widths h_n. As $z \approx 0$ at T, the influence of the transformation function $\mathbf{f}\left(z\right)$ in Equation (3) vanishes and the system stays at the goal position $\mathbf{g}$. Note that the degrees of freedom (DoF) are usually all modeled independently in the second system in Equation (2). All DoFs are synchronous as the dynamical systems for all DoFs start at the same time, have the same duration and the shape of the movement is generated using the transformation $\mathbf{f}\left(z\right)$ in Equation (3), which is learned as a function of the shared canonical system in Equation (1).

One of the biggest advantages of this motor primitive framework [1], [2] is that the second system in Equation (2), is linear in the shape parameters $\boldsymbol{\theta}$. Therefore, these parameters can be obtained efficiently, and the resulting framework is well-suited for imitation [1] and reinforcement learning [6]. The resulting policy is invariant under transformations of the initial position $\mathbf{x}_1^0$, the goal g, the amplitude $\mathbf{A}$ and the duration T [1]. These four modification parameters can be used as the meta-parameters γ of the movement. Obviously, we can make more use of the motor primitive framework by adjusting the meta-parameters γ depending on the current situation or state s according to a meta-parameter function $\gamma(\mathbf{s})$. The state s can for example contain the current position, velocity and acceleration of the robot and external objects, as well as the target to be achieved. This paper focuses on learning the meta-parameter function $\gamma(\mathbf{s})$ by episodic reinforcement learning.

Illustration of the Learning Problem: As an illustration of the meta-parameter learning problem, we take a 2D dart throwing task with a dart on a launcher which is illustrated in Figure 1 (in Section III-B, we will expand this example to a robot application). Here, the desired skill is to hit a specified point on a wall with a dart. The dart is placed on the launcher and held there by friction. The motor primitive corresponds to the throwing of the dart. When modeling a single dart's movement with dynamical-systems motor primitives [1], the combination of retracting and throwing motions would be represented by one movement primitive and can be learned by determining the movement parameters $\boldsymbol{\theta}$. These parameters can either be estimated by imitation learning or acquired by reinforcement learning. The dart's impact position can be adapted to a desired target by changing the velocity and the angle at which the dart leaves the launcher. These variables can be influenced by changing the meta-parameters of the motor primitive such as the final position of the launcher and the duration of the throw. The state consists of the current position of the hand and the desired position on the target. If the thrower is always at the same distance from the wall the two positions can be equivalently expressed as the vertical distance. The meta-parameter function $\gamma(\mathbf{s})$ maps the state (the relative height) to the meta-parameters γ (the final position g and the duration of the motor primitive T).

The approach presented in this paper is applicable to any movement representation that has meta-parameters, i.e., a small set of parameters that allows to modify the movement. In contrast to [16], [17], [18] our approach does not require explicit (re)planning of the motion.

In the next sections, we derive and apply an appropriate reinforcement learning algorithm.

B. Problem Statement: Meta-Parameter Self-Improvement

The problem of meta-parameter learning is to find a stochastic policy $\pi(\gamma|\mathbf{x}) = p(\gamma|\mathbf{s})$ that maximizes the expected return

$$J(\pi) = \int_{\mathbb{S}} p(\mathbf{s}) \int_{\mathbb{G}} \pi(\gamma|\mathbf{s}) R(\mathbf{s},\gamma) d\gamma\, d\mathbf{s}, \tag{5}$$

where $R(\mathbf{s},\gamma)$ denotes all the rewards following the selection of the meta-parameter γ according to a situation described by state s. The return of an episode is $R(\mathbf{s},\gamma) = T^{-1}\sum_{t=0}^{T} r^t$ with number of steps T and rewards r^t. For a parametrized policy π with parameters $\mathbf{w}$ it is natural to first try a policy gradient approach such as finite-difference methods, vanilla

Algorithm 1: Meta-Parameter Learning
Preparation steps:
Learn one or more motor primitives by imitation and/or reinforcement learning (yields shape parameters $\boldsymbol{\theta}$).
Determine initial state $\mathbf{s}^0$, meta-parameters γ^0, and cost C^0 corresponding to the initial motor primitive.
Initialize the corresponding matrices $\mathbf{S}, \boldsymbol{\Gamma}, \mathbf{C}$.
Choose a kernel $\mathbf{k}, \mathbf{K}$.
Set a scaling parameter λ.
For all iterations j**:**
Determine the state $\mathbf{s}^j$ specifying the situation.
Calculate the meta-parameters γ^j by:
Determine the mean of each meta-parameter i
$\gamma_i(\mathbf{s}^j) = \mathbf{k}(\mathbf{s}^j)^{\mathrm{T}} (\mathbf{K} + \lambda\mathbf{C})^{-1} \boldsymbol{\Gamma}_i$,
Determine the variance
$\sigma^2(\mathbf{s}^j) = k(\mathbf{s}^j, \mathbf{s}^j) - \mathbf{k}(\mathbf{s}^j)^{\mathrm{T}} (\mathbf{K} + \lambda\mathbf{C})^{-1} \mathbf{k}(\mathbf{s}^j)$,
Draw the meta-parameters from a Gaussian distribution
$\gamma^j \sim \mathcal{N}(\gamma \vert \gamma(\mathbf{s}^j), \sigma^2(\mathbf{s}^j)\mathbf{I})$.
Execute the motor primitive using the new meta-parameters.
Calculate the cost c^j at the end of the episode.
Update $\mathbf{S}, \boldsymbol{\Gamma}, \mathbf{C}$ according to the achieved result.

policy gradient approaches and natural gradients[3]. Reinforcement learning of the meta-parameter function $\gamma(\mathbf{s})$ is not straightforward as only few examples can be generated on the real system and trials are often quite expensive. The credit assignment problem is non-trivial as the whole movement is affected by every change in the meta-parameter function. Early attempts using policy gradient approaches resulted in tens of thousands of trials even for simple toy problems, which is not feasible on a real system.

Dayan & Hinton [19] showed that an immediate reward can be maximized by instead minimizing the Kullback-Leibler divergence $D(\pi(\gamma|\mathbf{s})R(\mathbf{s},\gamma)||\pi'(\gamma|\mathbf{s}))$ between the reward-weighted policy $\pi(\gamma|\mathbf{s})$ and the new policy $\pi'(\gamma|\mathbf{s})$. Williams [20] suggested to use a particular policy in this context; i.e., the policy

$$\pi(\gamma|\mathbf{s}) = \mathcal{N}(\gamma|\gamma(\mathbf{s}), \sigma^2(\mathbf{s})\mathbf{I}),$$

where we have the deterministic mean policy $\gamma(\mathbf{s}) = \boldsymbol{\phi}(\mathbf{s})^{\mathrm{T}}\mathbf{w}$ with basis functions $\boldsymbol{\phi}(\mathbf{s})$ and parameters $\mathbf{w}$ as well as the variance $\sigma^2(\mathbf{s})$ that determines the exploration $\epsilon \sim \mathcal{N}(\mathbf{0}, \sigma^2(\mathbf{s})\mathbf{I})$. The parameters $\mathbf{w}$ can then be adapted by reward-weighted regression in an immediate reward [11] or episodic reinforcement learning scenario [6]. The reasoning behind this reward-weighted regression is that the reward can be treated as an improper probability distribution over indicator variables determining whether the action is optimal or not.

C. A Task-Appropriate Reinforcement Learning Algorithm

Designing good basis functions is challenging, a nonparametric representation is better suited in this context. There is an intuitive way of turning the reward-weighted regression into a Cost-regularized Kernel Regression. The kernelization of the reward-weighted regression can be done straightforwardly (similar to Section 6.1 of [21] for regular supervised learning). Inserting the reward-weighted regression solution $\mathbf{w} = (\boldsymbol{\Phi}^{\mathrm{T}}\mathbf{R}\boldsymbol{\Phi} + \lambda\mathbf{I})^{-1}\boldsymbol{\Phi}^{\mathrm{T}}\mathbf{R}\boldsymbol{\Gamma}_i$, and using the Woodbury formula $(\boldsymbol{\Phi}^{\mathrm{T}}\mathbf{R}\boldsymbol{\Phi} + \lambda\mathbf{I})\boldsymbol{\Phi}^{\mathrm{T}} = \boldsymbol{\Phi}^{\mathrm{T}}\mathbf{R}(\boldsymbol{\Phi}\boldsymbol{\Phi}^{\mathrm{T}} + \lambda\mathbf{R}^{-1})$, we

[3] While we will denote the shape parameters by $\boldsymbol{\theta}$, we denote the parameters of the meta-parameter function by $\mathbf{w}$.

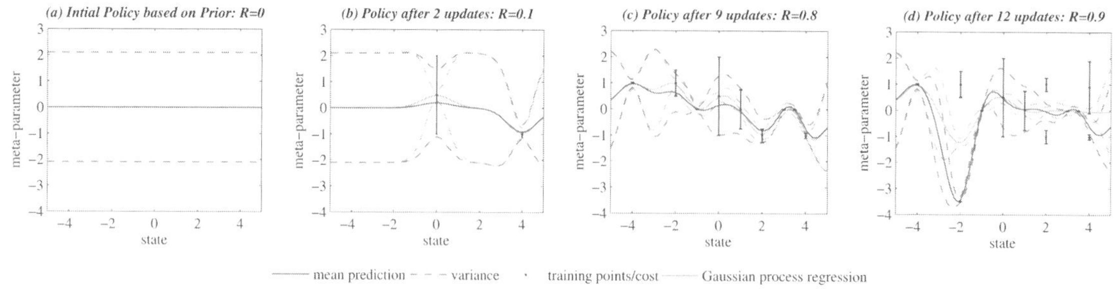

Figure 2: This figure illustrates the meaning of policy improvements with Cost-regularized Kernel Regression. Each sample consists of a state, a meta-parameter and a cost where the cost is indicated the blue error bars. The red line represents the improved mean policy, the dashed green lines indicate the exploration/variance of the new policy. For comparison, the gray lines show standard Gaussian process regression. As the cost of a data point is equivalent to having more noise, pairs of states and meta-parameter with low cost are more likely to be reproduced than others with high costs.

transform reward-weighted regression into a Cost-regularized Kernel Regression

$$\bar{\gamma}_i = \boldsymbol{\phi}(\mathbf{s})^{\mathrm{T}}\mathbf{w} = \boldsymbol{\phi}(\mathbf{s})^{\mathrm{T}} \left(\boldsymbol{\Phi}^{\mathrm{T}}\mathbf{R}\boldsymbol{\Phi} + \lambda\mathbf{I}\right)^{-1} \boldsymbol{\Phi}^{\mathrm{T}}\mathbf{R}\boldsymbol{\Gamma}_i$$

$$= \boldsymbol{\phi}(\mathbf{s})^{\mathrm{T}}\boldsymbol{\Phi}^{\mathrm{T}} \left(\boldsymbol{\Phi}\boldsymbol{\Phi}^{\mathrm{T}} + \lambda\mathbf{R}^{-1}\right)^{-1} \boldsymbol{\Gamma}_i, \quad (6)$$

where the rows of $\boldsymbol{\Phi}$ correspond to the basis functions $\boldsymbol{\phi}(\mathbf{s}_i) = \boldsymbol{\Phi}_i$ of the training examples, $\boldsymbol{\Gamma}_i$ is a vector containing the training examples for meta-parameter component γ_i, and λ is a ridge factor. Next, we assume that the accumulated rewards R_k are strictly positive $R_k > 0$ and can be transformed into costs by $c_k = 1/R_k$. Hence, we have a cost matrix $\mathbf{C} = \mathbf{R}^{-1} = \mathrm{diag}(R_1^{-1}, \ldots, R_n^{-1})$ with the cost of all n data points. After replacing $\mathbf{k}(\mathbf{s}) = \boldsymbol{\phi}(\mathbf{s})^{\mathrm{T}}\boldsymbol{\Phi}^{\mathrm{T}}$ and $\mathbf{K} = \boldsymbol{\Phi}\boldsymbol{\Phi}^{\mathrm{T}}$, we obtain the Cost-regularized Kernel Regression

$$\bar{\gamma}_i = \gamma_i(\mathbf{s}) = \mathbf{k}(\mathbf{s})^{\mathrm{T}} \left(\mathbf{K} + \lambda\mathbf{C}\right)^{-1} \boldsymbol{\Gamma}_i,$$

which gives us a deterministic policy. Here, costs correspond to the uncertainty about the training examples. Thus, a high cost is incurred for being further away from the desired optimal solution at a point. In our formulation, a high cost therefore corresponds to a high uncertainty of the prediction at this point.

In order to incorporate exploration, we need to have a stochastic policy and, hence, we need a predictive distribution. This distribution can be obtained by performing the policy update with a Gaussian process regression and we directly see from the kernel ridge regression

$$\sigma^2(\mathbf{s}) = k(\mathbf{s}, \mathbf{s}) + \lambda - \mathbf{k}(\mathbf{s})^{\mathrm{T}} \left(\mathbf{K} + \lambda\mathbf{C}\right)^{-1} \mathbf{k}(\mathbf{s}),$$

where $k(\mathbf{s}, \mathbf{s}) = \boldsymbol{\phi}(\mathbf{s})^{\mathrm{T}}\boldsymbol{\phi}(\mathbf{s})$ is the distance of a point to itself. We call this algorithm Cost-regularized Kernel Regression.

The algorithm corresponds to a Gaussian process regression where the costs on the diagonal are input-dependent noise priors. Gaussian processes have been used previously for reinforcement learning [22] in value function based approaches while here we use them to learn the policy.

If several sets of meta-parameters have similarly low costs the algorithm's convergence depends on the order of samples. The cost function should be designed to avoid this behavior and to favor a single set. The exploration has to be restricted to safe meta-parameters.

D. Meta-Parameter Learning by Reinforcement Learning

As a result of Section II-C, we have a framework of motor primitives as introduced in Section II-A that we can use for reinforcement learning of meta-parameters as outlined in Section II-B. We have generalized the reward-weighted regression policy update to instead become a Cost-regularized Kernel Regression (CrKR) update where the predictive variance is used for exploration. In Algorithm 1, we show the complete algorithm resulting from these steps.

The algorithm receives three inputs, i.e., (i) a motor primitive that has associated meta-parameters $\boldsymbol{\gamma}$, (ii) an initial example containing state $\mathbf{s}^0$, meta-parameter $\boldsymbol{\gamma}^0$ and cost C^0, as well as (iii) a scaling parameter λ. The initial motor primitive can be obtained by imitation learning [1] and, subsequently, improved by parametrized reinforcement learning algorithms such as policy gradients [3] or Policy learning by Weighting Exploration with the Returns (PoWER) [6]. The demonstration also yields the initial example needed for meta-parameter learning. While the scaling parameter is an open parameter, it is reasonable to choose it as a fraction of the average cost and the output noise parameter (note that output noise and other possible hyper-parameters of the kernel can also be obtained by approximating the unweighted meta-parameter function).

Illustration of the Algorithm: In order to illustrate this algorithm, we will use the example of the 2D dart throwing task introduced in Section II-A. Here, the robot should throw darts accurately while not destroying its mechanics. Hence, the cost corresponds to the error between desired goal and the impact point, as well as the absolute velocity of the end-effector. The initial policy is based on a prior, illustrated in Figure 2(a), that has a variance for initial exploration (it often makes sense to start with a uniform prior). This variance is used to enforce exploration. To throw a dart, we sample the meta-parameters from the policy based on the current state[4]. After the trial the cost is determined and, in conjunction with

[4]In the dart setting, we could choose the next target and thus employ CrKR as an active learning approach by picking states with large variances. However, often the state is determined by the environment, e.g., the ball trajectory in the table tennis experiment (Section III-C) depends on the opponent.

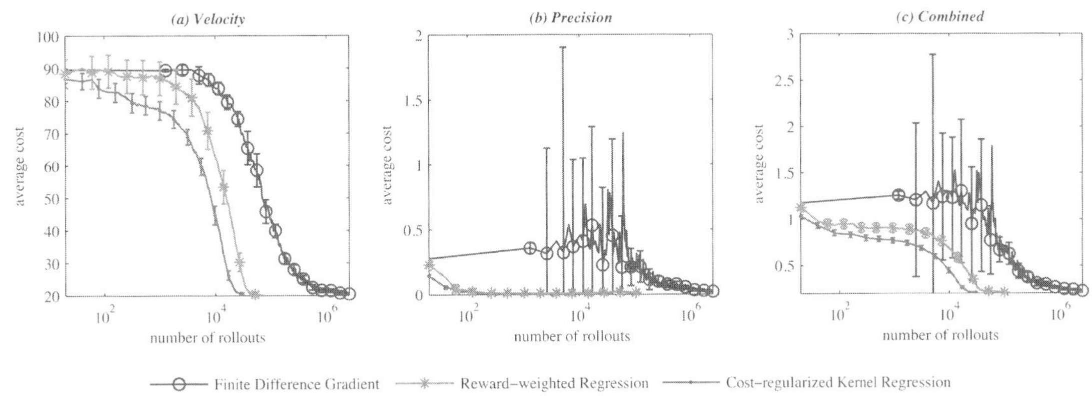

Figure 3: This figure shows the performance of the compared algorithms averaged over 10 complete learning runs. Cost-regularized Kernel Regression finds solutions with the same final performance two orders of magnitude faster than the finite difference gradient (FD) approach and twice as fast as the reward-weighted regression. At the beginning FD often is highly unstable due to our attempts of keeping the overall learning speed as high as possible to make it a stronger competitor. The lines show the median and error bars indicate standard deviation. The initialization and the initial costs are identical for all approaches. However, the omission of the first twenty rollouts was necessary to cope with the logarithmic rollout axis.

the employed meta-parameters, used to update the policy[5]. If the cost is large (for example the impact is far from the target), the variance of the policy is large as it may still be improved and therefore needs exploration. Furthermore, the mean of the policy is shifted only slightly towards the observed example as we are uncertain about the optimality of this action. If the cost is small, we know that we are close to an optimal policy and only have to search in a small region around the observed trial. The effects of the cost on the mean and the variance are illustrated in Figure 2(b). Each additional sample refines the policy and the overall performance improves (see Figure 2(c)). If a state is visited several times and different meta-parameters are sampled, the policy update must favor the meta-parameters with lower costs. Algorithm 1 exhibits this behavior as illustrated in Figure 2(d).

III. Evaluation

In Section II, we have introduced both a framework for meta-parameter self-improvement as well as an appropriate reinforcement learning algorithm used in this framework. In this section, we will first show that the presented reinforcement learning algorithm yields higher performance than off-the shelf approaches. Hence, we compare it on a simple planar cannon shooting problem [23] with the preceding reward-weighted regression and an off-the-shelf finite difference policy gradient approach.

[5] In the dart throwing example we have a correspondence between the state and the outcome similar to a regression problem. However, the mapping between the state and the meta-parameter is not unique. The same height can be achieved by different combinations of velocities and angles. Averaging these combinations is likely to generate inconsistent solutions. The regression must hence favor the meta-parameters with the lower costs. CrKR can be employed as a regularized regression method in this setting.

The proposed reinforcement learning method only requires a cost associated with the outcome of the trial. In the table tennis experiment (Section III-C), the state corresponds to the position and velocity of the ball over the net. We only observe the cost related to how well we hit the ball. After a table tennis trial, we do not know which state would have matched the employed meta-parameters, as would be required in a regression setting.

The resulting meta-parameter learning framework can be used in a variety of settings in robotics. We consider two scenarios here, i.e., (i) dart throwing with a simulated robot arm, a real Barrett WAM and the JST-ICORP/SARCOS humanoid robot CBi, and (ii) table tennis with a simulated robot arm and a real Barrett WAM. Some of the real-robot experiments are still partially work in progress.

A. Benchmark Comparison

In the first task, we only consider a simple simulated planar cannon shooting where we benchmark our Reinforcement Learning by Cost-regularized Kernel Regression approach against a finite difference gradient estimator and the reward-weighted regression. Here, we want to learn an optimal policy for a 2D toy cannon environment similar to [23].

The setup is given as follows: A toy cannon is at a fixed location $[0.0, 0.1]\,m$. The meta-parameters are the angle with respect to the ground and the speed of the cannon ball. In this benchmark we do not employ the motor primitives but set the parameters directly. The flight of the canon ball is simulated as ballistic flight of a point mass with Stokes's drag as wind model. The cannon ball is supposed to hit the ground at a desired distance. The desired distance $[1..3]\,m$ and the wind speed $[0..1]\,m/s$, which is always horizontal, are used as input parameters, the velocities in horizontal and vertical directions are predicted (which influences the angle and the speed of the ball leaving the cannon). Lower speed can be compensated by a larger angle. Thus, there are different possible policies for hitting a target; we intend to learn the one which is optimal for a given cost function. This cost function consists of the sum of the squared distance between the desired and the actual impact point and one hundredth of the squared norm of the velocity at impact of the cannon ball. It corresponds to maximizing the precision while minimizing the employed energy according to the chosen weighting. All approaches performed well in this setting, first driving the position error to zero and, subsequently, optimizing the impact

(a) The dart is placed on the launcher. (b) The arm moves back. (c) The arm moves forward on an arc. (d) The arm stops. (e) The dart is carried on by its momentum. (f) The dart hits the board.

Figure 4: This figure shows a dart throw in a physically realistic simulation.

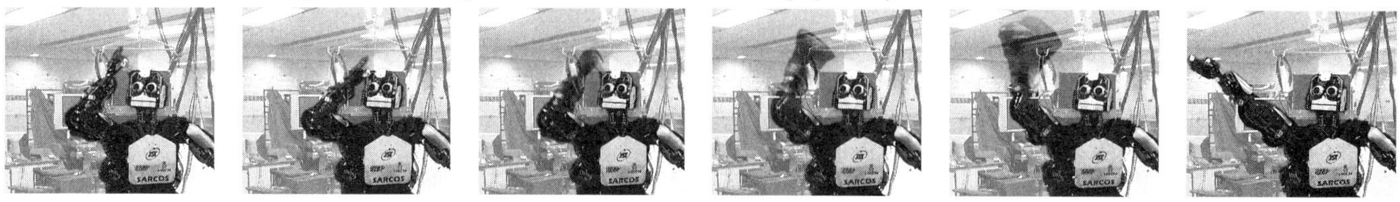

(a) The dart is placed in the hand. (b) The arm moves back. (c) The arm moves forward on an arc. (d) The arm continues moving. (e) The dart is released and the arm follows through. (f) The arm stops and the dart hits the board.

Figure 5: This figure shows a dart throw on the real JST-ICORP/SARCOS humanoid robot CBi.

velocity. The experiment was initialized with $[1, 10]\, m/s$ as initial ball velocities and $1\, m/s$ as wind velocity. This setting corresponds to a very high parabola, which is far from optimal. For plots, we evaluate the policy on a test set of 25 uniformly randomly chosen points that remain the same throughout of the experiment and are never used in the learning process but only to generate Figure 3.

We compare our novel algorithm to a finite difference policy gradient (FD) method [3] and to the reward-weighted regression (RWR) [11]. The FD method uses a parametric policy that employs radial basis functions in order to represent the policy and adds Gaussian exploration. The learning rate as well as the magnitude of the perturbations were tuned for best performance. We used 51 sets of uniformly perturbed parameters for each update step. The FD algorithm converges after approximately 2000 batch gradient evaluations, which corresponds to $2,550,000$ shots with the toy cannon.

The RWR method uses the same parametric policy as the finite difference gradient method. Exploration is achieved by adding Gaussian noise to the mean policy . All open parameters were tuned for best performance. The RWR algorithm converges after approximately $40,000$ shots with the toy cannon. For the Cost-regularized Kernel Regression (CrKR) the inputs are chosen randomly from a uniform distribution. We use Gaussian kernels and the open parameters were optimized by cross-validation on a small test set prior to the experiment. Each trial is added as a new training point if it landed in the desired distance range. The CrKR algorithm converges after approximately $20,000$ shots with the toy cannon.

After convergence, the costs of CrKR are the same as for RWR and slightly lower than those of the FD method. The CrKR method needs two orders of magnitude fewer shots than the FD method. The RWR approach requires twice the shots of CrKR demonstrating that a non-parametric policy, as employed by CrKR, is better adapted to this class of problems than a parametric policy. The squared error between the actual and desired impact is approximately 5 times higher for the finite difference gradient method, see Figure 3.

B. Dart-Throwing

Now, we turn towards the complete framework, i.e., we intend to learn the meta-parameters for motor primitives in discrete movements. We compare the Cost-regularized Kernel Regression (CrKR) algorithm to the reward-weighted regression (RWR). As a sufficiently complex scenario, we chose a robot dart throwing task inspired by [23]. However, we take a more complicated scenario and choose dart games such as *Around the Clock* [13] instead of simple throwing at a fixed location. Hence, it will have an additional parameter in the state depending on the location on the dartboard that should come next in the sequence. The acquisition of a basic motor primitive is achieved using previous work on imitation learning [1]. Only the meta-parameter function is learned using CrKR or RWR.

The dart is placed on a launcher attached to the end-effector and held there by stiction. We use the Barrett WAM robot arm in order to achieve the high accelerations needed to overcome the stiction. See Figure 4, for a complete throwing movement. The motor primitive is trained by imitation learning with kinesthetic teach-in. We use the Cartesian coordinates with respect to the center of the dart board as inputs. The parameter for the final position, the duration of the motor primitive and the angle around the vertical axis are the meta-parameters. The popular dart game *Around the Clock* requires the player to hit the numbers in ascending order, then the bulls-eye. As energy is lost overcoming the stiction of the launching sled, the darts fly lower and we placed the dartboard lower than official rules require. The cost function is the sum of ten times the squared error on impact and the velocity of the motion. After approximately 1000 throws the algorithms have converged but CrKR yields a high performance already much earlier (see Figure 6). We again used a parametric policy with radial basis functions for RWR. Designing a good parametric policy proved very difficult in this setting as is reflected by the poor performance of RWR.

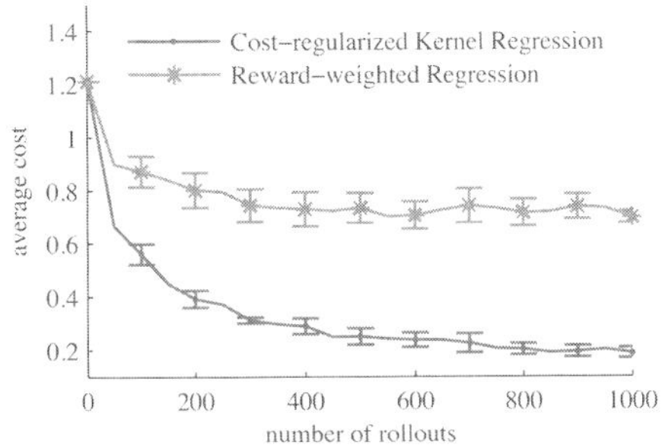

Figure 6: This figure shows the cost function of the dart-throwing task for a whole game *Around the Clock* in each rollout. The costs are averaged over 10 runs with the error-bars indicating standard deviation.

This experiment is also being carried out on two real, physical robots, i.e., a Barrett WAM and the humanoid robot CBi (JST-ICORP/SARCOS). CBi was developed within the framework of the JST-ICORP Computational Brain Project at ATR Computational Neuroscience Labs. The hardware of the robot was developed by the American robotic development company SARCOS. CBi can open and close the fingers which helps for more human-like throwing instead of the launcher employed by the Barrett WAM. See Figure 5 for a throwing movement. Parts of these experiments are still in-progress.

C. Table Tennis

In the second evaluation of the complete framework, we use it for hitting a table tennis ball in the air. The setup consists of a ball gun that serves to the forehand of the robot, a Barrett WAM and a standard sized table. The movement of the robot has three phases. The robot is in a rest posture and starts to swing back when the ball is launched. During this swing-back phase, the open parameters for the stroke are predicted. The second phase is the hitting phase which ends with the contact of the ball and racket. In the final phase the robot gradually ends the stroking motion and returns to the rest posture. See Figure 8 for an illustration of a complete episode. The movements in the three phases are represented by motor primitives obtained by imitation learning.

The meta-parameters are the joint positions and velocities for all seven degrees of freedom at the end of the second phase (the instant of hitting the ball) and a timing parameter that controls when the swing back phase is transitioning to the hitting phase. We learn these 15 meta-parameters as a function of the ball positions and velocities when it is over the net. We employed a Gaussian kernel and optimized the open parameters according to typical values for the input and output. As cost function we employ the metric distance between the center of the paddle and the center of the ball at the hitting time. The policy is evaluated every 50 episodes with 25 ball launches picked randomly at the beginning of the learning. We initialize the behavior with five successful strokes observed from another player. After initializing the meta-parameter function with only these five initial examples, the robot misses ca. 95% of the balls as shown in Figure 7. Trials are only used

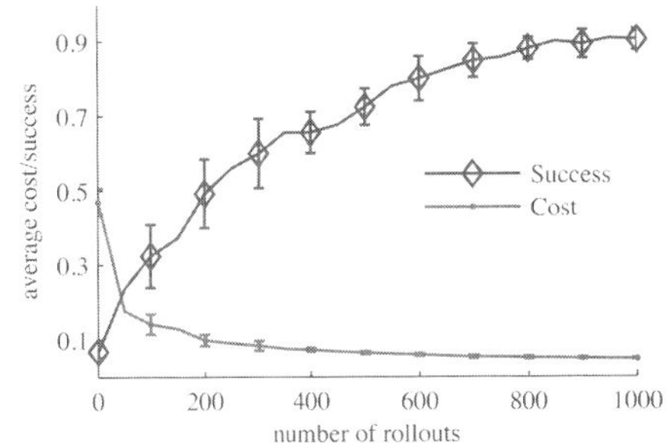

Figure 7: This figure shows the cost function of the table tennis task averaged over 10 runs with the error-bars indicating standard deviation. The red line represents the percentage of successful hits and the blue line the average cost. At the beginning the robot misses the ball 95% of the episodes and on average by $50\,cm$. At the end of the learning the robot hits almost all balls.

to update the policy if the robot has successfully hit the ball. Figure 9 illustrates different positions of the ball the policy is capable of dealing with after the learning. Figure 7 illustrates the costs over all episodes. Preliminary results suggest that the resulting policy performs well both in simulation and for the real system. We are currently in the process of executing this experiment also on the real Barrett WAM.

IV. Conclusion & Future Work

In this paper, we have studied the problem of meta-parameter learning for motor primitives. It is an essential step towards applying motor primitives for learning complex motor skills in robotics more flexibly. We have discussed an appropriate reinforcement learning algorithm for mapping situations to meta-parameters.

We show that the necessary mapping from situation to meta-parameter can be learned using a Cost-regularized Kernel Regression (CrKR) while the parameters of the motor primitive can still be acquired through traditional approaches. The predictive variance of CrKR is used for exploration in on-policy meta-parameter reinforcement learning. We compare the resulting algorithm in a toy scenario to a policy gradient algorithm with a well-tuned policy representation and the reward-weighted regression. We show that our CrKR algorithm can significantly outperform these preceding methods. To demonstrate the system in a complex scenario, we have chosen the *Around the Clock* dart throwing game and table tennis implemented both on simulated and real robots.

Adapting movements to situations is also discussed in [16] in a supervised learning setting. Their approach is based on predicting a trajectory from a previously demonstrated set and refining it by motion planning. The authors note that kernel ridge regression performed poorly for the prediction if the new situation is far from previously seen ones as the algorithm yields the global mean. In our approach we employ a cost weighted mean that overcomes this problem. If the situation is far from previously seen ones, large exploration will help to find a solution.

(a) The robot is the rest posture. (b) The arm swings back. (c) The arm strikes the ball. (d) The arm follows through and decelerates. (e) The arm returns to the rest posture.

Figure 8: This figure shows a table tennis stroke on the real Barrett WAM.

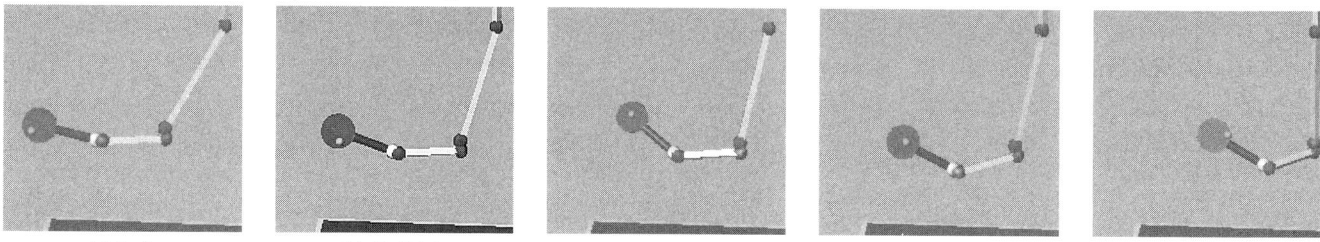

(a) Left. (b) Half left. (c) Center high. (d) Center low. (e) Right.

Figure 9: This figure shows samples of the learned forehands. Note that this figure only illustrates the learned meta-parameter function in this context but cannot show timing and velocity and it requires a careful observer to note the important configuration differences resulting from the meta-parameters.

Future work will require to sequence different motor primitives by a supervisory layer. This supervisory layer would for example in a table tennis task decide between a forehand motor primitive and a backhand motor primitive, the spatial meta-parameter and the timing of the motor primitive would be adapted according to the incoming ball, and the motor primitive would generate the trajectory. This supervisory layer could be learned by an hierarchical reinforcement learning approach [24] (as introduced in the early work by [25]). In this framework, the motor primitives with meta-parameter functions could be seen as robotics counterpart of options [9] or macro-actions [26].

References

[1] A. J. Ijspeert, J. Nakanishi, and S. Schaal, "Learning attractor landscapes for learning motor primitives," in *Advances in Neural Information Processing Systems 16*, 2003.

[2] S. Schaal, P. Mohajerian, and A. J. Ijspeert, "Dynamics systems vs. optimal control — a unifying view," *Progress in Brain Research*, vol. 165, no. 1, pp. 425–445, 2007.

[3] J. Peters and S. Schaal, "Policy gradient methods for robotics," in *Proc. Int. Conf. Intelligent Robots and Systems*, 2006.

[4] D. Pongas, A. Billard, and S. Schaal, "Rapid synchronization and accurate phase-locking of rhythmic motor primitives," in *Proc. Int. Conf. Intelligent Robots and Systems*, 2005.

[5] J. Nakanishi, J. Morimoto, G. Endo, G. Cheng, S. Schaal, and M. Kawato, "Learning from demonstration and adaptation of biped locomotion," *Robotics and Autonomous Systems*, vol. 47, no. 2-3, pp. 79–91, 2004.

[6] J. Kober and J. Peters, "Policy search for motor primitives in robotics," in *Advances in Neural Information Processing Systems 22*, 2009.

[7] H. Urbanek, A. Albu-Schäffer, and P. van der Smagt, "Learning from demonstration repetitive movements for autonomous service robotics," in *Proc. Int. Conf. Intelligent Robots and Systems*, 2004.

[8] R. Caruana, "Multitask learning," *Machine Learning*, vol. 28, pp. 41–75, 1997.

[9] A. McGovern and A. G. Barto, "Automatic discovery of subgoals in reinforcement learning using diverse density," in *Proc. Int. Conf. Machine Learning*, 2001.

[10] K. Mülling, "Motor control and learning in table tennis," Master's thesis, University of Tübingen, 2009.

[11] J. Peters and S. Schaal, "Reinforcement learning by reward-weighted regression for operational space control," in *Proc. Int. Conf. Machine Learning*, 2007.

[12] C. E. Rasmussen and C. K. Williams, *Gaussian Processes for Machine Learning*. MIT Press, 2006.

[13] Masters Games Ltd., "The rules of darts," online *http://www.mastersgames.com/rules/darts-rules.htm*, July 2010.

[14] G. Wulf, *Attention and motor skill learning*. Champaign, IL: Human Kinetics, 2007.

[15] D.-H. Park, H. Hoffmann, P. Pastor, and S. Schaal, "Movement reproduction and obstacle avoidance with dynamic movement primitives and potential fields," in *Proc. Int. Conf. Humanoid Robots*, 2008.

[16] N. Jetchev and M. Toussaint, "Trajectory prediction: learning to map situations to robot trajectories," in *Proc. Int. Conf. Machine Learning*, 2009.

[17] D. B. Grimes and R. P. N. Rao, "Learning nonparametric policies by imitation," in *Proc. Int. Conf. Intelligent Robots and System*, 2008.

[18] D. C. Bentivegna, A. Ude, C. G. Atkeson, and G. Cheng, "Learning to act from observation and practice," *Int. Journal of Humanoid Robotics*, vol. 1, no. 4, pp. 585–611, 2004.

[19] P. Dayan and G. E. Hinton, "Using expectation-maximization for reinforcement learning," *Neural Computation*, vol. 9, no. 2, pp. 271–278, 1997.

[20] R. J. Williams, "Simple statistical gradient-following algorithms for connectionist reinforcement learning," *Machine Learning*, vol. 8, pp. 229–256, 1992.

[21] C. M. Bishop, *Pattern Recognition and Machine Learning*. Springer Verlag, 2006.

[22] Y. Engel, S. Mannor, and R. Meir, "Reinforcement learning with gaussian processes," in *Proc. Int. Conf. Machine Learning*, 2005.

[23] G. Lawrence, N. Cowan, and S. Russell, "Efficient gradient estimation for motor control learning," in *Proc. Int. Conf. Uncertainty in Artificial Intelligence*, 2003.

[24] A. Barto and S. Mahadevan, "Recent advances in hierarchical reinforcement learning," *Discrete Event Dynamic Systems*, vol. 13, no. 4, pp. 341 – 379, 2003.

[25] M. Huber and R. Grupen, "Learning robot control using control policies as abstract actions," in *NIPS'98 Workshop: Abstraction and Hierarchy in Reinforcement Learning*, 1998.

[26] A. McGovern, R. S. Sutton, and A. H. Fagg, "Roles of macro-actions in accelerating reinforcement learning," in *Grace Hopper Celebration of Women in Computing*, 1997.

Analysis and Control of a Dissipative Spring-Mass Hopper with Torque Actuation

M. Mert Ankaralı
Dept. of Electrical and Electronics Eng.,
Middle East Technical University,
06531 Ankara, Turkey
ankarali@eee.metu.edu.tr

Uluç Saranlı
Dept. of Computer Engineering,
Bilkent University,
06800 Ankara, Turkey
saranli@cs.bilkent.edu.tr

***Abstract*— It has long been established that simple spring-mass models can accurately represent the dynamics of legged locomotion. Existing work in this domain, however, almost exclusively focuses on the idealized Spring-Loaded Inverted Pendulum (SLIP) model and neglects passive dissipative effects unavoidable in any physical robot or animal. In this paper, we extend on a recently proposed analytic approximation to the stance trajectories of a dissipative SLIP model to analyze stability properties of a planar hopper with a single rotary actuator at the hip. We first describe how a suitably chosen torque controller can compensate for damping losses, maintaining the same energy level across strides and hence reducing the return map to a single dimension. We then identify and characterize equilibrium points for this return map under a fixed leg placement policy and show that "uncontrolled" asymptotic stability is feasible for this energy-regulated system. Subsequent presentation of simulation evidence establishes that the predictions of this approximate model are consistent with the exact plant model. The paper concludes with the application of our energy-regulation scheme to the design of a task-level gait controller that uses explicit leg placement commands in conjunction with the hip torque.**

I. Introduction

Long term practical utility of mobile robots in unstructured environments critically depends on their locomotory aptitude. In this context, the performance of ground mobility that can ultimately be achieved by legged platforms is superior to any other alternative as evidenced by numerous examples in nature as well as a number of very successful dynamically stable autonomous legged robots that have been built to date [10, 25, 26, 30, 37]. Unfortunately, even on flat ground, legged morphologies do not enjoy the simplicity of models supported by the conveniently constrained and continuous modes of ground interaction observed in wheeled and, to some extent, tracked vehicles. Even the most basic legged behaviors such as walking and running require hybrid dynamic models whose analysis and control involve difficult challenges [14, 20, 23]. In the world of quasi-static locomotion with multi-legged robots, one can recover some of this simplicity through active or structural suppression of second order dynamics [39], but these methods are not directly applicable to dynamically dexterous modes of locomotion such as running.

One of the most significant discoveries in this context was most likely the recognition of similar center of mass (COM) movement patterns in running animals of widely different sizes and morphologies [1, 6, 7, 9, 24]. This led to the development of the simple yet accurate Spring-Loaded Inverted Pendulum (SLIP) model to describe such behaviors [21, 34]. Significant research effort was devoted to both the use of this model as a basis for the design of fast and efficient legged robots [10, 19, 27, 30] as well as its analysis to reveal fundamental aspects of associated locomotory behaviors [20]. The present paper falls into the latter category and contributes by investigating the previously unaddressed question of how the presence of passive damping affects the behavioral characteristics of running with the SLIP model.

Our treatment of this question is based on the use of analytic approximations to the otherwise non-integrable stance dynamics of the model. A number of such approximations have already been proposed in the literature. In particular, [35] uses a Hamiltonian formulation of the SLIP dynamics with an iterative application of the mean-value theorem to obtain an accurate return map for symmetric steps. More recently, [16] presents an analytically simple approximation to the stance dynamics of a conservative SLIP equipped with a linear spring. This approach is based on a linearization of gravity around mid-stance, similar in form to the solution proposed in [28] but involving a much more carefully formulated derivation. Inaccuracies of these approximations in the presence of non-symmetric gravitational effects were partially addressed in [5] using explicit corrections. In contrast to the lossless models adopted in all these approaches, a new return map for the lossy SLIP model with viscous damping in the leg was proposed in [4], providing a basis for our present inquiry.

Despite the availability of methods to analyze stability properties of locomotory behaviors in the absence of closed-form expressions for a Poincaré map [2, 3], a number of different possible approaches become available once a sufficiently accurate analytic return map is available. For example, [18] investigates in depth stability properties of a SLIP model attached to a rigid body by neglecting the effects of gravity, which allows for the derivation of suitable closed-form expressions for stride trajectories. A similar but less rigorous stability analysis was provided in [16] with comparisons to previous numerical results in [36] as well as biological data. In contrast to the lack of feedback control in these characterizations, [33] studies the stability of the SLIP model under a novel leg placement control strategy that only relies on easily obtainable state measurements. Another inquiry on how to achieve effective control of the SLIP model, now extended to a spatial setting, is provided in [8], focusing on lateral motions of the model. In conjunction with these studies primarily focused on running behaviors, similar analytically motivated contributions were also made to the structurally different walking behaviors [15, 22, 40], providing evidence that the same dynamic model can provide a unified description for both walking and running [17].

Our contributions in the present paper have a number of important differences from existing work. Firstly, our plant model is dissipative, impairing the accuracy of most existing analytic approximations and associated predictions. Secondly, in contrast to the usual energy regulation mechanisms in the literature through adjustments of the leg length or changing stiffness, our model uses only a single torque actuator at the hip relative to a virtual body with fixed orientation to compensate for energy losses. These changes are motivated by being much more realistic from an implementation point of view, as evidenced by the successful use of similar actuation mechanisms in the Scout quadrupeds [26] and the RHex hexapod [30] as well as a number of other monopedal platforms [12, 32]. Finally, our approximate solutions to the return map also take into account the effect of gravity on the angular momentum for steps that are non-symmetric with respect to the gravitational vertical, increasing the practical applicability of associated stability results.

II. The Torque-Actuated Dissipative SLIP Model

A. System Dynamics and the Apex Return Map

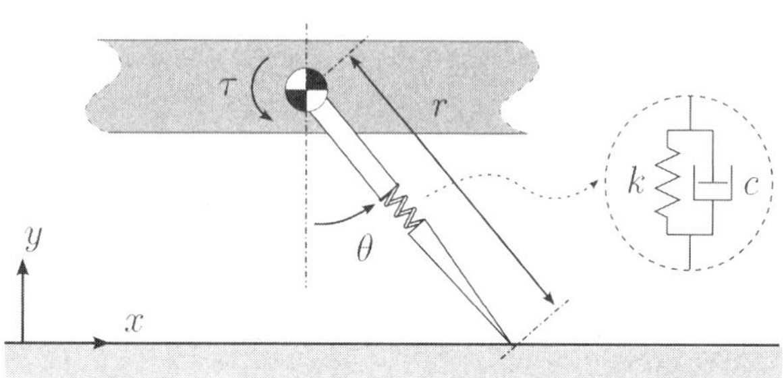

Fig. 1. TD-SLIP : Dissipative spring-mass hopper with rotary hip actuation

Fig. 1 illustrates the Torque-actuated Dissipative Spring-Loaded Inverted Pendulum (TD-SLIP) plant we investigate in this paper. It consists of a fixed orientation (2-DOF) planar rigid body with mass m, connected to a massless, fully passive leg with linear compliance k, rest length r_0 and linear viscous damping c, through an actuated rotary joint with torque τ. The system alternates between *stance* and *flight* phases during running, with the flight phase further divided into the *ascent* and *descent* subphases. Fig. 2 illustrates the three important events that define transitions between these phases: *touchdown*, where the leg comes into contact with the ground, *liftoff*, where the toe takes off from the ground and finally *apex*, where the body reaches its maximum height during flight with $\dot{y} = 0$. Another important event, not illustrated in the figure, is *bottom*, corresponding to the point of maximal leg compression during stance. Table I details the notation used throughout the paper.

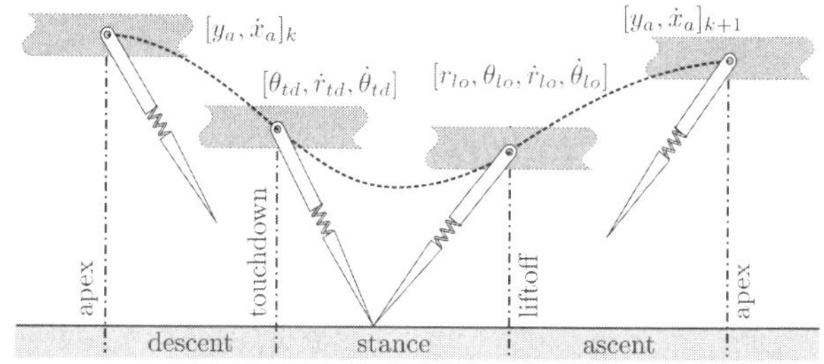

Fig. 2. A single TD-SLIP stride with definitions of transition states

TABLE I

NOTATION USED THROUGHOUT THE PAPER

System States, Event States and Control Inputs	
$x, y, \dot{x}, \dot{y}$	Cartesian body position and velocities
$r, \theta, \dot{r}, \dot{\theta}$	Leg length, leg angle and velocities
τ	Hip torque command during stance
$y_a, \dot{x}_a$	Apex height and velocity
$\theta_{td}, \dot{r}_{td}, \dot{\theta}_{td}$	Touchdown leg angle, polar velocities
t_b, r_b, θ_b	Bottom time, leg length and angle
$t_{lo}, r_{lo}, \theta_{lo}, \dot{r}_{lo}, \dot{\theta}_{lo}$	Liftoff time, leg length, angle and velocities
p_θ	Angular momentum around the toe
Kinematic and Dynamic Parameters	
m, g	Body mass and gravitational acceleration
k, r_0, c	Leg stiffness, rest length and damping

During flight, the body obeys ballistic flight dynamics

$$\begin{bmatrix} \ddot{x} \\ \ddot{y} \end{bmatrix} = \begin{bmatrix} 0 \\ -g \end{bmatrix}$$

and the massless leg can be arbitrarily positioned. In contrast, during stance, the toe remains stationary on the ground while the body mass feels forces generated by both the passive spring-damper pair and the hip torque. The stance dynamics of the planar SLIP model in polar leg coordinates with respect to the toe location take the form

$$\frac{d}{dt}\begin{bmatrix} m\dot{r} \\ mr^2\dot{\theta} \end{bmatrix} = \begin{bmatrix} mr\dot{\theta}^2 - mg\cos\theta - k(r - r_0) - c\dot{r} \\ mgr\sin\theta + \tau \end{bmatrix}. \quad (1)$$

A very useful abstraction for the analysis and control of cyclic TD-SLIP trajectories is provided by the apex return map, defined as a Poincaré map from one apex point to the next. In the following sections, we will use this map to study stability properties of TD-SLIP, and later adopt it as a task-level gait representation for a closed-loop running controller.

The apex return map can be formulated as $P := P_a \circ P_s \circ P_d$ by composing three individual submaps P_d, P_s, P_a for the descent, stance and ascent phases, respectively. The descent and ascent maps are trivial and are given by

$$P_d : \begin{bmatrix} \dot{r}_{td} \\ r_0\dot{\theta}_{td} \end{bmatrix} = R(\pi/2 - \theta_{td}) \begin{bmatrix} -\dot{x}_a \\ \sqrt{2g(y_a - r_0\cos\theta_{td})} \end{bmatrix} \quad (2)$$

$$P_a : \begin{bmatrix} y_a \\ \dot{x}_a \end{bmatrix} = \begin{bmatrix} r_{lo}\cos\theta_{lo} + \dot{y}_{lo}^2/(2g) \\ \dot{x}_{lo} \end{bmatrix} \quad (3)$$

where $\dot{x}_{lo}$ and $\dot{y}_{lo}$ are liftoff velocities in Cartesian coordinates and R denotes the standard 2D rotation matrix. Unfortunately, the dynamics of (1) are not integrable in closed form. Consequently, we will use an analytical approximation for the stance map, which we describe in the next section.

B. An Approximate Stance Map for the Unforced TD-SLIP

A new analytical approximation to the dynamics of a dissipative SLIP model was proposed in [4]. However, this method assumes the presence of radial leg actuation, either in the form of a controllable leg stiffness, or the regulation of touchdown and liftoff leg lengths. In this section, we briefly review their method and extend it to support the hip torque actuation of our model.

The approximation proposed in [4], which, in turn, is based on the methods described in [16], relies on two key assumptions: 1). The angular travel throughout stance is relatively small and remains close to the vertical, allowing linearization of the gravitational potential in the Lagrangian with subsequent conservation of the angular momentum $p_\theta := mr^2\dot{\theta}$ and 2). the radial compression is small with $r_0 - r \ll r_0$, allowing a truncated Taylor expansion of related terms.

As described in [4], under these conditions and assuming, for now, that $\tau = 0$, the radial component of (1) reduces to

$$\ddot{r} + (c/m)\dot{r} + (\omega_0^2 + 3\omega^2)r = -g + r_0\omega_0^2 + 4r_0\omega^2 \,, \quad (4)$$

where we define $\omega_0 := \sqrt{k/m}$ and $\omega := p_\theta/(mr_0^2)$. Solutions to this simple second-order ODE can be found as

$$r(t) = e^{-\zeta\hat{\omega}_0 t}(A\cos(\omega_d t) + B\sin(\omega_d t)) + F/\hat{\omega}_0^2 \,, \quad (5)$$

where we have $\hat{\omega}_0 := \sqrt{\omega_0^2 + 3\omega^2}$, $\zeta := c/(2m\hat{\omega}_0)$, $\omega_d := \hat{\omega}_0\sqrt{1-\zeta^2}$, $F := -g + r_0\omega_0^2 + 4r_0\omega^2$ and A and B determined by touchdown states as

$$\begin{aligned} A &:= r_0 - F/\hat{\omega}_0^2 \,, \\ B &:= (\dot{r}_{td} + \zeta\hat{\omega}_0 A)/\omega_d \,. \end{aligned}$$

Simple differentiation and further simplification yields radial TD-SLIP trajectories as

$$\begin{aligned} r(t) &= M\, e^{-\zeta\hat{\omega}_0 t}\cos(\omega_d t + \phi) + F/\hat{\omega}_0^2 \,, & (6) \\ \dot{r}(t) &= -M\hat{\omega}_0\, e^{-\zeta\hat{\omega}_0 t}\cos(\omega_d t + \phi + \phi_2) \,, & (7) \end{aligned}$$

with M, ϕ and ϕ_2 determined through trigonometric identities. At this point, the angular trajectories can be determined using the constant angular momentum. An additional linearization of the term $1/r^2$ leads to an analytical solution for the rate of change of the leg angle as

$$\begin{aligned} \dot{\theta}(t) = \; & 3\omega - 2\omega F/(r_0\hat{\omega}_0^2) - \quad (8) \\ & 2\omega M e^{-\zeta\hat{\omega}_0 t}\cos(\omega_d t + \phi)/r_0 \,, \end{aligned}$$

integrated to yield the angular trajectory

$$\begin{aligned} \theta(t) = \; & \theta_{td} + X\,t + \quad (9) \\ & Y(e^{-\zeta\hat{\omega}_0 t}\cos(\omega_d t + \phi + \phi_3) - \cos(\phi + \phi_3)). \end{aligned}$$

with X, Y and ϕ_3 computed accordingly as in [4].

The final step in completing the stance map requires finding the time of liftoff. Only one of the two liftoff conditions described in [4] is applicable in the context of the present paper since we do not allow control of the liftoff leg length. Consequently, the liftoff time is solely determined by the solution to the equation $k(r_0 - r(t_{lo})) - c\;\dot{r}(t_{lo}) = 0$, for which a sufficiently accurate analytical approximation can be found by approximating the exponential coefficient in the radial solution of (6) by its value at a specific instant during decompression. In particular, noting that the compression and decompression times are roughly equal, we use $e^{-\zeta\hat{\omega}_0 t} \approx e^{-\zeta\hat{\omega}_0 2t_b}$, where t_b denotes the *bottom time*, easily found by solving (7). Under this assumption, we have

$$\begin{aligned} t_{lo} \approx \; & (2\pi - \arccos(k(r_0 - F/\hat{\omega}_0^2)/(\overline{M}Me^{-\zeta\hat{\omega}_0\gamma t_b})) \\ & - \phi - \phi_4)/\omega_d \,, \quad (10) \end{aligned}$$

which yields the stance map as

$$P_s : \begin{bmatrix} r_{lo} \\ \theta_{lo} \\ \dot{r}_{lo} \\ \dot{\theta}_{lo} \end{bmatrix} = \begin{bmatrix} r(t_{lo}) \\ \theta(t_{lo}) \\ \dot{r}(t_{lo}) \\ \dot{\theta}(t_{lo}) \end{bmatrix} . \quad (11)$$

where the right hand side is a function of touchdown states.

Note, however, that these derivations completely ignore the presence of the hip torque. In the next section, we propose a new method to incorporate the effects of the hip torque through a fixed correction on the angular momentum value p_θ in a way similar to the one used in [5] for gravity corrections.

C. Stance Map for the Torque Controlled TD-SLIP

Hip actuation in legged systems can serve a number of different purposes. Among both biological [1] and robotic [13, 19, 27] systems, its most common uses involve retraction of legs in flight and control of body posture with legs in stance. Interestingly, the use of hip actuation to provide thrust has not been studied as extensively in the robotics literature. In addition to a few direct experimental inquiries [12, 32] and indirect uses in multi-legged platforms [26, 30], it has received limited attention in [2] in the form of an active spring.

In the present paper, we propose an open-loop hip actuation regime that enforces the ramp torque profile

$$\tau(t) = \begin{cases} \tau_0(1 - \frac{t}{t_f}) & \text{if } 0 \le t \le t_f \\ 0 & \text{if } t > t_f \end{cases} \quad (12)$$

during stance, with τ_0 and t_f chosen prior to touchdown. This open-loop profile has three important advantages. Firstly, its simple functional dependence on time allows us to easily incorporate its effects into the derivations of the previous section. Second, if we choose t_f to be the liftoff time, we have $\tau(t_{lo}) = 0$, which prevents premature leg liftoff due to the action of the hip and ensures a structural match to the trajectories of the unforced system. Finally, its unidirectional action ensures that no negative work is done during stance.

Inspection of the TD-SLIP dynamics of (1) shows that the hip torque directly acts on the angular dynamics and only indirectly effects radial motion. Consequently, we hypothesize that an average correction to the constant angular momentum p_θ of Section II-B can capture the effects of the hip torque on system trajectories. Normally, the angular momentum during stance can be formulated as

$$p_\theta(t) = p_\theta(0) + \int_0^t \tau(\eta)d\eta + \int_0^t mgr(\eta)\sin\theta(\eta)d\eta, \quad (13)$$

by integrating the angular dynamics. Adopting the method proposed in [5], we compute a corrected angular momentum

$$\hat{p_\theta} = p_\theta(0) + \Delta p_\tau + \Delta p_g. \quad (14)$$

where Δp_τ and Δp_g incorporate the time averaged effects of the leg torque and gravitational acceleration, respectively. Assuming $t_f = t_{lo}$, we have

$$\Delta p_\tau := \frac{1}{t_{lo}} \int_0^{t_{lo}} \left(\int_0^{\eta_1} \tau(\eta_2) d\eta_2 \right) d\eta_1 = \tau_0 \frac{t_{lo}}{3} . \quad (15)$$

However, even with available analytic approximations, derivation of an exact closed-form expression for Δp_g is not feasible. Instead, we use a linear approximation to the integrand $r(\eta)\sin\theta(\eta)$ using its values at the touchdown and liftoff, resulting in

$$\Delta p_g := \frac{mgt_{lo}}{6}(2r_0 \sin\theta_{td} + r_{lo}\sin\theta_{lo}) . \quad (16)$$

Estimated values for the liftoff time t_{lo}, leg angle θ_{lo} and leg length r_{lo} are provided by the unforced approximations of the previous section. Substituting $\hat{p_\theta}$ for the constant angular momentum in all derivations of Section II-B, we obtain a new approximation that takes into account the effects of both the hip torque and gravity on the stance trajectories.

Note that the corrections we propose have an iterative character since both (15) and (16) use prior estimates of t_{lo} and θ_{lo}. Consequently, starting from the unforced approximations, it is possible to iteratively apply these corrections to obtain more accurate predictions at the expense of analytic simplicity. Our simulations show that more than a single iteration is only needed for extreme conditions such as the angle of attack being very close to the touchdown leg angle, causing a bounce-back.

III. Stability of an Energy-Regulated TD-SLIP

A. Compensation of Damping Losses

Unlike previous stability studies of lossless spring-mass hoppers, fully passive self-stabilization with a fixed touchdown angle and no active control is not possible with the TD-SLIP model since damping losses will eventually drain out all energy in the system. Consequently, active hip thrust must be employed to sustain locomotion.

Recall that our choice of the hip torque in (12) incorporates two parameters: τ_0 and t_f. We have already shown that choosing $t_f = t_{lo}$ is advantageous in preventing early liftoff and ensuring structural correspondence of system trajectories to our analytical approximation. The simplest possible strategy for the remaining parameter τ_0, very close in spirit to the radial actuation strategy adopted by Raibert's runners [27] and its subsequent analysis in [21], is to choose a particular value and keep it constant across subsequent strides. The underlying idea is that since damping losses are proportional to the total energy level of the system, constant energy input will give rise to trajectories at a stable energy level. Unfortunately, in our 2-DOF model, such a strategy results in a two dimensional return map, for which, analytical solution and characterization of equilibrium points is not feasible.

Another possibility, which we adopt in the present paper, is to use the hip torque to compensate for all dissipative effects within a single step, ensuring conservation of energy in the apex return map and hence reducing its dimension by one. Note that the total energy dissipated within a single TD-SLIP step is given by

$$E_{loss} = E_c + E_k , \quad (17)$$

where E_c represents damping losses with

$$E_c := \int_0^{t_{lo}} c\dot{r}^2(\eta)\, d\eta , \quad (18)$$

and $E_k := (r_{lo} - r_0)^2/2$ captures the leftover energy in the leg spring when it lifts off before it is fully extended due to damping. Fortunately, our analytic approximations provide closed form expressions for both of these components. In particular, damping losses can be approximately computed as

$$\begin{aligned} E_c &= \frac{-c/M^2\hat{\omega}_0}{4\zeta}(\zeta\cos(2(\phi+\phi_2)+\phi_3)+1 \quad (19) \\ &- e^{-2\zeta\hat{\omega}_0 t_{lo}}(\zeta\cos(2\omega_d t_{lo} + 2(\phi+\phi_2)+\phi_3)+1)) , \end{aligned}$$

while E_k only depends on the previously computed r_{lo}.

In contrast, the energy supplied by the hip torque is

$$E_\tau = \tau_0 \int_0^{t_{lo}} (1 - \frac{t}{t_{lo}})\, \dot{\theta}(t) dt , \quad (20)$$

for which our analytical approximations can also be used to obtain closed-form expressions. We omit the details here for space considerations. Since both (17) and (20) can be obtained in closed form as a function of initial conditions and the choice of touchdown angle θ_{td}, we can easily find the desired torque magnitude τ_0 by solving

$$E_\tau = E_{loss} . \quad (21)$$

As noted above, this choice of torque results in successive apex states having the same energy, at least while working within our approximate apex return map. Naturally, additional corrections would be needed to apply these ideas to the exact plant model since inaccuracies of our approximations would invalidate this conservation. Nevertheless, we use this active compensation regime to reduce the dimension of our analytic apex return map, allowing us to easily identify its equilibrium points and characterize their stability.

B. Equilibrium Points with a Fixed Leg Placement Policy

In this section, we use our analytic approximations to identify and characterize equilibrium points of the one dimensional "energy-regulated" return map on the apex height y_a arising from the use of a fixed touchdown angle policy with

$\theta_{td} = \beta$ and the energy-regulating hip torque described in Section III-A. All results in this section will be presented in non-dimensional versions of relevant variables, defined as

$$\begin{aligned}
\bar{y}_a &:= y_a/r_0 \\
\bar{\dot{x}}_a &:= \dot{x}_a/\sqrt{gr_0} \\
\bar{E}_a &:= E_a/(mgr_0) \\
\bar{k} &:= kr_0/(mg) \\
\zeta_0 &:= c/(2\sqrt{mk})\ .
\end{aligned}$$

Finally, in order to facilitate comparison with earlier studies, we use kinematic and dynamic parameters that roughly match those of an average human with $m = 80kg$ and $r_0 = 1m$.

Fig. 3 shows two families of return maps for $\beta = 20°$ and $\beta = 32°$, respectively, together with the dependence of equilibrium points on the energy level of the system. These results show that the TD-SLIP exhibits asymptotically stable behavior under the fixed touchdown angle, energy-regulated regime, with the location of the equilibrium point depending on the chosen energy level. We can also observe that as the fixed touchdown angle β increases, the energy range for which stable fixed points exist increases as well. This is rather natural since the torque actuation at the hip can only supply energy through the angular momentum, which directly increases the angular span during stance. Increasing the touchdown angle admits a larger angular span for stance, allowing stable fixed points to form at higher energy levels as well.

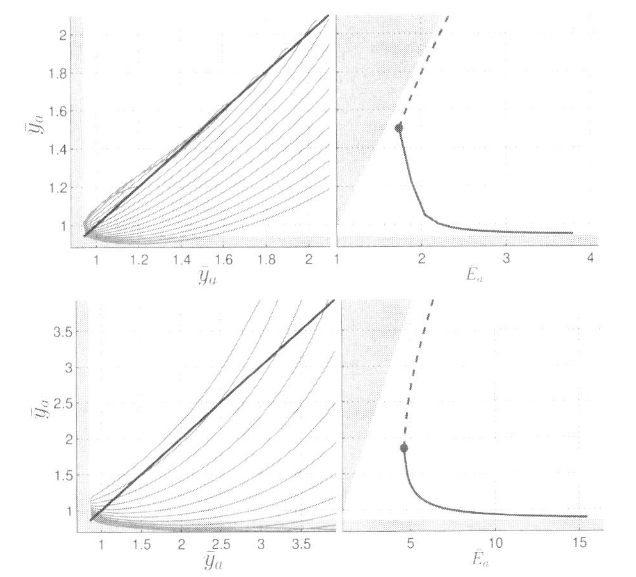

Fig. 3. Apex height return map (left) and associated equilibrium points (right) for the TD-SLIP model as a function of different (dimensionless) energy levels, generated with the proposed analytical approximations. The top plots are obtained with $\beta = 20°$, $\bar{k} = 40$, $\zeta_0 = 0.07$, whereas the bottom plots use $\beta = 32°$, $\bar{k} = 40$, $\zeta_0 = 0.07$ in dimensionless coordinates. Solid and dashed lines in the right figure indicate stable and unstable equilibrium points respectively. Shaded regions correspond to kinematically infeasible configurations.

Having established the presence of stable equilibrium points for the torque-controlled SLIP model, Fig. 4 shows a comparison of fixed points predicted by our analytic approximations,

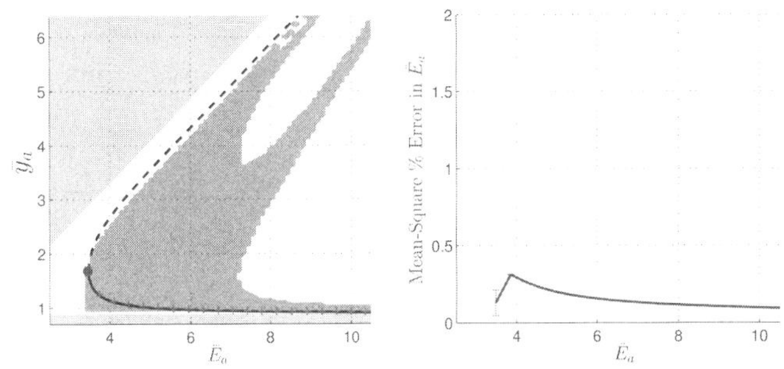

Fig. 4. Left: Comparison of stable equilibrium points predicted by our analytic approximation (solid line) with those obtained by numeric simulation of TD-SLIP dynamics (plus signs) for $\beta = 28°$ and different apex energy levels in the range $\bar{E}_a \in [2, 10]$. Shaded region in the middle illustrates the stable domain of attraction for the simulated plant model. Right: Percentage mean-square error between initial and steady-state dimensionless energy levels for the simulated plant.

with those that arise within simulations of the exact TD-SLIP model. In order to make direct comparisons possible, we started TD-SLIP simulations from a large range of initial y_a and E_a values, with a fixed touchdown angle and an energy regulation controller similar to the one presented Section III-A, but now taking the energy level of the very first step as an overall regulation goal. This modification was necessary since using the approximations to locally enforce energy conservation at every step would slowly cause prediction errors to accumulate, either draining all energy out of the system, or causing it to diverge. We then checked whether the system converges to a stable equilibrium point in apex coordinates after 25 steps up to a tolerance of 10^{-4}. Shaded region in the middle of the left plot of Fig. 4 illustrates the resulting domain of attraction, while the plus marks in the same plot illustrate the associated set of fixed points.

Note, also, that the domain of attraction exhibited by the simulation almost exactly covers the region between the unstable and stable fixed points predicted by our approximations. There is also an almost exact match between the fixed points predicted by our approximations and those obtained from simulation. The cavities to the right of the region of attraction arise from the presence of the "gap" region in the return map, resulting from kinematic constraints that require the apex height to be sufficiently large to allow leg placement. The reason for this can be clearly seen in the bottom right plot of Fig. 3, where parts of the return map overlap with the kinematically infeasible gray region on the bottom. This means that some initial conditions at high energy levels will lead to apex states for which leg placement at an angle of β is impossible. This gap was also observed by previous studies [16], and is reproduced by both our analytical approximations, and the simulated plant.

The right plot in Fig. 4 shows the mean and standard deviations of the percentage mean-square energy difference between the initial and steady state apex points for the simulated plant. The fact that this difference is consistently below 0.3% shows that our approximations are capable of very accurately modeling energy losses and successfully predict fixed points of the exact TD-SLIP plant. It is worth noting, also, that accuracy also increases significantly with increasing energy levels.

C. Parameter Dependence of Equilibrium Points

Equilibrium points that arise from our fixed touchdown angle, energy-regulated regime naturally depend on the kinematic and dynamic parameter choices. Fig. 5 illustrates the dependence of stable fixed points on each individual parameter (the touchdown angle β, the dimensionless leg stiffness $\bar{k}$ or leg damping ζ_0) with the remaining two parameters kept constant. The leftmost figure mirrors our observations in the previous section, namely that the range of stable energy levels increase with larger touchdown angles.

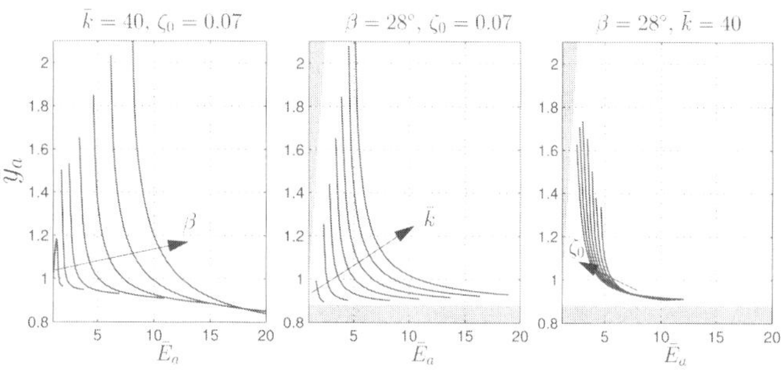

Fig. 5. Dependence of stable equilibrium points on variations of the touchdown angle β (left), leg spring stiffness $\bar{k}$ (middle) and leg damping ratio ζ_0 (right). Arrows indicate increasing directions for each varied parameter.

The dependence of equilibrium points on the leg stiffness, illustrated in the middle figure shows that increasing spring constants cause an increase in the range of stable energy levels. This is also natural since an increased stiffness corresponds to shorter stance times, resulting in decreased damping losses and a corresponding decrease in the necessary torque input. Finally, we observe that the impact of the damping coefficients on the equilibrium points is not as pronounced, providing evidence that our compensation strategy successfully balances damping losses. Nevertheless, increasing the amount of damping causes a slight decrease in the range of stable energy levels.

D. Correspondence of the Model to Biological Data

A recent quantitative comparison of ground reaction force data from a variety of running animals to those predicted by a simple, passive spring-mass model shows that despite the very good correspondence of vertical force components between biological data and the idealized SLIP model, there are some discrepancies in how well horizontal forces can be predicted [38]. In this section, we report on an interesting property of the torque-actuated TD-SLIP morphology: It seems to be capable of qualitatively reproducing ground reaction force profiles very similar to those observed in biological systems.

Fig. 6 illustrates the body trajectory for a single stride of steady-state running with the TD-SLIP model, together with a depiction of "virtual footfalls" in the direction of instantaneous ground reaction force vectors throughout the stance phase. As a result of the ramp torque profile we use for supplying energy to the system, large backward horizontal forces are

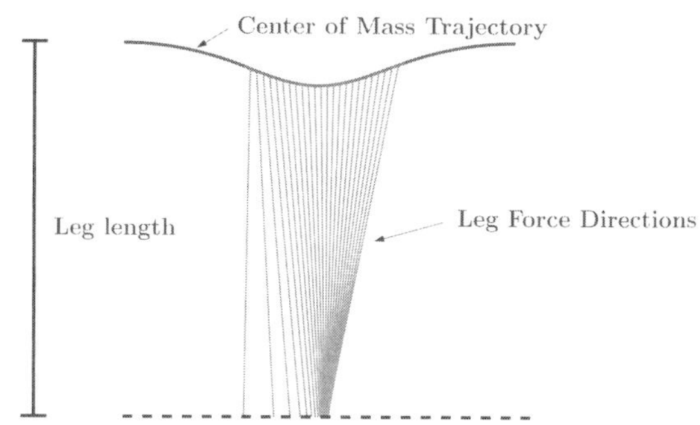

Fig. 6. COM trajectory for a single steady-state stride of the TD-SLIP model with $\bar{k} = 40$, $\zeta_0 = 0.07$, running at approximately 3m/s (0.96 in dimensionless units). Comparison of ground reaction force directions during stance to biological data presented in [38] reveals a remarkable qualitative match.

introduced in the beginning of the stance phase, resulting in associated virtual footfalls appearing behind the actual toe location. Towards the end of the stance phase, the hip torque approaches zero and brings the virtual footfall and actual toe locations together.

This qualitative structure is observed for all steady-state trajectories of the TD-SLIP model and is remarkably consistent with biological data presented in [38]. Even though we do not yet have any quantitative basis in which any predictive claims can be made, we think that this correspondence may provide evidence towards the use of hip torque as an additional source of energy used by biological runners, improving the predictive accuracy and utility of dynamic models of running.

IV. Feedback Control of TD-SLIP Running

A. Deadbeat Control by Inversion of the Apex Return Map

The presence of a sufficiently accurate analytic formulation of the apex return map naturally motivates its inversion to obtain a controller for stabilizing the system around a desired operating point $[y_a^*, \dot{x}_a^*]$ in apex state coordinates. A similar approach was adopted in a number of studies [8, 29, 31], but never in the context of a lossy model or torque actuation. In this section, we describe a deadbeat gait controller for TD-SLIP as an application of our approximations, and show that it is capable of very accurately regulating the apex states of a running TD-SLIP and improves on both the accuracy and stability of previous attempts to control a similar, torque-actuated model in [12].

An explicitly specified desired apex state will require a nonzero change in the energy level of the system. Using a strategy similar to the energy-conserving torque controller of Section III-A, we will use the hip torque to supply the requested energy input to the system in a single step. Similar to (21), this energy is given by

$$E_\tau = \frac{1}{2}m((\dot{x}_a^*)^2 - \dot{x}_a^2) + mg(y_a^* - y_a) + E_{loss}, \quad (22)$$

which can easily be solved to determine the ramp torque magnitude τ_0, assuming, once again, that $t_f = t_{lo}$.

Once the desired torque profile is determined, the return map has only one remaining degree of control freedom: the touchdown angle θ_{td}. A deadbeat controller can be formulated as a one dimensional minimization problem in the form

$$\theta_{td} = \underset{\frac{-\pi}{2} < \theta < \frac{-\pi}{2}}{\text{argmin}} \quad (\dot{x}_a^* - (\ \pi_{\dot{x}_a} \circ P(\theta_{td}, [y_a, \dot{x}_a]_k)\ ))^2 \ , \quad (23)$$

whose numerical solution is trivial due to the availability of our analytic approximation for the return map P. This yields an effective, one-step deadbeat controller for the regulation of forward speed and hopping height for the TD-SLIP model.

B. Controller Performance and Comparison

As noted before, there are very few in depth studies of how hip torque actuation can be used to achieve stable locomotion. Among notable exceptions is recent work on locomotion over mildly rough terrain [12] where the authors use TD-SLIP equations of motion to derive an approximate energy controller to regulate hopping height, and a PD-based torque policy to regulate forward speed. In this section, we present a comparison of the new controller we described in Section IV-A, with the control strategy proposed in [12]. In order to maintain consistency with our previous stability results, we use the same kinematic and dynamic parameters with Section III-B, roughly corresponding to an average human morphology. Note that parameters used in [12] are not substantially different from ours when converted to dimensionless units. All simulations were run in Matlab using a fourth order Runge-Kutta integrator together with accurate detection of transition events. Each run consisted of 25 steps, at the end of which we determined whether there was convergence to a fixed point in apex coordinates.

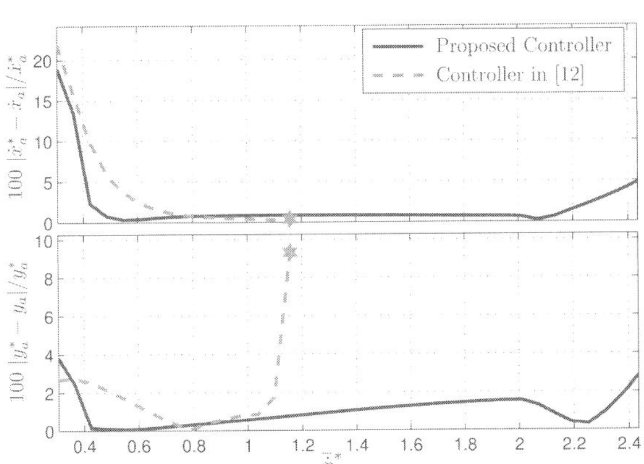

Fig. 7. Comparison of tracking performance for apex speed (top) and apex height (bottom) between the proposed controller (solid) and the PD control policy described in [12] (dashed) as a function of (dimensionless) desired velocity $\dot{x}_a^*$. Markers indicate where the controller of [12] loses stability. Vertical axes are percentage errors.

Fig. 7 illustrates tracking performances of both controllers for apex speed and height variables in terms of normalized percentage error measures. Note that our controller based on an accurate analytic model for the dynamics of TD-SLIP significantly increases the range of velocity goals that can be achieved without losing stability. Moreover, improvements can be observed in the tracking accuracy for both the apex speed and height variables. Finally, our controller does not require any feedback or sensory measurements during stance, but relies only on accurate measurement of apex states. This makes practical implementations much more feasible compared to the active PD control strategy since high-bandwidth feedback is usually very challenging for fast legged robots.

V. Conclusions and Future Work

In this paper, we presented a novel method to obtain analytical approximations to the stance trajectories of a dissipative, torque actuated planar spring-mass hopper. We have successfully used our approximations to design an energy-regulation controller for the hip torque that can accurately compensate for the effects of damping within a stride, allowing us to obtain a one-dimensional return map under a fixed angle leg placement policy. Consequently, we were able to analyze stability properties of this model, identifying and characterizing its equilibrium points. The predictive accuracy of our analytical approximations was confirmed by a very close match to fixed points and their domains of attraction obtained through numerical simulations of the exact plant model. We have also demonstrated the utility of our approximations through their use in designing a gait controller.

It is important to note that neither the energy regulating hip torque controller, nor the subsequent stability analysis would have been possible in the absence of our analytical approximations. Consequently, we believe this paper presents the first careful study of stability properties of running in the presence of non-negligible damping. In this context, we believe that the incorporation of damping as a significant component in the dynamical model substantially increases the applicability of associated analytical tools and controllers to practical robot platforms in which dissipative effects will always be present and may sometimes be a dominant factor particularly if compliance is achieved through composite materials. In the future, we hope to demonstrate the practical utility of our approximations by experimental verification of their predictions with respect to a physical monopedal runner.

Our choice of hip torque as the primary source of energy input to the system was motivated by the difficulty of implementing radial actuation in physical robot platforms, and the simplicity and success of existing robot platforms with similar actuation mechanisms [26, 30]. We have also further simplified our model by assuming a fixed body angle that may be justified by morphologies in which additional legs on the front and back of the body provide a stabilizing effect, or where the body link is explicitly constrained by an experimental setup [12, 32]. In this context, we discovered an interesting correspondence between the ground reaction force profiles resulting from the use of a hip torque and biological data presented in [38], leading to a possible explanation for the inability of the original SLIP model in reproducing horizontal force components during running and a very preliminary hypothesis that hip torque may be playing a previously unaddressed important role in the control of legged locomotion. In the future, we hope to generalize our results to a freely rotating body link, making the results applicable to less constrained morphologies such as bipeds. For example, one of the interesting possibilities is how forward-bending body posture and the resulting gravitational torque can be used to balance the torque input from the hip, making it possible to both have a freely rotating body, while using the hip torque to provide thrust. This seems to be one of the ways in which ideas similar to those used for passive dynamic walking can be applied to efficient bipedal running and we hope to extend our results in this paper to such scenarios.

VI. Acknowledgments

M. Mert Ankaralı was partially supported by the National Scientific and Technological Research Council of Turkey (TUBITAK). This project was also partially supported by TUBITAK project 109E032. Finally, we thank Afsar Saranli for inspiring discussions and his support for this work.

References

[1] R. Alexander. *Principles of Legged Locomotion*. Princeton University Press, 2006.

[2] R. Altendorfer, D. E. Koditschek, and P. Holmes. Stability Analysis of a Clock-Driven Rigid-Body SLIP Model for RHex. *Int. Journal of Robotics Research*, 23(10-11):1001–1012, 2004.

[3] R. Altendorfer, D. E. Koditschek, and P. Holmes. Stability Analysis of Legged Locomotion Models by Symmetry-Factored Return Maps. *Int. Journal of Robotics Research*, 23(10-11):979–999, 2004.

[4] M. M. Ankarali, O. Arslan, and U. Saranli. An analytical solution to the stance dynamics of passive spring-loaded inverted pendulum with damping. In *12th Int. Conf. on Climbing and Walking Robots and The Support Technologies for Mobile Machines (CLAWAR'09)*, Istanbul, Turkey, September 2009.

[5] O. Arslan, U. Saranli, and O. Morgul. An aproximate stance map of the spring mass hopper with gravity correction for nonsymmetric locomotions. In *Proc. of the IEEE Int. Conf. on Robotics and Automation*, Kobe, Japan, May 2009.

[6] R. Blickhan. The spring-mass model for running and hopping. *J. of Biomechanics*, 22:1217–1227, 1989.

[7] R. Blickhan and R. J. Full. Similarity in multilegged locomotion: Bouncing like a monopode. *J. of Comparative Physiology A: Neuroethology, Sensory, Neural, and Behavioral Physiology*, 173(5):509–517, Nov. 1993.

[8] S. G. Carver, N. J. Cowan, and J. M. Guckenheimer. Lateral stability of the spring-mass hopper suggests a two-step control strategy for running. *Chaos: An Interdisciplinary Journal of Nonlinear Science*, 19(2):026106, 2009.

[9] G. A. Cavagna, F. P. Saibene, and R. Margaria. Mechanical work in running. *J. of Applied Physiology*, 19(2):249–256, 1964.

[10] J. G. Cham, S. A. Bailey, J. E. Clark, R. J. Full, and M. R. Cutkosky. Fast and robust: Hexapedal robots via shape deposition manufacturing. *Int. Journal of Robotics Research*, 21(10):869–882, 2002.

[11] N. Cherouvim and E. Papadopoulos. Speed and height control for a special class of running quadruped robots. In *Proceedings of the IEEE Int. Conf. on Robotics and Automation*, pages 825–830, May 2008.

[12] N. Cherouvim and E. Papadopoulos. Control of hopping speed and height over unkown rough terrain using a single actuator. In *Proceedings of the IEEE Int. Conf. on Robotics and Automation*, Kobe, Japan, 2009.

[13] C. Chevallereau, E. R. Westervelt, and J. W. Grizzle. Asymptotically Stable Running for a Five-Link, Four-Actuator, Planar B ipedal Robot. *Int. Journal of Robotics Research*, 24(6):431–464, 2005.

[14] S. Collins, A. Ruina, R. Tedrake, and M. Wisse. Efficient Bipedal Robots Based on Passive-Dynamic Walkers. *Science*, 307(5712):1082–1085, 2005.

[15] M. Garcia, A. Chatterjee, A. Ruina, and M. Coleman. The simplest walking model: Stability, complexity, and scaling. *J. of Biomechanical Engineering*, 120(2):281–288, 1998.

[16] H. Geyer, A. Seyfarth, and R. Blickhan. Spring-mass running: simple approximate solution and application to gait stability. *J. of Theoretical Biology*, 232(3):315–328, Feb. 2005.

[17] H. Geyer, A. Seyfarth, and R. Blickhan. Compliant leg behaviour explains basic dynamics of walking and running. *Proceedings of the Royal Society B: Biological Sciences*, 273(1603):2861–2867, 2006.

[18] R. M. Ghigliazza, R. Altendorfer, P. Holmes, and D. Koditschek. A simply stabilized running model. *SIAM Journal on Applied Dynamical Systems*, 2:187–218, 2003.

[19] P. Gregorio, M. Ahmadi, and M. Buehler. Design, control, and energetics of an electrically actuated legged robot. *Transactions on Systems, Man, and Cybernetics*, 27(4):626–634, August 1997.

[20] P. Holmes, R. Full, D. Koditschek, and J. Guckenheimer. The dynamics of legged locomotion: Models, analyses, and challenges. *SIAM Review*, 48(2):207–304, 2006.

[21] D. E. Koditschek and M. Buehler. Analysis of a simplified hopping robot. *Int. Journal of Robotics Research*, 10(6):587–605, 1991.

[22] A. D. Kuo. Stabilization of lateral motion in passive dynamic walking. *Int. Journal of Robotics Research*, 18(9):917–930, September 1999.

[23] T. McGeer. Passive dynamic walking. *Int. Journal of Robotics Research*, 9(2):62–82, 1990.

[24] T. A. McMahon and G. C. Cheng. The mechanics of running: How does stiffness couple with speed. *J. of Biomechanics*, 23(Suppl. 1):65–78, 1990.

[25] R. Playter, M. Buehler, and M. Raibert. BigDog. In *Society of Photo-Optical Instrumentation Engineers (SPIE) Conference Series*, volume 6230 of *Presented at the Society of Photo-Optical Instrumentation Engineers (SPIE) Conference*, June 2006.

[26] I. Poulakakis, J. A. Smith, and M. Buehler. Modeling and Experiments of Untethered Quadrupedal Running with a Bounding Gait: The Scout II Robot. *Int. Journal of Robotics Research*, 24(4):239–256, 2005.

[27] M. Raibert. *Legged robots that balance*. MIT Press series in artificial intelligence. MIT Press, Boston, 1986.

[28] J. J. Robilliard and A. M. Wilson. Prediction of kinetics and kinematics of running animals using an analytical approximation to the planar spring-mass system. *J Exp Biol*, 208(23):4377–4389, 2005.

[29] U. Saranli. *Dynamic Locomotion with a Hexapod Robot*. PhD thesis, The University of Michigan, Ann Arbor, MI, September 2002.

[30] U. Saranli, M. Buehler, and D. E. Koditschek. RHex: A simple and highly mobile robot. *Int. Journal of Robotics Research*, 20(7):616–631, July 2001.

[31] U. Saranli, W. J. Schwind, and D. E. Koditschek. Toward the control of a multi-jointed, monoped runner. In *Proc. of the IEEE Int. Conf. On Robotics and Automation*, volume 3, pages 2676–82, New York, 1998.

[32] A. Sato and M. Buehler. A planar hopping robot with one actuator: design, simulation, and experimental results. In *Proceedings of the IEEE/RSJ Int. Conf. on Intelligent Robots and Systems*, volume 4, pages 3540–3545, Sept.-2 Oct. 2004.

[33] J. Schmitt. A simple stabilizing control for sagittal plane locomotion. *J. of Computational and Nonlinear Dynamics*, 1(4):348–357, 2006.

[34] W. J. Schwind. *Spring Loaded Inverted Pendulum Running: A Plant Model*. Phd, University of Michigan, 1998.

[35] W. J. Schwind and D. E. Koditschek. Approximating the stance map of a 2-dof monoped runner. *J. of Nonlinear Science*, 10:533–568, 2000.

[36] A. Seyfarth, H. Geyer, M. Gunther, and R. Blickhan. A movement criterion for running. *J. of Biomechanics*, 35(5):649–655, May 2002.

[37] M. J. Spenko, G. C. Haynes, J. A. Saunders, M. R. Cutkosky, A. A. Rizzi, R. J. Full, and D. E. Koditschek. Biologically inspired climbing with a hexapedal robot. *J. of Field Robotics*, 25(4-5):223–242, 2008.

[38] M. Srinivasan and P. Holmes. How well can spring-mass-like telescoping leg models fit multi-pedal sagittal-plane locomotion data? *J. of Theoretical Biology*, 255(1):1 – 7, 2008.

[39] M. Vukobratovic and B. Borovac. Zero-moment point-thirty five years of its life. *Int. Journal of Humanoid Robotics*, 1(1):157–173, 2004.

[40] E. Westervelt, J. Grizzle, and D. Koditschek. Hybrid zero dynamics of planar biped walkers. *IEEE Transactions on Automatic Control*, 48(1):42–56, January 2003.

On Motion and Force Control of Grasping Hands with Postural Synergies

D. Prattichizzo[1,3], M. Malvezzi[1], A. Bicchi[2,3]

[1] Dept. of Information Engineering, University of Siena, Siena, Italy
[2] Interdept. Research Center "E. Piaggo", University of Pisa, Pisa, Italy
[3] IIT - Istituto Italiano di Tecnologia, Genova, Italy

Abstract— **The complexity of robotic hands is needed to adapt devices to the many kinds of tasks, but the large number of motors needed to fully actuate the DoFs comes at the cost of size, complexity and weight of devices. A possible approach to solve this problem consists of reducing the number of actuators thus resulting more efficient, simpler and reliable than their fully actuated alternatives. Reducing control inputs seems to inspire also biological systems and in particular motor control of human hands, which share with robotic hands the large number of DoFs. Recent studies demonstrated that a few control variables, named postural synergies, are able to account for most of the variance in the patterns of hand movements and configurations of hands. This paper focuses on hands with postural synergies. Reducing the number of control inputs, from fully actuated joints to few synergies, might reduce the dimension of the force and motion controllability subspaces thus compromising the dexterity of the grasp, however, this is not true in general but strongly depends on how synergies are distributed. The paper investigates to what extent a hand with many DoFs can exploit postural synergies to control force and motion of the grasped object.**

I. Introduction

Robotic hands have many degrees of freedom distributed among several kinematic chains, the fingers. The complexity of the mechanical design is needed to adapt hands to the many kinds of tasks required in unstructured environments. Roboticists over the years have attempted to imitate the human hand in terms of dexterity and adaption capabilities. Some remarkable example of robotic hand design are the DLR hand II [6] and the UTAH/MIT hand with 16 actuated joints, 4 per each finger [9]. One of the main issues in designing and controlling robotic hands is that a large number of motors is needed to fully actuate the degrees of freedom but this comes at the cost of size, complexity and weight of the device. This disadvantage could be overtaken if the robotic hands were actuated and controlled by a reduced number of inputs, thus resulting more efficient, simpler and reliable than their fully actuated alternatives as shown in [3, 4].

A simplified control seems to inspire also biological systems and in particular motor control of human hands, which share with robotic ones the large number of degrees of freedom. Recent studies in neuroscience [13, 14] demonstrated that a limited set of input variables, named postural synergies, are able to describe most of the variance in hand movements and configurations in manipulation and grasping tasks.

Recently, these studies on human hands inspired new researches on design and control strategies for robotic hands whose main issue is to achieve a trade-off between simplicity, gained through synergy based control, and its versatility [5, 8]. In [8] the synergy idea concept has been applied to control different hand models: a simple gripper, the Barrett hand, the DLR hand, the Robonaut hand and the human hand model. In [5] authors proposed a robotic hand design able to match postural synergies mechanically coupling motion of the single joints.

Postural synergies in robotic hands allow to control the whole device through a lower dimension set of actions. Postural synergies can be easily derived for a given robotic hand, designed with postural synergies, while in human hands it can be evaluated performing a Principal Component Analysis (PCA) of hand postures during grasping operations [13].

To the best of our knowledge, the research in this field lacks of some basic and structural results which relates for instance the number and types of synergies to the possibility of controlling contact forces and object motion in grasping and manipulation tasks. This is the main focus of this paper.

This paper builds upon previous contributions by the authors [10] and extends results to hands controlled by postural synergies.

II. Modeling hands with synergies

A. Kinematics of the grasp

Consider a robotic hand that grasps an object as in Fig. 1. Let $\{N\}$ represent the inertial frame fixed in the workspace and let frame $\{B\}$ be fixed to the object. Let n_c be the number of contact points between the object and the grasp. Contacts may occur at any place of the robotic hand. At contact point i, though as fixed to the object, the frame $\{C_i^o\}$ is defined, with axes $\{\hat{n}_i^o, \hat{t}_i^o, \hat{o}_i^o\}$. The unit vector $\hat{n}_i^o$ is normal to the contact tangent plane, and directed toward the object. The other two unit vectors are orthogonal and lie in the tangent plane of the contact. Similarly we can define the frame $\{C_i^h\}$, fixed to the contact point thought as fixed to the hand. Let $u \in \mathbb{R}^6$ denote the vector describing the position and orientation of $\{B\}$ relative to $\{N\}$. Vector $\tilde{c}_i^o \in \mathbb{R}^6$ ($\tilde{c}_i^h \in \mathbb{R}^6$) describes the position and orientation of the i-th contact reference frame

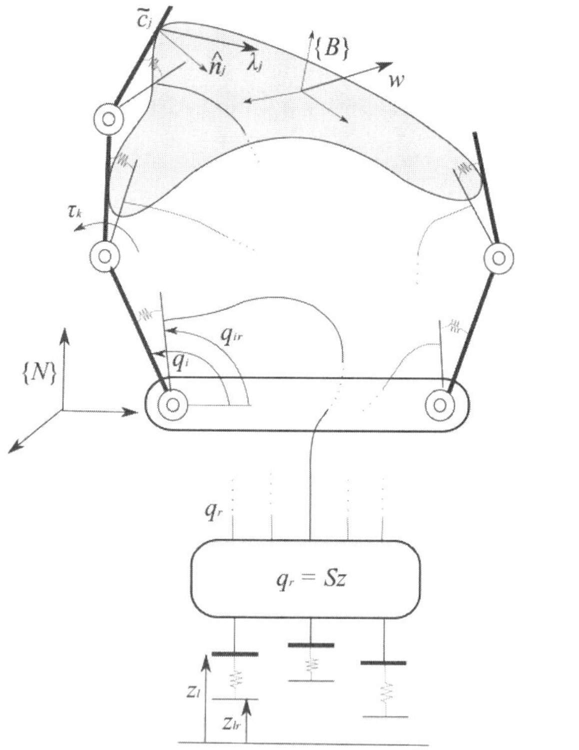

Fig. 1: Hand-object grasp with postural synergies: main quantities.

Notation	Definition
$u \in \mathbb{R}^6$	position and orientation of object
$w \in \mathbb{R}^6$	external wrench applied to the grasped object
n_c	number of contact ponits
C_i^o	reference system at the i-th contact point on the object
$\tilde{c}_i^o \in \mathbb{R}^6$	position and orientation of reference frame C_i^o
C_i^h	reference system at the i-th contact point on the hand
$\tilde{c}_i^h \in \mathbb{R}^6$	position and orientation of reference frame C_i^h
λ_i	vector of forces (and moments) at the contact i
n_q	number of joints
$q \in \mathbb{R}^{n_q}$	actual joint variables
$q_r \in \mathbb{R}^{n_q}$	reference joint variables
τ	vector of joint forces and torques
n_z	number of postural synergies
$z \in \mathbb{R}^{n_z}$	actual synergy variables
$z_r \in \mathbb{R}^{n_z}$	reference synergy variables
σ	generalized forces along synergies
$G \in \mathbb{R}^{6 \times n_l}$	grasp matrix
$J \in \mathbb{R}^{n_l \times n_q}$	hand jacobian matrix
$X \in \mathbb{R}^{n_q \times n_q}$	joint displacement gain matrix ($\Delta q = X \Delta q_r$)
$Y \in \mathbb{R}^{n_z \times n_z}$	postural synergy gain matrix ($\Delta z = Y \Delta z_r$)
$S \in \mathbb{R}^{n_q \times n_z}$	synergy matrix ($\Delta q_r = S \Delta z$)

TABLE I: Primary notation for grasp analysis.

$\{C_i\}$, thought as fixed to the object (hand), relative to $\{N\}$. Group all these vectors in the overall contact vector $\tilde{c}^o = [\tilde{c}_1^{o\mathrm{T}}, \cdot, \tilde{c}_{nc}^{o\mathrm{T}}]^\mathrm{T}$. Let $q = [q_1 \ \cdots \ q_{n_q}]^\mathrm{T} \in \mathbb{R}^{n_q}$ define the vector of actual joint displacements, and let $\tau \in \mathbb{R}^{n_q}$ represent joint loads (forces in prismatic joins and torques in revolute joints). Let $f \in \mathbb{R}^3$ be the force applied to the object at the point p and let $m_u \in \mathbb{R}^3$ be the applied moment. These are combined into the object load, or wrench, denoted by $w = \left[f^\mathrm{T} \ m^\mathrm{T}\right]^\mathrm{T} \in \mathbb{R}^6$.

For *small* displacements some relevant linear relationships can be defined between the vectors of the many reference systems, such as the contact reference frames, and other relevant variables such as joint displacements. In the following small variations from a reference configuration will be indicated with the prefix Δ. Two matrices are of the utmost importance in grasp kinematic analysis: the Grasp Matrix G and the Hand Jacobian J Following definitions reviewed in [12], let us introduce matrices $\tilde{G}$ and $\tilde{J}$: the transpose of the complete Grasp Matrix $\tilde{G}^\mathrm{T} \in \mathbb{R}^{6n_c \times 6}$ maps the object displacement Δu to the displacements of all the n_c contact frames $\Delta \tilde{c}^o$

$$\Delta \tilde{c}^o = \tilde{G}^\mathrm{T} \Delta u \tag{1}$$

while the complete Hand Jacobian Matrix $\tilde{J} \in \mathbb{R}^{6n_c \times n_q}$ relates the joint displacement variation to the displacements of the contact frame fixed to the hand structure:

$$\Delta \tilde{c}^h = \tilde{J}^\mathrm{T} \Delta q \tag{2}$$

In order to define the kinematic constraint and the contact forces imposed by the contact between the hand and the object, a suitable contact model has to be introduced. Two contact models have been considered in this work: hard-finger (HF), and soft-finger (SF) [12]. These models select components of the contact velocities to be constrained by the contact model. This is done by equating a subset of n_l components of the hand and object contact velocities. The corresponding components of the contact force and moment are considered without constraints imposed by the friction model. The HF model constraints the three components ($n_l = 3$) of the linear velocity while the SF model constraints also the angular velocity about the normal at the contact ($n_l = 4$).

The constrained velocities components are coded in the Selection Matrix $H \in \mathbb{R}^{n_l \times 6n_c}$ [12] which selects n_l components of the relative contact twists for all the contacts and sets them to zero:

$$H(\Delta \tilde{c}^h - \Delta \tilde{c}^o) = 0. \tag{3}$$

Finally, from (1) and (2) into (3) the following constraint equation is obtained:

$$\begin{bmatrix} J & -G^\mathrm{T} \end{bmatrix} \begin{bmatrix} \Delta q \\ \Delta u \end{bmatrix} = 0 \tag{4}$$

where the Grasp Matrix and Hand Jacobian are:

$$\begin{aligned} G^\mathrm{T} = H\tilde{G}^\mathrm{T} & \in \mathbb{R}^{n_l \times 6} \\ J = H\tilde{J} & \in \mathbb{R}^{n_l \times n_q} \end{aligned} \tag{5}$$

For more details on the construction of H, G, and J readers are referred to [12] and therein references.

B. Quasistatic model of the grasp

The forces and torques acting on the system composed of the hand and the grasped object are the object external wrench w, the forces/torques applied to the joints τ and the contact forces λ exchanged between the hand and the object at the contact points. The static equilibrium of the hand and of the object is given by

$$\tau = J^{\mathrm{T}}\lambda \tag{6}$$

$$w = -G\lambda \tag{7}$$

The general solution of eq. (7), assuming that w is in the column space of G, $\mathcal{R}(G)$, is:

$$\lambda = -G^{+}w + A\xi \tag{8}$$

where G^{+} is a generic right–inverse of the grasp matrix and $A \in \mathbb{R}^{n_l \times n_h}$ is a matrix whose columns form a basis of the nullspace of G, $\mathcal{N}(G)$, and the vector $\xi \in \mathbb{R}^{n_h}$ parametrizes the homogeneous part of the solution. The term $A\xi$ represents the solution to (7) when no external load w is applied and are usually referred to as internal forces. The control of internal forces is paramount to stabilize the grasp since these allow to satisfy the contact constraints. In previous works [1] it has been shown that for general grasp kinematics, e.g. in wholearm grasps where contacts are allowed with inner part of the fingers or the palm [2], controlling internal forces is not straightforward since the number of internal forces directions, i.e. the dimension of the subspace $\mathcal{N}(G)$, turns to be larger than the number of controlled joint actions. The problem of controlling internal forces becomes even worse in hands controlled with synergies since the number of synergy actions is even lower than the number of joints.

C. Hands controlled with postural synergies

We suppose that the hand is actuated using a number of inputs whose dimension is lower than the number of hand joints and we define it as *synergies*. The synergies are then collected in a vector $z \in \mathbb{R}^{n_z}$.

Recent results on the organization of the human hand in grasping and manipulation have inspired many research activities in robotics and more in general in hand design and control. In [13] authors proved that, notwithstanding the complexity of the human hand, a few variables are able to account for most of the variance in the patterns of human hands configuration and movement. These conclusions were based on the results of experimental tests in which subjects were asked to perform grasping actions on a wide variety of objects. Data were recorded by means of data gloves and were analyzed with principal component analysis techniques. In [7], the authors deal with robotic grasping with synergies and refer to the single principal component, or synergy, as an eigengrasp which can also be thought of as a special direction of motion in the joint space of the hand. Robotic hand design has been inspired by the synergy organization of the human hand [5]. The numerical example section of this paper deals with the Barrett Hand, a robotic hand with 8 joints controlled through 4 postural synergies corresponding to 4 motors. It is worth noting that the number of synergies is always lower than the number of joints and in general, according to the performed grasping tasks more than two synergies can be engaged.

This paper refers to postural synergies no matters what type of grasp, human or robotic, is considered. Differently from other approaches where the actual joint variables is a linear combination of synergies [5, 7], in this paper we define the postural synergies as a joint displacement aggregation corresponding to a reduced dimension representation of hand movements according to a compliant model of joint torques.

Definition 1: The reference vector q_r for joint variables is a linear combination of postural synergies $z \in \mathbb{R}^{n_z}$ with $n_z \leq n_q$

$$q_r = Sz \tag{9}$$

through the *synergy matrix* $S \in \mathbb{R}^{n_q \times n_z}$, whose columns describes the shapes, or directions, of each synergy i the joint space.

To map the postural synergies Sz to the actual joint variables, a compliant model for joint torques has been chosen:

$$q_r - q = C_q\tau \tag{10}$$

where $C_q \in \Re^{n_q \times n_q}$ is the compliance matrix that takes into account the static gains of the joint torque control and possibly the hand link compliance, while τ are the generalized force/torque applied to the joints (Fig. 1).

From (6) and (9), eq. (10) can be rewritten as

$$Sz - q = C_q J^{\mathrm{T}}\lambda. \tag{11}$$

A compliant model of the synergy actuation is here assumed (Fig. 1). The synergy actuator generalized forces σ are proportional to the difference between the reference and the actual synergy values:

$$\sigma = C_z^{-1}(z_r - z) \tag{12}$$

where $C_z \in \mathbb{R}^{n_z \times n_z}$ is a matrix whose elements are the reciprocal of the static gains of the synergy motor control.

D. Forces and object displacements controlled by synergies

Consider an equilibrium configuration where an object with an external wrench w_0 is grasped by a hand whose synergy values are z_0 and the corresponding joint displacements are q_0. The contact forces in this reference equilibrium are λ_0. Starting from this equilibrium configuration, we consider a variation of the input synergy reference values Δz_r, which leads to an actual variation of the postural synergies Δz, to a variation of the joint displacement Δq and a variation of contact forces $\Delta\lambda$ for the new equilibrium configuration of the quasi-static model. We suppose that the object wrench w_0 is kept constant. Lets furthermore assume that the variation of the jacobian matrix J due to the joint displacement is disregardable. In the new equilibrium configuration, the object is in a static equilibrium and all the contact forces $\Delta\lambda$ result to be self-balanced

$$G\Delta\lambda = 0. \tag{13}$$

In other terms, the activation of postural synergy displacements Δz_e induces a variation of contact force $\Delta\lambda \in \mathcal{N}(G)$, the internal forces subspace.

As discussed in [12], when the dimension of the contact force subspace is larger than the dimension of the control inputs, the grasp can results to be statically indeterminate, or hyperstatic and this typically happens for robotic hands controlled with a few postural synergies. In this case a possible way to analyze the grasp, and in particular to study the contact forces, is to relax the rigid contact hypothesis and consider elastic contacts: the kinematic constraint imposed by (4) is no longer satisfied and, as discussed in [1], the contact force $\Delta\lambda$ is modeled as

$$\begin{bmatrix} J & -G^{\mathrm{T}} \end{bmatrix} \begin{bmatrix} \Delta q \\ \Delta u \end{bmatrix} = C_s \Delta\lambda \tag{14}$$

where $C_s \in \mathbb{R}^{n_l \times n_l}$ denotes the contact compliance matrix. From (11), the variation $\Delta\lambda$ corresponding to the above described variation from the reference configuration can be written as

$$JS\Delta z - J\Delta q = JC_q J^{\mathrm{T}} \Delta\lambda. \tag{15}$$

Then summing up (14) and (15) we get

$$JS\Delta z - G^{\mathrm{T}}\Delta u = \left(C_s + JC_q J^{\mathrm{T}}\right)\Delta\lambda. \tag{16}$$

from which we can compute the contact forces as a function of synergy and object posture displacements

$$\Delta\lambda = K\left(JS\Delta z - G^{\mathrm{T}}\Delta u\right) \tag{17}$$

where the total stiffness matrix K takes into account the contact compliance C_s and the joint control gains C_q

$$K = \left(C_s + JC_q J^{\mathrm{T}}\right)^{-1} \tag{18}$$

Note that eq. (17) relates contact force displacements $\Delta\lambda$ to both synergy and object displacements Δz and Δu, but object and synergy displacements are not independent variables. Inspired by [1], we will use the Principle of Virtual Works to evaluate the quasi-static changes $\Delta\lambda$ generated, or controlled, by the postural synergy displacement Δz only, and similarly we will compute the quasi-static changes Δu generated by the postural synergy displacement Δz only.

It is worth underlying that this work studies the effect on the grasp due to changes of the postural synergies which play the role of controlled variables. No other actions is considered on the grasp. For example we assume that no change on the external wrench occurs.

The Principle of Virtual Works, applied to the configuration obtained applying a synergy variation Δz to a reference condition, leads to

$$\Delta\lambda^{\mathrm{T}} G^{\mathrm{T}} \delta u = 0 \tag{19}$$

where δu indicates the *virtual displacement* of the object reference frame, i.e. an infinitesimal displacement compatible with the contact constraints imposed by the hand. From (17) we obtain

$$\Delta z^{\mathrm{T}} S^{\mathrm{T}} J^{\mathrm{T}} K^{\mathrm{T}} G^{\mathrm{T}} \delta u = \Delta u^{\mathrm{T}} G^{\mathrm{T}} K^{\mathrm{T}} G \delta u \tag{20}$$

which, holding true for any arbitrary object virtual displacement δu, can be simplified as

$$GKJS\Delta z = GKG^{\mathrm{T}}\Delta u \tag{21}$$

Thus object displacement from one equilibrium configuration to another, generated by a synergy change Δz is

$$\Delta u = \left(GKG^{\mathrm{T}}\right)^{-1} GKJS\Delta z \tag{22}$$

and from (17) one gets the contact force changes $\Delta\lambda$ generated by Δz as

$$\Delta\lambda = \left(I - G_K^{+} G\right) KJS\Delta z \tag{23}$$

where G_K^{+} is the pseudoninverse of grasp matrix G weighted with the stiffness matrix K (18). From (23) and (15) we can then express the variation of joint displacement Δq due to an activation Δz of synergies as

$$\Delta q = X\Delta q_r = XS\Delta z \tag{24}$$

with $X = I - C_q J^{\mathrm{T}}\left(I - G_K^{+}G\right)KJ\right)$. We will refer to matrix X as the joint displacement gain matrix.

Remark 1: Defining the postural synergies as in Definition 1 with compliance models allows to still find a linear relationship between joint variables and synergies but trough matrix XS as in (24) and not through the synergy matrix S only as in [5, 7].

The linear map (24) between Δq and Δz is yet not complete. In fact, according to the compliance model pictorially described in Fig. 1, we still need to relate the reference changes for synergies Δz_r to joint displacements Δq. After some algebra one gets that

$$\Delta z = Y\Delta z_r \tag{25}$$

with $Y = \left(S^{\mathrm{T}} X^{\mathrm{T}} C_q^{-1}\left(I - X\right) SC_z + I\right)^{-1}$. Matrix Y will be referred to as the postural synergy gain matrix.

Finally, from (25) and Definition 1, we can express the joint displacement variation due to a variation of the reference postural synergies as

$$\Delta q = XSY\Delta z_r \tag{26}$$

Consequently the object displacement variation Δu and the internal force variation $\Delta\lambda$ can be evaluated as a function of the synergy reference variation Δz_r as

$$\Delta u = \left(GKG^{\mathrm{T}}\right)^{-1} GKJSY\Delta z_r \tag{27}$$

$$\Delta\lambda = \left(I - G_K^{+}G\right) KJSY\Delta z_r \tag{28}$$

In (28) the first matrix on the right side is a projector onto the nullspace of matrix G. Consequently $\Delta\lambda$ belongs to the nullspace of the grasp matrix. In other terms the quasistatically controllable contact forces are internal forces which do not affect the motion of the object. This is an intuitive result since the quasi-static model moves from one equilibrium configuration to another where the object does not move.

The control of internal forces is paramount in robotic grasping [12]. It allows to steer the contact forces to satisfy the constraints imposed by friction models at the contacts thus guarantying to not loose the contact with the object which would compromise the whole grasp.

From (28) , define a basis matrix E_s for the subspace of controllable internal forces by postural synergies as

$$\mathcal{R}(E_s) = \mathcal{R}\left(\left(I - G_K^+ G\right) KJSY\right) \quad (29)$$

Concluding all internal forces controllable by synergy actions can be parametrized through a free vector as $E_s\alpha$.

Remark 2: Starting from a reference configuration and acting on synergies, the joint displacements depends both on the synergy matrix S and on the whole system compliance, and then in the more general case $\delta q \neq S\Delta z_r$.

If the synergy actuation is perfectly stiff, i.e. if $C_z = 0$, it is clear from (25) that $Y = I$ and thus $\Delta z = \Delta z_r$. Furthermore if the links are perfectly stiff and the joint control gains are infinite, i.e. $C_q = 0$, from (24) it results that $X = I$ and $\Delta q = S\Delta z$.

Summarizing in case of $C_z = 0$ and $C_q = 0$ one gets a simplified version of eq. (26)

$$\Delta q = S\Delta z = \Delta z_r \quad (30)$$

which is similar to the definition of synergy control given in [5, 7]. Note that (30) holds true for contacts that are not perfectly indeformable, i.e. $C_s \neq 0$.

E. Control of rigid body object motion

Eq. (27) shows how the object displacements Δu are controlled from one equilibrum configuration to another by synergy small variations Δz_r.

Among all the possible motions of the grasped objects, rigid-body motion is perhaps the most interesting since it does not involve visco-elastic deformations in the contact points. Rigid-body motion can be regarded as low energy motions, in other words they represent the natural way to change the posture of the grasped object. Rigid-body motion controllable by synergies has to be compatible with kinematic contact constraint (4) and (26) which relates controlled postural synergies and joint displacements.

Thus a description of this motion can be obtained computing $\mathcal{N}\left[JXSY \quad -G^{\mathrm{T}}\right]$. Let's then define a matrix Γ, whose columns form a basis of such subspace. Under the hypothesis that the object motion is not indeterminate [12], i.e. $\mathcal{N}(G^{\mathrm{T}}) \neq 0$, i.e. the object is completely restrained by contacts, matrix Γ can be expressed as

$$\Gamma = \mathcal{N}\left[\begin{array}{cc} JXSY & -G^T \end{array}\right] = \left[\begin{array}{cc} \Gamma_{zr} & \Gamma_{zcs} \\ 0 & \Gamma_{ucs} \end{array}\right] \quad (31)$$

where Γ_{zr} is a basis matrix of the subspace of redundant motions $\mathcal{N}(JXSY)$, and Γ_{zcs} and Γ_{ucs} are conformal partitions of a complementary basis matrix. The image spaces of Γ_{zcs} and Γ_{ucs} consist of coordinated rigid–body motions of the mechanism, for the postural synergy references and the object position and orientation, respectively.

Coming back to the rigid-body object motion controlled by synergies, as already discussed it is easy to show that

$$\mathcal{R}(\Gamma_{ucs}) \subseteq \mathcal{R}((GKG^{\mathrm{T}})^{-1}GKJSY). \quad (32)$$

i.e. rigid-body motions of the object are not all the possible motions of the object controlled by synergies as in (27). The subspace of all synergy controlled object motion $\mathcal{R}((GKG^T)^{-1}GKJSY)$ also contains motions due to deformations of elastic elements in the model.

Concluding all rigid-body displacements of the object can be parametrized through a free vector as $\Gamma_{ucs}\beta$.

III. Main Result

In grasps by hands controlled with synergies it is possible that not all the object motions and contact forces result controllable by synergy actions. According to (31) and (32), desired quasi-static rigid-body object motions Δu_{des} can be performed if they remain within subspace $\mathcal{R}(\Gamma_{ucs})$ defined in (31) and analogously, according to (28) and (29), arbitrary quasi-static contact force displacements $\Delta\lambda_{des}$ can be performed if they evolve within subspace $\mathcal{R}(E_s)$ defined in (29). It is worth noting that with the results obtained up to here, we can arbitrarily control motions in $\mathcal{R}(\Gamma_{ucs})$ or contact forces in $\mathcal{R}(E_s)$ when we do not want to jointly control them. In other terms we are not guaranteed that in coupled motion and force control, we can jointly control two vectors lying on these subspaces.

However in grasping, due to the presence of unilateral, conic contact constraints, task specifications can not be given disjointly in terms of either object positions or contact forces. Therefore conditions $\Delta u_{des} \in \mathcal{R}(\Gamma_{ucs})$ and $\Delta\lambda_{des} \in \mathcal{R}(E_s)$ are only necessary, but no longer sufficient, for joint control of object motions and contact forces. Moreover, specifications of jointly controllable object motions and contact forces may not exhaust the control capabilities of synergy actions for the given grasp due to the presence of synergy redundancy.

Our goal is therefore to define a set of controlled outputs for a grasp with synergies that is guaranteed to be feasible with synergy actions, that fully exploits the control inputs and that is convenient for the specification of the tasks. The first requirement implies that the output vector of forces and motions can be controlled by synergies, the second that controlled output vector has the same dimension n_z of the synergy vector z_r and the third that the output vector consider the typical approach of a grasping task:

- contact forces that can be controlled so as to avoid violation of contact constraints;
- object trajectories that can be accommodated for by the grasp with synergies;
- reconfiguration of limbs in presence of redundancy in synergies.

The following theorem proposes a set of outputs for grasps with synergies.

Theorem 1: Under the technical assumption that the grasp is not indeterminate ($\mathcal{N}(G^{\mathrm{T}}) = 0$), and the symplifying assumption of Remark 2, i.e. $C_z = 0$ and $C_q = 0$ ($X = I$ and $Y = I$) consider the quasi–static model of any grasp with synergies described in (27) and (28). It is always possible to control, jointly but independently, the controllable internal forces, the rigid–body object motions and redundancy with the control input as synergy displacement Δz_r. Algebraically, this corresponds to state that for any α, β and γ, there always exists a Δz_r solving the linear system of equations

$$\begin{bmatrix} E_s\alpha \\ \Gamma_{ucs}\beta \\ \Gamma_{zr}\gamma \end{bmatrix} = \begin{bmatrix} (I - KG^T(GKG^T)^{-1}G)KJS \\ (GKG^T)^{-1}GKJS \\ I \end{bmatrix} \Delta z_r \tag{33}$$

where Γ_{ucs} and Γ_{zr} have been defined in (31), and E_s has been defined in (29).

Moreover, solution for Δz_r is unique and the number of synergies n_z is equal to the sum of the dimensions of the controlled output subspaces:

$$n_z = \dim(E_s) + \dim(\Gamma_{ucs}) + \dim(\Gamma_{zr}) \tag{34}$$

Proof:

Linear system (33) can be rewritten as

$$\begin{bmatrix} \alpha \\ \beta \\ \gamma \end{bmatrix} = \begin{bmatrix} E_s^+(I - KG^T(GKG^T)^{-1}G)KJS \\ \Gamma_{ucs}^+(GKG^T)^{-1}GKJS \\ \Gamma_{zr}^+ \end{bmatrix} \Delta z_r \tag{35}$$

where $B^+ = (B^{\mathrm{T}}B)^{-1}B^{\mathrm{T}}$ denotes the pseudoinverse of a basis matrix B.

The linear system is square if the number $\#(\Gamma_{ucs}) + \#(E_s) + \#(\Gamma_{zr}) = n_z$ where $\#(N)$ denotes the number of colums of matrix N,

Since Γ_{zr}, Γ_{ucs}, and E_s are full column rank by definition, from (31) we have $\#(\Gamma_{ucs})+\#(\Gamma_{zr}) = \dim(\mathcal{N}[JS \quad -G^T]) - \dim(\mathcal{N}(G^T))$. Observing that $\mathcal{N}(I-G_K^+G) = \mathcal{R}(KG^T)$, from (29) we obtain

$$\begin{aligned} \#(E_s) &= \#(JS) - \dim(\mathcal{N}(JS)) - \dim(\mathcal{R}(JS) \cap \mathcal{R}(G^T)) = \\ &= q - \dim(\mathcal{N}(JS)) - \big[\dim(\mathcal{N}([JS \quad -G^T])) \\ &\quad -\dim(\mathcal{N}(JS)) - \dim(\mathcal{N}(G^T))\big] = q - \#(\Gamma_{zr}) - \#(\Gamma_{ucs}); \end{aligned}$$

To complete the proof it suffices to show that coefficient matrix in (35) is full row rank which is equivalent to prove that

$$\mathcal{N}\left(\begin{bmatrix} E_s^+(I - G_K^+G)KJS \\ \Gamma_{ucs}^+(GKG^T)^{-1}GKJS \\ \Gamma_{zr}^+ \end{bmatrix}^{\mathrm{T}}\right) = \{0\}. \tag{36}$$

Observe that each block of the matrix in the equation above is full column rank, in fact

i: E_s is a basis for $\mathcal{R}((I - G_K^+G)KJS)$ (cf. (29));

ii: $\mathcal{R}(\Gamma_{ucs}) \subseteq \mathcal{R}((GKG^T)^{-1}GKJS)$, directly from (32);

iii: Γ_{zr} is a basis matrix for $\mathcal{N}(JS)$;

Hence, to prove (36) it is sufficient to show that the raw spaces of the three blocks are also mutually linearly independent and this directly follows from these observations.

iv: The columns of the third block span $\mathcal{N}(JS)$, while the column space of the first two blocks lies within $\mathcal{R}(S^{\mathrm{T}}J^{\mathrm{T}})$;

v: $\mathcal{R}(G_K^+\Gamma_{ucs})$ and $\mathcal{R}(I - G_K^+G)KE_s)$ are disjoint, then so are the spans of the columns of the first and second blocks.

□

Remark 3: The result in (34) deals with dimensions of subspaces and is numerical in nature. It states a very basic structural property of grasp analysis with postural synergies: if n_z control inputs are available, one cannot control, jointly and independently, more than n_z variables among internal forces, object motion directions and kinematic redundancy. If one wants to control more than n_z output variables between motions and forces, it happens that the grasp can still accomplish the requirements but in this case desired object motions and internal forces cannot be independently assigned. This means for instance that one cannot find contact forces satisfying friction constraints to track a desired object trajectory o vice versa.

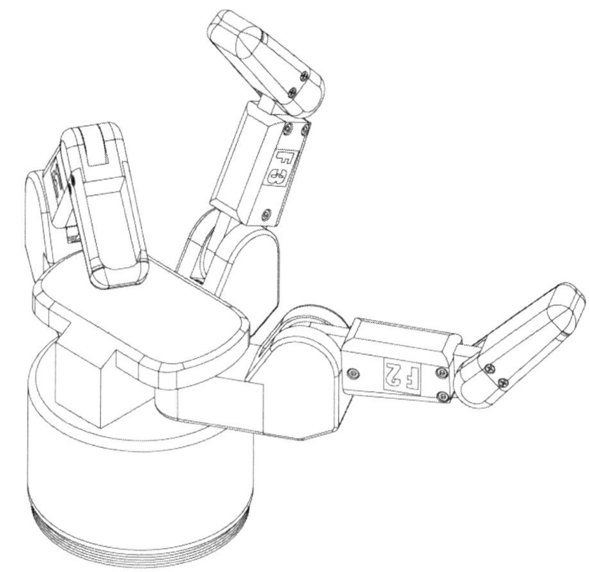

Fig. 2: The Barrett Hand.

Remark 4: The result in (33) deals with grasp control with postural synergies. When the mechanical structure is complex, with many joints, but the control inputs are few, it is not easy to understand which synergy one needs to activate to accomplish a given tasks. The solution of linear system (33) allows to simply compute the control variables, the synergy references, according to a task to be performed.

The theorem is useful also to find the minimal design requirements in terms of number of synergies to be used to accomplish the given task.

It is worth underlying that the motion of the object considered in this paper are motions performed with respect to the palm of the hand. In other terms we are considering fine motion control of grasped objects more than the large displacements which can be performed with moving the wrist with a robot arm, and not with the hand joints.

Finally, note that results presented in this paper still work for fully actuated robotic hands provided that matrix S is substituted with the identity matrix.

IV. Numerical Examples

Results on the motion and force control of grasps with synergies have been applied to the Barrett Hand, shown in Figure 2 : it is a three finger, eight-axis mechanical hand, in which each finger has two joints. One of the fingers, referred to as 1, is stationary, while the other two can spread synchronously up to 180 degrees about the palm. Although there are eight axes, the hand is actuated by four motors: each finger has an actuated inner link and a coupled outer link that moves at a fixed rate with the inner link. An additional motor control the synchronous spread of the two fingers about the palm. A clutch mechanism allows the outer link to continue to move even if the inner link motion is obstructed, however this feature has not been considered in the presented analysis.

Let us refer with $\theta_{i,1}$ $(i = 1,..,3)$ the rotation of the inner link with respect to the palm, with $\theta_{i,2}$ $(i = 1,..,3)$ the rotation of the outer link with respect to the inner one and with $\theta_{i,0}$ $(i = 2,3)$ the spread of the two fingers about the palm. Thus the configuration vector can be defined as: $q = [\theta_{1,1}, \theta_{1,2}, \theta_{2,0}, \theta_{2,1}, \theta_{2,2}, \theta_{3,0}, \theta_{3,1}, \theta_{3,2}]^T$.

The mechanical couplings between the joints are expressed by the following relations:

$$\begin{aligned} \theta_{2,0} &= -\theta_{3,0} = z_1 \\ \theta_{2,i} &= \alpha_i \theta_{1,i} = z_{i+1} \quad i = 1,..,3 \end{aligned}$$

where α_i represents the ratio between the outer and the inner angle for the i-th link. The joint angles are controlled acting on four parameters, collected in the vector $z = [z_1, ..., z_4]^T$. Accordingly, the synergy matrix can be defined as

$$S = \begin{bmatrix} 0 & 1 & 0 & 0 \\ 0 & \alpha_1 & 0 & 0 \\ 1 & 0 & 0 & 0 \\ 0 & 0 & 1 & 0 \\ 0 & 0 & \alpha_2 & 0 \\ -1 & 0 & 0 & 0 \\ 0 & 0 & 0 & 1 \\ 0 & 0 & 0 & \alpha_3 \end{bmatrix}. \qquad (37)$$

A reference value for the input parameters z_r has then been selected. By means of the synergy matrix and the forward kinematic analysis the hand configuration has been defined and the grasping has been analysed. The contact points between the hand and the grasped object were located on the three finger tips and the normal directions at the contact points have been thought as oriented towards the center of the object. Both HF and SF contact models have been considered.

Matrix G dimensions are 6×9 in case for HF and 6×12 in case of SF contact models, while dimensions of $\mathcal{N}(G)$ are 3 and 6, respectively for the same cases. The contact compliance matrix C_s has been chosen as a n_l–dimensional identity matrix where $n_l = 9$ for the HF model and $n_l = 12$ for the SF model. The compliance matrices C_q and C_s have been considered zero, so that, according to remark 2, X and Y are equal to identity matrices with dimensions n_q and n_z respectively.

According to the previously described analysis, the dimension of the controllable internal forces and object motions have been evaluated with both the hypothesis that the hand is controlled with the four synergies, as in the real case, and considering a case when all the eight joints are actuated. Table II summarizes the obtained results, in particular the dimensions of the controllable internal forces, rigid body motions and hand

	HF			SF		
	E	Γ_{ucs}	Γ_{zr}	E	Γ_{ucs}	Γ_{zr}
synergies (4 inputs)	3	0	1	4	0	0
fully actuated (8 joints)	3	2	3	6	1	1

TABLE II: Barrett Hand: controllable internal forces and allowable movements for the synergy actuated (4 degrees of freedom) and fully actuated (8 degrees of freedom) hand.

redundancy subspaces. We can observe that in all the cases equation (34) on subspace dimensions is verified, since the sum between the dimensions of E_s, Γ_{ucs} and Γ_{zr} is equal to the number of synergies or to the number of joints.

Figure 3 shows, for the first three synergies, the hand motion (first row) and the set of internal contact forces generated acting through each single synergy, evaluated by means of equation (28) (second row). The results shown in the figure have been obtained considering the HF contact model. Only three synergies have been represented since the activation of the fourth synergy does not produce internal forces, this result can be justified by the observation that the nullspace of JS, whose basis is the matrix Γ_{zr} is not empty thus the system is redundant.

It is worth noting that in the given grasping configuration the four synergies of the Barret Hand are not able to control any rigid-body motion of the object, since Γ_{ucs} is zero as shown in Table II.

V. Conclusions

In grasping hands with n_z postural synergies, a structural relationship exists between the dimension of controllable internal forces and object motion subspaces and the number synergy control inputs. We proved that it is not possible to control, jointly and independently, more than n_z variables among internal forces, object motion directions and kinematic redundancy. This paper provides control strategies to perform complex manipulation tasks, involving control of motion and forces, through very few synergies as control inputs. Furthermore tools for design requirements of complex robotic hands in terms of number of synergies to accomplish manipulation tasks are provided. We believe that providing structural and basic results like the controllability of forces and motions in hand grasps with postural synergies will allow to better understand and exploit the synergies in both robotics and human studies. A numerical example relative to the Barrett Hand is showed, further analysis on the human hand are in progress.

Acknowledgment

This work has been partially supported by the European Commission with the Collaborative Project no. 248587, “THE Hand Embodied”, within the FP7-ICT-2009-4-2-1 program “Cognitive Systems and Robotics”, and by the Ministero Italiano dell’Università e Ricerca with the project “Innovative micro-actuators for advanced haptic and grasping devices” within the PRIN 2008 research program.

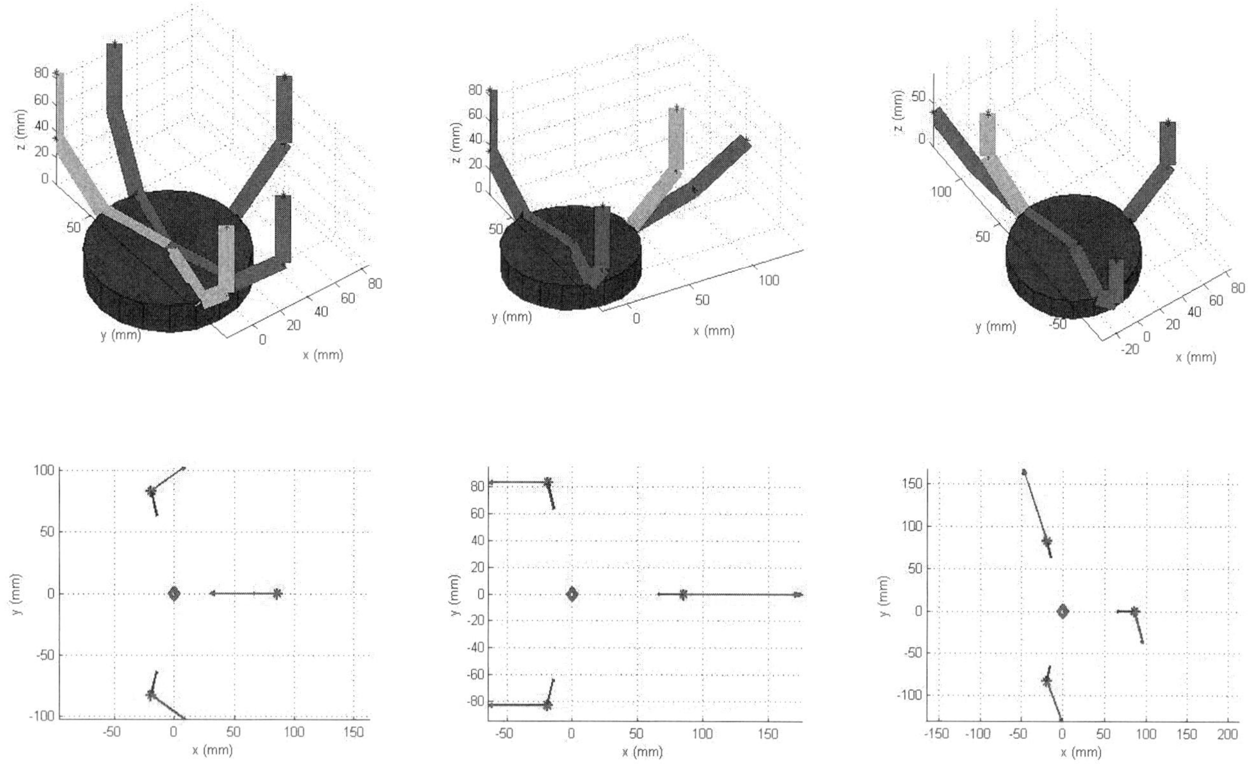

Fig. 3: Synergies in the Barrett hand. First row: hand configuration obtained acting on each synergy; second row: contact points, contact normals and internal forces generated activating each synergy.

References

[1] A. Bicchi. Force distribution in multiple whole-limb manipulation. In *Proc. IEEE Int. Conf. Robotics and Automation*, pages 196–201, Atlanta, 1993.

[2] A. Bicchi. On the problem of decomposing grasp and manipulation forces in multiple whole-limb manipulation. *Int. Journal of Robotics and Autonomous Systems*, 13:127–147, 1994.

[3] A. Bicchi. Hands for dextrous manipulation and robust grasping: a difficult road towards simplicity. *IEEE Trans. on Robotics and Automation*, 16(6):652–662, December 2000.

[4] L. Birglen, T. Laliberté, and C. Gosselin. *Underactuated robotic hands*, volume 40 of *Springer Tracts in Advanced Robotics*. Springer Verlag, 2008.

[5] C. Y. Brown and H. H. Asada. Inter-Finger Coordination and Postural Synergies in Robot Hands via Mechanical Implementation of Principal Components Analysis. In *2007 IEEE/RSJ International Conference on Intelligent Robots and System*, pages 2877–2882, 2007.

[6] J. Butterfass, M. Grebenstein, H. Liu, and G. Hirzinger. DLR-hand II: next generation of a dextrous robot hand. In *Robotics and Automation, 2001. Proceedings 2001 ICRA. IEEE International Conference on*, volume 1, pages 109–114, 2001.

[7] Matei Ciocarlie, Corey Goldfeder, and Peter Allen. Dimensionality reduction for hand-independent dexterous robotic grasping. In *Intelligent Robots and Systems, 2007. IROS 2007. IEEE/RSJ International Conference on*, pages 3270–3275, 2007.

[8] Matei T. Ciocarlie and Peter K. Allen. On-line interactive dexterous grasping. In Manuel Ferre, editor, *Haptics: Perception, Devices and Scenarios*, volume 5024 of *LNCS*, pages 104–113. Springer, 2008.

[9] SC Jacobsen, JE Wood, DF Knutti, and KB Biggers. The Utah/MIT dextrous hand: work in progress. *The International Journal of Robotics Research*, 3(4):21, 1984.

[10] D. Prattichizzo and A. Bicchi. Consistent specification of manipulation tasks for defective mechanical systems. *ASME Jour. Dynam. Systems, Measurement, and Control*, 119:767–777, December 1997.

[11] D. Prattichizzo and A. Bicchi. Dynamic analysis of mobility and graspability of general manipulation systems. *IEEE Trans. on Robotics and Automation*, 14(2):241–258, April 1998.

[12] D. Prattichizzo and J. Trinkle. Grasping. In B. Siciliano and O. Kathib, editors, *Handbook on Robotics*, pages 671–700. Springer, 2008.

[13] Marco Santello, Martha Flanders, and John F. Soechting. Postural hand synergies for tool use. *The Journal of Neuroscience*, 18(23):10105–10115, December 1998.

[14] Marco Santello and John F. Soechting. Force synergies for multifingered grasping. *Experimental Brain Research*, 133(4):457–467, August 2000.

Back-drivable and Inherently Safe Mechanism for Artificial Finger

Koichi Koganezawa

***Abstract*—This paper shows mechanisms for artificial finger based on a planetary gear system (PGS). Using the PGS as a transmitter provides an under-actuated system for driving three joints of a finger with back-drivability that is crucial characteristics for fingers as an end-effector when it interacts with external environment. This paper also shows the artificial finger employed with the originally developed mechanism called "double planetary gear system" (DPGS). The DPGS provides not only back-drivable and under-actuated flexion-extension of the three joints of a finger, which is identical to the former, but also adduction-abduction of the MP joint. Both of the above finger mechanisms are *inherently safe* due to being back-drivable with no electric device or sensor in the finger part. They are also rigorously solvable in kinematics and kinetics as shown in this paper.**

I. Introduction

Artificial finger has been strenuously studied in recent two decades since an anthropomorphic robot or spacecraft manipulator requires end-effectors to achieve dexterous task. However it is tremendously difficult to develop a hand having multiple fingers with a lot of DOF, which is feasible and practical not only on a mechanical point of view but also on their control. Some studies have been pursued to produce a hand of which fingers have full of DOF; each one is controlled by tiny DC-motor assembled in the finger [1][2] or by a remote control via wires[3][4][5]. This approach is interested in developing a research platform to study various control methods of multi-DOF hand when it handles objects, but it may not be oriented for practical usage as a hand for anthropomorphic robots or forearm prostheses.

In the meanwhile practical usage for robot hands or forearm prostheses requires reducing DOF. There have been some studies to control three joints of one finger (MP, PIP, DIP joints) by one motor via pulleys and wire [6] [7] or closed linkage system. It is a matter of course that the motion of the finger developed in these studies is restricted in one typical motion such as open-and-grip.

The study of this paper aims to develop finger mechanisms and a hand with multiple-fingers that can achieve dexterous handling tasks while sustaining feasibility for practical usage. The author considers an artificial finger or hand should have following properties to take into consideration for practical usage.

Manuscript received February 1, 2010. This work was supported in part by Japan Society for the Promotion of Science, via Grants-in-Aid for Scientific Research: No. 20560242 '*Development of Artificial Hand using the Double Planetary Gear System and the Handling of Objects*'.

Koichi Koganezawa is the Professor of the Department of Mechanical Engineering, Tokai University (e-mail: kogane@keyaki.cc.u-tokai.ac.jp)

***Back-drivability* (BD)**: The hand with multiple fingers is an end-effecter that always interacts with external environment, of which effect to the fingers is usually not expected to be predictable. This fact requires a hand to have some "resiliency" to the external interaction, which is interpreted as "back-drivability" of the joints of the hand in the technical point of view. This property have been pursued in some studies to develop tendon-driven hand [6][7] .

***Inherently safe design* (ISD)**: This property is much related to the BD in the meaning that the hand never harmful to external environment (especially to human). In addition to the BD, ISD requires some items to a controller of the fingers as follows.

(a) Any electric/electronic devices or sensors should not be equipped in the finger part. Because of the trait of the fingers as an end-effecter, it has to cope with severe external environments to be possible, such as electromagnetic or radio-active fields.

(b) According to (a) the controller can not much rely on the feedback from sensors such as tactile or pressure sensors.

ISD does not completely deny using the sensory feedback, which might be helpful to accomplish some dexterous and precise motion of the hand if a measure against noise is satisfactory. Therefore ISD requires that some primitive motions of the fingers should be provided without sensory feedback, in other words, mechanically, but it allows using the sensory feedback as subsidiary measures.

***All-in-one design* (AOD)**: All of the actuators should be mounted in a hand (distal to the wrist). A lot of works have equipped the actuators at the place away from the hand [3][4][5], the forearm part for example, and taken the wire driven as a force/torque transmission. But it will hamper to compose a wrist joint that has three DOF if trying to mimic a human. Moreover the wire-driven have some drawbacks; friction loss, control delay, unpredictable non-linear behavior due to the friction and the elastic element of the wire, all of which make the control precision worse. The place for mounting the actuator is therefore severely restricted due to AOD and ISD requirements. Only the palm is remained for it. Therefore the AOD requires reducing the number of actuators as many as possible by employing some under-actuation mechanical systems.

***Simple Control* (SC)**: The control of the multiple fingers of the hand should be as simple as possible. This requirement will be achieved along with pursuing the above BD and ISD properties. BD will reduce the control burden to determine the fingers' shape when they interact with external objects. ISD also provides the simple control that does not much rely on the sensory feedback.

Hence the purpose of the study is to develop a finger and a hand with multiple fingers that meet or have a potential to meet the above properties, BD, ISD, AOD and SC.

This paper is organized as follow. In the following section it shows a finger using a planetary gear system, the principle of its motion, shape-fitting property, kinematics and kinetics to show the back-drivability. In the third section it shows the advanced finger mechanism using the double planetary gear system. The last section is devoted to the conclusive remarks.

II. Finger Using Planetary Gear System[8]

A. Structure

Fig.1 shows the structure of the finger. It has three joint (DIP, PIP and MP joints) that are connected with a compound four-bar linkage system and a planetary gear system. The link 1 is directly attached to the inner gear to rotate with it. The link 2 is connected to one of the planetary gear axis. The size is almost the same as an adult index finger. The compound four-bar

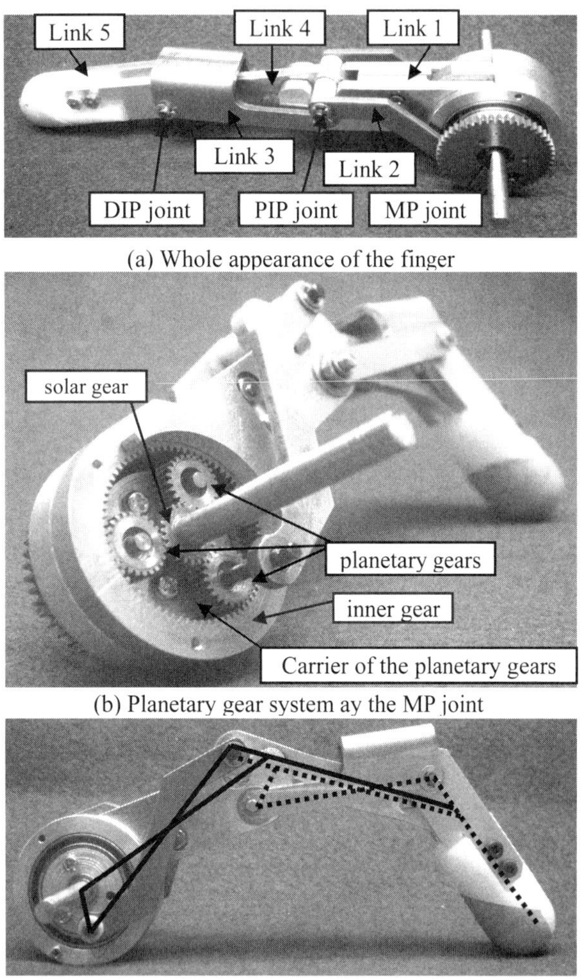

(a) Whole appearance of the finger

(b) Planetary gear system ay the MP joint

(c) Compound four-bar linkages

Fig.1 The finger using the planetary gear system

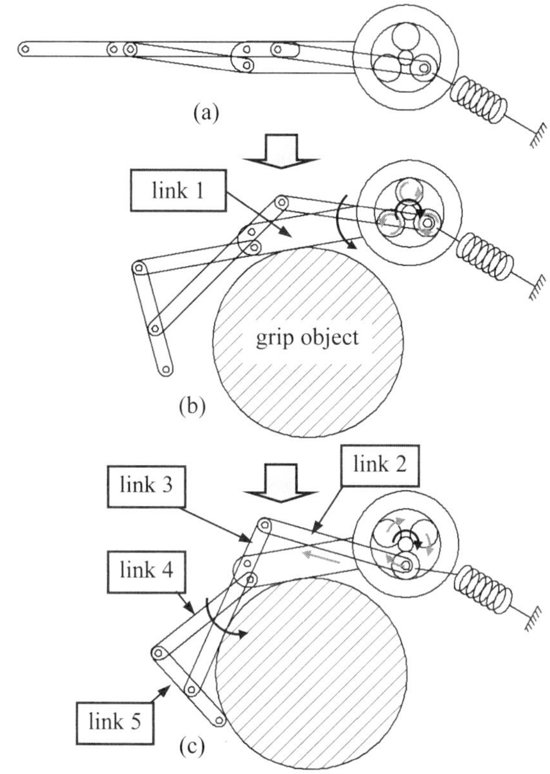

Fig.2 Sequence of the synergetic gripping motion

linkage system developed in this study is in part similar to the parallel mechanism developed in Laval University[9]. However our mechanism can be achieved by lesser number of parts in a smaller volume.

B. Synergetic gripping motion

The new mechanism allows the three joints of one finger (DIP, PIP and MP joints) to synergically wrap an object with no sensory feedback.

The key point of the mechanism resides in the fact that the planetary gear system is a transmitter that receives two inputs; the solar gear receives active torque from an actuator, while the carrier of the planetary gear receives a passive torque from some elastic element or some dissipative element. This amalgamation of active and passive torques provides synergetic motion of the three joints according to the shape of a gripping object under a simple control.

Fig.2 illustrates the motion sequence of gripping an object. The spring pulls the carrier of the planetary gears of which force is strong enough to sustain the weight of the finger and not to have the carrier rotate under no external force lording to the finger (a).

The clock-wise rotation of the solar gear gives rise to the counter clock-wise rotation of the inner gear via the rotation of the planetary gears on their axes (b).

Once the motion of the link 1 is hampered by touching the object and the rotation of the inner gear is stopped, further clock-wise rotation of the solar gears give rise to the orbital

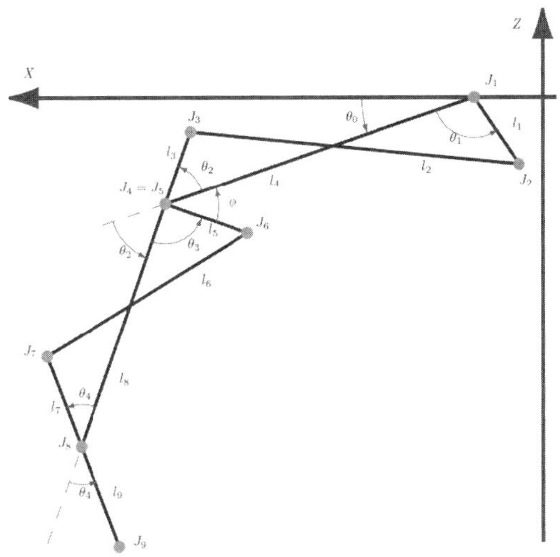

Fig.3 Finger configuration

rotation of the planetary gears around the solar gear accompanied by the rotation of the carrier. The rotation of the carrier push the link 2 forward, which leads the wrapping motion of the object by the link 3, 4 and 5 (c).

It is noteworthy that the line of motions described above can be carried out only by one-way rotation of the solar gears with no sensory feedback. Therefore this mechanism will meet the entire requirement of BD, ISD, AOD and SC defined in the former section.

C. Kinematics and kinetics

According to the assignment of the joint angles and the links shown in Fig.3, we can derive the kinematics and the kinetics of the finger. First we find that the angles θ_0 and θ_1 are related to the solar gear angle θ_S and the carrier angle θ_P described by,

$$\begin{bmatrix}\theta_0\\ \theta_1\end{bmatrix}=\begin{bmatrix}-Z_S/Z_I & (Z_S+Z_I)/Z_I\\ Z_S/Z_I & -Z_S/Z_I\end{bmatrix}\begin{bmatrix}\theta_S\\ \theta_P\end{bmatrix}=\mathbf{A}\begin{bmatrix}\theta_S\\ \theta_P\end{bmatrix} \tag{1}$$

where, Z_S and Z_I are the number of teeth of the solar gear and the inner gear respectively.

Next we can derive the two constraint equations for determining the dependent variables θ_2 and θ_4 (θ_3 is equal to $\pi-\phi-\theta_2$ with the constant angle ϕ) as,

$$\mathbf{\Phi}=\begin{bmatrix}(\mathbf{J}_2-\mathbf{J}_3)^T(\mathbf{J}_2-\mathbf{J}_3)-l_2^2\\ (\mathbf{J}_6-\mathbf{J}_7)^T(\mathbf{J}_6-\mathbf{J}_7)-l_6^2\end{bmatrix}=\mathbf{0}_2 \tag{2}$$

Partial differentiation of (2) with respect to θ_0 θ_1, θ_2 and θ_4 provides,

$$\mathbf{B}[\Delta\theta_2 \quad \Delta\theta_4]^T+\mathbf{C}[\Delta\theta_0 \quad \Delta\theta_1]^T=\mathbf{0}_2 \tag{3}$$

with

$$\mathbf{B}=[\partial\mathbf{\Phi}_i/\partial\theta_2 \quad \partial\mathbf{\Phi}_i/\partial\theta_4],\ \mathbf{C}=[\partial\mathbf{\Phi}_i/\partial\theta_0 \quad \partial\mathbf{\Phi}_i/\partial\theta_1]$$

It follows by combining (3) with (1),

$$[\Delta\theta_2 \quad \Delta\theta_4]^T=-\mathbf{B}^{-1}\mathbf{C}\mathbf{A}[\Delta\theta_S \quad \Delta\theta_P]^T \tag{4}$$

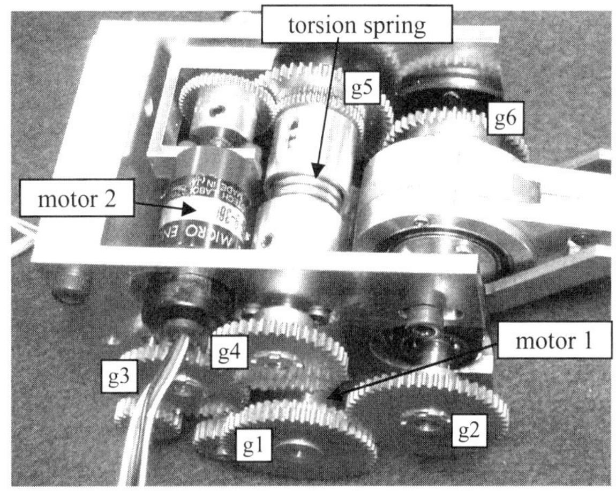

Fig.4 The part of the prime mover

By using (1) and (4) we can determine all of the dependent angles from the infinitesimal variations of θ_S and θ_P.

The infinitesimal variation of the finger tip position $\mathbf{J}_9=[J_{9z} \ J_{9x}]^T$ is calculated by,

$$[\Delta J_{9z} \quad \Delta J_{9x}]^T=\mathbf{D}[\Delta\theta_2 \quad \Delta\theta_4]^T+\mathbf{E}[\Delta\theta_0 \quad \Delta\theta_1]^T \tag{5}$$

with

$$\mathbf{D}=[\partial\mathbf{J}_9/\partial\theta_2 \quad \partial\mathbf{J}_9/\partial\theta_4],\ \mathbf{E}=[\partial\mathbf{J}_9/\partial\theta_0 \quad \partial\mathbf{J}_9/\partial\theta_1]$$

It follows the final result by substituting (1) and (4) into (5),

$$[\Delta J_{9z} \quad \Delta J_{9x}]^T=(\mathbf{E}-\mathbf{D}\mathbf{B}^{-1}\mathbf{C})\mathbf{A}[\Delta\theta_S \quad \Delta\theta_P]^T \tag{6}$$

and we can identify the Jacobian of the finger,

$$\mathbf{\Lambda}=(\mathbf{E}-\mathbf{D}\mathbf{B}^{-1}\mathbf{C})\mathbf{A} \tag{7}$$

Once the Jacobian is identified we can establish the kinetics that relates the force at the finger tip (J_9) $\mathbf{f}=[f_z \quad f_x]^T$ and the torque of the solar gear and the carrier of the planetary gears $\boldsymbol{\tau}=[\tau_S \quad \tau_P]^T$ as based on the virtual work principle,

$$\boldsymbol{\tau}=\mathbf{\Lambda}^T\mathbf{f} \tag{8}$$

D. Control of the force of the finger tip

The force exerted at the finger tip can be controlled if one extra motor is equipped to operate the spring tension. Fig.4 shows the part of the prime mover of the finger. The torque transmission passes are as follows.

Motor 1 $\rightarrow$ g1 $\rightarrow$ g2 $\rightarrow$ solar gear

Motor 2 $\rightarrow$ g3 $\rightarrow$ g4 $\rightarrow$ torsion spring $\rightarrow$ g5 $\rightarrow$ g6 $\rightarrow$ carrier

The carrier is basically rotated passively accompanied by twisting the torsion spring. The force exerted at the finger tip is determined by a torque of the solar gear and a passive torque of the torsion spring. Therefore if the motor 2 adjusts the torsion angle of the spring, the force at the finger tip can be controlled .according to the eq.(8).

E. Pinching an object by two fingers

The developed finger mechanism shown above allows us to scheme a very simple control when the two fingers pinch an object and handle it. Basically the motor 1 that directly drives the solar gear has a role to handle the object position, while the

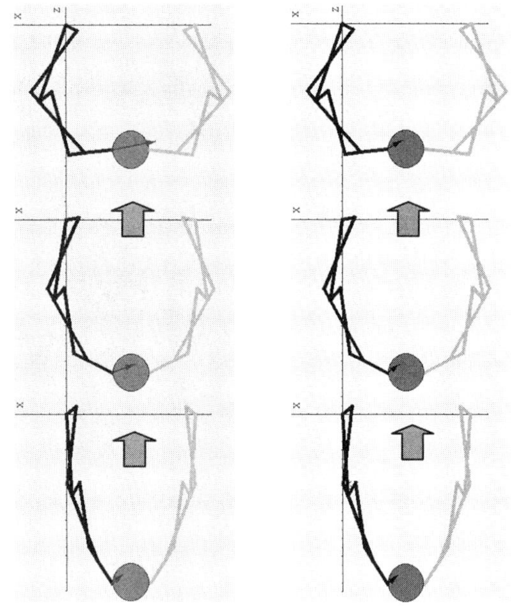

(a) No torsion spring control (b) Torsion spring control

Fig.5 Kinetic simulation of the object lifting-up by two fingers

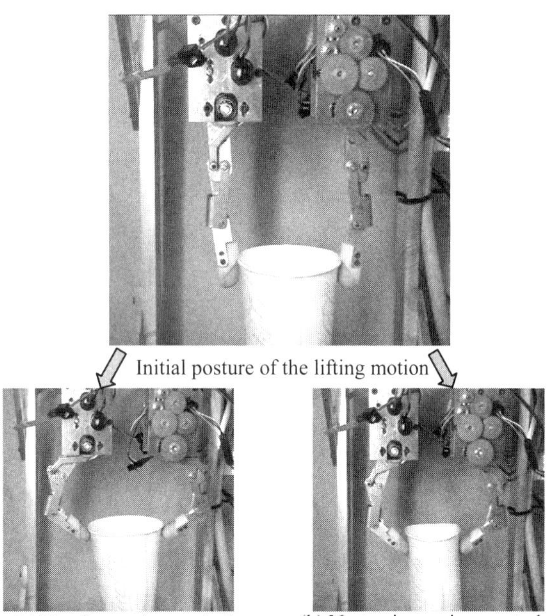

(a) Torsion spring control (b) No torsion spring control

Fig.6 Lifting experiment by two fingers

motor 2 that twists the torsion spring has a role to adjust the force exerted at the finger tip, although they are not rigorously discriminated.

For example the simple lifting-up of an object by two fingers can be achieved by just controlling the motor 1. Fig.5 shows the kinetic simulation of lifting the 0.1 kg object. Fig.5 (a) is the results when the motor 1 exerts the necessary torque τ_S to hold the lifting posture and the torsion spring exerts the carrier torque τ_p that is passively determined by the rotation angle of the carrier. The arrows shown at the finger tip designate the force vector (only those of the left finger are shown). As shown, the force exerted at the finger tip increases according to the lifting height because the torsion spring is more twisted as proceeding the lifting. Fig.5 (a), on the other hand, is the result that adds the torsion spring control, in which the motor 2 controls the torsion angle of the spring to be always constant. As shown in the force vector at the finger tip, the pinching force is sustained to be almost constant during the lifting. Fig.6 is the experimental result, of which control method is identical to the simulation. It shows the paper cup gets rid of being crushed when applying the torsion angle control (Fig.6 (a)), in which the torsion angle of the spring is monitored and fed back. In other words, the torsion spring has a role as a sensor that monitors the pinching force.

III. Finger Using Double Planetary Gear System

In parallel with developing the finger mechanism shown in the former section we have been developing an advanced mechanism that covers all of the functions the former mechanism has and has one more DOF; abduction/adduction of the MP joint [10]. It mainly consists of two parts: the *double planetary gear system* (DPGS) and the compound four-bar linkages as shown in Fig.7. The appearance of the assembled finger, the model developed in 2008, is shown in Fig.10.

A. DPGS

Fig.8 describes the structure of the DPGS. It consists of one inner gear in which two sets of a solar gear and three planetary gears are embedded face to face. The left-hand and the right-hand solar gears are driven by individual motors (hereafter we call both of them as "motor 1"). The planetary gears located around each solar gear rotate on their own axes and transmit the rotation torque to the inner gear, or take an orbital rotation around the solar gear accompanied by rotating an individual carrier.

B. Abduction/adduction of the MP joint by DPGS

Abduction/adduction of the MP joint is carried out as follows. A link 2-p (see Fig.9) is connected to the axle of one planetary gear on the carrier. The rotation of the left side and right side solar gears in opposite direction gives rise to the orbital rotation of the planetary gears (the rotation of the carriers), which is also opposite between the left-hand side and the right-hand side. Then the link 2-p of both sides moves forward and backward as shown in the right-hand figure in Fig.9, which induces the abduction/adduction of the MP joint.

C. Synergic gripping motion

The new mechanism can achieve the completely identical gripping motion to the former mechanism. Three joints of one finger (DIP, PIP and MP joints) concurrently grip the object if both sides of the solar gears are rotated in the same direction, which provides the identical motion as schematically shown in Fig.2. Fig.11 shows the actual gripping motion of the finger. The sequential view of the left hand side is the case that there is no object to be gripped, in which simultaneous flexion of MP,

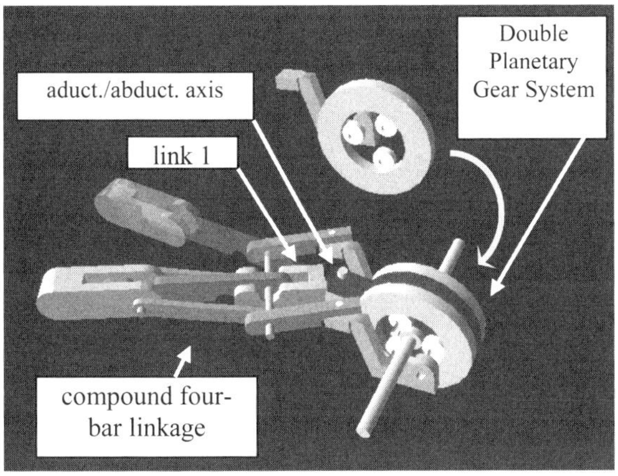

Fig.7 Configuration of the new mechanism

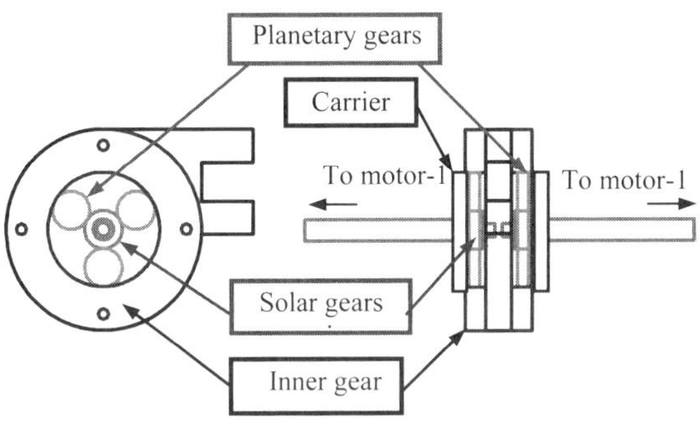

Fig.8 Double planetary gear system (DPGS)

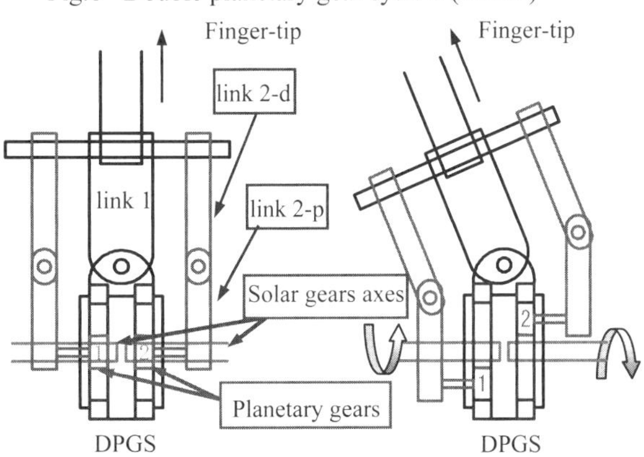

Fig.9 Abduct./adduct. by the DPGS

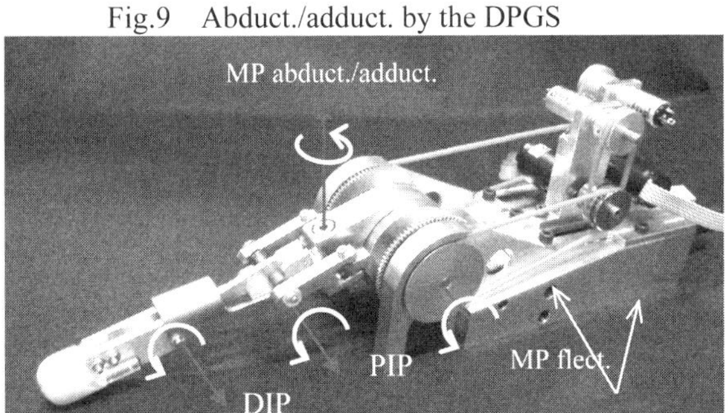

Fig.10 Overview of the finger developed in 2008

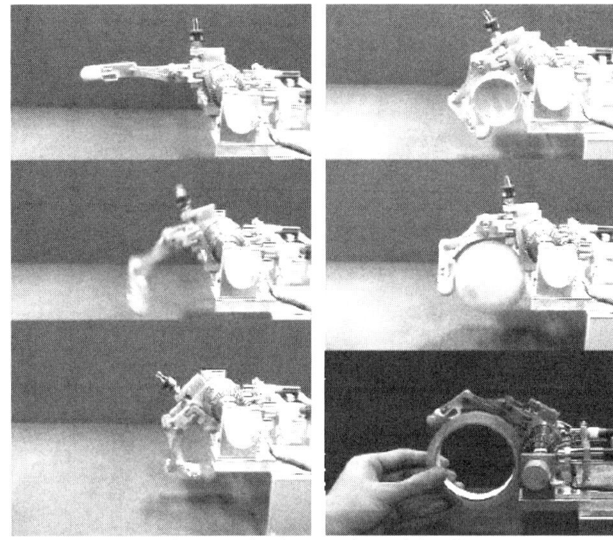

Fig.11 Gripping motion of the finger

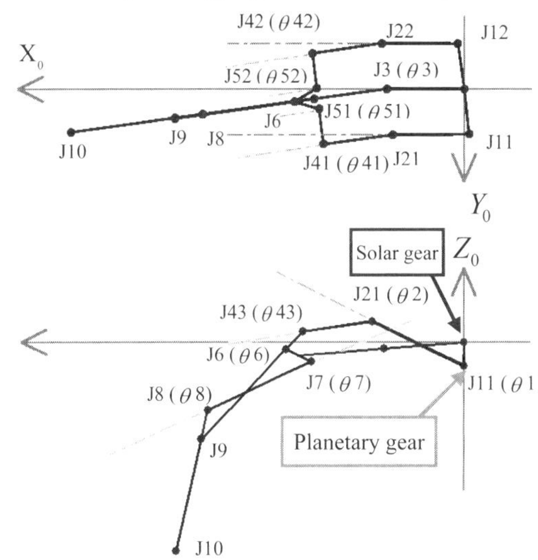

Fig.12 Finger configuration for simulation

PIP and DIP joints occurs by the single rotation of the solar gear. The right hand side shows to grip objects in various size.

It is noteworthy that these gripping motions occur in just one way rotation of the solar gear with no sensory feedback to satisfy the BD, ISD and SC requirements described in the introduction.

D. Kinematics and kinetics

According to the assignment of the joint angles and the links shown in Fig.12, we can derive the kinematics and the kinetics of the finger. In the case of the finger mechanism based on the DPGS we have two solar gears angles θ_{S1}, θ_{S2}, two carrier angles θ_{P1}, θ_{P2} and one inner gear angle θ_I in the DPGS. First we define two angles as,

$$\theta_{SD} \equiv (\theta_{S1} - \theta_{S2})/2, \quad \theta_{SS} \equiv (\theta_{S1} + \theta_{S2})/2 \tag{9}$$

Then we can establish the following relationship of the DPGS angles,

$$\begin{bmatrix} \Delta\theta_I \\ \Delta\theta_{P1} \\ \Delta\theta_{P2} \end{bmatrix} = \begin{bmatrix} -1/\rho_{SI} & 0 \\ 0 & 1/(1+\rho_{SI}) \\ 0 & -1/(1+\rho_{SI}) \end{bmatrix} \begin{bmatrix} \Delta\theta_{SS} \\ \Delta\theta_{SD} \end{bmatrix} = \mathbf{A} \begin{bmatrix} \Delta\theta_{SS} \\ \Delta\theta_{SD} \end{bmatrix} \quad (10)$$

where, $\rho_{SI} = Z_I / Z_S$ with Z_S *and* Z_I being the number of teeth of the solar gear and the inner gear respectively.

Next we find the following eight kinematical constraints existing in the closed link system,

$$f_{C1} \equiv J_{61,x} - J_{62,x} = 0,\quad f_{C2} \equiv J_{61,y} - J_{62,y} = 0,$$
$$f_{C3} \equiv J_{61,z} - J_{62,z} = 0,\quad f_{C4} \equiv J_{61,x} - J_{63,x} = 0,$$
$$f_{C5} \equiv J_{61,y} - J_{63,y} = 0,\quad f_{C6} \equiv J_{61,z} - J_{63,z} = 0,$$
$$f_{C7} \equiv J_{91,x} - J_{92,x} = 0,\quad f_{C8} \equiv J_{91,z} - J_{92,z} = 0$$

Partial differentiation with respect to angular variables provides the following equation to derive the infinitesimal variations of eight dependent variables $\boldsymbol{\theta}_d = [\theta_{11}\ \theta_{121}\ \theta_3\ \theta_{41}\ \theta_{42}\ \theta_{43}\ \theta_7\ \theta_8]^T$,

$$\Delta\boldsymbol{\theta}_d = -\mathbf{B}^{-1}\mathbf{C}\mathbf{A}[\Delta\theta_{SS}\ \Delta\theta_{SD}]^T \quad (11)$$

where, $\mathbf{B} = \partial\mathbf{f}_C / \partial\boldsymbol{\theta}_d \in \Re^{8\times8}$ and $\mathbf{C} = \partial\mathbf{f}_C / \partial\boldsymbol{\theta}_{ind} \in \Re^{8\times3}$ with $\mathbf{f}_C = \{ f_{Ci} \mid i = 1,\cdots,8\}$ and $\boldsymbol{\theta}_{ind} = [\theta_I\ \theta_{P1}\ \theta_{P2}]^T$.

At the same time the infinitesimal variation of the finger tip position $\mathbf{J}_{10} = [J_{10x}\ J_{10y}\ J_{10z}]^T$ will be,

$$\Delta\mathbf{J}_{10} = \left(\mathbf{E} - \mathbf{D}\mathbf{B}^{-1}\mathbf{C}\right)\mathbf{A}[\Delta\theta_{SS}\ \Delta\theta_{SD}]^T = \boldsymbol{\Lambda}[\Delta\theta_{SS}\ \Delta\theta_{SD}]^T \quad (12)$$

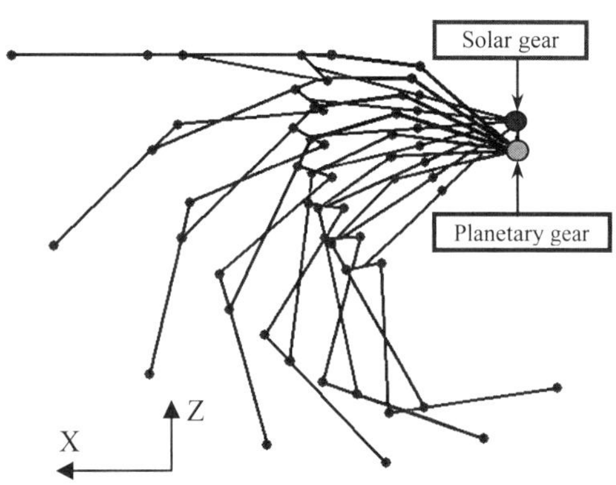

(a) Finger motion when two solar gears rotate in the same direction

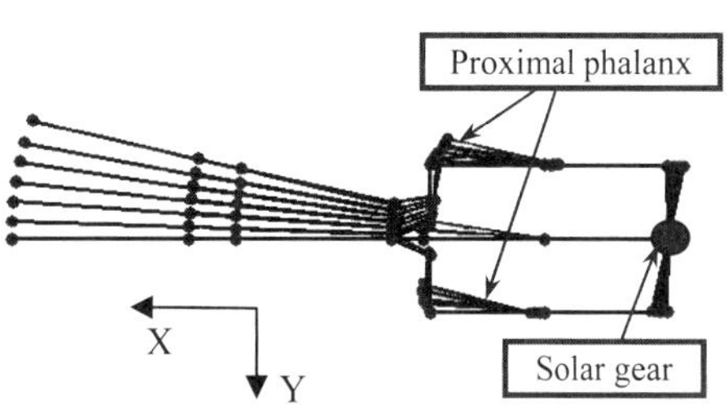

(b) Finger motion when two solar gears rotate in the opposite direction

Fig.13 Kinematical simulation of the finger motion

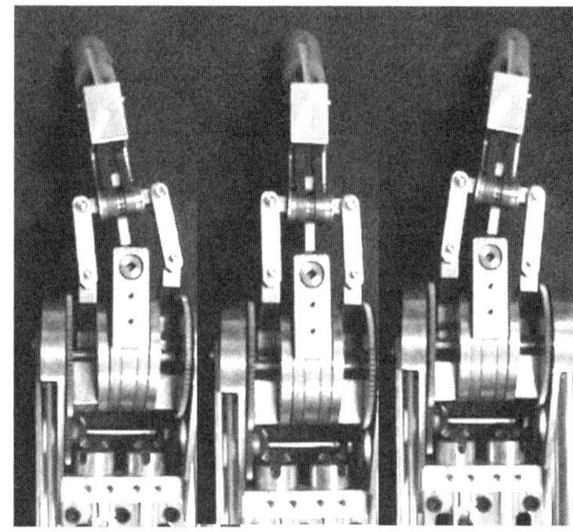

Fig.14 The abduct./adduct. motion of the finger

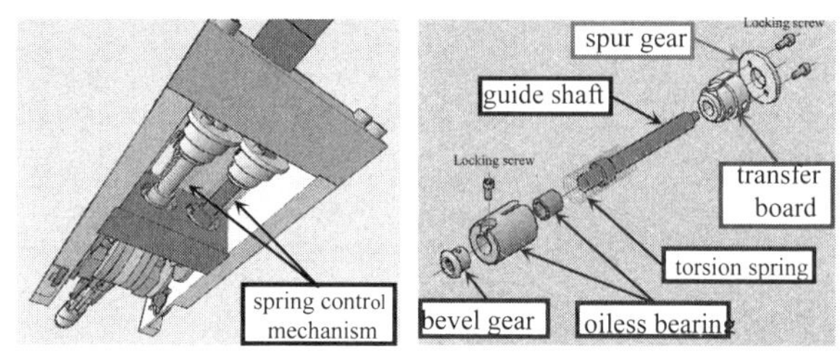

Underside view of the base　　Spring control mechanism

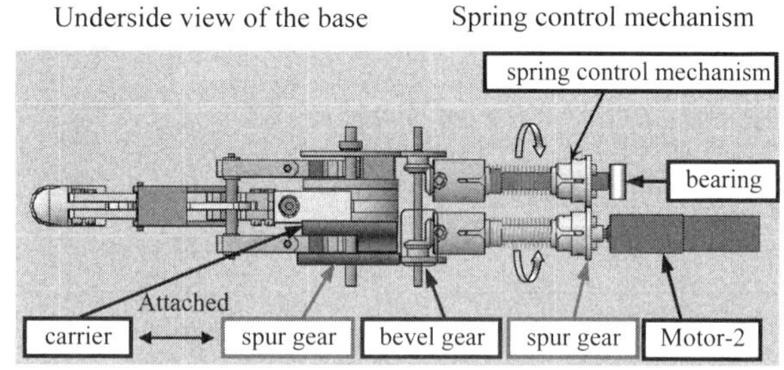

Fig.15 The parts of the spring controller

with $\mathbf{D} = \partial\mathbf{J}_{10} / \partial\boldsymbol{\theta}_d \in \Re^{3\times8}$, $\mathbf{E} = \partial\mathbf{J}_{10} / \partial\boldsymbol{\theta}_{ind} \in \Re^{3\times3}$ and the Jacobian $\boldsymbol{\Lambda} \equiv \left(\mathbf{E} - \mathbf{D}\mathbf{B}^{-1}\mathbf{C}\right)\mathbf{A}$.

Fig.13 shows the kinematical motion of the finger. Fig.13 (a) is the result when two solar gears rotate in the same direction and Fig.13 (b) is those for the opposite direction. Both results are assumed that the carriers do not rotate. As shown, the solar gears' rotation in the same direction gives rise to the synergetic flexion of MP, PIP and DIP joints. On the other hand their rotation in the opposite direction gives rise to the abduction/adduction of the MP joint about $\pm 25°$ that is almost the same range as those of a human finger. Fig.14 shows the actual abduction/adduction of the machine.

E. Pinching force control

It also needs springs for the finger to allow a back-drivable motion that is identical to the finger based on the PGS. Fig.15 shows the assembling view of the spring controller. We need two torsion springs for the carriers of both sides of DPGS. But

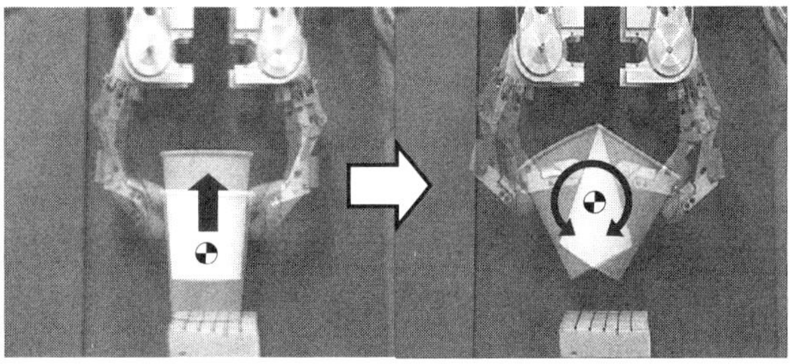

(a) Lifting and rolling an object

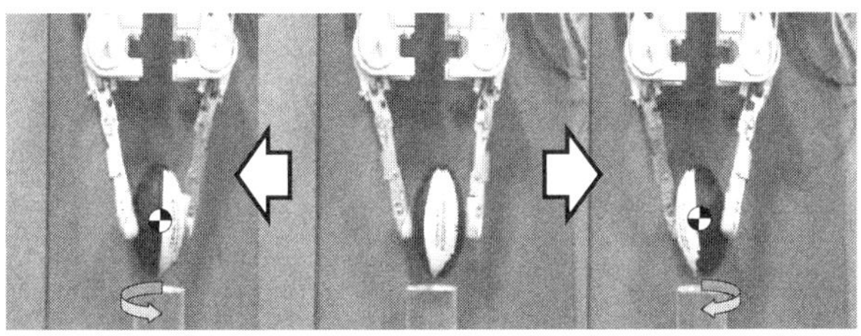

(b) Yawing an object

Fig.16 Object handling by two fingers

the developed spring controller allows for one motor (Motor 2) to twist both of the torsion springs simultaneously.

F. Pinching an object by two fingers

The mechanism based on the DPGS allows two fingers to handle the object with simple control schemes as same as the mechanism based on the PGS shown in the former section. Fig.16 (a) is the example for lifting and rolling the paper cup. In this motion two solar gears in one finger rotates identically and the torsion springs that connect to the left-hand and right-hand carriers of the DPGS are controlled by the motor 2 to keep the constant torsion angle of the spring. Therefore this motion can be achieved by the mechanism based on the PGS, but the motion shown in Fig.16 (b) can not be. It is an object yawing, which can be achieved by fingers having a DOF of abduction/ adduction. During the motion two solar gears rotates in the opposite direction.

IV. Conclusions

This paper shows two finger mechanisms; one is based on the planetary gear system (PGS) and the other is based on the originally-devised double planetary gear system (DPGS). Both mechanisms will satisfy the requirements of BD (back-drivability), ISD (inherently safe design) and SC (simple control), all of which are indispensable properties for robot hand or prosthesis. The finger with PGS will also satisfy the AOD (all-in-one design) for developing a multiple finger hand. Moreover the mechanisms have the following properties;

(a) Its kinematics, kinetics and dynamics can be solved in reliable level because they have no tendon wire.

(b) Although they have no sensors in their finger portion for the sake of ISD, the spring equipped in the PGS or DPGS has a role of force sensor.

The above facts will enhance to use the mechanisms into applications that require more precise and dexterous works.

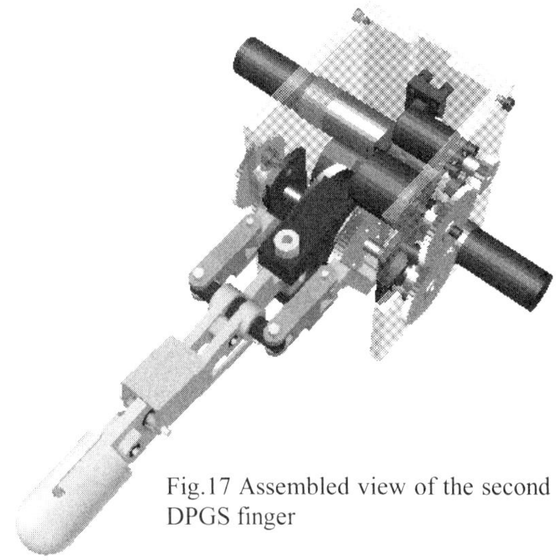

Fig.17 Assembled view of the second DPGS finger

We are now developing the finger based on the DPGS with a new reduced volume design in the prime mover part as shown in Fig.17 in order to compose a hand with multiple fingers that will satisfy the AOD (all-in-one design).

Acknowledgment

The authors thank the Japan Society for the Promotion of Science, via Grants-in-Aid for Scientific Research: No. 20560242 '*Development of Artificial Hand using the Double Planetary Gear System and the Handling of Objects*', for their support in the pursuance of this work.

References

[1] Kawasaki, H., et. al., Mechanism of Anthropomorphic Robot Hand: Gifu Hand I, J. Robot. And Mech., 11,4, pp.269-273, 1999.

[2] Namiki, A., Y. Imai, M. Kaneko and M. Ishikawa, Development of a High-speed Multifingered Hand System, Proc. of Int. Conf. On Intelligent Manipulation and Grasping, Genova-Itary, pp.85-90, 2004.

[3] Salisbury, K. and B. Roth, Kinematics and force analysis of articulate mechanical hands, ASME J. Mechanism, Transmissions and Autom. In Design, Vol.105, pp.35-41, 1983.

[4] Maekawa, H., et al., Development of Three-Fingered Robot Hand with Stiffness Control Capability, Mechanism, Vol.2, No.5, pp.483-494, 1992.

[5] Lovchik, C.S. and M.A. Diftler, The Robonaut Hand:A Dextrous Robot Hand for Space, Proc. of the 1999 IEEE Int. Conf. On Robotics and Autom, pp.907-912, 1999.

[6] Lee, Y-T, H-R Choi, W-K Chung and Y. Youm, Stiffness Control of a Coupled Tendon-Driven Robot Hand, IEEE Control Systems, pp.10-19, 1994.

[7] Harada Electric Industry Co., Myo-electric controlled forearm prosthesis SH-1, URL http://www.h-e-i.co.jp.

[8] Koganezawa, K., Artificial Finger with Shape-Fitting Mechanism, Proceedings of the International Conference on Intelligent Manipulation and Grasping, pp. 103-109, Genova, Italy, 2004

[9] Martin, E. and T. L. C. Gosselin, SARAH Hand Used for Space Operations on STVF Robot, Proceedings of the International Conference on Intelligent Manipulation and Grasping, pp. 279-284, Genova, Italy, 2004

[10] Koganezawa, K. and Y. Ishizuka, Novel Mechanism of Artificial Finger using Double Planetary Gear System, Proceedings of the 2008 IEEE/RSJ International Conference on Intelligent Robots and Systems, pp. 3184-3191, Nice, France, Sept 22-26, 2008.

Segmentation and Unsupervised Part-based Discovery of Repetitive Objects

Rudolph Triebel Jiwon Shin Roland Siegwart
Autonomous Systems Lab, ETH Zurich
Tannenstrasse 3, 8092 Zurich, Switzerland
email: {rudolph.triebel,jiwon.shin}@mavt.ethz.ch, rsiegwart@ethz.ch

Abstract— **In this paper, we present an unsupervised technique to segment and detect objects in indoor environments. The main idea of this work is to identify object instances whenever there is evidence for at least one other occurence of an object of the same kind. In contrast to former approaches, we do not assume any given segmentation of the data, but instead estimate the segmentation and the existence of object instances concurrently. We apply graph-based clustering in feature and in geometric space to presegmented input data. Each segment is treated as a potential object part, and the inter-dependence of object labels assigned to part clusters are modeled using a Conditional Random Field (CRF) named the "parts graph". Another CRF is then applied to the scene graph to smooth the class labels using the distributions obtained from the parts graph. First results on indoor 3D laser range data are evaluated and presented.**

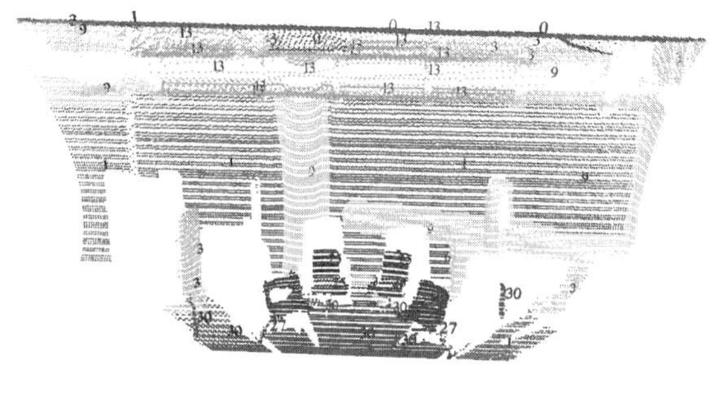

Fig. 1: 3D range scan of an indoor scene. Two different types of chairs are detected by exploiting the fact that particular constellations between back rests and seats occur more than once in the scene.

I. INTRODUCTION

The ability for a robot to learn and discover objects without any human guidance enhances its autonomy and makes it more independent. Such a robot requires no prior training and can more easily adapt to new, unknown environments. It is also able to autonomously draw conclusions about the structure of its environment. This functionality is useful when robots operate fully autonomously without human interaction. But also when robots live with humans, a high-level semantic analysis of the environment helps the robot to communicate with a human. As an example, a robot which can detect similarities of objects encountered in the environment, no longer requires a human to first label all occurences of objects in a previously acquired data set. Instead, it may ask the human, "I discovered several instances of something that looks like an interesting object. What is the name of the object?" In this paper, we take a first step in this direction.

We propose an approach to segment and discover objects of multiple occurrences without supervision, where an object is defined as a constellation of object parts. We segment input point clouds and treat each segment as an instance of a potential object part. In our work, object parts are determined by grouping similar segments together using clustering in a predefined feature space. In addition, the segments are clustered in the geometric space and the number of resulting connected components is used as an upper bound on the number of potential object classes. Then, two major reasonings are used to determine a class label for each segment: First, different object parts that often occur close to each other are more likely to correspond to the same object class. For example, the fact that a back rest of a chair and a chair seat are frequently observed in close vicinity to each other rises the evidence that there is an object class for which more than one instance appears in the scene (see Fig. 1). Second, an instance of an object part may appear to correspond to some object class, but given its physical context it is more likely to be part of another class. For example, a segment may appear to be a chair leg, but if surrounded by table parts it is more likely to be a part of a table. Both ideas are implemented using probabilistic reasoning based on Conditional Random Fields (CRFs) [1].

II. RELATED WORK

Most work on repetition detection has been in the field of image analysis. Detection of regularly repeating patterns has been the focus of many researchers [2, 3] with some recent work by Loy and Eklundh [4] on grouping of features based on symmetry, and by Wenzel *et al.* [5] on using symmetry to detect repetitive structures in facade images. Zeng and van Gool [6] employ point-wise repetition to improve segmentation results using mutual information. In 3D, discovery and utilization of repetition has been adressed in computer aided design and other synthetic models [7, 8, 9]. The work of Bokeloh *et al.* [10] is more closely related to this work. The authors proposed an algorithm for detecting structural redundancy by matching symmetric constellations of feature lines. In terms of repetition detection, the main challenge of our work lies in the lack of a repetition unit as we do not assume any regularity or symmetry of the repetition pattern.

In this work, we employ clustering to group similar segments in feature and geometric space. Clustering has received a considerable amount of attention by machine learning and pattern recognition communities. Some classic methods such as the expectation-maximization algorithm and k-means clustering assume that data can be modeled by a simple distribution, while other methods such as agglomerative clustering are sensitive to noise and outliers. To overcome these challenges, alternative approaches have been proposed. Ng, Jordan, and Weiss [11] presented a spectral clustering algorithm, which uses eigenvectors of the data matrix to group points together, and demonstrate how well the algorithm clusters even challenging data. Another work of interest is affinity propogation proposed by Frey and Dueck [12]. Affinity propagation clusters data by finding a subset of exemplars, which are cluster centers selected from data. This method avoids the pitfalls of bad initialization and does not require the number of clusters to be prespecified. In this work, we use affinity propagation to cluster segments in feature space.

Conditional Random Fields (CRFs) [1] are discriminative models, that have also been applied to object recognition problems [13, 14]. Notably, Quattoni, Collins, and Darrell [15] presented a part-based approach for object class recognition using a CRF. We also take a parts-based approach for objects. Ma and Grimson [16] proposed a coupled CRF to allow for interaction between contour and texture in image data. While our work does not explicitly use coupled CRFs, the interaction between part labels and class labels play a critical role in the success. The main idea in our work which has not been adressed previously is the use of Conditional Random Fields without any training set. Instead, we infer from clustering results the possible object labels.

In unsupervised object detection, several authors have proposed adaptation of text analysis methods in image analysis. For example, Liu and Chen [17] proposed a modified probablilistic latent semantic analysis (pLSA) method to detect foreground objects from images. Sivic *et al.* [18] compare pLSA and Latent Dirichlet Allocation (LDA) to discover object categories from sets of images. Also, Endres *et al.* [19], use LDA to discover object classes from range data without supervision. While this approach can classify objects of multiple classes, it assumes that a ground plane and walls are extracted a priori and the objects are spatially disconnected. In our work, we do not make such assumptions. We consider every segment as a potential object part and test them to determine if they belong to an object. Lastly, this work is similar to our previous work [20], which also discovers objects without supervision, but it does not explicitly label the segments as we do in this work.

III. SEGMENTATION AND CLUSTERING

Object detection using unsupervised learning is significantly different from using supervised learning. In fact, objects can not be *detected*; that is, it is not possible to identify some part of the input data by matching it against an instance of a previously known object class as there is no such known object class. Instead, objects can only be *discovered* by hypothesizing the existence of an object based on some sort of repetition or pattern found in the data. Thus, to discover objects, we need to find similarities in the data. We do this by extracting features (see Sec. III-B) and comparing them by a clustering algorithm in the feature space, which is described in Sec. III-C.

Algorithm: SSCGF

data : Point Cloud $\mathcal{P} = \{\mathbf{p}_1, \ldots, \mathbf{p}_N\}$

input :

- Segmentation parameters κ and τ
- Cluster parameters ϑ_f and ϑ_g

output:

- low-level segmentation $\mathcal{S} = \{\mathbf{s}_1, \ldots, \mathbf{s}_M\}$, $\mathbf{s}_i \subset \mathcal{P}$
- feature-space clusters of segments $\mathcal{F}_1, \ldots, \mathcal{F}_C$
- geometric clusters of segments $\mathcal{G}_1, \ldots, \mathcal{G}_K$
- class label distributions $\mathbf{d}_1, \ldots, \mathbf{d}_M$, $\mathbf{d}_i \in [0, 1]^K$

procedure:

$\mathcal{S} \leftarrow$ `SuperPixelSegmentation`$(\mathcal{P}, \kappa, \tau)$
$\mathbf{f}_1, \ldots, \mathbf{f}_M \leftarrow$ `FeatureExtraction`$(\mathcal{S})$
$\mathcal{F}_1, \ldots, \mathcal{F}_C \leftarrow$ `ClusterInFeatureSpace`$(\mathcal{S}, \vartheta_f, \{\mathbf{f}_i\})$
$\mathcal{G}_1, \ldots, \mathcal{G}_K \leftarrow$ `ClusterInGeometricSpace`$(\mathcal{S}, \vartheta_g)$
$\mathfrak{P} \leftarrow$ `MakePartsGraph`$(\{\mathcal{F}_i\}, \{\mathcal{G}_i\}, \mathcal{S})$
$\mathfrak{P} \leftarrow$ `SmoothPartsGraph`$(\mathfrak{P}, K)$
$\mathfrak{S} \leftarrow$ `MakeSceneGraph`$(\mathfrak{P}, \{\mathcal{G}_i\}, \mathcal{S})$
$\mathfrak{S} \leftarrow$ `SmoothSceneGraph`$(\mathfrak{S}, K)$
$\mathbf{d}_1, \ldots, \mathbf{d}_M \leftarrow$ `ReadFromGraphNodes`$(\mathfrak{S})$

Alg. 1: Segmentation and smoothed clustering in geometric and in feature space (SSCGF). Note that the number of segments M, as well as the numbers C and K of clusters are computed inside the particular subroutines (see text).

Another problem which arises here is that the *segmentation* of the data is unknown, i.e. we do not know where the boundaries of the objects are. The segmentation problem is tightly bound to the detection problem because a perfect segmentation would make object detection very easy – a simple comparison of the segmented object instances would suffice. To tackle this problem we perform three steps in our algorithm: first, we apply a low-level segmentation as described in Sec. III-A. Then, we obtain a coarser data segmentation by clustering in the geometric space as presented in Sec. III-C. Finally, we obtain a further improved segmentation by reasoning on parts of objects that occur in a similar constellation in different instances of the scene. The details of this are described in Sec. IV and Sec. V. An overview of the entire algorithm is shown in Alg. 1.

A. Low-level Segmentation

The first step in our algorithm is the segmentation of the data using the graph-based segmentation algorithm of Felzenszwalb and Huttenlocher [21], adapted to range image data. In the modified algorithm, we create a graph $\mathfrak{G} = \{\mathcal{V}, \mathcal{E}\}$ of vertices $\mathcal{V}$ and edges $\mathcal{E}$, where each point $\mathbf{p}$ in a given point cloud $\mathcal{P}$

corresponds to a vertex and an edge connects adjacent points. Here, adjacency of two points is determined from a triangular mesh built on the point cloud. Every edge $(\mathbf{p}_i, \mathbf{p}_j)$ has an associated weight w_{ij}, which is equal to the dissimilarity Δ of the connected points $\mathbf{p}_i$ and $\mathbf{p}_j$. In the case of a camera image, this can be the difference of the pixel intensities; in our case, we define $\Delta(\mathbf{p}_i, \mathbf{p}_j)$ as the dot product of the normal vectors $\mathbf{n}_i$ and $\mathbf{n}_j$ computed at $\mathbf{p}_i$ and $\mathbf{p}_j$. This yields for smooth surfaces, e.g. a plane or a sphere, being grouped as one segment, while surfaces with sharp edges, e.g. between two sides of a box, are grouped into two segments.

The algorithm begins with each vertex in its own segment. The edges are processed by increasing weights, and the two segments $\mathbf{s}_i$ and $\mathbf{s}_j$ connected by a given edge are merged whenever

$$w_{ij} \leq \min\left(d(\mathbf{s}_i) + \frac{\kappa}{|\mathbf{s}_i|}, d(\mathbf{s}_j) + \frac{\kappa}{|\mathbf{s}_j|}\right),$$

where $d(\mathbf{s})$ is the *internal difference* function defined by the maximal edge weight of all edges in the minimum spanning tree of the segment $\mathbf{s} \subseteq \mathcal{V}$, and κ is a consistency parameter that influences the granularity of the segmentation: a low value of κ requires segments to be more consistent and thus produces more but smaller segments.

In addition to introducing a 3D extension, we make other modifications to the original algorithm. First, as the normal at points with an insufficient number of neighboring points is ill-defined, we do not force every point to be in a segment. No vertices are generated in the graph for these points, and thus no segments are created containing them. Second, as a post-processing step, we remove segments that contain fewer points than a given minimal value τ, which are often caused by sensor imperfections or occlusions. In our experiments, a good choice of the segmentation parameters turned out to be $\kappa = 9$ and $\tau = 100$.

B. Feature Extraction

As shape descriptors, we use spin images [22], shape distributions [23], and shape factors [24]. A *spin image* for a given point $\mathbf{p}$ with normal vector $\mathbf{n}$ is defined as a 2D histogram H oriented along the line l through $\mathbf{p}$ with direction $\mathbf{n}$. Each bin of H counts the points with a certain distance to l and the plane through $\mathbf{p}$ with normal vector $\mathbf{n}$. For the spin image descriptor of a segment $\mathbf{s}$, we form vectors $\mathbf{h}_i$ of stacked lines of the histograms H_i for all points $\mathbf{p}_i \in \mathbf{s}$ and compute the average $\mathbf{h}^s$ over all $\mathbf{h}_i$.

A *shape distribution* is defined as a histogram of values of a predefined function $f : \mathcal{P}^r \to \mathbb{R}$, where r is the arity of f and is usually a value between 1 and 4. In our implementation, we use two binary functions $f_d(\mathbf{p}_i, \mathbf{p}_j)$ and $f_a(\mathbf{p}_i, \mathbf{p}_j)$, namely the Euclidean distance between $\mathbf{p}_i$ and $\mathbf{p}_j$ and the dissimilarity $\Delta(\mathbf{p}_i, \mathbf{p}_j)$ as defined above. The resulting histogram vectors $\mathbf{h}^d$ and $\mathbf{h}^a$ are computed by evaluating f_d and f_a on all pairs of points in a segment $\mathbf{s}$. To make the feature vectors invariant with respect to the sample density, we normalize the histograms $\mathbf{h}^d$ and $\mathbf{h}^a$ by the total number of bin entries. As we consider scale as a feature of an object, we do not perform normalization with respect to the maximum distance encountered in a segment.

Lastly, we compute *shape factors* per segment, i.e. the normalized eigenvalues of the covariance matrix C_i of all points in segment $\mathbf{s}_i$, collected in a vector $\mathbf{h}^f$. For each individual descriptor, we compute a PCA to reduce the dimensionality, and the results are combined into a feature vector $\mathbf{f}$.

C. Clustering in Feature Space

To find similar segments, we apply a clustering algorithm in the feature space. The number of existing clustering algorithms is large, and they include agglomerative clustering, k-means clustering [25], mean-shift estimation [26], spectral clustering [27, 11], and, more recently, affinity propagation (AP) [12]. We explored some of these clustering methods and decided for AP clustering because of its robustness and its ability to estimate the number of clusters implicitly. The basic principle is to determine *exemplars* out of all given points that are well-suited to explain the remaining data points. The application of the algorithm to our case is sketched as follows:

First a similarity value ς_{ij} is computed for all pairs $(\mathbf{f}_i, \mathbf{f}_j)$ of feature vectors. In our implementation, we use the negative squared Euclidean distance between $\mathbf{f}_i$ and $\mathbf{f}_j$. Then, in an iterative manner, two functions, namely the *responsibility* $r(i, j)$ and the *availability* $a(i, j)$ are computed for each vector pair, where r expresses how well-suited $\mathbf{f}_j$ is to serve as an exemplar for $\mathbf{f}_i$ and a expresses how appropriate it would be for $\mathbf{f}_i$ if $\mathbf{f}_j$ were its exemplar. These functions are defined as

$$r(i, j) = \varsigma_{ij} - \max_{j' s.t. j' \neq j} \left\{a(i, j') + \varsigma_{ij'}\right\} \quad (1)$$

$$a(i, j) = \min\left\{0, r(j, j) + \sum_{i' s.t. i' \notin \{i, j\}} \max\{0, r(i', j)\}\right\}, \quad (2)$$

where Eq. (2) is applied only if $i \neq j$. Initially, $a(i, j)$ is set to zero for all i and j. For the special case of "self-availability" the rule

$$a(i, i) = \sum_{i' s.t. i' \neq i} \max\{0, r(i', i)\} \quad (3)$$

is used. In each iteration, responsibilities and availabilities are computed and then, for each $\mathbf{f}_i$, an $\mathbf{f}_j$ is determined so that the sum $a(i, j) + r(i, j)$ is maximized. If the resulting j equals i, then $\mathbf{f}_i$ is identified as an exemplar, otherwise $\mathbf{f}_j$ serves as an exemplar for $\mathbf{f}_i$. The stopping criterion of the iteration is met when the assignments of points to exemplars do not change over a fixed number of iterations or a given number of maximum iterations is reached. The only parameter ϑ_f of the algorithm is the *self-similarity*, which can be specified either for each data point individually or commonly for all data points. This value influences the number of resulting clusters: a high value results in more clusters, a low value in fewer clusters. In our experiments, a good value turned out to be -0.2, specified equally for all data points.

As a result, we obtain C clusters $\mathcal{F}_1, \ldots, \mathcal{F}_C$ of similar segment instances, where each cluster defines a potential object part.

D. Clustering in Geometric Space

From the clustering in feature space we obtain a grouping of segments into object parts. However, we also want to reason on the object level, where objects are considered to consist of several parts. To accomplish this, we also perform a clustering in the geometric space where segments are represented by their center of gravity (COG). Unfortunately, affinity propagation (AP) is not a good choice to perform the clustering in the geometric space because it produces "star-like" clusters, i.e. all points in a cluster are connected directly to the cluster exemplar. This restricts the type of objects that can be detected, as for many objects such an exemplar part can not be found. Also, in AP clustering, the distances between points are not explicitly bounded, which often results in counter-intuitive clustering results.

Therefore, we apply a different, much simpler strategy to cluster the segments. We define a distance threshold ϑ_g and connect only those pairs of segments $(\mathbf{s}_i, \mathbf{s}_j)$ with an edge, for which the COGs are closer to each other than ϑ_g. As a result, all connected components will be farther away from each other than ϑ_g, which rises the evidence that they correspond to different object instances. This will be of importance later on.

IV. SCENE GRAPH AND PARTS GRAPH

One important aspect of the work presented here is the reasoning about object *instances* solely based on the extraction of potential object *parts*, represented as segments. The challenge here is to find a proper definition of an object class as we do not know of how many parts an object consists and whether or not all of its parts are visible. In addition, the number of observed objects is unknown – we only know the number of object parts. Two intuitions and one assumption help us to reason on parts to discover objects: First, we exploit the fact that segments which occur physically close to each other are more likely to correspond to the same object. Second, segments of one type which occur often in the vicinity to segments of another type give evidence that several instances from the same object class exist. Referring back to the introductory example, we can say: if we find many backrests of a chair that are all close to chair seats, then there is probably an object class which consists of at least these two parts. Furthermore, we work under the assumption that the number of possible object classes is bounded by the number of connected components in the graph which results from clustering the geometric space using the distance threshold ϑ_g. We call this the *scene graph* and give details in Sec. IV-A. Our assumption implicitly states that two objects are always supposed to be farther away from each other than ϑ_g. This may seem very restrictive, but in fact, it limits the applicability of the algorithm not as much as it appears. The reason is that the number of connected components only bounds the number of object *classes*, not the number of actual object *instances*. Thus, even if two different objects are closer to each other than ϑ_g, there are usually enough other connected components with at least two objects of the same class, so that the number of connected components exceeds the number of object classes.

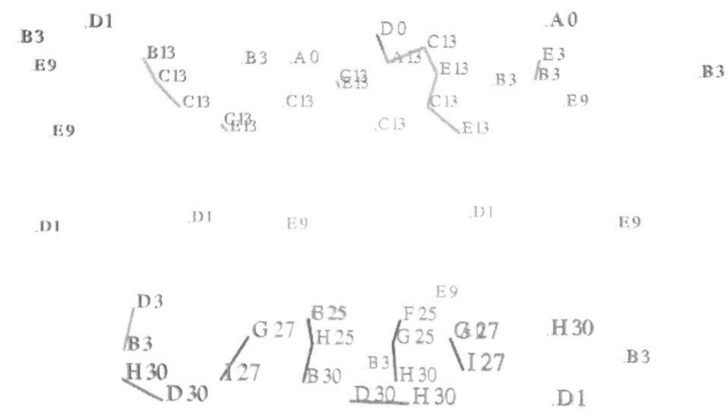

Fig. 2: Example of the scene graph obtained from the range scan shown in Fig. 1. Notice that segments with the label G are given the class labels 25 and 27, depending on their neighboring segments. The figure shows the scene graph after smoothing as described in Sec. V-B.

To analyse the multiple occurence of segments representing the same object part in constellation with segments that represent a different part, we define another, simpler graph structure named the *parts graph*. The nodes in the parts graph correspond to clusters in feature space as described in Sec. III-C and the edges represent connections in the scene graph. Details on the parts graph are given in Sec. IV-B.

A. The Scene Graph

After clustering the segments in the geometric space, we obtain K subsets of $\mathcal{S}$, namely $\mathcal{G}_1, \ldots, \mathcal{G}_K$. From our assumption, the number of potential object classes is bounded by K. To encode the fact that neighboring segments are more likely to correspond to the same object class, we define the scene graph $\mathfrak{S}$, which consists of the node set $\mathcal{V}_s = \mathcal{S}$ and the edge set $\mathcal{E}_s = \{(\mathbf{s}_i, \mathbf{s}_j) \mid \exists \mathcal{G}_i : (\mathbf{s}_i, \mathbf{s}_j) \subset \mathcal{G}_i\}$. Thus, $\mathcal{E}_s$ consists of all connections between segments that are closer to each other than ϑ_g. Furthermore, to each node in scene graph we assign a *class label* $y \in \{1, \ldots, K\}$. And finally, as each node $\mathbf{s}$ of $\mathfrak{S}$ corresponds to an instance of an object part, we can associate it with a *part label* $x \in \{1, \ldots, C\}$. At first sight, x should be fixed to the index of the feature space cluster $\mathcal{F}$ to which $\mathbf{s}$ belongs. However, as we formulate our problem in a probabilistic framework, we say that this index is only the *most likely* part label for $\mathbf{s}$ and all others are still possible. Details of this are explained in Sec. V.

An example of a scene graph is shown in Fig. 2. Here, part labels are represented as capital letters and class labels as numbers. We can see that many of the connected components only consist of one segment, as the segments are mostly far apart from each other. Also observe that there are segments with the same part label x, but with different class labels y. This stems from the fact that they occur in different contexts.

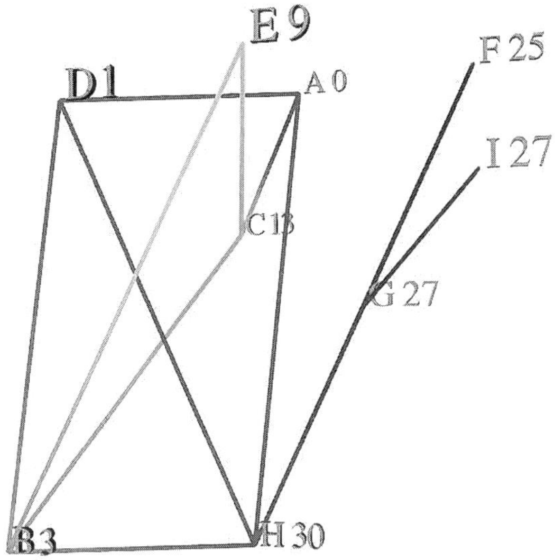

Fig. 3: Example of the parts graph obtained from the scan shown in Fig. 1. The positions of the nodes are determined at random, as only the topology of the graph is important. Again, part labels are represented as capital letters and class labels as numbers.

B. The Parts Graph

As mentioned before, apart from the class labels of specific segment instances, we also want to reason on the interaction among different object parts in general. We do this using the parts graph $\mathfrak{P}$. The node set $\mathcal{V}_p$ of $\mathfrak{P}$ corresponds to all clusters $\mathcal{F}_1, \ldots, \mathcal{F}_C$ of the feature space, and the set of edges $\mathcal{E}_p$ is defined as

$$\mathcal{E}_p = \{(\mathcal{F}_i, \mathcal{F}_j) \mid i \neq j, \exists (\mathbf{s}_k, \mathbf{s}_l) \in \mathcal{E}_s : \mathbf{s}_k \in \mathcal{F}_i \wedge \mathbf{s}_l \in \mathcal{F}_j\}.$$

This means, whenever there are two segments connected in the scene graph, there is a connection between the corresponding segment types in the parts graph. As in the scene graph, each node of the parts graph has an assigned part label x and a class label y, although with a slightly different interpretation: An assignment of a class label y to a node $\mathbf{s}$ in $\mathfrak{S}$ means that the particular segment instance $\mathbf{s}$ is most probably part of an object of class y, whereas the same assignment to a node $\mathcal{F}$ in $\mathfrak{P}$ means that in general segments that are elements of the cluster $\mathcal{F}$ are primarily more likely to be of class y. In contrast to the scene graph, each node of $\mathfrak{P}$ has a unique most-likely part label x, because these are directly determined by the feature space clustering.

In Fig. 3, we see an example of a parts graph. The positions of the nodes have been defined randomly as only the topology of the graph is important. Note that this concrete example of a parts graph is non-planar, although there is an equivalent planar representation. In general however, parts graphs may not be representable as planar graphs, for example in the case of a full connectivity.

V. SMOOTHING

So far, we described the construction of the two graph structures "scene graph" and "parts graph", but we did not specify how these are used to determine class labels for the segments. We do this using probabilistic reasoning: the nodes in both graphs are interpreted as random variables and the edges are used to model conditional dependencies between adjacent nodes. For the scene graph, this means that the class label y_i of a given segment $\mathbf{s}_i$ not only depends on the local evidence of the node i, i.e. the features $\mathbf{f}_i$ extracted from $\mathbf{s}_i$, but also on the class labels y_j of all neighbors $\mathbf{s}_j$ in the scene graph. Intuitively, a strong evidence for $\mathbf{s}_i$ to belong to a certain class may "outvote" the weaker evidence for $\mathbf{s}_j$ being of a different class. Similarly, a class label y_i for a given node $\mathcal{F}_i$ in the parts graph may be so strong that it *propagates* to the class labels of the neighbors of $\mathcal{F}_i$. The reasoning here is that if a certain type of segment $\mathcal{F}_i$ is very likely to be of a given class, then all segment types that occur frequently in the vicinity of $\mathcal{F}_i$ are also more likely to be of the same class. Applying this strategy to a graph leads to a *smoothed* graph because it tends to remove sudden changes of class labels between adjacent graph nodes.

In our implementation, we use Conditional Random Fields (CRFs) [1] to perform the smoothing. Our CRF models the conditional distribution

$$p(\mathbf{y} \mid \mathbf{f}) = \frac{1}{Z(\mathbf{f})} \prod_{\mathcal{V}} \varphi(\mathbf{f}_i, y_i) \prod_{(i,j)\in\mathcal{E}} \psi(\mathbf{f}_i, \mathbf{f}_j, y_i, y_j), \qquad (4)$$

where $Z(\mathbf{f}) = \sum_{\mathbf{y}'} \prod_{i=1}^{N} \varphi(\mathbf{f}_i, y_i') \prod_{(ij)\in\mathcal{E}} \psi(\mathbf{f}_i, \mathbf{f}_j, y_i', y_j')$ is the *partition function*, $\mathcal{V}$ is either $\mathcal{V}_s$ or $\mathcal{V}_p$ and $\mathcal{E}$ is $\mathcal{E}_s$ or $\mathcal{E}_p$. The dinstinction between the scene graph and the parts graph is made using different definitions for the *node potential* φ and the *edge potential* ψ. This will be described next.

However, before we proceed with the definition of the potentials, we mention some aspects in our formulation of the CRFs that are slightly different from others found in the literature. First, we recall that node and edge potentials are usually defined using the log-linear model so that, for the case of the node potential

$$\log \varphi(\mathbf{f}_i, y_i) = w_n \cdot f_n(\mathbf{f}_i, y_i), \qquad (5)$$

where w_n is a weight vector and f_n a feature function, which is high if $\mathbf{f}_i$ and y_i in some sense match well. For example, it can be defined as the outcome of a local classification. For our case of an unsupervised setting, we let f_n be the conditional probability $p(y_i \mid \mathbf{f}_i)$, i.e. a scalar value that is high if y_i matches well to $\mathbf{f}_i$. The same is also done for the edge potentials. As a result, the feature functions range between 0 and 1. This simplifies the weighting between node and edge potentials and turns the weight vectors w_n and w_e into scalars as well. In contrast to supervised learning with CRFs, we can not learn the node and edge weights w_n and w_e, as there is no training data available. Instead, we have to determine them manually. We do this using an appropriate evaluation measure on a validation set. This is described in Sec. VI.

A. Smoothing the Parts Graph

After creating the parts graph, the next step in our algorithm is to run the inference step using the CRF that corresponds to $\mathfrak{P}$. First, we note that the node features of this CRF are not equal to the feature vectors $\mathbf{f}_1, \ldots, \mathbf{f}_M$ extracted from the segments, because a node in $\mathfrak{P}$ actually represents a cluster in feature space and not a single segment. Instead, we define the node features in $\mathfrak{P}$ to be the mean $\bar{\mathbf{f}}$ of all feature vectors inside the corresponding cluster.

Furthermore, we observe that the dependence between feature vectors $\mathbf{f}$ and labels y is only implicit as we cannot model it directly. We can, however, model conditional probabilities between segment labels x and class labels y, as well as between feature vectors $\mathbf{f}$ and segment labels x. Following the definition of the node feature of $\mathfrak{P}$ as a conditional probability, we have

$$p(y_i \mid \bar{\mathbf{f}}_i) = \sum_{x=1}^{C} p(y_i \mid x) p(x \mid \bar{\mathbf{f}}_i), \tag{6}$$

where we assume a conditional independence of the labels and the features given the segment types. To obtain the two terms in the sum, we proceed as follows.

The class label posterior $p(y_i \mid x)$ is computed using Bayes' rule with the assumption of a uniform prior of the labels:

$$p(y_i \mid x) = \frac{p(x \mid y_i) p(y_i)}{\sum_{y'} p(x \mid y_i') p(y_i')} = \frac{p(x \mid y_i)}{\sum_{y'} p(x \mid y_i')} \tag{7}$$

The class likelihoods $p(x \mid y_i)$ are determined by counting the occurence of segments of type x inside the geometric cluster y_i and dividing by the number of all segments in y_i. The posteriors $p(y_i \mid x)$ can be computed at creation time of $\mathfrak{P}$ and collected in a $K \times C$ matrix, as they do not depend on the node features.

For the segment type posterior $p(x \mid \bar{\mathbf{f}}_i)$, we perform a nearest-neighbor search in feature space inside a sphere of radius ρ around $\bar{\mathbf{f}}_i$. Denoting the number of feature vectors in the sphere with type x as ν_x and the number of all elements in the sphere as ν, we approximate $p(x \mid \bar{\mathbf{f}}_i)$ with ν_x/ν.

Similar to Eqn. (6), we define the edge feature of $\mathfrak{P}$ as

$$p(y_i, y_j \mid \bar{\mathbf{f}}_i, \bar{\mathbf{f}}_j) = \sum_{x_i=1}^{C} \sum_{x_j=1}^{C} p(y_i, y_j \mid x_i, x_j) p(x_i, x_j \mid \bar{\mathbf{f}}_i, \bar{\mathbf{f}}_j). \tag{8}$$

As is common in literature related to CRFs, the edge features are designed to be zero whenever the labels y_i and y_j of the adjacent nodes are different ("generalized Potts model", see [28, 29]). The rationale of this is that only edges between equally labeled nodes should propagate the belief between the nodes. Thus, the first term in the sum of Eqn. (8) can be simplified to $p(y_{ij} \mid x_i, x_j)$, where y_{ij} is the common label of the nodes. We can compute this expression by counting the occurences of edges between x_i and x_j in cluster y_{ij} and applying Bayes rule as in Eq. (7).

For the second term in the sum of Eqn. (8), we apply the formulation

$$\begin{aligned} p(x_i, x_j \mid \bar{\mathbf{f}}_i, \bar{\mathbf{f}}_j) &= p(x_i \mid \bar{\mathbf{f}}_i, \bar{\mathbf{f}}_j) p(x_j \mid \bar{\mathbf{f}}_i, \bar{\mathbf{f}}_j) \\ &= p(x_i \mid \bar{\mathbf{f}}_i) p(x_j \mid \bar{\mathbf{f}}_j), \end{aligned} \tag{9}$$

which results from the conditional independence assumptions on x_i and x_j, and from those for x_i and $\mathbf{f}_j$, as well as x_j and $\mathbf{f}_i$. The resulting terms $p(x_i \mid \bar{\mathbf{f}}_i)$ and $p(x_j \mid \bar{\mathbf{f}}_j)$ are again computed using nearest-neighbor.

Once the edge and node potentials are defined, we can do inference on the parts graph. We do this using max-product loopy belief propagation. This is an approximate algorithm that returns labels $\mathbf{y}$ that maximize the conditional probability given in Eqn. (4). However, it also returns distributions over the class labels at each node. These will be used later.

B. Smoothing the Scene Graph

From the output of the inference step run on $\mathfrak{P}$, we obtain at each node of $\mathfrak{P}$ a distribution over class labels y. As the nodes of $\mathfrak{P}$ uniquely represent the segment types x, we can use that information to read the probability $p(y \mid x)$ directly from the parts graph. Thus, when creating the scene graph, we can again determine values for $p(y \mid x)$, as we did this for the parts graph, with the difference that now $p(y \mid x)$ also reveals the conditional dependencies between labels of segment types x, that have been observed in a close distance. This means, that we again compute a matrix for $p(y \mid x)$, but now we simply read the class label distributions off the nodes of $\mathfrak{P}$, as they result from belief propagation. In accordance to the node feature of $\mathfrak{P}$, we define the node feature of $\mathfrak{S}$ as the conditional probability

$$p(y_i \mid \mathbf{f}_i) = \sum_{x=1}^{C} p(y_i \mid x) p(x \mid \mathbf{f}_i). \tag{10}$$

Note that now the feature vectors $\mathbf{f}_i$ correspond to those that are actually extracted for each segment, i.e. at each node of $\mathfrak{S}$. As before, the term $p(x \mid \mathbf{f}_i)$ is computed with a nearest-neighbor search in feature space.

Finally, we define the edge feature of $\mathfrak{S}$ as

$$p(y_i, y_j \mid \mathbf{f}_i, \mathbf{f}_j) = \sum_{x_i=1}^{C} \sum_{x_j=1}^{C} p(y_{ij} \mid x_i, x_j) p(x_i, x_j \mid \mathbf{f}_i, \mathbf{f}_j), \tag{11}$$

where we denote the common class label with y_{ij} as above and the right term in the sum is computed according to Eqn. (9). Unfortunately, the computation of $p(y_{ij} \mid x_i, x_j)$ is not straightfoward, as we can not simply read this value from the edges of $\mathfrak{P}$. Instead, we use the following strategy: The only interesting case is when the class labels of x_i and x_j are equal. Thus, we can interpret the common label y_{ij} as an *edge label* that is determined by one of the adjacent node labels. In a sense, this expresses which of the nodes was responsible for the common edge label y_{ij}. We can formulate that using a binary variable c that is true if node x_i is responsible for y_{ij} and false otherwise. Using this, we can estimate $p(y_{ij} \mid x_i, x_j)$ by marginalizing out over c:

$$\begin{aligned} p(y_{ij} \mid x_i, x_j) &= \sum_{c} p(y_{ij} \mid x_i, x_j, c) p(c, \mid x_i, x_j) \\ &= p(y_{ij} \mid x_i, x_j, c) p(c) + p(y_{ij} \mid x_i, x_j, \neg c) p(\neg c) \\ &= 0.5 * p(y_{ij} \mid x_i) + 0.5 * p(y_{ij} \mid x_j) \end{aligned} \tag{12}$$

Here, we assumed that c is independent on x_i and x_j and that its prior probability is 0.5. Using Eqn. (12), we can compute $p(y_{ij} \mid x_i, x_j)$ by reading the values for $p(y_{ij} \mid x_i)$ and $p(y_{ij} \mid x_j)$ from the parts graph. As for the parts graph, we use max-product loopy belief propagation for the inference in the scene graph.

C. Obtaining the Class Label Distributions

Once the inference step on the scene graph is performed, we read the class label distributions $\mathbf{d}_1, \ldots, \mathbf{d}_M$ from all nodes of $\mathfrak{S}$. This is possible, as mentioned before, because belief propagation stores these distributions for each node. For the final discovery result, we report the most likely class label at each node, but it is important to note that the distributions may be useful for later reference, for example in an online application, where several data sets are acquired subsequently and the discovery of similar objects is done across the data sets.

VI. EXPERIMENTAL RESULTS

We tested the algorithm on data acquired from real-world scenes using a nodding SICK laser scanner with a horizontal opening angle of 100 degrees and a nodding range of 90 degrees. Each set was captured at a horizontal resolution of 0.25 degrees and a vertical resolution of 0.2 degrees. We evaluated 50 data sets from four different rooms, each room containing some number of chairs, trash cans, flip charts, plants, etc. Objects were placed up to 90 degrees of rotation from each other. Most scenes contained two or three objects of the same type, but some scenes contained up to four objects of three different kinds.

A. Qualitative Evaluation

In addition to the result shown in Fig. 1, Fig. 4 shows some more results of our object dicovery algorithm. All points that belong to the same object are depicted with the same color, and the numbers represent the class label to which each segment belongs. For instance, the scene in Fig. 1 contains four chairs of two different kinds, and they are correctly labeled as 25 (blue) and 27 (violet). Furthermore, we see that also most of the background segments, e.g. on the floor, ceiling and walls, have plausible labels. For many of them, only the local class label evidence was relevant, as they are not connected in the scene graph.

B. Quantitative Evaluation

In contrast to supervised learning algorithms, the performance of an unsupervised object discovery method is difficult to evaluate, as there is no real ground truth. The major problem here is that humans tend to be focused and might miss similarities in the data that are irrelevant for them. However, a good way to still evaluate unsupervised object discovery methods has been recently proposed by Tuytelaars *et al.* [30]. This method uses the *conditional entropy* of the "ground truth" class labels $\mathbf{y}^*$ given the class labels $\mathbf{y}$ that resulted from the discovery algorithm. Applied to our case, the conditional entropy is computed as

$$H(Y^* \mid Y) = \sum_{\sigma(y)=1}^{K^*} p(\sigma(y)) \sum_{y^*=1}^{K^*} p(y^* \mid \sigma(y)) \log \frac{1}{p(y^* \mid \sigma(y))},$$

where K^* is the number of ground truth classes and σ is an *oracle mapping* from the set of discovered classes to the set of ground truth classes, which is determined from a *tuning set* (for details see [30]). Intuitively, $H(Y^* \mid Y)$ determines the number of ground truth labels a segment can have once its discovered object label is known. The smaller this number is, the better is the result of the discovery algorithm. In the best case it is zero, which means that each discovered class label directly implies a ground truth label.

We created hand labeled ground truth data consisting of the five classes "ceiling", "floor", "wall", "chair", and "other", and evaluated the performance of our algorithm for different values of the parameters ϑ_g, ρ, and the node and edge weights (w_n, w_e). The results are shown in Fig. 5. Two conclusions can be drawn from these graphs: First, with values around 1, the results are very good compared to the results of similar algorithms described in [30]. And second, our algorithm is relatvely robust against small changes in the choice of the parameters, especially for the distance threshold ϑ_g, which is responsible for the clustering in geometric space.

VII. CONCLUSION AND OUTLOOK

We presented a fully unsupervised approach to segment 3D range scan data and to discover objects of a similar type that occur more than once in the scene. Our approach uses the only assumption that the number of actually existing object classes is not higher than the number of connected components in the scene, which holds in most cases. We applied probablistic reasoning based on Conditional Random Fields to model conditional dependencies of object part labels that are close to each other. In experiments on real data we showed that our algorithm is able to discover objects such as chairs in an indoor environment. In the future, we plan to use this technique in an online framework where the evidence of existing object classes is accumulated over time and across different data sets.

ACKNOWLEDGMENTS

This work was partially supported by the EC in the project robots@home under contract number FP6-IST-045350, in the project BACS under contract number FP6-IST-027140, and in the project EUROPA under contract number EUROPA-FP7-231888.

REFERENCES

[1] J. Lafferty, A. McCallum, and F. Pereira, "Conditional random fields: Probabilistic models for segmenting and labeling sequence data," in *Proceedings of International Conference on Machine Learning*, 2001.

[2] T. Tuytelaars, A. Turina, and L. van Gool, "Noncombinatorial detection of regular repetitions under perspective skew," *IEEE Trans. Pattern Analysis and Machine Intelligence*, vol. 25, no. 4, pp. 418–432, 2003.

[3] Y. Liu, R. T. Collins, and Y. Tsin, "A computational model for periodic pattern perception based on frieze and wallpaper groups," *IEEE Trans. on Pattern Anal. and Machine Intell.*, vol. 26, pp. 354–371, 2004.

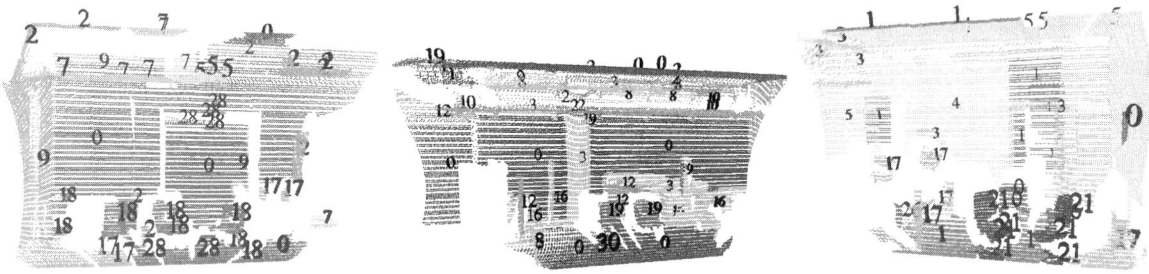

Fig. 4: Three example scenes observed with our nodding SICK laser scanner. Objects that are discovered through the algorithm are colored, where all points which belong to the same object class are assigned to one color. For visualization when viewed in greyscale, the discovered class labels are also shown.

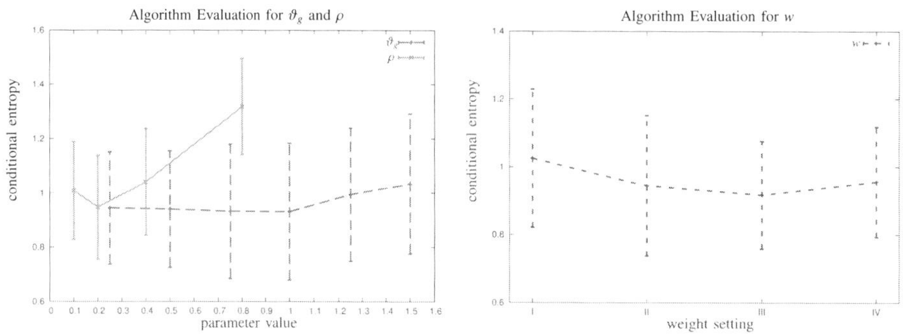

Fig. 5: Evaluation of our discovery algorithm using the conditional entropy with its standard deviation over 50 data sets. Left: Results for different values of ϑ_g and ρ. As it can be seen, ρ has a stronger influence on the results as ϑ_g. Right: Results for different settings of the node and edge weights (w_n, w_e). Here, 'I' corresponds to the pair $(0.5, 1.5)$, 'II' to $(1, 1)$, 'III' to $(1.5, 0.5)$ and 'IV' to $(2.0, 0)$. The best result is obtained for $w_n = 1.5$, $w_e = 0.5$.

[4] G. Loy and J.-O. Eklundh, "Detecting symmetry and symmetric constellations of features," in *ECCV*, 2006, pp. 508–521.

[5] S. Wenzel, M. Drauschke, and W. Förstner, "Detection of repeated structures in facade images," *Pattern Recognition and Image Analysis*, vol. 18, no. 3, pp. 406–411, 2008.

[6] G. Zeng and L. van Gool, "Multi-label image segmentation via point-wise repetition," in *IEEE Conf. on Comp. Vis. and Pat. Recog. (CVPR)*, Alaska, USA, June 2008.

[7] D. Shikhare, S. Bhakar, and S. Mudur, "Compression of large 3d engineering models using automatic discovery of repeating geometric features," in *Vision, Modeling, and Visualization*, 2001.

[8] J. Podolak, P. Shilane, A. Golovinskiy, S. Rusinkiewicz, and T. Funkhouser, "A planar-reflective symmetry transform for 3d shapes," in *SIGGRAPH*. New York, NY, USA: ACM, 2006, pp. 549–559.

[9] M. Pauly, N. J. Mitra, J. Wallner, H. Pottmann, and L. J. Guibas, "Discovering structural regularity in 3d geometry," *ACM Trans. Graph.*, vol. 27, no. 3, pp. 1–11, 2008.

[10] M. Bokeloh, A. Berner, M. Wand, H. Seidel, and A. Schilling, "Symmetry detection using feature lines," *Computer Graphics Forum (Proceedings of Eurographics)*, vol. 28, no. 2, pp. 697–706(10), April 2009.

[11] A. Y. Ng, M. I. Jordan, and Y. Weiss, "On spectral clustering: Analysis and an algorithm," in *Adv. in Neural Information Processing Systems*, T. Dietterich, S. Becker, and Z. Ghahramani, Eds., 2002.

[12] B. J. Frey and D. Dueck, "Clustering by passing messages between data points," *Science*, vol. 315, no. 5814, pp. 972–976, February 2007.

[13] A. Rabinovich, A. Vedaldi, C. Galleguillos, E. Wiewiora, and S. Belongie, "Objects in context," in *IEEE Conf. on Computer Vision*, 2007.

[14] C. Galleguillos, A. Rabinovich, and S. Belongie, "Object categorization using co-occurrence, location and appearance," in *IEEE Conference on Computer Vision and Pattern Recognition*, 2008.

[15] A. Quattoni, M. Collins, and T. Darrell, "Conditional random fields for object recognition," in *Adv. in Neural Inform. Proc. Systems*, L. Saul, Y. Weiss, and L. Bottou, Eds., 2005, pp. 1097–1104.

[16] X. Ma and W. E. L. Grimson, "Learning coupled conditional random field for image decomposition with application on object categorization," in *IEEE Conference on Computer Vision and Pattern Recognition*, 2008.

[17] D. Liu and T. Chen, "Semantic-shift for unsupervised object detection," in *Proc. of the Conf. on Comp. Vision and Pattern Rec. Workshop*, 2006.

[18] J. Sivic, B. Russell, A. Efros, A. Zisserman, and W. Freeman, "Discovering object categories in image collections," in *IEEE Conf. on Comp. Vis. and Pat. Recog. (CVPR)*, 2005.

[19] F. Endres, C. Plagemann, C. Stachniss, and W. Burgard, "Unsupervised discovery of object classes from range data using latent dirichlet allocation," in *Proc. of Robotics: Science and Systems*, 2009.

[20] J. Shin, R. Triebel, and R. Siegwart, "Unsupervised discovery of repetitive objects," in *IEEE Int. Conf. Robotics and Automation*, 2010.

[21] P. F. Felzenszwalb and D. P. Huttenlocher, "Efficient graph-based image segmentation," *Int. J. Comput. Vision*, vol. 59, no. 2, pp. 167–181, 2004.

[22] A. Johnson, "Spin-images: A representation for 3-d surface matching," Ph.D. dissertation, Robotics Institute, Carnegie Mellon Univ., 1997.

[23] R. Osada, T. Funkhouser, B. Chazelle, and D. Dobkin, "Shape distributions," *ACM Trans. on Graphics*, vol. 21, no. 4, pp. 807–832, 2002.

[24] C.-F. Westin, S. Peled, H. Gudbjartsson, R. Kikinis, and F. A. Jolesz, "Geometrical diffusion measures for MRI from tensor basis analysis," in *ISMRM '97*, Vancouver Canada, April 1997, p. 1742.

[25] S. P. Lloyd, "Least squares quantization in pcm," *IEEE Transactions on Information Theory*, vol. 28, no. 2, pp. 129–137, 1982.

[26] D. Comaniciu, P. Meer, and S. Member, "Mean shift: A robust approach toward feature space analysis," *IEEE Transactions on Pattern Analysis and Machine Intelligence*, vol. 24, pp. 603–619, 2002.

[27] J. Shi and J. Malik, "Normalized cuts and image segmentation," *IEEE Transactions on Pattern Analysis and Machine Intelligence*, vol. 22, no. 8, pp. 888–905, 2000.

[28] D. Anguelov, B. Taskar, V. Chatalbashev, D. Koller, D. Gupta, G. Heitz, and A. Ng, "Discriminative learning of markov random fields for segmentation of 3d scan data," in *IEEE Conf. on Comp. Vis. and Pat. Recog. (CVPR)*, 2005, pp. 169–176.

[29] R. B. Potts, "Some generalized order-disorder transformations," *Proc. Cambridge Phil Soc.*, vol. 48, 1952.

[30] T. Tuytelaars, C. H. Lampert, M. B. Blaschko, and W. Buntine, "Unsupervised object discovery: A comparison," *Int. Journal of Computer Vision*, vol. 88, no. 2, pp. 284–302, 2009.

Scale Drift-Aware Large Scale Monocular SLAM

Hauke Strasdat
Department of Computing,
Imperial College London, UK
Email: strasdat@doc.ic.ac.uk

J.M.M. Montiel
Instituto de Investigación
en Ingeniería de Aragón (I3A),
Universidad de Zaragoza, Spain
Email: josemari@unizar.es

Andrew J. Davison
Department of Computing,
Imperial College London, UK
Email: ajd@doc.ic.ac.uk

***Abstract*— State of the art visual SLAM systems have recently been presented which are capable of accurate, large-scale and real-time performance, but most of these require stereo vision. Important application areas in robotics and beyond open up if similar performance can be demonstrated using monocular vision, since a single camera will always be cheaper, more compact and easier to calibrate than a multi-camera rig.**

With high quality estimation, a single camera moving through a static scene of course effectively provides its own stereo geometry via frames distributed over time. However, a classic issue with monocular visual SLAM is that due to the purely projective nature of a single camera, motion estimates and map structure can only be recovered up to scale. Without the known inter-camera distance of a stereo rig to serve as an anchor, the scale of locally constructed map portions and the corresponding motion estimates is therefore liable to drift over time.

In this paper we describe a new near real-time visual SLAM system which adopts the continuous keyframe optimisation approach of the best current stereo systems, but accounts for the additional challenges presented by monocular input. In particular, we present a new pose-graph optimisation technique which allows for the efficient correction of rotation, translation and scale drift at loop closures. Especially, we describe the Lie group of similarity transformations and its relation to the corresponding Lie algebra. We also present in detail the system's new image processing front-end which is able accurately to track hundreds of features per frame, and a filter-based approach for feature initialisation within keyframe-based SLAM. Our approach is proven via large-scale simulation and real-world experiments where a camera completes large looped trajectories.

I. Introduction

Visual SLAM with a single camera is more challenging than when stereo vision can be used, but successful solutions have the potential to make a much wider impact because of the wealth of application domains in robotics and beyond where a single camera can more cheaply, compactly and conveniently be installed. Monocular SLAM research, which naturally overlaps with the highly developed Structure from Motion (SfM) field in computer vision, has seen great progress over the last ten years, and several impressive real-time systems have emerged as researchers have managed to transfer the best sequential SLAM techniques to the single camera domain. However, the successful sequential systems presented have mostly either been limited to operation in a local workspace (e.g. [7, 14, 9, 2]) or, more rarely, have been designed for open-loop visual odometry exploration of larger areas [21, 3].

Meanwhile, in stereo vision-based SLAM there have now been systems presented which offer end to end solutions to real-time large scale SLAM. Probably the most complete is that of Mei *et al.* [18], which combines robust and accurate local visual odometry with constant-time large-scale mapping (enabled by a continuous relative map representation), high-performance appearance-based loop closure detection, and global metric map optimisation if required. Another similar system is that presented by Konolige and Agrawal [16].

It has proven more difficult to achieve real-time large-scale mapping with a monocular camera, due to its nature as a purely projective sensor. Geometry does not just 'pop out' of the data from a moving camera, but must be inferred over time from multiple images. Difficulties had to be overcome before a probabilistic sequential approach could successfully be implemented for monocular SLAM due to the fact that image features must be observed from multiple viewpoints with parallax before fully constrained 3D landmark estimates can be made. Special attention was given to feature initialisation schemes which permitted sequential probabilistic filtering of the joint camera and map state [7, 28, 20]. Beyond this issue, the fact that a single camera does not measure metric scale means that monocular mapping must either proceed with scale as an undetermined factor or that scale has to be introduced by an additional information source such as a calibration object as in [7] or even by exploiting nonholonomic motion constraints [27]. These issues, and the fact that monocular maps are just generally less well constrained than those built by metric sensors, meant that it was more difficult than in SLAM systems with other sensors to build algorithms which could tackle large areas by composing local fragments.

One of the first attempts to put a whole large scale monocular SLAM together was by Clemente *et al.* [4], who took a hierarchical mapping approach where locally filtered submaps were built and joined by an upper level joint map. Map matching based on feature correspondences between submaps allowed global correction of motion and scale drift around a large loop. In [24] a more accurate later evolution of this work based on conditional independent local maps was presented. Eade and Drummond [10, 8] presented a system which had several similarities to these approaches, but also many improvements, in that it was able to operate automatically in true real-time, incorporating effective loop-closing and re-localisation methods, as well as making use of non-linear optimisation techniques at various different levels to improve accuracy. At its core, though, it built a graph of local filtered submaps, each with a state vector and covariance, and with the similarity transforms between neighbouring submaps estimated via feature correspondences.

A. SLAM via Optimisation

Recently, Strasdat *et al.* [29] presented results which indicate, on the basis of a systematic investigation into the accuracy of local map components versus computational cost, that the keyframes + optimisation approach of Klein and Murray [14] — based ultimately on well known bundle adjustment methods [30] — is strongly advantageous compared to methods like those of Eade [8] or Clemente [4] whose building blocks are locally filtered submaps. Essentially, the accuracy analysis in [29] shows that it is the ability to harvest measurements from large numbers of image features per frame that plays the major role in increasing accuracy in local monocular motion estimation, while incorporating information from a large number of closely spaced camera frames provides less information for a given computation budget. This plays into the hands of the optimisation methods which cope much more efficiently with large numbers of features than filters can.

In light of this analysis, the previous large scale monocular systems [4] and [8] can be seen as somewhat unsatisfactory approaches, which combine filtering at a local level with optimisation at the global level. For instance, when loop closures are imposed, the relative positions of their filtered submaps changes, but any drift within the local maps themselves cannot be resolved. In this paper, we have therefore resolved to take a keyframe optimisation approach from top to bottom, aiming for maximum accuracy while taking into full account the special character of SLAM with monocular vision.

B. Gauge Freedoms, Monocular SLAM and Scale Drift

Metric SLAM systems aim to build coherent maps, in a single coordinate frame, of the areas that a robot moves through. But they must normally do this based on purely relative measurements of the locations of scene entities observable by their on-board sensors. There will therefore always be certain degrees of *gauge freedom* in the maps that they create, even when the best possible job is done of estimation. These gauge freedoms are degrees of transformation freedom through which the whole map, consisting of feature and robot position estimates taken together, can be transformed without affecting the values of the sensor measurements. In SLAM performed by a robot moving on a ground plane and equipped with a range-bearing sensor, there are three degrees of gauge freedom, since the location of the entire map with regard to translations and rotation in the plane is undetermined by the sensor measurements. In SLAM by a robot moving in 3D and equipped with a sensor like calibrated stereo vision or a 3D laser range-finder, there are six degrees of gauge freedom, since the whole map could experience a rigid body transformation in 3D space. In monocular SLAM, however, there are fundamentally seven degrees of gauge freedom [30], since the overall scale of the map, as well as a 6 DoF (degrees of freedom) rigid transformation, is undetermined (scale and a rigid transformation taken together are often known as a similarity transformation).

It is the number of gauge degrees of freedom in a particular type of SLAM which therefore determines the ways in which drift will inevitably occur between different fragments of a map. Consider two distant fragments in a large map built continuously by a single robot: local measurements in each of the fragments have no effect in pulling either towards a particular location in the degrees of gauge freedom. If they are not too distant from each other, they will share some coherence in these degrees of freedom, but only via compounded local measurements along the chain of fragments connecting them. The amount of drift in each of these degrees of freedom will grow depending on the distance between the fragments, and the distribution of the potential drift can be calculated if needed by uncertainty propagation along the chain.

It is very well known that planar maps built by a ground-based robot drift in three degrees of freedom; so maps built by a monocular camera with no additional information drift in seven degrees of freedom. It is through these degrees of freedom therefore which loop closure optimisations must adjust local map fragments (poses or submaps in a graph).

II. General Framework for State Estimation using Optimisation

In this section, we give a short review of state estimation using optimisation. The goals of this are first to define common notation and terminology for the rest of the paper, and second to clarify the natural close relationship between the different types of optimisation our approach uses, such as bundle adjustment and pose-graph optimisation.

In a general estimation problem, we would like to estimate a vector of parameters $\mathbf{p}$ given a vector of measurements $\mathbf{f}$, where we know the form of the likelihood function $p(\mathbf{f}|\mathbf{p})$. The most probable solution is the set of values $\mathbf{p}$ which maximises this likelihood, which is equivalent to minimising the negative log-likelihood:

$$\arg\max_{\mathbf{p}} p(\mathbf{f}|\mathbf{p}) = \arg\min_{\mathbf{p}}(-\log p(\mathbf{f}|\mathbf{p})) \ . \qquad (1)$$

In many optimisation problems, negative log-likelihood is known as *energy* and the goal is to minimise it, which is fully equivalent to maximising the probability of the solution.

A. Least Square Problems and Gauss-Newton Optimisation

We now speak in more specific but still very widely applicable terms, and assume that the likelihood distribution $p(\mathbf{f}|\mathbf{p})$ is jointly Gaussian, and thus has the distribution (up to a constant of proportionality):

$$p(\mathbf{f}|\mathbf{p}) \propto \exp((\mathbf{f} - \hat{\mathbf{f}}(\mathbf{p}))^\top \Lambda_{\mathbf{f}}(\mathbf{f} - \hat{\mathbf{f}}(\mathbf{p}))) \ , \qquad (2)$$

where $\Lambda_{\mathbf{f}} = \Sigma_{\mathbf{f}}^{-1}$, the *information matrix* or inverse of the *measurement covariance matrix* $\Sigma_{\mathbf{f}}$ of the likelihood distribution, and $\hat{\mathbf{f}}(\mathbf{p})$ is the 'measurement function' which computes (predicts) the distribution of measurements $\mathbf{f}$ given a set of parameters $\mathbf{p}$.

The negative log-likelihood, or energy, $\chi^2(\mathbf{p}) = -\log p(\mathbf{f}|\mathbf{p})$ therefore has the quadratic form:

$$\chi^2(\mathbf{p}) = (\mathbf{f} - \hat{\mathbf{f}}(\mathbf{p}))^\top \Lambda_{\mathbf{f}} (\mathbf{f} - \hat{\mathbf{f}}(\mathbf{p})) \ . \tag{3}$$

If $\Lambda_{\mathbf{f}}$ is block-diagonal, $\chi^2(\mathbf{p})$ simplifies to:

$$\chi^2(\mathbf{p}) = \sum_i (\mathbf{f}_i - \hat{\mathbf{f}}_i(\mathbf{p}_i))^\top \Lambda_{\mathbf{f},i} (\mathbf{f}_i - \hat{\mathbf{f}}_i(\mathbf{p}_i)) \ . \tag{4}$$

This is the case if $\mathbf{f}$ consists of a stack of measurements which are independent, which can very commonly be assumed when many measurements are made from a calibrated sensor — of particular interest here, where each measurement is the image location of a scene feature in one frame taken by a camera.

Further, and again in our case, $\Lambda_{\mathbf{f}}$ can very often be taken to equal the identity matrix. This implies that the uncertainty in each individual parameter of every measurement has exactly the same value, and that they are all independent. The absolute magnitude of the values on the diagonal of $\Lambda_{\mathbf{f}}$ does not affect the optimisation result so they can be set to one. Now, $\chi^2(\mathbf{p})$ is simply a sum of squares, and minimising it is called *non-linear least squares optimisation.*

A common technique for non-linear least squares optimisation is the Gauss-Newton method, which is an approximation of the Newton method. This consists of iteratively updating an initial estimate of the parameter vector $\mathbf{p}$ by the rule

$$\mathbf{p}_{(i+1)} = \mathbf{p}_{(i)} + \delta \ , \tag{5}$$

where at each step the update vector δ is found by solving the *normal equation*:

$$\left(\mathtt{J}_{\mathbf{p}}^\top \Lambda_{\mathbf{f}} \mathtt{J}_{\mathbf{p}}\right) \delta = -\mathtt{J}_{\mathbf{p}}^\top \Lambda_{\mathbf{f}} \mathbf{r} \ . \tag{6}$$

Here, $\mathtt{J}_{\mathbf{p}} = \frac{\partial \mathbf{r}}{\partial \mathbf{p}}$ and $\mathbf{r} = \mathbf{f} - \hat{\mathbf{f}}(\mathbf{p})$ is the *residual error.*

All estimation described in this paper is done using a variant of Gauss-Newton called Levenberg-Marquardt (LM), which alters the normal equation as follows:

$$\left(\mathtt{J}_{\mathbf{p}}^\top \Lambda_{\mathbf{f}} \mathtt{J}_{\mathbf{p}} + \mu \mathtt{I}\right) \delta = -\mathtt{J}_{\mathbf{p}}^\top \Lambda_{\mathbf{f}} \mathbf{r} \ . \tag{7}$$

The parameter μ rotates the update vector δ towards the direction of the steepest descent. Thus, if $\mu \to 0$ pure Gauss-Newton is preformed, whereas if $\mu \to \infty$, *gradient descent* is used. In LM, we only perform an update $\mathbf{p}_{(i)} + \delta$ if it will be successful in significantly reducing the residual error. After a successful update, μ is decreased; otherwise μ is increased. The motivation behind this heuristic is that Gauss-Newton can approach a quadratic rate of convergence, but gradient descent behaves much better far from the minimum.

B. Covariance Back-propagation

So far, we have considered only how to find the single most probable set of parameters $\mathbf{p}$. If there is no gauge freedom in the minimisation $\arg\min_{\mathbf{p}} \chi^2(\mathbf{p})$, then $\mathtt{J}_{\mathbf{p}}^\top \Lambda_{\mathbf{f}} \mathtt{J}_{\mathbf{p}}$ has full rank and we can determine a full distribution for $\mathbf{p}$. A first-order approximation of the covariance $\Sigma_{\mathbf{p}}$ can be calculated using covariance back-propagation:

$$\Sigma_{\mathbf{p}} = (\mathtt{J}_{\mathbf{p}}^\top \Lambda_{\mathbf{f}} \mathtt{J}_{\mathbf{p}})^{-1} \ . \tag{8}$$

Note that if $\mathbf{p}$ is high-dimensional, this calculation is very expensive because of the large matrix inversion required.

C. Optimising Rigid Body Transformations in $\mathbb{R}^3$

In optimisation, especially in 3D SLAM, the correct treatment of angles consistently causes confusion. On one hand, a minimal parametrisation is desired, but also singularities should be avoided. Interpolation of angles is not straightforward, since the group of rotation is only locally Euclidean.

Probably the most elegant way to represent rigid body transformations in optimisation is using a Lie group/algebra representation. It is well known that the Newton method has a quadratic convergence rate given we optimise over a Euclidean space. However, this result can be generalised to general Lie groups [17]. A rigid body transformation in $\mathbb{R}^n$ can be expressed as an $(n+1) \times (n+1)$ matrix which can be applied to homogeneous position vectors:

$$\mathtt{T} = \begin{bmatrix} \mathtt{R} & \mathbf{t} \\ \mathtt{0} & 1 \end{bmatrix} \quad \text{with} \quad \mathtt{R} \in \mathrm{SO}(n) \ , \quad \mathbf{t} \in \mathbb{R}^n \ , \tag{9}$$

and $\mathrm{SO}(n)$ being the Lie group of rotation matrices. The rigid body transformations in $\mathbb{R}^n$ form a smooth manifold and therefore a Lie group — the *Special Euclidean* group $\mathrm{SE}(n)$ [11]. The group operator is the matrix multiplication.

A minimal representation of this transformation is defined by the corresponding Lie algebra $\mathfrak{se}(n)$ which is the tangent space of $\mathrm{SE}(n)$ at the identity. In $\mathbb{R}^3$, the algebra elements are 6-vectors $(\omega, \upsilon)^\top$: $\omega = (\omega_1, \omega_2, \omega_3)$ is the axis-angle representation for rotation, and υ is a rotated version of the translation $\mathbf{t}$. Elements of the $\mathfrak{se}(3)$ algebra can be mapped to the $\mathrm{SE}(3)$ group using the exponential mapping $\exp_{\mathrm{SE}(3)}$:

$$\exp_{\mathrm{SE}(3)}(\omega, \upsilon) := \begin{bmatrix} \exp_{\mathrm{SO}(3)}(\omega) & \mathtt{V}\upsilon \\ \mathtt{0} & 1 \end{bmatrix} = \begin{bmatrix} \mathtt{R} & \mathbf{t} \\ \mathtt{0} & 1 \end{bmatrix} . \tag{10}$$

Here, $\exp_{\mathrm{SO}(3)}(\omega) = \mathtt{I} + \frac{\sin(\theta)}{\theta}(\omega)_\times + \frac{1-\cos(\theta)}{\theta^2}(\omega)_\times^2$ is the well-known *Rodrigues' formula*, $\mathtt{V} = \mathtt{I} + \frac{1-\cos(\theta)}{\theta^2}(\omega)_\times + \frac{\theta - \sin(\theta)}{\theta^3}(\omega)_\times^2$, $\theta = ||\omega||_2$, and $(\cdot)_\times$ is an operator which maps a 3-vector to its skew-symmetric matrix. Since $\exp_{\mathrm{SE}(3)}$ is surjective, there exists also an inverse relation $\log_{\mathrm{SE}(3)}$.

During optimisation, incremental updates δ are calculated in the tangent space around the identity $\mathfrak{se}(3)$ and mapped back onto the manifold $\mathrm{SE}(3)$, leading to a modified version of (5):

$$\mathtt{T}_{(i+1)} = \exp_{\mathrm{SE}(3)}(\delta) \cdot \mathtt{T}_{(i)} \ . \tag{11}$$

In this way, we avoid singularities — since δ is always close to the identity — while ensuring a minimal representation during optimisation. The Jacobian of (11) is linear in $\mathtt{T}$,

$$\left.\frac{\partial \exp_{\mathrm{SE}(3)}(\delta) \cdot \mathtt{T}}{\partial \delta}\right|_{\delta=0} = \begin{pmatrix} -(\mathbf{r}_1)_\times & \mathtt{0}_{3\times3} \\ -(\mathbf{r}_2)_\times & \mathtt{0}_{3\times3} \\ -(\mathbf{r}_3)_\times & \mathtt{0}_{3\times3} \\ -(\mathbf{t})_\times & \mathtt{I}_{3\times3} \end{pmatrix} , \tag{12}$$

where $(\mathbf{r}_1, \mathbf{r}_2, \mathbf{r}_3) := \mathtt{R}$. A general Jacobian $\frac{\partial \mathbf{f}(\exp_{\mathrm{SE}(3)}(\delta) \cdot \mathtt{T})}{\partial \delta}$ can be easily determined using the chain rule.

III. A Keyframe Optimisation System for Large Scale Monocular SLAM

We will now describe the elements of our complete system for sequential monocular SLAM. Similar to Klein and Murray's PTAM system [14], our SLAM framework consists of two parts: a front-end which tracks features in the image and estimates the pose of each live frame given a set of known 3D points, and a back-end which performs joint bundle adjustment over a set of keyframes. In PTAM which focuses on accurate and fast mapping in a small environment, the front-end only performs pose and feature tracking, while the optimisation back-end does everything else such as feature initialisation and dropping of new keyframes. Since our approach focusses on mapping large scenes, we use the opposite emphasis. The optimisation back-end does only bundle adjustment, but the tracking front-end takes over all important decisions such as when to drop new keyframes and which feature to initialise. This design puts a higher computational burden on the tracking front-end, but facilitates exploration since the stability of the tracking front-end does not depend on the optimisation back-end to drop new keyframes.

A. The Camera Projection Function and Camera Poses

Points in the world $\mathbf{x}_j \in \mathbb{R}^3$ are mapped to the camera image using the observation function $\hat{\mathbf{z}}$:

$$\hat{\mathbf{z}}(\mathtt{T}_i, \mathbf{x}_j) = \mathbf{proj}(\mathtt{K} \cdot \mathtt{T}_i \cdot \mathbf{x}_j) \ . \tag{13}$$

Here, $\mathbf{x}_j$ is a homogeneous point, $\mathtt{T}_i \in \mathrm{SE}(3)$, $\mathtt{K}$ is the camera calibration matrix (which we assume is known from prior calibration) and $\mathbf{proj}$ is the 3D-2D projection:

$$\mathbf{proj}(\mathbf{a}) := \frac{1}{a_3}(a_1, a_2)^\top \quad \text{for} \quad \mathbf{a} \in \mathbb{R}^3 \ . \tag{14}$$

We represent the pose of the camera at a given time-step i by the rigid body transformation $\mathtt{T}_i$. Note that the position of the camera center is *not* explicitly represented, but can be recovered as $-\mathtt{R}_i^\top \mathbf{t}_i$ [13, pp.155].

B. Optimisation Back-end

In the back-end, we perform an optimisation jointly over a set of keyframes $\mathtt{T}_{\text{key}:i}$ and large set of points $\mathbf{x}_j$ called bundle adjustment (BA) [30]. In order to be constant-time, the number of keyframes included in BA is restricted. Since we focus on exploration, we optimise over a sliding window of keyframes. At loop closure, different measures have to be taken (see Sec. IV).

In general in BA, not all points are visible in every frame. Let $\mathtt{Z}$ be a sparse matrix where an entry is either an observation $\mathbf{z}_{i,j}$ if point $\mathbf{x}_j$ is visible in frame $\mathtt{T}_{\text{key}:i}$ or zero otherwise. Let I be the sequence of non-zero row indices of $\mathtt{Z}$: $\mathtt{I} = \{(i) : \mathtt{Z}_{i,j} \neq 0\}$. The column index list J is defined similarly. Now, it is obvious how to define the *stacked projection function*:

$$\hat{\mathbf{z}}(\mathtt{T}_{\text{key}:1}, \ldots, \mathtt{T}_{\text{key}:m}, \mathbf{x}_1, \ldots, \mathbf{x}_n, \mathtt{I}, \mathtt{J}) \ . \tag{15}$$

We solve BA conventionally using LM by minimising

$$\chi^2(\mathbf{p}) = \sum_{(i,j)\in I\times J} (\mathbf{z}_{i,j} - \hat{\mathbf{z}}(\mathtt{T}_{\text{key}:1}, \ldots, \mathtt{T}_{\text{key}:m}, \mathbf{x}_1, \ldots, \mathbf{x}_n, \mathtt{I}, \mathtt{J}))^2 \tag{16}$$

with respect to the parameter vector $\mathbf{p} = (\mathtt{T}_{\text{key}:3}, ..\mathtt{T}_{\text{key}:m}, \mathbf{x}_1, \ldots, \mathbf{x}_n)$. As a sliding window BA is applied, the optimisation window has to be anchored to the previous map. Thus, at least 7 DoF (rotation, translation and scale) should be fixed. Because of this, $\mathtt{T}_{\text{key}:1}, \mathtt{T}_{\text{key}:2}$ are excluded from the optimisation. In order to guard against spurious matches, we use the pseudo-Huber cost function [13, p.619] as a robust kernel.

The primary sparseness pattern in BA — that there are only links between points and frames, but no point-point or frame-frame constraints — is exploited in the usual way using the Schur complement [30]. However, our BA implementation also exploits the secondary sparseness pattern — that not every point is visible in every frame. This is done using the CSparse library [6] which performs efficient symbolic Cholesky decomposition using elimination trees.

C. Feature Tracking and Pose Optimisation

The main task of the tracking front-end is to estimate the current camera pose $\mathtt{T}_i$. However, it also deals with feature and keyframe initialisation. It takes as input a set of keyframe poses $\mathtt{T}_{\text{key}:i}$ and 3D points $\mathbf{x}_j \in \mathcal{X}$ which are associated with 2D measurements $\mathbf{z}_{\text{key}:i,j}$. The poses $\mathtt{T}_{\text{key}:i}$ and points $\mathbf{x}_j$ were already jointly estimated in the optimisation back-end (see Sec. III-B).

For feature tracking and pose optimisation, we use a mixture of bottom-up (for every pixel) and top-down approaches (guided techniques). First, we estimate the relative pose $\Delta\mathtt{T}_i = \mathtt{T}_{i-1} \cdot \mathtt{T}_i$ between the previous and current frames. This is done by the help of dense variational optical flow implemented very efficiently on the GPU [25]. Given the pose estimate of the previous frame $\mathtt{T}_{i-1}$, we can calculate an initial pose estimate for the current frame: $\mathtt{T}_i^{[0]} = \Delta\mathtt{T}_i \cdot \mathtt{T}_{i-1}^{-1}$. Feature tracking is not done using frame to frame matching. Whereas frame to frame matching is very reliable for pure visual odometry, it prevents mini-loop closures.

Given our initial pose guess $\mathtt{T}_i^{[0]}$, we estimate the subset $\mathcal{X}' \subset \mathcal{X}$ of landmarks which might be visible in the current frame. Based on the projection (13), we perform active search [7] for each point $\mathbf{x}_{j'} \in \mathcal{X}'$, where the centre of the search region is defined by the prediction $\hat{\mathbf{z}}(\mathtt{T}_i^{[0]}, \mathbf{x}_{j'})$. Generally, the uncertainty of the projection and thus the size and shape of the search region depends on the uncertainty of $\mathbf{x}_{j'}$ and $\mathtt{T}_i$. Given that all points in $\mathcal{X}$ are already optimised using keyframe bundle adjustment, and that pose $\mathtt{T}_{i-1}$ is already optimised with respect to the $\mathcal{X}$, then apart from the measurement noise $\Sigma_{\mathbf{z}}$, the only crucial uncertainty originates from the motion between the previous and the current frame. Let, $\Sigma_{\Delta\mathtt{T}_i}$ be the uncertainty of the relative motion. Then the

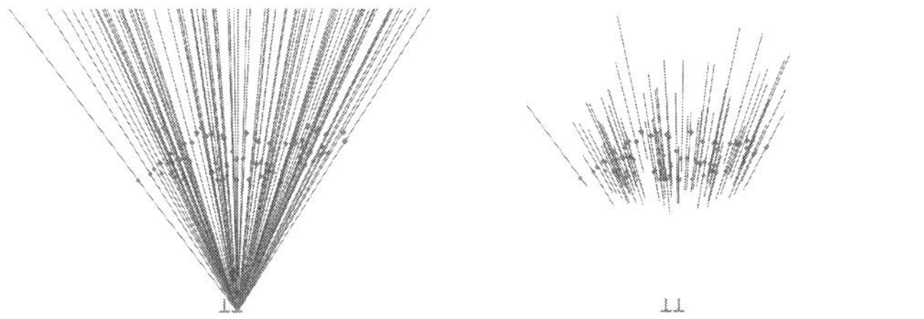

a) Feature initialisation b) After a single update

Fig. 1. Illustration of the feature initialisation process. First features are initialised as inverse depth points with infinite uncertainty on the depth (a). A single update leads to a significant reduction of the depth uncertainty (b). The inverse depth features are plotted using a 99.7 percent confidence interval.

search area is constrained by the innovation covariance

$$\left(\frac{\partial\hat{\mathbf{z}}(\mathtt{T}',\mathbf{x}_{j'})}{\partial\delta}\right)^{\top}\Sigma_{\Delta\mathtt{T}_i}\left(\frac{\partial\hat{\mathbf{z}}(\mathtt{T}',\mathbf{x}_{j'})}{\partial\delta}\right)+\Sigma_{\mathbf{z}}\ , \qquad (17)$$

with $\mathtt{T}' = \exp_{\mathrm{SE}(3)}(\delta)\cdot\mathtt{T}_i^{[0]}$. The search within this region is based on template matching using normalised cross-correlation. Using the initial guess $\mathtt{T}_i^{[0]}$, the templates are perspectively warped. In order to speed-up the search, we do not apply the matching at every single pixel in the search region, but only there where FAST features [26] are detected.

As a result, we receive a set of 2D-3D matches between image locations $\mathbf{z}_k$ and feature points $\mathbf{x}_k$. Finally, we optimize the pose $\mathtt{T}_i$ using motion-only BA. Thus, we minimize $\chi^2(\mathtt{T}_i) = \sum_k(\mathbf{z}_k - \hat{\mathbf{z}}_k(\mathtt{T}_i,\mathbf{x}_j))^2$ wrt. $\mathtt{T}_i$ using LM. Again, we use the pseudo-Huber cost function as a robust kernel.

D. Feature Initialisation

In monocular SLAM approaches using filtering, no special treatment for feature initialisation is needed if an inverse depth representation [20] is used. New features are jointly estimated together with all other parameters, but at a cost of $O(n^3)$ where n is the number of features visible.

In PTAM [14], new features are triangulated between keyframes. In order to restrict the matching, a strong depth prior is used (restricting the PTAM to using features relatively close to the camera). We suggest a feature initialisation method for keyframe-based SLAM systems based on a set of three dimensional information filters which can estimate the position of arbitrarily distant features. Very recently, a similar method was briefly described by Klein and Murray [15]. Compared to our approach, their filters are applied on every frame, are one-dimensional and are only used for data association.

Ultimately, we would like to update a set of partially initialised 3D points $\mathbf{x}_{new:j}$ given the current camera pose $\mathtt{T}_i$. If $\mathtt{T}_i$ is known, the features $\mathbf{x}_{new:n}$ become independent:

$$p(\mathbf{x}_{new:1},...,\mathbf{x}_{new:j},..|\mathtt{T}_i) = p(\mathbf{x}_{new:1})\cdots p(\mathbf{x}_{new:n})\ . \qquad (18)$$

Since the current camera $\mathtt{T}_i$ is well-optimised wrt. the set of known points $\mathcal{X}$ (as described in the previous section), equation (18) is approximately true. Thus, our method employs a set of information filters. Each filter estimates the position of the a single landmark given the current pose estimate. In this sense, our approach is similar to FastSLAM [19]. The difference to FastSLAM is that the partially initialised features $\mathbf{x}_{new:j}$ are not used for state estimation immediately. New features are only used for pose estimation after they are jointly bundle adjusted and added to $\mathcal{X}$ (see Sec. III-E).

The design of a single information filter is greatly inspired by Eade's filtering framework [8]. Features are represented using inverse depth coordinates $\mathbf{y} = (u, v, q)^{\top}$ wrt. the origin frame $\mathtt{T}_{\mathrm{org}}$ in which they were seen first. They can be mapped to a Euclidean point $\mathbf{x}_{\mathrm{org}} = \frac{1}{q}(u, v, 1)^{\top}$ in this local coordinate frame $\mathtt{T}_{\mathrm{org}}$. The uncertainty of an inverse depth feature is represented using the inverse covariance matrix $\Lambda_{\mathbf{y}} = \Sigma_{\mathbf{y}}^{-1}$. New features are jointly initialised at keyframes where FAST [26] produces candidate locations. In order to ensure an equal distribution of feature locations over the image, we employ a quadtree similar to Mei *et al.* [18]. Given the FAST feature location $\mathbf{z} = (u, v)^{\top}$, we set $\mathbf{y} = (u, v, q)$ with $q \in \mathbb{R}^+$. The uncertainty $\Lambda_{\mathbf{y}}^{(0)}$ is set to $\mathrm{diag}(\frac{1}{\sigma_{\mathbf{z}}^2}, \frac{1}{\sigma_{\mathbf{z}}^2}, 0)$. Note that initially there is an infinite uncertainty along the feature depth, so we do not enforce any depth prior no matter which start value we assign for q. In order to save computation and to avoid divergence, inverse depth features are only updated on selected frames with sufficient parallax.

We filter $\mathbf{y}$ by minimising:

$$\chi^2(\mathbf{y}) = (\mathbf{y} - \hat{\mathbf{y}})^{\top}\Lambda_{\mathbf{y}}(\mathbf{y} - \hat{\mathbf{y}}) + (\Delta\mathbf{z})^{\top}\Lambda_{\mathbf{z}}(\Delta\mathbf{z}) \qquad (19)$$

wrt. $\mathbf{y}$ using LM. Here, $\hat{\mathbf{y}}$ is set to the initial guess $\mathbf{y}_{[0]}$ and $\Delta\mathbf{z} = \mathbf{z} - \hat{\mathbf{z}}(\mathtt{T}_i, \mathtt{T}_{\mathrm{org}}^{-1}\mathbf{x}_{\mathrm{org}})$. The first term in $\chi^2(\mathbf{y})$ ensures that the optimisation of $\mathbf{y}$ is based on its prior distribution $\langle\hat{\mathbf{y}}, \Lambda_{\mathbf{y}}\rangle$. The second term takes care that the projection error wrt. the current frame is reduced. In other words, the point $\mathbf{y}$ is moved along its uncertain depth rather than via its certain u, v coordinate in order to reduce the projection error in the current image. We also update the uncertainty using covariance back-propagation (see Fig. 1):

$$\Lambda_{\mathbf{y}}^{(k+1)} = \Lambda_{\mathbf{y}}^{(k)} + \left(\frac{\partial\Delta\mathbf{z}}{\partial\mathbf{y}}\right)^{\top}\Lambda_{\mathbf{z}}^{(k)}\left(\frac{\partial\Delta\mathbf{z}}{\partial\mathbf{y}}\right)\ . \qquad (20)$$

E. Keyframe Dropping

A new keyframe is dropped as soon the distance of the current camera from the closest keyframe exceeds a threshold. Only inverse depth features $\mathbf{y}$ which could be reliably tracked and whose position is certain enough are considered for adding to $\mathcal{X}$. Before, they are bundle adjusted with respect to a small number (e.g. three) of surrounding keyframes. Only those features which do not produce a large projection error in any of the surrounding keyframes are added to $\mathcal{X}$.

IV. Loop Closure

A. Loop Closure Detection

It is well known that loop closures can be detected effectively using appearance information only [1, 5]. These methods often rely on visual bags of words based on SIFT or SURF features. Given that we have a loop closure detection between two frames each associated with a set of SURF features, the standard method would apply RANSAC in conjunction with the 5-point method [22] in order to estimate the epipolar geometry. Then the relative Euclidean motion up to an unknown scale in translation can be recovered.

However, we can exploit the fact that in our SLAM system each frame is associated with a large set of three-dimensional feature points. First, we create a candidate set of SURF feature pairs by matching features between the current frame and the loop frame based on their descriptors. Then, we create a dense surface model using the three-dimensional feature points visible in the loop frame. Next, the unknown depth of the candidate SURF features of the loop frame is calculated using this dense surface model.[1] The computationally very efficient k-nearest neighbour regression algorithm proved to be sufficient for our needs. In other words, we calculate the depth of SURF features by simply interpolating the depth of nearby 3D points.

Finally, a 6 DoF loop constraint $\mathtt{T}_{\text{loop}}$ is calculated based on the 2D-3D SURF correspondences using the P4P-method plus RANSAC. Given this initial pose estimate, more matches can be found using perspective-warped image patches, and the pose is refined using robust optimisation. However the relative change in scale s_{loop} due to scale drift has to be computed as well, so we also calculate the depth of the SURF features visible in the current frame using a dense surface model, giving a set of 3D-3D correspondences. We estimate s_{loop} robustly as the median scale difference over all correspondences.

B. Loop Closure Correction

Let us consider a loop closure scenario in a large-scale map. After the loop closure is detected and matches are found between the current frame and the loop frame, the loop closing problem can be stated as a large BA problem. We have to optimize over all frames and points in the map. However, optimising over a large number of frames is computationally demanding. More seriously, since BA is not a convex problem, and we are far away from the global minimum, it is likely that BA will get stuck in a local minimum.

One solution would be to optimise over relative constraints between poses using pose-graph optimisation [12][23]. The relative constraints between two poses $\mathtt{T}_i$ and $\mathtt{T}_j$ can be calculated easily: $\Delta\mathtt{T}_{i,j} = \mathtt{T}_j \cdot \mathtt{T}_i^{-1}$. But while an optimisation over the 6 DoF constraints would efficiently correct translational and rotational drift, it would not deal with scale drift, and would lead to an unsatisfactory overall result as we will show experimentally in Section V. Therefore, we perform optimisation based on 7 DoF constraints $\mathtt{S}$, which are elements of the group of similarity transformations Sim(3):

$$\mathtt{S} = \begin{bmatrix} s\mathtt{R} & \mathbf{t} \\ \mathtt{0} & 1 \end{bmatrix} \tag{21}$$

[1] The underlying assumption is that the scene structure is smooth enough. However, this assumption is not only vital if dense surface models are used, but always if we aim to calculate the 3D position of a blob feature no matter which method is used. In other words, the 3D position of a blob is only properly defined if its 'carrying' surface is smooth enough.

where $s \in \mathbb{R}^+$ is a positive scale factor. A minimal representation $\mathfrak{sim}(3)$ can be defined be the following 7-vector $(\omega, \sigma, \upsilon)^\top$ with $s = e^\sigma$. Indeed, it can be shown that Sim(3) is a Lie group, $\mathfrak{sim}(3)$ is the corresponding Lie algebra and their exponential map is:

$$\exp_{\mathsf{Sim}(3)}\begin{pmatrix} \omega \\ \sigma \\ \upsilon \end{pmatrix} = \begin{bmatrix} e^\sigma \exp_{\mathsf{SO}(3)}(\omega) & \mathtt{W}\upsilon \\ \mathtt{0} & 1 \end{bmatrix} = \begin{bmatrix} s\mathtt{R} & \mathbf{t} \\ \mathtt{0} & 1 \end{bmatrix} \tag{22}$$

with $\mathtt{W} = \frac{a\sigma+(1-b)\theta}{\theta(\sigma^2+\theta^2)}(\omega)_\times + \left(c - \frac{(b-1)\sigma+a\theta}{\sigma^2+\theta^2}\right)\frac{(\omega)^2_\times}{\theta^2} + c\mathtt{I}$, $a = e^\sigma \sin(\theta)$, $b = e^\sigma \cos(\theta)$ and $c = \frac{e^\sigma - 1}{\sigma}$. Furthermore, $\exp_{\mathsf{Sim}}(3)$ is surjective, so an inverse relation $\log_{\mathsf{Sim}(3)}$ exists.

In order to prepare for the 7 DoF optimisation, we transform each absolute pose $\mathtt{T}_i$ to an absolute similarity $\mathtt{S}_i$, and each relative pose constraint $\Delta\mathtt{T}_{i,j}$ to a relative similarity constraint $\Delta\mathtt{S}_{i,j}$ by leaving rotation and translation unchanged and setting the scale $s = 1$. Only for the loop constraint $\Delta\mathtt{S}_{\text{loop}}$ do we set the scale to $s_{\text{loop}} \neq 1$ (using the estimate obtained as explained in the previous section). We define the residual $\mathbf{r}_{i,j}$ between two transformations $\mathtt{S}_i$ and $\mathtt{S}_j$ minimally in the tangent space $\mathfrak{sim}(3)$:

$$\mathbf{r}_{i,j} = \log_{\mathsf{Sim}(3)}(\Delta\mathtt{S}_{i,j} \cdot \mathtt{S}_i \cdot \mathtt{S}_j^{-1}) \ . \tag{23}$$

The graph of similarity constraints is optimised by minimising:

$$\chi^2(\mathtt{S}_2, ..., \mathtt{S}_m) = \sum_{i,j} \mathbf{r}_{i,j}^\top \Lambda_{\Delta\mathtt{S}i,j} \mathbf{r}_{i,j} \tag{24}$$

using LM. The first transform $\mathtt{S}_1$ is fixed and defines the coordinate frame. The corresponding Jacobian is sparse. We exploit the sparseness pattern using the CSparse library.

After the similarities $\mathtt{S}_i^{\text{cor}}$ are corrected, we also need to correct the set of points. For each point $\mathbf{x}_j$, a frame $\mathtt{T}_i$ is selected in which it is visible. Now we can map each point relative to its corrected frame: $\mathbf{x}_j^{\text{cor}} = (\mathtt{S}_i^{\text{cor}})^{-1}(\mathtt{T}_i \mathbf{x}_j)$. Afterwards, each similarity transform $\mathtt{S}_i^{\text{cor}}$ is transformed back to a rigid body transform $\mathtt{T}_i^{\text{cor}}$ by setting the translation to $s\mathbf{t}$ and leaving the rotation unchanged. Finally, the whole map can be further optimised using structure-only or full BA.

V. Experiments

We performed the evaluation of our SLAM system using the Keble College data set [4] — a sequence where a handheld sideways-facing camera completes a circuit around a large outdoor square. Images were captured using a low cost IEEE Unibrain camera with resolution 320×240 at 30 FPS, and using a lens with 80 degree horizontal field of view. Our SLAM framework performed almost in real-time. For the 5952 images in the sequence, the run-time was 465 seconds which corresponds to frame rate of more than 12 FPS. The computation was performed on a desktop computer with an Intel Core 2 Duo processor and an NVIDIA 8800 GT GPU which were used for dense optical flow computation.

The trajectory before loop closure is shown in Fig. 2 (a). It consists of 766 keyframes, 11885 points and 84820 observations. In the optimisation back-end, BA with a sliding window

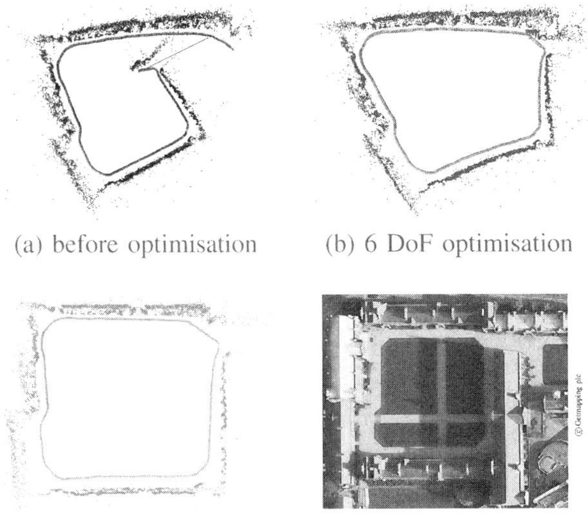

(a) before optimisation (b) 6 DoF optimisation

(c) 7 DoF optimisation (d) aerial photo

Fig. 2. Keble college data set

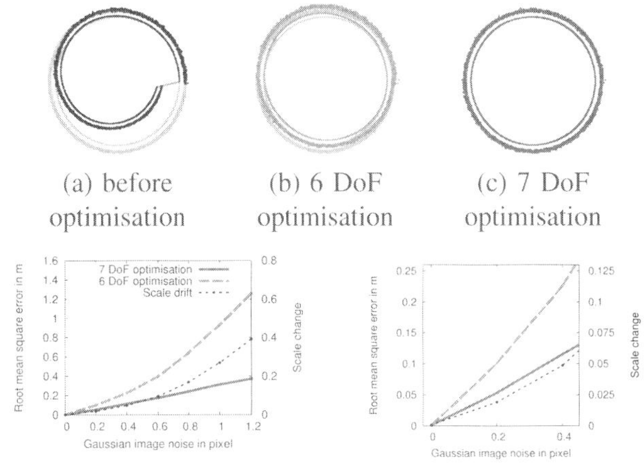

(a) before optimisation (b) 6 DoF optimisation (c) 7 DoF optimisation

(d) whole plot (e) close-up

Fig. 3. Simulation experiments

of 10 keyframes was used. A loop closure constraint ΔT_{loop} was successfully calculated and refined using the method described in Sec.IV. A relative scale change of $1 : 0.37$ was detected. The large amount of drift can be partially explained by the fact that the visible scene is always very local and that only a rough intrinsic calibration was available. Also, future improvements in our visual odometry implementation could lead to a reduction of drift during exploration. Nevertheless, a certain amount of drift during exploration is unavoidable and our main focus is how to deal with drift when it occurs. A traditional 6 DoF graph optimisation such as [12] closes the loop but leaves the scale drift unchanged which leads to a deformed trajectory as shown in Fig. 2 (b). However, if we perform graph optimisation using the similarity transform, the result looks significantly better (see Fig. 2 (c)). The graph optimisation converged after a single iterations, taking 0.57 seconds. After graph optimisation, ten iterations of structure-only bundle adjustment were performed, taking 0.25 seconds. The application of full BA over all 766 keyframes did not lead to a notable further improvement (not shown here).

In addition to this real-world experiment, we also performed a series of simulation experiments in full 3D space. A simulated camera was moved on a circular trajectory with radius 10m. The camera is directed outwards. A set of 5000 points was drawn from a ring shaped distribution with radius 11m. The camera trajectory consists of 720 poses. In this simulation environment, our full SLAM system was applied including feature initialisation and fixed-window bundle adjustment with size 10. Only the camera was not simulated and data association was given. The difference between the true trajectory and the estimated one is shown in Fig. 3 (a). In this particular example, we simulated Gaussian image noise with a standard deviation of one pixel. Fig. 3 (b) and (c) show loop closure correction using $\mathrm{SE}(3)$ and $\mathrm{Sim}(3)$.

We varied the amount of image noise from 0 to 1.2 pixels and performed 10 simulation runs for each setting. An important result of our experiment is that there is a clear relation between scale drift and image noise (see Fig. 3 (d), thin curve). This indicates the correctness of our characterisation of scale as a parameter in SLAM which drifts during exploration in a similar way to rotation and translation. If we define the first pose as our origin, there is still a scale ambiguity of possible maps. Therefore we define a measure between the corrected poses T_i^{cor} and the true poses T_i^{true} using the minimum of the sum of square differences over the scale s: $M = \min_s \sum_i (\mathbf{t}_i^{\text{true}} - s\mathbf{t}_i^{\text{cor}})^2$. By dividing M by the number of frames and taking the square root, we obtain the root mean square error $RMSE = \sqrt{\frac{M}{720}}$. The average RMSE over the ten simulation runs is shown in Fig. 3 (d). One can see that $\mathrm{Sim}(3)$ optimisation (red curve) outperforms $\mathrm{SE}(3)$ optimisation (green curve) by a large amount, particularly if the scale change is large. But interestingly, even for small scale changes of one to four percent the $\mathrm{Sim}(3)$ optimisation performs significantly better than $\mathrm{SE}(3)$ optimisation (see Fig. 3 (e)).

Finally we did an experiment to illustrate that our optimisation framework naturally extends to multiple loop closures. Imagine an aircraft flying over a sphere and performing monocular SLAM using a downward directed camera (see Fig. 4 (a)). First we perform visual odometry, and afterwards we compute a set of loop closure constraints (shown as blue line segments in Fig. 4 (b)). In this particular example, $\mathrm{Sim}(3)$ optimisation leads to a small RMSE of only 0.328 (Fig. 4 (c)), whereas $\mathrm{SE}(3)$ optimisation results in an RMSE of 2.187. Videos illustrating the experiments[2] as well as large parts of the source code[3] are available online.

VI. Conclusions

We have presented a new monocular SLAM algorithm, based consistently on non-linear optimisation, which we be-

[2] http://www.doc.ic.ac.uk/~strasdat/rss2010videos
[3] http://www.openslam.org/robotvision

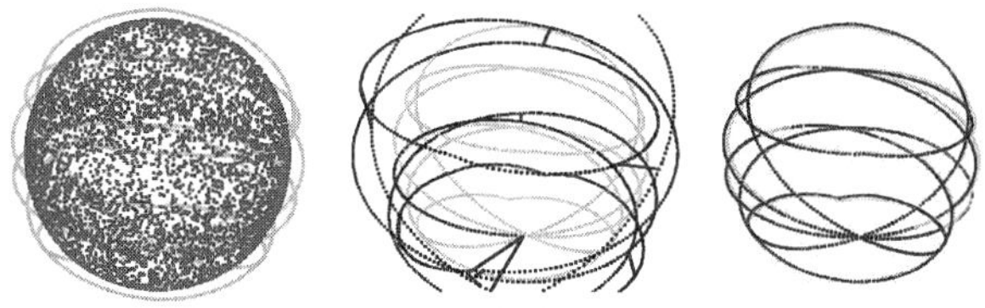

Fig. 4. Multi-loop-closure example: An aircraft is flying over a sphere and doing SLAM using a single downward-directed camera.

lieve takes advantage of the best current knowledge of the hard problem of large scale monocular SLAM to achieve state of the art results. Our algorithm is suitable for sequential operation in large domains, currently at near real-time rates. Importantly, our approach explicitly acknowledges the issue of scale drift at all stages, and offers a practical way to resolve this drift effectively upon loop closures. We show experimentally, both in simulation and using a real outdoor sequence, that a certain amount of scale drift is unavoidable during exploration and this must be taken into account during loop closure to achieve the best results. Also, we have shown that our optimisation approach naturally extends to multiple loop closures.

BA over all frames and scene points is the gold standard for SfM problems where camera networks are typically strong. However, a lightweight approach of 7 DoF graph-optimisation plus structure-only BA seems to be sufficient to accurately close large loops. It is unclear and worth further investigating whether full BA leads to significant better results for the weak camera networks which occur in monocular SLAM.

Acknowledgments

This research was supported by the European Research Council Starting Grant 210346, the Spanish MEC Grant DPI2009-07130 and EU FP7-ICT-248942 RoboEarth. We are very grateful to colleagues at Imperial College London and the University of Zaragoza for discussions and software collaboration. Special thanks to Giorgio Grisetti and Ethan Eade for the insightful discussions about pose-graph optimisation and Lie theory respectively.

References

[1] A. Angeli, D. Filliat, S. Doncieux, and J.-A. Meyer. Fast and incremental method for loop-closure detection using bags of visual words. *IEEE Transactions on Robotics (T-RO)*, 24(5):1027–1037, 2008.

[2] D. Chekhlov, M. Pupilli, W. W. Mayol, and A. Calway. Real-time and robust monocular SLAM using predictive multi-resolution descriptors. In *Proceedings of the 2nd International Symposium on Visual Computing*, 2006.

[3] J. Civera, O. G. Grasa, A. J. Davison, and J. M. M. Montiel. 1-point RANSAC for EKF-based structure from motion. In *Proceedings of the IEEE/RSJ Conference on Intelligent Robots and Systems (IROS)*, 2009.

[4] L. A. Clemente, A. J. Davison, I. Reid, J. Neira, and J. D. Tardós. Mapping large loops with a single hand-held camera. In *Proceedings of Robotics: Science and Systems (RSS)*, 2007.

[5] M. Cummins and P. Newman. Highly scalable appearance-only SLAM — FAB-MAP 2.0. In *Proceedings of Robotics: Science and Systems (RSS)*, 2009.

[6] T. A: Davis. *Direct Methods for Sparse Linear Systems*. SIAM, 2006.

[7] A. J. Davison. Real-time simultaneous localisation and mapping with a single camera. In *Proceedings of the International Conference on Computer Vision (ICCV)*, 2003.

[8] E. Eade. *Monocular Simultaneous Localisation and Mapping*. PhD thesis, University of Cambridge, 2008.

[9] E. Eade and T. Drummond. Scalable monocular SLAM. In *Proceedings of the IEEE Conference on Computer Vision and Pattern Recognition (CVPR)*, 2006.

[10] E. Eade and T. Drummond. Monocular SLAM as a graph of coalesced observations. In *Proceedings of the International Conference on Computer Vision (ICCV)*, 2007.

[11] J. Gallier. *Geometric Methods and Applications for Computer Science and Engineering*. Springer-Verlag, 2001.

[12] G. Grisetti, R. Kümmerle, C. Stachniss, U. Frese, and C. Hertzberg. Hierarchical optimization on manifolds for online 2D and 3D mapping. In *Proceedings of the IEEE International Conference on Robotics and Automation (ICRA)*, 2010.

[13] R. Hartley and A. Zisserman. *Multiple View Geometry in Computer Vision*. Cambridge University Press, second edition, 2004.

[14] G. Klein and D. W. Murray. Parallel tracking and mapping for small AR workspaces. In *Proceedings of the International Symposium on Mixed and Augmented Reality (ISMAR)*, 2007.

[15] G. Klein and D. W. Murray. Parallel tracking and mapping on a camera phone. In *Proceedings of the International Symposium on Mixed and Augmented Reality (ISMAR)*, 2009.

[16] K. Konolige and M. Agrawal. FrameSLAM: From bundle adjustment to real-time visual mapping. *IEEE Transactions on Robotics (T-RO)*, 24:1066–1077, 2008.

[17] R. Mahony and J.H. Manton. The geometry of the Newton method on non-compact Lie groups. *Journal of Global Optimization*, 23(3):309–327, 2002.

[18] C. Mei, G. Sibley, M. Cummins, P. Newman, and I. Reid. A constant time efficient stereo SLAM system. In *Proceedings of the British Machine Vision Conference (BMVC)*, 2009.

[19] M. Montemerlo and S. Thrun. Simultaneous localization and mapping with unknown data association using FastSLAM. In *Proceedings of the IEEE International Conference on Robotics and Automation (ICRA)*, 2003.

[20] J. M. M. Montiel, J. Civera, and A. J. Davison. Unified inverse depth parametrization for monocular SLAM. In *Proceedings of Robotics: Science and Systems (RSS)*, 2006.

[21] E. Mouragnon, M. Lhuillier, M. Dhome, F. Dekeyser, and P. Sayd. Real-time localization and 3D reconstruction. In *Proceedings of the IEEE Conference on Computer Vision and Pattern Recognition (CVPR)*, 2006.

[22] D. Nistér. An efficient solution to the five-point relative pose problem. *IEEE Trans. Pattern Anal. Mach. Intell.*, 26(6):756–777, 2004.

[23] E. Olson, J. J. Leonard, and S. Teller. Fast iterative alignment of pose graphs with poor initial estimates. In *Proceedings of the IEEE International Conference on Robotics and Automation (ICRA)*, 2006.

[24] P. Pinies and J. D. Tardós. Large scale SLAM building conditionally independent local maps: Application to monocular vision. *IEEE Transactions on Robotics (T-RO)*, 24(5):1094–1106, 2008.

[25] T. Pock, M. Unger, D. Cremers, and H. Bischof. Fast and exact solution of total variation models on the GPU. In *Proceedings of the CVPR Workshop on Visual Computer Vision on GPU's*, 2008.

[26] E. Rosten and T. Drummond. Fusing points and lines for high performance tracking. In *Proceedings of the International Conference on Computer Vision (ICCV)*, 2005.

[27] D. Scaramuzza, F. Fraundorfer, M. Pollefeys, and R. Siegwart. Absolute scale in structure from motion from a single vehicle mounted camera by exploiting nonholonomic constraints. In *Proceedings of the International Conference on Computer Vision (ICCV)*, 2009.

[28] J. Solà, M. Devy, A. Monin, and T. Lemaire. Undelayed initialization in bearing only SLAM. In *Proceedings of the IEEE/RSJ Conference on Intelligent Robots and Systems (IROS)*, 2005.

[29] H. Strasdat, J. M. M. Montiel, and A. J. Davison. Real-time monocular SLAM: Why filter? In *Proceedings of the IEEE International Conference on Robotics and Automation (ICRA)*, 2010.

[30] B. Triggs, P. McLauchlan, R. Hartley, and A. Fitzgibbon. Bundle adjustment — a modern synthesis. In *Vision Algorithms: Theory and Practice*, volume 1883 of *Lecture Notes in Computer Science*, pages 298–372. Springer-Verlag, 2000.

Preliminary Results in Decentralized Estimation for Single-Beacon Acoustic Underwater Navigation

Sarah E. Webster
Dept. of Mechanical Engineering
Johns Hopkins University
Baltimore, MD 21218
Email: swebster@jhu.edu

Louis L. Whitcomb
Dept. of Mechanical Engineering
Johns Hopkins University
Baltimore, MD 21218
Email: llw@jhu.edu

Ryan M. Eustice
Dept. of Naval Arch. and Marine Eng.
University of Michigan
Ann Arbor, MI, 48109
Email: eustice@umich.edu

Abstract— **We report a decentralized, vehicle-based extended information filter designed to enable single-beacon acoustic navigation of multiple underwater vehicles. In single-beacon navigation, ranges from a single reference beacon to a moving underwater vehicle are used in addition to the vehicle's inertial navigation sensors to perform absolute (as opposed to relative) localization and navigation. For this implementation we assume a moving reference beacon from which range measurements are calculated using asynchronous acoustic data broadcasts that also contain information about the reference beacon's position and recent sensor measurements. We assume that other than these acoustic data transmissions the vehicle has no knowledge of the beacon's position or sensor measurements. This implementation allows the simultaneous navigation of multiple vehicles within acoustic range of the reference beacon. Within this framework we show that the decentralized information filter yields identical results to a centralized extended Kalman filter at the instant of each range measurement; in addition we show that between range measurements the results from the two filters differ only by linearization errors. We compare the state estimation results of the decentralized information filter to that of a centralized extended Kalman filter using a simulated data set.**

I. Introduction

This paper reports the derivation of a decentralized extended information filter (DEIF), for single-beacon one-way-travel-time (OWTT) navigation [6], [7], [26]. This extended information filter is designed to run locally on a submerged vehicle with real-time access to the vehicle's navigation sensors and infrequent, asynchronous access to acoustic broadcasts from a moving reference beacon. The DEIF does *not* have access to real-time global positioning system (GPS) measurements from the reference beacon or any other information except information that is received acoustically.

Our goal is to enable high-precision absolute navigation of multiple underwater vehicles over length scales of $\mathcal{O}$(1-100km). Conventional "dead-reckoning" navigation solutions rely on strap-down sensors such as Doppler velocity logs (DVLs) and inertial measurement units (IMUs) to measure the vehicle attitude, linear and angular velocity, and acceleration. These measurements can be integrated to estimate relative change in vehicle position but yield an estimate of local displacement with errors that are unbounded over time. In order to achieve bounded-error navigation, additional navigation information is required from an absolute georeferenced source. Traditional methods such as ultra short baseline navigation (USBL) and some implementations of long baseline navigation (LBL) suffer from a lack of scalability where the rate at which multiple vehicles can receive navigation updates decreases linearly as the number of vehicles in the water increases [13]. In addition, LBL navigation requires external, fixed reference beacons that limit the vehicle's navigable range to 5-10 km from the beacon. In contrast, the data broadcasts used in OWTT navigation enable the navigation of any vehicle within acoustic range and the use of a single, moving reference beacon eliminates the need for multiple, fixed beacons and their associated cost and range limitations.

Within the context of single-beacon navigation, the decentralized approach provides a flexible, scalable solution for vehicle navigation. Navigation algorithms that rely on a centralized observer suffer from the severely limited bandwidth and high latency associated with underwater acoustic communication in comparison to typical land-based radio frequency communication networks [14]. Given the speed of sound in water (~1500 m/s), transmitting acoustic data over length scales on the order of kilometers results in latency on the order of seconds. And though the bandwidth of acoustic modem technology has increased dramatically in recent years, achieving throughput of up to 2400 bps [22], operationally the average throughput is on the order of 10-50 bps due to the low duty cycle with which these messages are typically transmitted during at-sea operation. The proposed information filter lends itself to single-beacon navigation because the sparse structure and additive nature of measurements updates in the EIF, described in Section III, requires a minimal quantity of information to be transmitted from the beacon to the vehicle.

This paper is organized as follows: Section II describes previous work in the area of single-beacon navigation and the use of information filters in decentralized estimation. Section III presents the derivation of the decentralized extended information filter and shows that it produces identical results to that of a centralized extended Kalman filter immediately following each range measurement. Section IV presents the results of the simulation used to validate the DEIF. Section V concludes.

II. Previous Work

Single-beacon navigation relies on range measurements from a single, georeferenced beacon to provide an absolute position reference. Previous work in the area of single-beacon navigation is extensively reviewed in [26]. This section reviews some of the references most relevant to this paper. Navigation with a single, fixed beacon whose position is known *a priori* has been reported using several different estimation techniques—a least squares approach in [21] and [1], and a vehicle-based EKF in [16] and [11]. Navigation with respect to a moving beacon whose position is not known *a priori* is reported in [17] using a nonlinear least mean square method; in [7] using a maximum likelihood method; and in [26] using a centralized EKF. However, each of these methods, as reported, is only structurally tractable for post-processing, though the authors of [17] suggest improvements that would allow for real-time implementations.

Decentralized estimation in the context of underwater communication and navigation faces unique constraints in terms of low bandwidth and high latency, which renders many of the decentralized estimation solutions from land-based applications unsuitable. Until recently little research has been done on the topic of decentralized estimators and multi-vehicle navigation in the field of underwater robotics. However, as the cost of vehicles has decreased and their reliability improved, the increased interest in multi-vehicle operations within the ocean science community has precipitated new research in this area. In [2] the author presents a method that allows a vehicle running a Bayes estimator to use range and position information broadcast from one or more moving beacons. This work expands on the moving long baseline concept in [24] to encompass multiple range sources and real-time operation. The authors of [18] address a similar concept to moving long baseline and compare the use of the Kalman filter to the particle filter on the vehicle's localization performance.

The estimation algorithm that most closely resembles our work is [8], where the authors rely on a single moving georeferenced beacon to support the localization of multiple vehicles through asynchronous acoustic broadcasts. The main difference between the algorithm presented in [8] and the algorithm presented herein, is that [8] employs a vehicle-based EKF and performs range measurement updates using the absolute position and covariance broadcast from the reference beacon. The benefit of this formulation is that the algorithm is trivially robust to packet loss, however excluding the reference beacon position from the state vector of the filter makes the results difficult or impossible to compare analytically to a centralized filter. In contrast, the decentralized algorithm reported herein recreates the results of a centralized EKF that has access to measurements from both the vehicle and the beacon's navigation sensors. In addition, by keeping both vehicle and ship states in the filter, range measurement updates do not risk over confidence in and correlation of observations. In [2], the work upon which [8] is based, a multi-hypothesis strategy is employed to avoid over-confidence by preventing measurement data from being incorporated multiple times.

Finally, we look specifically at the use of the information filter for navigation in the field of robotics. Derived in detail in [19], the extended information filter has been employed for vehicle navigation [3] and SLAM algorithms [20], [23] in the context of land-based robotics. In the context of underwater vehicles, the EIF is most widely used for coordinated control, but there are a few examples of the EIF being employed in SLAM algorithms [4], [5]. To the best knowledge of the authors, the present paper is the first to formulate and test an extended information filter in the context of decentralized single-beacon navigation for underwater vehicles.

III. Derivation of the Decentralized Extended Information Filter

In this section we present a brief review of the update and measurement equations for the extended information filter followed by a derivation of the vehicle-based decentralized extended information filter (DEIF). The implementation of the DEIF relies on two separate filters, both of which process sensor data causally and asynchronously. The information filter on the ship has access to ship sensor data but not range measurements (we will use the term ship to refer to the reference beacon for the remainder of this derivation). The ship-based filter is used to calculate the change in the ship's information vector and information matrix between acoustic broadcasts, and this delta information is acoustically transmitted to the vehicle. As described in [25], the ship-to-vehicle range is measured using the one-way-travel-time of the acoustic broadcast. The DEIF on the vehicle has real-time access to vehicle sensor data and the asynchronous acoustic broadcasts from the ship, but does *not* have access to the ship's sensor measurements apart from the acoustic broadcasts. A centralized EKF (CEKF), which has simultaneous, real-time access to both vehicle and ship sensor data, provides the benchmark for the performance of the DEIF. Figure 1 shows a schematic of the delta ship information transmitted from the ship to the vehicle.

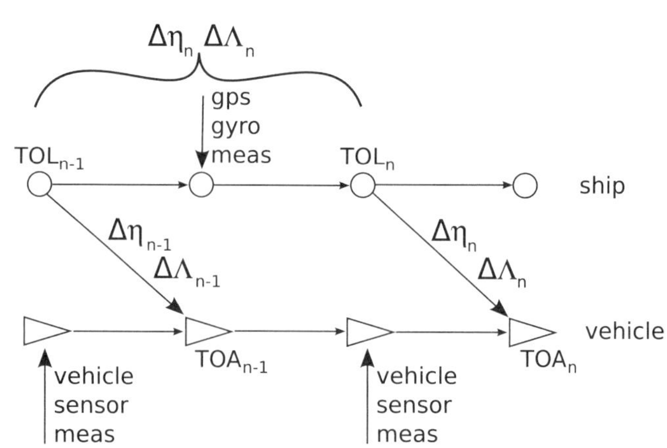

Fig. 1. A schematic of the information contained in the range packet acoustically transmitted from the ship to the vehicle.

A. Extended Information Filter Equations

The EIF is characterized by the information matrix, $\boldsymbol{\Lambda}$, and the information vector, $\boldsymbol{\eta}$, which can be defined in terms of the mean, $\boldsymbol{\mu}$, and covariance, $\boldsymbol{\Sigma}$, of the state vector, $\boldsymbol{x}$, [12], [19], as

$$\boldsymbol{\Lambda} = \boldsymbol{\Sigma}^{-1} \tag{1}$$

$$\boldsymbol{\eta} = \boldsymbol{\Lambda}\boldsymbol{\mu} \tag{2}$$

where

$$\boldsymbol{\mu} = E[\boldsymbol{x}] \tag{3}$$

$$\boldsymbol{\Sigma} = E\left[(\boldsymbol{x}-\boldsymbol{\mu})(\boldsymbol{x}-\boldsymbol{\mu})^\top\right]. \tag{4}$$

1) Process Prediction: For the general process prediction equations we consider a state vector with two states in it, the current state $\boldsymbol{x}_k$ and a previous state $\boldsymbol{x}_p$, so that

$$\boldsymbol{x}_{k|k} = \begin{bmatrix} \boldsymbol{x}_k \\ \boldsymbol{x}_p \end{bmatrix} \tag{5}$$

and

$$\boldsymbol{\Lambda}_{k|k} = \begin{bmatrix} \boldsymbol{\Lambda}_{kk} & \boldsymbol{\Lambda}_{kp} \\ \boldsymbol{\Lambda}_{pk} & \boldsymbol{\Lambda}_{pp} \end{bmatrix} \tag{6}$$

$$\boldsymbol{\eta}_{k|k} = \begin{bmatrix} \boldsymbol{\eta}_k \\ \boldsymbol{\eta}_p \end{bmatrix}. \tag{7}$$

The process prediction equations for the information filter, which represent the distribution $p(\boldsymbol{x}_{k+1}, \boldsymbol{x}_p|\mathbf{Z}^{1:k}, \mathbf{U}^{1:k+1})$, are

$$\boldsymbol{\Lambda}_{k+1|k} = \begin{bmatrix} \Psi_k & \boldsymbol{Q}_k^{-1}\boldsymbol{F}_k\boldsymbol{\Omega}_k^{-1}\boldsymbol{\Lambda}_{kp} \\ \boldsymbol{\Lambda}_{kp}\boldsymbol{\Omega}_k^{-1}\boldsymbol{F}_k^\top\boldsymbol{Q}_k^{-1} & \boldsymbol{\Lambda}_{pp} - \boldsymbol{\Lambda}_{pk}\boldsymbol{\Omega}_k^{-1}\boldsymbol{\Lambda}_{kp} \end{bmatrix} \tag{8}$$

$$\boldsymbol{\eta}_{k+1|k} = \begin{bmatrix} \boldsymbol{Q}_k^{-1}\boldsymbol{F}_k\boldsymbol{\Omega}_k^{-1}\boldsymbol{\eta}_k + \Psi_k(\boldsymbol{f}(\boldsymbol{\mu}_{k|k}, \boldsymbol{u}_{k+1}) - \boldsymbol{F}_k\boldsymbol{\mu}_{k|k}) \\ \boldsymbol{\eta}_p - \boldsymbol{\Lambda}_{pk}\boldsymbol{\Omega}_k^{-1}\boldsymbol{\eta}_k^* \end{bmatrix} \tag{9}$$

where $\mathbf{Z}^{1:k}$ are the vehicle sensor measurements up to time k and $\mathbf{U}^{1:k+1}$ are the control inputs up to time $k+1$, $\boldsymbol{f}(\cdot)$ is the nonlinear process model, $\boldsymbol{F}_k$ is the Jacobian, $\boldsymbol{Q}_k$ is the covariance of the discretized process model noise, and

$$\Psi_k = (\boldsymbol{Q}_k + \boldsymbol{F}_k\boldsymbol{\Lambda}_{kk}^{-1}\boldsymbol{F}_k^\top)^{-1} \tag{10}$$

$$\boldsymbol{\Omega}_k = \boldsymbol{\Lambda}_{kk} + \boldsymbol{F}_k^\top\boldsymbol{Q}_k^{-1}\boldsymbol{F}_k \tag{11}$$

$$\boldsymbol{\eta}_k^* = \boldsymbol{\eta}_k - \boldsymbol{F}_k^\top\boldsymbol{Q}_k^{-1}(\boldsymbol{f}(\boldsymbol{\mu}_{k|k}, \boldsymbol{u}_{k+1}) - \boldsymbol{F}_k\boldsymbol{\mu}_{k|k}) \tag{12}$$

as derived in [5], [23].

2) Process Prediction with Augmentation: In equations (8) and (9), the current state at time k is propagated to time $k+1$, so that

$$\boldsymbol{x}_{k+1|k} = \begin{bmatrix} \boldsymbol{x}_{k+1} \\ \boldsymbol{x}_p \end{bmatrix}. \tag{13}$$

If instead we augment the state vector to include the state at time $k+1$ in addition to the original states,

$$\boldsymbol{x}_{k+1|k} = \begin{bmatrix} \boldsymbol{x}_{k+1} \\ \boldsymbol{x}_k \\ \boldsymbol{x}_p \end{bmatrix}, \tag{14}$$

the process prediction equations, which represent the distribution $p(\boldsymbol{x}_{k+1}, \boldsymbol{x}_k, \boldsymbol{x}_p|\mathbf{Z}^{1:k}, \mathbf{U}^{1:k+1})$, have a very different structure, [5],

$$\boldsymbol{\Lambda}_{k+1|k} = \begin{bmatrix} \boldsymbol{Q}_k^{-1} & -\boldsymbol{Q}_k^{-1}\boldsymbol{F}_k & \mathbf{0} \\ -\boldsymbol{F}_k^\top\boldsymbol{Q}_k^{-1} & \boldsymbol{F}_k^\top\boldsymbol{Q}_k^{-1}\boldsymbol{F}_k + \boldsymbol{\Lambda}_{kk} & \boldsymbol{\Lambda}_{kp} \\ \mathbf{0} & \boldsymbol{\Lambda}_{pk} & \boldsymbol{\Lambda}_{pp} \end{bmatrix} \tag{15}$$

$$\boldsymbol{\eta}_{k+1|k} = \begin{bmatrix} \boldsymbol{Q}_k^{-1}(\boldsymbol{f}(\boldsymbol{\mu}_{k|k}, \boldsymbol{u}_{k+1}) - \boldsymbol{F}_k\boldsymbol{\mu}_{k|k}) \\ \boldsymbol{\eta}_k - \boldsymbol{F}_k^\top\boldsymbol{Q}_k^{-1}(\boldsymbol{f}(\boldsymbol{\mu}_{k|k}, \boldsymbol{u}_{k+1}) - \boldsymbol{F}_k\boldsymbol{\mu}_{k|k}) \\ \boldsymbol{\eta}_p \end{bmatrix} \tag{16}$$

using the same definitions as in (8) and (9). Equation (15), can be written as the sum of two matrices—one containing the process prediction information and the other containing the previous information matrix.

$$\boldsymbol{\Lambda}_{k+1|k} = \begin{bmatrix} \boldsymbol{Q}_k^{-1} & -\boldsymbol{Q}_k^{-1}\boldsymbol{F}_k & \mathbf{0} \\ -\boldsymbol{F}_k^\top\boldsymbol{Q}_k^{-1} & \boldsymbol{F}_k^\top\boldsymbol{Q}_k^{-1}\boldsymbol{F}_k & \mathbf{0} \\ \mathbf{0} & \mathbf{0} & \mathbf{0} \end{bmatrix} + \begin{bmatrix} \mathbf{0} & \mathbf{0} & \mathbf{0} \\ \mathbf{0} & \boldsymbol{\Lambda}_{kk} & \boldsymbol{\Lambda}_{kp} \\ \mathbf{0} & \boldsymbol{\Lambda}_{pk} & \boldsymbol{\Lambda}_{pp} \end{bmatrix} \tag{17}$$

As was first noted in [5], this results in $\boldsymbol{\Lambda}$ having a sparse, block-tridiagonal structure. The sparsity of $\boldsymbol{\Lambda}$ is important in the context of acoustic navigation because it bounds the amount of information that must be acoustically transmitted in order to fully reconstruct the ship's state (see Section III-B).

3) Measurement Update: The measurement update equations for the extended information filter are

$$\boldsymbol{\Lambda}_{k|k} = \boldsymbol{\Lambda}_{k|k-1} + \boldsymbol{H}_k^\top\boldsymbol{R}_k^{-1}\boldsymbol{H}_k \tag{18}$$

$$\boldsymbol{\eta}_{k|k} = \boldsymbol{\eta}_{k|k-1} + \cdots \quad \boldsymbol{H}_k^\top\boldsymbol{R}_k^{-1}(\boldsymbol{z}_k - \boldsymbol{h}(\boldsymbol{\mu}_{k|k-1}) + \boldsymbol{H}_k\boldsymbol{\mu}_{k|k-1}) \tag{19}$$

where $\boldsymbol{z}_k$ is the measurement, $\boldsymbol{R}_k$ is the covariance matrix of the measurement noise, $\boldsymbol{h}(\cdot)$ is the measurement model, $\boldsymbol{H}_k$ is the Jacobian of the measurement with respect to $\boldsymbol{x}$

$$\boldsymbol{H}_k = \left.\frac{\partial \boldsymbol{h}(\boldsymbol{x})}{\partial \boldsymbol{x}}\right|_{\boldsymbol{x}=\boldsymbol{\mu}_{k|k-1}} \tag{20}$$

and $\boldsymbol{\mu}_{k|k-1}$ is the mean of the state vector [12], [19], [23].

B. Ship-Based Information Filter

The DEIF, running on the vehicle, relies on acoustic data transmissions from the ship to make range measurements. The acoustic transmission contains a *range packet* with ship state information from an independent ship-based information filter that is used to maintain both current and historic ship state estimates. This filter is independent of the vehicle state, vehicle sensor measurements, and range measurements.

We assume a constant-velocity *linear* kinematic process model for the ship, which, in the experience of the authors, is a reasonable assumption for ship motion in the context of AUV survey operations. For the ship's process model the state vector contains the ship's xy-position, heading, and the respective velocities,

$$\boldsymbol{x}_s = [x_s, y_s, \theta_s, \dot{x}_s, \dot{y}_s, \dot{\theta}_s]^\top. \tag{21}$$

The ship process model is identical to that reported in [26], to which the reader is referred for further details.

1) State Vector Augmentation: The independent ship filter maintains an estimate of the current ship state as well as copies of historic ship states. Each time a new acoustic range packet is broadcast, i.e. at the time-of-launch (TOL), a copy of the current ship state is appended onto the ship state vector. This results in a state vector, after n range packets have been transmitted, of the form

$$\boldsymbol{x}_{s_{k|k}} = \begin{bmatrix} \boldsymbol{x}_{s_k} \\ \boldsymbol{x}_{sTOL_n} \\ \vdots \\ \boldsymbol{x}_{sTOL_2} \\ \boldsymbol{x}_{sTOL_1} \end{bmatrix} \tag{22}$$

where we have adopted the following convention: $\boldsymbol{x}_{s_{k|k}}$ denotes the entire ship state vector at time k; $\boldsymbol{x}_{s_k}$ is the current ship state; and $\boldsymbol{x}_{sTOL_n}$ is the ship state when the nth range packet was broadcast. Note that we are using $^\top$ for the matrix transpose operator. Because in practice it is undesirable, and unnecessary, for the state vector to grow without bound, we marginalize out [5] the oldest historic ship states in order to maintain a fixed-length state vector.

2) Acoustic Data Packets: What differentiates the DEIF from other estimators used in decentralized single-beacon navigation is the information that is transmitted with the range measurements and how that information is incorporated into the decentrialized vehicle navigation filter in conjunction with the range measurement. To initiate a range measurement, the ship broadcasts an acoustic data packet, or range packet, containing information about the ship state. In other formulations of single-beacon navigation, such as [2] and [8], the range packet contains the mean and covariance of the ship's current x-y position, which is used by the filter on the vehicle to perform a range measurement update. This approach, while certainly reasonable, does not contain enough information to allow the vehicle to recreate the global state of the system, including full state recovery of both the vehicle and the ship.

In this formulation, the range packet contains the *change* in $\boldsymbol{\Lambda}_s$ and $\boldsymbol{\eta}_s$ between the time of the current acoustic broadcast, TOL$_n$, and the time of the previous acoustic broadcast, TOL$_{n-1}$,

$$\Delta\boldsymbol{\Lambda}_{TOL_n} = \boldsymbol{\Lambda}_{sTOL_n} - \boldsymbol{\Lambda}_{sTOL_{n-1}} \tag{23}$$

$$\Delta\boldsymbol{\eta}_{TOL_n} = \boldsymbol{\eta}_{sTOL_n} - \boldsymbol{\eta}_{sTOL_{n-1}} \tag{24}$$

where, for conformability, $\boldsymbol{\Lambda}_{sTOL_{n-1}}$ and $\boldsymbol{\eta}_{sTOL_{n-1}}$ have been padded with zeros to match the size of $\boldsymbol{\Lambda}_{sTOL_n}$ and $\boldsymbol{\eta}_{sTOL_n}$ respectively. These range packets are reassembled subsea in the DEIF.

The contribution to $\Delta\boldsymbol{\Lambda}_{TOL_n}$ and $\Delta\boldsymbol{\eta}_{TOL_n}$ from the ship sensor measurement updates, (19), and prediction with augmentation, (17), are independent of other measurements, because of the additive nature of the respective operations and the linear ship process model. As a result, the $\Delta\boldsymbol{\Lambda}_{TOL_n}$ and $\Delta\boldsymbol{\eta}_{TOL_n}$ calculated from the ship-based filter, which has no knowledge of the vehicle and is not subject to range measurement updates, is *identical* to what the DEIF would have calculated if it was performing the ship process predictions and ship sensor measurement updates *even though the ship states in the DEIF have been modified by prior range measurements.* This is discussed in more detail in Section III-C.

C. Vehicle-Based Decentralized Extended Information Filter

The DEIF uses a constant velocity nonlinear process model with a 12 degree-of-freedom (DOF) state vector,

$$\boldsymbol{x}_v = [\boldsymbol{s}^\top, \boldsymbol{\varphi}^\top, \boldsymbol{v}^\top, \boldsymbol{\omega}^\top]^\top \tag{25}$$

where $\boldsymbol{s}$ is the local-level vehicle pose in the local frame, $\boldsymbol{\varphi}$ is the local-level vehicle attitude (Euler roll, pitch, heading), $\boldsymbol{v}$ is the body-frame linear velocity, and $\boldsymbol{\omega}$ is the body-frame angular velocity. The vehicle process model is identical to that reported in [26], to which the reader is referred for further details.

1) DEIF State Vector: In addition to the current vehicle state, the DEIF maintains a copy of the historic *ship* states reconstructed from the delta information acoustically broadcast from the ship. As a result, the DEIF state vector consists of two parts, the current vehicle state and the historic ship states,

$$\boldsymbol{x}_{v_{k|k}} = \begin{bmatrix} x_{v_k} \\ x_{sTOL_n} \\ \vdots \\ x_{sTOL_1} \end{bmatrix} \tag{26}$$

where we adopt the convention that $\boldsymbol{x}_{v_{k|k}}$ denotes the entire vehicle state vector at time k; x_{v_k} is the current vehicle state; and x_{sTOL_n} is the ship state when the nth range packet was broadcast. Note that by definition, because acoustic broadcasts have a non-negligible time delay associated with them, the DEIF state vector cannot contain information about the *current* ship state, it can only contain information about *historic* ship states.

The reconstructed ship states are not subjected to process predictions or measurement updates other than range measurements, because the DEIF has no knowledge of the ship process model or measurement from ship sensors. The ship states remain unchanged over time except for range measurement updates and when new information from acoustic broadcasts is incorporated.

2) Incorporating Delta Ship Information: At the time-of-arrival (TOA) of a range packet at the vehicle, the delta ship information included in each range packet is incorporated in the DEIF, and then the range measurement update is performed. The delta ship information is incorporated into the DEIF simply by addition, in the analogous operation to (23) and (24):

$$\boldsymbol{\Lambda}_{vTOA_n} = \bar{\boldsymbol{\Lambda}}_{vTOA_n} + \Delta\boldsymbol{\Lambda}_{TOL_n} \quad (27)$$

$$\boldsymbol{\eta}_{vTOA_n} = \bar{\boldsymbol{\eta}}_{vTOA_n} + \Delta\boldsymbol{\eta}_{TOL_n} \quad (28)$$

where $\bar{\boldsymbol{\Lambda}}_{vTOA_n}$ is the information matrix before the delta ship information is incorporated and $\boldsymbol{\Lambda}_{vTOA_n}$ is the information matrix after the delta ship information is incorporated, accounting for conformability. As noted in Section III-B.2, the delta ship information encapsulates all of the information that the filter has gained about the ship state since the last range packet was transmitted. The simplicity of this computation is one of the advantages of the information filter.

3) Range Measurement Updates: At the time-of-arrival (TOA) of the range packet, after the delta ship information is incorporated, the range measurement update (19) is performed. The observation model for the range measurement is

$$z_k = \sqrt{(\boldsymbol{x}_{v_{xyz}} - \boldsymbol{x}_{s_{xyz}})^\top (\boldsymbol{x}_{v_{xyz}} - \boldsymbol{x}_{s_{xyz}})} + v_k \quad (29)$$

where $\boldsymbol{x}_{v_{xyz}}$ is the vehicle pose at the TOA, $\boldsymbol{x}_{s_{xyz}}$ is the ship pose at the TOL, and $v_k \sim \mathcal{N}(0, R_k)$.

D. CEKF Delayed Measurement Update

Immediately after performing a range update, the current state of the vehicle recovered by the DEIF is identical to the vehicle state recovered from the CEKF. There are several subtleties in this observation that we address here.

At the TOA of the range packet, the range measurement is made between $\boldsymbol{x}_{v_{TOA}}$ and $\boldsymbol{x}_{s_{TOL}}$. Comparing the probability distributions of the two filters immediately after the range measurement update, we find that they are *not* identical:

$$\text{DEIF: } p\left(\boldsymbol{x}_{v_k}, \boldsymbol{x}_{s_{TOL}} | z_{rng}^{TOA}, \mathbf{Z}_v^{1:TOA}, \mathbf{U}_v^{1:TOA}, \boxed{\mathbf{Z}_s^{1:TOL}}\right) \quad (30)$$

$$\text{CEKF: } p\left(\boldsymbol{x}_{v_k}, \boldsymbol{x}_{s_{TOL}} | z_{rng}^{TOA}, \mathbf{Z}_v^{1:TOA}, \mathbf{U}_v^{1:TOA}, \boxed{\mathbf{Z}_s^{1:TOA}}\right) \quad (31)$$

where z_{rng}^{TOA} is the most recent range measurement, $\mathbf{Z}_v^{1:TOA}$ are the vehicle sensor measurements up to the TOA, $\mathbf{U}_v^{1:TOA}$ are the vehicle control inputs up to the TOA, and $\mathbf{Z}_s^{1:TOL}$ are the ship sensor measurements up to the TOL. Note that the CEKF has access to ship sensor measurements between the TOL and the TOA of which the DEIF has no knowledge. This is illustrated in Figure 2 where the CEKF has access to the GPS and gyrocompass measurements made at the ship between the TOL and TOA, while the DEIF does not.

The ramifications of this are that the DEIF performs a range measurement between the current vehicle state and the the best estimate of the ship's state at the TOL given ship sensor measurements *only up to the TOL*. In contrast, the CEKF performs a range measurement between the current vehicle state and the best estimate of the ship's state at the TOL given ship sensor measurement *up to the TOA*. The CEKF is essentially performing a smoothing operation on the

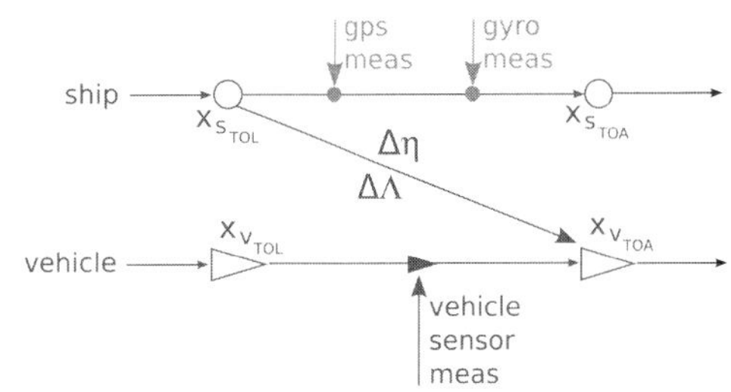

Fig. 2. At the TOA, the CEKF has access to the GPS and gyrocompass measurements made at the ship after the TOL, while the DEIF does not.

ship's state at the TOL, because it has access to additional information from the ship's sensors after the data packet was broadcast at the TOL.

To address this discrepancy we use a two-step delayed measurement update in the CEKF, first performing a measurement update for the range measurement with only ship measurements up until the TOL, (32), and then performing another measurement update for the ship measurements that happened between the TOL and the TOA, (33):

$$p(\boldsymbol{x}_{v_k}, \boldsymbol{x}_{s_{TOL}} | z_{rng}^{TOA}, \mathbf{Z}_v^{1:TOA}, \mathbf{U}_v^{1:TOA}, \boxed{\mathbf{Z}_s^{1:TOL}}) \quad (32)$$

$$p(\boldsymbol{x}_{v_k}, \boldsymbol{x}_{s_{TOL}} | z_{rng}^{TOA}, \mathbf{Z}_v^{1:TOA}, \mathbf{U}_v^{1:TOA}, \mathbf{Z}_s^{1:TOL}, \boxed{\mathbf{Z}_s^{TOL:TOA}}) \quad (33)$$

so that now the DEIF distribution in (30) is identical to the CEKF distribution in (32), without compromising the CEKF's final distribution, i.e. (33) is identical to (31).

Between range measurements, the CEKF and DEIF estimates of the vehicle's state will not be identical because of linearization errors, as seen in (30) versus (31). However, at the instant of *each* range measurement, through the two-step delayed update, the filters estimates will be identical.

Due to the cumulative nature of this navigation technique, packet loss is an operational concern for real-time implementation. The authors are currently investigating several viable options for addressing the effects of packet loss, including broadcasting redundant range packets or reformulating the delta information in order to mitigate the effect of packet loss.

A final note on this derivation, mentioned briefly in Section III-B.2, is that the ship process model must be linear in order for the delta ship information calculated in the independent ship filter to be identical to a centralized filter. A linear process model guarantees that the process prediction and measurement updates are independent of the ship's current state. This is essential because the independent ship filter will have a different estimate of the ship's current state than a centralized filter, because the centralized filter's estimate is conditioned on previous range measurements. This presents a non-trivial challenge for future work where we would like to incorporate range measurements from other vehicles that will not necessarily have linear process models.

TABLE I

SIMULATED NAVIGATION SENSORS: SAMPLING FREQUENCY AND NOISE

Vehicle Sensors	Frequency	Noise
OCTANS [a]	3 Hz	h: 0.1° rp: 0.01° $\dot{h}$: 0.5°/s $\dot{r}\dot{p}$: 0.25°/s
depth sensor	0.9 Hz	5 cm
DVL	3.0 Hz	1 cm/s
modem	every 2.5 min	4 m

[a] h, p, and r are heading, pitch, and roll respectively; $\dot{h}$, $\dot{p}$, and $\dot{r}$ are heading, pitch, and roll rates.

Ship Sensors	Frequency	Noise
GPS	1.0 Hz	0.5 m
gyrocompass	2.0 Hz	0.1°

IV. SIMULATION

The DEIF is tested using a simulated 6 hours survey at 3800m depth. As described in detail in Section III, the DEIF has real-time access to vehicle sensor data and asynchronous acoustic broadcasts from the ship that are used to make range measurements. The DEIF does *not* have access to the ship's sensor measurements apart from these acoustic broadcasts. To test the validity of the filter, we compare the DEIF estimation results to those obtained with a CEKF, reported in [26], at every time step.

A. Simulation Setup

For comparison purposes this simulation is designed to mimic the experimental setup of the deep water survey [26]. In the simulated mission presented here, the vehicle drives ten 700 m tracklines spaced 80 m apart at a velocity of 0.35 m/s. The vehicle's depth is constant at 3800 m. The vehicle takes approximately 6 hours to complete the survey, during which time the ship drives around the vehicle's survey area in a diamond pattern at 0.5 m/s, broadcasting acoustic data packets every 2.5 minutes.

We assume that the ship is equipped with a differential global positioning system (DGPS) receiver and a gyrocompass to measure heading. The vehicle has an OCTANS fiber-optic gyrocompass to measure attitude and attitude rates; a Paroscientific pressure sensor to measure depth; and an RDI Doppler velocity log (DVL) to measure bottom-referenced velocities. Acoustic modems are used to measure the range between the ship and the vehicle. The vehicle and ship navigation sensors, their sampling frequencies, and the noise statistics for each sensor are given in Table I.

B. Results

The results of the simulation are shown in Figures 3, 4 and 5. Figure 3 shows the estimated vehicle trajectory overlaid with the 3-sigma covariance of the vehicle position as estimated by the DEIF. The GPS-reported position of the ship as it moves around the vehicle survey area is also shown. Figure 4 shows the difference between the vehicle's true position and the estimate from the DEIF of the vehicle's position over the course of the simulated dive. The error at the end of the dive between the DEIF's estimate of the vehicle position and the true vehicle position is 3.7 m cross-track and 0.2 m along-track both with 3.1 m standard deviation. For comparison, had the vehicle relied solely on dead reckoning throughout the dive with no range measurements, the error at the end of the dive between the estimated and true vehicle position would have been 8.8 m cross-track and 5.6 m along-track with a 7.8 m standard deviation. The 3-sigma error bars are included to show that the filter maintains consistency over the course of the dive.

Comparing the mean of the vehicle's 12 degree-of-freedom (DOF) state vector as estimated by the DEIF versus the CEKF, Figure 5 shows the norm of the difference over the course of the simulation. The lower plot highlights the norm of the difference immediately after a range measurement, as marked by the asterisks. Note that the y-axis on the lower plot has been scaled down by two orders of magnitude to show the precision with which the DEIF is able to reproduce the results of the CEKF. The average difference between the filters across the entire dive is 5.68e-3 (5.7 mm) in x-y position and 3.35e-8 in the other state elements. The average difference immediately after a range measurement is 8.27e-5 m in x-y position and 1.70e-10 in the other vehicle states.

C. Discussion

As discussed in Section III-D, we expect the DEIF to produce state estimates that are comparable to the CEKF: *immediately after* each range update the results should be identical; *between* range updates, the results should differ only due

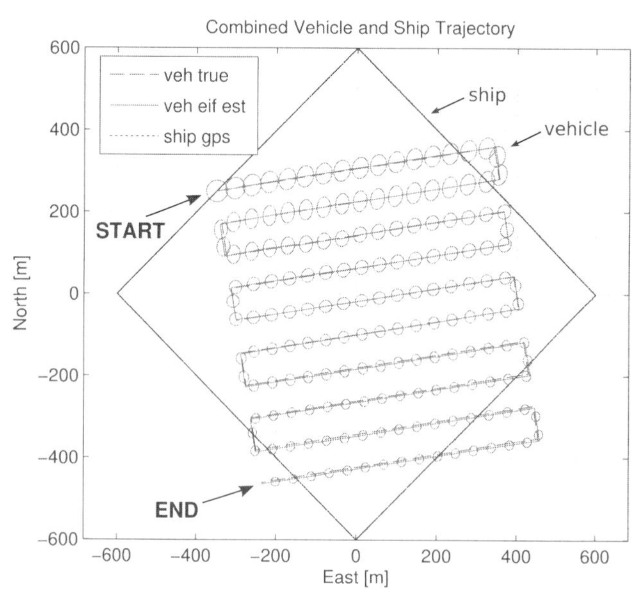

Fig. 3. Ship and vehicle trajectories. The vehicle started in the northwest and proceeded east, ending at the southwest corner of the survey. The ship moved counter-clockwise around the diamond starting at the eastern-most apex.

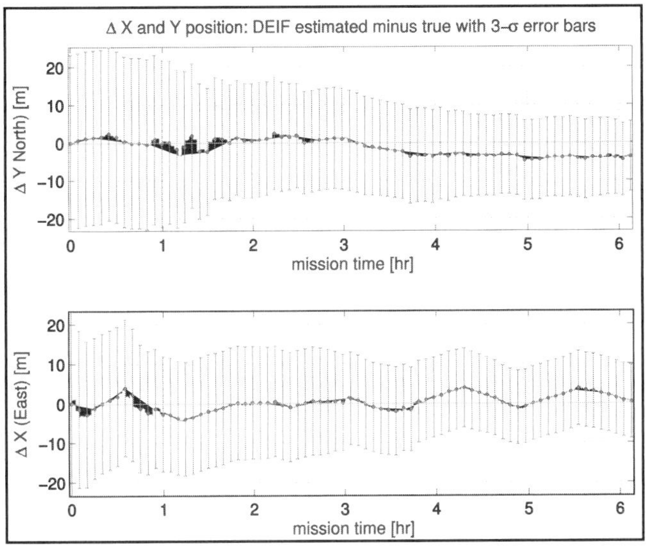

Fig. 4. The difference between the true vehicle position and estimate from the DEIF over time. The error bars are 3 times the standard deviation in each degree of freedom.

to linearization errors. The results from the simulation shown in Figure 5 support this within the tolerance of numerical precision. Over the course of the 6 hour simulated dive the difference in x-y position between the DEIF and the EKF is, on average, 8.27e-5 m immediately after each range update. In addition, the difference between the filters due to linearization errors (averaged over the entire dive) is 5.7 mm on average with a maximum difference of 4.9 cm.

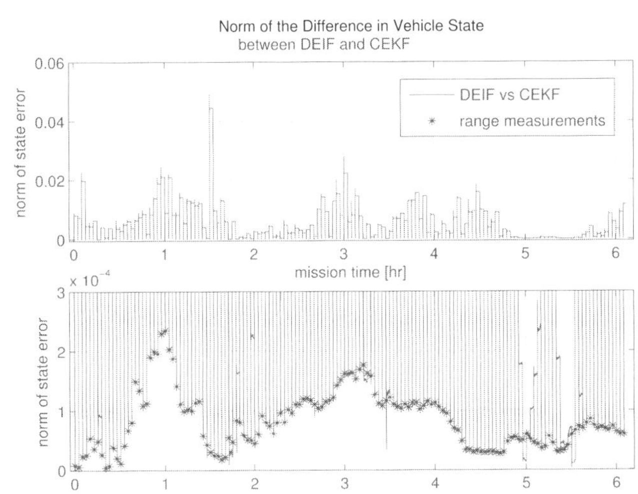

Fig. 5. The sum of the squared error between the mean vehicle position as estimated by the DEIF versus the CEKF.

Because these results are based on a simulated data set, there are several possible discrepancies compared to experimental data. The assumed noise characteristics of the navigation sensors in Table I are used both in the simulation of noisy sensor data and in the measurement models in the DEIF and CEKF. As a result the measurement models exactly predict the performance of the navigation sensors. In addition, the noise model of every sensor is assumed to be Gaussian. While these assumptions may be reasonable for common vehicle navigation sensors that have been tested extensively in the field [15], acoustic range measurements suffer from highly-variable, non-Gaussian noise sources including multi-path and ray-bending errors. In an attempt to account for this, we use a large assumed variance for the range measurements. In a real-world context outlier filtering of the range data would be necessary.

V. Conclusions and Future Work

The structure of the information filter makes it a natural choice for a decentralized implementation. Delayed ship updates are simply additive and require a minimal amount of information to be acoustically transmitted that is well within the function limits of available acoustic modems [9], [10]. In this paper we have derived a vehicle-based extended information filter that is able to estimate a vehicle's state, including x-y position, using only vehicle-based inertial navigation sensors and asynchronous acoustic broadcasts from a single, moving, georeferenced beacon. The DEIF is able to locally recreate vehicle state estimates that are commensurate with the results from a centralized extended Kalman filter within a margin of numerical error, and did so over the course of a simulation that is representative of an actual, deep-water survey in both physical scale and the frequency of measurements. In addition, the filter in its current form could be used on multiple underwater vehicles where each vehicle simultaneously receives acoustic data broadcasts from the reference beacon. Given the favorable results in simulation of the DEIF, we look forward to experimentally validating this algorithm and continuing to work towards a full multi-vehicle implementation.

In the future, the natural expansion of this algorithm is to incorporate acoustic broadcasts from *other vehicles* in addition to broadcasts from the reference beacon. The addition of vehicle-based acoustic broadcasts would generate inter-vehicle range information that could be used to further constrain each receiving vehicle's navigation solution. Incorporating inter-vehicle ranges presents a number of challenges for continued research, including the nonlinearity of the process models of the vehicles initiating the acoustic broadcast, and the problem of over confidence associated with double counting information passed between the vehicles.

Acknowledgments

This work was supported by the National Science Foundation under NSF Awards ATM-0427220, ATM-0428122, and IIS-0746455.

References

[1] P. Baccou and B. Jouvencel, "Homing and navigation using one transponder for AUV, postprocessing comparisons results with long base-line navigation," in *Proc. IEEE Int. Conf. Robot. Auto. (ICRA)*, vol. 4, 2002, pp. 4004–4009.

[2] A. Bahr, "Cooperative localization for autonomous underwater vehicles," Ph.D. dissertation, Massachusetts Institute of Technology, Cambridge, MA, USA, February 2009.

[3] M. Bozorg, E. Nebot, and H. Durrant-Whyte, "A decentralised navigation architecture," in *Proc. IEEE Int. Conf. Robot. Auto. (ICRA)*, vol. 4, May 1998, pp. 3413–3418 vol.4.

[4] J. Diosdado and I. Ruiz, "Decentralised simultaneous localisation and mapping for AUVs," in *Proceedings of the IEEE OCEANS-Europe Conference and Exhibition*, June 2007, pp. 1–6.

[5] R. M. Eustice, H. Singh, and J. J. Leonard, "Exactly sparse delayed-state filters for view-based SLAM," *IEEE Trans. Robot.*, vol. 22, no. 6, pp. 1100–1114, Dec. 2006.

[6] R. M. Eustice, L. L. Whitcomb, H. Singh, and M. Grund, "Recent advances in synchronous-clock one-way-travel-time acoustic navigation," in *Proc. IEEE/MTS OCEANS Conf. Exhib.*, Boston, MA, USA, Sep. 2006, pp. 1–6.

[7] ——, "Experimental results in synchronous-clock one-way-travel-time acoustic navigation for autonomous underwater vehicles," in *Proc. IEEE Int. Conf. Robot. Auto. (ICRA)*, Rome, Italy, Apr. 2007, pp. 4257–4264.

[8] M. F. Fallon, G. Papadopoulos, and J. J. Leonard, "Cooperative AUV navigation using a single surface craft," in *Proc. of Field and Service Robotics*, Cambridge, MA, Jul 2009, pp. 1–10.

[9] L. Freitag, M. Grund, J. Partan, K. Ball, S. Singh, and P. Koski, "Multi-band acoustic modem for the communications and navigation aid AUV," in *Proc. IEEE/MTS OCEANS Conf. Exhib.*, Washington, D.C., Sep. 2005, pp. 1080–1085.

[10] L. Freitag, M. Grund, S. Singh, J. Partan, P. Koski, and K. Ball, "The WHOI micro-modem: an acoustic communications and navigation system for multiple platforms," in *Proc. IEEE/MTS OCEANS Conf. Exhib.*, Washington, D.C., Sep. 2005, pp. 1086–1092.

[11] A. Gadre and D. Stilwell, "A complete solution to underwater navigation in the presence of unknown currents based on range measurements from a single location," in *Proc. IEEE/RSJ Intl. Conf. Intell. Robots Systems (IROS)*, Edmonton AB, Canada, 2005, pp. 1420–1425.

[12] A. Gelb, Ed., *Applied optimal estimation*. Cambridge, MA: MIT Press, 1982.

[13] M. Hunt, W. Marquet, D. Moller, K. Peal, W. Smith, and R. Spindel, "An acoustic navigation system," Woods Hole Oceanographic Institution, Tech. Rep. WHOI-74-6, Dec. 1974.

[14] D. Kilfoyle and A. Baggeroer, "The state of the art in underwater acoustic telemetry," *Oceanic Engineering, IEEE Journal of*, vol. 25, no. 1, pp. 4–27, Jan 2000.

[15] J. C. Kinsey, R. M. Eustice, and L. L. Whitcomb, "A survey of underwater vehicle navigation: recent advances and new challenges," in *IFAC Conf. on Manoeuvring and Control of Marine Craft*, Lisbon, Portugal, Sep. 2006.

[16] M. B. Larsen, "Autonomous navigation of underwater vehicles," Ph.D. dissertation, Technical University of Denmark, Denmark, Feb. 2001.

[17] S. McPhail and M. Pebody, "Range-only positioning of a deep-diving autonomous underwater vehicle from a surface ship," *Oceanic Engineering, IEEE Journal of*, vol. 34, no. 4, pp. 669–677, Oct. 2009.

[18] D. Mirabello, A. C. Sanderson, and D. R. Blidberg, "Comparing Kalman and particle filter approaches to coordinated multi-vehicle navigation," in *Proc. Intl. Symp. Unmanned Untethered Subm. Tech. (UUST)*, Aug. 2007.

[19] A. G. O. Mutambara, *Decentralized Estimation and Control for Multi-sensor Systems*. Boca Raton, FL, USA: CRC Press, Inc., 1998.

[20] S. Reece and S. Roberts, "Robust, low-bandwidth, multi-vehicle mapping," in *Information Fusion, 2005 8th International Conference on*, vol. 2, July 2005, pp. 1319–1326.

[21] A. Scherbatyuk, "The AUV positioning using ranges from one transponder LBL," in *Proc. IEEE/MTS OCEANS Conf. Exhib.*, vol. 3, San Diego, California, 1995, pp. 1620–1623.

[22] S. Singh, S. E. Webster, L. Freitag, L. L. Whitcomb, K. Ball, J. Bailey, and C. Taylor, "Acoustic communication performance of the WHOI micro-modem in sea trials of the *Nereus* vehicle to 11,000 m depth," in *Proc. IEEE/MTS OCEANS Conf. Exhib.*, Biloxi, MS, Oct 2009, pp. 1–6.

[23] S. Thrun, Y. Liu, D. Koller, A. Y. Ng, Z. Ghahramani, and H. Durrant-Whyte, "Simultaneous localization and mapping with sparse extended information filters," *International Journal of Robotics Research*, vol. 23, pp. 693–716, 2004.

[24] J. Vaganay, J. Leonard, J. Curcio, and J. Willcox, "Experimental validation of the moving long base-line navigation concept," in *Autonomous Underwater Vehicles, 2004 IEEE/OES*, June 2004, pp. 59–65.

[25] S. E. Webster, R. M. Eustice, C. Murphy, H. Singh, and L. L. Whitcomb, "Toward a platform-independent acoustic communications and navigation system for underwater vehicles," in *Proc. IEEE/MTS OCEANS Conf. Exhib.*, Biloxi, MS, Oct 2009, pp. 1–7.

[26] S. E. Webster, R. M. Eustice, H. Singh, and L. L. Whitcomb, "Preliminary deep water results in single-beacon one-way-travel-time acoustic navigation for underwater vehicles," in *Proc. IEEE/RSJ Intl. Conf. Intell. Robots Systems (IROS)*, St. Louis, MO, Oct. 2009, pp. 2053–2060.

A Non-invasive, Real-Time Method for Measuring Variable Stiffness

Giorgio Grioli
Centro Interdipartimentale
"E. Piaggio"
Università di Pisa
Pisa, Italy 56125
Email: g.grioli@centropiaggio.unipi.it

Antonio Bicchi
Centro Interdipartimentale
"E. Piaggio"
Università di Pisa
Pisa, Italy 56125
Email: bicchi@centropiaggio.unipi.it

Abstract— **The need for adaptability to the environment, energy conservation, and safety during physical interaction with humans of many advanced robotic applications has prompted the development of a number of Variable Stiffness Actuators (VSA). These have been implemented in a variety of ways, using different transduction technologies (electromechanical, pneumatic, hydraulic, but also piezoelectric, active polymeric, etc.) and arrangements with elastic elements. All designs share a fundamentally unavoidable nonlinear behavior. The control schemes proposed for these actuators typically aim at independently controlling the position (or force) of the link, and its stiffness with respect to external disturbances. Although effective feedback control schemes using position and force sensors are commonplace in robotics, control of stiffness is at present completely open–loop. In practice, instead of measuring stiffness, it is inferred from the mathematical model of the actuator. Being this in most cases only roughly known, model mismatches affect severely stiffness control, undermining its utility. It should be noticed that, while for constant stiffness elements an accurate calibration of the model is possible, the same approach is hardly viable for variable stiffness systems.**

We propose a method for estimating stiffness while it is varied, either intentionally or not, hence without knowledge of the command inputs. The method uses instantaneous measurements of force and position at one of the ends of the compliant elements in the system, and derives a measure that asymptotically converges to the current value of stiffness, up to an error which can be bounded by an arbitrarily small value. Simulation and experimental results are provided, which illustrate the performance of the proposed measurement method.

I. Introduction

To fully characterize the motion of human limbs, not only knowledge of their position and velocity, but also of their physical behavior in interaction with the environment, i.e. their mechanical impedance is needed [1, 2, 3]. Analogously, many modern robots are capable of changing their mechanical impedance to better perform a task and adapt to an environment. Variable impedance is obtained in these system by either intrinsic physical properties of the actuators (e.g. muscles in humans, or Variable Stiffness Actuators [4] in robots) or by low-level control (neural reflexes or impedance control).

Because of its ubiquitous importance, accurate measurements of the mechanical impedance of limbs are very important. Unfortunately, impedance is a rather elusive object to measure, as it is not, strictly speaking, a physical quantity

(a) robotic application (b) biometric application

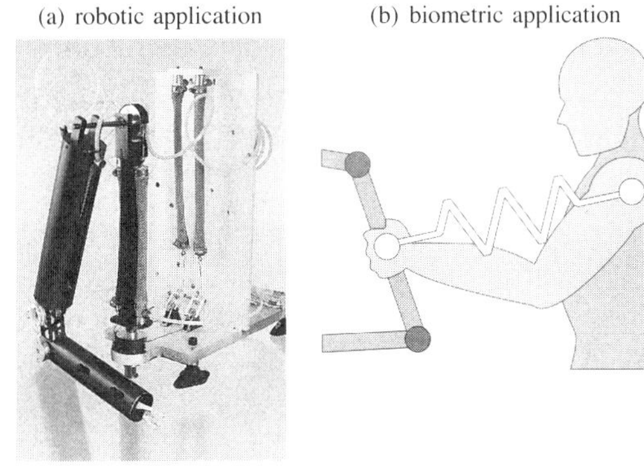

Fig. 1. Two possible applications for Non-invasive real-time stiffness measurments: control of a Variable Impedance Actuated robot (a) and identification of human limb impedance (b) [6].

per se (where by physical quantity it is meant "a property of a phenomenon, body, or substance, where the property has a magnitude that can be expressed as a number and a reference" [5]. Indeed, impedance is rather a differential operator relating the time course of physical quantities (forces and displacements). In full generality, therefore, the process of characterizing impedance of a system is more a problem of dynamical system identification than a direct measurement in a traditional sense.

Current protocols for identifying impedance in human motion typically measure the basic parameters of mass, damping, and stiffness, which concur in forming impedance, by experiments in which perturbations are purposefully injected in the system, and their effects are measured. These experiments are designed specifically so as to isolate the effects of different parameters of impedance, while at the same time minimizing perturbation of the task during which the measurement is needed [3]. In artificial robotic systems, impedance parameters are either calculated on the basis of a precise description of the model (wherever this is available), or obtained through accurate calibration procedures.

In both natural and artificial systems (e.g. see fig.1), it would be of great utility to have a method which could measure all impedance parameters in real time, without perturbing the normal execution of the task, and robust to inaccurate modeling and time-varying parameters which could alter calibration results. Of particular interest in robotics would be the application of such methods to the class of Variable Impedance Actuators (VIA), which have been recently introduced to address the need for adaptability to the environment, energy conservation, and safety during physical interaction with humans. These have been implemented in a variety of ways, using different transduction technologies (electromechanical, pneumatic, hydraulic, but also piezoelectric, active polymeric, etc.) and arrangements. All designs share a fundamentally unavoidable nonlinear behavior.

The control schemes proposed for these actuators typically aim at independently controlling the position (or force) of the link, and its impedance (most often just stiffness) with respect to external disturbances. Because of the lack of a real-time technique to measure impedance, or even just stiffness, controllers proposed so far (see e.g. [7]), are in reality open–loop, thus completely prone to model uncertainties.

Several works in the field of Admittance/Impedance robot control and haptic telemanipulation discuss the estimation of impedance. Online estimation of the impedance parameters of the contact dynamics is used to improve stability and overall performance of interaction and telemanipulation tasks. Most of these works assume a linear model for the environment dynamics (a simple spring, a spring-damper, or a mass-spring-damper system), and use methods such as adaptive control, recursive least squares, Extended Kalman filters and Active Observers to obtain an estimate of the parameters which fits best the assumed model best with the observed measurements. An extensive list of references about those methods can be found in [11]. Other works assume nonlinear impedance contact model, as Hunt-Crossley's, see in example [12].

Our work while drawing some inspiration from those aforementioned approaches, moves a step forward with the substantial improvement of considering the possibility of dealing with a time-varying non-linear spring, without assuming any a-priori description of the spring model. Notice that nonlinear and time-varying impedance parameter are ubiquitous and unavoidable in VIA.

Moreover, in this paper, we consider the problem of real-time impedance identification along motions that are not intentionally perturbed. After setting up and discussing the problem in general terms, we focus on stiffness identification, and propose a method that uses position and force sensors to provide a measurement without recurring to any a priori knowledge of the mechanical model of the system. Simulation and experimental results are provided that confirm the viability and practical applicability of our theoretical results.

II. Problem Statement

The simplest example of mechanical impedance is a *spring*, i.e. a sample of material which exhibits a relation between the applied force f and the steady–state displacement y given by

$$f = f(y).$$

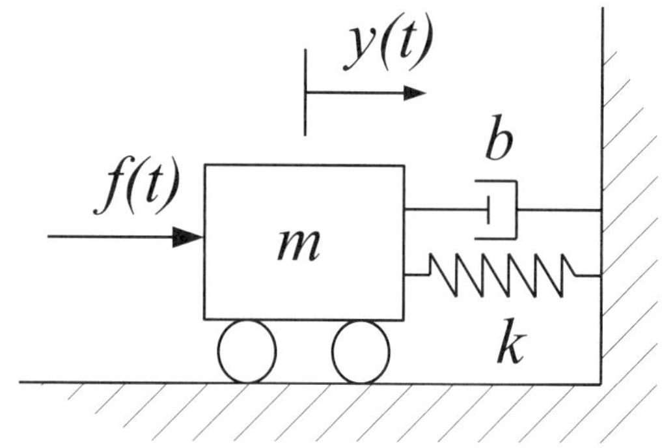

Fig. 2. A simple mass-spring-damper system.

In linear springs, $f(y) = ky$, where the constant k is the spring *stiffness*. In general, the force function f is nonlinear and time-varying, as it may depend on other parameters u (which represent internal states or inputs), which in turn may vary in time. Stiffness σ is in this case defined as

$$\sigma(y,u) = -\frac{\partial f(y,u)}{\partial y}.$$

To introduce impedance, consider first the paradigmatic example of a mass-spring-damper system (see fig. 2), described as a relation between the applied force $f(t)$ and position $y(t)$ through

$$f = m\ddot{y} + b\dot{y} + ky. \quad (1)$$

the three parameters m, b, k are constant, the O.D.E. (1) is linear and time-invariant. Introducing the Laplace transforms $F(s), Y(s)$ of force and position, one has immediately

$$F(s) = (ms^2 + bs + k)Y(s). \quad (2)$$

The operator $Z(s) := (ms^2 + bs + k)$ is called the mechanical *impedance* of the spring-damper-mass system [1]. The reciprocal operator of impedance, called *admittance* $A(s)$, generalizes compliance as it maps forces in displacements: $Y(s) = A(s)F(s)$. The admittance operator is causal (while impedance is not).

The above approach can be generalized to a nonlinear dynamic setting by considering the relation between forces, displacements, first and second derivatives of displacements, and internal states u, and its graph $G \subset F \times Y \times DY \times D^2Y \times U$, comprised of 5-tuples $d(t) := (f(t), y(t), \dot{y}(t), \ddot{y}(t), u)$ corresponding to an idealized, infinite set of experiments. If $G(f, y, \dot{y}, \ddot{y}, u) = 0$ is an analytical description of the graph, and d_0 is a regular point, then a *force function* $f(y, \dot{y}, \ddot{y}, u)$ is defined in a neighborhood of d_0. Defining *generalized stiffness* as

$$k(d) = -\left(\frac{\partial G(d)}{\partial f}\right)^{-1} \frac{\partial G(d)}{\partial y},$$

[1]It should be noted that in the literature, the term *impedance* is sometimes used to denote the relationship between velocity and force.

generalized damping as

$$b(d) = -\left(\frac{\partial G(d)}{\partial f}\right)^{-1} \frac{\partial G(d)}{\partial \dot{y}},$$

generalized mass as

$$m(d) = -\left(\frac{\partial G(d)}{\partial f}\right)^{-1} \frac{\partial G(d)}{\partial \ddot{y}},$$

and *generalized I/O characteristic* as

$$\nu(d) = -\left(\frac{\partial G(d)}{\partial f}\right)^{-1} \frac{\partial G(d)}{\partial u},$$

one can compute the Frèchet differential of the force function as

$$\delta f = m(d)\,\delta\ddot{y} + b(d)\,\delta\dot{y} + k(d)\,\delta y + \nu(d)\,\delta u\ .$$

From consideration of the positive definiteness of the kinetic energy, Rayleigh dissipation function, and elastic energy associated with the generalized inertia, damping and stiffness, it follows that m, b, and k are always greater than zero.

Alternatively, one can describe admittance along a given trajectory as the linear operator mapping small changes of the external force to changes in the resulting motion. To do this, consider the nonlinear ODE obtained by solving $G(f, y, \dot{y}, \ddot{y}, u, t) = 0$ at a regular point with respect to $\ddot{y}$ as

$$\ddot{y} = g(y, \dot{y}, f, u)$$

and its state space form, with $x \in \mathbb{R}^2$, $x_1 = y$, $x_2 = \dot{y}$, i.e.

$$\frac{d}{dt}\begin{bmatrix} x_1 \\ x_2 \end{bmatrix} = \begin{bmatrix} x_2 \\ g(x_1, x_2, f, u) \end{bmatrix}$$

For given initial conditions $x(0) = \bar{x}_0$ and a given course in time for force $\bar{f}(t)$, let $\bar{x}(t)$ be the trajectory obtained, i.e. the solution of the dynamics ODE. The first-order approximation of the dynamics of the perturbed motion $\tilde{x}(t) = x(t) - \bar{x}(t)$ corresponding to a change in force $\tilde{f}(t) = f(t) - \bar{f}(t)$ is the time-varying linear system

$$\dot{\tilde{x}} = \Gamma(t)\tilde{x} + \Theta(t)\begin{bmatrix} \tilde{f} \\ \tilde{u} \end{bmatrix},$$

where

$$\Gamma(t) = \begin{bmatrix} 0 & 1 \\ -\frac{k(d)}{m(d)} & -\frac{b(d)}{m(d)} \end{bmatrix}, \quad \Theta(t) = \begin{bmatrix} 0 & 0 \\ \frac{1}{m(d)} & \frac{\nu(d)}{m(d)} \end{bmatrix}.$$

Example: Consider the link in fig. 3, actuated by two antagonistic actuators (with a role vaguely similar to that of the biceps and triceps muscles at the elbow), with a quadratic damping and subject to gravity and to an external torque load τ_e. The system dynamics are

$$I\ddot{\theta} + \beta\dot{\theta}|\dot{\theta}| - \tau_b + \tau_t - mgl\sin\theta - \tau_e = 0$$

Assume first that the two actuators generate torques according to the model (cf. [8])

$$\begin{aligned} \tau_b &= (\tau_{max} - \alpha\theta_b)\,u_b \\ \tau_t &= (\tau_{max} - \alpha\theta_t)\,u_t \end{aligned} \qquad (3)$$

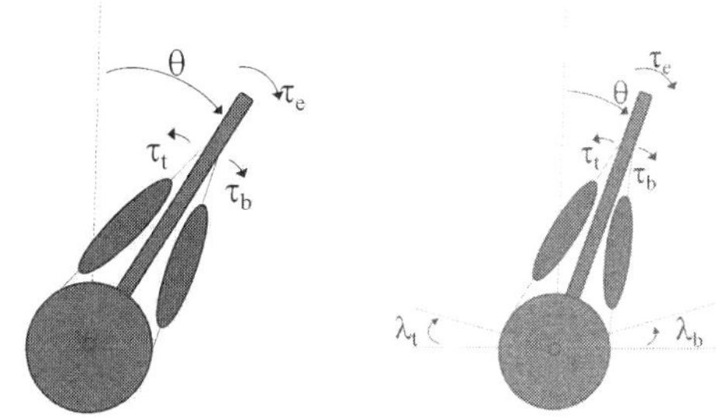

Fig. 3. A link subject to an external load and actuated by two antagonistic actuators. Nomenclature refers to actuator models used in (3) and (4), respectively.

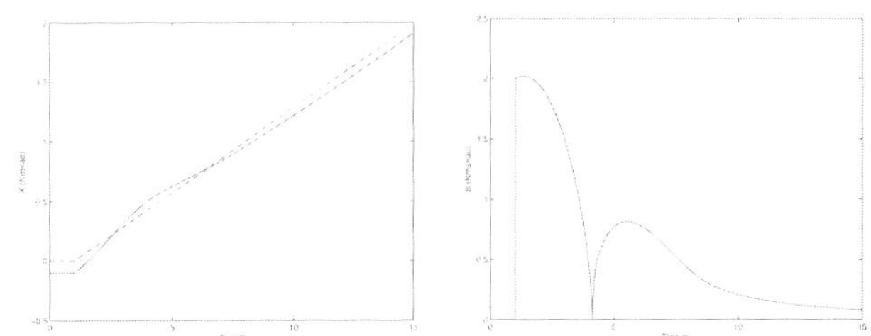

Fig. 4. Generalized stiffness (left - dashed is without gravity term) and generalized damping for the example with actuators as in (3), subject to a unit step in external torque at $T = 1$s, and with time-varying activation $u_b(t) = u_t(t)$ linearly increasing from 0 at $T = 1$s to 1 at $T = 15$s. Numerical values used in simulation: $I = 0.05$Nms^2, $mgl = 0.1$Nm, $\beta =$ 1Ns^{2}m, $alpha =$ 1Nm/rad, $tau_{max} =$ 2Nm.

where $\theta_b = (\pi/2 + \theta)$, $\theta_t = (\pi/2 - \theta)$, τ_{max} is the maximum isometric torque, u_b, u_t are the normalized contraction parameters ($0 \le u \le 1$, $u = u_b, u_t$), and α is a constant assumed to be equal for the two actuators. We easily obtain $m = I$ for the generalized mass, $b(\dot{\theta}) = 2\beta|\dot{\theta}|$ for the generalized damping, and

$$k(\theta, u) = \alpha(u_b + u_t) - mgl\cos(\theta)$$

for the generalized stiffness. In the latter expression, the role of a gravity-induced term and a co-activation stiffness term are apparent.

If a different actuator model is adopted, namely (cf. [9])

$$\begin{aligned} \tau_b &= -\alpha(\theta_b - \lambda_b)^2, \\ \tau_t &= -\alpha(\theta_t - \lambda_t)^2 \end{aligned} \qquad (4)$$

where λ_b, λ_t are interpreted as the rest lengths of the actuators, one has

$$k(\theta) = 2\alpha(\pi - \lambda_b - \lambda_t) - mgl\cos(\theta)$$

The values of stiffness and damping for the two examples above, corresponding to time-varying values of the control parameters, are reported in figs. 4 and 5, respectively. •

III. Impedance Observers

For the motivations above introduced, we seek a method for characterizing impedance using force and position measurements taken during the natural evolution of a system, as

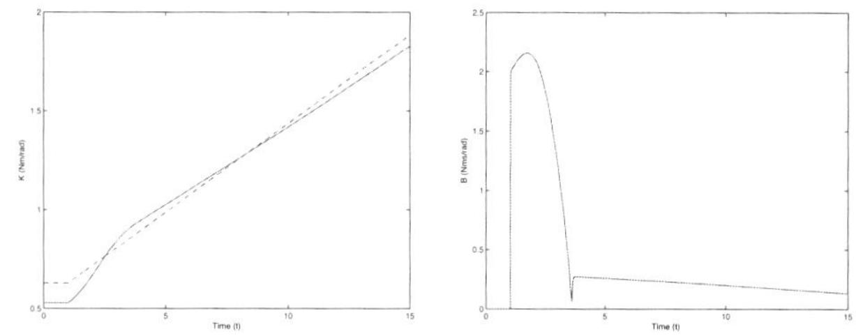

Fig. 5. Generalized stiffness (left - dashed is without gravity term) and generalized damping for the example with actuators as in (4), subject to a unit step in external torque at $T = 1$s, and with time-varying reference angle $\lambda_b(t) = \lambda_t(t)$ linearly decreasing from $\pi/3$ at $T = 1$s to 0 at $T = 15$s. Numerical values used in simulation as in fig. 4, except for $\alpha = 0.3$.

e.g. represented by the reference motion $\bar{x}(t)$ under a given reference excitation $\bar{f}(t)$, assuming that the design of the excitation $\bar{f}(t)$ is not at our disposal.

According to the description of impedance as a differential operator, its characterization can be cast as an identification problem of a dynamical system. This is done first for the case of linear impedance as follows.

A. Observers for linear impedance

Consider an extended state vector $z = \begin{bmatrix} y & \dot{y} & -\frac{k}{m} & -\frac{b}{m} & \frac{1}{m} \end{bmatrix}$, and rewrite the dynamics of (1) as

$$\begin{aligned} \dot{z} &= \begin{bmatrix} z_2 \\ z_1 z_3 + z_2 z_4 \\ 0 \\ 0 \\ 0 \end{bmatrix} + \begin{bmatrix} 0 \\ z_5 \\ 0 \\ 0 \\ 0 \end{bmatrix} f, \\ y &= h(z) = z_1. \end{aligned} \tag{5}$$

The identification of the impedance parameters can thus be cast as a nonlinear state estimation problem, i.e., from the measurement of the external force f and position y, estimate the initial state $z(0)$, and in particular its three last components which completely determine the linear impedance.

We preliminarily establish that the problem is well posed. Indeed, considering the observability codistribution for this system,

$$\Omega(z) = \begin{bmatrix} 1 & 0 & 0 & 0 & 0 \\ 0 & 1 & 0 & 0 & 0 \\ z_3 & z_4 & z_1 & z_2 & 0 \\ 0 & 0 & 0 & 0 & 1 \\ z_3 z_4 & z_3 + z_4^2 & z_2 + z_1 z_4 & z_1 z_3 + 2 z_2 z_4 & 0 \\ 0 & 0 & 0 & z_5 & z_4 \end{bmatrix}$$

it turns out that for $m, b, k > 0$, $\dim \Omega(z)^{\perp} = 0$, $\forall z$ except $z_1 = z_2 = 0$. Hence, if the system moves from the equilibrium, the three linear impedance parameters can be reconstructed from position and force measurements.

To actually estimate the impedance in this case, different methods can be adopted. These include standard off-line identification techniques (which exploit the linear nature of the regressor for the unknown parameters), such as e.g. in [10], or on-line nonlinear state observers (e.g. Extended Kalman Filters) applied to system (5).

B. Variable stiffness observers

Unfortunately, generalization of the above straightforward approach to the case when impedance is nonlinear and/or time varying is not trivial. To convince oneself, it is sufficient to consider the case of an unknown force function $s(y, u(t))$ replacing the linear spring term in (1), i.e.

$$f = m\ddot{y} + b\dot{y} + s(y, u), \tag{6}$$

when no information on the structure of f or on the variable $u(t)$ is available [2].

In the rest of this paper we describe a different approach to measure stiffness in a system such as (6). For simplicity's sake, let us assume for the time being that accurate measures of the applied force $f(t)$ and of the position $y(t)$ are available, and that numerical derivatives of these signals can be done. Assume also that both the mass and damping coefficients, m and b, are known (these strong assumptions will be discussed later on). No assumptions are made on the function $s(y, u)$ except that it is smooth in both arguments, with bounded derivatives of all orders.

We assume that the stiffness-regulating input $u(t)$ is bounded with its first derivative $\dot{u}(t)$. It should be noticed that, in building an observer that relies only on measurements of the position $y(t)$ corresponding to the external load $f(t)$, it is physically impossible to observe a stiffness which is changing in time ($\dot{u}(t) \neq 0$) while the system is at equilibrium ($\dot{y}(t) = 0$). More precisely then, we will make the assumption that the ratio between the stiffness regulation rate of change and the velocity of the measured trajectory is bounded, namely that, for all times t during the application of the observer, it holds

$$\frac{|\dot{u}(t)|}{|\dot{y}(t)|} < v \in \mathbb{R}, \quad \forall t.$$

Let

$$\frac{\partial f}{\partial y} = \frac{\partial s(y, u)}{\partial y} := \sigma(y, u(t))$$

denote the stiffness to be measured. Also let $\hat{\sigma}(t)$ denote its estimate at time t, and $\tilde{\sigma}(t) = \sigma(y, u(t)) - \hat{\sigma}(t)$ be the estimation error.

Differentiate (6) once with respect to time to get

$$\dot{f} = m\dddot{y} + b\ddot{y} + \sigma\dot{y} + s_u \dot{u},$$

where $s_u := \frac{\partial s(y,u)}{\partial u}$. Using the current estimate of stiffness and the assumptions stated above, a best-effort prediction for $\dot{f}$ can be written (in the absence of information on $s(y, u)$ and on $u(t)$) as

$$\dot{\hat{f}} = m\dddot{y} + b\ddot{y} + \hat{\sigma}\dot{y}$$

[2] In the case a parametric description of the force function is available, e.g. in terms of a finite polynomial expansion $f(y, u) = k_0(u) + k_1(u)y + k_2(u)y^2 + \ldots$, and u is constant, an observable finite dimensional nonlinear system can be built. However, the possibility to achieve good performance of the corresponding observer is dubious.

We will show that the update law

$$\dot{\hat{\sigma}} = \alpha \dot{\tilde{f}} \operatorname{sgn}(\dot{y}), \quad (7)$$

with $\alpha > 0$ and

$$\operatorname{sgn}(x) := \begin{cases} \frac{x}{\|x\|} & \text{if } \|x\| \neq 0 \\ 0 & \text{if } \|x\| = 0 \end{cases},$$

is such that $\hat{\sigma}(t)$ can be made to converge to the true stiffness value $\sigma(t)$ within an uniformly ultimately bounded error.

Indeed, consider the positive definite error function

$$V_\sigma := \frac{1}{2}\tilde{\sigma}^2$$

and its derivative along the trajectory defined in (7), i.e.

$$\dot{V}_\sigma = \tilde{\sigma}\dot{\tilde{\sigma}} = \tilde{\sigma}\dot{\sigma} - \tilde{\sigma}\dot{\hat{\sigma}} = \tilde{\sigma}\dot{\sigma} - \alpha\tilde{\sigma}s_u\,\dot{u}\operatorname{sgn}(\dot{y}) - \alpha\tilde{\sigma}^2|\dot{y}| \,. \quad (8)$$

While the first two terms in the rightmost sum in (8) are indefinite in sign, the third term is negative definite. Therefore, wherever the inequality holds

$$|\tilde{\sigma}| > |s_u|\frac{|\dot{u}|}{|\dot{y}|} + \frac{1}{\alpha}\frac{|\dot{\sigma}|}{|\dot{y}|}$$

the derivative of the error function $\dot{V}_\sigma$ is negative, hence the estimation error decreases. By writing $\dot{\sigma} = \sigma_y\dot{y} + \sigma_u\dot{u}$, and using the upper bound above assumed on the rate of stiffness change, we have that stiffness estimates converge to the true value within an ultimately uniformly bounded error given by

$$|\tilde{\sigma}| > \frac{|\sigma_y|}{\alpha} + \left(|s_u| + \frac{|\sigma_u|}{\alpha}\right)v \quad (9)$$

Remark: The assumption that the mass and damping are exactly known is not realistic. However, it is easy to verify that the analysis above carries over exactly even with no knowledge of m and b, provided that a force sensor directly placed on the elastic element provides a measure of the force $s(x, u)$. In case this is not available, then errors on the m and b parameters have the effect of making the ultimate error larger (this effect can be countered by increasing the observer gain α).

IV. Simulation Results

As a first illustration, the proposed stiffness observer is tested using numeric simulations in two systems.

A. Single spring

In this first test, the observer algorithm is used to track the stiffness value of a single spring with non-linear, time-invariant stiffness characteristic. The simulated elastic element is an exponential spring, whose force/displacement characteristic is

$$s(y) = ae^{by} \,. \quad (10)$$

Such springs are designed so that the stiffness at any operating point is proportional to the force the spring is exchanging with the external environment.

Figure 5(a) compares the observer estimate of the spring stiffness with its exact value $\sigma(y) = bae^{by}$, for three different values of the spring parameters. The springs were subject to a external force $f(t)$ given by a chirp in the frequency range from 0.1 to 10 Hz, with deformations $y(t)$ in the range $[0.25, 2.25]$cm. The observation error (shown in fig. 5(b)) decays to less than one percent in less than 0.3 s.

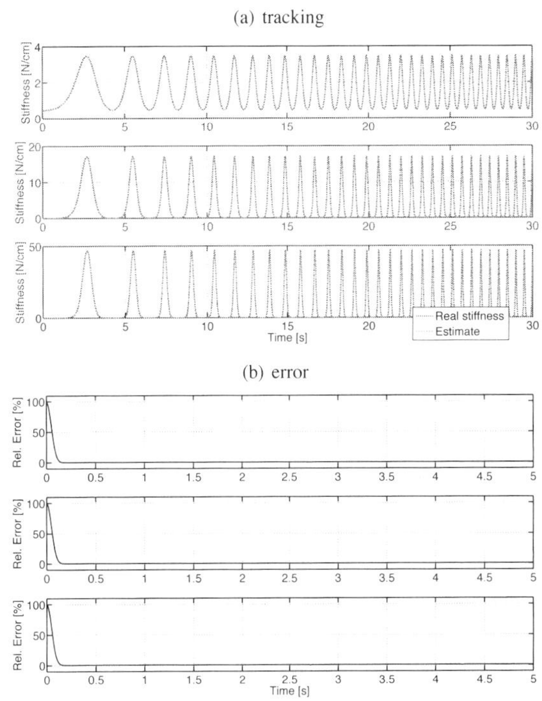

Fig. 6. Simulation results for the exponential spring test, with three different springs ($a = 2.7e^{-2}$, $b = 1$; $a = 8.1e^{-6}$, $b = 3$; $a = 13.5e^{-10}$, $b = 5$, respectively). Stiffness tracking is shown in panel (a), and relative estimation error in panel (b). Only the first 5s of the relative error are shown, to focus on the transient phase. Relative error remains under 1% for the rest of the time.

B. Antagonistic VSA systems

One of the simplest and most common examples of variable stiffness, both in natural systems and in robotics, is the agonist-antagonist arrangement on nonlinear actuators. To illustrate how our proposed stiffness observer applies to antagonist VSA systems, consider the examples reported above in fig. 3. Application of the stiffness observer in this case can be carried out in two ways: 1) the tendon tensions τ_b, τ_t are measured directly, or 2) the external torque $\tau_e(t)$ is measured, and estimates of the link inertia and damping are used. In all cases, a measurement of the link angle $\theta(t)$ is necessary. It should be noticed that, while the first method does not require any estimate of link parameters, it is more invasive in the system, and is inapplicable to e.g. stiffness measurement in a human elbow joint. On the opposite, the second method is easily applicable to this case, although its accuracy will be reduced if poor estimates of inertia and damping are available.

Simulation results for the antagonist arrangements of two muscle-like actuators as described in (3) and (4) are reported in fig. 7 a) and b), respectively. In both simulations, the external

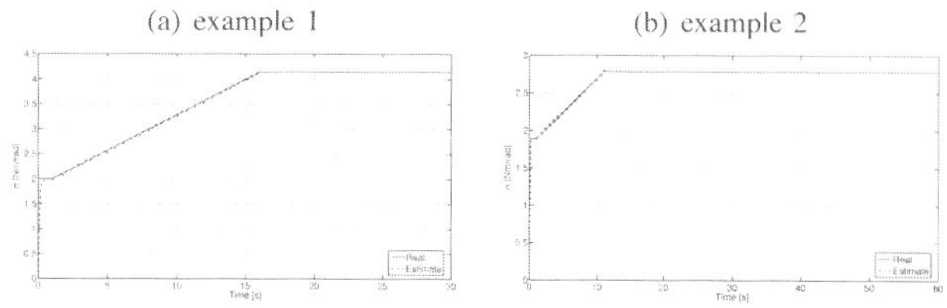

Fig. 7. Stiffness tracking for an antagonist VSA system realized adopting nonlinear muscle-like actuators as in equations 3 and 4 (panels (a) and (b) respectively).

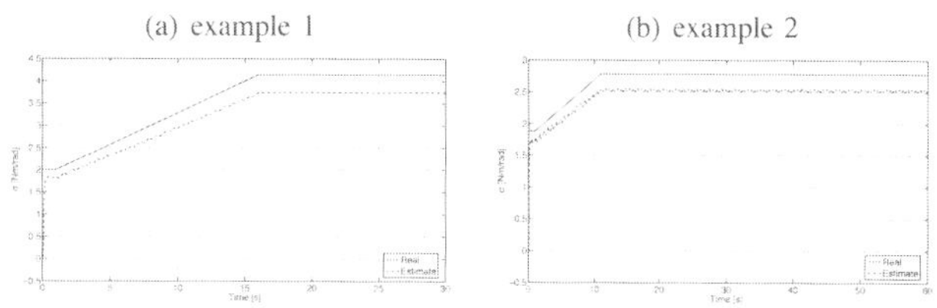

Fig. 8. Stiffness tracking of the antagonist VSA systems with muscle-like actuators as in (3) (a), and (4) (b), with a 10% error in the knowledge of parameters m and b.

force τ_e is a sinusoid with $\omega = 5$ rad/s and amplitude 0.02 Nm. Stiffness is varied during the simulation in a saturated ramp fashion. The ensuing joint motion $\theta(t)$ is in the range ± 0.01rad, while $\dot{\theta}$ varies from zero to 0.05 rad/s.

To assess how strongly the performance of the stiffness observer is affected by inertia and damping parameter mismatches (in the case that only external torques are measured), we performed simulations in the hypothesis that m and b were in error by 10% of their actual value. Results reported in fig. 8 indicate that, for both muscle models, the relative error on stiffness is of comparable magnitude.

Finally, a simulation is reported for the same link actuated by two exponential springs (9). In this case, the stiffness is exponentially increasing in time (by linearly varying the co-contraction of the antagonist springs). A sinusoidal external force is applied during the initial and final phases of the experiment, while it is set to zero in the time interval between 10s and 20s. Correspondingly, motion of the links stops ($\dot{\theta} = 0$), and the stiffness estimate is not updated in the interval. When motion resumes, the estimation recovers quickly to the exact value.

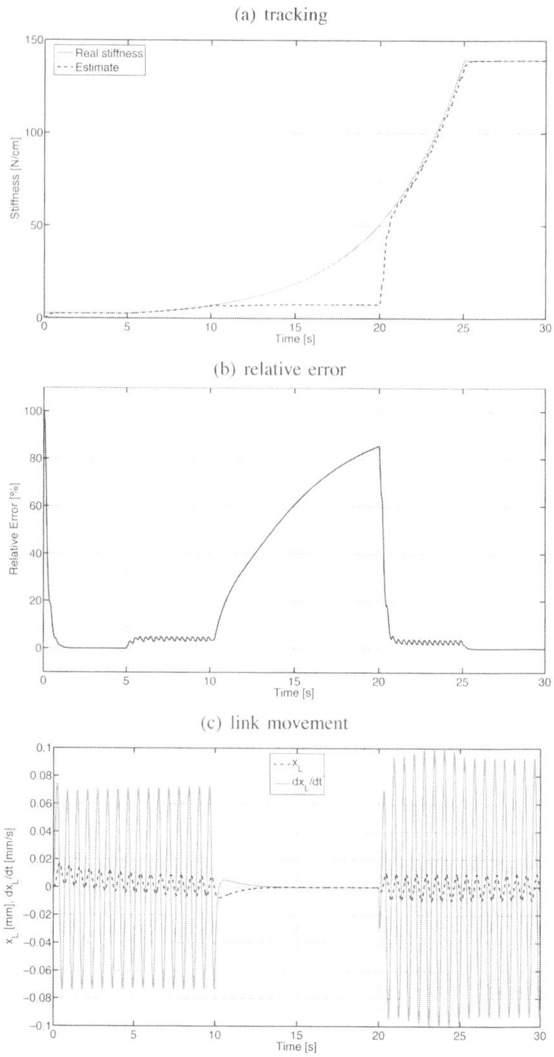

Fig. 9. Simulation results for the VSA system with exponential springs. a) Comparison of the link stiffness with its estimate; b) Relative estimation error; c) link motion during the simulated experiment.

V. Experimental Results

The algorithm has been tested on the experimental device shown in figure 10, implementing the antagonistic VSA device with exponential springs described previously. Two strain-gauge load cells were used to measure the tendon tensions directly, while positions of the link and of the tendon origin were measured using three HEDS-5540 encoders with a resolution of 2000 CPRs. Data were acquired using a National Instruments PCI6251 ADC board for the strain gauges, and an USB-PCI4e for the encoders. Data were sampled with sampling time $T_s = 0.015s$, and afterward filtered with a second–order filter with time constant of $0.02s$. Signal derivatives used in the algorithm were approximated by the numerical filter described by the transfer function

$$D(s) = \frac{s}{1 + 10^{-4}s} \,. \tag{11}$$

Despite the simplicity of such "Dirty Derivatives" technique, results were suitable for the purposes of the present work, proving the practical feasibility of the proposed method.

It should be pointed out that in this paper the impedance estimates are not used for control in a feedback loop, hence the effect of derivation noise can not destabilize the system. Future work will address this problem when closed loop control of impedance will be considered.

Input signals derivation could be avoided through use of sensors of the rate of change of desired quantities, e.g.

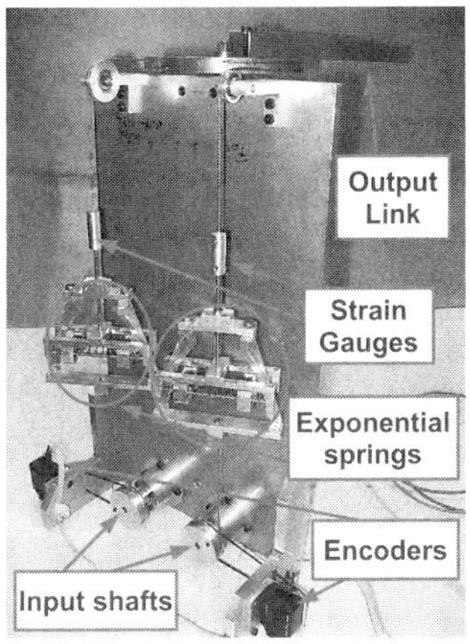

Fig. 10. The experimental setup consists of an antagonistic VSA system with exponential springs, realized using a linear spring forced to move on a suitable cam profile. Force sensors (strain gauges) are mounted on the tendons connecting the springs to the link. Position sensors (encoders) are mounted on the link and on two tendon pulleys coupled with the input levers.

inductive sensors for positions, and piezo-electric sensors for forces.

Both the external load and the torque actuating the tendon tensions were generated manually, and not measured.

To obtain ground-truth data, the force functions of the two springs were experimentally evaluated through careful preliminary calibration experiments. The calibration procedure consisted in collecting a large number of force-displacement pairs (x, f), translating them in semi-logarithmic coordinates space $(x, y = ln(f))$, finding the regression line in the semi-logarithmic space, such that $y = mx + q$, to finally go back to the original space and obtain $f = e^y = e^{mx+q} = e^q \cdot e^{mx}$, from which $a = m$ and $b = e^q$. From the mean square error of the regression SME, the relative error margin of the model can be easily evaluate as $r = 1 - e^{SME}$.

The numeric values of the exponential curves fitting our data are

$$a_1 = 0.999 \quad , \quad b_1 = 3.267 \; ,$$
$$a_2 = 0.950 \quad , \quad b_2 = 2.780 \; ,$$

where subscripts are relative to the two spring. Figure 11 shows the regressed curve alongside with experimental data for both the left (a) and right (b) springs. It is noticeable that, due to un-modeled friction in the mechanism implementing the exponential springs, a certain hysteresis is present, making the model correct only up to a relative error margin of about 25%. Anyway, this error only marginally affected the performance of the proposed observer.

Raw experimental data are reported in fig. 12. The estimate of stiffness reconstructed in real-time by the stiffness observer ($\alpha = 3$) is compared with the calibrated stiffness data in fig. 13

VI. Conclusion

In this paper we presented an algorithm which can be used to measure stiffness in real-time, using force and position sensors. The method's main advantages are that it does not require injecting any intentional perturbation in the system, and that uses no a piori knowledge on the model of the physical actuators. Simulations and experimental test shown that the method is practically applicable and robust to noisy data and uncertain parameters.

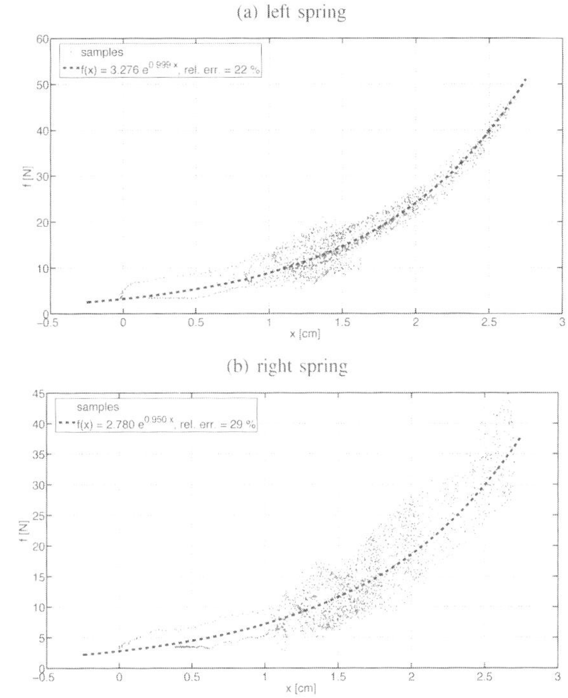

Fig. 11. Experimental characterization of the two exponential springs. Force and displacement pairs recorded during a calibration experiment, and regression curve are shown for each of the two springs of the experimental VSA system.

Acknowledgment

The authors wish to thank Alessandro De Luca and Riccardo Schiavi for the useful discussions, and Soumen Sen for the help on the realization of the experimental set-up. Further gratitude goes to the anonymous reviewers for their mindful comments and constructive critiques. This work was partially supported by the VIACTORS and THE projects, funded by the European Community under contract IST - 231554 2008 and ICT - 248587 - 2010, respectively.

References

[1] N. Hogan, "Impedance control: An approach to manipulation, part i - theory, part ii - implementation, part iii - applications," *Journal of Dynamic Systems, Measurement and Control*, vol. 107, pp. 1–24, 1985.

[2] M. Kawato, "Internal models for motor control and trajectory planning," in *Current Opinion in Neurobiology*. Elsevier Science Ltd., 1999, vol. 9, pp. 718–727.

[3] E. Burdet, R. Osu, D.W. Franklin, T. Yoshioka, T.E. Milner and M. Kawato, "A method for measuring endpoint stiffness during multi-joint arm movements," *Journal of Biomechanics*, vol. 33, no. 12, pp. 1705–1709(5), December 2000.

[4] B. Vanderborght K.W. Hollander R. Ham, T. Sugar and D. Lefeber, "Compliant actuator designs," *IEEE, Robotics and Automation Magazine*, vol. 16, no. 3, pp. 81–94, September 2009.

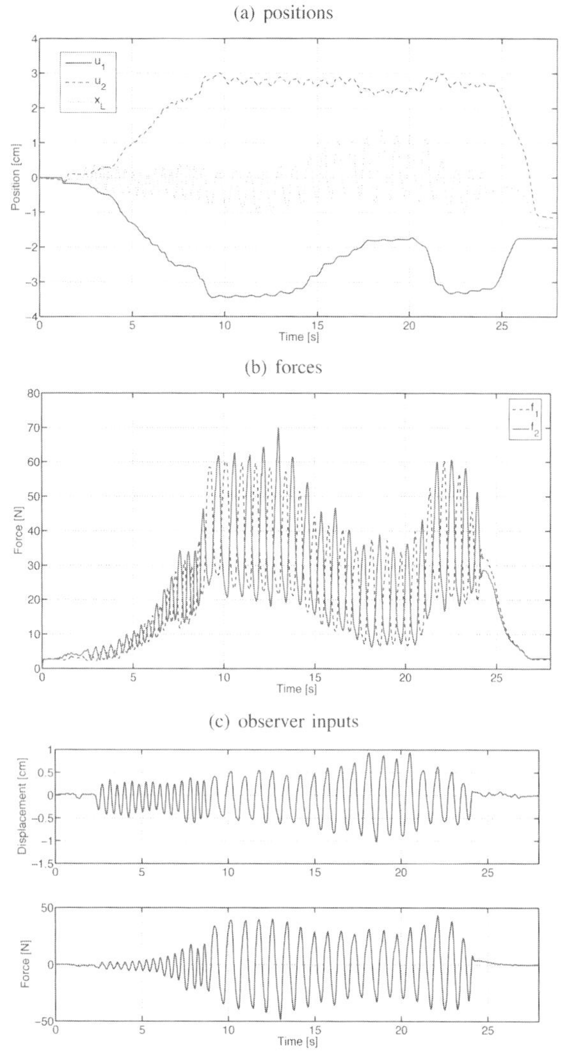

Fig. 12. Experimental data. Raw data recording from sensors is shown in the first two panels. The third panel shows the values of f and x actually fed to the observer during the experiment.

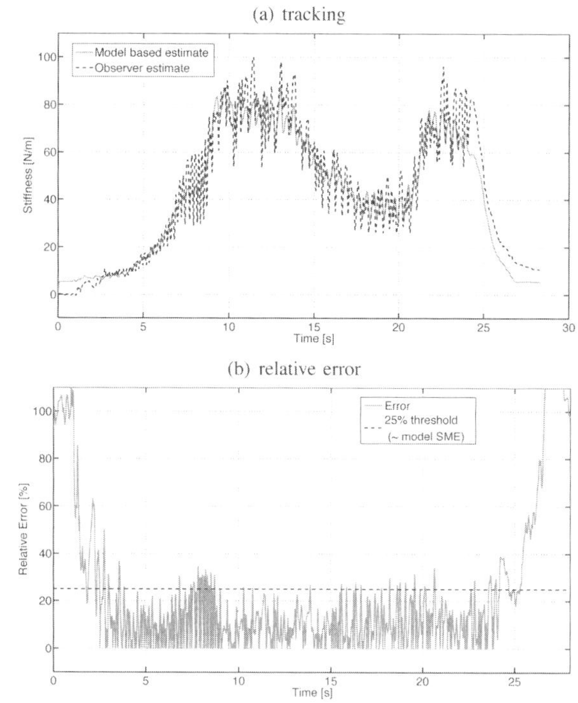

Fig. 13. Experimental results for the stiffness observer. The first panel shows a comparison between the stiffness values derived by the calibrated spring model and the estimate performed by the stiffness observer algorithm. The second panel shows the relative error (difference between estimator and model, normalized w.r.t. the model) compared with the relative error underlying the model itself.

[5] J. C. for Guides in Metrology (JCGM), Ed., *International Vocabulary of Metrology (VIM)*, iii ed. Pavillon de Breteuil, 2008, vol. Basic and General Concepts and Associated Terms, no. JCGM 200:2008, `http://www.bipm.org/utils/common/documents/jcgm/JCGM_200_2008.pdf`.

[6] B. Sepehri1, A. Esteki, E. Ebrahimi-Takamjani, G.-A. Shahidi, F. Khamseh, and M. Moinodin "Quantification of rigidity in Parkinson disease," in*Annals of Biomedical Eng.*, vol. 35, no. 12, pp. 2196-2203, December 2007.

[7] A. De Luca and P. Lucibello, "A general algorithm for dynamic feedback linearization of robots with elastic joints," in *IEEE Int. Conf. on Robotics and Automation*, 1998, pp. 504–510.

[8] N. Hogan, "Adaptive control of mechanical impedance by coactivation of antagonist muscles," *IEEE Trans. Aut. Cont.*, vol. AC-29, no. 8, pp. 681–690, August 1984.

[9] A.G. Feldman, "Functional tuning of the nervous system with control of movement or maintenance of a steady posture. ii: Controllable parameters of the muscle," *Biophys.*, vol. 11, pp. 565–578, 1966.

[10] K. Hashimoto, T. Kureha, Y. Nishimura, K. Okumura, and S. Muraoka, "Measurement of mechanical impedance using quartz resonator force sensor during the process of grasping," in *SICE 2004 Annual Conference*, vol. 1, pp. 722–726, August 2004.

[11] D. Verscheure, I. Sharf, H. Bruyninckx, J. Swevers, and J. De Schutter, "Identification of Contact Dynamics Parameters for Stiff Robotic Payloads," in *IEEE Trans. Robot.*, vol. 25, no. 2, pp. 240-252, April 2009.

[12] N. Diolaiti, C. Melchiorri, and S. Stramigioli, "Contact impedance estimation for robotic systems," in *IEEE Trans. Robot.*, vol. 21, no. 5, pp. 925-935, October 2005.

Consistent Data Association in Multi-robot Systems with Limited Communications

Rosario Aragüés Eduardo Montijano Carlos Sagüés
Departamento de Informática e Ingeniería de Sistemas, Instituto de Investigación en Ingeniería de Aragón,
Universidad de Zaragoza, María de Luna, 50018 Zaragoza, Spain
raragues@unizar.es, emonti@unizar.es, csagues@unizar.es

Abstract— **In this paper we address the data association problem of features observed by a robot team with limited communications. At every time instant, each robot can only exchange data with a subset of the robots, its neighbors. Initially, each robot solves a local data association with each of its neighbors. After that, the robots execute the proposed algorithm to agree on a data association between all their local observations which is globally consistent. One inconsistency appears when chains of local associations give rise to two features from one robot being associated among them. The contribution of this work is the decentralized detection and resolution of these inconsistencies. We provide a fully decentralized solution to the problem. This solution does not rely on any particular communication topology. Every robot plays the same role, making the system robust to individual failures. Information is exchanged exclusively between neighbors. In a finite number of iterations, the algorithm finishes with a data association which is free of inconsistent associations. In the experiments, we show the performance of the algorithm under two scenarios. In the first one, we apply the resolution and detection algorithm for a set of stochastic visual maps. In the second, we solve the feature matching between a set of images taken by a robotic team.**

I. Introduction

During the last years, there has been an increasing interest in multi robot systems. In these scenarios, a team of robots cooperatively perform some task in a more efficient way than a single robot would do. In this paper, we address the data association problem in multi robot scenarios. It consists of establishing correspondences between different measurements or estimates of a common element. It is of high interest in localization, mapping, exploration, and tracking applications [1]. The Nearest Neighbor (NN), and the Maximum Likelihood (ML), are widely used methods which associate each observation with its closest feature in terms of the Euclidean or the Mahalanobis distance [23], [15], [12]. Other popular method is the Joint Compatibility Branch and Bound (JCBB) [18], which considers the compatibility of many associations simultaneously. The Combined Constraint Data Association [2] builds a graph where the nodes are individually compatible associations and the edges relate binary compatible assignments. Over this graph, a Maximal Common Subgraph problem is solved for finding the maximum clique in the graph. Scan matching and Iterative Closest Point (ICP) [6] are popular methods for comparing two laser scans. Other methods, like the Multiple Hypothesis Tracking, and the Joint Probabilistic Data Association, maintain many association hypothesis instead of selecting one of them. And exist many variations of these techniques that combine RANSAC [9] for higher robustness.

In solutions based on submaps, one of them is usually transformed into an observation of another. The local submaps are merged with the global map following a sequence [22], or in a hierarchical binary tree fashion [5]. All the mentioned data association approaches, operate on elements from two sets. One set usually contains the current observations, and the other one consists of the feature estimates. These sets may be two images, two laser scans, or two probabilistic maps.

Lately, many localization, mapping, and exploration algorithms for multi robot systems have been presented. However, they have not fully addressed the problem of multi robot data association. Some solutions have been presented for merging two maps [21], [23] that do not consider a higher number of robots. Many approaches rely on broadcasting all controls and observations measured by the robots. Then, the data association is solved like in a single robot scenario, using scan matching and ICP for laser scans [14], [10], [16], [20], or NN, ML, and visual methods for feature-based maps [12], [17]. In these methods, the problem of inconsistent data associations is avoided by forcing a cycle-free merging order. This limitation has also been detected in the computer vision literature. In [8] they approach an inconsistent association problem for identifying equal regions in different views. They consider a centralized scenario, where each 2 views are compared among them in a 2-by-2 way. Then, their results are arranged on a graph where associations are propagated and conflicts are solved. The work in [7], from the target tracking literature, simultaneously considers the association of all local maps. It uses an expectation-maximization method for both computing the data association and the final global map. The main limitation of this work is that the data from all sensors needs to be processed together, what implies a centralized scheme, or a broadcast method.

All the previous methods rely on centralized schemes, full communication between the robots, or broadcasting methods. However, in multi robot systems, distributed approaches are more interesting. They present a natural robustness to individual failures since there are no central nodes. Besides, they do not rely on any particular communication scheme, and they are robust to changes in the topology. On the other hand, distributed algorithms introduce an additional level of complexity in the algorithm design. Although the robots make decisions based on their local data, the system must exhibit a global behavior.

In this paper, we address the data association problem for distributed robot systems. Each of our robots posses a local observation of the environment. Instead of forcing a specific order for associating their observations, we allow the robots compute its data association with each of its neighbors in the graph. Although this scenario is more flexible, it may lead to inconsistent global data associations in the presence of cycles in the communication graph. These inconsistencies are detected when chains of local associations give rise to two features from one robot being associated among them. These situations must be correctly identified and solved before merging the data. Otherwise, the merging process would be wrong and could not be undone. In this paper, we approach a distributed data association, under limited communications. Instead of comparing any 2 local observations among them, only the local observations of neighboring robots can be compared. Besides, there is no central agent that has knowledge of all the local associations and each robot exclusively knows the associations computed by itself. Then, each robot updates its local information by communicating with its neighbors. We present an algorithm where, finally, each robot is capable of detecting and solving any inconsistent association that involves any of its features.

II. Problem Description

We consider a robotic team composed of $n \in \mathbb{N}$ robots. The n robots have communication capabilities to exchange information with the other robots. However, these communications are limited. Let $\mathcal{G}_{com} = (\mathcal{R}_{com}, \mathcal{E}_{com})$ be the undirected communication graph. The nodes are the robots, $\mathcal{R}_{com} = \{1, \ldots, n\}$. If two robots i, j can exchange information then there is an edge between them, $(i, j) \in \mathcal{E}_{com}$. Let N_i be the set of neighbors of robot i,

$$N_i = \{j \mid (i, j) \in \mathcal{E}_{com}\}.$$

Each robot i has observed a set $\mathcal{S}_i$ of m_i features,

$$\mathcal{S}_i = \{f_1^i, \ldots, f_{m_i}^i\}. \tag{1}$$

It can compute the data association between its own set $\mathcal{S}_i$, and the sets of its neighbors $\mathcal{S}_j$, with $j \in N_i$. However, these data associations are not perfect. There may appear inconsistent data associations relating different features from the same set $\mathcal{S}_i$ (Fig. 1). If the robots merge their data as soon as they solve the local data association, inconsistent associations cannot be managed since the merging cannot be undone. The goal of our algorithm is to detect and resolve these inconsistent associations before executing the merging.

In order to make the reading easy, along the paper we will use the indices i, j and k to refer to robots and indices r, r', s, s', to refer to features. The rth feature observed by the ith robot will be denoted as f_r^i. Given a matrix A, the notation $[A]_{r,s}$ will correspond to the component (r, s) of the matrix whereas A_{ij} will denote the block (i, j) when the matrix is defined by blocks.

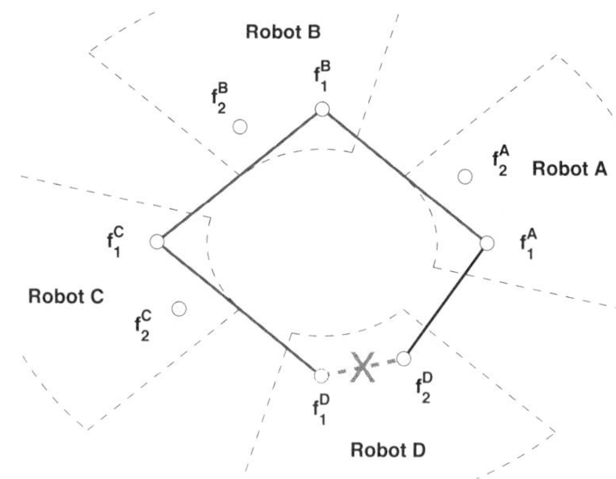

Fig. 1. Robots A, B, C and D associate their features comparing their maps in a two-by-two way. Robot A associates its feature f_1^A with f_2^D and with f_1^B; robot B associates f_1^B with f_1^C; robot C associates f_1^C with f_1^D (solid lines). As a result, there is a path (dashed line) between f_1^D and f_2^D. This is an inconsistent situation. Finding this path would require the knowledge of the whole association graph.

Let F be a function that computes the data association between any two sets of features, $\mathcal{S}_i$ and $\mathcal{S}_j$, and returns an association matrix $A_{ij} \in \mathbb{N}^{m_i \times m_j}$ where

$$[A_{ij}]_{r,s} = \begin{cases} 1 & \text{if } f_r^i \text{ and } f_s^j \text{ are associated,} \\ 0 & \text{otherwise,} \end{cases}$$

for $r = 1, \ldots, m_i$ and $s = 1, \ldots, m_j$. We assume that F satisfies the following conditions.

***Assumption** 2.1:* When F is applied to the same set $\mathcal{S}_i$, it returns the identity, $F(\mathcal{S}_i, \mathcal{S}_i) = A_{ii} = I$.

***Assumption** 2.2:* The returned association A_{ij} has the property that the features are associated in a one-to-one way,

$$\sum_{r=1}^{m_i} [A_{ij}]_{r,s} \leq 1 \text{ and } \sum_{s=1}^{m_j} [A_{ij}]_{r,s} \leq 1,$$

for all $r = 1, \ldots, m_i$ and $s = 1, \ldots, m_j$.

***Assumption** 2.3:* Robots i and j associate their features in the same way. Given two sets $\mathcal{S}_i$ and $\mathcal{S}_j$ it holds that $F(\mathcal{S}_i, \mathcal{S}_j) = A_{ij} = A_{ji}^T = (F(\mathcal{S}_j, \mathcal{S}_i))^T$.

Most of the existing data association methods satisfy these assumptions.

If F is applied to all pairs of sets of features $\mathcal{S}_i$, $\mathcal{S}_j$, for $i, j \in \{1, \ldots, n\}$, its results can be represented by an undirected graph $G_{map} = (V_{map}, E_{map})$. Each node in V_{map} is a feature f_r^i, for $i = 1, \ldots, n$, $r = 1, \ldots, m_i$. There is an edge between two features f_r^i, f_s^j iff $[A_{ij}]_{r,s} = 1$. If F were providing the ground truth data association, then G_{map}^{gt} would exclusively contain disjoint cliques, identifying features observed by multiple robots [8]. Since F is not perfect, G_{map} is a perturbed version of G_{map}^{gt} that includes additional spurious edges, while it misses others.

In addition, due to communication restrictions, the data associations available to our robots are just a subset of G_{map}. We name $G = (V, E)$ this available association graph. It has the same set of nodes than G_{map}, $V = V_{map}$. It has an edge between two features f_r^i, f_s^j only if the edge exists in G_{map} and the robots i and j are neighbors in the communication graph. The goal of our algorithm is to detect and resolve inconsistencies in this graph $G \subseteq G_{map}$ in a decentralized fashion.

***Definition** 2.4:* An *association set* is a set of features such that they form a connected component in G. Such set is an *inconsistent association* or a *conflictive set* if there exists a path in G between two or more features from the same robot. A feature is *inconsistent* or *conflictive* if it belongs to an inconsistent association.

Along this paper, we name m_{sum} the number of features in G, $m_{sum} = \sum_{i=1}^{n} m_i$. We name d the diameter of G, the length of the longest path between any two nodes in G. The diameter satisfies $d \leq m_{sum}$. We name $W \in \mathbb{N}^{m_{sum} \times m_{sum}}$ the adjacency matrix of G,

$$W = \begin{bmatrix} W_{11} & \dots & W_{1n} \\ \vdots & \ddots & \vdots \\ W_{n1} & \dots & W_{nn} \end{bmatrix}, \qquad (2)$$

where W_{ij} is the block within W for the associations between robots i and j,

$$W_{ij} = \begin{cases} A_{ij} & \text{if } j \in \{N_i \cup i\}, \\ \mathbf{0} & \text{otherwise.} \end{cases} \qquad (3)$$

III. Decentralized Detection of Inconsistent Associations

Considering definition 2.4 we observe that in order to detect an inconsistent association it is required to compute the paths that exist among the elements in G. As the following lemma states [4], given a graph G, the powers of its adjacency matrix contains the information about the number of paths existing between the nodes of G:

***Lemma** 3.1 (Lemma 1.32 [4]):* Let G be a weighted graph of order $|V|$ with un-weighted adjacency matrix $W \in \{0,1\}^{|V| \times |V|}$, and possibly with self loops. For all $i, j \in \{1, \dots, |V|\}$ and $t \in \mathbb{N}$ the (i, j) entry of the tth power of W, W^t, equals the number of paths of length t (including paths with self loops) from node i to node j.

The computation of the powers of W requires, a priori, the information about the whole matrix. We show now that this computation can also be done in a decentralized manner. Let each robot $i \in \mathcal{R}_{com}$ maintain the blocks within W^t associated to its own features, $X_{ij}(t) \in \mathbb{N}^{m_i \times m_j}$, $j = 1, \dots, n$, $t \geq 0$,, which are initialized as

$$X_{ij}(0) = \begin{cases} I, & j = i, \\ \mathbf{0}, & j \neq i, \end{cases} \qquad (4)$$

and are updated, at each time step, with the following algorithm

$$X_{ij}(t+1) = \sum_{k \in \{N_i \cup i\}} W_{ik} X_{kj}(t), \qquad (5)$$

with W_{ik} as defined in (3). It is observed that the algorithm is fully distributed because the nodes only use information about its direct neighbors in the communication graph.

***Theorem** 3.2:* Let $[W^t]_{ij} \in \mathbb{N}^{m_i \times m_j}$ be the block within W^t related to the associations between robot i and robot j. The matrices $X_{ij}(t)$ computed by each robot i using the decentralized algorithm (5) are exactly the sub-matrices $[W^t]_{ij}$,

$$X_{ij}(t) = [W^t]_{ij}, \qquad (6)$$

for all $i, j \in \{1, \dots, n\}$ and all $t \in \mathbb{N}$.

Proof: The proof is done using induction. First we show that eq. (6) is satisfied for $t = 0$. In this case we have that $W^0 = I$, thus for all $i, j \in \{1, \dots, n\}$, $[W^0]_{ii} = I$ and $[W^0]_{ij} = \mathbf{0}$, which is exactly the initial value of the variables X_{ij} (eq. (4)).

Now we have that for any $t > 0$,

$$[W^t]_{ij} = \sum_{k=1}^{n} W_{ik}[W^{t-1}]_{kj} = \sum_{k \in \{N_i \cup i\}} W_{ik}[W^{t-1}]_{kj},$$

because $W_{ik} = \mathbf{0}$ for $k \notin \{N_i \cup i\}$,

Assuming that for all $i, j \in \{1, \dots, n\}$ and a given $t > 0$, $X_{ij}(t-1) = [W^{t-1}]_{ij}$ is true, then

$$\begin{aligned} X_{ij}(t) &= \sum_{k \in \{N_i \cup i\}} W_{ik} X_{kj}(t-1) = \\ &= \sum_{k \in \{N_i \cup i\}} W_{ik}[W^{t-1}]_{kj} = [W^t]_{ij}. \end{aligned} \qquad (7)$$

Then, by induction, $X_{ij}(t) = [W^t]_{ij}$ is true for all $t > 0$. ∎

***Corollary** 3.3:* The variables $X_{ij}(t)$ contain the information about all the paths of length t between features observed by robots i and j.

Proof: By direct application of Lemma 3.1. ∎

Analyzing the previous algorithm the first issue to deal with is how to simplify the computation of the matrices in order to avoid high powers of W. In the case we are studying it is just required to know if there is a path between two elements in G and not how many paths are. This means that in this situation it is enough that $[X_{ij}(t)]_{r,s} > 0$ in order to know that features f_r^i and f_s^j are connected by a path. Another issue is to decide when the algorithm in (5) must stop. Since the maximum length of a path between any two nodes in a graph is its diameter, then after d iterations the algorithm should stop. However, in general situations the robots will not know neither d nor m_{sum}, which makes this decision hard to be made a priori.

***Definition** 3.4:* We will say that two matrices W and $\bar{W}$ of the same dimensions are equivalent, $W \sim \bar{W}$, if for all r and s it holds

$$[W]_{r,s} > 0 \Leftrightarrow [\bar{W}]_{r,s} > 0 \text{ and } [W]_{r,s} = 0 \Leftrightarrow [\bar{W}]_{r,s} = 0.$$

In practice any equivalent matrix to the $X_{ij}(t)$ will provide the required information, which allows us to simplify the computations simply by changing any positive value in the matrices by 1. Moreover, the equivalency is also used to find a criterion to stop the algorithm:

***Proposition** 3.5:* For a robot i, let t_i be the first time instant, t, such that $X_{ij}(t) \sim X_{ij}(t-1)$ for all $j = 1, \dots, n$. Then robot i can stop to execute the algorithm at time t_i.

Proof: Let $\bar{X}_{ij}(t)$ be the components in $X_{ij}(t)$, such that $[X_{ij}(t-1)]_{r,s} = 0$ and $[X_{ij}(t)]_{r,s} > 0$. The cardinal, $|\bar{X}_{ij}(t)|$, represents the number of features $f_s^j \in \mathcal{S}_j$ such that the minimum path length in G between them and one feature $f_r^i \in \mathcal{S}_i$ is t. At time t_i, $X_{ij}(t_i) \sim X_{ij}(t_i - 1) \quad \forall j$ for the first time, and then $\sum_{j=1}^{n} |\bar{X}_{ij}(t_i)| = 0$ because no component has changed its value from zero to a positive. This means that there is no path of minimum distance t_i linking any feature f_r^i with any other feature in G. By the physical properties of a path, it is obvious that if there are no features at minimum distance t_i, it will be impossible that a feature is at minimum distance $t_i + 1$ and all the paths that connect features of robot i with any other feature have been found. ■

***Corollary** 3.6:* All the robots end the execution of the iteration rule (5) in at most in $d + 1$ iterations.

Proof: Recalling that the maximum distance between two nodes in G is the diameter of the graph, denoted by d, then $\sum_{j=1}^{n} |\bar{X}_{ij}(d+1)| = 0$ for all $i = 1, \ldots, n$. ■

If a robot j at time t does not receive the information $X_{ij}(t)$ from robot i then it will use the last matrix received, because robot i has already finished computing its paths and $X_{ij}(t) \sim X_{ij}(t-1)$.

It remains to analyze which features are conflictive and which are not. Each robot has the information about all the association paths of its features and the features of the rest of the robots in the network in the different variables $X_{ij}(t_i)$. The robots detect all the conflictive features using two simple rules. A feature f_r^i is conflictive if and only if one of the following conditions are satisfied:

(i) There exists other feature $f_{r'}^i$, with $r \neq r'$, such that

$$[X_{ii}(t_i)]_{r,r'} > 0; \tag{8}$$

(ii) There exist features f_s^j and $f_{s'}^j$, $s \neq s'$, such that

$$[X_{ij}(t_i)]_{r,s} > 0 \text{ and } [X_{ij}(t_i)]_{r,s'} > 0. \tag{9}$$

In conclusion, the proposed algorithm will be able to find all the inconsistencies in a finite number of iterations. The algorithm is decentralized and it is based only on local interactions between the robots. Each robot only needs to know its local data associations. It updates its information based on the data exchanged with its neighbors. When the algorithm finishes, each robot i can extract from its own matrices $X_{ij}(t_i)$ all the information of any conflict that involves any of its features. If the robot has any conflictive feature, it also knows the rest of features that belong to the conflictive set independently of the robot that observed such features. In the next section we present an additional technique to solve the inconsistencies found by the algorithm.

IV. Decentralized Resolution of Inconsistent Associations

The resolution of inconsistent associations consists of deleting edges from G so that the resulting graph is conflict-free.

***Definition** 4.1:* Let c denote the number of conflictive sets in G. We say a conflictive set C is *detectable* by a robot i if there exists a $r \in \{1, \ldots, m_i\}$ such that $f_r^i \in C$. The set of robots that detect a conflictive set C is $R \subseteq \mathcal{R}_{com}$. The number of features from each robot $i \in R$ involved in C is $\tilde{m}_i$. We say G is *conflict-free* if $c = 0$.

All the edges whose deletion transforms G into a conflict-free graph, belong to any of the c conflictive sets of G. Since the conflictive sets are disjoint, they can be considered separately. From now on, we focus on the resolution of one of the conflictive sets C. The other conflictive sets are managed in the same way. The resolution problem consists of partitioning C into a set of disjoint conflict-free components C_q such that

$$\cup_q C_q = C, \text{ and } C_q \cap C_{q'} = \emptyset,$$

for all $q, q' = 1, 2, \ldots$. The number of such conflict-free components is a priori unknown and it will be discussed later in this section.

Obtaining an optimal partition that minimizes the number of deleted edges is complicated. If there were only two inconsistent features f_r^i, $f_{r'}^i$, it could be approached as a max-flow min-cut problem [19]. However, in general there will be more inconsistent features, $\tilde{m}_i \geq 2$, within C associated to a robot $i \in R$. Besides, there may also be $\tilde{m}_j \geq 2$ inconsistent features belonging to a different robot $j \in R$. The application of [19] separately to any pair of inconsistent features does not necessarily produce an optimal partition. It may happen that a single edge deletion simultaneously resolves more than one inconsistent association. Therefore, an optimal solution should consider multiple combinations of edge deletions, what makes the problem computationally intractable, and imposes a centralized scheme. We propose a resolution algorithm that is not optimal but is efficient and is proven to be correct. Besides, it allows a decentralized computation.

***Proposition** 4.2:* Let R be the set of robots that detect C. Let $i_\star$ be the robot with the most features involved in C,

$$i_\star = \arg\max_{i \in R} \tilde{m}_i. \tag{10}$$

The number of conflict-free components in which C can be decomposed is lower bounded by $\tilde{m}_{i_\star}$.

Proof: Each conflict-free component can contain, at most, one feature from a robot $i \in R$. Then there must be, at least, $\max_{i \in R} \tilde{m}_i = \tilde{m}_{i_\star}$ components. ■

The resolution algorithm constructs $\tilde{m}_{i_\star}$ conflict-free components using a strategy close to a BFS tree construction. Initially, each robot i detects the conflictive sets for which it is the root using its local information $X_{i1}(t_i), \ldots, X_{in}(t_i)$. The root robot for a conflictive set is the one with the most inconsistent features involved. In case two robots have the same number of inconsistent features, the one with the lowest robot id is selected. Then, each robot executes the resolution algorithm (Algorithm 1).

The root robot creates $\tilde{m}_{i_\star}$ components and initializes each component C_q with one of its features $f^{i_\star} \in C$. Then, it tries to add to each component C_q the features directly associated to $f^{i_\star} \in C_q$. Let us consider that f_s^j has been assigned to C_q. For all f_r^i such that $[W_{ij}]_{r,s} = 1$, robot j sends a component

Algorithm 1 Decentralized resolution algorithm - Robot i

– *Initialization*
for each conflictive set C for which i is root ($i = i_\star$) **do**
    create $\tilde{m}_{i_\star}$ components
    assign each inconsistent feature $f_r^{i_\star} \in C$ to a different component C_q
    send component request to all its neighboring features
end for
– *Algorithm*
for each component request from f_s^j to f_r^i **do**
    if (b) or (c) **then**
        $[W_{ij}]_{r,s} = 0$
        send reject message to j
    else if (d) **then**
        assign f_r^i to the component
        send component request to all its neighboring features
    end if
end for
for each component reject from f_s^j to f_r^i **do**
    $[W_{ij}]_{r,s} = 0$
end for

request message to robot i. When robot i receives it, it may happen that

(a) f_r^i is already assigned to C_q;
(b) f_r^i is assigned to a different component;
(c) other feature $f_{r'}^i$ is already assigned to C_q;
(d) f_r^i is unassigned and no feature in i is assigned to C_q.

In case (a), f_r^i already belongs to the component C_q and robot i does nothing. In cases (b) and (c), f_r^i cannot be added to C_q; robot i deletes the edge $[W_{ij}]_{r,s}$ and replies with a reject message to robot j; when j receives the reject message, it deletes the equivalent edge $[W_{ji}]_{s,r}$. In case (d), robot i assigns its feature f_r^i to the component C_q and the process is repeated.

***Theorem** 4.3:* Let us consider that each robot $i \in \mathcal{R}_{com}$ executes the decentralized resolution algorithm (Algorithm 1) on G, obtaining G',

(i) after $t = n$ iterations no new features are added to any component C_q and the algorithm finishes;
(ii) each obtained C_q is a connected component in G';
(iii) C_q is conflict free;
(iv) C_q contains at least two features;

for all $q \in \{1, \ldots, \tilde{m}_{i_\star}\}$ and all conflictive sets.

Proof: (i) The maximal depth of a conflict-free component is n since, if there were more features, at least two of them would belong to the same robot. Then, after at most n iterations of this algorithm, no more features are added to any component C_q and the algorithm finishes.

(ii) There is a path in G between any two features belonging to a conflictive set C. Therefore, there is also a path in G between any two features assigned to the same component C_q. Since the algorithm does not delete edges from G within a component (case (a)), then C_q it is also connected in G'. Since none feature can be assigned to more than one component (case (b)), the components are disjoint. Therefore, C_q is a connected component in G'.

(iii) By construction, two features from the same robot are never assigned to the same component C_q (case (c)). Therefore, each component is conflict-free.

(iv) Each conflictive set has more than one feature. Because of Assumptions 2.1 and 2.2, each feature and its neighbors are conflict free. Therefore, each component C_q contains, at least, its originating feature, and a neighboring feature. Thus, it has at least two features. ■

***Corollary** 4.4:* After executing Algorithm 1, the size of each conflict set C is reduced by at least $2\ \tilde{m}_{i_\star}$, where $\tilde{m}_{i_\star} \geq 2$.

When the algorithm finishes, each original conflictive set C has been partitioned into $\tilde{m}_{i_\star}$ conflict-free components. It may happen that a subset of features remains unassigned. These features may still be conflictive in G'. The detection algorithm can be executed on the subgraph defined by this smaller subset of features.

***Proposition** 4.5:* Consider each robot i iteratively executes the detection (Section III) and the resolution (Section IV) algorithms. Then, in a finite number of iterations, all conflictive sets disappear.

Proof: After each execution of the resolution algorithm, the size of each conflict set C is reduced by, at least, $2\ \tilde{m}_{i_\star} \geq 4$ (Corollary 4.4). Then, in a finite number of iterations, it happens that $|C| < 4$. A set with 3 features $f_r^i, f_{r'}^i, f_s^j$ cannot be conflictive; this would require the existence of edges (f_r^i, f_s^j) and $(f_{r'}^i, f_s^j)$, what is impossible (Assumption 2.2). A set with 2 features cannot be conflictive (Assumptions 2.1 and 2.2), and a set with a single feature cannot be inconsistent by definition. Therefore, there will be no remaining inconsistencies or conflictive sets. ■

The main interest of the presented resolution algorithm, is that it is fully decentralized and it works on local information. Each robot uses its own $X_{ij}(t_i)$ for detecting the root robot of each conflictive set. During the resolution algorithm, the decisions and actions taken by each robot are based on its local associations W_{ij}, and the components assigned to its local features. Moreover, each robot is responsible of deleting the edges from its local association matrices W_{ij}, with $j \in \{1, \ldots, n\}$. In addition, the presented algorithm works in finite time. Let us note that although we presented the algorithm for a single conflictive set, all conflictive sets are managed in parallel.

V. Experiments

A. Data association of stochastic maps

The presented algorithm is of high interest in multi-robot exploration scenarios with limited communications. Each robot has explored a section of the environment and it has built a stochastic map. When the exploration finishes, the local maps are merged into a global map of the environment. Each robot can compute the data association between its local map, and the maps of its neighbors. However, it does not know the data associations obtained by other robots in the team. If they start the merging process using only its local associations, and there is any inconsistency, at some point a robot will be forced to fuse two or more of its features into a single one. To avoid this situation, they execute the presented algorithm for resolving any inconsistent association before merging the maps.

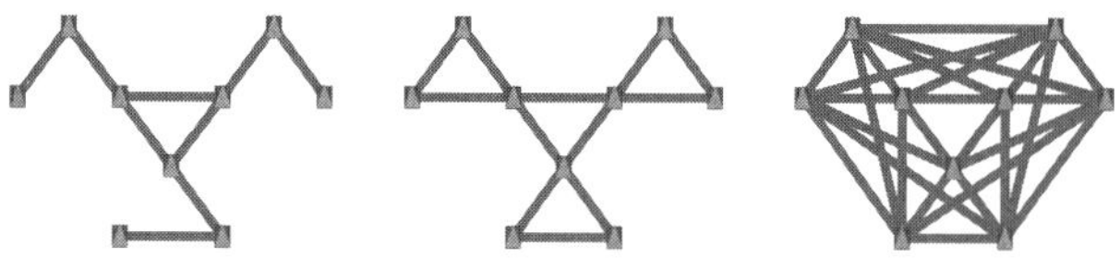

(a) One cycle (b) Four cycles (c) Complete

Fig. 2. Communication graphs between the 9 robots after the exploration.

The behavior of the algorithm is analyzed under real data. We use a data set [11] with bearing information obtained with vision (Sony EVI-371DG). The landmarks are vertical lines extracted from the images. The measurements are labeled so that we can compare our results with the ground-truth data association. We select 9 subsections of the whole path for the operation of 9 different robots (Fig. 3 (a)). A separate SLAM is executed on each subsection, producing the 9 local maps (Fig. 3 (b)). The local data associations are computed using the JCBB [18] since it is very convenient for clutter situations like the considered scenario (Fig. 3 (b)). The JCBB is applied to the local maps of any pair of neighboring robots. We analyze the performance of the algorithm under 3 communication graphs (Fig. 2).

Table I gives statistics about the number of inconsistencies found considering the different network topologies in Fig. 2. We show the obtained associations compared to the ground truth results. The number of association sets is the number of connected components of W^t. The number of good links (true positives) are obtained associations between 2 features which are true (ground truth). The missing links (false negatives) are associations that are in the ground truth information, but have been not detected. And spurious links (false positives) are associations found between features that are different according to the ground truth. The sixth row, c, is the number of conflictive sets. The next row in the table shows the total number of features which have been associated to any other feature from other local map. The last row gives information about how many of those features are conflictive. The amount

TABLE I

INITIAL ASSOCIATIONS BETWEEN THE 9 LOCAL MAPS

Comm. graph	(a)	(b)	(c)
Association sets (ground truth)	242	284	400
Association sets	182	218	290
Good links (true positives)	160	190	228
Missing links (false negatives)	82	94	172
Spurious links (false positives)	22	28	62
Conflictive sets (c)	3	5	8
Number of features m_{sum}	138	144	154
Conflictive features	16	24	51

of missing and spurious associations obtained is very high for the three network topologies. This is the expected result for many real scenarios, where the landmarks are close to each other, and where the only available information are their cartesian coordinates. As a result, the conflictive features are more than a 10% of the total. In communication graphs with more cycles (Fig. 2 (b)(c)), there are more conflictive features. In the three cases, after a single execution of the detection and the resolution algorithms, all the inconsistencies are solved (Table II, 1st row). An interesting result is that, although our algorithm cannot distinguish between good and spurious edges, in practice a high number of the deleted edges (last row) are spurious ones.

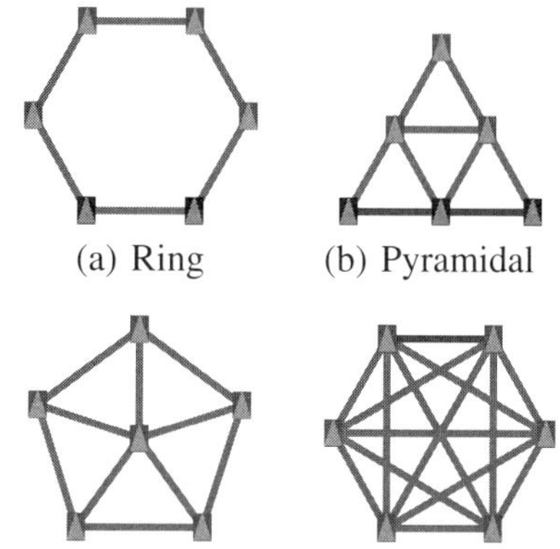

(a) Ring (b) Pyramidal

(c) Star-Ring (d) Complete

Fig. 4. Formations used in the experiments

TABLE II

MANAGEMENT OF THE INCONSISTENCIES

Comm. graph	(a)	(b)	(c)
Iterations	1	1	1
Initial conflictive sets	3	5	8
Deleted links	6	10	34
Good deleted links (true positives)	2	2	12
Spurious deleted links (false positives)	4	8	22

B. Matching sets of images

We have also tested the performance of our proposal with a set of images. A team of robots equipped with cameras and limited communication capabilities is a typical situation in which not all the images are available to execute a global matching. A solid set of matches is required independently of the task the team is performing (visual SLAM, formation control, etc.). In the proposed experiment 6 robots moving in formation are considered. Each robot acquires one image with its onboard camera and extracts SURF features [3]. The local matching is only applied to pairs of images which are connected in the communication graph. For the local matching the epipolar constraint combined with RANSAC is imposed [13]. The detection and resolution of inconsistencies is analyzed for four different typical communication graphs (Fig. 4).

Although the epipolar constraint discards most part of the wrong matches, some spurious associations are still found.

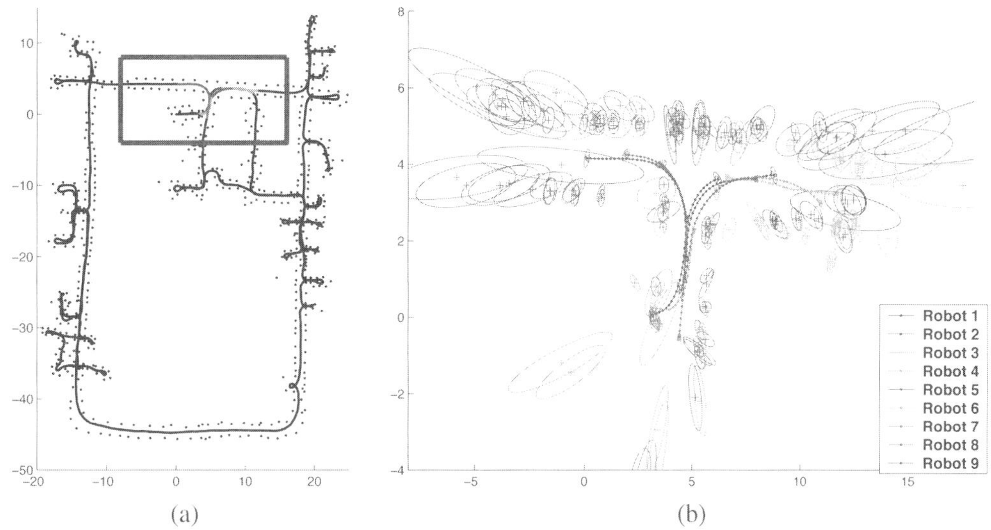

Fig. 3. (a) Section of the dataset used in the experiments. (b) Local maps acquired by 9 robots exploring the region in (a).

Figure 5 shows an example of an inconsistency found by our algorithm in the case of the formation (d). The figure also shows how the Algorithm 1 solves such inconsistency; the result is a subset of the initial matches with more connected components than before but without any conflict.

More general results about the experiment can be seen in Table III. Since for this experiment there is no ground truth available, the number of missing and spurious links found in the local matching cannot be provided. Some interesting conclusions are extracted with the obtained results. First of all, the number of association sets and the number of features involved m_{sum} are increased with the number of edges in $\mathcal{G}_{com}$. The more local associations there are, the more matches the algorithm will have available. With respect to the inconsistencies, they grow with the number of cycles in $\mathcal{G}_{com}$ because each cycle can generate inconsistencies independently of the rest of the communication network. The size of the cycles also influence on the conflicts, cycles of smaller length will cause more inconsistencies because the number of local associations required to find a conflict is also smaller.

TABLE III

INITIAL ASSOCIATIONS FOR THE DIFFERENT FORMATIONS

Formation	(a)	(b)	(c)	(d)
Association sets	528	590	605	632
Conflictive sets (c)	2	22	31	72
Number of features m_{sum}	1335	1643	1743	2069
Conflictive features	14	144	204	521

Another important aspect to analyze is how the resolution of the conflicts is performed. We have observed that some of the inconsistencies are caused by features that are very close to each other in one image and due to the RANSAC threshold the miss-match appears. These inconsistencies are introduced because of the extractor parameters but the errors they can cause are negligible. However, there are also some other inconsistencies which can generate problems to the vision system because they include features from the same image which are far away from each other. Table IV shows the statistics on how are the solutions to the conflicts provided by our algorithm. Although ground truth is not available in this experiment it is easily observable when the inconsistent matches have been separated correctly in the used images and when not (Fig. 5). We have manually classified the conflicts as well solved and wrongly solved. We consider a conflict as well solved when it does not contain any spurious edge in any of the associations that the algorithm returns. The spurious and the good links that the algorithm deletes, also manually counted, are shown in fifth and sixth rows of table IV respectively.

TABLE IV

MANAGEMENT OF THE INCONSISTENCIES

Formation	(a)	(b)	(c)	(d)
Initial conflictive sets	2	22	31	72
Inconsistencies well solved	1	20	24	55
Inconsistencies wrongly solved	1	2	7	17
Deleted links	2	26	46	131
Good deleted links (true positives)	1	2	19	43
Spurious deleted links (false positives)	1	24	27	88

VI. CONCLUSIONS

We have presented a new technique to match several sets of features observed by a team of robots in a consistent way under limited communications. Local associations are found only within robots that are neighbors in the communication graph. After that a fully decentralized method to compute all the paths between local associations is carried out, allowing the robots to detect all the inconsistencies related with their observations. For every conflictive set detected, in a second step the method is able to delete local associations to break the conflict using only local communications. The whole method is proved to finish in a finite amount of time finding and solving all the inconsistent associations. Experimental results show the performance of the method in two different scenarios with

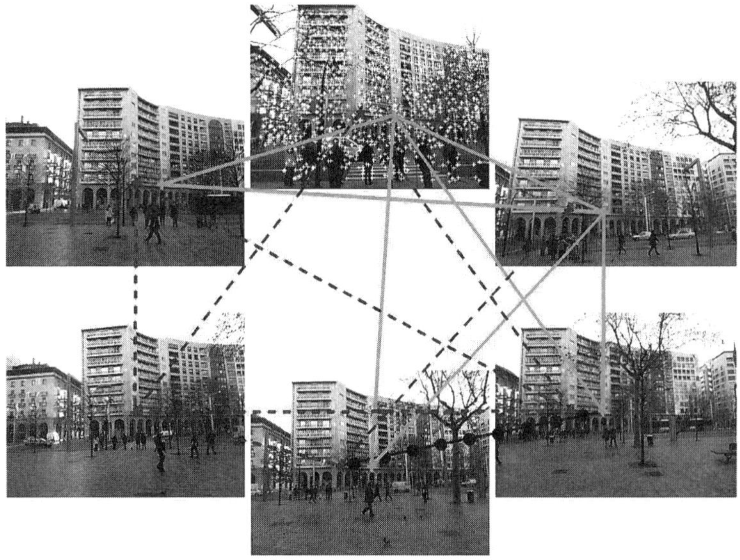

Fig. 5. Example of one inconsistency found by the algorithm. The inconsistency is represented by the whole set of depicted links. It is observed that if all the links are considered, features of the same image are matched. After executing Algorithm 1 the inconsistency is solved. In this example the root of the inconsistency is the top-middle image. The dark line with circles is the link deleted by the algorithm. Solid lines represent one of the conflict free components and dashed lines the second. For clarity, the rest of the SURF features are only shown in the top middle image.

great interest in robotic tasks. To the best of our knowledge this is the first algorithm that is able to do it.

Acknowledgment

This work was supported by project Ministerio de Ciencia e Innovacion DPI2009-08126, and grants MEC BES-2007-14772 and AP2007-03282. The data set used in one of the experiments was provided by U. Frese and J. Kurlbaum. The authors gratefully acknowledge them for providing data and support.

References

[1] T. Bailey and H. Durrant-Whyte. Simultaneous localization and mapping: part II. *IEEE Robotics & Automation Magazine*, 13(3):108–117, 2006.

[2] T. Bailey, E. M. Nebot, J. K. Rosenblatt, and H. F. Durrant-Whyte. Data association for mobile robot navigation: a graph theoretic approach. In *IEEE Int. Conf. on Robotics and Automation*, pages 2512–2517, San Francisco, USA, April 2000.

[3] H. Bay, T. Tuytelaars, and L. Van Gool. Surf: Speeded up robust features. In *European Conference on Computer Vision*, pages 404–417, 2006.

[4] F. Bullo, J. Cortés, and S. Martínez. *Distributed Control of Robotic Networks*. Applied Mathematics Series. Princeton University Press, 2009. Electronically available at http://coordinationbook.info.

[5] C. Cadena, F. Ramos, and J. Neira. Efficient large scale SLAM including data association using the Combined Filter. In *European Conference on Mobile Robotics, ECMR*, pages 217–222, Mlini/Dubrovnik, Croatia, September 2009.

[6] A. Censi. An accurate closed-form estimate of ICP's covariance. In *IEEE Int. Conf. on Robotics and Automation*, pages 3167–3172, Roma, Italy, April 2007.

[7] R. W. Deming and L. I. Perlovsky. Concurrent multi-target localization, data association, and navigation for a swarm of flying sensors. *Information Fusion*, 8(3):316 – 330, 2007.

[8] V. Ferrari, T. Tuytelaars, and L. Van Gool. Wide-baseline multiple-view correspondences. In *IEEE Computer Society Conference on Computer Vision and Pattern Recognition*, volume 1, pages 718–725, Madison, USA, June 2003.

[9] M. A. Fischler and R. C. Bolles. Random sample consensus: A paradigm for model fitting with applications to image analysis and automated cartography. *Communications of the ACM*, pages 381–395, 1981.

[10] D. Fox, J. Ko, K. Konolige, B. Limketkai, D. Schulz, and B. Stewart. Distributed multirobot exploration and mapping. In *Proceedings of the IEEE*, volume 94, pages 1325–1339, July 2006.

[11] U. Frese and J. Kurlbaum. A data set for data association, June 2008. http://www.sfbtr8.spatial-cognition.de/insidedataassociation/.

[12] A. Gil, O. Reinoso, M. Ballesta, and M. Julia. Multi-robot visual SLAM using a rao-blackwellized particle filter. *Robotics and Autonomous Systems*, 58(1):68–80, 2009.

[13] R. Hartley and A. Zisserman. *Multiple View Geometry in Computer Vision*. Cambridge University Press, Cambridge, 2000.

[14] A. Howard. Multi-robot simultaneous localization and mapping using particle filters. *International Journal of Robotics Research*, 25(12):1243–1256, 2006.

[15] M. Kaess and F. Dellaert. Covariance recovery from a square root information matrix for data association. *Robotics and Autonomous Systems*, 57(12):1198 – 1210, 2009.

[16] K. Konolige, J. Gutmann, and B. Limketkai. Distributed map-making. In *Workshop on Reasoning with Uncertainty in Robotics, Int. Joint Conf. on Artificial Intelligence*, Acapulco, Mexico, August 2003.

[17] H. S. Lee and K. M. Lee. Multi-robot SLAM using ceiling vision. In *IEEE/RSJ Int. Conf. on Intelligent Robots and Systems*, pages 912–917, St. Louis, USA, October 2009.

[18] J. Neira and J. D. Tardós. Data association in stochastic mapping using the joint compatibility test. *IEEE Transactions on Robotics and Automation*, 17(6):890–897, 2001.

[19] C. H. Papadimitriou and K. Steiglitz. *Combinatorial Optimization: Algorithms and Complexity*, chapter 6.1 The Max-Flow, Min-Cut Theorem, pages 120–128. Dover, 1998.

[20] M. Pfingsthorn, B. Slamet, and A. Visser. A scalable hybrid multi-robot SLAM method for highly detailed maps. In U. Visser, F. Ribeiro, T. Ohashi, and F. Dellaert, editors, *RoboCup 2007: Robot Soccer World Cup XI*, volume 5001 of *Lecture Notes in Artificial Intelligence*, pages 457–464. Springer Verlag, Berlin Heidelberg New York, 2008.

[21] S. Thrun and Y. Liu. Multi-robot SLAM with sparse extended information filers. In *Int. Symposium of Robotics Research*, pages 254–266, Sienna, Italy, October 2003.

[22] S. B. Williams and H. Durrant-Whyte. Towards multi-vehicle simultaneous localisation and mapping. In *IEEE Int. Conf. on Robotics and Automation*, pages 2743–2748, Washington, DC, USA, May 2002.

[23] X. S. Zhou and S. I. Roumeliotis. Multi-robot SLAM with unknown initial correspondence: The robot rendezvous case. In *IEEE/RSJ Int. Conf. on Intelligent Robots and Systems*, pages 1785–1792, Beijing, China, October 2006.

Singularity-invariant Leg Rearrangements in Doubly-planar Stewart-Gough Platforms

Júlia Borràs, Federico Thomas, and Carme Torras
Institut de Robòtica i Informàtica Industrial (CSIC-UPC), Barcelona, Spain
E-mails: {jborras, fthomas, ctorras}@iri.upc.edu

Abstract— **In general, rearranging the legs of a Stewart-Gough platform, *i.e.*, changing the locations of its leg attachments, modifies the platform singularity locus in a rather unexpected way. Nevertheless, some leg rearrangements have been recently found to leave singularities invariant but, unfortunately, these rearrangements are only valid for Stewart-Gough platforms containing rigid components. In this work, the authors go a step further presenting singularity-invariant leg rearrangements that can be applied to any Stewart-Gough platform whose base and platform attachments are coplanar. The practical consequences of the presented theoretical results are illustrated with several examples including well-known architectures.**

I. Introduction

Even when there is no known solution to a given mathematical problem, it is always possible to try to find the set of transformations to the problem that leave its solution invariant. Although this does not solve the problem itself, it provides a lot of insight into its nature. This way of thinking is at the root of the development of Group Theory and it is the one applied herein for the characterization of the singularity loci of Stewart-Gough platforms.

The Stewart-Gough platform is defined as a 6-DoF parallel mechanism with six identical S$\underline{P}$S legs [14], [6]. It triggered the research on parallel manipulators, and it has remained one of the most widely studied because, despite its geometric simplicity, its analysis translates into challenging mathematical problems. One important part of this analysis corresponds to the characterization of its singularities.

The geometric and topological characterization of the singularity locus of a given Stewart-Gough platform in its six-dimensional configuration space is, in general, a huge task which has only been completely solved for some specializations —*i.e.*, designs in which some spherical joints in the platform, the base, or both, coalesce to form multiple spherical joints [2], [1]. In this context, it seems reasonable to find leg rearrangements in a given Stewart-Gough platform that leave its singularity locus invariant for two main reasons:

(a) If the singularity locus of the platform at hand has already been characterized, it could be interesting to modify the location of its legs to optimize some other platform characteristics without altering such locus.

(b) If the singularity locus of the analyzed platform has not been characterized yet, it could be of interest to simplify the platform's geometry by changing the location of its legs, thus easing the task of obtaining this characterization.

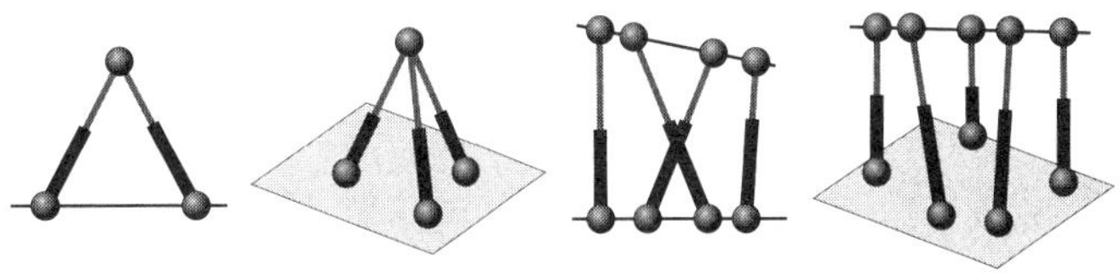

Fig. 1. The four possible rigid components involving linear geometric elements in Stewart-Gough platforms.

Let us suppose that we want to apply a singularity-invariant leg rearrangement limited to a subset of legs. Clearly, this is only possible if this subset of legs defines a rigid subassembly. Kong and Gosselin refer to these subassemblies as components [12]. The simplest component arises when two legs share an attachment. The result is called the Point-Line component. Similarly, the three other components involving linear geometric entities (points, lines and planes) are the Point-Plane, Line-Line and Line-Plane components (Fig.1). The singularity-invariant leg rearrangements for each of these four components have already been fully characterized [4], [3], [5]. In this paper, the authors go a step further presenting the rules for the leg rearrangements that can be applied to any Stewart-Gough platform whose base and platform attachments are coplanar (*i.e.*, the equivalent to a Plane-Plane component).

We will show that, for a leg rearrangement to be singularity-invariant, it is necessary and sufficient that the linear actuators' velocities, before and after the rearrangement, are linearly related. By integrating this differential condition, the above statement can be reformulated as follows: a leg rearrangement is singularity-invariant if the squared leg lengths, before and after the rearrangement, are affinely related. It is important to realize that, if this condition is satisfied, a one-to-one correspondence between the elements of the platform forward kinematics solution sets, before and after the rearrangement, arises. Actually, the invariance in the singularities and the assembly modes of a parallel platform are two faces of the same coin. These ideas are closely related to those that made possible the development of kinematic substitutions [8]. They are general in the sense that they can be applied to any kind of mechanism, not only parallel platforms. We will also show that their application to well-studied platforms leads to interesting new results. For example, we will see that it is not necessary that a platform has collinear attachments to behave like a Griffis-Duffy type II manipulator.

This paper is organized as follows. In Section II, a necessary and sufficient condition that must be satisfied by any singularity-invariant leg rearrangement in a Stewart-Gough platform is presented. Then, the challenge becomes that of finding the geometric transformations that satisfy this condition. This is discussed in Sections III and IV for any doubly-planar Stewart-Gough platform. Section V presents some cases that exemplify the potentialities of the obtained transformation. Finally, Section VI summarizes the main results.

II. General condition for a leg rearrangement to be singularity-invariant

In general, if we change the location of the leg attachments in a Stewart-Gough platform, its singularity locus is modified. However, it is shown below that the singularity locus remains invariant if, and only if, the squared lengths of the legs in their new location can be expressed in terms of those in the original location through an affine relation (for any arbitrary pose of the platform with respect to the base).

Let us consider a general Stewart-Gough platform, *i.e.*, a 6-DoF parallel mechanism with six identical S$\underline{P}$S legs [14], [6]. For this kind of platform, the linear actuators' velocities, $\dot{l}_1, \dot{l}_2, \ldots, \dot{l}_6$, can be expressed in terms of the platform velocity vector $(\mathbf{v}, \mathbf{\Omega})$ as follows:

$$\mathrm{diag}(l_1,\ldots,l_6)\begin{pmatrix}\dot{l}_1\\ \dot{l}_2\\ \vdots\\ \dot{l}_6\end{pmatrix} = \mathbf{J}\begin{pmatrix}\mathbf{v}\\ \mathbf{\Omega}\end{pmatrix}, \quad (1)$$

where $\mathbf{J}$ is the matrix of normalized Plücker coordinates of the six leg lines [14]. The parallel singularities of the platform are those configurations in which $\det(\mathbf{J}) = 0$ [13].

Now, let us change the location of the leg attachments so that the lengths of the legs in their new locations, say $d_1, d_2, \ldots, d_6$, are related to those of the original legs, $l_1, l_2, \ldots, l_6$, through the relation:

$$\begin{pmatrix}d_1^2\\ d_2^2\\ \vdots\\ d_6^2\end{pmatrix} = \mathbf{A}\begin{pmatrix}l_1^2\\ l_2^2\\ \vdots\\ l_6^2\end{pmatrix} + \mathbf{b}, \quad (2)$$

where $\mathbf{A}$ and $\mathbf{b}$ are a constant matrix and a constant vector, respectively. Differentiating this equation with respect to time and substituting (1) in the result, we get

$$\mathrm{diag}(d_1,\ldots,d_6)\begin{pmatrix}\dot{d}_1\\ \dot{d}_2\\ \vdots\\ \dot{d}_6\end{pmatrix} = \mathbf{AJ}\begin{pmatrix}\mathbf{v}\\ \mathbf{\Omega}\end{pmatrix}. \quad (3)$$

Then, the singularities of the platform after the leg rearrangement leading to (2) are those configurations in which $\det(\mathbf{AJ}) = \det(\mathbf{A})\det(\mathbf{J}) = 0$. If $\det(\mathbf{A}) \neq 0$, the leg rearrangement is said to be *singularity-invariant*. If $\det(\mathbf{A}) = 0$, the rearrangement introduces an *architectural singularity*, *i.e.*, the platform is always in a singularity irrespective of its leg lengths.

Since lengths are assumed to be positive magnitudes, equation (2) defines a one-to-one relationship between leg lengths before and after a singularity-invariant leg rearrangement. As a consequence, this kind of transformations leaves not only the singularities of the platform unaltered, but also the nature and number of its assembly modes.

III. Leg rearrangements

It has been shown in the preceding section that any singularity-invariant leg rearrangement results in an affine relation between the leg lengths before and after the rearrangement. In this section this idea is exploited to define the singularity-invariant conditions for the doubly-planar general Stewart-Gough platform.

In a doubly-planar Stewart-Gough platform, the i-th leg joins a base attachment with coordinates $\mathbf{a}_i = (x_i, y_i, 0)^T$ with a platform attachment whose local coordinates are $\widehat{\mathbf{b}}_i = (z_i, t_i, 0)^T$. Given the position $\mathbf{p} = (p_x, p_y, p_z)^T$ and the rotation

$$\mathbf{R} = (\mathbf{i}, \mathbf{j}, \mathbf{k}) = \begin{pmatrix} i_x & j_x & k_x\\ i_y & j_y & k_y\\ i_z & j_z & k_z\end{pmatrix}, \quad (4)$$

the coordinates of the platform attachments in the base reference frame are $\mathbf{b}_i = \mathbf{p} + \mathbf{R}(z_i, t_i, 0)^T$, for $i = 1\ldots, 6$.

Let l_i be the length of the original i-th leg, then $l_i^2 = \|\mathbf{b}_i - \mathbf{a}_i\|^2$, for $i = 1, \ldots, 6$.

We define a leg rearrangement of a single leg, *i.e.*, the substitution of any of the legs by another one going from the base attachment located at $\mathbf{a} = (x, y, 0)^T$ to the platform attachment at $\mathbf{b} = \mathbf{p} + \mathbf{R}(z, t, 0)^T$. We will call it *leg rearrangement towards* (x, y, z, t). Now we proceed to compute the length of that newly introduced leg $d^2 = \|\mathbf{b} - \mathbf{a}\|^2$.

Subtracting from the expressions for d^2 and $l_i^2, i = 1, \ldots, 6$, the equations $i_x^2 + i_y^2 + i_z^2 = 1$ and $j_x^2 + j_y^2 + j_z^2 = 1$, and using the relation $\mathbf{i} \cdot \mathbf{j} = i_x j_x + i_y j_y + i_z j_z = 0$, quadratic terms in the rotation variables cancel out, yielding

$$-\frac{1}{2}(p_x^2 + p_y^2 + p_z^2) - z_i u - t_i v + x_i p_x + y_i p_y + x_i z_i i_x + y_i z_i i_y + x_i t_i j_x + y_i t_i j_y - k_i = 0 \quad (5)$$

$$-\frac{1}{2}(p_x^2 + p_y^2 + p_z^2) - zu - tv + xp_x + yp_y + xzi_x + yzi_y + xtj_x + ytj_y - k + d^2/2 = 0 \quad (6)$$

where $u = \mathbf{p} \cdot \mathbf{i}$, $v = \mathbf{p} \cdot \mathbf{j}$, and the constant factors $k_i = \frac{1}{2}(x_i^2 + y_i^2 + z_i^2 + t_i^2 - l_i^2)$ and $k = \frac{1}{2}(x^2 + y^2 + z^2 + t^2)$.

If we subtract the first equation from the others, quadratic terms in p_x, p_y and p_z cancel too, yielding six linear equations in the 9 unknowns p_x, p_y, u, v, i_x, i_y , j_x, j_y and d^2 [equation (7)]. Note that k_i depends on l_i.

From the matrix $\mathbf{Q}$ of the system (7), let us call $\mathbf{Q}_{ijk}$ the square matrix obtained from $\mathbf{Q}$ after deleting columns i, j and k, and Q_{ijk} its determinant. The system can be solved if we

$$
\underbrace{\begin{pmatrix}
z_1-z_2 & t_1-t_2 & x_2-x_1 & y_2-y_1 & x_2z_2-x_1z_1 & y_2z_2-y_1z_1 & x_2t_2-x_1t_1 & y_2t_2-y_1t_1 & 0\\
z_1-z_3 & t_1-t_3 & x_3-x_1 & y_3-y_1 & x_3z_3-x_1z_1 & y_3z_3-y_1z_1 & x_3t_3-x_1t_1 & y_3t_3-y_1t_1 & 0\\
z_1-z_4 & t_1-t_4 & x_4-x_1 & y_4-y_1 & x_4z_4-x_1z_1 & y_4z_4-y_1z_1 & x_4t_4-x_1t_1 & y_4t_4-y_1t_1 & 0\\
z_1-z_5 & t_1-t_5 & x_5-x_1 & y_5-y_1 & x_5z_5-x_1z_1 & y_5z_5-y_1z_1 & x_5t_5-x_1t_1 & y_5t_5-y_1t_1 & 0\\
z_1-z_6 & t_1-t_6 & x_6-x_1 & y_6-y_1 & x_6z_6-x_1z_1 & y_6z_6-y_1z_1 & x_6t_6-x_1t_1 & y_6t_6-y_1t_1 & 0\\
z_1-z & t_1-t & x-x_1 & y-y_1 & xz-x_1z_1 & yz-y_1z_1 & xt-x_1t_1 & yt-y_1t_1 & \frac{1}{2}
\end{pmatrix}}_{\mathbf{Q}}
\begin{pmatrix} u\\ v\\ p_x\\ p_y\\ i_x\\ i_y\\ j_x\\ j_y\\ d^2 \end{pmatrix}
=
\begin{pmatrix} k_2-k_1\\ k_3-k_1\\ k_4-k_1\\ k_5-k_1\\ k_6-k_1\\ k-k_1 \end{pmatrix}
\tag{7}
$$

take three unknowns as parameters, for example u, v, p_x. The resulting linear system is:

$$
\mathbf{Q}_{123}\begin{pmatrix} p_y\\ i_x\\ i_y\\ j_x\\ j_y\\ d^2 \end{pmatrix}
=
\begin{pmatrix}
(z_2-z_1)u+(t_2-t_1)v-(x_2-x_1)p_x+k_2-k_1\\
(z_3-z_1)u+(t_3-t_1)v-(x_3-x_1)p_x+k_3-k_1\\
(z_4-z_1)u+(t_4-t_1)v-(x_4-x_1)p_x+k_4-k_1\\
(z_5-z_1)u+(t_5-t_1)v-(x_5-x_1)p_x+k_5-k_1\\
(z_6-z_1)u+(t_6-t_1)v-(x_6-x_1)p_x+k_6-k_1\\
(z-z_1)u+(t-t_1)v-(x-x_1)p_x+k-k_1
\end{pmatrix}.
$$

Solving the system for d^2 using Crammer's rule and then applying multi-linear properties of determinants to split the determinant of the resulting matrix into 4 determinants yields

$$
d^2=\frac{Q_{239}u+Q_{139}v+Q_{129}p_x+Q^*_{123}}{\frac{1}{2}Q_{123}}, \tag{8}
$$

where Q^*_{123} is the determinant of $\mathbf{Q}_{123}$ except for the last column that contains the elements k_i-k_1 for $i=2,\ldots,6$ and $k-k_1$. As a result, if we impose $Q_{239}=Q_{139}=Q_{129}=0$, then equation (8) becomes affine in $l_1^2,\ldots,l_6^2$. Indeed, expanding Q^*_{123} leads to an expression of the form

$$
d^2=c_1l_1^2+c_2l_2^2+c_3l_3^2+c_4l_4^2+c_5l_5^2+c_6l_6^2+c_0,
$$

where all coefficients are known constants. Then, following Section II, any leg rearrangement satisfying $Q_{239}=Q_{139}=Q_{129}=0$ leaves singularities invariant.

In other words, if we substitute any leg by a new leg with base attachment located at $\mathbf{a}=(x,y,0)^T$ and platform attachment at $\mathbf{b}=\mathbf{p}+\mathbf{R}(z,t,0)^T$, the singularities will remain invariant as long as (x,y,z,t) satisfies the system $Q_{239}=Q_{139}=Q_{129}=0$, where Q_{ijk} can be simplified into a 7×7 determinant using simple row/column operations, yielding the following system:

$$
Q_{239}=\begin{vmatrix}
y_1 & x_1z_1 & y_1z_1 & x_1t_1 & y_1t_1 & z_1 & 1\\
y_2 & x_2z_2 & y_2z_2 & x_2t_2 & y_2t_2 & z_2 & 1\\
y_3 & x_3z_3 & y_3z_3 & x_3t_3 & y_3t_3 & z_3 & 1\\
y_4 & x_4z_4 & y_4z_4 & x_4t_4 & y_4t_4 & z_4 & 1\\
y_5 & x_5z_5 & y_5z_5 & x_5t_5 & y_5t_5 & z_5 & 1\\
y_6 & x_6z_6 & y_6z_6 & x_6t_6 & y_6t_6 & z_6 & 1\\
y & xz & yz & xt & yt & z & 1
\end{vmatrix}=0, \tag{9}
$$

$$
Q_{139}=\begin{vmatrix}
y_1 & x_1z_1 & y_1z_1 & x_1t_1 & y_1t_1 & t_1 & 1\\
y_2 & x_2z_2 & y_2z_2 & x_2t_2 & y_2t_2 & t_2 & 1\\
y_3 & x_3z_3 & y_3z_3 & x_3t_3 & y_3t_3 & t_3 & 1\\
y_4 & x_4z_4 & y_4z_4 & x_4t_4 & y_4t_4 & t_4 & 1\\
y_5 & x_5z_5 & y_5z_5 & x_5t_5 & y_5t_5 & t_5 & 1\\
y_6 & x_6z_6 & y_6z_6 & x_6t_6 & y_6t_6 & t_6 & 1\\
y & xz & yz & xt & yt & t & 1
\end{vmatrix}=0, \tag{10}
$$

$$
Q_{129}=\begin{vmatrix}
y_1 & x_1z_1 & y_1z_1 & x_1t_1 & y_1t_1 & -x_1 & 1\\
y_2 & x_2z_2 & y_2z_2 & x_2t_2 & y_2t_2 & -x_2 & 1\\
y_3 & x_3z_3 & y_3z_3 & x_3t_3 & y_3t_3 & -x_3 & 1\\
y_4 & x_4z_4 & y_4z_4 & x_4t_4 & y_4t_4 & -x_4 & 1\\
y_5 & x_5z_5 & y_5z_5 & x_5t_5 & y_5t_5 & -x_5 & 1\\
y_6 & x_6z_6 & y_6z_6 & x_6t_6 & y_6t_6 & -x_6 & 1\\
y & xz & yz & xt & yt & -x & 1
\end{vmatrix}=0, \tag{11}
$$

A. Generalizing and simplifying the condition

The above reasoning fails if $Q_{123}=0$ but, for a non-architecturally singular manipulator, a 6×6 matrix $\mathbf{Q}_{ijk}$ with non-zero determinant can always be found (otherwise, the 6 leg length equations would be linearly dependent). However, this may change the expression of the singularity-invariant leg rearrangement condition in equations (9)-(11).

To avoid such ambiguity, we can reformulate the condition in terms of rank deficiency of the matrix $\mathbf{Q}_9$ (that is, the matrix $\mathbf{Q}$ in equation (7) without the last column). The 5 first rows of $\mathbf{Q}_9$ are full rank for any non-architecturally singular manipulator. Furthermore, $\mathbf{Q}_9$ is rank defective if, and only if, all its submatrices have null determinant. However, it is only necessary to check 3 of its submatrix determinants. Thus, the condition in equations (9)-(11) is equivalent to the rank deficiency of $\mathbf{Q}_9$. The advantage of this formulation is that any set of 3 submatrices could be used instead of the three determinants in (9)-(11).

To simplify the notation, we consider the following simpler matrix:

$$
\mathbf{P}=\begin{pmatrix}
-z_1 & -t_1 & x_1 & y_1 & x_1z_1 & y_1z_1 & x_1t_1 & y_1t_1 & 1\\
-z_2 & -t_2 & x_2 & y_2 & x_2z_2 & y_2z_2 & x_2t_2 & y_2t_2 & 1\\
-z_3 & -t_3 & x_3 & y_3 & x_3z_3 & y_3z_3 & x_3t_3 & y_3t_3 & 1\\
-z_4 & -t_4 & x_4 & y_4 & x_4z_4 & y_4z_4 & x_4t_4 & y_4t_4 & 1\\
-z_5 & -t_5 & x_5 & y_5 & x_5z_5 & y_5z_5 & x_5t_5 & y_5t_5 & 1\\
-z_6 & -t_6 & x_6 & y_6 & x_6z_6 & y_6z_6 & x_6t_6 & y_6t_6 & 1\\
-z & -t & x & y & xz & yz & xt & yt & 1
\end{pmatrix}. \tag{12}
$$

Let us denote by P_{ij} the determinant of the submatrix obtained from $\mathbf{P}$ after deleting columns i and j, and P_{ijk} the determinant of the submatrix formed by the first 6 rows of $\mathbf{P}$ after deleting columns i, j and k.

Note that $P_{ij}=Q_{ij9}$ for $i,j\neq 9$ and $P_{ijk}=\frac{1}{2}Q_{ijk}$ for $k\neq 9$. Using these relations it can be proved that $\mathbf{Q}_9$ is rank defective if, and only if, $\mathbf{P}$ is also rank defective. Thus, a much simpler condition can now be stated: a leg rearrangement towards (x, y, z, t) leaves singularities invariant as long as the matrix $\mathbf{P}$ is rank defective.

One practical methodology to check rank deficiency is to apply Gaussian elimination on $\mathbf{P}$. The last row of the resulting matrix has 3 nonzero terms dependent on x, y, z and t. The corresponding 3 equations are equivalent to the system $\{(9), (10), (11)\}$. Different equations arise depending on the order of the columns. For example, Gaussian elimination on matrix $\mathbf{P}$ as it appears in equation (12) leads to a matrix whose last row is

$$\frac{1}{P_{789}}\begin{pmatrix} 0 & 0 & 0 & 0 & 0 & 0 & P_{89} & P_{79} & P_{78} \end{pmatrix}.$$

Then, as long as $P_{789} \neq 0$, the singularity-invariant leg rearrangements are defined by the system

$$\{P_{89} = 0, P_{79} = 0, P_{78} = 0\}. \quad (13)$$

Alternatively, if the matrix $\mathbf{P}$ columns are sorted as $[y, xz, yz, xt, yt, 1, -z, -t, x]$, then the corresponding system is $\{(9), (10), (11)\}$ and P_{123} should be non-zero.

IV. Geometric Interpretation of the Condition

Note that any equation consisting of a submatrix determinant P_{ij} equated to zero will be bilinear in the unknowns, but with different monomials. Let us consider the system of equations (13), which after cofactor expansion, leads to

$$\left.\begin{aligned} -P_{891}z + P_{892}t + P_{893}x - P_{894}y + P_{895}xz \\ -P_{896}yz + P_{897}xt = 0 \\ -P_{791}z + P_{792}t + P_{793}x - P_{794}y + P_{795}xz \\ -P_{796}yz + P_{798}yt = 0 \\ -P_{781}z + P_{782}t + P_{783}x - P_{784}y + P_{785}xz \\ -P_{786}yz + P_{789} = 0 \end{aligned}\right\} \quad (14)$$

As the system is linear both in (x, y) and in (z, t), it can be rewritten in matrix form as

$$\mathbf{S}_b \begin{pmatrix} z \\ t \\ 1 \end{pmatrix} = \begin{pmatrix} 0 \\ 0 \\ 0 \end{pmatrix}, \quad (15)$$

where $\mathbf{S}_b$ is the matrix

$$\begin{pmatrix} P_{895}x - P_{896}y - P_{891} & P_{892} + P_{897}x & P_{893}x - P_{894}y \\ P_{795}x - P_{796}y - P_{791} & P_{792} + P_{798}y & P_{793}x - P_{794}y \\ P_{785}x - P_{786}y - P_{781} & P_{782} & P_{783}x - P_{784}y + P_{789} \end{pmatrix},$$

that only depends on x and y (the b refers to $base$, as x and y are the coordinates of the base plane). The other way round, the system can also be written as

$$\mathbf{S}_p \begin{pmatrix} x \\ y \\ 1 \end{pmatrix} = \begin{pmatrix} 0 \\ 0 \\ 0 \end{pmatrix}, \quad (16)$$

where now matrix $\mathbf{S}_p$ is

$$\begin{pmatrix} P_{893} + P_{895}z + P_{897}t & -P_{894} - P_{896}z & P_{892}t - P_{891}z \\ P_{793} + P_{795}z & P_{798}t - P_{794} - P_{796}z & P_{792}t - P_{791}z \\ P_{783} + P_{785}z & -P_{784} - P_{786}z & P_{782}t - P_{781}z + P_{789} \end{pmatrix},$$

that only depends on z and t (and p refers to $platform$, as z and t are the coordinates of the platform plane).

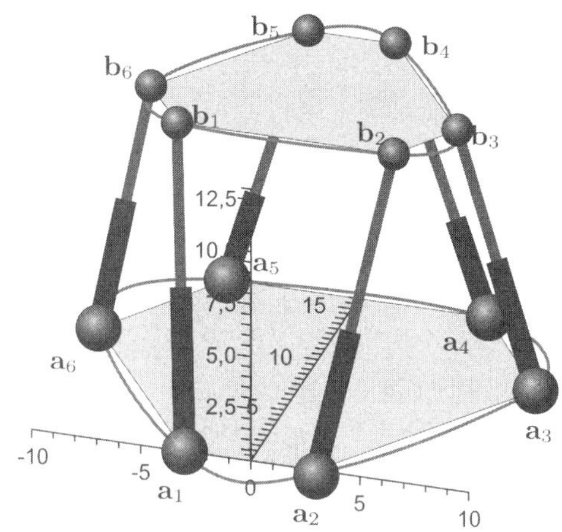

Fig. 2. Scheme of the platform described in Table I.

From equation (15) it is clear that the system has a solution on (z, t) only for those (x, y) that satisfy $\det(\mathbf{S}_b) = 0$, and this solution is unique (assuming that the matrix $\mathbf{S}_b$ has rank 2). In the same way, there exists a solution on (x, y) only for those (z, t) that make $\det(\mathbf{S}_p) = 0$. Both determinants define cubic curves on the base and platform planes, respectively. In other words, the system (14) defines a one-to-one correspondence between points on the two cubic curves. However, the correspondence may be not one-to-one for singular points on the cubics, as will be seen in the example of Section V-B.

Depending on the geometric placement of the attachments, these curves can be generic curves of degree 3, or a line and a conic, or even 3 lines crossing 2 by 2. This will be exemplified in the following section.

V. Examples

A. Classic Stewart-Gough platform

In [9] Husty *et al.* analyzed the classic Stewart-Gough platform, searching where additional legs could be placed without changing the forward kinematics solution, to obtain a redundant manipulator. The same example is analyzed here. The local coordinates of the attachments are listed in Table I.

TABLE I

Coordinates of the attachments $\mathbf{a}_i = (x_i, y_i, 0)$ and $\mathbf{b}_i = \mathbf{p} + \mathbf{R}(z_i, t_i, 0)^T$ for the analyzed robot

i	x_i	y_i	z_i	t_i
1	−3	0	−5	0
2	3	0	5	0
3	10	10	7	3
4	6	16	2	10
5	−6	16	−2	10
6	−10	10	−7	3

After substituting the corresponding numerical values, the

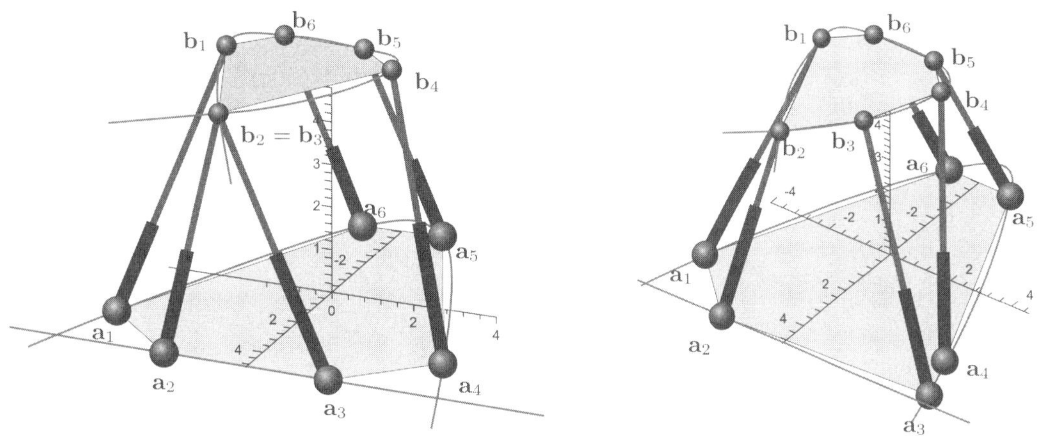

Fig. 3. Scheme of the platform described in Table II (left), and the equivalent platform after moving the 3rd leg (right).

system of equations (14) results in:

$$\left.\begin{aligned} 2430z - 4050x + 255yz + 188xt &= 0 \\ -280t + 45y + 13yt &= 0 \\ -70t + 43y - 4xz + 60 &= 0 \end{aligned}\right\}$$

and, thus, matrices $\mathbf{S}_b$ and $\mathbf{S}_p$ are:

$$\mathbf{S}_b = \begin{pmatrix} 2430 + 255y & 188x & -4050x \\ 0 & 13y - 280 & 45y \\ -4x & -70 & 60 + 43y \end{pmatrix},$$

$$\mathbf{S}_p = \begin{pmatrix} 188t - 4050 & 255z & 2430z \\ 0 & 13t + 45 & -280t \\ -4z & 43 & 60 - 70t \end{pmatrix},$$

whose determinants equated to zero give the two cubic curves

$$\begin{aligned} &-16296x^2y + 9503y^3 + 302400x^2 \\ &\quad - 47312y^2 - 1599420y - 2721600 = 0, \\ &20598z^2t - 8554t^3 + 21870z^2 \\ &\quad + 275173t^2 - 1932795t - 546750 = 0; \end{aligned} \tag{17}$$

plotted in Fig. 4. The cubic in the base coincides with the one appearing in [9], whereas the cubic in the platform is not given explicitly there.

By definition, all attachment coordinates are solutions of the system, and therefore belong to the curves, as shown in Fig.4.

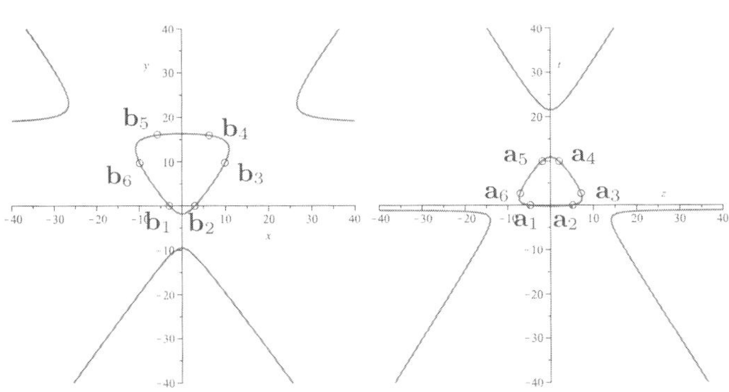

Fig. 4. The base and the platform curves defined in (17).

In [9], the authors propose to add additional legs to obtain redundant manipulators. Instead, here we propose to substitute any leg by another leg satisfying the one-to-one correspondence between the base and platform cubics defined by (17). The singularity locus will remain unchanged, and other performance indices can be improved, such as stiffness, or maneuverability, or the workspace can be enlarged by reducing the risk of leg collisions.

B. Degenerate cubic curves

Interesting cases appear when one or both of the curves are degenerate. Consider the example given in Table II, where two of the attachments on the platform are made coincident, $\mathbf{b}_2 = \mathbf{b}_3$ [Fig. 3-(left)]. The two legs sharing an endpoint form a Point-Line component, and it was proved in [4] that the base attachments $\mathbf{a}_2$ and $\mathbf{a}_3$ can be rearranged on any point on the line $\mathbf{a}_2\mathbf{a}_3$ without modifying the singularity locus.

TABLE II

COORDINATES OF THE ATTACHMENTS $\mathbf{a}_i = (x_i, y_i, 0)$ AND $\mathbf{b}_i = \mathbf{p} + \mathbf{R}(z_i, t_i, 0)^T$ FOR THE ANALYZED ROBOT

i	x_i	y_i	z_i	t_i
1	3	−4	−2	−2
2	5	−2	2	−1/2
3	5	2	2	−1/2
4	3	4	−2	2
5	−4	1	−3	1
6	−4	−1	−3	−1

After applying Gaussian elimination to the matrix $\mathbf{P}$ given in equation (12) corresponding to this example, the following equations are obtained:

$$\left.\begin{aligned} &372z + 18988t + 1302x - 5656y + 527xz \\ &\qquad +2828yz + 1212xt = 0 \\ &5172z + 808t - 2502x + 404y + 257xz \\ &\qquad +404yz + 2424yt = 0 \\ &74z - 44x - 13xz + 202 = 0 \end{aligned}\right\} \tag{18}$$

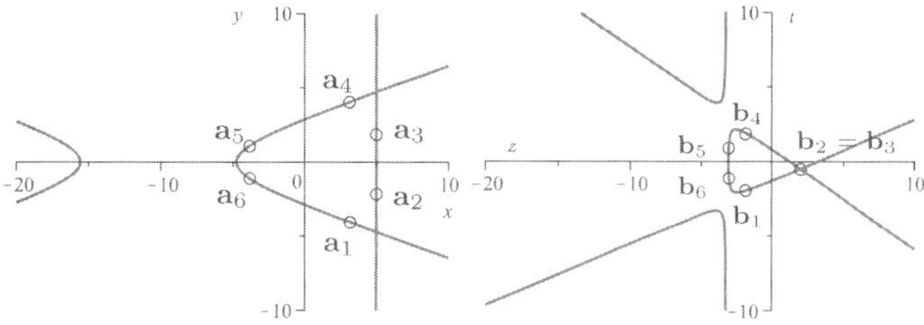

Fig. 5. The base and the platform curves defined in (19).

and the corresponding cubic curves are defined by

$$
\begin{aligned}
&1713768(x-5)\\
&\qquad(31x^2-280y^2+631x+2308)=0,\\
&1713768(-132z^3+124z^2t+476zt^2+191z^2\\
&\qquad+620zt+1528t^2+1259z+744t-1606)=0.
\end{aligned}
\tag{19}
$$

In other words, the cubic curve in the base factorizes into a conic (an hyperbola) and a line, while the platform curve remains a cubic, but with a singular point (called node) on the vertex of the Point-Line component, $\mathbf{b}_2=(2,-1/2)$. Fig. 5 shows the plot of these two curves and the corresponding location of the attachments.

The correspondence between the base line and the platform cubic curve can be derived by solving system (18) for any point on the base line $\mathbf{a}_2\mathbf{a}_3$ (*i.e.*, $x=5$), leading to

$$
\left.\begin{aligned}
3007z+25048t+6510-5656y+2828yz&=0\\
+6457z+808t-12510+404y+404yz+2424yt&=0\\
9z-18&=0
\end{aligned}\right\}
$$

From the last equation, $z=2$. Substituting this value into the first two equations and factorizing the result yields

$$
\begin{aligned}
12524(1+2t)&=0,\\
404(1+3y)(1+2t)&=0.
\end{aligned}
$$

The solution of the system is $\{x=5, y=y, z=2, t=-1/2\}$, that is, any point on the line $x=5$ corresponds to the vertex $\mathbf{b}_2=\mathbf{b}_3$ in accordance with previous results [4].

For the rest of the points on the cubic curve, the correspondence can be written in terms of a single parameter z. Given a point on the platform cubic

$$
\left(z, \frac{-31z^2-155z-186\pm\sqrt{\Delta}}{2(119z+382)}\right),
$$

the corresponding point on the base hyperbola is

$$
\left(\frac{2(37z+101)}{44+13z},\right.
$$
$$
\left.\frac{3(z-2)(16669z^2+103981z+162022)\mp(26z+88)\sqrt{\Delta}}{4(2-z)(44+13z)^2\pm 2(39z+132)\sqrt{\Delta}}\right),
\tag{20}
$$

where the discriminant $\Delta=(16669z^2+103981z+162022)(z-2)^2$ determines whether points are real or complex. Real points on the platform always correspond to real points on the base, and vice versa. Observe that, on the singular point $z=2$, (20) is undefined. However, terms $(z-2)$ can be simplified and then the resulting point gives the intersection between the line and the hyperbola. In other words, this parametrization represents the one-to-one correspondence between points on the platform cubic and the base hyperbola, except for the singular platform point $(2,-1/2)$, a double point that corresponds to two points on the hyperbola (the two intersections of the line with the conic).

To avoid multiple spherical joints, here the Point-Line component can be split by substituting any of its legs by another leg going from the conic to the base cubic. For example, take the point on the platform cubic given by $z=0$ and $t=\frac{-93+\sqrt{162022}}{382}$, and solve system (18) after evaluating it on this point, or equivalently, evaluate expression (20) for $z=0$. The result is:

$$
x=\frac{101}{22} \text{ and } y=\frac{243033-44\sqrt{162022}}{-3872+132\sqrt{162022}}.
$$

Hence, we can substitute the 3rd leg by a new leg going from the base point $(\frac{101}{22}, \frac{243033-44\sqrt{162022}}{-3872+132\sqrt{162022}}, 0)$ to the platform attachment with local coordinates $(0, \frac{-93+\sqrt{162022}}{382}, 0)$ and the resulting Jacobian is the same as before, multiplied by a constant $\frac{15990+93\sqrt{162022}}{67232}=0.794$, and with no coincident attachments (Fig. 3-right).

C. *Griffis-Duffy platforms*

In 1993, Griffis and Duffy patented two manipulators named thereafter Griffis-Duffy type I and II platforms [7]. Both platforms have their attachments distributed on triangles, three attachments on the vertices and three on the midpoints of the edges. Type I platforms are formed by joining the attachments on the midpoints on the base to the vertices on the platform, and the vertices on the base to midpoints on the platform (Fig. 6-left). Type II join midpoints to midpoints and vertices to vertices (Fig. 7-left).

A type I Griffis-Duffy platform is shown to be singularity equivalent to the octahedral manipulator [4]. In [11], type II Griffis-Duffy manipulators are shown to be always non-architecturally singular.

Consider the two examples specified in Table III, where the same triangles define two manipulators of type I and type II, respectively.

TABLE III

COORDINATES OF THE ATTACHMENTS $\mathbf{a}_i=(x_i,y_i,0)$ AND $\mathbf{b}_i=\mathbf{p}+\mathbf{R}(z_i,t_i,0)^T$ FOR THE ANALYZED ROBOT

i	z_i	t_i
1	1	0
2	1/2	0
3	−1	0
4	−1/2	$\sqrt{3}/2$
5	0	$\sqrt{3}$
6	1/2	$\sqrt{3}/2$

i Type I	i Type II	x_i	y_i
2	1	2	0
3	2	2/3	0
4	3	−2	0
5	4	−2/3	$(4/3)\sqrt{3}$
6	5	0	$2\sqrt{3}$
1	6	1	$\sqrt{3}$

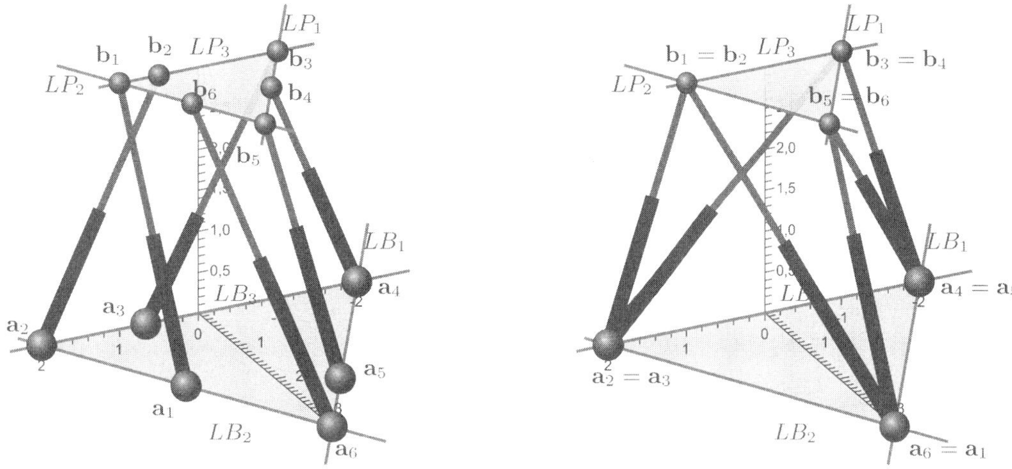

Fig. 6. Scheme of the platform described in Table III of a Griffis-Duffy type I platform (left), and its equivalent octahedral manipulator after applying a leg rearrangement (right).

The computation of the base and platform curves factorizes into the 3 same lines for both type I and type II platforms:

$$\underbrace{(\sqrt{3}z - t + \sqrt{3})}_{LP_1}\underbrace{(\sqrt{3}z + t - \sqrt{3})}_{LP_2}\underbrace{t}_{LP_3} = 0,$$

$$\underbrace{(-3x + \sqrt{3}y - 6)}_{LB_1}\underbrace{(3x + \sqrt{3}y - 6)}_{LB_2}\underbrace{y}_{LB_3} = 0;$$

but the system obtained by applying Gaussian elimination on matrix $\mathbf{P}$ results in different equations. The system corresponding to the type I platform is:

$$\left.\begin{aligned} 2t - y + yz + xt = 0 \\ (\sqrt{3}z + t - \sqrt{3})y = 0 \\ -2\sqrt{3}z + 4t + \sqrt{3}x - y + \sqrt{3}xz + 3yz - 2\sqrt{3} = 0 \end{aligned}\right\}$$

whereas for the type II platform is:

$$\left.\begin{aligned} 2t - y + xt - yz = 0 \\ 3\sqrt{3}y - 8\sqrt{3}t + \sqrt{3}yz + yt = 0 \\ 10\sqrt{3}z - 16t - 5\sqrt{3}x + 9y + \sqrt{3}xz + yz - 2\sqrt{3} = 0 \end{aligned}\right\} \quad (21)$$

The resolution of these systems gives correspondences between base and platform attachments that leave the singularities invariant.

For the type I platform, the correspondence is between points and lines (in accordance with results in [4]), that is, to each vertex of the base (platform) triangle corresponds a line on the platform (base) triangle, in the same way as in the preceding section the vertex of the Point-Line component was in correspondence with a line on the base. Thus, by moving the six midpoint attachments along their supporting lines, the manipulator can be rearranged into the equivalent octahedral manipulator depicted in Fig. 6-right (a result concordant with that in [4]).

On the other hand, for the type II platform, there is a one-to-one correspondence between points on line LP_i and LB_i, for $i = 1, 2, 3$. That is, the same geometrical elements determine the invariance of the singularity locus, but in the first case the correspondence is between points and lines and in the second case it is point-to-point between lines.

The legs of the type II manipulator can be rearranged following the correspondence $LB_i \leftrightarrow LP_i$. But some rearrangements must be avoided, for example, placing four legs in the same line-line correspondence leads to an architecturally singular manipulator (as it contains a Line-Line component in projective correspondence [4, 10]).

An interesting rearrangement consists in removing all collinearities from the type II manipulator. As a result, an equivalent platform such as that shown in Fig. 7-right is obtained.

To remove collinearities, all legs from vertex to vertex need to be rearranged. The 1st leg can be placed going from a point on LB_2 to the corresponding point on LP_2. In other words, take a point on the line LB_2, substitute the values on system (21) and the solution gives a point on the line LP_2:

$$\underbrace{\begin{array}{c} x = 1/2 \\ y = (3/2)\sqrt{3} \end{array}}_{\text{on } LB_2} \quad \begin{array}{c}\text{Substitute on (21)} \\ \longrightarrow \\ \text{and solve}\end{array} \quad \underbrace{\begin{array}{c} z = 1/4 \\ t = (3/4)\sqrt{3} \end{array}}_{\text{on } LP_2}$$

The same can be done to substitute the 3rd leg by a leg going from LB_3 to LP_3:

$$\underbrace{\begin{array}{c} x = -2/3 \\ y = 0 \end{array}}_{\text{on } LB_3} \quad \begin{array}{c}\text{Substitute on (21)} \\ \longrightarrow \\ \text{and solve}\end{array} \quad \underbrace{\begin{array}{c} z = -1/7 \\ t = 0 \end{array}}_{\text{on } LP_3}$$

and finally, the 5th leg is substituted by a leg going from a point on LB_1 to a point on LP_1

$$\underbrace{\begin{array}{c} x = -3/2 \\ y = (1/2)\sqrt{3} \end{array}}_{\text{on } LB_1} \quad \begin{array}{c}\text{Substitute on (21)} \\ \longrightarrow \\ \text{and solve}\end{array} \quad \underbrace{\begin{array}{c} z = -6/7 \\ t = (1/7)\sqrt{3} \end{array}}_{\text{on } LP_1}$$

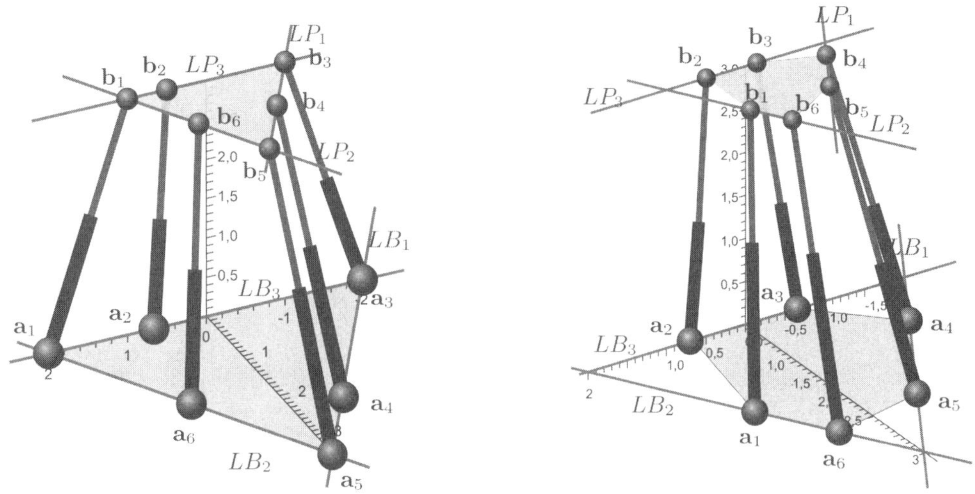

Fig. 7. Scheme of the platforms described in Table III of a Griffis-Duffy type II platform (left), and its equivalent platform after removing all collinearities (right).

The resulting manipulator depicted in Fig. 7-right is equivalent to the one in Fig.7-left, as regards to both its kinematics and its singularity locus.

In conclusion, we have seen that it is not necessary that a platform has collinear attachments to behave like a Griffis-Duffy type II manipulator, and what is more important, we have found a general doubly-planar Stewart-Gough platform equivalent to a Griffis-Duffy type II manipulator.

VI. Conclusions and Future Work

In this work we have started to analyze the singularity-invariant leg rearrangements that can be applied to Stewart-Gough platforms with coplanar attachments on the base and on the platform. It has been shown that the resulting attachments need to obey a one-to-one correspondence between two cubic curves, one on the base and another on the platform. The presented approach generalizes previous results.

Many promising possibilities open up for future research. There exist 15 different types of non-degenerate cubic curves, so we can classify all Stewart-Gough platforms according to the type of cubic curves they generate. Furthermore, many interesting cases can be studied that have degenerate cubic curves. In addition, efficient parameterizations of cubic curves available in literature would allow us to define the attachments using less parameters, thus simplifying the analysis.

On the other hand, we are working on the relation between the coefficients of the Jacobian polynomial and the minors of matrix $\mathbf{P}$ in equation (12), similarly as it was done in previous works for particular cases [3]. Based on the analysis of the $\mathbf{P}$ matrix, the number of assembly modes of the general Stewart-Gough platform can be lowered by imposing simple relations between the attachments, as it was done in [5].

The usefulness of the proposed method has been shown with three examples. The derived leg rearrangements permit reducing the number of multiple spherical joints as well as devising reconfigurable robots that maintain the same singularity locus at each reconfiguration state.

Acknowledgment

This work has been partially supported by the Generalitat de Catalunya through the VALTEC program, cofinanced by FEDER funds, and the Spanish Ministry of Science and Innovation, under the I+D project DPI2007-60858.

References

[1] M. Alberich-Carramiñana, M. Garolera, F. Thomas, and C. Torras, "Partially-flagged parallel manipulators: Singularity charting and avoidance," *IEEE Transactions on Robotics*, vol. 25, no. 4, pp. 771–784, 2009.

[2] M. Alberich-Carramiñana, F. Thomas, and C. Torras, "Flagged parallel manipulators," *IEEE Transactions on Robotics*, vol. 23, no. 5, pp. 1013–1023, 2007.

[3] J. Borràs and F. Thomas, "Kinematics of the line-plane subassembly in Stewart platforms," in *IEEE International Conference on Robotics and Automation*, 2009, pp. 4094–4099.

[4] J. Borràs, F. Thomas, and C. Torras, "On Δ-transforms," *IEEE Transactions on Robotics*, vol. 25, no. 6, pp. 1225–1236, 2009.

[5] ——, "A family of quadratically-solvable 5-UPS parallel robots," in *IEEE International Conference on Robotics and Automation*, 2010, pp. 4703–4708.

[6] B. Dasguptaa and T. Mruthyunjayab, "The Stewart platform manipulator: a review," *Mechanism and Machine Theory*, vol. 35, pp. 15–40, 2000.

[7] M. Griffis and J. Duffy, "Method and apparatus for controlling geometrically simple parallel mechanisms with distinctive connections," *US Patent 5,179,525*, 1993.

[8] K. Hunt, *Kinematic Geometry of Mechanisms*. Oxford University Press, 1978, p. 323.

[9] M. Husty, S. Mielczarek, and M. Hiller, "A redundant spatial Stewart-Gough platform with a maximal forward kinematics solution set," in *International Symposium on Advances in Robot Kinematics*, 2002, pp. 147–154.

[10] M. Husty and A. Karger, "Architecture singular parallel manipulators and their self-motions," in *International Symposium on Advances in Robot Kinematics*, 2000, pp. 355–364.

[11] ——, "Self-motions of Griffis-Duffy type parallel manipulators," in *IEEE International Conference on Robotics and Automation*, 2000, pp. 7–12.

[12] X. Kong and C. Gosselin, "Classification of 6-SPS parallel manipulators according to their components," in *Proc. of ASME Design Engineering Technical Conferences*, 2000, pp. DETC2000/MECH–14 105.

[13] O. Ma and J. Angeles, "Architecture singularities of platform manipulators," in *IEEE International Conference on Robotics and Automation*, vol. 2, 1991, pp. 1542–1547.

[14] J.-P. Merlet, *Parallel Robots*. Springer, 2000.

On the Kinematic Design of Exoskeletons and Their Fixations with a Human Member

Nathanaël Jarrassé
Guillaume Morel

University P. & M. Curie, ISIR (Institut des Systèmes Intelligents et de Robotique) CNRS - UMR 7222
4 place Jussieu, 75005 Paris - France
Emails : jarrasse@isir.fr, guillaume.morel@upmc.fr

***Abstract*— A crucial problem in developing robotic exoskeletons lies in the design of physical connexions between the device and the human limb it is connected to. Indeed, because in general the human limb kinematics and the exoskeleton kinematics differ, using an embedment at each connection point leads to hyperstaticity. Therefore, uncontrollable forces can appear at the interaction port. To cope with this problem, literature suggests to add passive mechanisms at the fixation points. However, empirical solutions proposed so far suffer from a lack of proper analysis and generality.**

In this paper, we study the general problem of connecting two similar kinematic chains through multiple passive mechanisms. We derive a constructive method that allows to determine all the possible repartitions of freed DoFs across the different fixation mechanisms. It also provides formal proofs of global isostaticity. Practical usefulness is illustrated through an example with conclusive experimental results.

I. Introduction

More and more exoskeletons are being designed by researchers for a growing number of applications, ranging from military applications [1] to rehabilitation [2].

For years, research has mainly focused on technological aspects (actuators, embedment, energy...) and followed a paradigm defined in [3]: *"an exoskeleton is an external structural mechanism with joints and links corresponding to those of the human body"* . In other words, designing the kinematics of an exoskeleton generally consists of trying to replicate the human limb kinematics. This brings a number of advantages: similarity of the workspaces, singularity avoidance [4], one-to-one mapping of joint force capabilities over the workspace. The major drawback of this paradigm is that, in fact, human kinematics is impossible to precisely replicate with a robot. Indeed two problems occur: morphology drastically varies between subjects and, for a given subject, the joints kinematics is very complex and cannot be imitated by conventional robot joints [5]. In fact, it is impossible to find any consensual model of the human kinematics in the biomechanics literature due to complex geometry of bones interacting surfaces. For example, different models are used for the shoulder-scapula-clavicle group [6].

Discrepancies between the two kinematic chains thus seem unavoidable. Because of the connexions between multiple loops, it generates kinematic compatibility problems. Indeed, when connecting two-by-two the links of two *kinematically similar* chains that are not perfectly identical, hyperstaticity occurs. This phenomenon leads, if rigid models are used, to the impossibility of moving and the appearance of non-controllable (possibly infinite) internal forces. In practice, though, rigidity is not infinite and mobility can be obtained thanks to deformations. When a robotic exoskeleton and a human limb are connected, most likely, these deformations occur at the interface between the two kinematic chains, caused by the low stiffness of human skin and tissues surrounding the bones [7].

Solutions found in the literature to cope with this problem are of two kinds. In a first approach the exoskeleton design can be thought in such a way that adaptation to human limb kinematics is maximized. Robotic segments with adjustable length were thus developed, and pneumatic systems were added to introduce elasticity in the robot fixations and adaptability to variable limb section [8]. This minimizes the kinematic differences, but drastically increases the complexity of the device, leading to weight increase, stiffness limitations, etc. Furthermore, again, it seems that perfect matching at any instant is yet out of reach.

The second approach consists in adding passive DoFs to connect the two kinematic chains. This was proposed back in the 1970s in the context of passive orthoses, [9], [10]. The same principle was recently proposed for a one degree of freedom device in [7], but the force transmission is analyzed only in a plane, and relies on explicit equations derived for a particular planar mechanism. It thus suffers from a lack of generality and the author neglects all the off-plane forces that unavoidably arise from the unmodeled lack of parallelism between the human limb plane and the exoskeleton plane.

Rather, the constructive method proposed here applies to a general spatial problem, which is properly formalized and then solved thanks to a set of necessary and sufficient conditions for global isostaticity (Section II). In Section III, the method is applied to ABLE, a given active 4DoF arm exoskeleton. In Section IV, experimental results illustrate the practical interest of the approach.

II. General Methodology

The main question addressed in this paper is: given a proposed exoskeleton structure designed to (approximately) replicate a human limb kinematic model, how to connect it to the human limb while avoiding the appearance of uncontrollable forces at the interface? The answer takes the form of a set of passive frictionless mechanisms used to connect the robot and the subject's limb that allows to avoid hyperstaticity.

A. Problem formulation

We consider two different serial chains with multiple couplings as illustrated in Fig. 1. One represents a human limb **H** and the other the robot structure **R**.

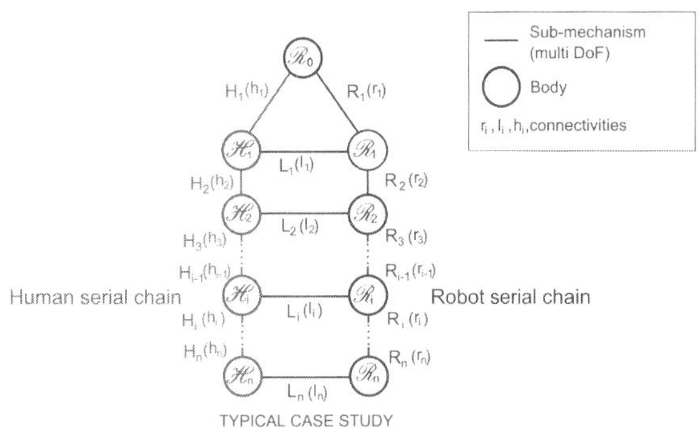

Fig. 1. Schematic of two serial chains parallel coupling

The base body of the exoskeleton is supposed to be attached to a body of the human subject. This common body is denoted $\mathscr{R}_0 \equiv \mathscr{H}_0$. The robot and the limbs are supposed to be connected through n fixations. Each fixation is a mechanism $\mathbf{L}_i$ for $i \in \{1,..,n\}$ consisting in a passive kinematic chain which connects a human body $\mathscr{H}_i$ to a robot body $\mathscr{R}_i$. Mechanisms $\mathbf{L}_i$ are supposed to possess a connectivity l_i. Recall that connectivity is the minimum and necessary number of joint scalar variables that determine the geometric configuration of the $\mathbf{L}_i$ chain [11]. Typically, $\mathbf{L}_i$ will be a nonsingular serial combination of l_i one DoF joints. The fixation can be an embedment ($l_i = 0$) or can release several DoFs, such that:

$$\forall i \in \{1,..,n\}, \quad 0 \leq l_i \leq 5 \quad . \tag{1}$$

Indeed choosing $l_i \geq 6$ would correspond to complete freedom between $\mathscr{H}_i$ and $\mathscr{R}_i$ which would not make any practical sense in the considered application where force transmission is required.

Between $\mathscr{R}_{i-1}$ and $\mathscr{R}_i$, on the robot side, there is an active mechanism $\mathbf{R}_i$ which connectivity is denoted r_i. Similarly, between $\mathscr{H}_{i-1}$ and $\mathscr{H}_i$ on the human side, there is a mechanism $\mathbf{H}_i$ of connectivity h_i. Note that, due to the complexity of human kinematic h_i is not always exactly known, and literature from biomechanics provides controversial data on this point. For example, the elbow is often modeled as a one DoF joint, but in reality a residual second DoF can be observed [12].

Our goal is to design mechanisms $\mathbf{L}_i$ with $i \in \{1,..,n\}$ in such a way that on one side, all the forces generated by the exoskeleton on the human limb are controllable and on the other side, there is no possible motion for the exoskeleton when the human limb is still. We shall thus consider in the next that the human limbs are virtually attached to the base body $\mathscr{R}_0$. This represents the case, when the subject does not move at all. The resulting mechanism, depicted in Fig. 2, is denoted S_n.

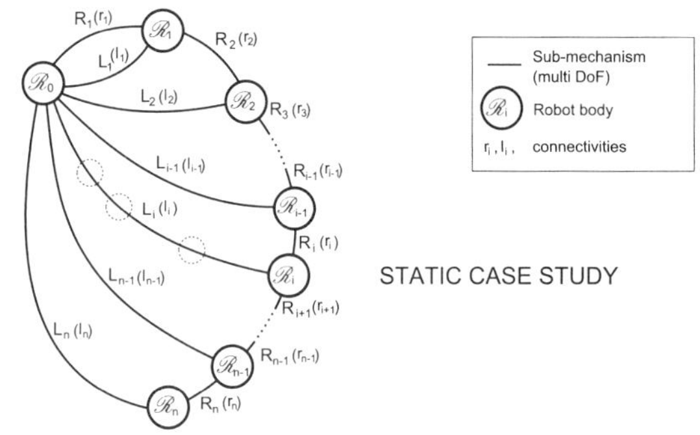

Fig. 2. Studied problem with a fixed human limb

A proper design for the passive mechanisms $\mathbf{L}_i$ shall guarantee that, in the absence of any external forces, both:

$$\forall i \in 1 \cdots n, \quad {}^{\mathbf{S}_n}T_i = \{0\} \quad and \tag{2a}$$

$$\forall i \in 1 \cdots n, \quad {}^{\mathbf{S}_n}W_{\mathbf{L}_i \to 0} = \{0\} \quad , \tag{2b}$$

where ${}^{\mathbf{S}_n}T_i$ is the space of twists describing the velocities of robot body $\mathscr{R}_i$ relative to $\mathscr{R}_0$ when the whole mechanism $\mathbf{S}_n$ is considered and ${}^{\mathbf{S}_n}W_{\mathbf{L}_i \to 0}$ is the space of wrenches (forces and moments) statically admissible transmitted through the $\mathbf{L}_i$ chain on the reference body $\mathscr{R}_0$, when the whole mechanism $\mathbf{S}_n$ is considered.

Equation (2a) expresses the fact that the mobility of any robot body connected to a human limb should be null, which is required since the human member is supposed here to be still. Moreover, Eq. (2b) imposes that, considering the whole mechanism, there can be no forces of any kind exerted on the human limb. Indeed, since the actuators are supposed to apply null generalized forces, the presence of any force at the connection ports would be an uncontrollable force due to hyperstaticity. In the next Eq. (2) is referred as the *global isostaticity condition*.

B. Conditions on the twist space ranks

At first, one can notice the recursive structure of the considered system: if we name $\mathbf{S}_i$ the sub-mechanism constituted by the bodies $\mathscr{R}_0$ to $\mathscr{R}_i$, the chains $\mathbf{R}_0$ to $\mathbf{R}_i$ and $\mathbf{L}_0$ to $\mathbf{L}_i$, we can represent $\mathbf{S}_i$ recursively from $\mathbf{S}_{i-1}$, as in Fig. 3, where m_{i-1}

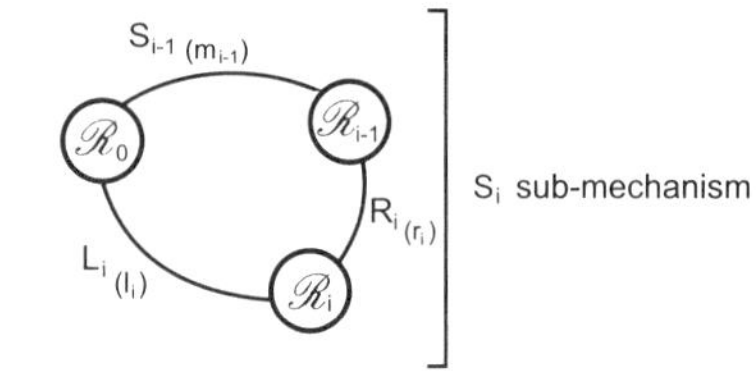

Fig. 3. Recursive structure $\mathbf{S}_i$ of the system

is the connectivity of $\mathbf{S}_{i-1}$. In this convention, $\mathbf{S}_0$ represents a zero DoF mechanism. Using this recursive representation one can establish the following proposition:

Proposition 1: The conditions (2) are equivalent to :

$$\forall i \in 1 \cdots n, \quad dim(T_{\mathbf{S}_{i-1}} + T_{\mathbf{R}_i} + T_{\mathbf{L}_i}) = 6 \quad and \tag{3a}$$

$$\forall i \in 1 \cdots n, \quad dim(T_{\mathbf{S}_{i-1}} \cap T_{\mathbf{R}_i}) = 0 \quad and \tag{3b}$$

$$dim(T_{\mathbf{S}_n}) = 0 \ , \tag{3c}$$

where $T_{\mathbf{S}_j} = {}^{\mathbf{S}_j}T_j$ is the space of twists describing the velocities of $\mathscr{R}_j$ relative to $\mathscr{R}_0$, when $\mathbf{S}_j$ is considered isolated from the rest of the mechanism (then it is different from ${}^{\mathbf{S}_n}T_j$), $T_{\mathbf{R}_i}$ is the space of twists produced by $\mathbf{R}_i$ – *i.e.* the space of twists of $\mathscr{R}_i$ relative to $\mathscr{R}_{i-1}$ if they were only connected through $\mathbf{R}_i$, $T_{\mathbf{L}_i}$ is the space of twists produced by $\mathbf{L}_i$ *i.e.* the space of twists of $\mathscr{R}_i$ relative to $\mathscr{R}_0$ if they were only connected through $\mathbf{L}_i$. ∎

The demonstration can be found in Appendix A.

Remarkably, conditions (3) involve the space of twists generated by $\mathbf{R}_i$ and $\mathbf{L}_i$ when taken isolated, which is of great help for design purposes. In the next, we convert these conditions into constraints on the connectivities $r_i = dim(T_{\mathbf{R}_i})$ and $l_i = dim(T_{\mathbf{L}_i})$. To do so, we suppose that kinematic singularities are avoided. In other words, summing the subspaces of twists will always lead to a subspace of maximum dimension given the dimensions of individual summed subspaces. This hypothesis will lead to determine how many DoFs shall be included in the passive fixation mechanisms $\mathbf{L}_i$. Of course as it is usual in mechanism design, when a particular design is finally proposed, it will be necessary to verify *a posteriori* the singularity avoidance condition.

C. Conditions on connectivities

At first, let's compute the connectivity of $\mathbf{S}_i$. One has:

$$T_{\mathbf{S}_i} = T_{\mathbf{L}_i} \cap (T_{\mathbf{R}_i} + T_{\mathbf{S}_{i-1}}) \ , \tag{4}$$

which directly results from the space sum law for serial chains and the intersection law for parallel chains (see [13]). Furthermore, since for any vector subspaces $\mathbf{A}$ and $\mathbf{B}$, $dim(\mathbf{A}) + dim(\mathbf{B}) = dim(\mathbf{A}+\mathbf{B}) + dim(\mathbf{A} \cap \mathbf{B})$, one gets:

$$\begin{aligned} m_i &= dim(T_{\mathbf{L}_i}) + dim(T_{\mathbf{R}_i} + T_{\mathbf{S}_{i-1}}) - dim(T_{\mathbf{L}_i} + T_{\mathbf{R}_i} + T_{\mathbf{S}_{i-1}}) \\ &= dim(T_{\mathbf{L}_i}) + dim(T_{\mathbf{R}_i}) + dim(T_{\mathbf{S}_{i-1}}) - dim(T_{\mathbf{R}_i} \cap T_{\mathbf{S}_{i-1}}) \\ &\quad - dim(T_{\mathbf{L}_i} + T_{\mathbf{R}_i} + T_{\mathbf{S}_{i-1}}). \end{aligned}$$

If condition (3) is respected and under full rank asumption, one gets:

$$m_i = l_i + r_i + m_{i-1} - 6 \tag{5}$$

Finally, using $m_0 = 0$, this recursive equation simplifies to:

$$m_i = \sum_{j=1}^{i} (l_j + r_j) - 6.i \ . \tag{6}$$

Therefore, from Eq. (3a), noticing that any vector subspaces $\mathbf{A}$,$\mathbf{B}$ and $\mathbf{C}$ of a vector space $\mathbf{E}$, $dim(\mathbf{A}+\mathbf{B}+\mathbf{C}) \leq dim(\mathbf{A}) + dim(\mathbf{B}) + dim(\mathbf{C})$, it is necessary that:

$$\forall i \in 1 \cdots n, \quad m_{i-1} + r_i + l_i \geq 6, \quad \text{or :} \quad \sum_{j=1}^{i} (l_j + r_j) \geq 6.i \tag{7}$$

Moreover, since if $\mathbf{A}$ and $\mathbf{B}$ are two vector subspaces of $\mathbf{E}$ and $dim(\mathbf{A}) + dim(\mathbf{B}) > dim(\mathbf{E})$, then $\mathbf{A} \cap \mathbf{B} \neq \{0\}$, Eq. (3b) imposes that:

$$\forall i \in 1 \cdots n, \quad m_{i-1} + r_i \leq 6 \text{ or : } \sum_{j=1}^{i-1} (l_j + r_j) + r_i \leq 6.i \tag{8}$$

Finally, the last condition (3c) leads to:

$$m_n = 0 \text{ or : } \sum_{j=1}^{n} (l_j + r_j) = 6.n \tag{9}$$

Notice that (9) provides the total number of DoFs to be freed for the mechanism $\mathbf{S}_n$, while (7) gives the minimal value (to prevent from hyperstaticity in the sub-mechanisms $\mathbf{S}_j$) for l_j and (8) provides the maximal one (to prevent from internal mobility in $\mathbf{S}_j$).

Thanks to these three last necessary conditions, we are able to calculate the different possible solutions for distributing the additional passive DoFs over the structure:

- the possible choices for l_1 are such that $5 \geq l_1 \geq 6 - r_1$.
- for each choice of l_1, the possible choices for l_2 are such that $5 \geq l_2 \geq 12 - r_1 - r_2 - l_1$.

This leads to a tree that groups all the admissible combinations for l_i, as illustrated in Fig (4).

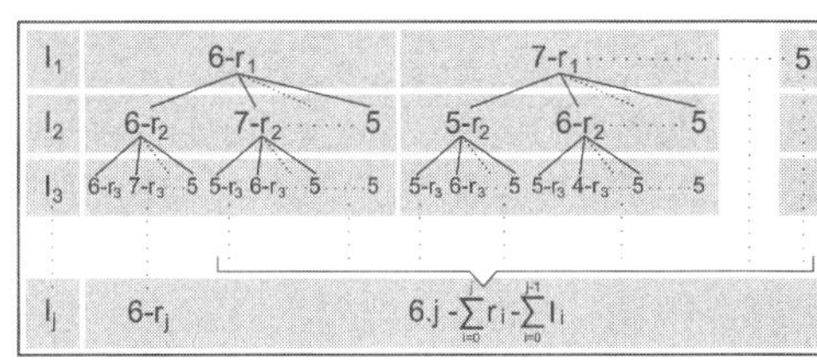

Fig. 4. Tree of possible solutions for the number of passive DoFs to add at every fixation point

Out of this tree, all the possible combinations of connectivities for the fixations are given. Of course, the selection among these solutions is to be made depending on the exoskeleton kinematics. Generally speaking, an important aspect to be considered is the force transmission: through any linear or rotational DoF that is not freed by the fixation mechanism, a force or a moment will be transmitted to the human limb, which is surrounded by soft tissues. Therefore, typically, transmitting moments around P_i would lead to locally deform the tissues which in turn can generate discomfort. The next section illustrates, on a concrete spatial example involving two fixations, how to integrate this kind of considerations in the design of fixation mechanisms.

III. Application to a given exoskeleton

A. ABLE: an upper limb exoskeleton for rehabilitation

ABLE (see Fig. 5) is a 4 axis exoskeleton that has been designed by CEA-LIST [14] on the basis of an innovative screw-and-cable actuation technology ([15]). Its kinematics is composed of a shoulder spherical joint composed of 3 coincident pivots and a 1 DoF pivot elbow. The forearm,

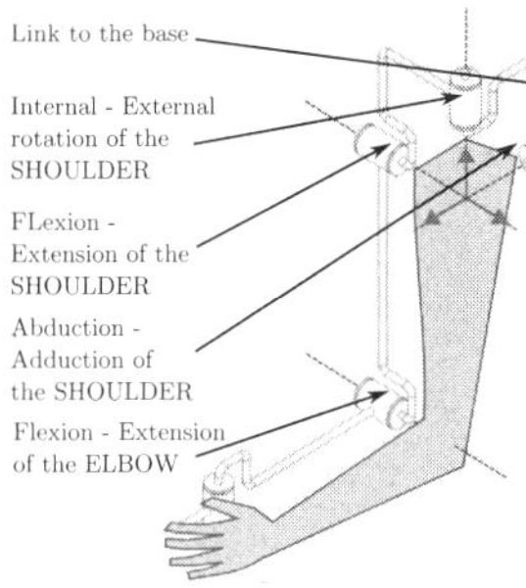

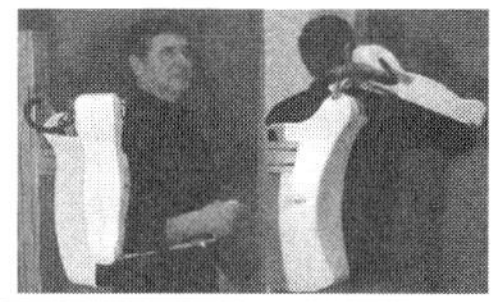

Joint	α_{i-1}	θ_i	a_{i-1}	d_i
0→1	0	θ_{1mes}	0	0
1→2	90°	$-\theta_{2mes}$-90°	0	0
2→3	-90°	θ_{3mes}-90°	0	0
3→4	0	θ_{4mes}	357mm	0

Fig. 5. Kinematics of ABLE

terminated by a handle, is not actuated. Details on this robot can be found in [14].

B. *Fixations design for ABLE*

In this section, we apply general method proposed in Sec. II to ABLE. We proceed in three steps:

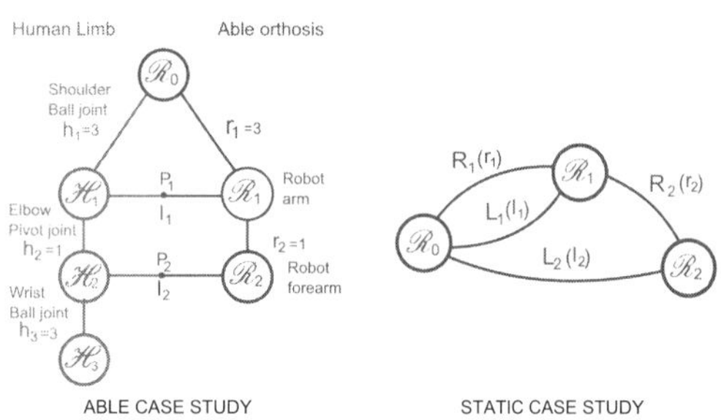

Fig. 6. Schematic of the ABLE and human arm coupling

- compute the tree of possible values for l_i
- choose among them a preferred solution by examining force transmission properties and kinematic complementarity
- verify the full kinematic rank which is reported in Appendix B.

Firstly, since ABLE comprises an upper arm and a forearm, we shall use two fixations (See Fig 6). The total number of passive DoF to be added is given by Eq. (9):

$$\sum_{j=1}^{n=2} l_j = 12 - \sum_{j=1}^{n=2} r_j = 12-(3+1) \quad \Rightarrow l_1 + l_2 = 8 \qquad (10)$$

Moreover, for the first fixation, Eq. (7) and (8) give:

$$6 - r_1 \leq l_1 \leq 6 \quad \Rightarrow \quad 3 \leq l_1 \leq 5 \ .$$

Since the total number of DoFs is fixed, the tree of possible solutions consists here of three parallel branches where l_1 is chosen between 3 and 5 and $l_2 = 8 - l_1$. Possible couples for (l_1, l_2) are (3,5), (4,4) and (5,3). Hereafter, these three options are analyzed in order to choose a preferred design among them.

• *Case a:* $l_1 = 3$ *and* $l_2 = 5$. In this case, both $\mathbf{S}_1$ taken isolate and $\mathbf{S}_2$ are isostatic, which corresponds to the most intuitive way of achieving global isostaticity. Degrees of Freedom for $\mathbf{L}_1$ have to be chosen complementary to those of $\mathbf{R}_1$ in order to satisfy the full rank assumption. Since $\mathbf{R}_1$ is a ball joint that generates three independent rotational velocities around its center M_1, $\mathbf{L}_1$ must generate three independent velocities at point M_1. For example, three non coplanar translations could be used for $\mathbf{L}_1$. However, in this case, the fixation would transmit a null force, *i.e.* a pure couple. This seems undesirable due to the torsion of the soft tissues that it would create around P_1 at the level of the attachment to the limb. One could thus think of using for $\mathbf{L}_1$ a ball joint around P_1, but in this case, the full rank condition would not be respected, because $\mathbf{R}_1$ and $\mathbf{L}_1$ would both generate the same rotation around $\vec{z}_1 = \frac{1}{\|\overrightarrow{M_1P_1}\|}\overrightarrow{M_1P_1}$. Finally, a preferred solution is to choose for $\mathbf{L}_1$ two pivot joints perpendicular to the arm main axis $\vec{z}_{arm}$, and one translation joint collinear $\vec{z}_{arm}$ (see 8 for further definition of P_i, $\vec{z}_{arm}$ and $\vec{z}_{forearm}$). In this case, two forces perpendicular to $\vec{z}_{arm}$ and one moment around $\vec{z}_{arm}$ can be exchanged between the exoskeleton and the arm through $\mathbf{L}_1$. Moreover, since $\mathbf{S}_1$ is isostatic, one has $m_1 = 0$. Therefore $\mathbf{L}_2$ needs to be designed in order to be kinematically complementary to $\mathbf{R}_2$, which is a pivot of axis $(M_2, \vec{z}_2)$ (See Appendix for $\vec{z}_2$ definition). In other words, $\mathbf{L}_2$ must generate independently 2 rotations perpendicular to $\vec{z}_2$ and 3 velocities at point M_2. A natural solution is to choose a ball joint around P_2 and two translations in a plane perpendicular to $\vec{z}_2$. The resulting overall design is noted (a) and represented in Fig. 7.

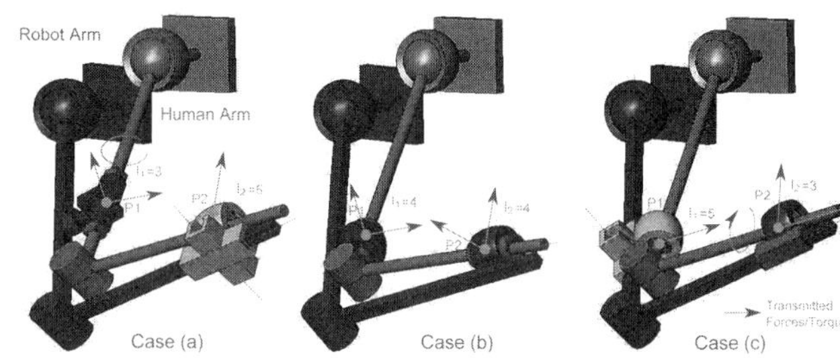

Fig. 7. Considered possibilities for coupling ABLE to an human arm. Case (a): ball joint alone at P_1 and ball joint + 2 slides at P_2; case (b): Universal joint + 1slide at P_1 and ball joint + 2 slides at P_2; case (c) Ball joints with slide at both P_1 and P_2.

• *Case b:* $l_1 = 4$ *and* $l_2 = 4$. Note that in this case, $\mathbf{S}_1$ taken isolate is a 1 DoF mechanism, while only $\mathbf{S}_2$ is isostatic. We consider solution (a), for which one DoF must be added to $\mathbf{L}_1$ and one must be removed from $\mathbf{L}_2$. Concerning $\mathbf{L}_1$, keeping freed the 3 DoF liberated for the isostatic solution (a), it seems preferable to choose, for the extra freed DoF, the rotation around z_1. Indeed, this will cancel the local tissue torsion due to moment transmission around $\vec{z}_1$. As a result, $\mathbf{S}_1$ is now a 1 DoF mechanism consisting of a pivot around $(M_1, \vec{z}_1)$. Concerning $\mathbf{L}_2$, the DoF to be removed from the solution (a) shall not degrade the dimension of $T_{\mathbf{S}_1} + T_{\mathbf{R}_2} + T_{\mathbf{L}_2}$. It seems preferable to keep the freed three rotations around P_2 and only one translation along the forearm axis $\vec{z}_{forearm}$. Indeed, again, this choice avoids any torsion around P_2. Furthermore, it is shown in Appendix B that singular configurations of this solution, noted (b) and represented in Fig. 7 are easily identifiable and far away from nominal conditions of operation.

• *Case c:* $l_1 = 5$ *and* $l_2 = 3$. Similarly to solution (a), this combination will necessary lead to transmit at least one torsion moment around $\vec{z}_{forearm}$, as illustrated in Fig. 7 (solution (c)). Therefore, the finally preferred solution is (b).

Note that with solution (b), generating a moment to the human upper arm around $\vec{z}_{arm}$ is obtained by applying opposite pure forces perpendicular to $\vec{z}_{arm}$ at P_1 and to $\vec{z}_{forearm}$ at P_2 (see Fig. 8). Interestingly, this reproduces the method used by physical therapists to assist patients in generating internal rotations of the shoulder without torsion to the tissue. As a price, the full extension configuration, when M_1, P_1 and P_2 are aligned, is singular, as detailed in the Appendix B. This configuration can be easily avoided by limiting the range of the elbow extension.

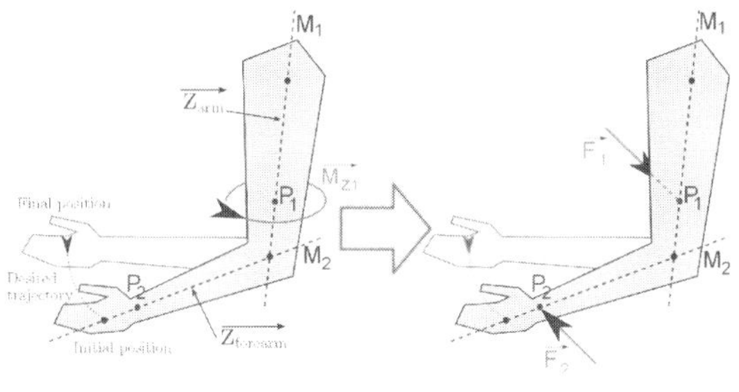

Fig. 8. Transmitting a moment around the upper arm axis with solution (b) (left) and (c) (right)

C. Fixations realization

The two fixations mechanisms are finally identical. They shall generate three independent rotations and one translation along the limb. The mechanism used to realize this function consists of three successive pivot joints which axis coincide and one slider whose axis is parrallel to human limb (see Fig 9).

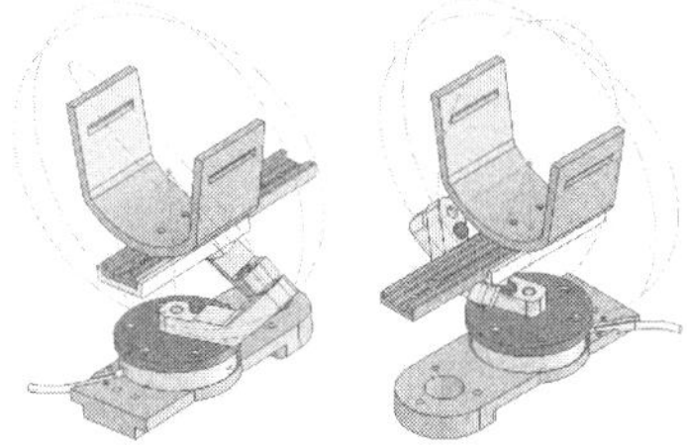

Fig. 9. Fixation simplification and realization (rear and front)

The fixations were dimensionned differently: one to allow forearm pronosupination and the other not to collide with arm tissues. As a result, possible motions left by the passive fixations have the ranges defined in Table I.

These fixations were both fitted with one force sensor placed on the base (ATI Nano43 6-axis Force/Torque sensor) allowing us to reconstruct the three forces and three torques components at P_1 and P_2 respectively.

DoF	Arm Fixation	Forearm Fixation
Rotation1 (⊥ to the limb axis)	360°	360°
Rotation2 (⊥ to the limb axis)	90°	90°
Rotation3 (around the limb axis)	110°	110°
Translation	100mm	100mm

TABLE I

For the experiments presented in the next section, in order to compare the forces involved with and without DoF liberation, the fixations were also equipped with removable metallic pins, allowing us to quickly lock the passive DoF without detaching the subject from the exoskeleton. These fixations were mounted on the 4 DoF ABLE exoskeleton. Arm fixation is placed near the elbow, just under the triceps. Forearm fixation is placed near the wrist. Thermoformable materials were also used to create two splints perfectly adapted to human morphology. These splints are serially connected to the last fixation body. Wrist splint was specifically created to lock the wrist flexions which are not studied here. Only passive pronosupination is allowed.

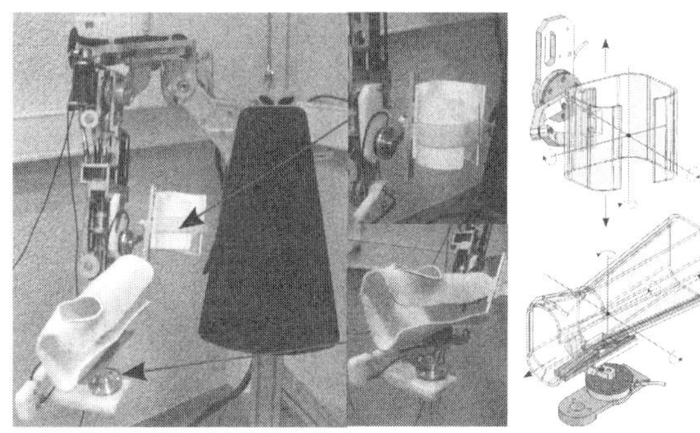

Fig. 10. The two fixations on the exoskeleton

IV. Experimental results

A. Experimental setup

An articulated mannequin was used for the experiment. Its arms possess 5 passive DoFs (a ball joint shoulder, a pivot elbow and a pronosupination). Analyzing the interaction force and torque variations at the interfaces during the same movement with isostatic fixations and without (locked case) will allow us to evaluate their impact on preventing the appearance of uncontrolled forces but also to quantify them roughly.

The mannequin was thus placed in the exoskeleton and attached through the two fixations. The thermoformable splints allow to avoid any looseness in the fixation and increase the contact stiffness (no foam needed).

During the experiments, the exoskeleton imposes a controlled trajectory, with a constant speed, to the mannequin arm. The experiment consists in a serie of six simple point-to-point movements (with a limited range of motion) to the same target but reached with different joints movements (thanks

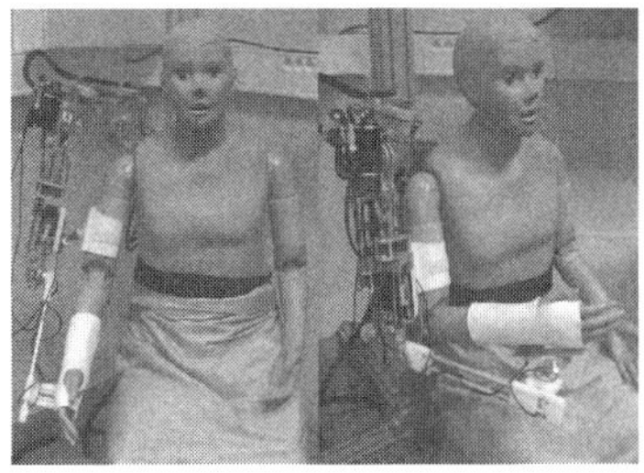

Fig. 11. Mannequin connected to the ABLE exoskeleton

to arm redundancy). Target was reached at constant and low speed (0.05 m/s) to limit inertial forces. Due to the rigidity of the mannequin surface, the movement amplitude on every exoskeleton joint was limited to a 15 ° maximum range of motion in order not to destroy force sensors whose maximum allowable value is about 36N. Indeed when exoskeleton is connected to human limb, thanks to skin and muscles deformations, the hyperstatic force level applied on the human kinematic structure (the bones) is reduced.
The use of a mannequin controlled by exoskeleton allows to obtain a perfect repeatability during the experiments. This is really representative of co-manipulation cases where the robot generates a controlled motion by applying forces, as during robotic rehabilitation or movement assistance for impaired people.

B. Results and discussions

Principal results are presented below. In Fig. 12, we plotted the undesired force absolute value and mean moment averaged norm during the experiments, for the two sensors, averaged across the six movements (moments are computed at the rotation center of the fixation). We can observe on the arm

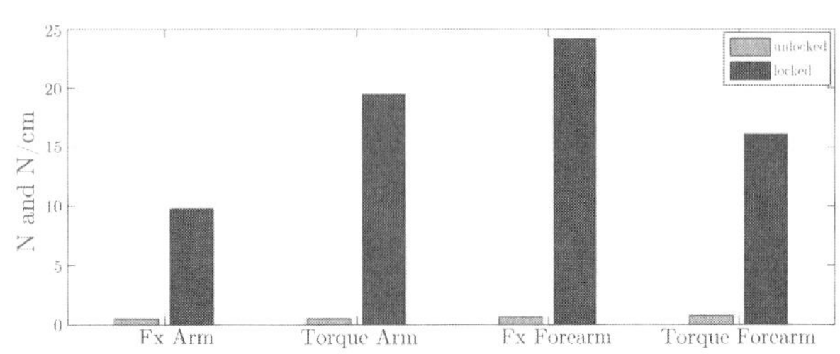

Fig. 12. Averaged absolute value of the undesired force $|F_x|$ and moments norm $\sqrt{(M_x^2+M_y^2+M_z^2)}$ on the two fixations (mean for the six movements)

fixation a decrease in the undesired force (F_x) level and the undesired torques level by approximatively 95%. For the forearm fixation, an approximative 96% decrease can be observed for the undesired force and moment components. In Fig. 13, the norm of the components (F_y and F_z)corresponding to the components transmitted by the passive fixations is presented. The exoskeleton ability to transmit forces to the subject is not altered. Note that the percentage of hyperstatic force level decrease achieved by the fixations resulting from our methodology and the obtained numerical value of the

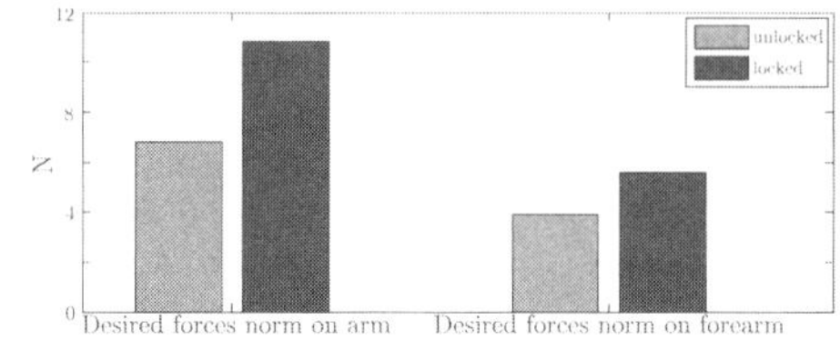

Fig. 13. Allowed forces ($\sqrt{(F_x^2+F_y^2)}$) norm on the two fixation (mean for the six movements)

hyperstatic forces have to be interpreted. Indeed, due to the mannequin arm smallness (see Fig. 11) and it body suface rigidity, hyperstatic force level is higher than the one occuring during a comanipulation between the exoskeleton and a human subject.

V. Conclusion

In this paper we presented a methodology aimed at designing the kinematics of fixations between an exoskeleton and a human member. Thanks to this method, we realized isostatic fixations for a 4 DoF exoskeleton and experimentally verified their benefit on minimizing uncontrollable hyperstatic forces at the human robot interface and thus on a fine control of the interaction forces. These results show that the provided solution avoids hyperstaticity but also adapts to large variations of the human limb geometry without requiring a complex adaptable robot structure.

Appendix

A. *Demonstration of Proposition 1*

1) Conditions (3) are sufficient: $[(3) \Rightarrow (2)]$.
We here suppose that conditions (3) are verified.
Because in $\mathbf{S}_n$, $\mathscr{R}_{i-1}$ is connected on one side to $\mathscr{R}_0$ through $\mathbf{S}_{i-1}$ and on the other side to $\mathscr{R}_i$ through $\mathbf{R}_i$ (see Fig. 3), one has:

$$\forall i \in \{1 \ldots n\}, \quad {}^{\mathbf{S}_n}T_{i-1} = {}^{\mathbf{S}_{i-1}}T_{i-1} \cap \left[T_{\mathbf{R}_i} + {}^{\mathbf{S}_n}T_i\right] \quad , \tag{11}$$

which is a recursive relationship for ${}^{\mathbf{S}_n}T_i$. Recalling that, by assumption, ${}^{\mathbf{S}_n}T_{\mathbf{S}_n} = \{0\}$ (condition 3c) and $T_{\mathbf{S}_{i-1}} \cap T_{\mathbf{R}_i} = \{0\}$ (condition 3b), this recursive law trivially leads to (2a).
Furthermore, the kinemato-static duality principle applied to the loop $(\mathscr{R}_0 \to \mathscr{R}_{i-1} \to \mathscr{R}_i \to \mathscr{R}_0)$ in Fig. 3 writes:

$$\forall i \in \{1 \ldots n\}, \quad dim({}^{\mathbf{S}_i}W_{\mathbf{L}_i \to 0}) + dim(T_{\mathbf{S}_{i-1}} + T_{\mathbf{R}_i} + T_{\mathbf{L}_i}) = 6 \quad . \tag{12}$$

Thanks to condition (3a), this leads to:

$$\forall i \in \{1 \ldots n\}, \quad {}^{\mathbf{S}_i}W_{\mathbf{L}_i \to 0} = \{0\} \quad . \tag{13}$$

Considering again the system $\mathbf{S}_i$ depicted in Fig. 3, and recalling that $\mathbf{L}_i$ and $\mathbf{R}_i$ are serial chains, one has, $\forall i \in \{1 \ldots n\}$:

$${}^{\mathbf{S}_i}W_{\mathbf{L}_i \to 0} = {}^{\mathbf{S}_i}W_{\mathbf{L}_i \to i} = {}^{\mathbf{S}_i}W_{\mathbf{R}_i \to i} = {}^{\mathbf{S}_i}W_{\mathbf{R}_i \to i-1} = \{0\} \quad . \tag{14}$$

Therefore, statically speaking, the multi-loop system $\mathbf{S}_{i-1}$ is in the same state when included in $\mathbf{S}_i$ than when isolated from the rest of the mechanism.

$$\forall i \in \{2 \ldots n\}, \quad {}^{\mathbf{S}_i}W_{\mathbf{L}_{i-1} \to 0} = {}^{\mathbf{S}_{i-1}}W_{\mathbf{L}_{i-1} \to 0} \; ,$$

which, together with (13) recursively leads to condition (2b).

2) Conditions (3) are necessary : $\left[\overline{(3)} \Rightarrow \overline{(2)}\right]$.
Firstly, if condition (3c) is not verified, then ${}^{\mathbf{S}_n}T_n = T_{\mathbf{S}_n} \neq \{0\}$. In this case, (2a) is not satisfied.
Secondly, if (3b) is not verified, then $\exists i,\ (T_{\mathbf{R}_i} \cap T_{\mathbf{S}_{i-1}}) \neq \{0\}$. Thanks to Eq. (11), this leads to:

$$\exists i \in \{1 \cdots n\},\quad {}^{\mathbf{S}_n}T_{i-1} \neq \{0\} \quad , \tag{15}$$

which directly contradicts (2a).
Thirdly, if (3a) is not verified, i.e.:

$$\exists i,\ dim(T_{\mathbf{S}_{i-1}} + T_{\mathbf{R}_i} + T_{\mathbf{L}_i}) \leq 6 \quad , \tag{16}$$

then $\exists i,\ {}^{\mathbf{S}_i}W_{\mathbf{L}_{i\to 0}} \neq \{0\}$, meaning that $\mathbf{S}_i$ taken isolate is hyperstatic. Obviously, adding the rest of the mechanism to build $\mathbf{S}_n$, which consists of adding a parallel branch to $\mathbf{S}_i$ between $\mathscr{R}_0$ and $\mathscr{R}_i$ will not decrease the degree of hyperstaticity. Therefore $\exists i,\ {}^{\mathbf{S}_n}W_{\mathbf{L}_{i\to 0}} \neq \{0\}$, which contradicts condition (2b).

B. *Singularity analysis for ABLE and the two proposed fixation mechanisms*

We study the mechanism depicted in Fig. 14: $\mathbf{R}_1$ is a ball joint which center is M_1; $\mathbf{L}_1$ is composed of a ball joint which center is P_1 (with $\overrightarrow{M_1P_1} = l_1.\overrightarrow{z_1}$ and $l_1 \neq 0$) and a slide along $(P_1, \overrightarrow{z_{arm}})$; $\mathbf{R}_2$ is a pivot joint which axis is $(M_2, \overrightarrow{x_2})$; $\mathbf{L}_2$ is composed of a ball joint which center is P_2 (with $\overrightarrow{M_2P_2} = l_2.\overrightarrow{z_2}$ and $l_2 \neq 0$) and a slide along $(P_2, \overrightarrow{z_{forearm}})$.
In order to find the singular configurations of this system, we use the necessary and sufficient conditions (3).

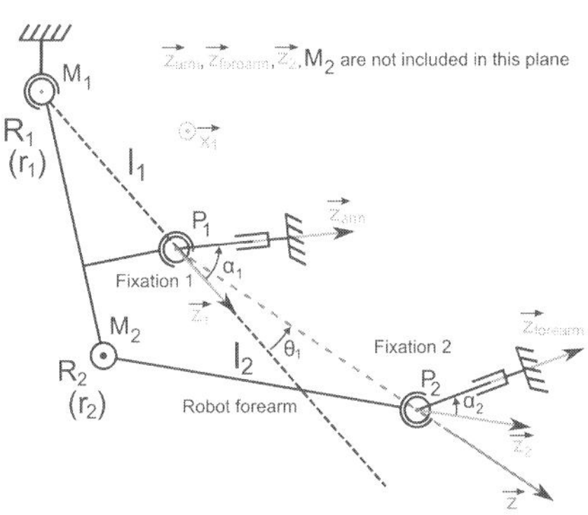

Fig. 14. Kinematics of ABLE + its fixations. The plane of the figure, perpendicular to $\vec{x}_1$, is defined by M_1, P_1 and P_2 while M_2 is off the plane.

1) Examination of Condition (3a)

- For $i = 1$, (3a) writes $dim(T_{\mathbf{R}_1} + T_{\mathbf{L}_1}) = 6$.

At point P_1, velocities allowed by $\mathbf{L}_1$ belong to the vector subspace $T_{\mathbf{L}_1} = span\{t_1, t_2, t_3, t_4\}$ and the velocities allowed by $\mathbf{R}_1$ belong to $T_{\mathbf{R}_1} = span\{t_5, t_6, t_3\}$, with

$$t_1 = (x_1{}^{\mathrm{T}}\ 0_3{}^{\mathrm{T}})^{\mathrm{T}},\quad t_3 = (z_1{}^{\mathrm{T}}\ 0_3{}^{\mathrm{T}})^{\mathrm{T}},\quad t_5 = (x_1{}^{\mathrm{T}}\ -l_1.y_1{}^{\mathrm{T}}\ )^{\mathrm{T}}$$
$$t_2 = (y_1{}^{\mathrm{T}}\ 0_3{}^{\mathrm{T}})^{\mathrm{T}},\quad t_4 = (0_3{}^{\mathrm{T}}\ z_{arm}{}^{\mathrm{T}}\ )^{\mathrm{T}},\quad t_6 = (y_1{}^{\mathrm{T}}\ l_1.x_1{}^{\mathrm{T}}\ )^{\mathrm{T}}$$

Thus $T_{\mathbf{R}_1} + T_{\mathbf{L}_1} = span\{t_1, ..., t_6\}$. Defining

$$t'_5 = \frac{(t_6 - t_2)}{l_1} = (0_3{}^{\mathrm{T}}\ x_1{}^{\mathrm{T}}\ )^{\mathrm{T}} \text{ and } t'_6 = \frac{(t_1 - t_5)}{l_1} = (0_3{}^{\mathrm{T}}\ y_1{}^{\mathrm{T}}\ )^{\mathrm{T}},$$

we can easily show that

$$[t_1\ t_2\ t_3\ t_4\ t'_5\ t'_6] = A\ [t_1\ t_2\ t_3\ t_4\ t_5\ t_6]$$

with $det(A) = \frac{1}{l_1^2}$. Since $l_1 \neq 0$, $\tau_1 = \{t_1, .., t_6\}$ is a basis of $\mathbb{R}^6$ if and only if $\tau_2 = \{t_1, .., t_4, t'_5, t'_6\}$ is a basis of $\mathbb{R}^6$. Let's consider now $a_i \in \mathbb{R}$, $i \in \{1, .., 6\}$ such that:

$$a_1t_1 + a_2t_2 + a_3t_3 + a_4t_4 + a_5t'_5 + a_6t'_6 = 0 \tag{17}$$

It is trivial to show that $a_1 = a_2 = a_3 = 0$, $a_4d_z = 0$, $a_6 + a_4d_y = 0$ and $a_5 + a_4d_x = 0$ where $\overrightarrow{z_{arm}} = d_x\overrightarrow{x_1} + d_y\overrightarrow{y_1} + d_z\overrightarrow{z_1}$. If $d_z \neq 0$ then $a_6 = a_5 = a_4 = 0$. Therefore, τ_2 and τ_1 are bases of $\mathbb{R}^6$. Otherwise, there exists a non null combination of a_i that verifies (17) which means that τ_2 and τ_1 are not free anymore. Condition (3a) is thus verified for $i = 1$ if and only if $\overrightarrow{z_{arm}}.\overrightarrow{z_1} \neq 0$. This is equivalent to $\alpha_1 \neq \pm\frac{\pi}{2}$ and this is a singular value to be avoided. In the rest of the study we will thus consider that $\overrightarrow{z_{arm}}.\overrightarrow{z_1} \neq 0$.

- For $i = 2$, (3a) writes $dim(T_{\mathbf{S}_1} + T_{\mathbf{R}_2} + T_{\mathbf{L}_2}) = 6$.

We know that $T_{\mathbf{S}_1} = T_{\mathbf{R}_1} \cap T_{\mathbf{L}_1}$. Let's consider $t \in T_{\mathbf{L}_1}$ and $t' \in T_{\mathbf{R}_1}$. One has:

$$\exists(\alpha_1, \alpha_2, \alpha_3, \alpha_4) \quad \text{such that} \quad t = \sum_{i=1}^{4} \alpha_i\ t_i \tag{18}$$
$$\exists(\alpha'_1, \alpha'_2, \alpha'_3,) \quad \text{such that} \quad t' = \alpha'_1\ t_5 + \alpha'_2\ t_6 + \alpha'_3\ t_3 \tag{19}$$

Using $\overrightarrow{z_{arm}}.\overrightarrow{z_1} \neq 0$, one easily gets:

$$t = t' \Leftrightarrow \alpha_1 = \alpha_2 = \alpha_4 = \alpha'_1 = \alpha'_2 = 0 \quad . \tag{20}$$

or:

$$t = t' \Leftrightarrow t = \alpha_3\ t_3 = \alpha'_3\ t_3 \quad . \tag{21}$$

In other words, at point P_1:

$$T_{\mathbf{S}_1} = T_{\mathbf{R}_1} \cap T_{\mathbf{L}_1} = span(\{t_3\}) = span(\{(z_1{}^{\mathrm{T}}\ 0_3{}^{\mathrm{T}})^{\mathrm{T}}\}) \quad . \tag{22}$$

We know write twists at point P_2. We get: $T_{\mathbf{S}_1} = span(\{t_7\})$, $T_{\mathbf{R}_2} = span(\{t_8\})$ and $T_{\mathbf{L}_2} = span(\{t_9\ t_{10}\ t_{11}\ t_{12}\})$, with:

$$t_7 = (z_1{}^{\mathrm{T}}\ l\sin\theta_1 x_1{}^{\mathrm{T}})^{\mathrm{T}} \quad , \quad t_8 = (x_2{}^{\mathrm{T}}\ -l_2\ y_2{}^{\mathrm{T}})^{\mathrm{T}} \quad , \quad t_9 = (x_2{}^{\mathrm{T}}\ 0^{\mathrm{T}})^{\mathrm{T}}$$
$$t_{10} = (y_2{}^{\mathrm{T}}\ 0^{\mathrm{T}})^{\mathrm{T}} \quad , \quad t_{11} = (z_2{}^{\mathrm{T}}\ 0^{\mathrm{T}})^{\mathrm{T}} \quad , \quad t_{12} = (0^{\mathrm{T}}\ z_{forearm}{}^{\mathrm{T}})^{\mathrm{T}},$$

where $\overrightarrow{P_1P_2} =: l\vec{z}$ and $\theta_1 := \left(\widehat{\overrightarrow{z_1}, \overrightarrow{z}}\right)$ measured around $\vec{x}_1$. Thus $T_{\mathbf{S}_1} + T_{\mathbf{R}_2} + T_{\mathbf{L}_2} = span(\{t_7, t_8, t_9, t_{10}, t_{11}, t_{12}\})$.
Suppose first that $\sin\theta_1 = 0$. Then, denoting $\overrightarrow{z_1} = z_{1x}.\overrightarrow{x_2} + z_{1y}.\overrightarrow{y_2} + z_{1z}.\overrightarrow{z_2}$, one gets:

$$t_7 = z_{1x}t_9\ +\ z_{1y}t_{10}\ +\ z_{1z}t_{12} \tag{23}$$

In this particular case, $\{t_7\ ..\ t_{12}\}$ is not a basis, which identifies a second singular configuration, when M_1, P_1 and P_2 are aligned. In the rest of the study we will thus assume that this singular configuration is also avoided, that is: $\sin\theta_1 \neq 0$. Defining

$$t'_7 = \frac{(t_7 - z_{1x}t_9 - z_{1y}t_{10} - z_{1z}t_{12})}{l\sin\theta_1} = (0^{\mathrm{T}}\ x_1{}^{\mathrm{T}})^{\mathrm{T}} \quad \text{,and}$$
$$t'_8 = \frac{(t_{10} - t_8)}{l_2} = (0^{\mathrm{T}}\ y_2{}^{\mathrm{T}})^{\mathrm{T}} \quad ,$$

we get $[t'_7\ t'_8\ t'_9\ ..\ t'_{12}] = B.[t_7\ t_8\ ..\ t_{12}]$ with $det(B) = \frac{-1}{l_2\sin\theta_1} \neq 0$. Thus $\tau_3 = \{t_7\ ..\ t_{12}\}$ is a basis of $\mathbb{R}^6$ if and only if $\tau_4 = \{t'_7\ ..\ t'_{12}\}$ is a basis of $\mathbb{R}^6$. Lets's consider $b_i \in \mathbb{R},\ i \in \{1,..,6\}$ such that:

$$b_1t'_7 + b_2t'_8 + b_3t_9 + b_4t_{10} + b_5t_{11} + b_6t_{12} = 0\ . \qquad (24)$$

It comes easily that $b_3 = b_4 = b_5 = 0$ and $b_1t'_7 + b_2t'_8 + b_6t'_{12} = 0$ which is equivalent to $b_1\overrightarrow{x_1} + b_2\overrightarrow{y_2} + b_6\overrightarrow{z_{forearm}} = \overrightarrow{0}$. The necessary and sufficient conditions to have a non-null triplet b_1, b_2, b_6 verifying the previous equation is that $\overrightarrow{x_1}, \overrightarrow{y_2}, \overrightarrow{z_{forearm}}$ are coplanar. This identifies a third singularity, which, again, is supposed to be avoided in the rest of the study.

2) Examination of the condition (3b)

- For $i = 1$, since $T_{\mathbf{S}_0} = \{0\}$, one directly gets $dim(T_{\mathbf{S}_0} \cap T_{\mathbf{L}_1}) = 0$.
- For $i = 2$, it is necessary to verify that $dim(T_{\mathbf{S}_1} \cap T_{\mathbf{L}_2}) = 0$.

Let's consider $t \in T_{\mathbf{S}_1}$ and $t' \in T_{\mathbf{L}_2}$. One has:

$$\exists\alpha_1 \in \mathbb{R} \quad / \quad t = \alpha_1 t_7$$
$$\exists\alpha'_1, \alpha'_2, \alpha'_3, \alpha'_4 \in \mathbb{R} \quad / \quad t' = \alpha'_1 t_9 + \alpha'_2 t_{10} + \alpha'_3 t_{11} + \alpha'_4 t_{12}\ .$$

One easily shows that $t = t'$ is equivalent to:

$$\begin{cases} \alpha_1 l\sin\theta_1\overrightarrow{x_1} + \alpha'_4\overrightarrow{z_{forearm}} = \overrightarrow{0} \\ (\alpha_1 z_{1x} + \alpha'_1)\overrightarrow{x_2} + (\alpha_1 z_{1y} + \alpha'_2)\overrightarrow{y_2} + (\alpha_1 z_{1z} + \alpha'_3)\overrightarrow{z_2} = \overrightarrow{0} \end{cases}$$

Since $\overrightarrow{x_1}$ is not colinear to $\overrightarrow{z_{forearm}}$, the first equation leads to $\alpha_1 = \alpha'_4 = 0$. Similarly, since $\{\overrightarrow{x_2}, \overrightarrow{y_2}, \overrightarrow{z_2}\}$ forms a basis, $\alpha'_1 = \alpha'_2 = \alpha'_3 = 0$. In conclusion, $dim(T_{\mathbf{S}_1} \cap T_{\mathbf{L}_2}) = \{0\}$.

3) Examination of the condition (3c)

For the considered example, $n = 2$ and condition (3c) writes $dim(T_{\mathbf{S}_2}) = 0$. Since $T_{\mathbf{S}_2} = (T_{\mathbf{S}_1} + T_{\mathbf{R}_2}) \cap T_{\mathbf{L}_2}$, we need to verify that any vector that belongs to both $(T_{\mathbf{S}_1} + T_{\mathbf{R}_2})$ and $T_{\mathbf{L}_2}$ is null. Let's consider $t \in (T_{\mathbf{S}_1} + T_{\mathbf{R}_2})$ and $t' \in T_{\mathbf{L}_2}$. One has:

$$\exists\ \alpha_1, \alpha_2 \in \mathbb{R} \quad / \quad t = \alpha_1 t_7 + \alpha_2 t_8$$
$$\exists\ \alpha'_1, .., \alpha'_4 \in \mathbb{R} \quad / \quad t' = \alpha'_1 t_9 + \alpha'_2 t_1 0 + \alpha'_3 t_1 1 + \alpha'_4 t_1 2$$

Therefore $t = t'$ is equivalent to:

$$\begin{cases} \alpha_1 l\sin\theta_1\overrightarrow{x_1} - \alpha_2 l_2\overrightarrow{y_2} + \alpha'_4\overrightarrow{z_{forearm}} = \overrightarrow{0} \\ (\alpha_1 z_{1x} + \alpha'_1 + \alpha_2)\overrightarrow{x_2} + (\alpha_1 z_{1y} + \alpha'_2)\overrightarrow{y_2} + (\alpha_1 z_{1z} + \alpha'_3)\overrightarrow{z_2} = \overrightarrow{0} \end{cases}$$

The first of these two equations leads to $\alpha_1 = \alpha_2 = \alpha'_4 = 0$ since it is supposed that $\overrightarrow{x_1}, \overrightarrow{y_2}$ and $\overrightarrow{z_{forearm}}$ are not coplanar in order to avoid the third singularity, and $\sin\theta_1 \neq 0$ in order to avoid the second singularity. Therefore, the second equation leads to $\alpha_1 = \alpha_2 = \alpha'_4 = 0$ because $\{\overrightarrow{x_2}, \overrightarrow{y_2}, \overrightarrow{z_2}\}$ forms a basis. In conclusion, $t = t' \Rightarrow t = 0$, or $dim(T_{\mathbf{S}_2}) = 0$.

4) Summary.

In conclusion, we identified three singularities:

1) $\overrightarrow{z_{arm}}.\overrightarrow{z_1} = 0$ representing the case where the passive slide, mounted parallel to the upper arm axis, is perpendicular to the robot upper limb axis. This case will never appear since the angle between $\overrightarrow{z_{arm}}$ and $\overrightarrow{z_1}$ reflects small discrepancies between the exoskeleton and human kinematics, and remains smaller than a few degrees.
2) $\sin\theta_1 = 0$ representing the case where M_1, P_1 and P_2 are aligned. This singular configuration can be avoided by limiting the range of motion for the robot elbow to a few degrees before full extension.
3) $\overrightarrow{x_1}, \overrightarrow{y_2}$ and $\overrightarrow{z_{forearm}}$ coplanar. This configuration does not appear in practice, since in the nominal configuration, $\overrightarrow{x_1}$ is perpendicular to the plane generated by $\overrightarrow{y_2}$ and $\overrightarrow{z_{forearm}}$.

Therefore, under normal conditions of operation, the ABLE exoskeleton with its two fixations never falls into a singular configuration.

REFERENCES

[1] A.B. Zoss, H. Kazerooni, and A. Chu. Biomechanical design of the berkeley lower extremity exoskeleton (bleex). *Mechatronics, IEEE/ASME Transactions on*, 11(2):128 –138, april 2006.

[2] M. Mihelj, T. Nef, and R. Riener. Armin ii - 7 dof rehabilitation robot: mechanics and kinematics. In *Robotics and Automation, 2007 IEEE International Conference on*, pages 4120 –4125, 10-14 2007.

[3] J.C. Perry, J. Rosen, and S. Burns. Upper-limb powered exoskeleton design. *Mechatronics, IEEE/ASME Transactions on*, 12(4):408 –417, aug. 2007.

[4] Jos L. Pons. *Wearable Robots: Biomechatronic Exoskeletons*. Wiley, April 2008.

[5] Stephen H. Scott and David A. Winter. Biomechanical model of the human foot: Kinematics and kinetics during the stance phase of walking. *Journal of Biomechanics*, 26(9):1091–1104, September 1993.

[6] F.C.T. Van der Helm, H.E.J. Veeger, G.M. Pronk, L.H.V. Van der Woude, and R.H. Rozendal. Geometry parameters for musculoskeletal modelling of the shoulder system. *Journal of Biomechanics*, 25(2):129–144, February 1992.

[7] A. Schiele. An explicit model to predict and interpret constraint force creation in phri with exoskeletons. In *Robotics and Automation, 2008. ICRA 2008. IEEE International Conference on*, pages 1324 –1330, 19-23 2008.

[8] A. Schiele and F.C.T. van der Helm. Kinematic design to improve ergonomics in human machine interaction. *Neural Systems and Rehabilitation Engineering, IEEE Transactions on*, 14(4):456 –469, dec. 2006.

[9] L W Lamoreux. Kinematic measurements in the study of human walking. *Bull Prosthet Res.*, 10(15):3–84, 1971. PMID: 5131748.

[10] KL Markolf, JS Mensch, and HC Amstutz. Stiffness and laxity of the knee–the contributions of the supporting structures. a quantitative in vitro study. *J Bone Joint Surg Am*, 58(5):583–594, 1976.

[11] C. Diez-Martnez, J. Rico, J. Cervantes-Snchez, and J. Gallardo. Mobility and connectivity in multiloop linkages. In *Advances in Robot Kinematics*, pages 455–464. 2006.

[12] M. Stokdijk, C. G. M. Meskers, H. E. J. Veeger, Y. A. de Boer, and P. M. Rozing. Determination of the optimal elbow axis for evaluation of placement of prostheses. *Clinical Biomechanics*, 14(3):177–184, March 1999.

[13] K. J. Waldron. The constraint analysis of mechanisms. *Journal of Mechanisms*, 1(2):101–114, 1966.

[14] P. Garrec, J.P. Friconneau, Y. Measson, and Y. Perrot. Able, an innovative transparent exoskeleton for the upper-limb. *Intelligent Robots and Systems, 2008. IROS 2008. IEEE/RSJ International Conference on*, pages 1483–1488, Sept. 2008.

[15] Garrec. P. French patent: Transmission vis, ecrou et cable attache a la vis - fr0101630, 2000 (eur 01938347.0-2421 and us 10/296,740 (screw and nut transmission and cable). 2000.

Assessing Optimal Assignment under Uncertainty: An Interval-based Algorithm

Lantao Liu
Computer Science and Engineering Department
Texas A&M University
College Station, USA
Email: lantao@cse.tamu.edu

Dylan A. Shell
Computer Science and Engineering Department
Texas A&M University
College Station, USA
Email: dshell@cse.tamu.edu

***Abstract*— We consider the problem of multi-robot task-allocation when robots have to deal with uncertain utility estimates. Typically an allocation is performed to maximize expected utility; we consider a means for measuring the robustness of a given optimal allocation when robots have some measure of the uncertainty (e.g., a probability distribution, or moments of such distributions). We introduce a new $O(n^4)$ algorithm, the Interval Hungarian algorithm, that extends the classic Kuhn-Munkres Hungarian algorithm to compute the maximum interval of deviation (for each entry in the assignment matrix) which will retain the same optimal assignment. This provides an efficient measurement of the tolerance of the allocation to the uncertainties, for both a specific interval and a set of interrelated intervals. We conduct experiments both in simulation and with physical robots to validate the approach and to gain insight into the effect of location uncertainty on allocations for multi-robot multi-target navigation tasks.**

I. Introduction

Task-allocation mechanisms are among the most successful non-domain-specific means for coordinating the actions of multiple robots. These involve treating the work to be performed as a set of tasks and the robots themselves as workers to be assigned to particular tasks. By estimating the expected utility of a particular robot's performance of a particular task, algorithms can optimize the allocation of robots to tasks (or vice versa) in order to maximize expected collective performance. The complexity of the allocation problem depends on the particular capabilities required to achieve the task, the capabilities of the robots, and whether the allocation must consider temporal scheduling aspects [1]. Generally the appropriateness of the task-allocation approach and the difficulty of the allocation problem both depend on the degree to which each of the robots and each of the tasks can be considered independent.

We consider the archetype multi-robot task-allocation problem which involves performing an instantaneous assignment of single-task robots to single-robot tasks (following the taxonomic characterization in [1]). This reduces to an instance of the well-studied Optimal Assignment Problem (OAP) for which the Kuhn-Munkres Hungarian algorithm, which was first proposed by H. W. Kuhn [2] in 1955 and improved by J. Munkres [3] in 1957, is a solution.

For multi-robot task-allocation, however, outstanding issues remain. The Hungarian algorithm maximizes the utility for the team because it is provided with an estimate of each robot's expected utility. Calculating this estimate is costly because every robot must provide estimates for each task, and the optimality of the resultant allocation is only meaningful when these estimates are accurate. Furthermore, the robots each have to deal with uncertainty about the state of the world in constructing these estimates. Even if the robots maintain a representation of this uncertainty (*e.g.,* a distribution over potential states) the expected utility is only the first moment of the utility distribution given a particular robot-task assignment pair. Important questions are: (1) How much effort should the robots invest in constructing the utility estimates? For n robots and n tasks, n^2 estimates are provided—but only n elements make up the optimum assignment; not all utility estimates need to be known with equal fidelity. (2) Once an allocation is computed, how stable is that allocation with respect to changes in the matrix of utility estimates? (3) If these utility estimates arise from an underlying probability distribution, what is the likelihood that the assignment is sub-optimal?

This paper makes the following contributions toward addressing these questions:

- Identification of properties of intervals of utility estimates, and proofs of these properties.
- Introduction of a new algorithm, the Interval Hungarian Algorithm, ideally suited to multi-robot systems, although broadly applicable to any OAP where the matrix of utility estimates is subject to uncertainty.
- Introduction of an efficient method for quantifying the effects of uncertainty on an allocation. This includes analysis for instances with a single specific estimate and multiple interrelated estimates.
- Analysis of the impact of uncertainty with concrete examples, using standard localization methods to produce real utility distributions with physical robots and in simulation.

II. Related Work

It is worthwhile drawing a distinction between multi-robot coordination strategies that employ assignment methods in which expected utilities for each robot's task performance is provided, and those that model the task performance itself in greater detail. The former depend on an independence assumption, because a linear function (like the sum) of the group's utilities is optimized. Other effects, like interference or inter-robot synergy, can only be captured in the way they effect particular utilities in the expectation. Centralized and distributed algorithms for performing allocations have been developed, including greedy allocations [4], optimization techniques [1, 5] and auction [6, 7] and market-based approaches [8, 9]. It is within this framework that the present study falls. It uses little information about the domain-specific aspects that lead to the structure of the coordination problem, or even the source of the uncertainty. However, aspects like interrelated utilities (or in the present study, more generally, interrelated uncertainty in the utilities) are not explicitly captured. If the effect of these higher-order interactions on utility values can be calculated, then the presented algorithm can still be of use.

Schemes that use a richer model of agent and task in order to construct a probabilistic model, for example, stochastic games/decentralized-MDPs [10], factored-MDPs,[11], POMDPs [12], permit one to address explicitly the question of when to perform particular actions (movement, sensing, communication) in order to reduce uncertainty if doing so is beneficial for the task performance. However, these problems do not admit polynomial-time solutions, and often factorization or independence assumptions are introduction in order to make the problem tractable.

Algorithm II.1 The Hungarian Algorithm

Input:

A valid $n \times n$ assignment matrix represented as the equivalent complete weighted bipartite graph $G = (X, Y, E)$, where $|X| = |Y| = n$.

Output:

A perfect matching, M.

1: Generate an initial labelling l and matching M in G_e.

2: If M perfect, terminate algorithm. Otherwise, randomly pick an exposed vertex $u \in X$. Set $S = \{u\}$, $T = \varnothing$.

3: If $N(S) = T$, update labels:

$$\delta = min_{x \in S, y \in Y-T}\{l(x) + l(y) - w(x, y)\}$$

$$l'(v) = \begin{cases} l(v) - \delta & if\ v \in S \\ l(v) + \delta & if\ v \in T \\ l(v) & otherwise \end{cases}$$

4: If $N(S) \neq T$, pick $y \in N(S) - T$.

(a) If y exposed, $u \rightarrow y$ is augmenting path. Augment M and go to step 2.

(b) If y matched, say to z, extend Hungarian tree: $S = S \bigcup \{z\}$, $T = T \bigcup \{y\}$, and go to step 3.

$*$ Definitions:

- Equality graph $G_e = \{e(x, y) : l(x) + l(y) = w(x, y)\}$;
- Neighbor $N(u)$ of vertex $u \in X$: $N(u) = \{v : e(u, v) \in G_e\}$.

III. HUNGARIAN ALGORITHM

The Hungarian Algorithm treats the optimization OAP as a combinatorial problem in order to efficiently solve an $n \times n$ task assignment problem in $O(n^3)$ time. The utility estimates become edge weights in a complete bipartite graph in which each robot and task becomes a vertex. The Hungarian Algorithm (Algorithm II.1) searches for a perfect matching in a sub-graph of the complete bipartite graph, where the perfect matching is exactly the optimal assignment solution. In step 4 the search process either increases the matching size, or the so-called *equality graph* in which the matching resides. (Graph and matching definitions and notation are adopted from [13].)

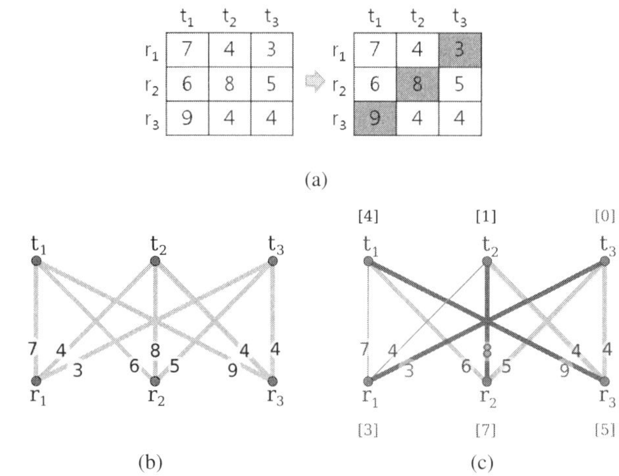

Fig. 1. Hungarian algorithm solves the OAP using a complete bipartite graph. (a) An assignment matrix and solution; (b) A complete bipartite graph; (c) The perfect matching for assignment solution. The state of the graph after an assignment, we term the *resultant bipartite graph*.

Figure 1 shows an example assignment problem and the corresponding perfect matching in form of the associated bipartite graph. In Figure 1(a), the task assignment problem is described as an assignment matrix (which need not be square in general). An element u_{ij} is an estimated utility, meaning that robot r_i has an expected utility when assigned to perform task t_j. The algorithm generates the maximal allocation, shown as shaded cells in the matrix. Figs. 1(b) and 1(c) show the same information but in bipartite graph form. Edges that are used during the calculation are differentiated by 3 colors: red (bold dark) edges comprise the matching, green (thin dark) edges are unmatched but they are in the equality graph G_e, grey (bold light) edges are also unmatched but do not appear the equality graph. Edges within G_e are termed admissible (both red and green edges) and will have weights that satisfy $w_{ij} = l(r_i) + l(t_j)$. Bracketed integers beside each vertex represent the labeling value $l(\cdot)$. Edge set M and scalar m represent a specific perfect matching solution and the corresponding optimum value, respectively.

Since the development of the Hungarian Algorithm many variations have been proposed (see [14] for a recent survey). Two relate directly to this paper: [15] provide an incremental Hungarian Method, which is an $O(n^2)$ technique for inserting a pair of new vertices in the bipartite graph resulting from a previous assignment. An extension to this in [16] also permits deletions, which enables one to solve assignment problems with k utility changes in $O(kn^2)$ time. The current work computes a description of an assignment problem (in $O(n^4)$)

which permits subsequent $O(1)$ queries of whether or not the assignment has changed. When assignments are being recomputed frequently with small absolute changes in value (*e.g.*, frequent replanning of distances as a robot moves) this can be a considerable saving.

IV. Interval Hungarian Algorithm

For each utility value, we compute the interval in which the utility may be (independently) perturbed before the optimality of the computed assignment is violated. Thus, given an input matrix of utilities, the algorithm characterizes a set of inputs which yield the same output. The intervals are computed based on the three categories of edges described above.

A. Interval Analysis for Matched Edges

The allowable intervals for matched edge weights is analyzed as follows: for any such edge $e_m(r_\alpha, t_\beta)$, the interval can be described as $[w_{m\alpha\beta} - \varepsilon_m, +\infty)$, where $w_{m\alpha\beta}$ is the edge weight of $e_m(r_\alpha, t_\beta)$ and ε_m is the *tolerance margin* that the weight can decrease without violating the optimality of the current matching solution. It is safe to increase the weight as this is a maximization problem. We say a matched edge is *hidden* if its weight has decreased so as to no longer form part of a matching solution.

Lemma 4.1: With the resultant matching solution M_0 and bipartite graph of Hungarian algorithm, if a matched edge $e_m(r_\alpha, t_\beta)$ is hidden, then the Hungarian algorithm can be completed with one iteration rooted at exposed node r_α. When a new perfect matching solution M' exists, the labeling reduction of the root r_α satisfies $l(r_\alpha) - l'(r_\alpha) = m_0 - m'$.

Proof: The proof is based on the Hungarian algorithm solution method. If one matched edge $e_m(r_\alpha, t_\beta)$ is hidden, the bipartite graph remains feasible, but the equality graph G_e loses one matched edge. Hiding $e_m(r_\alpha, t_\beta)$ exposes r_α and t_β, requiring an iteration to complete the Hungarian algorithm. All other unexposed vertices are on corresponding matched edges and, thus, are included in G_e, and moreover the sum of their labels is constant for all subsequent iterations. To grow G_e in bridging a new augmenting path[1], $l(r_\alpha)$ may decrease, but $l(t_\beta)$ remains constant because it is the end of augmenting path and will be reached last. Therefore, $l(r_\alpha)$ is the only variable that can reduce the optimum. This proves that the reduction of $l(r_\alpha)$ is exactly the reduction of optimum from M_0 to M', namely, $l(r_\alpha) - l'(r_\alpha) = m_0 - m'$. ■

*Theorem 4.2 (**Matched Edge Interval**):* Hiding a matched edge from the Hungarian solution leads to a new solution, and the labeling reduction ε_m at the root of the Hungarian tree is the tolerance margin for this element, *i.e.*, the safe interval for matched edge $e_m(r_\alpha, t_\beta)$ is $[w_{m\alpha\beta} - \varepsilon_m, +\infty)$.

Proof: Assigning weight $w = w_{m\alpha\beta} - \varepsilon_m$ to edge $e_m(r_\alpha, t_\beta)$ results in optimum $m_0 - \varepsilon_m$ for the original matching M_0. Any new matching M' which excludes $e_m(r_\alpha, t_\beta)$ also has an optimum of $m' = m_0 - \varepsilon_m$, therefore both M_0 and M' are optimal matching solutions. Whenever the weight of edge $e_m(r_\alpha, t_\beta)$ satisfies $w \geq w_{m\alpha\beta} - \varepsilon_m$, we have $m_0 \geq m'$ implying that M_0 is optimal, since otherwise M_0 would be substituted by M'. Thus, the maximum allowable interval for matched edge $e_m(r_\alpha, t_\beta)$ is $[w_{m\alpha\beta} - \varepsilon_m, +\infty)$. ■

[1]Although the definition of *augmenting path* is not necessary to understand Algorithm II.1, we note that here we deviate from [13, pg. xxxii], where they use the term M-augmenting path.

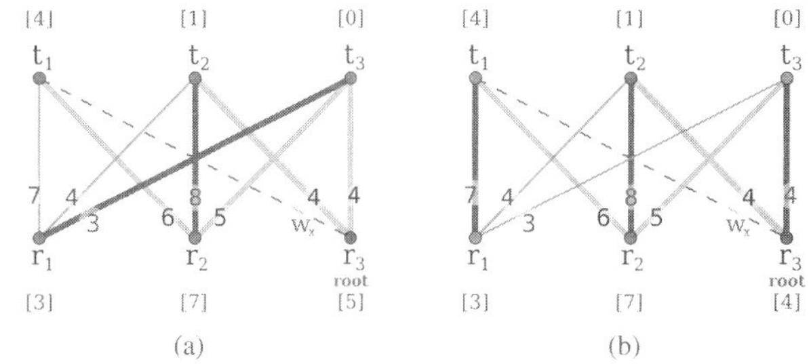

Fig. 2. Interval analysis for a matched edge. (a) Hide an objective matched edge and assign it with weight w_x; (b) New matching solution without the hidden edge.

Algorithm IV.1 Intervals of Matched Edges

Input:

A matched edge $e_m(r_\alpha, t_\beta)$ and the corresponding resultant bipartite graph.

Output:

Interval (lower bound ξ_m) for $e_m(r_\alpha, t_\beta)$.

1: Hide $e_m(r_\alpha, t_\beta)$ by assigning it an unknown weight w_x. Set $S = \{r_\alpha\}$, $T = \varnothing$.

2: If $N(S) = T$, update labels:
$\delta = min_{x \in S, y \in Y - T, e(x,y) \neq e_m(r_\alpha, t_\beta)}\{l(x) + l(y) - w(x, y)\}$

$$l'(v) = \begin{cases} l(v) - \delta & if\ v \in S \\ l(v) + \delta & if\ v \in T \\ l(v) & otherwise \end{cases}$$

$l'(r_\alpha) + l'(t_\beta) - w_x > \delta \Rightarrow w_x < l'(r_\alpha) + l'(t_\beta) - \delta$.
Update $\xi_m = l'(r_\alpha) + l'(t_\beta) - \delta$.

3: If $N(S) \neq T$, pick $y \in N(S) - T$.
(a) If $y = t_\beta$, there must be an augmenting path $r_\alpha \longrightarrow t_\beta$. Augment matching and terminate algorithm.
(b) If y matched, say to z, extend Hungarian tree: $S = S \bigcup \{z\}$, $T = T \bigcup \{y\}$. Go to step 2.

Lemma 4.1 and Theorem 4.2 permit computation of the interval of a matched edge $e_m(r_\alpha, t_\beta)$ in the following way: first hide $e_m(r_\alpha, t_\beta)$ from the bipartite graph and assign it an undecided weight w_x that satisfies the constraint: $w_x < l(r_\alpha) + l(t_\beta)$. Next, let exposed vertex r_α be the root of a Hungarian tree and construct an augmenting path excluding $e_m(r_\alpha, t_\beta)$. The algorithm terminates when such a path is found that generates a perfect matching. Because $l(t_\beta)$ stays unchanged but $l(r_\alpha)$ is decreased, w_x will decrease per iteration; the lower bound $w_{m\alpha\beta} - \varepsilon_m$ occurs moment when the new perfect matching exists.

Figure 2 illustrates an example. Hiding matched edge $e(r_3, t_1)$ requires construction of a Hungarian tree rooted at newly exposed vertex r_3. Here $l(r_3)$ decreases while searching for an augmenting path, and a new matching solution

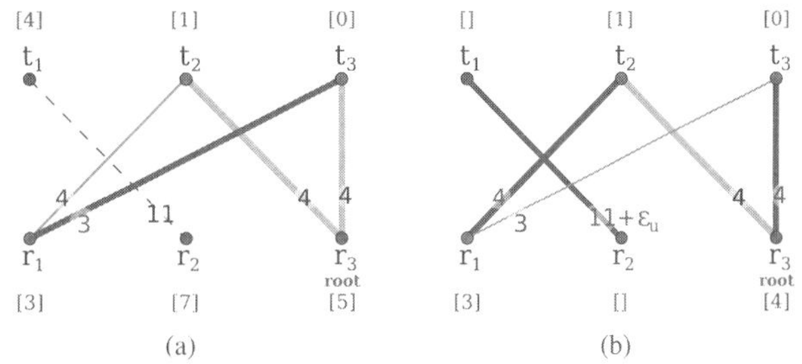

Fig. 3. Interval analysis for an unmatched edge. (a) Hide all associated edges of objective unmatched edge; (b) New matching solution formed with the objective edge and matching solution in auxiliary bipartite graph.

$\langle e(r_1,t_1), e(r_2,t_2), e(r_3,t_3)\rangle$ replaces the original one in augmenting path $r_3 \rightarrow t_3 \rightarrow r_1 \rightarrow t_1$. The reduction of labeling for r_3 is $5-4=1=\varepsilon_m$, and the interval for $e(r_3,t_1)$ is $[8,+\infty)$.

B. Interval Analysis for Unmatched Edges

An unmatched edge $e_u(r_\alpha,t_\beta)$ has an interval $(-\infty,\xi_u]$, where the upper bound ξ_u reflects the maximum value of the utility for it to remain an unmatched robot and task pair.

Lemma 4.3: In the resultant bipartite graph of the Hungarian algorithm, the weight of any unmatched edge $e_u(r_x,t_y)$ can be increased to the sum of two associated labeling values $l(r_x)+l(t_y)$ without affecting the assignment optimum.

Proof: An increment of this form does not violate the feasibility of the bipartite graph, nor does the modification remove any admissible edges from G_e. Therefore, the original matching in G_e remains perfect. ■

However, the upper bound in Lemma 4.3 is not tight. For example, $e(r_2,t_1)$ in Figure 1(c), the weight can safely increase to 12 rather than $4+7=11$ (12 is the upper bound because greater weights result in optimal matching $\langle e(r_1,t_2), e(r_2,t_1), e(r_3,t_3)\rangle$), and there is a tolerance margin of 1. Next we show how to find the tolerance margin ε_u and further improve the upper bound. At present we redefine the interval as $(-\infty, l(r_\alpha)+l(t_\beta)+\varepsilon_u]$. Note that this also shows that all unmatched edges, whether in G_e or not, can be treated uniformly once they become admissible.

To obtain ε_u for unmatched edge $e_u(r_\alpha,t_\beta)$, hide $e_u(r_\alpha,t_\beta)$ and all other edges incident to vertices r_α and t_β from the resultant bipartite graph. This yields a bipartite graph with $n-1$ vertices in each partition. We term this new bipartite graph the *auxiliary* bipartite graph G_a. Notice that auxiliary bipartite graph is associated with a particular edge, and that the auxiliary bipartite graph has only $n-2$ matched edges. It therefore requires the addition of one edge for a matching solution.

*Theorem 4.4 (**Unmatched Edge Interval**):* Any unmatched edge $e_u(r_\alpha,t_\beta)$ in the Hungarian resultant bipartite graph, has interval tolerance margin $\varepsilon_u = m_0 - (m_a + l(r_\alpha) + l(t_\beta))$, where m_0 is the optimum of the original solution, and m_a is the optimum of the auxiliary bipartite graph associated with $e_u(r_\alpha,t_\beta)$. The allowable interval for edge $e_u(r_\alpha,t_\beta)$ is $(-\infty, m_0-m_a]$.

Proof: For an arbitrary unmatched edge $e_u(r_\alpha,t_\beta)$ and its associated auxiliary bipartite graph G_a of size $n-1$, we add $e_u(r_\alpha,t_\beta)$ to the matching solution M_a of G_a. This forms a new matching M' of size n. If the weight of edge $e_u(r_\alpha,t_\beta)$ satisfies $w_{u\alpha\beta} > m_0 - m_a$, then the matching M' containing $e_u(r_\alpha,t_\beta)$ must satisfy $m' = w_{u\alpha\beta} + m_a > m_0$. But this contradicts the fact that the original matching M_0 was perfect. This proves that the upper bound for unmatched edge $e_u(r_\alpha,t_\beta)$ is m_0-m_a, and that the allowable interval is $(-\infty, m_0-m_a]$. ■

Algorithm IV.2 Intervals of Unmatched Edges

Input:

An unmatched edge $e_u(r_\alpha,t_\beta)$ and the corresponding resultant bipartite graph.

Output:

Interval (upper bound ξ_u) for $e_u(r_\alpha,t_\beta)$.

1: Assume $e(r_\alpha, mate(r_\alpha))$, $e(t_\beta, mate(t_\beta))$ are matched edges, then set $S=\{mate(t_\beta)\}$, $T=\varnothing$.

2: Hide $e_u(r_\alpha,t_\beta)$ and all other edges incident to vertices r_α and t_β, and obtain the auxiliary bipartite graph G_a.

3: In G_a, if $N(S)=T$, update labels:
$\delta = min_{x\in S, y\in Y-T}\{l(x)+l(y)-w(x,y)\}$

$$l'(v) = \begin{cases} l(v)-\delta & if\ v\in S \\ l(v)+\delta & if\ v\in T \\ l(v) & otherwise \end{cases}$$

4: In G_a, if $N(S)\neq T$, pick $y\in N(S)-T$.
(a) If $y = mate(r_\alpha)$, there must be an augmenting path $mate(t_\beta) \rightarrow mate(r_\alpha)$. Augment matching and go to step 5.
(b) If y matched, say to z, extend Hungarian tree: $S = S\bigcup\{z\}$, $T = T\bigcup\{y\}$. Go to step 3.

5: $\xi_u = m_0 - m_a$.

∗ Definitions:

- $mate(v)$ is the other ending vertex with regard to vertex v;
- m_0 is optimum of the original solution, m_a is optimum of G_a.

Consider Figure 3 as an example. Hiding $e(r_2,t_1)$ and all edges incident to r_2 and t_1 leaves auxiliary bipartite graph G_a containing vertices r_1, r_3, t_2, t_3 and associated edges (see Figure 3(a)). The Hungarian tree is created rooted at newly exposed vertex r_3, and finally the matching solution M_a of G_a is $\langle e(r_1,t_2), e(r_3,t_3)\rangle$, as shown in Figure 3(b). The tolerance margin and allowable interval for $e(r_2,t_1)$ are $\varepsilon_u = m_0 - (m_a + l(r_2) + l(t_1)) = 1$ and $(-\infty, 12]$, respectively.

C. Interval Hungarian Algorithm

Combining the interval analysis of matched and unmatched edges, we have the Interval Hungarian algorithm described in Algorithm IV.3. Figure 4 shows the corresponding intervals for assignment matrix in Figure 1(a).

D. Computational Complexity

The algorithm has worst-case time complexity $O(n^4)$. The Hungarian algorithm has a computational complexity of $O(n^3)$. Obtain the intervals for n matched edges, needs an

Algorithm IV.3 Interval Hungarian Algorithm

Input:

A resultant bipartite graph from running Algorithm II.1.

Output:

An interval matrix $mx_{itv}(n, n)$ storing all intervals.

```
mx_itv(n, n) = NULL.
for all edges e(i, j) in bipartite graph
  if e(i, j) is matched
    compute interval I(i, j) with Algorithm IV.1.
    mx_itv(i, j) = I(i, j).
  else
    compute interval I(i, j) with Algorithm IV.2.
    mx_itv(i, j) = I(i, j).
return mx_itv.
```

	t_1	t_2	t_3
r_1	$(-\infty,8]$	$(-\infty,6]$	$[2,+\infty)$
r_2	$(-\infty,12]$	$[6,+\infty)$	$(-\infty,7]$
r_3	$[8,+\infty)$	$(-\infty,8]$	$(-\infty,5]$

Fig. 4. An example of an interval matrix

extra $O(n \times n^2) = O(n^3)$ operations since for each edge requires construction and search of the Hungarian tree costing $O(n^2)$. The computation of all the unmatched edges has a worst case cost of $O(n^4)$ since, for each unmatched edge, we do the same searching iterations in G_a, which has a size of $n-1$ and thus needs a computational complexity of $O((n-1)^2)$. There are $n^2 - n$ unmatched edges, yielding a total of $O((n^2-n) \times (n-1)^2) = O(n^4)$.

V. Quantifying the Effect of Uncertainty

When the Hungarian algorithm is applied to a matrix of expected utilities calculated from uncertain data (*e.g.*, using the mean of a utility distribution) one has little idea of the impact the uncertainty has on the resultant assignment. The output from the Interval Hungarian algorithm can be used to analyze the changes in optimal allocation as changes are made to particular utility values. This can be used to evaluate likelihood that the calculated assignment will be sub-optimal.

A. Uncertainty Measurement for a Single Utility

*Theorem 5.1 (**Uncertainty of a Single Interval**):* With regard to any specific single utility value, assuming other utilities are certain, the perfect matching solutions are identical if and only if any specific utility is within its allowable interval.

Proof: Assuming other utilities in the assignment matrix are certain, taken together Theorems 4.2 and 4.4 prove that if the matching solution remains the same, then any specific utility must be within its allowable interval. To prove the converse: suppose the matching solutions are not identical, then there must be a new perfect matching M' that replaces the original one M_0 *i.e.*, $m' > m_0$. However, this cannot happen since any value within interval must produce a matching solution with weight sum less than or equal to m_0. ■

To analyze the effect of uncertainty on a specific utility in the assignment matrix, we assume the other values are certain. Given a probability density function $f(x)$ for this specific expected utility, and associated interval I as output from the algorithm, the probability of a sub-optimal assignment is:

$$P_I = \begin{cases} \int_{\xi_m}^{+\infty} f(x), & \text{when } I = [\xi_m, +\infty) \\ \int_{-\infty}^{\xi_u} f(x), & \text{when } I = (-\infty, \xi_u]. \end{cases} \tag{1}$$

For applications in which robots are actively estimating quantities involved in producing $f(x)$, one may set some threshold T, such that robots only commit to an assignment if $P_I \geq T$, and instead invest resources in reducing the uncertainty in the estimate if it is likely to have a major bearing on the particular assignment. High values of T will ensure the robots only commit to allocations that are robust to errors in estimates of the expected utility.

B. Uncertainty Measurement for Interrelated Utilities

The previous subsection gives an approach for quantifying the effect of uncertainty on the robot-to-task assignment when only one utility was uncertain. Most often, however, multiple utilities are uncertain, and they may all be related if they involve inference over the same underlying state variables. For example, a row in the assignment matrix represents all relevant expected utilities for a specific robot. A change that effects the performance of the robot (*e.g.,* low battery) effects all the entries in the row. Here we use the term *interrelated edges* to represent all directly related utilities in a single row or column. For the same assignment to be preserved, despite n interrelated edges, there must be one and only one edge that remains matched, and all the others should be unmatched.

*Theorem 5.2 (**Uncertainty of Interrelated Intervals**):* Given a set of n interrelated edges, assume e_m is the matched edge with interval $[w_m - \varepsilon_m, +\infty)$, and e_{ui} are unmatched edges with intervals $(-\infty, w_{ui} + \varepsilon_{ui}]$, $(i = 1, 2, ..., n-1)$, then for any $\varepsilon' \leq \varepsilon_m$, the weight of e_m can be safely substituted with $w_m - \varepsilon'$, and the interval for e_{ui} becomes $(-\infty, w_{ui} + \varepsilon_{ui} - \varepsilon']$, $(i = 1, 2, ..., n-1)$.

Proof: $\varepsilon' \leq \varepsilon_m$ indicates that new weight $w'_m = w_m - \varepsilon'$ is within the interval associated with edge e_m, thus a substitution of w'_m will not violate the matching solution. To prove the interval form for e_{ui} $(i = 1, 2, ..., n-1)$: suppose m'_0 is the solution optimum with substituted weight w'_m, then we have $m'_0 = m_0 - \varepsilon'$. From Theorem 4.4, the interval for e_{ui} is $(-\infty, m'_0 - m_a]$, which can be substituted with $(-\infty, m_0 - \varepsilon' - m_a]$. Because $m_0 - m_a = w_{ui} + \varepsilon_{ui}$, the interval for e_{ui} becomes $(-\infty, w_{ui} + \varepsilon_{ui} - \varepsilon']$. ■

Notice that Theorem 5.2 exploits the mutual exclusion property of interrelated unmatched edges: at any time one and only one interrelated unmatched edge can possibly become matched. This means that as the matched edge's weight is decreased, one unmatched edge moves closer to it's interval's upper-bound. However, as the matched edge's value decreases, the unmatched edge's bounds decrease too. To allow the simultaneous occurrence of interrelated matched edge and

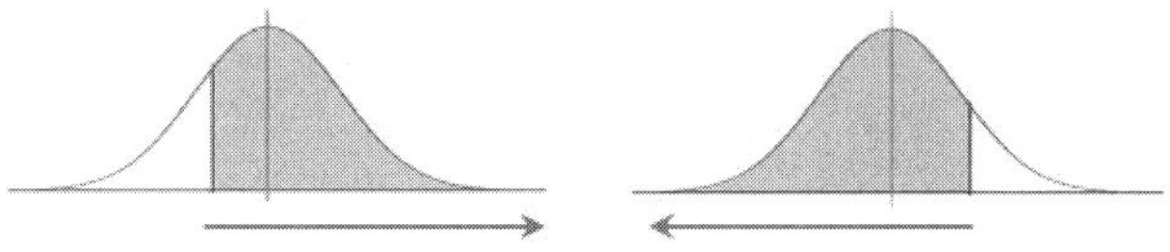

Fig. 5. Reliability levels (shaded area) with Gaussian distribution density. The horizontal axis represents the uncertain utility estimates.

unmatched edges, we compromise between them: intervals of interrelated unmatched edges shrinks by ε', while interval for interrelated matched edge shrinks by $\varepsilon_m - \varepsilon'$. We design the method for measuring the uncertainties of interrelated edges in the following steps:

1) Determine ε_{min} from all interrelated edges:
$\varepsilon_{min} = min(\varepsilon_m, \varepsilon_{ui}),\ (i = 1, 2, ..., n-1)$
2) Determine each interrelated interval I_i:
$$I_i = \begin{cases} [w_m - k \cdot \varepsilon_{min}, +\infty) \\ (-\infty, w_{ui} + \varepsilon_{ui} - k \cdot \varepsilon_{min}], (i = 1, 2, ..., n-1) \end{cases}$$
$*I_0$ represents interval for the matched edge. k is an empirical coefficient and $k \in [0, 1]$ which effects the degree to which the matched and unmatched interval's are scaled.
3) Determine probability:
$$P_{Ii} = \begin{cases} \int_{\xi'_m}^{+\infty} f(x),\ (\xi'_m = w_m - k \cdot \varepsilon_{min}) \\ \int_{-\infty}^{\xi'_{ui}} f(x),\ (\xi'_{ui} = w_{ui} + \varepsilon_{ui} - k \cdot \varepsilon_{min}) \end{cases} \quad (2)$$
4) Determine reliability level:
The assignment is reliable when $P_{Ii} \geq T$, and unreliable otherwise.

This approach utilizes a parameter k to balance the shrinking intervals (generally, $k \to 0$ when the number of robots is larger, which controls the loss of compromised ranges). In our experiments involving 3 robots, we use $k = 0.5$ to guarantee a margin of at least $0.5\varepsilon_{min}$ from each estimated utility). This is most effective for Gaussian-like distributions in which mean and medians of the distributions are close. Figure 5 illustrates the reliability levels (shaded area) under a Gaussian distribution density for the matched and unmatched edges, respectively. This method can measure the uncertainties for a horizontal row or a vertical column in the assignment matrix, assuming other non-directly interrelated rows or columns are known with certainty. The uncertainty involving multiple rows or columns is complex and beyond the scope of this paper. In practical applications, however, a conservative approach is simply to raise the threshold T so that the inaccuracy arising from multiple interrelated rows or columns can be compensated for.

VI. Experiments and Results

We demonstrate, through simulation and physical robot experiments, that the Interval Hungarian algorithm permits the effect of uncertainty on the allocation to be quantified. We consider the problem of dispatching a group of homogeneous

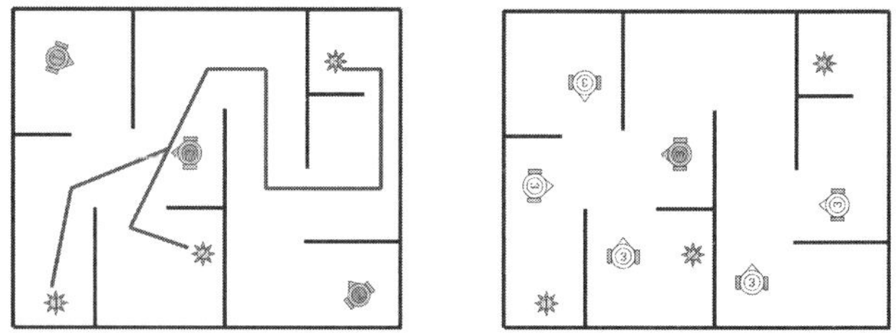

(a) (b)

Fig. 6. Multi-robot task assignment using localization. (a) Planned path with wavefront driver; (b) Possible localization results of robot #3.

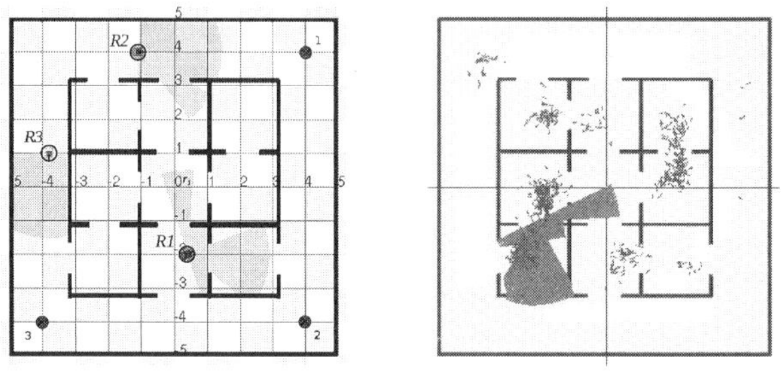

(a) (b)

Fig. 7. (a) Multi-robot task assignment and commitment in a maze; (b) Uncertain hypothesis for uncertainty-robot (robot $R1$ in red).

robots to a set of destination locations. We selected this task because it (and variations to it) have been used in the literature for the purpose of evaluating task assignment methods (*e.g.,* [7] and subsequent paper by Koenig's group). Additionally, the source of uncertainty and the means for actively estimating this has been well-studied in the last decade. Other tasks (*e.g.,* Parker's [4] Cooperative Multi-robot Observation of Multiple Moving Targets (CMOMMT) task) have different forms of uncertainty, but once a utility matrix is constructed for purposes of robot task-allocation, the method applies directly.

The assignment matrix is a negative estimation of the distance from current location to goal location and the uncertainty in the assignment problem comes from localization error. The robot attempts to localize itself in the environment (with a given map) by employing an particle filter-based approach [17]. Particles are clustered and the weighted means of these clusters taken as pose hypotheses. A planner is used to estimate the cost from estimated location to a given goal location. Figure 6 illustrates three robots being dispatched to three task locations in a maze-like room. The line segments connecting to the task locations in Figure 6(a) are the planned paths, and the path cost is the total length of each path. Figure 6(b) shows the possible poses of robot #3.

A. Simulation

1) Experimental Setup: The simulation environment is designed as in Figure 7(a) within the Stage simulator [18]. The maze was designed to be somewhat symmetric so as to provide multiple uncertain poses for the localization method. Three robots were randomly positioned within the maze and dispatched to the three task locations shown as big dots in the corners of Figure 7(a). Of the three robots, only one

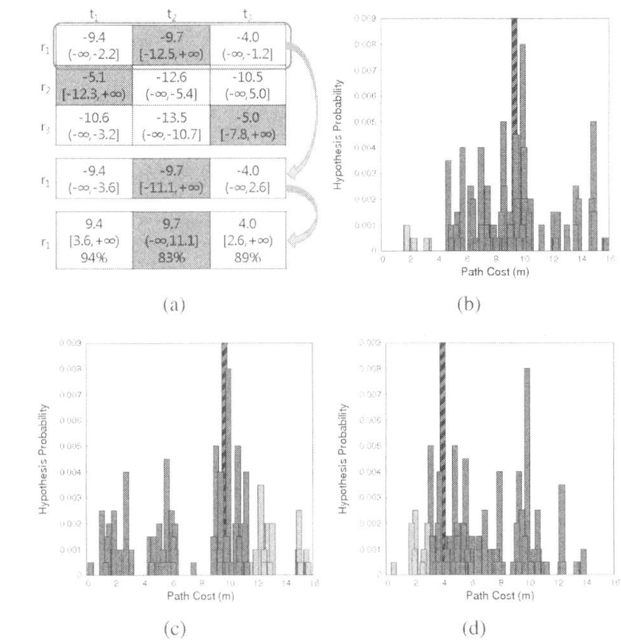

(a) (b) (c) (d)

Fig. 8. (a) Uncertainty analysis of interrelated intervals for robot $R1$; (b)—(d) Hypothesis distributions of interrelated utilities captured at 105th second.

(Robot $R1$ in red color, named *uncertainty-robot*) is using the localization algorithm and thus has uncertain poses, the other two ($R2$ in green and $R3$ in yellow, called *localized-robots*) obtain ground-truth poses from the simulator. This facilitates the verification of uncertainty measurement methods provided in Section V, which assume other non-interrelated utilities are known (actually non-interrelated uncertainties are allowed and also can be measured so long as not more than one set of interrelated uncertainties appear concurrently). The uncertainty-robot uses a simulated scanning laser ranging sensor and a map identical to the simulation environment. Figure 7(b) shows the particles for uncertainty-robot.

A path cost distribution shows the effect of uncertainty on a robot's utility estimates; it is obtained by computing path lengths from all available hypothesis of uncertainty-robot to a specific task location.

To capture the uncertainty, we set a fixed period t and empirical reliability threshold T, for example, $t = 5s$ and $T = 80\%$. After every period t, the following was performed: check the number of hypothesis n, if $n = 1$, the the robot is completely localized. If $n > 1$ then check the maximal probability p_{max} among all hypothesis, and if $p_{max} > T$, it indicates the robot is sufficiently certain of its location, otherwise, trigger the uncertainty measurement mechanism provided in Section V.

2) Results and Analysis: Figure 8 is the corresponding captured localization uncertainties at $t = 105s$. Figure 8(a) is the assignment matrix with optimal solution (shaded cells) and allowable intervals for all utilities. Figures 8(b)—8(d) are the path cost distributions for the uncertainty robot. In each sub-figure, the bar with highest probability (colored with stripe) is the path cost for the current most-likely localization result and thus is used in the assignment matrix. Other bars around it are the path costs for all other uncertain hypotheses. Remember that here only Robot $R1$ has imperfect localization, and the intervals for the interrelated utilities (the first row Figure 8(a)) are $\langle(-\infty, -2.2], [-12.5, +\infty), (-\infty, -1.2]\rangle$, thus we can compute its tolerance margin $\varepsilon_m = -9.7 - (-12.5) = 2.8$. Using the method discussed in Section V, we get the allowable interrelated intervals $\langle(-\infty, -3.6], [-11.1, +\infty), (-\infty, -2.6]\rangle$. The intervals are flipped from negative to positive to fit the real path cost distributions: $\langle[3.6, +\infty), (-\infty, 11.1], [2.6, +\infty)\rangle$.

In Figures 8(b)—8(d), the bars within interrelated intervals are filled red (dark), and the others are in pink (light). The calculated results of P_I are $\langle 94\%, 83\%, 89\%\rangle$, and we use the reliability threshold $T = 80\%$, therefore all utilities are considered reliable and we can trust the assignment solution.

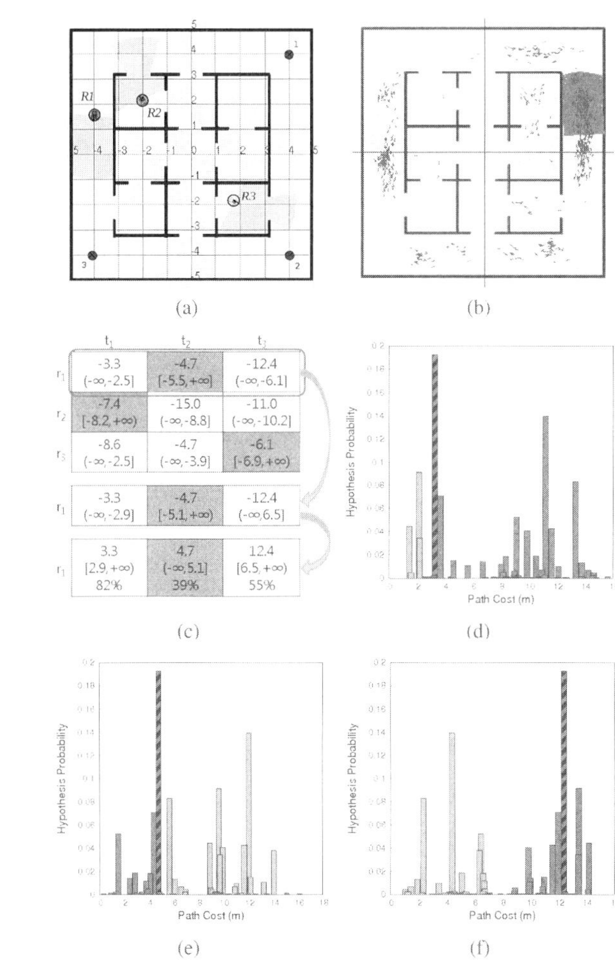

(a) (b) (c) (d) (e) (f)

Fig. 9. (a)—(b) Multi-robot assignment simulation captured at 76th second; (c)—(f) Uncertainty analysis of interrelated utilities.

Figure 9 is another example that illustrates the unreliable assignment solution. The reliability levels of interrelated utilities are $\langle 82\%, 39\%, 55\%\rangle$, as shown in Figure 9(c). There are two utilities below the reliability threshold, thus the assignment solution is considered unreliable. In fact, from Figure 9(a) and 9(b) we see that, if the localization of Robot $R1$ is at its true location, then the assignment solution should be

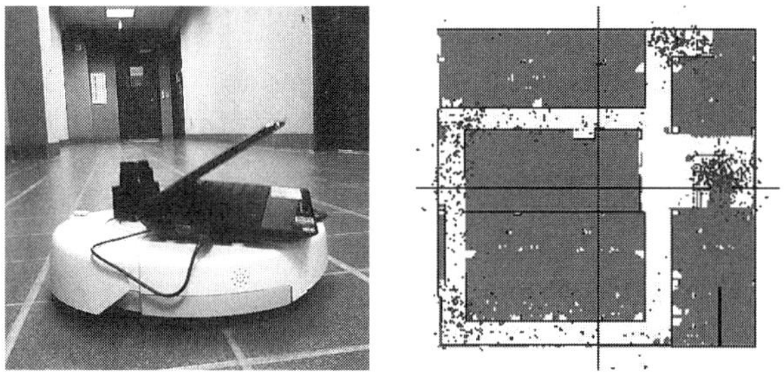

(a) (b)

Fig. 10. Physical robot localizing in a floor legend

$r_1 \rightarrow t_3$ (Robot $R1$ is assigned to the task location #3), $r_2 \rightarrow t_1$ and $r_3 \rightarrow t_2$. The case illustrates the sensitivity of the assignment solution to uncertainty, the algorithm's output suggests refining of the localization before committing to an unreliable assignment.

B. Simulation vs. Real Robots

1) Experimental Setup: We also conducted experiments with physical robot in our research building (Figure 10(b) shows the floor plan, as drawn by hand, with pose estimates.) A single iRobot create robot with Hokuyo URG-04LX-UG01 laser sensor, as shown in Figure 10(a), was used to collect pose estimates, which were in turn used to compute path cost distributions. To best compare the distributions we used the same parameters for the localization parameters and randomly placed the robot in a corridor with the same initial poses and goal positions at the same locations.

2) Results and Analysis: Compared with simulation, the physical robot shows greater certainty and converges more slowly. Figure 11(a) is a typical simulation result of the hypothesis distribution captured at $t = 117s$, and Figure 11(b) is the result from physical robot captured at $t = 319s$.

Qualitatively both distributions are clustered Gaussian-distributed bars. The simulation data have fewer clusters than the physical robot data. The greater uncertainty and slower convergence suggests that using the Interval Hungarian Method can be particularly important for physical robots with significant measurement error.

VII. Conclusion

This paper presents the Interval Hungarian algorithm and an approach for measuring the effect of uncertainty on the optimality and stability of the output from the OAP with direct application to task-allocation in multi-robot systems. The Interval Hungarian algorithm is based on the bipartite matching variant of the Hungarian algorithm. Given an input utility matrix it outputs an assignment matrix along with intervals which are associated with each utility in the input. Each interval conveys a tolerance of the optimal assignment to perturbations in the utility value. We illustrated how uncertainties in multi-robot assignment problems can be quantified when probability distributions describe the utility estimates. Data from simulated and physical robot implementations were presented and compared.

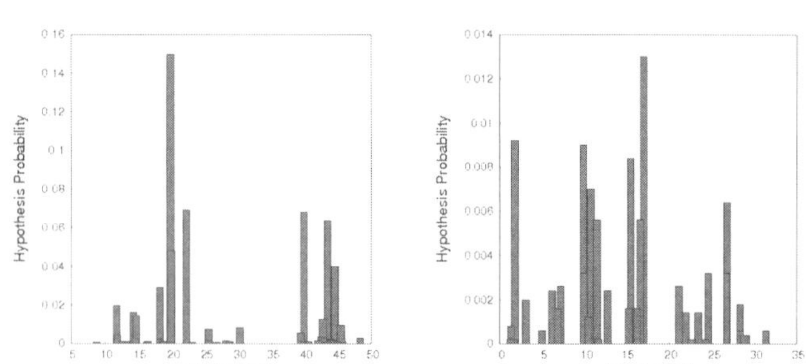

(a) Path cost distribution from the simulated robot after 117 seconds.

(b) Path cost distribution from the physical robot after 319 seconds.

Fig. 11. Comparison of hypothesis distributions between simulation (left) and physical robot (right)

References

[1] B. P. Gerkey and M. J. Matarić, "A formal analysis and taxonomy of task allocation in multi-robot systems," *International Journal of Robotics Research*, vol. 23, no. 9, pp. 939–954, Sept. 2004.

[2] H. W. Kuhn, "The hungarian method for the assignment problem," *Naval Research Logistic Quarterly*, 1955.

[3] J. Munkres, "Algorithms for the assignment and transportation problems," *Journal of the Soc. for Industrial and Applied Math.*, March 1957.

[4] L. E. Parker, "Alliance: An architecture for fault-tolerant multi-robot cooperation," *IEEE Transactions on Robotics and Automation*, vol. 14, no. 2, pp. 220–240, 1998.

[5] N. Atay and B. Bayazit, "Mixed-Integer Linear Programming Solution to Multi-Robot Task Allocation Problem," Department of Computer Science & Engineering, Washington University, Tech. Rep. WUCSE-2006-54, 2006.

[6] D. P. Bertsekas, "The auction algorithm for assignment and other network flow problems: A tutorial," *Interfaces*, 1990.

[7] M. Berhault, H. Huang, P. Keskinocak, S. Koenig, W. Elmaghraby, P. Griffin, and A. J. Kleywegt, "Robot Exploration with Combinatorial Auctions," in *Proceedings of the IEEE/RSJ International Conference on Intelligent Robots and Systems (IROS'03)*, Las Vegas, NV, U.S.A., Oct. 2003, pp. 1957–1962.

[8] M. B. Dias and A. T. Stentz, "A market approach to multirobot coordination," Robotics Institute, Pittsburgh, PA, Tech. Rep. CMU-RI -TR-01-26, August 2001.

[9] M. B. Dias, R. Zlot, N. Kalra, and A. Stentz, "Market-based multirobot coordination: a survey and analysis," *Proceedings of the IEEE—Special Issue on Multi-robot Systems*, vol. 94, no. 7, pp. 1257–1270, July 2006.

[10] D. S. Bernstein, S. Zilberstein, and N. Immerman, "The complexity of decentralized control of Markov Decision Processes," in *Conference on Uncertainty in Artificial Intelligence*, June 2000.

[11] C. Guestrin, S. Venkataraman, and D. Koller, "Context-Specific Multi-agent Coordination and Planning with Factored MDPs," in *Proceedings of the eighteenth AAAI National Conference on Artificial Intelligence (AAAI'02)*, Edmonton, Alberta, Canada, July 2002, pp. 253–259.

[12] M. Roth, R. Simmons, and M. Veloso, "What to Communicate? Execution-Time Decision in Multi-agent POMDPs," in *Proceedings of the eighth International Conference on Distributed Autonomous Robotic Systems (DARS'06)*, Minneapolis, MN, U.S.A., July 2006, pp. 177–186.

[13] L. Lovász and M. D. Plummer, *Matching Theory.* North-Holland, 1986.

[14] D. W. Pentico, "Assignment problems: A golden anniversary survey," *European J. of Op. Research*, vol. 176, no. 2, pp. 774–793, 2007.

[15] I. H. Toroslu and G. Üçoluk, "Incremental assignment problem," *Information Sciences*, March 2007.

[16] G. A. Mills-Tettey, A. T. Stentz, and M. B. Dias, "The dynamic hungarian algorithm for the assignment problem with changing costs," Robotics Institute, Pittsburgh, PA, Tech. Rep. CMU-RI-TR-07-27, July 2007.

[17] D. Fox, "Kld-sampling: Adaptive particle filters," *In Advances in Neural Information Processing Systems*, 2001.

[18] B. Gerkey, R. T. Vaughan, and A. Howard, "The player/stage project: Tools for multi-robot and distributed sensor systems," *Proceedings of the 11th International Conference on Advanced Robotics*, June 2003.

LQG-MP: Optimized Path Planning for Robots with Motion Uncertainty and Imperfect State Information

Jur van den Berg Pieter Abbeel Ken Goldberg

Abstract— **This paper presents LQG-MP (linear-quadratic Gaussian motion planning), a new approach to robot motion planning that takes into account the sensors and the controller that will be used during execution of the robot's path. LQG-MP is based on the linear-quadratic controller with Gaussian models of uncertainty, and explicitly characterizes in advance (i.e., before execution) the a-priori probability distributions of the state of the robot along its path. These distributions can be used to assess the quality of the path, for instance by computing the probability of avoiding collisions. Many methods can be used to generate the needed ensemble of candidate paths from which the best path is selected; in this paper we report results using the RRT-algorithm. We study the performance of LQG-MP with simulation experiments in three scenarios involving a kinodynamic car-like robot, multi-robot planning with differential-drive robots, and a 6-DOF manipulator.**

I. Introduction

Motion uncertainty, i.e. the fact that the motion of the robot unpredictably deviates from what a dynamics model predicts, and imperfect state information due to partial or noisy measurements of the robot's state, arise in many real-world robotic tasks ranging from guiding mobile robots over uneven terrain to performing robotic surgery with high-DOF manipulators. The amount of motion and sensing uncertainty may depend on the particular motion that is executed and the state the robot is in, so different paths for the robot will have different uncertainties associated with them. Because safety and accuracy are of critical importance for many robotic tasks, these uncertainties will have significant influence on which path is best for the task at hand. The challenge we discuss in this paper is to precisely quantify these uncertainties in advance, such that the best path can be selected.

Many traditional path planners assume deterministic motion and full knowledge of the state [18], [13], and leave issues of uncertainty to the *control* phase in which the path may be executed using a feedback controller [15]. Planning and control are related but distinct fields. While recent work on path planning has addressed motion and/or sensing uncertainty (see Section II), most planning methods do not account for control during execution and most control methods take the path as given. LQG-MP builds a bridge between these disciplines and draws from results in both.

This work was supported in part by NSF Award 0905344 and NIH Award 1R01EB-006435-01A1.

The authors are with the University of California at Berkeley, Berkeley, CA, USA. E-mail: {berg, pabbeel, goldberg}@berkeley.edu.

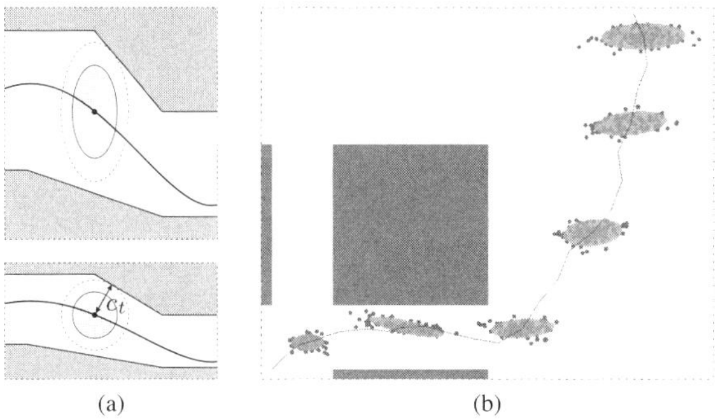

Fig. 1. (a) The maximum factor c_t by which the ellipse containing the positions within one standard deviation can be scaled before it intersects obstacles gives an indication of the probability that collisions will be avoided (top). c_t is computed as the Euclidean distance to the nearest obstacle in the environment transformed such that the ellipse becomes a unit disc (bottom). (b) The ellipses show the a-priori distributions as computed by LQG-MP along the best among the 1000 candidate paths for Scenario A. The samples result from performing 100 simulations.

The key insight of LQG-MP is that the a-priori knowledge of the sensors and controller that will be used during the execution of the path can be used to optimize the path in the planning phase. We base our approach on the linear-quadratic controller (LQG-controller) with Gaussian models of the motion and sensing uncertainty, as it provides *optimal* control for guiding a robot along a planned path [4]. We will show that for a given stochastic model of the motion dynamics, and a stochastic model of the sensor measurements obtained during execution, it is possible to derive *in advance* (i.e. before execution) the a-priori probability distributions of the states and the control inputs of the robot along a given path (see Fig. 1). These distributions can be used to compute, for example, the probability that collisions will be avoided, the likelihood that the robot will arrive at the goal, or any other measure defining the quality of the path. We can then use any motion planning method to generate a large set of candidate paths, and select the path that is best with respect to the chosen planning objective.

Our approach is generally applicable to both holonomic and non-holonomic robots with state spaces of arbitrary dimension and kinematics and dynamics constraints. We assume that the stochastic dynamics model of the robot and the stochastic observation model are given explicitly, and that their stochasticity can be modeled by Gaussian noise.

Our approach is designed for linear models, but can also be applied to non-linear models if they are locally well approximated by their linearizations.

We implemented our approach using the RRT motion planning algorithm [18] for representative path planning problems, and validated our approach using simulation experiments. We will show that the quality of candidate paths can differ starkly based on the uncertainty, even if traditional planning criteria such as path length or clearance from obstacles are similar, and that the type of sensors used during execution of the path has a significant influence on which path is best. A path planner that is unaware of the sensors, the controller and their uncertainties would not be able to make this distinction, and may produce sub-optimal paths.

The remainder of this paper is organized as follows. We start by discussing related work in Section II. We formally define the problem addressed in this paper in Section III. In Section IV we show how LQG-MP computes the a-priori probability distributions for a given path. In Section V, we discuss application examples and simulation results of LQG-MP for several motion and sensing models and planning objectives. We conclude in Section VI.

II. Related Work

A substantial body of work has addressed uncertainty in motion planning. The uncertainty typically originates from three sources: (i) motion uncertainty, (ii) sensing uncertainty and partial observations, and (iii) uncertainty about the environment. Our approach focuses on the first two, but is to some extent also applicable to the latter, as we will show in one of our experiments.

Planners that specifically take into account motion uncertainty include [14], [21], [29]. These planners plan paths that avoid rough terrain, but do not consider partial observability and sensing uncertainty. In [10], the probability of collisions is minimized for the specific case of a manipulator with base pose uncertainty. The sensing uncertainty is taken into account in the planner of [28], which aims to optimize the information content along a path. Planners in [5], [7], [20] assume that landmark regions exist in the environment where the accumulated motion uncertainty can be "reset".

Other approaches blend planning and control by defining a global control policy over the entire environment. MDPs, for instance, can be used with motion uncertainty to optimize probability of success [1], [30]. However, they require discretization of the state and control input spaces. The MDP concept can be extended to POMDPs to also include sensing uncertainty [12], [16], [26], but these suffer from issues of scalability [24]. The method of [17] also provides a global control policy in case of motion and sensing uncertainty.

Another class of planners considers the uncertainty about the environment and obstacles, rather than motion and sensing uncertainty [6], [9], [22], [23]. They typically aim to plan paths for which the probability of collisions is minimal.

Existing planners that are most directly related to LQG-MP take into account the available sensing capability to maximize the probability of arriving at the goal or to minimize expected cost [8], [11], [25], [27]. However, these algorithms implicitly assume to receive maximum-likelihood measurements from the sensors, which does not result in the true probability distributions of the state of the robot, but rather a measure of how well one will be able to infer the state. Besides the sensors, LQG-MP also takes into account the controller that will be used for executing the path, and computes the true a-priori probability distributions of the state of the robot along its future path.

III. Problem Definition

Let $\mathcal{X} = \mathbb{R}^n$ be the *state space* of the robot, and let $\mathcal{U} = \mathbb{R}^m$ be the *control input space* of the robot. We assume that time is discretized into stages of equal duration, and that applying a control input $\mathbf{u}_t \in \mathcal{U}$ at stage t brings the robot from state $\mathbf{x}_t \in \mathcal{X}$ at stage t to state $\mathbf{x}_{t+1} \in \mathcal{X}$ at stage $t+1$ according to a given stochastic dynamics model:

$$\mathbf{x}_t = f(\mathbf{x}_{t-1}, \mathbf{u}_{t-1}, \mathbf{m}_t), \qquad \mathbf{m}_t \sim \mathcal{N}(\mathbf{0}, M_t), \quad (1)$$

where $\mathbf{m}_t$ is the process noise at stage t drawn from a zero-mean Gaussian distribution with variance M_t that models the motion uncertainty. We assume that the function f is either linear or locally well approximated by its linearization.

Let us be given a start state $\mathbf{x}^{\text{start}} \in \mathcal{X}$ where the robot begins and a goal region $\mathcal{X}^{\text{goal}} \subset \mathcal{X}$ where the robot needs to go. A *path* Π for the robot is defined as a series of states and control inputs $(\mathbf{x}^\star_0, \mathbf{u}^\star_0, \ldots, \mathbf{x}^\star_\ell, \mathbf{u}^\star_\ell)$, such that $\mathbf{x}^\star_0 = \mathbf{x}^{\text{start}}$, $\mathbf{x}^\star_\ell \in \mathcal{X}^{\text{goal}}$, and $\mathbf{x}^\star_t = f(\mathbf{x}^\star_{t-1}, \mathbf{u}^\star_{t-1}, \mathbf{0})$ for $0 < t \leq \ell$, where ℓ is the number of stages of the path. That is, a path connects the start state and the goal region, and is consistent with the dynamics model if there were no process noise.

During execution of the path, the robot will deviate from the path due to motion uncertainty. To compensate for unexpected motions, we assume that the path will be executed using a feedback controller that aims to keep the robot close to the path by minimizing the cost function

$$\mathbb{E}\Big(\sum_{t=0}^{\ell}((\mathbf{x}_t - \mathbf{x}^\star_t)^T C (\mathbf{x}_t - \mathbf{x}^\star_t) + (\mathbf{u}_t - \mathbf{u}^\star_t)^T D (\mathbf{u}_t - \mathbf{u}^\star_t))\Big), \quad (2)$$

which quadratically penalizes deviations from the path. C and D are given positive-definite weight matrices.

We assume that noisy sensors provide us with partial information about the state during execution of the path according to a given stochastic observation model:

$$\mathbf{z}_t = h(\mathbf{x}_t, \mathbf{n}_t), \qquad \mathbf{n}_t \sim \mathcal{N}(\mathbf{0}, N_t), \quad (3)$$

where $\mathbf{z}_t$ is the measurement obtained at stage t that relates to state $\mathbf{x}_t$ through function h, and $\mathbf{n}_t$ is the measurement noise drawn from a zero-mean Gaussian with variance N_t. We assume that the function h is either linear or locally well approximated by its linearization.

We define our problem in two parts; (i) given the stochastic dynamics model, the stochastic observation model, and the cost function, compute the a-priori distributions of the state and control input along a given path, and (ii) given a planning objective based on the probability distributions, select the best path among a large set of candidates.

IV. A-PRIORI PROBABILITY DISTRIBUTIONS

In this section we describe how to compute the a-priori probability distributions of the state and control input of the robot along a given path Π. For this, we use the fact that we know in advance what controller will be used to execute the path: for linear dynamics and observation models with Gaussian noise and a quadratic cost function, the optimal approach for executing the path is to use an LQR feedback controller in parallel with a Kalman filter for state estimation, which is called linear-quadratic Gaussian (LQG) control [4]. A Kalman filter provides the optimal estimate of the state given previous state estimates, measurements and control inputs, and an LQR controller provides the optimal control input given the estimate of the state.

We will first discuss how to linearize the dynamics and observation model, and then review the Kalman filter and LQR controller. From these, we compute the a-priori probability distributions of the states and the control inputs of the robot along the path.

A. Linear(ized) Dynamics and Observation Model

In principle, our approach applies to linear dynamics and observation models f and h. However, since the robot is controlled to stay close to the path during execution, we can approximate non-linear models with local linearizations (i.e. first-order Taylor expansions) around the path Π. This gives the following linear(ized) stochastic models:

$$\mathbf{x}_t = f(\mathbf{x}^\star_{t-1}, \mathbf{u}^\star_{t-1}, \mathbf{0}) + A_t(\mathbf{x}_{t-1} - \mathbf{x}^\star_{t-1}) + B_t(\mathbf{u}_{t-1} - \mathbf{u}^\star_{t-1}) + V_t\mathbf{m}_t, \quad (4)$$

$$\mathbf{z}_t = h(\mathbf{x}^\star_t, \mathbf{0}) + H_t(\mathbf{x}_t - \mathbf{x}^\star_t) + W_t\mathbf{n}_t, \quad (5)$$

where $A_t = \frac{\partial f}{\partial \mathbf{x}}(\mathbf{x}^\star_{t-1}, \mathbf{u}^\star_{t-1}, \mathbf{0})$, $B_t = \frac{\partial f}{\partial \mathbf{u}}(\mathbf{x}^\star_{t-1}, \mathbf{u}^\star_{t-1}, \mathbf{0})$, $V_t = \frac{\partial f}{\partial \mathbf{m}}(\mathbf{x}^\star_{t-1}, \mathbf{u}^\star_{t-1}, \mathbf{0})$, $H_t = \frac{\partial h}{\partial \mathbf{x}}(\mathbf{x}^\star_t, \mathbf{0})$, $W_t = \frac{\partial h}{\partial \mathbf{n}}(\mathbf{x}^\star_t, \mathbf{0})$ are the Jacobian matrices of f and h along path Π.

It is convenient to express the control problem in terms of the *deviation* from the path. By defining

$$\bar{\mathbf{x}}_t = \mathbf{x}_t - \mathbf{x}^\star_t, \quad \bar{\mathbf{u}}_t = \mathbf{u}_t - \mathbf{u}^\star_t, \quad \bar{\mathbf{z}}_t = \mathbf{z}_t - h(\mathbf{x}^\star_t, \mathbf{0}), \quad (6)$$

as the state deviation, control input deviation and measurement deviation, respectively, we can formulate the dynamics and observation model of Equations (4) and (5) as

$$\bar{\mathbf{x}}_t = A_t\bar{\mathbf{x}}_{t-1} + B_t\bar{\mathbf{u}}_{t-1} + V_t\mathbf{m}_t, \quad \mathbf{m}_t \sim \mathcal{N}(\mathbf{0}, M_t), \quad (7)$$

$$\bar{\mathbf{z}}_t = H_t\bar{\mathbf{x}}_t + W_t\mathbf{n}_t, \quad \mathbf{n}_t \sim \mathcal{N}(\mathbf{0}, N_t), \quad (8)$$

and the cost function of Equation (2) as

$$\mathbb{E}\Big(\sum_{t=0}^{\ell}(\bar{\mathbf{x}}_t^T C \bar{\mathbf{x}}_t + \bar{\mathbf{u}}_t^T D \bar{\mathbf{u}}_t)\Big). \quad (9)$$

This is the standard formulation of an LQG-control problem.

B. Kalman Filter for Optimal State Estimation

The Kalman filter keeps track of the estimate $\tilde{\mathbf{x}}_t$ and variance P_t of the true state $\bar{\mathbf{x}}_t$ during the execution of the path. It continually performs two steps; a process update to propagate the applied control input $\bar{\mathbf{u}}_t$, and a measurement update to incorporate the obtained measurement $\bar{\mathbf{z}}_t$:

Process update:

$$\tilde{\mathbf{x}}^-_t = A_t\tilde{\mathbf{x}}_{t-1} + B_t\bar{\mathbf{u}}_{t-1} \quad (10)$$

$$P^-_t = A_t P_{t-1} A_t^T + V_t M_t V_t^T, \quad (11)$$

Measurement update:

$$K_t = P^-_t H_t^T (H_t P^-_t H_t^T + W_t N_t W_t^T)^{-1} \quad (12)$$

$$\tilde{\mathbf{x}}_t = \tilde{\mathbf{x}}^-_t + K_t(\bar{\mathbf{z}}_t - H_t\tilde{\mathbf{x}}^-_t) \quad (13)$$

$$P_t = (I - K_t H_t) P^-_t. \quad (14)$$

These are the standard Kalman filter equations for optimal estimation given the dynamics and observation model of Equations (7) and (8) [31]. Note that the Kalman-gain matrices K_t can be computed *in advance* (i.e. before execution) given the initial variance P_0, without knowledge of the actual control inputs $\bar{\mathbf{u}}_t$ and measurements $\bar{\mathbf{z}}_t$.

C. LQR for Optimal Control

The control inputs $\bar{\mathbf{u}}_t$ that are optimal to apply during execution of the path are determined by the control policy that minimizes the cost function of Equation (9). For the dynamics model of Equation (7), the cost function is minimal when $\bar{\mathbf{u}}_t = L_t\bar{\mathbf{x}}_t$, where L_t is the *feedback matrix*, which is computed in advance for all $t \in 0, \ldots, \ell - 1$ using:

$$S_\ell = C \quad (15)$$

$$L_t = -(B_{t+1}^T S_{t+1} B_{t+1} + D)^{-1} B_{t+1}^T S_{t+1} A_{t+1} \quad (16)$$

$$S_t = C + A_{t+1}^T S_{t+1} A_{t+1} + A_{t+1}^T S_{t+1} B_{t+1} L_t. \quad (17)$$

These are the standard equations for a finite-horizon discrete-time LQR controller [4].

As the true state $\bar{\mathbf{x}}_t$ is unknown, the estimate $\tilde{\mathbf{x}}_t$ of the state which is obtained from the Kalman filter is used to determine the control input $\bar{\mathbf{u}}_t$ at each stage t during execution of the path. Hence, the control policy is:

$$\bar{\mathbf{u}}_t = L_t\tilde{\mathbf{x}}_t. \quad (18)$$

After application of the control input, the Kalman filter produces the estimate of the next state from which in turn a new control input is determined. This cycle repeats until the execution of the path is complete.

D. A-priori Distributions of State and Control Input

Given the LQR control policy and the Kalman filter, we can analyze in advance how the true state $\bar{\mathbf{x}}_t$ and the estimated state $\tilde{\mathbf{x}}_t$ will evolve during execution of the path as functions of each other. The evolution of the true state $\bar{\mathbf{x}}_t$ is dependent on the estimated state through the LQR control policy (Equation (18)) and the evolution of the estimated state $\tilde{\mathbf{x}}_t$ is dependent on the true state through the measurement obtained in the Kalman filter (Equation (13)). This gives the following equations:

$$\bar{\mathbf{x}}_t = A_t\bar{\mathbf{x}}_{t-1} + B_t L_{t-1}\tilde{\mathbf{x}}_{t-1} + V_t\mathbf{m}_t, \quad (19)$$

$$\tilde{\mathbf{x}}_t = A_t\tilde{\mathbf{x}}_{t-1} + B_t L_{t-1}\tilde{\mathbf{x}}_{t-1} + K_t\big(\bar{\mathbf{z}}_t - H_t(A_t\tilde{\mathbf{x}}_{t-1} + B_t L_{t-1}\tilde{\mathbf{x}}_{t-1})\big) \quad (20)$$

$$
\begin{aligned}
&= A_t\tilde{\mathbf{x}}_{t-1} + B_tL_{t-1}\tilde{\mathbf{x}}_{t-1} + K_t\big(H_t\bar{\mathbf{x}}_t + W_t\mathbf{n}_t - \\
&\quad H_t(A_t\tilde{\mathbf{x}}_{t-1} + B_tL_{t-1}\tilde{\mathbf{x}}_{t-1})\big) \\
&= A_t\tilde{\mathbf{x}}_{t-1} + B_tL_{t-1}\tilde{\mathbf{x}}_{t-1} + K_t\big(\\
&\quad H_t(A_t\bar{\mathbf{x}}_{t-1} + B_tL_{t-1}\tilde{\mathbf{x}}_{t-1} + V_t\mathbf{m}_t) + W_t\mathbf{n}_t - \\
&\quad H_t(A_t\tilde{\mathbf{x}}_{t-1} + B_tL_{t-1}\tilde{\mathbf{x}}_{t-1})\big) \\
&= A_t\tilde{\mathbf{x}}_{t-1} + B_tL_{t-1}\tilde{\mathbf{x}}_{t-1} + K_tH_tA_t\bar{\mathbf{x}}_{t-1} + \\
&\quad K_tH_tV_t\mathbf{m}_t + K_tW_t\mathbf{n}_t - K_tH_tA_t\tilde{\mathbf{x}}_{t-1},
\end{aligned}
$$

Equation (19) follows from substituting Equation (18) into Equation (7). The first equality of (20) follows from substituting Equation (18) into Equation (10) and Equation (10) into Equation (13); the second and third equalities follow after substituting Equations (8) and (19), respectively, and the fourth equality follows after expanding the terms.

Combining Equations (19) and (20) gives the matrix form:

$$
\begin{bmatrix}\bar{\mathbf{x}}_t \\ \tilde{\mathbf{x}}_t\end{bmatrix} = \begin{bmatrix} A_t & B_tL_{t-1} \\ K_tH_tA_t & A_t + B_tL_{t-1} - K_tH_tA_t \end{bmatrix}\begin{bmatrix}\bar{\mathbf{x}}_{t-1} \\ \tilde{\mathbf{x}}_{t-1}\end{bmatrix} + \begin{bmatrix} V_t & 0 \\ K_tH_tV_t & K_tW_t\end{bmatrix}\begin{bmatrix}\mathbf{m}_t \\ \mathbf{n}_t\end{bmatrix}, \quad \begin{bmatrix}\mathbf{m}_t \\ \mathbf{n}_t\end{bmatrix} \sim \mathcal{N}(\mathbf{0}, \begin{bmatrix} M_t & 0 \\ 0 & N_t\end{bmatrix}),
$$

which we write shorthand (for the appropriate definitions of $\mathbf{y}_t$, $\mathbf{q}_t$, F_t, G_t and Q_t) as:

$$
\mathbf{y}_t = F_t\mathbf{y}_{t-1} + G_t\mathbf{q}_t, \qquad \mathbf{q}_t \sim \mathcal{N}(\mathbf{0}, Q_t). \tag{21}
$$

From this, we can compute the mean $\hat{\mathbf{y}}_t$ and the variance R_t of $\mathbf{y}_t = \left[\begin{smallmatrix}\bar{\mathbf{x}}_t \\ \tilde{\mathbf{x}}_t\end{smallmatrix}\right]$ for any stage t of the execution of the path:

$$
\hat{\mathbf{y}}_t = F_t\hat{\mathbf{y}}_{t-1}, \qquad \hat{\mathbf{y}}_0 = \mathbf{0}, \tag{22}
$$

$$
R_t = F_tR_{t-1}F_t^T + G_tQ_tG_t^T, \qquad R_0 = \begin{bmatrix} P_0 & 0 \\ 0 & 0\end{bmatrix}. \tag{23}
$$

Note that the mean $\hat{\mathbf{y}}_t$ is zero for all stages t. Hence, $\left[\begin{smallmatrix}\bar{\mathbf{x}}_t \\ \tilde{\mathbf{x}}_t\end{smallmatrix}\right] \sim \mathcal{N}(\mathbf{0}, R_t)$. As it follows from Equations (18) and (6) that

$$
\begin{bmatrix}\mathbf{x}_t \\ \mathbf{u}_t\end{bmatrix} = \begin{bmatrix} I & 0 \\ 0 & L_t\end{bmatrix}\begin{bmatrix}\bar{\mathbf{x}}_t \\ \tilde{\mathbf{x}}_t\end{bmatrix} + \begin{bmatrix}\mathbf{x}_t^\star \\ \mathbf{u}_t^\star\end{bmatrix}, \tag{24}
$$

the a-priori distribution of the state $\mathbf{x}_t$ and the control input $\mathbf{u}_t$ at stage t of the execution of the path is:

$$
\begin{bmatrix}\mathbf{x}_t \\ \mathbf{u}_t\end{bmatrix} \sim \mathcal{N}(\begin{bmatrix}\mathbf{x}_t^\star \\ \mathbf{u}_t^\star\end{bmatrix}, \Lambda_tR_t\Lambda_t^T), \qquad \Lambda_t = \begin{bmatrix} I & 0 \\ 0 & L_t\end{bmatrix}. \tag{25}
$$

The covariance between $\left[\begin{smallmatrix}\mathbf{x}_i \\ \mathbf{u}_i\end{smallmatrix}\right]$ and $\left[\begin{smallmatrix}\mathbf{x}_j \\ \mathbf{u}_j\end{smallmatrix}\right]$ is given by:

$$
\mathrm{cov}(\begin{bmatrix}\mathbf{x}_i \\ \mathbf{u}_i\end{bmatrix}, \begin{bmatrix}\mathbf{x}_j \\ \mathbf{u}_j\end{bmatrix}) = \Lambda_iR_iF_{i+1}^TF_{i+2}^T\cdots F_j^T\Lambda_j^T, \quad i < j. \tag{26}
$$

Using the a-priori distributions, the quality of path Π can be computed with respect to the chosen planning objective. We can then use any motion planner to generate a large set of candidate paths, from which the best one is selected.

V. Example Applications and Results

In this section, we report simulation results for three scenarios in which LQG-MP is used to select a path. In each of the three scenarios, we use a different dynamics model, observation model and planning objective, and provide comparative analysis with a brute-force approach. We report results for an Intel P7350 2GHz with 4GB RAM.

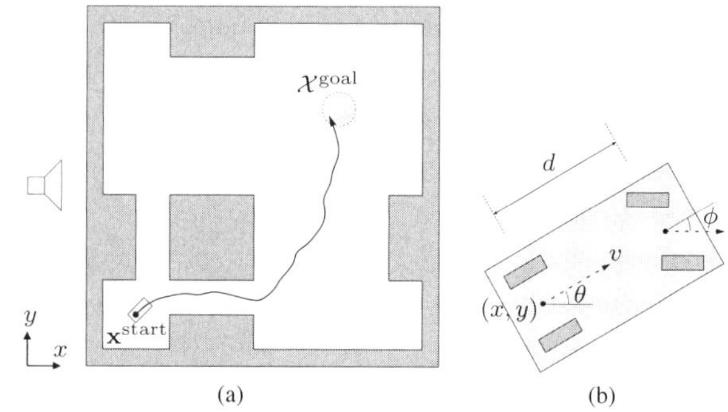

Fig. 2. (a) The environment of Scenario A, in which a car-like robot has to move between a start state and a goal region without colliding with obstacles. Sensors can only measure the y-coordinate of the position of the robot. The best path according to LQG-MP among the 1000 generated by RRT is shown. (b) The state $\mathbf{x}$ of a car-like robot.

For each scenario, we use the random rapidly-exploring tree (RRT) algorithm [18] to generate a large set of candidate paths. The RRT algorithm is well suited for our context as it can handle any dynamics model (without process noise) of the form of Equation (1) well. Even though it only plans a single path between the start state and the goal region, the path is generated randomly and will thus be different each time the algorithm is run. Hence, to generate multiple different paths, we run the RRT algorithm multiple times.

A. Car-Like Robot

In the first scenario, we apply LQG-MP to a non-holonomic car-like robot with 2nd-order dynamics in a 2-D environment with obstacles. The robot needs to move from a start state $\mathbf{x}^{\text{start}}$ to a goal region $\mathcal{X}^{\text{goal}}$ without colliding with the obstacles in the environment (see Fig. 2(a)).

1) Dynamics model: The state $\mathbf{x} = (x, y, \theta, v)$ of the robot is a 4-D vector consisting of its position (x, y), its orientation θ, and its speed v (see Fig. 2(b)). Its control input $\mathbf{u} = (a, \phi)$ is a 2-D vector consisting of an acceleration a and the steering wheel angle ϕ, corrupted by process noise $\mathbf{m} = (\tilde{a}, \tilde{\phi}) \sim \mathcal{N}(\mathbf{0}, \left[\begin{smallmatrix}\sigma_a^2 & 0 \\ 0 & \sigma_\phi^2\end{smallmatrix}\right])$. This gives the following non-linear dynamics model:

$$
f(\mathbf{x}, \mathbf{u}, \mathbf{m}) = \begin{bmatrix} x + \tau v\cos\theta \\ y + \tau v\sin\theta \\ \theta + \tau v\tan(\phi + \tilde{\phi})/d \\ v + \tau(a + \tilde{a})\end{bmatrix}, \tag{27}
$$

where τ is the duration of a stage (time step), and d the distance between the front and rear axle of the car [19].

2) Observation model: To show the effect of partial sensing, the robot only receives feedback on the y-coordinate of its position. Hence, the measurement vector $\mathbf{z}$ is univariate and consists of a measurement of the y-coordinate of the robot corrupted by measurement noise $\mathbf{n} = \tilde{y} \sim \mathcal{N}(0, \sigma_y^2)$. This gives the following linear observation model:

$$
h(\mathbf{x}, \mathbf{n}) = y + \tilde{y}. \tag{28}
$$

Even though the sensor feedback is very partial, information about the other variables is still obtained through the interplay with the dynamics model.

3) Planning objective: We aim to find the path for the robot with a minimal probability of colliding with obstacles. Instead of computing this probability exactly, we will use an approximation that can be computed efficiently given the probability distributions along the path. To this end, we look at the number of standard deviations that one can deviate from the path before the robot may collide with an obstacle. Let this number be denoted c_t for stage t along the path. For a multivariate Gaussian distribution of dimension n, the probability that a sample is within c_t standard deviations is given by $\Gamma(n/2, c_t^2/2)$, where Γ is the regularized Gamma function [32]. It provides a lower bound of the probability of avoiding collisions at stage t. We now define the quality of a path Π as:

$$\prod_{t=0}^{\ell} \Gamma(n/2, c_t^2/2), \tag{29}$$

which is indicative of the probability that collisions will be avoided during execution. It is the planning objective to find a path for which this measure is maximal.

The value of c_t for stage t is computed as follows. For simplicity, we approximate the geometry of the car by a bounding disc, such that its orientation has no influence on whether or not the car is colliding. Also its speed does not influence its collision status. Hence, c_t is determined by the distribution $\mathcal{N}(\mathbf{p}_t, \Sigma_t)$ of the position of the car (i.e. $n = 2$), which is the marginal distribution of the first two variables of $\mathcal{N}(\left[\begin{smallmatrix}\mathbf{x}_t^\star \\ \mathbf{u}_t^\star\end{smallmatrix}\right], \Lambda_t R_t \Lambda_t^T)$ as computed in Equation (25). Let U_t be a matrix such that $U_t U_t^T = \Sigma_t$. The set of positions within one standard deviation is then an ellipse centered at the mean $\mathbf{p}_t$ obtained by transforming a unit disc by U_t, and c_t is the maximum factor by which the ellipse can be scaled such that it does not intersect with obstacles (see Fig. 1(a)).

Computing c_t can efficiently be implemented using a collision-checker that is capable of performing distance calculations and linear transformations on the geometry, for instance SOLID [3]. Transforming the environment (including the robot) by U_t^{-1} (such that the uncertainty ellipse becomes a unit disc, see Fig. 1(a)), and calculating the Euclidean distance between the robot and the nearest obstacle in the transformed environment gives the value of c_t for stage t.

4) Results: We randomly generated 1000 paths using the RRT algorithm, which took 56.8 seconds. For each of the paths, we computed the a-priori probability distributions and the measure of Equation (29), which took in total 2.67 seconds. The best path among the 1000 is shown in Fig. 2(a). It can be seen that the "lower-right" passage is chosen to get to the goal. This can be explained as the uncertainty will mainly be in the x-coordinate given that the sensors only provide feedback on the y-coordinate. The geometry of the lower-right passage allows for more deviation in the x-direction than the upper-left passage. Indeed, changing the observation model such that only the x-coordinate is measured results in a path that takes the upper-left passage.

To validate our results, we used a brute-force approach to estimate for each path the "ground-truth" probability that it will be executed without collisions. We performed 10,000 simulations of executions of the path using the LQR-controller and an extended Kalman Filter with artificially generated process and measurement noise, and counted the number of collision-free executions. This took in total 10440 seconds, which is almost 4000 times as much as the time needed by LQG-MP to evaluate the paths. It turns out that the path selected by LQG-MP has a 99% probability of success. The average probability of success over the 1000 paths is 61%, and the worst path has a probability of success of 13%. This is an indication of the typical and worst-case success rate of paths planned by a planner unaware of the uncertainties. Among the paths taking the upper-left passage, the best one has a success rate of 88% (versus 99% for the best path overall). This shows that the type of sensors used during execution has a significant influence on which path is optimal, even as the environment is symmetric.

In Fig. 1(b) the samples of 100 simulations are shown for the best among the 1000 paths, along with the uncertainty ellipses of the a-priori probability distributions as computed by LQG-MP. As can be seen, the samples indeed follow the a-priori distributions computed by LQG-MP. This shows that any error introduced into LQG-MP by the linearization of the dynamics model is insignificant for this example.

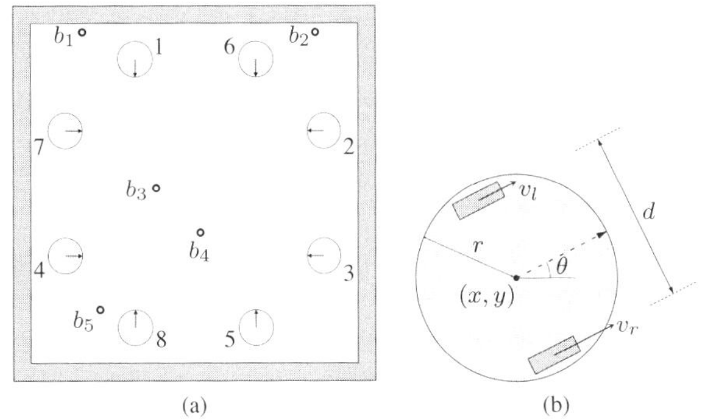

Fig. 3. (a) The environment of Scenario B, in which eight robots have to move to their antipodal position in the environment without mutual collisions. The numbers indicate the priority rank assigned to each robot. Five beacons $b_1, \ldots, b_5$ send out a signal whose strength decays quadratically with distance. (b) The state $\mathbf{x}$ of the differential-drive robot.

B. Multi-Robot Planning with Differential-Drive Robots

In the second experiment, we apply LQG-MP to multi-robot motion planning with disc-shaped differential-drive robots (e.g. Roomba vacuum cleaners). Eight robots need to move simultaneously to their antipodal position in the environment without mutual collisions (see Fig. 3(a)). We use a prioritized approach to the multi-robot planning problem: the robots are planned for one by one in order of a priority assigned to them, and aim to avoid collisions with robots of higher priority, which are treated as moving obstacles [2]. This means that for each robot we apply LQG-MP to a dynamic environment in which not only the robot itself is subject to uncertainty, but also the obstacles (i.e. the robots of higher priority).

1) Dynamics model: The state $\mathbf{x} = (x, y, \theta)$ of each robot is a 3-D vector consisting of its position (x, y) and its orientation θ (see Fig. 3(b)). Its control input $\mathbf{u} = (v_l, v_r)$ is a 2-D vector consisting of the speeds of the left and right wheel, respectively, corrupted by process noise $\mathbf{m} = (\tilde{v}_l, \tilde{v}_r) \sim \mathcal{N}(\mathbf{0}, \sigma_v^2 I)$. This gives the following non-linear dynamics model:

$$f(\mathbf{x}, \mathbf{u}, \mathbf{m}) = \begin{bmatrix} x + \frac{1}{2}\tau(v_l + \tilde{v}_l + v_r + \tilde{v}_r)\cos\theta \\ y + \frac{1}{2}\tau(v_l + \tilde{v}_l + v_r + \tilde{v}_r)\sin\theta \\ \theta + \tau(v_r + \tilde{v}_r - v_l - \tilde{v}_l)/d \end{bmatrix}, \quad (30)$$

where τ is the time step and d the distance between the left and right wheel of the robot [19].

2) Observation model: The robots receive feedback on their state from five beacons $b_1, \ldots, b_5$ scattered around the environment that each send out an identifiable signal of unit strength that decays quadratically with the distance to the beacon. Each beacon b_i has a known location $(\check{x}_i, \check{y}_i, 1)$. Hence, the measurement vector $\mathbf{z}$ consists of five readings of signal strengths, one from each beacon, corrupted by measurement noise $\mathbf{n} = (\tilde{b}_1, \ldots, \tilde{b}_5) \sim \mathcal{N}(\mathbf{0}, \sigma_b^2 I)$. This gives the following non-linear observation model:

$$h(\mathbf{x}, \mathbf{n}) = \begin{bmatrix} 1/((x - \check{x}_1)^2 + (y - \check{y}_1)^2 + 1) + \tilde{b}_1 \\ \vdots \\ 1/((x - \check{x}_5)^2 + (y - \check{y}_5)^2 + 1) + \tilde{b}_5 \end{bmatrix}. \quad (31)$$

3) Planning objective: For each robot, we aim to minimize the probability that it will collide with a robot of higher priority along its path. In this experiment, we approximate this probability more directly than we did for the first scenario. Let us assume we are planning for robot j, and that a path has already been planned for robots $1, \ldots, j-1$. As the robots are disc-shaped, only their position influences whether or not they collide. Let $\mathcal{N}(\mathbf{p}_t^i, \Sigma_t^i)$ be the marginal probability distribution of the position of robot i at stage t along i's path as computed by LQG-MP. Then, the distribution of the *relative* position of robot j and robot i (for $i \in 1, \ldots, j-1$) at stage t is $\mathcal{N}(\mathbf{p}_t^i - \mathbf{p}_t^j, \Sigma_t^i + \Sigma_t^j)$. The probability $\mathbb{P}_t(i \otimes j)$ that robot j collides with robot i at stage t is then given by:

$$\int_{\|\mathbf{p}\|<2r} \frac{\exp(-\frac{1}{2}(\mathbf{p}-\mathbf{p}_t^{ij})^T(\Sigma_t^i+\Sigma_t^j)^{-1}(\mathbf{p}-\mathbf{p}_t^{ij}))}{2\pi \det(\Sigma_t^i + \Sigma_t^j)^{1/2}} \, d\mathbf{p}, \quad (32)$$

where $\mathbf{p}_t^{ij} = \mathbf{p}_t^i - \mathbf{p}_t^j$. This is the integral over the set of relative positions $\mathbf{p}$ for which the robots collide (that is when $\|\mathbf{p}\| < 2r$, where r is the radius of the robots) of the probability density function of the distribution of relative positions, and can be evaluated numerically. It follows that the probability that robot j does not collide with any robot at any stage along its path is:[1]

$$\prod_{t=0}^{\ell} \prod_{i=1}^{j-1} (1 - \mathbb{P}_t(i \otimes j)). \quad (33)$$

It is the planning objective for robot j to maximize this probability.

[1]Note that we assume here that the probabilities of avoiding collisions at different stages along the path are independent. This is not the case, but it will for practical purposes be a reasonable assumption.

TABLE I
RESULTS FOR SCENARIO B (1000 PATHS PER ROBOT)

	Computation time		Success rate	
robot	RRT	LQG-MP	Best path	Avg. path
1	22.3s	0.23s	100%	100%
2	28.2s	0.99s	100%	70.3%
3	29.5s	1.75s	100%	69.2%
4	30.5s	2.79s	100%	60.9%
5	57.0s	2.92s	99.2%	10.6%
6	49.8s	3.90s	99.8%	21.0%
7	39.2s	5.26s	99.9%	24.8%
8	77.8s	6.85s	99.7%	13.0%
total	334s	24.7s	98.6%	2.13%

As a secondary objective, we aim to minimize the uncertainty around the robot's path to leave maximal "space" for the other robots. That is, in case of equal probabilities of success, we aim to minimize the function $\sum_{t=0}^{\ell} \operatorname{tr}(\Sigma_t^j)$. This is equivalent to maximizing the likelihood that the robot will exactly follow the path Π during execution. The robot with the highest priority does not need to avoid other robots, so it will select its path purely based on the secondary objective.

4) Results: For each of the robots in turn, we planned 1000 paths using the RRT algorithm and selected the path that is best according to the planning objective. Note that the paths were planned such that, if there were no uncertainty, they are collision-free with respect to the robots of higher priority for which a path has already been selected. The result is shown in Fig. 4, along with the uncertainty ellipses of the a-priori probability distributions along the paths. It can be seen that the robots need to get close to the beacons to be able to estimate their position accurately. Almost all of the robots move through the region around the central beacons b_3 and b_4. At the same time, the robots aim to stay far away from each other, in order to minimize the probability of collisions. Robot 2, for instance, makes a wide detour around robot 1. Robot 3 first avoids robot 1 and then robot 2, causing its path to have a wide S-shape.

The quantitative results are given in Table I. The second column shows the time needed to plan 1000 paths for each robot, and the third column shows the time needed by LQG-MP to compute the probabilities of success for all paths. It shows that these probabilities can be computed efficiently. Per path, it takes an order of magnitude less time than planning the path itself. The third column shows the probability of success of the best path among the 1000 paths. This is the path that LQG-MP selects for the particular robot. The fourth column shows the average probability of success of the 1000 paths. This provides an indication of what an uncertainty-unaware planner would typically achieve. The probability that all eight robots successfully reach their goal is the product of the robot's individual probabilities of success, and is shown in the bottom row. This is 98.6% for LQG-MP, whereas an uncertainty-unaware planner would on average only have a 2.13% probability of success.

C. 6-DOF Manipulator

In the third experiment, we apply LQG-MP to a holonomic 6-DOF articulated robot in a 3-D environment. The robot needs to move from its initial state $\mathbf{x}^{\text{start}}$ to a configuration

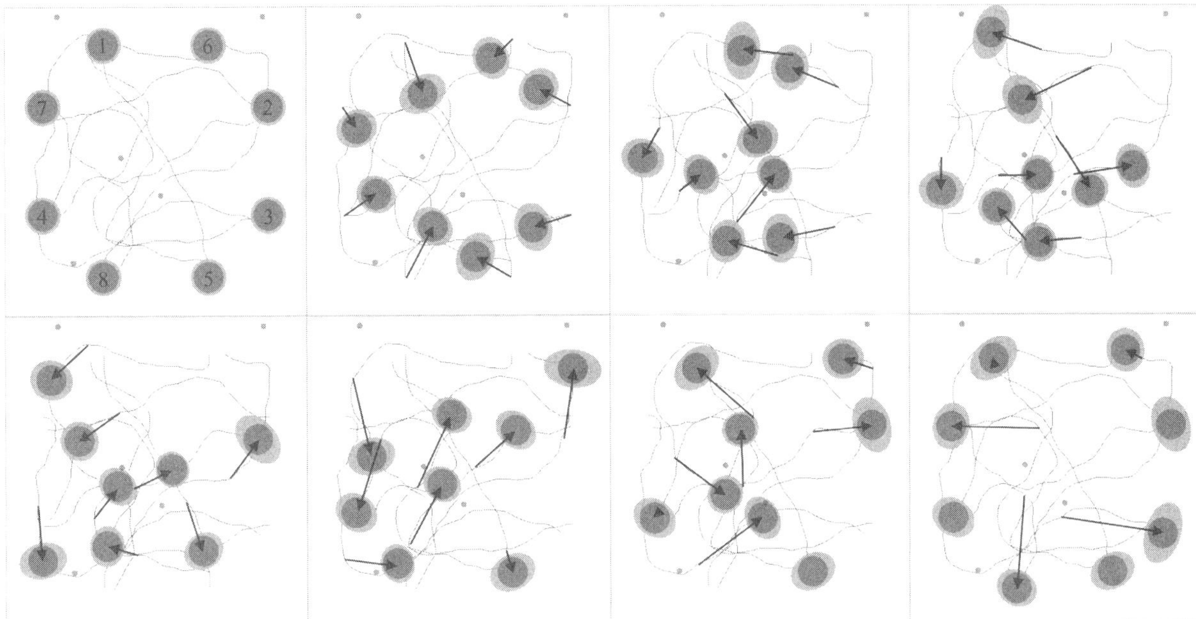

Fig. 4. The paths resulting from consecutively applying LQG-MP to each of the robots in Scenario B (snapshots at $t = 0, 3, 6, 9, 12, 16, 20, 28$). The numbers in the top-left image indicate the priority rank of the robots. The arrows show the movement with respect to the previous image. The robots enlarged by the uncertainty ellipses of their a-priori probability distributions are shown in green.

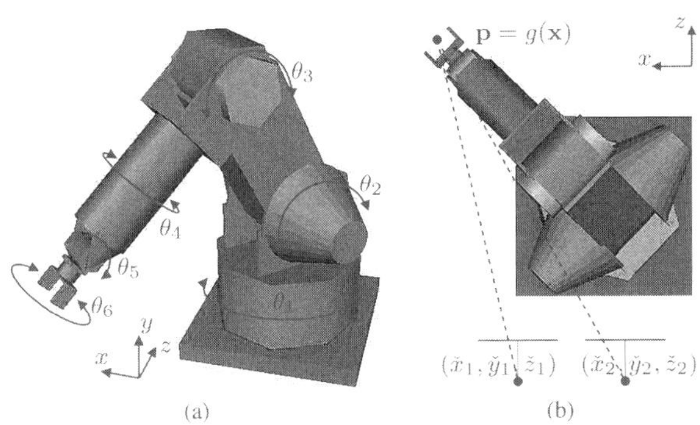

Fig. 5. (a) The state $\mathbf{x}$ of the articulated robot of Scenario C. (b) A stereo camera tracks the position $\mathbf{p}$ of the end-effector of the robot.

in which the end-effector is inside a goal region on the other side of the environment.

1) Dynamics model: The state $\mathbf{x} = (\theta_1, \ldots, \theta_6)$ of the robot is a 6-D vector consisting of the angles of rotation at each of the joints (see Fig. 5(a)). The control input $\mathbf{u} = (\omega_1, \ldots, \omega_6)$ is a 6-D vector consisting of the angular speeds at each of the joints, corrupted by process noise $\mathbf{m} = (\tilde{\omega}_1, \ldots, \tilde{\omega}_6) \sim \mathcal{N}(\mathbf{0}, \sigma_\omega^2 I)$. Ignoring higher order dynamics, this results in the following linear dynamics model:

$$f(\mathbf{x}, \mathbf{u}, \mathbf{m}) = \begin{bmatrix} \theta_1 + \tau(\omega_1 + \tilde{\omega}_1) \\ \vdots \\ \theta_6 + \tau(\omega_6 + \tilde{\omega}_6) \end{bmatrix}. \quad (34)$$

2) Observation model: The robot receives feedback from a stereo camera that tracks the position of the end-effector of the robot. Let $\mathbf{p} = g(\mathbf{x})$ be the function relating the set of joint angles of the state $\mathbf{x}$ to the position $\mathbf{p} \in \mathbb{R}^3$ of the end-effector. This point is projected on the imaging plane of each camera i, which has a unit focal distance and a known location $(\check{x}_i, \check{y}_i, \check{z}_i)$ (see Fig. 5(b)). Hence, the measurement $\mathbf{z}$ is a 4-D vector consisting of the pixel coordinates of the end-effector on the imaging planes of both cameras, corrupted by measurement noise $\mathbf{n} \sim \mathcal{N}(\mathbf{0}, \sigma_n^2 I)$. Ignoring occlusions, this gives the following non-linear observation model:

$$h(\mathbf{x}, \mathbf{n}) = \begin{bmatrix} (g_x(\mathbf{x}) - \check{x}_1)/(g_z(\mathbf{x}) - \check{z}_1) \\ (g_y(\mathbf{x}) - \check{y}_1)/(g_z(\mathbf{x}) - \check{z}_1) \\ (g_x(\mathbf{x}) - \check{x}_2)/(g_z(\mathbf{x}) - \check{z}_2) \\ (g_y(\mathbf{x}) - \check{y}_2)/(g_z(\mathbf{x}) - \check{z}_2) \end{bmatrix} + \mathbf{n}. \quad (35)$$

3) Planning objective: We aim to maximize the likelihood that the end-effector arrives at its goal position. Let $\mathcal{N}(\mathbf{p}_\ell, \Sigma_\ell)$ be the distribution of the position of the end-effector at the last stage of the path, then this likelihood is maximal when $\mathrm{tr}(\Sigma_\ell)$ is minimal. Σ_ℓ can be approximated from the variance X_ℓ of the state $\mathbf{x}_\ell$ computed by LQG-MP as $\Sigma_\ell = T_\ell X_\ell T_\ell^T$, where $T_\ell = \frac{\partial g}{\partial \mathbf{x}}(\mathbf{x}_\ell^\star)$, i.e. the Jacobian matrix of function g at the goal position.

4) Results: We planned 1000 paths for the robot using the RRT algorithm, and computed for each the likelihood of arriving at the goal. Constructing the paths took 192 seconds, and evaluating them using LQG-MP took 1.16 seconds. The path found best is shown in Fig. 6(a). Interestingly, the robot chooses to move in the plane perpendicular to the viewing direction of the camera while being fully stretched out. In such configurations, the position of the end-effector contains most information about the angles at the joints. This apparently outweighs the benefit of more precise positioning when the end-effector is closer to the camera. Indeed, an experiment in which the camera is placed above the robot

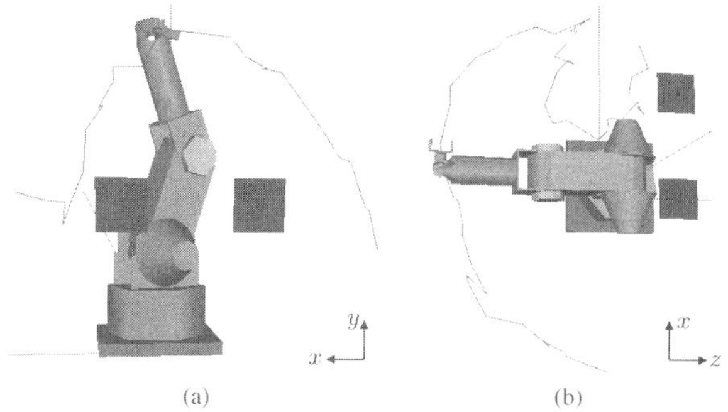

Fig. 6. The best path among the candidates for Scenario C when the cameras (blue squares) are placed (a) next to the robot and (b) above the robot. The jaggedness of the paths is due to the random nature of the RRT algorithm.

results in a path with similar characteristics (see Fig. 6(b)).

VI. Conclusion and Future Work

We have presented LQG-MP, a new approach to evaluate paths in motion planning for robots subject to motion and sensing uncertainty. LQG-MP precisely characterizes the a-priori probability distributions of the state of the robot along a given path, based on which the path can be optimized for the particular task. We have shown that this considerably increases the probability of a successful execution when compared to uncertainty-unaware planners. The key of LQG-MP is that it takes into account the a-priori knowledge of both the sensors and controller in the planning phase.

In the experiments we performed, we have not used the a-priori distributions of the control input that LQG-MP also computes, nor the covariances between the states at different stages along the path. We envision that these could be used to compute the conditional distributions of the remainder of the path after each application of a control input during the execution. If the new distributions indicate that the quality has dropped below a threshold, we might opt to *replan*. Current planning times, though, do not allow for real-time application of LQG-MP. It is a major objective of future work to bring planning times down, for instance by devising a focused planner such that planning a large set of candidate paths is no longer required. Other limitations, such as the fact that the candidate paths may not constitute a representative sample in high-dimensional state spaces, and the jaggedness of the paths that RRT produces, might then also be resolved.

We plan to use LQG-MP in future work on optimizing accuracy and safety in challenging robotic applications, such as autonomous helicopter flight, needle steering for prostate brachytherapy, and robotic-assisted surgery.

References

[1] R. Alterovitz, T. Siméon, K. Goldberg. The stochastic motion roadmap: a sampling framework for planning with Markov motion uncertainty. *Proc. Robotics: Science and Systems*, 2007.

[2] J. van den Berg, M. Overmars. Prioritized motion planning for multiple robots. *Proc. IEEE/RSJ Int. Conf. on Intelligent Robots and Systems*, 2005.

[3] G. van den Bergen. *Collision detection in interactive 3D environments*. Morgan Kaufmann Publishers, 2004.

[4] D. Bertsekas. *Dynamic programming and optimal control*. Athena Scientific, 2001.

[5] B. Bouilly, T. Simeon, R. Alami. A numerical technique for planning motion strategies of a mobile robot in presence of uncertainty. *Proc. IEEE Int. Conf. on Robotics and Automation*, 1995.

[6] B. Burns, O. Brock. Sampling-based motion planning with sensing uncertainty. *Proc. IEEE Int. Conf. on Robotics and Automation*, 2007.

[7] T. Fraichard, R. Mermond. Path planning with uncertainty for car-like robots. *Proc. IEEE Int. Conf. on Robotics and Automation*, 1998.

[8] J. Gonzalez, A. Stentz. Using linear landmarks for path planning with uncertainty in outdoor environments. *Proc. IEEE/RSJ Int. Conf. on Intelligent Robots and Systems*, 2009.

[9] L. Guibas, D. Hsu, H. Kurniawati, E. Rehman. Bounded uncertainty roadmaps for path planning. *Proc. Workshop on Algorithmic Foundations of Robotics*, 2008.

[10] Y. Huang, K. Gupta. Collision-probability constrained PRM for a manipulator with base pose uncertainty. *Proc. IEEE/RSJ Int. Conf. on Intelligent Robots and Systems*, 2009.

[11] V. Huynh, N. Roy. icLQG: combining local and global optimization for control in information space. *Proc. IEEE Int. Conf. on Robotics and Automation*, 2009.

[12] L. Kaelbling, M. Littman, A. Cassandra. Planning and acting in partially observable stochastic domains. *Artificial Intelligence* 101(1-2):99–134, 1998.

[13] L. Kavraki, P. Svestka, J.-C. Latombe, M. Overmars. Probabilistic roadmaps for path planning in high dimensional configuration spaces. *IEEE Trans. on Robotics and Automation* 12:4(566–580), 1996.

[14] G. Kewlani, G. Ishigami, K. Iagnemma. Stochastic mobility-based path planning in uncertain environments. *Proc. IEEE/RSJ Int. Conf. on Intelligent Robots and Systems*, 2009.

[15] Y. Kuwata, J. Teo, S. Karaman, G. Fiore, E. Frazzoli, J. How. Motion planning in complex environments using closed-loop prediction. *Proc. AIAA Guidance, Navigation, and Control Conf. and Exhibit*, 2008.

[16] H. Kurniawati, D. Hsu, W. Lee. SARSOP: Efficient point-based POMDP planning by approximating optimally reachable belief spaces. *Proc. Robotics: Science and Systems*, 2008.

[17] S. LaValle, S. Hutchinson. An objective-based framework for motion planning under sensing and control uncertainties. *Int. J. of Robotics Research* 17(1):19–42, 1998.

[18] S. LaValle, J. Kuffner. Randomized kinodynamic planning. *Int. Journal on Robotics Research* 20(5):378–400, 2001.

[19] S. LaValle. *Planning algorithms*. Cambridge University Press, 2006.

[20] A. Lazanas, J. Latombe. Motion planning with uncertainty: a landmark approach. *Artificial Intelligence*, 76(1-2):285–317, 1995.

[21] N. Melchior, R. Simmons. Particle RRT for path planning with uncertainty. *Proc. IEEE Int. Conf. on Robotics and Automation*, 2007.

[22] P. Missiuro, N. Roy. Adapting probabilistic roadmaps to handle uncertain maps. *Proc. IEEE Int. Conf. on Robotics and Automation*, 2006.

[23] A. Nakhaei, F. Lamiraux. A framework for planning motions in stochastic maps. *Proc. Int Conf. on Control, Automation, Robotics and Vision*, 2008.

[24] C. Papadimitriou, J. Tsisiklis. The complexity of Markov decision processes. *Mathematics of Operations Research*, 12(3):441–450, 1987.

[25] R. Pepy, A. Lambert. Safe path planning in an uncertain-configuration space using RRT. *Proc. IEEE/RSJ Int. Conf. on Intelligent Robots and Systems*, 2006.

[26] J. Porta, N. Vlassis, M. Spaan, P. Poupart. Point-based value iteration for continuous POMDPs. *Journal of Machine Learning Research* 7:23292367, 2006.

[27] S. Prentice, N. Roy. The belief roadmap: efficient planning in linear POMDPs by factoring the covariance. *Proc. Int. Symp. of Robotics Research*, 2008.

[28] N. Roy, W. Burgard, D. Fox, S. Thrun. Coastal navigation - mobile robot navigation with uncertainty in dynamic environments. *Proc. IEEE Int. Conf. on Robotics and Automation*, 1999.

[29] R. Tedrake. LQR-trees: Feedback motion planning on sparse randomized trees. *Proc. Robotics: Science and Systems*, 2009.

[30] S. Thrun, W. Burgard, D. Fox. *Probabilistic Robotics*, MIT Press, 2005.

[31] G. Welch, G. Bishop. An introduction to the Kalman filter. *Tech. Report TR 95-041*, University of North Carolina at Chapel Hill, 2006.

[32] Wikipedia. Chi-square distribution. http://en.wikipedia.org/wiki/Chi_square, 2010.

The Smooth Curvature Flexure Model: An Accurate, Low-dimensional Approach for Robot Analysis

Lael U. Odhner and Aaron M. Dollar
Yale University
Email: {lael.odhner, aaron.dollar}@yale.edu

Abstract— **This paper presents a new and comprehensive method for modeling robots having highly flexible members such as flexure joints. An accurate model of large deformation bending is important for precisely describing the configuration of the flexible member. Additionally, the accuracy of the Jacobian and Hessian of the forward kinematics are critically important at large angles for predicting the deformation and the stiffness of the joint under load. The model introduced here is based on the assumption that the curvature of a beam in bending is smooth, and thus can be approximated by low-order polynomials. This produces a parameterized description of flexure motion that can be used as a joint model when expressed in Denavit-Hartenberg form, as a transformation from one rigid link to the next in a serial manipulator. We will show that with only three parameters, this model faithfully reproduces the elastic deformation of a flexure hinge predicted by the continuum model, even for large angles, without requiring numerical integration or many finite elements. It can also be used to compute the compressive buckling load of the flexure as predicted by the continuum model.**

I. INTRODUCTION

Highly flexible members have been frequently considered in the context of robotic hardware. A number of studies into the behavior of flexible link robots have been conducted, often for the purposes of controlling for undesirable dynamic effects related to working with long, thin links (e.g. [1-3]). A smaller number of efforts have dealt with the beneficial aspects of highly flexible links, such as providing a large number of degrees of freedom for manipulation tasks [4] or low stiffness for grasping and assembly purposes [5]. A related application is the use of highly flexible members as *joints* between rigid links, typically referred to as flexures. The contrast between these two applications is shown in Fig. 1. These features are commonly used to allow motion in monolithic structures, and have been used as joints in a number of different robotic mechanisms, particularly in compliant hands [6-8]. The benefits of flexure-based joints include having no sliding parts (and therefore no friction or stick-slip effects), no backlash, and are able to compliantly deform in response to unplanned collisions, making them ideal for robots that must operate in unstructured environments [9].

Another major benefit of flexures is the simplicity and potential lower cost compared to standard revolute joints, which require bearings for smooth, accurate motion. As robots become more common as commercial products, flexures are likely to be used with increasing frequency due to their compatibility with inexpensive polymeric fabrication processes such as multi-shot injection molding [10] and shape deposition manufacturing [11].

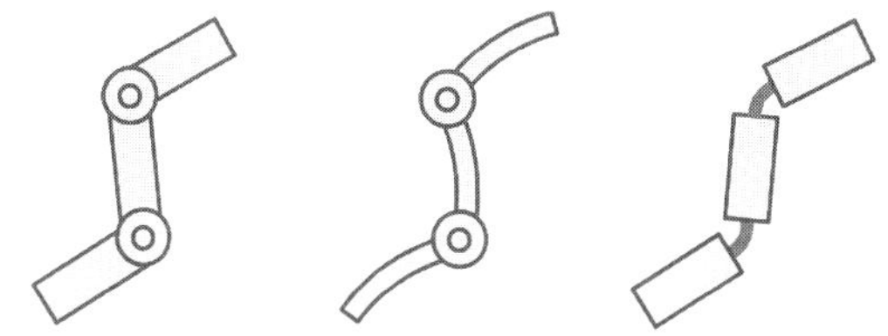

Figure 1. A comparison between traditional rigid robot manipulators (left), flexible link robots (center), and flexure-joint robots, in which flexible links act as hinges (right).

One drawback to flexure-based robot mechanisms is the complex mechanical behavior that they exhibit compared to pin joints. A pin joint has one degree of freedom, whereas the elastic deformation of a beam in bending has infinitely many degrees of freedom. Moreover, a flexure hinge in a robot often bends to angles up to 90 degrees or more, so classical small-deflection beam bending models are inapplicable. As a result, there is no canonical parametric model for planar flexure hinges suitable for robot analysis. This paper presents a model that can fill this role. In order to apply the rich set of tools available for serial manipulator design, control, and analysis, one must have a model of elastic behavior that is accurate and computationally simple. The design specification for a good flexure model can be broken into three functional requirements:

1. It should be possible to compute both the *shape* of the robot and the *elastic energy* associated with deformation as a function of a small set of generalized coordinates, as one might describe a jointed mechanism using the internal joint angles.

2. It should be possible to compute the *force* in generalized coordinates resulting from a force on the robot at any point using the Jacobian of that point's coordinates, as well as the *local equilibrium position* resulting from such a force.

3. It should be possible to compute the *stiffness* in generalized coordinates resulting from a force on the robot at a point by the Hessian of that point's coordinates, as well as any *buckling modes* the robot has (configurations/loads having zero stiffness in some direction).

Many models of flexible robot components meet some, but not all of these requirements. One common approach is to model flexure elements as having constant curvature [12]. Flexures have also been approximated as a single pin joint halfway between the ends of the flexure [13]. Both of these models capture the relative rotation between rigid bodies on the manipulator, making them useful for inverse kinematic computation and form closure grasp analysis. However, because these models have fewer degrees of freedom than a real flexure, they are too rigid and under-predict the deformation of a loaded flexure. Another family of models, called pseudo-rigid body models, consists of one or several joints placed to approximate the flexure's center of rotation, connected in parallel with nonlinear springs fit by regression to the exact force-deflection profile [14]. These models can be used to find flexure deflection under load, but they are unsuitable for manipulator analysis because the linkage geometries used to approximate the beam bending change based on the direction of applied force, and thus are not purely kinematic descriptions of flexure behavior. Another approach is based on assuming some set of superimposed deformational modes [15]. This technique has been applied to flexible links (e.g. [16]) and continuum manipulators [17]. Modal models for flexures have been proposed based on analytically calculated small-deformation solutions, as well as finite element solutions [18,19]. However, modal models are specific to the behavior they are designed to model, and none of the currently available models in the robotics literature accurately capture large-deformation flexure behavior.

The flexure model discussed in this paper is a modal model which approximates the curvature of a flexure using a polynomial basis. The polynomial coefficients define the relative position and orientation of two bodies connected by the flexure, as well as the elastic energy stored in the flexure itself. This model meets all three of the functional requirements introduced above, while avoiding the need to use numerical integration or to break the beam into many finite elements. It predicts not only the deflection of a flexure under load, but also second-order kinematic effects such as buckling and the change in flexure stiffness resulting from compressive or tensile loads. These second-order effects are particularly useful in the study of grasping and manipulation, where grasp stability may depend on the elastic stability of the manipulator itself [20]. Previous study of this model has dealt with stiffness prediction [21]. This paper is intended as a more general discussion of this model's applicability to robotic systems.

The remainder of this paper is divided into three sections. Section II is an overview of the flexure model, describing how the parameters define the shape and the energy function of a flexure hinge. Section III examines the shape of the flexure when an arbitrary load and moment is applied at one end. The results are compared to exact large deformation beam solutions. Additional results are shown comparing the smooth curvature model to finite element models of a sample mechanism. Section IV demonstrates the second-order kinematic accuracy of the model by comparing classical continuum buckling models to the discrete buckling predicted by the proposed parameterized model.

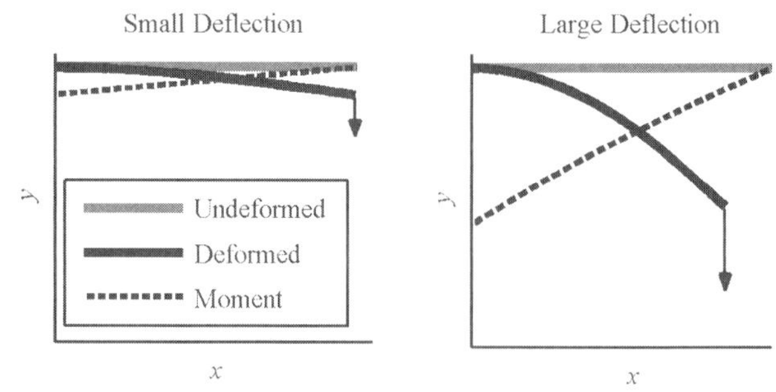

Figure 2. A comparison of the moment profile in small- and large-deflection cantilevered beam bending.

II. The Smooth Curvature Model

A. *Motivation*

In 1694, Jacob Bernoulli proposed (and solved) the problem of finding the shape of a pre-bent cantilevered beam of length L that would bend into a straight line when loaded with an arbitrarily large force P at the tip [22]. Today the curve is known as the clothoid or Euler spiral. The curvature $\kappa(s)$ of the clothoid curve varies linearly with the arc length s from the base of the cantilever,

$$\kappa(s) = \frac{P}{EI}\left(1 - \frac{s}{L}\right) \quad (1)$$

Here E is the elastic modulus and I is the constant planar moment of the beam area. The clothoid is also a passable approximate solution the more useful problem of finding the deformed shape of an *initially straight* beam loaded at one end with a large load. This can be seen by examining the nonlinear deformation of a cantilevered beam, as shown in Fig. 2. For a small end load (left), the bending moment will be almost exactly proportional to the distance from the tip of the flexure, as expected. A large end load (right) will produce a non-linear deformation profile, but the moment, plotted as a function of arc length, is still roughly linear. The curvature is directly proportional to bending moment in the beam, $\tau(s)$,

$$\kappa(s) = \frac{\tau(s)}{EI} \quad (2)$$

For some range of large loads, then, the curvature of a beam can be approximated as some constant plus a linear function of arc length. Horn discussed this approximation in the context of spline curves [23]. The accuracy of this model could be further improved by noting that while the curvature may not be exactly linear, it is certainly smooth, and so might be described with a basis of n Legendre polynomials, $G_0(s)_{\ldots}G_{n-1}(s)$. The curvature is expressed as a weighted sum of the bases,

$$\kappa(s, \underline{q}) = q_0 G_0(s) + q_1 G_1(s) + \ldots + q_{n-1} G_{n-1}(s) \quad (3)$$

The coefficients, $q_0 \ldots q_{n-1}$, used as a generalized coordinate vector $\underline{q}$ for describing the deformation of the flexure, are central to the proposed flexure model. Two particular cases will be considered here, corresponding to the models with 2 and 3 parameters, whose basis functions are Legendre polynomials, translated and scaled to be orthogonal on [0, L],

$$\kappa_2(s,\underline{q}) = q_0 \frac{1}{L} + q_1 \frac{2s-L}{L^2} \quad (4)$$

$$\kappa_3(s,\underline{q}) = q_0 \frac{1}{L} + q_1 \frac{2s-L}{L^2} + q_2 \frac{6s^2-6sL+L^2}{L^3} \quad (5)$$

Models of this type, which we will call smooth curvature models, can be used to predict the shape of the flexure in bending, as well as the elastic bending energy. These derivations follow in the next two sections.

B. *Flexure Shape*

Joints in a serial robot manipulator are often represented in Denavit-Hartenberg notation, that is, as a parameterized axial, radial and angular transformation from one joint to another. The analogous transformation for a pin joint is a rotation about the joint axis. In the case of a flexure, this transformation corresponds to the translation and rotation from one end of the flexure to the other, as shown in Fig. 3 [12]. This could be written as a matrix, for instance,

$$T = \begin{bmatrix} \cos(\varphi_{tip}) & -\sin(\varphi_{tip}) & x_{tip} \\ \sin(\varphi_{tip}) & \cos(\varphi_{tip}) & y_{tip} \\ 0 & 0 & 1 \end{bmatrix} \quad (6)$$

The three quantities characterizing this transformation are the flexure tip displacement (x_{tip}, y_{tip}), and the relative angle from the base of the flexure to the tip of the flexure, φ_{tip}. They can be written as functions of the polynomial coefficient vector, $\underline{q}$. The angular profile, $\varphi(s, \underline{q})$, is the integral of the curvature,

$$\varphi(s,\underline{q}) = \int \kappa(s,\underline{q})ds \quad (7)$$

$$\varphi_2(s,\underline{q}) = q_0 \frac{s}{L} + q_1 \frac{s^2-sL}{L^2} \quad (8)$$

$$\varphi_3(s,\underline{q}) = q_0 \frac{s}{L} + q_1 \frac{s^2-sL}{L^2} + q_2 \frac{2s^3-3s^2L+sL^2}{L^3} \quad (9)$$

At the end of the flexure (s=L), the tip angle is equivalent to q_0 irrespective of the model order. This is a happy side effect of using orthogonal polynomials, as all the higher, non-constant terms must integrate to zero:

$$\varphi_{tip} = \varphi(L,\underline{q}) = q_0 \quad (10)$$

The tip position of the flexure can be found by integrating the cosine and sine of the angular profile,

$$x_{tip} = \int_0^L \cos(\varphi(s,\underline{q}))ds \quad (11)$$

$$y_{tip} = \int_0^L \sin(\varphi(s,\underline{q}))ds \quad (12)$$

These expressions are transcendental. The second order solution can be solved in terms of Fresnel integrals by

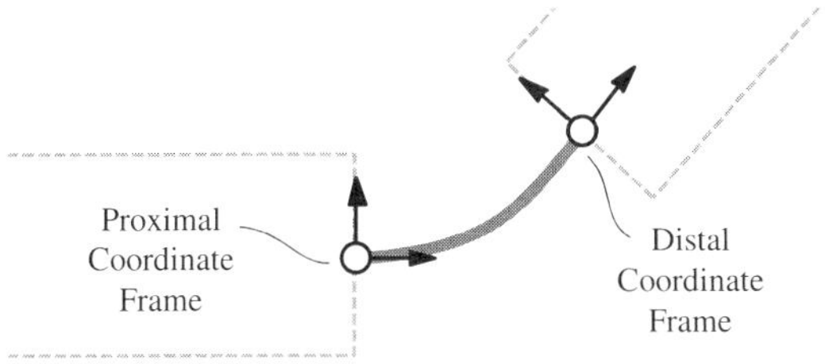

Figure 3. A kinematic model of a flexure consists of a transformation mapping the coordinate frame at the proximal side of the joint to the coordinate frame at the distal side of the joint.

completing the square and using trigonometric addition identities, but it contains discontinuities, and is not practically useful. Further, this strategy does not generalize to the 3 parameter model. Instead, an interpolating approximation was used, so that (11) and (12) can be analytically approximated within some reliable error bounds. In this paper, Chebyshev interpolation [24] was used to approximate the sine and cosine functions as polynomials. There is a trade-off between the domain of interpolation and the computational cost, so maximum flexure rotation was limited to be less than 108°. Alternatively, Gaussian quadrature could be used, essentially interpolating the entire integrand as a polynomial.

C. *Elastic Energy*

Having found the shape of the hinge as a function of $\underline{q}$, we now turn to finding an expression for the elastic energy in the flexure. The energy stored in an Euler-Bernoulli beam is proportional to the integral of the squared curvature [15]:

$$U(\underline{q}) = \frac{EI}{2} \int_0^L \kappa(s,\underline{q})^2 ds \quad (13)$$

If the 2 parameter curvature is used, $U(\underline{q})$ evaluates to a weighted sum of the squared parameters.

$$U_2(\underline{q}) = \frac{EI}{2} \int_0^L \left(\frac{q_0}{L} + \frac{4q_1(2s-L)}{L^2} \right)^2 ds = \frac{EI}{2L} \left(q_0^2 + \frac{q_1^2}{3} \right) \quad (14)$$

Because the polynomial basis is orthogonal under convolution over the interval [0, L], there are no cross-terms in this expression. The expression for energy given 3 parameters differs only in the addition of a single term:

$$U_3(\underline{q}) = \frac{EI}{2L} \left(q_0^2 + \frac{q_1^2}{3} + \frac{q_2^2}{5} \right) \quad (15)$$

D. *Summary*

The smooth curvature model for flexure hinges has been introduced, based on the observation that the curvature of a flexure can be approximated using a low-dimensional basis of orthogonal polynomials. The position and orientation of the flexure tip can be found relative to its base using only a few model parameters, and these can be used to represent the flexure as a joint in Denavit-Hartenberg form. The elastic energy stored in the bent beam is a weighted sum of the squared flexure parameters.

The remaining sections will demonstrate that this model can be used to satisfy the two other requirements for a flexure joint model, that is, that the model accurately predicts the equilibrium position of the flexure when an arbitrary force and moment are applied, and that the model accurately predicts variable stiffness effects and buckling due to compressive loads.

III. Deflection Under Load

A. *Jacobian Analysis of the Forces on a Robot*

Accurate descriptions of the force exerted on a robot and the resulting deflection are central to many problems in the control and analysis of robot manipulators. The net generalized force $\underline{F}$ on a manipulator experiencing a force $\underline{f}_p$ at some point $\underline{p}$ is given by the Jacobian of that point's coordinates, and the gradient of the potential energy function $U(\underline{q})$,

$$\underline{F} = \nabla_q(\underline{p})^T \underline{f}_p + \nabla_q(U(\underline{q})) \tag{16}$$

The generalized force balance equation is only realistic if the generalized coordinates faithfully represent all of the motions that the robot is capable of making. For a flexure hinge, it is important that the motion of flexure tip, as described in the previous section, is accurate, so that the forces and moments transmitted from one link to the next result in a physically realistic deformation of the flexure hinges.

This section considers two tests for benchmarking the ability of the smooth curvature flexure model to predict deformation under load. The first test compares the exact deflection of a cantilevered flexure (using numerically computed elastica curves [25]) to the tip position predicted by the smooth curvature model. The second test computes the deflection of a two-link finger from a tendon-driven elastic gripper developed by the authors. The two flexure hinges in the finger are modeled with finite element flexure models, and with the 3 parameter smooth curvature model, showing that the two models agree despite the vastly reduced parameter space of the smooth curvature model.

B. *Tip Deflection of a Loaded Flexure*

A straightforward method of examining the accuracy of a flexible beam is to clamp one end and examine the deflection of the other end when subject to an arbitrary moment M and force P exerted at an angle θ, as shown in Fig. 4. To ensure proper scaling of the results, a non-dimensional form of the beam bending equations should be used, based on these substitutions:

$$\tilde{P} = \frac{PL^2}{EI},\ \tilde{M} = \frac{ML}{EI},\ \tilde{\tau} = \frac{\tau L}{EI},\ \tilde{s} = \frac{s}{L},\ \tilde{x} = \frac{x}{L},\ \tilde{y} = \frac{y}{L} \tag{17}$$

The dimensionless beam bending equations are equivalent to modeling a beam as having length 1, and an elastic modulus and cross-sectional moment equal to 1.The only parameter that is unaffected by this scaling is the beam's angular profile, $\varphi(s)$. Results will be computed for the case when the tip angle, φ_{tip}, is equal to 90°, a prototypical test case in the study of large-deformation beam bending[1]. The equilibrium configurations were compared to the exact solution obtained by numerically integrating the large-deformation Euler-Bernoulli equation,

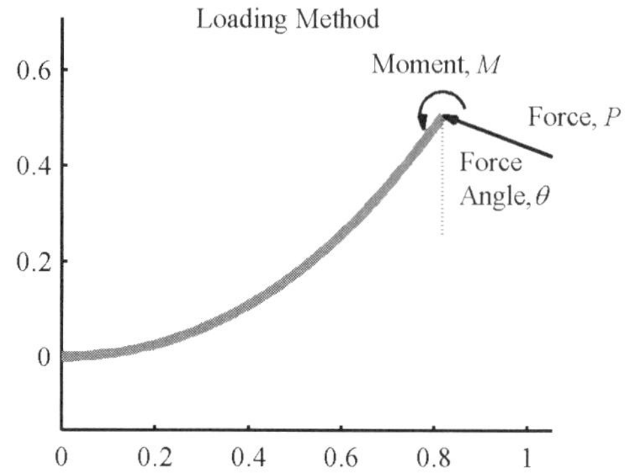

Figure 4. A flexure, loaded at the end by a force and a moment. This figure shows the direction of load, θ.

$$\frac{d}{d\tilde{s}}\begin{bmatrix}\tilde{\tau}(\tilde{s})\\ \varphi(\tilde{s})\\ \tilde{x}(\tilde{s})\\ \tilde{y}(\tilde{s})\end{bmatrix} = \begin{bmatrix}\tilde{P}(\cos(\theta)\cos(\varphi(\tilde{s})) - \sin(\theta)\sin(\varphi(\tilde{s})))\\ \tilde{\tau}(\tilde{s})\\ \cos(\varphi(\tilde{s}))\\ \sin(\varphi(\tilde{s}))\end{bmatrix} \tag{18}$$

This is a restatement of (2), (7), (11) and (12) in differential form, after applying the substitutions from (17). This equation was solved using a Runge-Kutta solver. The tip moment M, force angle θ, and tip angle φ_{tip}, were specified, and the integral from the tip of the flexure to the based was computed. The tip force, P, was found using a bisection search such that the boundary conditions at both ends of the flexure were simultaneously satisfied.

Each force-moment combination (P, θ, M) that was computed for the exact beam equation was applied to the tip of the smooth curvature flexure model, using the generalized force balance from (16). The Jacobian of the tip coordinates x_{tip}, y_{tip} and φ_{tip} was derived from (10), (11) and (12),

$$\frac{\partial \varphi_{tip}}{\partial q_i} = \begin{cases}1, & i = 0\\ 0, & i \neq 0\end{cases} \tag{19}$$

$$\frac{\partial x_{tip}}{\partial q_i} = -\int_0^L \sin(\varphi(s,\underline{q}))\frac{\partial \varphi_{tip}}{\partial q_i}\,ds \tag{20}$$

$$\frac{\partial y_{tip}}{\partial q_i} = \int_0^L \cos(\varphi(s,\underline{q}))\frac{\partial \varphi_{tip}}{\partial q_i}\,ds \tag{21}$$

As in (11) and (12), Chebyshev interpolation was used to produce analytical approximations of (20) and (21). The derivatives of the energy function are much simpler, and can be found from (15),

[1] This is a generalization of the rectangular elastica problem posed by Bernoulli, the problem of finding the shape of a cantilevered beam bent at a right angle by a force at the tip [25].

$$\nabla_q U(\underline{q}) = \frac{EI}{L}\begin{bmatrix} 1 & 0 & 0 \\ 0 & 1/3 & 0 \\ 0 & 0 & 1/5 \end{bmatrix}\begin{bmatrix} q_0 \\ q_1 \\ q_2 \end{bmatrix} \tag{22}$$

The force balance, computed from (19)-(22), was set to zero to form a system of nonlinear equations, which was solved numerically. Three force-moment combinations were used to compare the two models. These combinations, labeled A, B, and C, are shown in Fig. 5. In case A, a pure bending moment was applied, sufficient to bend the flexure to an angle of 90°. Case B was a pure force of a magnitude sufficient to bend the flexure to 90°. Case C was the most complex load, consisting of a moment equal and opposite to the moment applied in combination 1, counteracted by a force so that the flexure tip angle remained at 90°.

The prediction errors from the smooth curvature models were computed for values of θ ranging from 20 degrees to 105 degrees (as in Fig. 4). The lower bound of 20 degrees was chosen because the magnitude of the load required to bend a beam to 90° in case B has a vertical asymptote at θ=0. Thus, the behavior of the flexure becomes increasingly unrealistic in this case. The upper bound of 105 degrees was chosen because the elastica curves generated with the Runge-Kutta solver could not predict buckled configurations, and as the force on the flexure tip becomes increasingly compressive (i.e. θ > 90°), good reference comparisons could not be made. Instead, compressive loads were compared to finite element models in the following subsection.

The positional error of the flexure tip was found, that is, the norm of the vector from the predicted flexure tip to the tip of the numerically computed elastica curve. This is plotted in non-dimensional form, meaning that the error is given as a fraction of the flexure length. The angular error is also shown. The errors for the 2 parameter model are shown in Fig. 6, and the errors for the 3 parameter model are shown in Fig. 7. These plots show a number of significant results. First, the errors in case A (pure moment loading) were very small for both the 2 and 3 parameter smooth curvature models. The exact shape one would expect for a beam having a constant bending moment is an arc, a shape that can be exactly reproduced with both models. Thus, the error is correspondingly small. The errors observed in loading cases B and C indicate that the 3 parameter model is significantly more accurate, especially when the flexure is loaded by an opposed force and moment. Most importantly, in every case, the 3 parameter smooth curvature model was within a positional accuracy of 1 percent of the beam length, and an angular accuracy bound of 1°.

C. Finite Element Comparisons

One purpose of the smooth curvature model of particular interest to the authors is to enable efficient analysis of manipulators having multiple flexure joints. In previous work, the authors have developed robot hands incorporating polymeric elastic flexure joints [8,9]. These hands are made up of 2-link, tendon-driven fingers, represented in Fig. 8A. In order to evaluate the usefulness of the smooth curvature model for multi-link manipulators, a finite element model was constructed using an object-oriented Matlab library created by

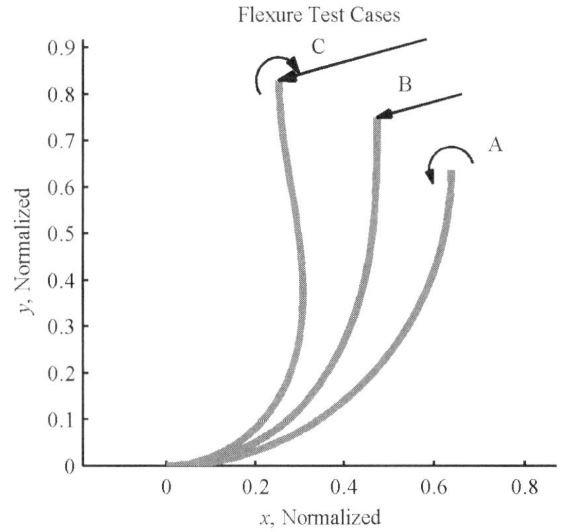

Figure 5. The three loading cases used to test the smooth curvature model.

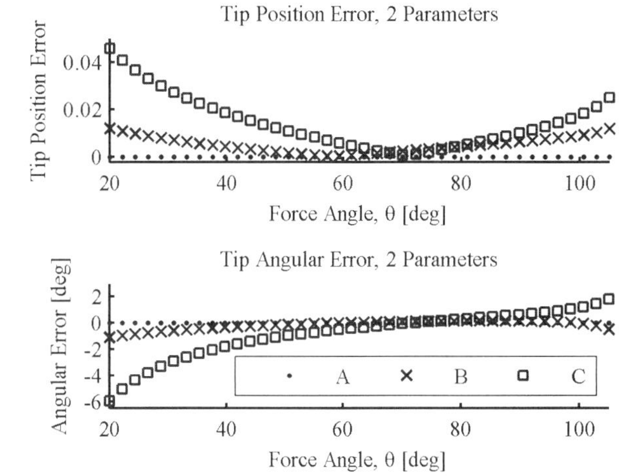

Figure 6. Prediction errors for the 2 parameter smooth curvature model. Cases A, B, and C correspond to the cases in Fig. 5.

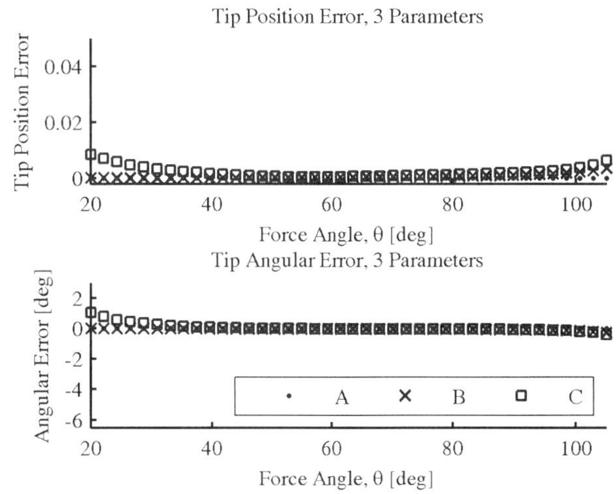

Figure 7. Prediction errors for the 3 parameter smooth curvature model. Cases A, B, and C correspond to the cases in Fig. 5.

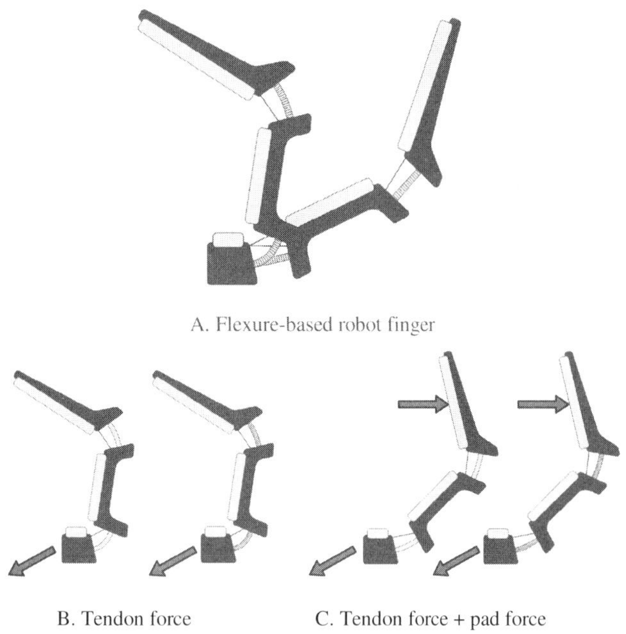

Figure 8. A comparison between FEM and smooth curvature (SC) models:
A. The FEM model, no tendon force vs 10 N tendon force.
B. SC model (i) vs FEM (ii), 10 N tendon force.
C. SC model (iii) vs. FEM (iv), 10 N force + 0.5 N force on distal link.

the authors [26]. This model represents each flexure in a fashion similar to the finite element model proposed in [1], consisting of 16 small rotational links. The transformation representing each joint in Denavit-Hartenberg form was the composition of the many resulting rotations and translations making up each finite element.

A model of the finger was also constructed using our 3 parameter smooth curvature flexures. This model uses the joint transformation from (6) and the energy function from (15) to describe the joint behavior in terms of three generalized coordinates per joint, for a total of 6. The finite element model, by comparison, had a total of 32 generalized coordinates. Both models were subjected to two different loading conditions: in the first condition, the finger was actuated with a single tendon connected to the distal link, as in Fig. 8B. The second condition, depicted in Fig. 8C, included the same tendon force and a horizontal force applied to the center of the pad on the distal link. In both cases, the generalized force balance was computed using (16), computing the kinematics of the tendon and the center of the distal pad using a composition of geometric joint and link transformations.

The results of the test are shown in Fig. 8 and Table I, which describes the agreement between the smooth curvature and finite element models as to the position and orientation of the center of the distal link. The results show that the displacement of the distal link pad as calculated by the smooth curvature model is within 0.1% of the position predicted by the finite element model. The angular agreement is similarly within a tenth of a degree in both cases. Attempts at visual comparison between the FEM and smooth curvature models by overlaying the two were unsuccessful, because they were almost indistinguishable to the eye.

TABLE I. FEM VS. SMOOTH CURVATURE FINGER MODEL RESULTS

Case	Pad x	Pad y	Pad angle
FEM, Tendon Force (i)	-0.0048	0.1012	147.6553
SC, Tendon Force (ii)	-0.0047	0.1012	147.5760
FEM, Tendon + Pad Force (iii)	0.0615	0.1046	103.0476
SC, Tendon + Pad Force (iv)	0.0615	0.1046	103.0168

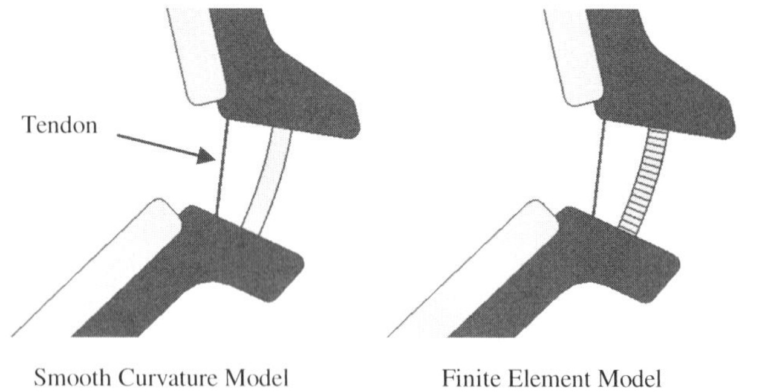

Figure 9. A single tendon-driven joint, modeled using the smooth curvature model (left) and finite rotational elements (right).

D. *Summary*

For both a single flexure hinge and a two-joint manipulator, the smooth curvature model has been shown to accurately predict the deformation of a manipulator under a wide range of forces and moments. This is interesting and new because most methods of accurately solving large-deflection beam bending problems involve numerical integration, or the breaking down of a beam into many finite elements. Because the smooth curvature model achieves a useful degree of accuracy with only three parameters per joint, calculation of dynamics and statics for control or motion planning is a much simpler process.

IV. STIFFNESS AND BUCKLING

A. *Stiffness of a Loaded Elastic Structure*

One major difference between the well-studied problem of flexible-link manipulators and the newer field of flexure-jointed manipulators is the relatively increased importance of buckling in flexure joints. The flexure joints shown in Fig. 9 show a typical tendon/flexure actuation scheme. The tendon exerts a force in tension, which is balanced by a compressive force in the flexure, unless a parallel load path exists. This compressive force is quite large, and can easily approach the Euler buckling load of a thin flexure (the load at which the lateral stiffness of the flexure is zero). Unlike structural columns, buckling does not represent a necessarily undesirable effect. The fact that the flexure is buckled just means that its rotational stiffness is very low. Most pin joints, for example, have zero rotational stiffness and this is not an impediment to their use in robots. However, this change in stiffness as a function of load is critically important in some robotic tasks. For example, a change in joint stiffness will affect computed torque control models. Additionally, some tasks such as grasping and manipulation rely on the elastic stability of the whole system; a buckling mode could be harmless, or it could correspond to configuration in which a grasped object twists out of its gripped position [20]. As a result of all these concerns, it is important that a flexure model provide a reasonable model of elastic buckling.

This section briefly describes the ability of the smooth curvature model to predict elastic buckling in a flexure using the Hessian of the flexure kinematics. As a proxy case for comparing the continuum behavior of a flexure to the smooth curvature model, the smooth curvature model will be used to predict buckling in compression by finding the smallest compressive load for which the generalized stiffness matrix is singular. This result can be compared to Euler's buckling load formula. The 2 and 3 parameter models produce successively better approximations of buckling.

B. *Continuum vs. Discrete Buckling*

A continuum structure is said to buckle when it has zero (or negative) stiffness in some direction, so that a small perturbation to the structure's shape is met by a destabilizing force, rather than a restoring force. The compressive load P_{crit} at which a clamped-free beam should buckle is given by Euler's well-known formula [27]:

$$\frac{P_{crit}L^2}{EI} = \left(\frac{\pi}{2}\right)^2 = 2.4674... \tag{23}$$

The concept analogous to elastic buckling in a generalized coordinate model has to do with the generalized stiffness matrix obtained by taking the gradient of the generalized force balance from (16) with respect to $\underline{q}$,

$$\mathbf{K} = \nabla_q \nabla_q (\underline{p})^T \underline{f}_p + \nabla_q \nabla_q (U(\underline{q})) \tag{24}$$

Stiffness is a function of the Hessian of the coordinates where force is applied, and of the Hessian of the energy, $U(\underline{q})$. When this stiffness matrix has an eigenvalue that is zero or negative, it buckles. In other words, there will exist some eigenvector $\delta\underline{q}$, which, when applied to the robot as a perturbation, will produce a destabilizing force. This could also be thought of as a test for the convexity of the total energy in the robot.

We will derive the generalized stiffness matrix for the smooth curvature model, when the flexure is loaded in the $-x$ direction with a force, P. Thus, the contact point $\underline{p}$ from (24) above is the scalar x_{tip}, as described in (8). The Hessian elements can be calculated from the Jacobian of x_{tip} in (20),

$$\frac{\partial^2 x_{tip}}{\partial q_i \partial q_j} = -\int_0^L \left[\cos(\phi(s))\frac{\partial\phi}{\partial q_i}\frac{\partial\phi}{\partial q_j} + \sin(\phi(s))\frac{\partial^2\phi}{\partial q_i \partial q_j}\right] ds \tag{25}$$

When the flexure is straight, $q_0=q_1=q_2=0$. In this configuration, (25) can be simplified, because the sine term disappears and the cosine term approaches one,

$$\frac{\partial^2 x_{tip}}{\partial q_i \partial q_j} = -\int_0^L \frac{\partial\phi}{\partial q_i}\frac{\partial\phi}{\partial q_j} ds \tag{26}$$

These are polynomials, which can be evaluated to compute the Hessian of x_{tip} with respect to the generalized coordinates. For the 2 parameter model, the Hessian is a 2 by 2 matrix,

$$\nabla_q \nabla_q (x_{tip}) = L\begin{bmatrix} 1/3 & -1/12 \\ -1/12 & 1/30 \end{bmatrix} \tag{27}$$

The stiffness due to potential energy can be found by taking the Hessian of the energy function derived in (14),

$$\nabla_q \nabla_q (U(\underline{q})) = \frac{EI}{L}\begin{bmatrix} 1 & 0 \\ 0 & 1/3 \end{bmatrix} \tag{28}$$

The coordinate Hessian and the energy Hessian can be substituted back into (24),

$$\mathbf{K} = \frac{EI}{L}\begin{bmatrix} 1 & 0 \\ 0 & 1/3 \end{bmatrix} - PL\begin{bmatrix} 1/3 & -1/12 \\ -1/12 & 1/30 \end{bmatrix} \tag{29}$$

The buckling load of the 2 parameter model, P_2, is the value of P for which the determinant of $\mathbf{K}$ is zero, indicating that the matrix has an eigenvalue equal to zero,

$$\begin{vmatrix} \frac{EI}{P_2L^2} - \frac{1}{3} & \frac{1}{12} \\ \frac{1}{12} & \frac{EI}{3P_2L^2} - \frac{1}{30} \end{vmatrix} = 0 \tag{30}$$

The smallest root of this polynomial is the most physically meaningful, as it represents the load at which the unconstrained flexure will buckle,

$$\frac{P_2L^2}{EI} = 2.4860... \tag{31}$$

This is only 0.75% larger than the true value reported by the continuum model in (23). This exercise can be repeated for the three parameter model, to find the predicted buckling load, P_3,

$$\begin{vmatrix} \frac{EI}{P_3L^2} - \frac{1}{3} & \frac{1}{12} & \frac{1}{60} \\ \frac{1}{12} & \frac{EI}{3P_3L^2} - \frac{1}{30} & 0 \\ \frac{1}{60} & 0 & \frac{EI}{5P_3L^2} - \frac{1}{210} \end{vmatrix} = 0 \tag{32}$$

The resulting buckling load prediction is within 0.02% of the value predicted by Euler's beam buckling formula,

$$\frac{P_3L^2}{EI} = 2.4677... \tag{33}$$

C. *Discussion*

The results of this brief study indicate that the smooth curvature model has no difficulty predicting the stiffness of a straightened flexure hinge as a function of load. A more general argument, too long to be presented here, would also cover the stiffness of a constrained flexure, such as one which is pinned or clamped at the distal end. Although the effect of constraints on the generalized stiffness is lengthier to describe, the 3 parameter model can successfully describe these cases. Finally, it is worth mentioning that although stiffness prediction of a flexure undergoing large deformation has not been presented here, the smooth curvature model will predict the stiffness matrix for large deformations, as we have shown in [21].

V. Conclusions

A. *In Summary*

In this paper, we have presented a model for flexible links that is accurate for large deformations, so that it can be used for the special case of flexible links as flexure hinges. These models are compatible with all of the standard tools used for manipulator analysis, because they are in a form where the shape of the joint and the elastic energy of the joint can be entirely described by a set of generalized coordinates. We have demonstrated that a flexure can be described to a high level of accuracy using only three parameters – arguably the minimum number of parameters capable of describing a flexure with three independent end conditions, x_{tip}, y_{tip} and φ_{tip}. This model is useful for "zeroth" order descriptions (shape and energy), first order descriptions (local deformation and force), and for second order descriptions (buckling configurations and stiffness) of mechanical behavior of flexible members undergoing large deformations under loads.

B. *Future Directions*

The smooth curvature flexure model presents several obvious directions for further study. First, this model does not take into account axial deformation, which is important in accurately analyzing parallel mechanisms. We are confident that this method can be extended to describe such mechanisms through the addition of one or two axial deformation modes.

Second, it is worth noting that although models of planar beam bending are quite useful, many flexures admit a great deal of out-of-plane motion. This behavior can be characterized by modal models similar to the one presented here. We are working on extensions of smooth curvature models to three dimensions, including possible techniques for coping with the greatly increased complexity of describing non-commutative spatial rotations.

References

[1] E. Bayo, "A finite-element approach to control the end-point motion of a single-link flexible robot," Journal of Robotic Systems, vol. 4(1), pp. 63-75, 1986.

[2] H. Asada, Z.-D. Ma, and H. Tokumaru, "Inverse Dynamics of Flexible Robot Arms: Modeling and Computation for Trajectory Control", J. Dynamic Systems, Measurement, and Control vol.112(2), pp. 177-186, 1990.

[3] Zheng-Hua Luo, "Direct strain feedback control of flexible robot arms: new theoretical and experimental results," IEEE Transactions on Automatic Control, vol.38(11), pp.1610-1622, 1993.

[4] A. Hemami, "Studies on a light weight and flexible robot manipulator," Robotics, vol. 1(1), pp. 27-36, 1985.

[5] S. Sur and R.M. Murray, "An Experimental Comparison of Tradeoffs in Using Compliant Manipulators for Robotic Grasping Tasks,"Proceedings of the 1997 IEEE International Conference on Robotics and Automation, pp. 1807-1814, 1997.

[6] F. Lotti and G. Vassura, "A Novel Approach to Mechanical Design of Articulated Fingers for Robotic Hands," Proceedings of the 2002 IEEE International Conference on Intelligent Robots and Systems, pp. 1687-1692, 2002.

[7] F. Lotti, P. Tiezzi, G. Vassura, L. Biagiotti, G. Palli, C. Melchiorri, "Development of UB Hand 3: Early Results," Proceedings of the 2005 IEEE International Conference on Robotics and Automation, pp. 4499 – 4504, 2005.

[8] A. Dollar and R. Howe, "The Highly Adaptive SDM Hand: Design and Performance Evaluation," Int. Journal of Robotics Research, v.29, pp. 585-597, 2010.

[9] A. Dollar and R. Howe, "Towards grasping in unstructured environments: Grasper compliance and configuration optimization," Advanced Robotics, v. 19, n. 5, pp. 523-543, 2005.

[10] R. Digiantonio and G. Lawrence, "Two-shot molding of thermoplastic elastomers," Proceedings of the 1992 SPE Annual Technology Conference, pp. 851-853, 1992

[11] R. Merz, F. Prinz, K. Ramaswami, M. Terk, L. Weiss, "Shape deposition manufacturing," Proceedings of the Solid Freeform Fabrication Symposium, v. 5, pp.1-8, 1994

[12] M. Hannan and I. Walker, "Kinematics and the implementation of an elephant's trunk manipulator and other continuum style robots," Journal of Robotics Systems, v. 20, n. 2, pp. 45-63, 2003

[13] M. Ciocarlie and P. Allen, "A Design and Analysis Tool for Underactuated Compliant Hands," Proceedings of the 2009 IEEE/RSJ Int. Conf. on Intelligent Robotics and Systems, pp. 5235-5239, 2009

[14] A. Midha, T. Norton, and L. Howell, "On the nomenclature, classification, and abstractions of compliant mechanisms," Trans. of the ASME, v. 116, n. 1, pp.270-280, 1994

[15] S. Timoshenko and D. Young, *Advanced Dynamics*, p. 204, McGraw-Hill, 1948.

[16] A. DeLuca and B. Siciliano, "Closed-form dynamic model of planar multilink lightweight robots," IEEE Trans. on Systems, Man and Cybernetics, v. 21, n. 4, pp. 826-839, 1991

[17] G. Chirikjian and J. Burdick, "A modal approach to hyper-redundant manipulator kinematics," IEEE Trans. on Robotics and Automation, v. 10, n. 3, pp. 343-354, 1994

[18] N. Lobontiu, "Distributed-Parameter dynamic model and optimized design of a four-link pendulum with flexure hinges," Mechanism and Machine Theory, v. 36, pp. 653-669, 2001

[19] N. Lobontiu and E. Garcia, "Two microcantilever designs: lumped-parameter model for static and modal analysis," Journal of Microelectromechanical Systems, v. 13, 2004

[20] H. Hanafusa and H. Asada, "Stable Prehension by a Robot Hand with Elastic Fingers," Proceedings of the 7th International Symposium on Industrial Robots, pp. 361-368, 1977.

[21] L. Odhner and A. Dollar, "Fast, accurate models for predicting the compliance of elastic flexure-jointed robots," ASME Internat. Design Engineering Technical Conf and Computers and Information in Engineering Conf., 2010.

[22] R. Levien, "The Euler spiral: a mathematical history," Technical Report No. UCB/EECS-2008-111, University of California at Berkeley, 2008

[23] B. Horn, "The Curve of Least Energy," ACM Trans. On Mathematical Software, vol. 9, n. 4, pp. 441-460, 1983

[24] W. Press, S. Teukolsky, W. Vetterling, and B. Flannery, *Numerical Recipes in C*, 2nd ed., p. 190, Cambridge University Press, 1992.

[25] R. Levien, "The elastica: a mathematical history," Technical Report No. UCB/EECS-2008-103, University of California at Berkeley, 2008

[26] L. Odhner and A. Dollar, "The Freeform Manipulator Analysis Tool" [Online]. Avaliable: http://www.eng.yale.edu/grablab/fmat/

[27] S. Crandall, C. Dahl and T. Lardner, *An Introduction to the Mechanics of Solids*, 2nd ed., p. 587, McGraw-Hill, 1999.

Multi-priority Cartesian Impedance Control

Robert Platt Jr.
Computer Science and Artificial Intelligence Laboratory
Massachusetts Institute of Technology
rplatt@csail.mit.edu

Muhammad Abdallah, Charles Wampler
Manufacturing Systems Research Lab
General Motors
{muhammad.abdallah,charles.w.wampler}@gm.com

Abstract— **Manipulator compliance is well known to be important to robot manipulation and assembly. Recently, this has been highlighted by the development of new higly-compliant robot manipulators such as the Barrett arm or the DLR lightweight manipulator [1, 2]. It is also clear that dexterous manipulation involves touching the environment at different locations simultaneously (perhaps at different points on the robot hand or fingers). In these situations, it is particularly attractive to control the system using a multi-priority strategy where several contact points are commanded in parallel. Multi-priority Cartesian impedance control is the natural combination of these two ideas. The system realizes several impedances with different reference positions at different points on the robot with a specified order of priority. We find a controller that minimizes an arbitrary quadratic norm on the second-priority impedance error subject to constraints deriving from the first priority impedance task. We also show that the locally optimal controller does not require force feedback in its implementation for passive desired inertias. The results are illustrated in simulation.**

I. Introduction[1]

Since the earliest days of robotics, researchers have been aware of the possibility of controlling second-priority objectives using degrees of freedom (DOFs) that are redundant with respect to a first-priority task [3, 4, 5]. However, the advent of high-DOF humanoid robots make it possible to consider executing more than two tasks simultaneously [6, 7]. This approach to robot control is particularly relevant to manipulation tasks where high-DOF hand-arm systems must be controlled so as to interact with an environment at a few desired contact regions. However, it is ironic that although end-effector compliance is important to robust manipulation, relatively little research exists that studies multi-priority impedance control. This paper addresses this deficit by proposing a locally-optimal dual-priority impedance law. First, we propose a general law that uses force feedback to realize an arbitrary dual-priority impedance. Then, we characterize the subset of impedances that can be realized using torque controlled manipulators without force sensors. In contrast to prior work, we consider the space of all possible control laws that realize the first-priority impedance and select the one that is optimal with respect to an arbitrary quadratic optimization criterion.

The multi-priority approach to robot manipulator control has been studied extensively. Many early approaches calculated joint velocities that attempt to achieve a second-priority objective (such as avoiding obstacles or manipulator singularities) while achieving a desired end-effector velocity [4, 5, 8]. Chiaverini considers a damped-least-squares version of the control law that is robust to algorithmic singularities [9]. Antonelli provides a Lyapunov analysis demonstrating that the basic approach is stable [10]. Related approaches have been applied in more general contexts [11, 12].

A significant body of work explores the general problem of impedance control in the context of redundant manipulators. Building on Hogan's early work [13], Natale *et. al.* propose a version of the impedance controller that correctly handles angular impedances [14]. Albu-Schaffer and Hirzinger propose a dual-priority impedance architecture where the end-effector impedance and null space joint impedance are controlled separately using a stiffness formulation [15] and Oh *et. al.* propose an impedance control formulation where task and redundant space tasks are dynamically decoupled [16]. Ott *et. al.* demonstrate stability even though the redundant space is non-integrable [17].

In contrast to the above, the current paper considers the question of simultaneously controlling the impedance at multiple end-effectors. Rather than stabilizing an arbitrary parameterization of the redundant space, this paper optimizes for an integrable operational space objective. To our knowledge, the most closely related work on the subject is the multi-priority framework of Sentis and Khatib [6, 18]. In their work, the multi-priority objectives are realized by projecting lower-priority objectives into the dynamically consistent null space. Although this formulation keeps the first priority task independent of the lower-priority tasks, it does not optimize for second-priority performance. The current paper explores this point. We provide a control law that minimizes quadratic error with respect to the desired lower-priority impedance while still ensuring independence of the first-priority objective. It turns out that for the optimal dual-priority controller, it is possible to identify the operational space directions where the second-priority task is independent of the first-priority task. These independent directions can be adjusted by changing the weighting matrix used by the optimization criterion.

II. Background

A. Generalized equation of motion

The dynamic motion of a robot arm with n revolute joints is typically understood in terms of the following generalized

[1]Portions of this publication have patents pending.

equation of motion:

$$M\ddot{q} + \eta = \tau_a + \tau,$$

where M is the $n \times n$ manipulator inertia matrix, $\ddot{q}$ is an n-vector of manipulator joint accelerations, τ is an n-vector of joint torques resulting from externally applied loads, τ_a is a vector of actuator torques, and η describes the sum of frictional, coriolis, centrifugal, and gravitational torques [19]. The dependence of these terms on manipulator configuration is implicitly assumed. In order to simplify notation, we introduce the following substitution:

$$\tau_a = u + \eta,$$

such that the equation of motion can be expressed in terms of a command vector, u:

$$M\ddot{q} = u + \tau. \tag{1}$$

B. Operational space

It is frequently useful to design controllers defined in operational space coordinates rather than in joint space. While the term "operational space" may refer to any coordinate system relevant to a robot task, it usually refers to the space of positions and orientations of the end-effector represented by a parameterization of $SE(3)$. In this paper, $SE(3)$ is parameterized using exponential coordinates whereby a Cartesian pose is encoded by a 6-vector with the first three numbers describing position and the last three numbers describing orientation using the axis-angle representation [20]. The Cartesian velocity of the end-effector will be represented as a twist and the acceleration as the derivative of twist. Similarly, loads in Cartesian space will be written as wrenches (six-vector that concatenates a force and a moment) [20]. The end-effector Jacobian, J, relates joint velocities, $\dot{q}$, to Cartesian twists at the end-effector, $\dot{x}$: $\dot{x} = J\dot{q}$.

III. Dual Priority Cartesian Space Impedance Control

Dual priority Cartesian impedance control is defined with respect to two points of reference (PORs). A POR is a reference frame attached to the manipulator that is the object of Cartesian control. In general, dual priority Cartesian impedance control realizes two different impedance laws defined with respect to two different PORs simultaneously. One impedance objective is first-priority while the other is second-priority. Dual-priority Cartesian impedance control minimizes error with respect to the first-priority impedance objective while also realizing the second-priority objective to the greatest extent possible.

A. Problem statement

Let x_1, $\dot{x}_1$, $\ddot{x}_1$, and f_1 be the pose, twist, acceleration, and externally applied wrench associated with the first-priority POR. Define x_2, $\dot{x}_2$, $\ddot{x}_2$, and f_2 similarly for the second-priority POR. The goal of dual-priority impedance control is to realize the first-priority impedance objective,

$$\Omega_1\ddot{x}_1 + f_1^* = f_1, \tag{2}$$

as completely as possible given the mechanics of the manipulator, where

$$f_1^* = B_1\dot{x}_1 + K_1\tilde{x}_1, \tag{3}$$

Ω_1 is the desired inertia, B_1 is the desired damping, K_1 is the desired stiffness, and $\tilde{x}_1$ is the pose error in the first-priority POR. In addition, dual-priority Cartesian impedance control must also minimize error with respect to the second-priority impedance objective,

$$\Omega_2\ddot{x}_2 + f_2^* = f_2, \tag{4}$$

to the greatest extent possible while not impacting performance with regard to the first-priority objective, where

$$f_2^* = B_2\dot{x}_2 + K_2\tilde{x}_2, \tag{5}$$

Ω_2 is the desired inertia, B_2 is the desired damping, K_2 is the desired stiffness, and $\tilde{x}_2$ is the pose error at the second-priority POR.

B. Optimization criterion

The possibility of realizing two Cartesian space impedances depends on the degree of mobility between the two PORs. In general, it is not possible to realize both Cartesian impedances concurrently because the two PORs are not necessarily sufficiently mobile with respect to each other. Therefore, the multiple-priority approach applies actuator torques such that the primary impedance objective is achieved while minimizing an optimization criterion associated with the second-priority impedance objective. We restrict our attention to the case where the optimization criterion is the weighted squared magnitude of the Cartesian acceleration at the second-priority POR:

$$\begin{aligned} \epsilon &= (\ddot{x}_2 - \ddot{x}_2')^T W (\ddot{x}_2 - \ddot{x}_2') \\ &= z^T z, \end{aligned} \tag{6}$$

where $\ddot{x}_2'$ is the desired Cartesian space acceleration at the second-priority POR and:

$$z = W^{\frac{1}{2}}(\ddot{x}_2 - \ddot{x}_2'). \tag{7}$$

The goal of dual-priority Cartesian impedance control is to minimize Equation 6 while realizing the first-priority impedance in Equation 2.

C. Control law

The first-priority Cartesian impedance constrains the space of joint accelerations that must be considered. Substituting $J_1\ddot{q} + \dot{J}_1\dot{q}$ for $\ddot{x}_1$ in Equation 2 and taking the pseudo-inverse, we have:

$$\ddot{q} = J_1^+\left[\Omega_1^{-1}(f_1 - f_1^*) - \dot{J}\dot{q}\right] + N_1\lambda, \tag{8}$$

where $N_1 = I - J_1^+J_1$ and λ is arbitrary. Using $\ddot{x}_2 = J_2\ddot{q} + \dot{J}_2\dot{q}$, the space of Cartesian accelerations at the second-priority POR consistent with the first-priority impedance is:

$$\ddot{x}_2 = J_2J_1^+\left[\Omega_1^{-1}(f_1 - f_1^*) - \dot{J}\dot{q}\right] + \dot{J}_2\dot{q} + J_2N_1\lambda. \tag{9}$$

Among those joint accelerations permitted by the constraint, we must find one that minimizes the optimization criterion. This can be achieved by solving Equation 9 for the value of λ that minimizes ϵ (Equation 6), or equivalently, z (Equation 7). Solving Equation 4 for the desired acceleration at the second POR, we have:

$$\ddot{x}_2' = \Omega_2^{-1}(f_2 - f_2^*).$$

Substituting Equation 9 into Equation 7 for $\ddot{x}_2$ and using $\ddot{x}_2'$ above, we have:

$$\begin{aligned} z &= W^{\frac{1}{2}}(\ddot{x}_2 - \ddot{x}_2') \\ &= W^{\frac{1}{2}}\Big[J_2 J_1^+\left(\Omega_1^{-1}(f_1 - f_1^*) - \dot{J}_1\dot{q}\right) + \dot{J}_2\dot{q} \\ &\quad -\Omega_2^{-1}(f_2 - f_2^*) + J_2 N_1 \lambda\Big]. \end{aligned} \tag{10}$$

The value of λ that minimizes $z^T z$ can be found using the pseudo-inverse:

$$\begin{aligned} \lambda &= \hat{J}_2\left[\Omega_2^{-1}(f_2 - f_2^*) - \dot{J}_2\dot{q}\right] \\ &\quad -\hat{J}_2 J_2 J_1^+\left[\Omega_1^{-1}(f_1 - f_1^*) - \dot{J}_1\dot{q}\right] \\ &\quad +(I - \hat{J}_2 J_2 N_1)\beta, \end{aligned} \tag{11}$$

where

$$\hat{J}_2 = (W^{\frac{1}{2}} J_2 N_1)^+ W^{\frac{1}{2}}. \tag{12}$$

The term $(I - \hat{J}_2 J_2 N_1)\beta$ (for arbitrary β) represents the additional freedoms left over in the case that the second-priority optimization criterion does not completely determine the motion of all joints. The matrix $(I - \hat{J}_2 J_2 N_1)$ is an orthogonal projection matrix that spans the joint accelerations irrelevant to either objective.

The joint accelerations that realize the first-priority impedance (Equation 2) while also minimizing the optimization criterion with respect to the second-priority impedance (Equation 4) can be found by substituting λ back into Equation 8:

$$\begin{aligned} \ddot{q} &= J_1^+\left[\Omega_1^{-1}(f_1 - f_1^*) - \dot{J}_1\dot{q}\right] \\ &\quad +N_1\hat{J}_2\left[\Omega_2^{-1}(f_2 - f_2^*) - \dot{J}_2\dot{q}\right] \\ &\quad -N_1\hat{J}_2 J_2 J_1^+\left[\Omega_1^{-1}(f_1 - f_1^*) - \dot{J}_1\dot{q}\right] \\ &\quad +N_1(I - \hat{J}_2 J_2 N_1)\beta. \end{aligned} \tag{13}$$

This equation can be simplified by using the fact that for any matrix, A, the following is true: $A^+ = A^T(AA^T)^+$. Therefore, we have that:

$$\begin{aligned} N_1\hat{J}_2 &= N_1(W^{\frac{1}{2}} J_2 N_1)^+ W^{\frac{1}{2}} \\ &= N_1(N_1 J_2^T W^{\frac{1}{2}})(W^{\frac{1}{2}} J_2 N_1 J_2^T W^{\frac{1}{2}})^+ W^{\frac{1}{2}} \\ &= (N_1 J_2^T W^{\frac{1}{2}})(W^{\frac{1}{2}} J_2 N_1 J_2^T W^{\frac{1}{2}})^+ W^{\frac{1}{2}} \\ &= \hat{J}_2. \end{aligned} \tag{14}$$

As a result, Equation 13 becomes:

$$\begin{aligned} \ddot{q} &= \hat{N}_2 J_1^+\left[\Omega_1^{-1}(f_1 - f_1^*) - \dot{J}_1\dot{q}\right] \\ &\quad +\hat{J}_2\left[\Omega_2^{-1}(f_2 - f_2^*) - \dot{J}_2\dot{q}\right] \\ &\quad +N_1\hat{N}_2 N_1\beta. \end{aligned} \tag{15}$$

where

$$\hat{N}_2 = (I - \hat{J}_2 J_2). \tag{16}$$

Substituting into the equation of motion (Equation 1) for $\ddot{q}$, we have:

$$\begin{aligned} u &= M\ddot{q} - \tau \\ &= M\hat{N}_2 J_1^+\left[\Omega_1^{-1}(f_1 - f_1^*) - \dot{J}_1\dot{q}\right] \\ &\quad +M\hat{J}_2\left[\Omega_2^{-1}(f_2 - f_2^*) - \dot{J}_2\dot{q}\right] \\ &\quad +MN_1\hat{N}_2 N_1\beta - \tau. \end{aligned} \tag{17}$$

In order to realize Equation 17 in practice, it is necessary to measure the externally applied loads. This might be accomplished by measuring the loads, f_1 and f_2, applied at the first- and second-priority PORs and assuming that no external loads are applied elsewhere on the manipulator. Accordingly, we have:

$$\tau = J_1^T f_1 + J_2^T f_2.$$

Substituting into Equation 17, we have:

$$\begin{aligned} u &= M\ddot{q} - \tau \\ &= M\hat{N}_2 J_1^+\left[\Omega_1^{-1}(f_1 - f_1^*) - \dot{J}_1\dot{q}\right] \\ &\quad +M\hat{J}_2\left[\Omega_2^{-1}(f_2 - f_2^*) - \dot{J}_2\dot{q}\right] \\ &\quad +MN_1\hat{N}_2 N_1\beta - J_1^T f_1 - J_2^T f_2. \end{aligned} \tag{18}$$

The control law in Equation 18 solves the problem proposed in Section III-A while minimizing the Cartesian acceleration of the second-priority POR and assuming that externally applied wrenches at each POR can be measured directly.

D. Analysis

The closed-loop behavior of the dual-priority Cartesian impedance controller (Equation 18) can be understood in terms of the Cartesian accelerations at the first- and second-priority PORs. First, note that:

$$\begin{aligned} J_1\hat{J}_2 &= J_1(W^{\frac{1}{2}} J_2 N_1)^+ W^{\frac{1}{2}} \\ &= J_1 N_1 J_2^T W^{\frac{1}{2}}(W^{\frac{1}{2}} J_2 N_1 J_2^T W^{\frac{1}{2}})^+ W^{\frac{1}{2}} \\ &= 0. \end{aligned}$$

Therefore, we have that:

$$\begin{aligned} J_1\hat{N}_2 &= J_1(I - \hat{J}_2 J_2) \\ &= J_1. \end{aligned} \tag{19}$$

Using the above results and substituting Equation 15 into $\ddot{x}_1 = J_1\ddot{q} + \dot{J}_1\dot{q}$, the Cartesian acceleration at the first POR is:

$$\ddot{x}_1 = \Omega_1^{-1}(f_1 - f_1^*),$$

thereby realizing the desired first-priority impedance.

The Cartesian acceleration of the second-priority POR can be calculated using $\ddot{x}_2 = J_2\ddot{q} + \dot{J}_2\dot{q}$:

$$\begin{aligned}\ddot{x}_2 &= J_2\hat{J}_2\left[\Omega_2^{-1}(f_2 - f_2^*) - \dot{J}_2\dot{q}\right] + \dot{J}_2\dot{q} \\ &\quad +(I - J_2\hat{J}_2)J_2J_1^+\left[\Omega_1^{-1}(f_1 - f_1^*) - \dot{J}_1\dot{q}\right] \\ &\quad +J_2N_1\hat{N}_2N_1\beta.\end{aligned}$$

The last term of this equation can be eliminated by noting:

$$\begin{aligned}J_2N_1\hat{N}_2N_1 &= J_2N_1(I - \hat{J}_2J_2)N_1 \\ &= J_2N_1 - J_2N_1\hat{J}_2J_2N_1 \\ &= J_2N_1 - J_2N_1(W^{\frac{1}{2}}J_2N_1)^+W^{\frac{1}{2}}J_2N_1 \\ &= W^{-\frac{1}{2}}\left[W^{\frac{1}{2}}J_2N_1\right. \\ &\quad \left.-W^{\frac{1}{2}}J_2N_1(W^{\frac{1}{2}}J_2N_1)^+W^{\frac{1}{2}}J_2N_1\right] \\ &= 0.\end{aligned}$$

Therefore, the resulting closed-loop impedance at x_2 is:

$$\begin{aligned}\ddot{x}_2 &= J_2\hat{J}_2\left[\Omega_2^{-1}(f_2 - f_2^*) - \dot{J}_2\dot{q}\right] + \dot{J}_2\dot{q} \\ &\quad +(I - J_2\hat{J}_2)J_2J_1^+\left[\Omega_1^{-1}(f_1 - f_1^*) - \dot{J}_1\dot{q}\right]. \quad (20)\end{aligned}$$

Equation 20 can be understood by recognizing that $J_2\hat{J}_2$ and $(I - J_2\hat{J}_2)$ are orthogonal weighted projection matrices: $J_2\hat{J}_2(I - J_2\hat{J}_2) = 0$. The first term is projected through $J_2\hat{J}_2$ and describes the contribution of the second-priority impedance to the closed-loop acceleration of x_2. The last term is projected through $(I - J_2\hat{J}_2)$ and describes the effect that the first-priority impedance objective has on the closed loop behavior at the second-priority POR. Notice that the second-priority impedance operates completely independently in the range space of $J_2\hat{J}_2$. Multiplying both sides of Equation 20 by $J_2\hat{J}_2$, we have:

$$J_2\hat{J}_2\ddot{x}_2 = J_2\hat{J}_2\Omega_2^{-1}(f_2 - f_2^*). \quad (21)$$

E. Comparison to Sentis-Khatib

These closed-loop dynamics may be compared to the dynamics of the multi-priority Sentis-Khatib control law. Based on [6], the Sentis-Khatib control law is:

$$\begin{aligned}\tau &= J_1^T\Lambda_1\left[\Omega_1^{-1}(f_1 - f_1^*) - \dot{J}_1\dot{q}\right] \\ &\quad +J_{2|1}^T\Lambda_{2|1}\left[\Omega_2^{-1}(f_2 - f_2^*) - \dot{J}_2\dot{q}\right]. \quad (22)\end{aligned}$$

In the above control law, we have used

$$\Lambda_{2|1} = (J_{2|1}M^{-1}J_{2|1}^T)^{-1}$$

and

$$J_{2|1} = J_2N_M,$$

where N_A is the inertia-weighted null space of the first priority objective:

$$N_M = I - M^{-1}J_1^T(J_1M^{-1}J_1^T)^{-1}J_1.$$

Substituting back into the equation of motion (Equation 1) and solving for $\ddot{x}_1$, we arrive at Equation 2, demonstrating that the second-priority objective does not influence the dynamics of the first-priority objective. However, substituting into the equation of motion and solving for $\ddot{x}_2$, we get:

$$\begin{aligned}\ddot{x}_2 &= J_2M^{-1}J_1^T\Lambda_1\Omega_1^{-1}(f_1 - f_1^*) \\ &\quad +J_2M^{-1}J_{2|1}^T\Lambda_{2|1}\Omega_2^{-1}(f_2 - f_2^*).\end{aligned}$$

It should be noted that the second term above is not orthogonal to the first term as is the case in Equation 20. This begs the question of whether this control law is optimal for any criterion of the form of Equation 6. This question can be answered using the Lagrange multiplier method. We are interested in control laws that minimize

$$f(\ddot{q}) = (J_2\ddot{q} - \beta)^TW(J_2\ddot{q} - \beta)$$

subject to

$$g(\ddot{q}) = J_1\ddot{q} - \alpha = 0,$$

where $\alpha = \Omega_1^{-1}(f_1 - f_1^*) - \dot{J}_1\dot{q}$ and $\beta = \Omega_2^{-1}(f_2 - f_2^*) - \dot{J}_2\dot{q}$. Differentiating f and g and constructing the Lagrangian, we have

$$h^TJ_1 + \ddot{q}^TJ_2^TWJ_2 - \beta^TWJ_2 = 0, \quad (23)$$

where h is the Lagrange multiplier. Equation 23 must be satisfied for some value of h at an optimum. Equivalently, we can require:

$$N_1\left[J_2^TWJ_2\ddot{q} - J_2^TW\beta\right] = 0.$$

Solving for $\ddot{q}$ by substituting Equation 22 into Equation 1 and substituting into the above, the condition becomes:

$$N_1\left[J_2^TWJ_2M^{-1}\left(J_1^T\Lambda_1\alpha + J_{2|1}^T\Lambda_{2|1}\beta\right) - J_2^TW\beta\right] = 0.$$

Since we must assume α to be arbitrary, the condition is only satisfied when

$$N_1J_2^TWJ_2M^{-1}J_1^T\Lambda_1\alpha = 0. \quad (24)$$

However, since there is no fixed W such that Equation 24 is true for arbitrary α, Equation 22 cannot be optimal.

IV. Impedance Control without Force Sensors

An important concern with multi-priority impedance control as it is expressed in Equation 18 is the need to sense externally applied loads at two different points on the manipulator. While this may be feasible in some cases, it is important to address the case where such feedback is not available.

A. Control law

Consider Equation 17, and suppose that the joint torque caused by all externally applied loads (τ) is substituted for each instance of $J_1^Tf_1$ or $J_2^Tf_2$. Let $\Lambda_1 = (J_1M^{-1}J_1^T)^{-1}$ be the passive manipulator inertia described in Cartesian coordinates at the first-priority POR. Let $\Lambda_2 = (J_2M^{-1}J_2^T)^{-1}$ be the similar quantity at the second-priority POR. Set the desired Cartesian inertias to these passive values: $\Omega_1 = \Lambda_1$ and $\Omega_2 = \Lambda_2$. Then, Equation 17 becomes:

$$\begin{aligned} u &= -M\hat{N}_2 J_1^+ \left[\Lambda_1^{-1} f_1^* + \dot{J}_1 \dot{q}\right] \\ &\quad -M\hat{J}_2 \left[\Lambda_2^{-1} f_2^* + \dot{J}_2 \dot{q}\right] \\ &\quad +M\hat{N}_2 J_1^+ J_1 M^{-1}\tau \\ &\quad +M\hat{J}_2 J_2 M^{-1}\tau \\ &\quad +MN_1\hat{N}_2 N_1\beta \\ &\quad -\tau. \end{aligned} \tag{25}$$

Since the value of β does not affect the closed-loop behavior of the control law with respect to the first- or second-priority impedance objectives, set $\beta = M^{-1}\tau + \gamma$, where γ is arbitrary. Then the last four terms of Equation 25 are:

$$M\left[\hat{N}_2 J_1^+ J_1 + \hat{J}_2 J_2 + N_1\hat{N}_2 N_1\right] M^{-1}\tau - \tau + N_1\hat{N}_2 N_1\gamma.$$

This expression can be simplified using Equation 14 to find that

$$\begin{aligned} N_1\hat{N}_2 N_1 &= N_1(I - \hat{J}_2 J_2)N_1 \\ &= (N_1 - N_1\hat{J}_2 J_2)N_1 \\ &= (N_1 - \hat{J}_2 J_2)N_1 \\ &= (I - \hat{J}_2 J_2)N_1 \\ &= \hat{N}_2 N_1. \end{aligned}$$

Therefore, the last four terms of Equation 25 can be simplified as follows:

$$\begin{aligned} & M\left[\hat{N}_2 J_1^+ J_1 + \hat{J}_2 J_2 + \hat{N}_2 N_1\right] M^{-1}\tau - \tau + \hat{N}_2 N_1\gamma \\ =\; & M\left[\hat{N}_2 + \hat{J}_2 J_2\right] M^{-1}\tau - \tau + \hat{N}_2 N_1\gamma \\ =\; & MM^{-1}\tau - \tau + \hat{N}_2 N_1\gamma \\ =\; & \hat{N}_2 N_1\gamma. \end{aligned}$$

As a result, Equation 25 can be re-written without the τ terms:

$$\begin{aligned} u &= -M\hat{N}_2 J_1^+ \left[\Lambda_1^{-1} f_1^* + \dot{J}_1 \dot{q}\right] \\ &\quad -M\hat{J}_2 \left[\Lambda_2^{-1} f_2^* + \dot{J}_2 \dot{q}\right] \\ &\quad +M\hat{N}_2 N_1\gamma. \end{aligned} \tag{26}$$

This control law has the same closed-loop impedance as does Equation 18 in the special case that $\Omega_1 = \Lambda_1$ and $\Omega_2 = \Lambda_2$. Since it does not require Cartesian force sensors to be mounted at the first- and second-priority PORs, this control law will be referred to as the "zero-force" dual-priority control law. (The forces are not zero, but we do not need to measure them.)

B. Analysis

The analysis of the zero-force dual-priority Cartesian impedance control law (Equation 26) is similar to the analysis performed in Section III-D. Substituting Equation 26 into the equation of motion (Equation 1) and using the fact that $\ddot{x}_1 = J_1\ddot{q} + \dot{J}_1\dot{q}$, the dynamics of the first-priority POR are:

$$\Lambda_1\ddot{x}_1 + f_1^* = \Lambda_1 J_1 M^{-1}\tau. \tag{27}$$

Note the similarities between the closed-loop impedance above and the desired impedance in Equation 2. When all external loads are applied to the first-priority POR such that $\tau = J_1^T f_1$, then:

$$\Lambda_1 J_1 M^{-1}\tau = f_1.$$

and Equation 27 is identical to Equation 2. When external loads are applied to both PORs such that $\tau = J_1^T f_1 + J_2^T f_2$, then loads applied to the second-priority POR, f_2, are projected through $\Lambda_1 J_1 M^{-1} J_2^T$ onto the first-priority POR:

$$\Lambda_1 J_1 M^{-1}\tau = f_1 + \Lambda_1 J_1 M^{-1} J_2^T f_2.$$

Substituting Equation 26 into the equation of motion (Equation 1) and using the fact that $\ddot{x}_2 = J_2\ddot{q} + \dot{J}_2\dot{q}$, the dynamics of the first-priority POR are:

$$\begin{aligned} \ddot{x}_2 &= J_2 M^{-1}\tau - J_2\hat{N}_2 J_1^+(\Lambda_1^{-1} f_1^* + \dot{J}_1\dot{q}) \\ &\quad -J_2\hat{J}_2(\Lambda_1^{-1} f_2^* + \dot{J}_2\dot{q}) + \dot{J}_2\dot{q} \\ &= J_2 M^{-1}\tau - (I - J_2\hat{J}_2)J_1^+(\Lambda_1^{-1} f_1^* + \dot{J}_1\dot{q}) \\ &\quad -J_2\hat{J}_2\Lambda_1^{-1} f_2^* + (I - J_2\hat{J}_2)\dot{J}_2\dot{q}. \end{aligned}$$

Multiplying the above by $J_2\hat{J}_2$, the resulting closed-loop behavior at the second-priority POR is:

$$J_2\hat{J}_2\ddot{x}_2 = J_2\hat{J}_2(J_2 M^{-1}\tau - \Lambda_2^{-1} f_2^*). \tag{28}$$

If all external loads are applied to the second-priority POR, then $\tau = J_2^T f_2$ and Equation 28 becomes Equation 21 for $\Omega_2 = \Lambda_2$. If loads are applied to both PORs, $\tau = J_1^T f_2 + J_2^T f_2$, then the load applied to the first-priority POR is projected onto the second-priority POR.

V. SIMULATION

The dual-priority approach proposed in this paper is compared to both a single-priority approach and the Sentis-Khatib dual-priority control law (Section III-E) in the context of a simple regrasp problem.

A. Overview

Figure 2 illustrates the simulation scenario. Both fingers are intially in contact with an object and touching it lightly (the entire manipulator is illustrated in Figure 1). The control problem is to move one finger while the other finger applies a small inward force resulting from a constant impedance. For example, this behavior could be important in an assembly application where one finger needs to maintain a constant force (perhaps in order to keep a part in place) while the other finger moves into a new grip position. What makes this a potentially difficult problem is that the first finger is not fully mobile with respect to the second finger. In view of Figure 1, notice that the two fingertips may only move independently in a plane; the wrist must swivel in order to move one of the fingers out of plane. It is not possible to move one finger to an arbitrary position while keeping *both* the position and orientation of the second finger fixed. If the position reference of the first finger cannot be fully reached while maintaining the pose and impedance of the second finger, then one or the other objective

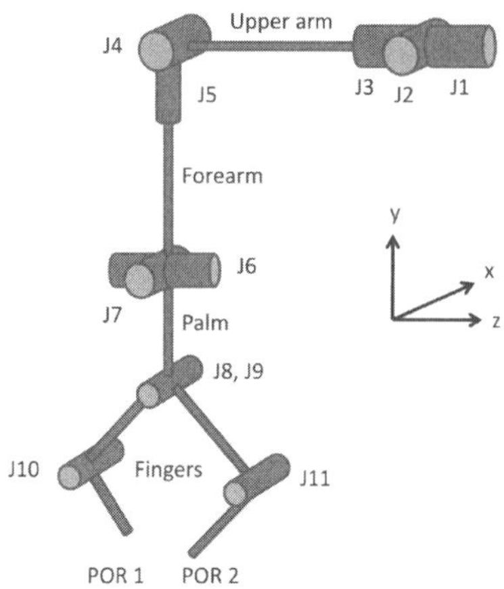

Fig. 1. The bifurcated manipulator (not drawn to scale) used in the simulations. A POR is defined at the end of each "finger."

Fig. 2. The manipulation scenario. The two-finger manipulator is initially contacting the object lightly. At the start of the simulation, the right finger impedance position reference changes so that the finger will move in the direction of the dashed arrow. The left finger impedance remains unchanged so that it will continue to apply a light inward force on the object. The dual-priority control law must move the right finger without changing the impedance of the left finger.

must be "sacrificed." In this application scenario, it is desirable to sacrifice those aspects of the first finger objective so that regardless of how the finger is commanded to move, the second finger is assured of maintaining a constant impedance.

The entire manipulator is illustrated in Figure 1. The three links proximal to the bifurcation will be referred to, in order proximal to distal, as: the "upper arm", the "forearm", and the "palm." The two branches will be referred to as the "fingers." The lengths, masses, and moments of inertia are given in Table I. Note that the finger masses and inertias are large enough to be of the same order as the arm. A POR is defined at the tip of each of the two fingers.

B. Comparison with a single-priority control law

For reference, the dual-priority control law proposed in this paper was compared to a single priority control law. Compared with the dual-priority control law, this is a simpler alternative and should only be discarded when significant advantages are offered by a dual-priority law.

1) Single-priority control law: The single-priority law is:

$$u = J^T f^* + MJ^+ \dot{J}\dot{q} + MNB_j\dot{q}. \quad (29)$$

This achieves a desired impedance,

$$\Lambda_s \ddot{x}_s + f_s^* = f_s,$$

where $\ddot{x}_s^T = (\ddot{p}_1, \dot{\omega}_1, \ddot{p}_2)$ is a generalized acceleration, Λ_s is the passive manipulator inertia projected into the corresponding operational space coordinates, and f_s is the generalized force. Both desired impedances are combined into a single impedance objective:

$$f_s^* = \begin{pmatrix} B_1^p & \mathbf{0} & \mathbf{0} \\ \mathbf{0} & B_1^r & \mathbf{0} \\ \mathbf{0} & \mathbf{0} & B_2^p \end{pmatrix} \dot{x}_s + \begin{pmatrix} K_1^p & \mathbf{0} & \mathbf{0} \\ \mathbf{0} & K_1^r & \mathbf{0} \\ \mathbf{0} & \mathbf{0} & K_2^p \end{pmatrix} \tilde{x}_s,$$

where $K_1^p = diag(45)$ (Newtons per Meter) and $B_1^p = diag(25)$ (Newton-seconds per Meter) are the desired translational stiffness and damping of POR 1, $K_1^r = diag(10)$ and $B_1^r = diag(5)$ are the desired angular stiffness and damping of POR 1, and $K_2^p = diag(20)$ and $B_2^p = diag(25)$ are the desired translational stiffness and damping of POR 2. The corresponding Jacobian is

$$J = \begin{pmatrix} J_1^p \\ J_1^r \\ J_2^p \end{pmatrix},$$

where J_1^p and J_1^r describe the Cartesian translational and angular velocities (respectively) at POR 1 and J_2^p is the translational Jacobian for POR 2.

2) Dual-priority control law: The single-priority control law above was compared to the dual priority control law of Equation 18. The first priority objective was to realize a translational and angular desired impedance at POR 1,

$$\Lambda_1 \ddot{x}_1 + f_1^* = f_1.$$

$\ddot{x}_1 = (\ddot{p}_1, \dot{\omega}_1)$ is the acceleration in $SE(3)$. Λ_1 is the corresponding passive operational space inertia, f_1^* is the desired damping and stiffness,

$$f_1^* = \begin{pmatrix} B_1^p & \mathbf{0} \\ \mathbf{0} & B_1^r \end{pmatrix} \dot{x}_1 + \begin{pmatrix} K_1^p & \mathbf{0} \\ \mathbf{0} & K_1^r \end{pmatrix} \tilde{x}_1, \quad (30)$$

and f_1 is the corresponding externally applied load. The second priority objective was to realize just a translational impedance at POR 2 – not an angular impedance,

$$\Lambda_2 \ddot{p}_2 + f_2^* = f_2,$$

where Λ_2 is the passive manipulator inertia projected into the translational Euclidean coordinates for POR 2, and f_2 is the externally applied force at POR 2. The desired POR 2 damping and stiffness is:

$$f_2^* = B_2^p \dot{p}_2 + K_2^p \tilde{p}_2. \quad (31)$$

Figure 3 compares the performance of the two control laws when the first-priority POR reference pose is the same as

Link	Length	Mass	I_{xx}	I_{yy}	I_{zz}
upper arm	0.38m	6.8Kg	0.082	0.082	0.0082
forearm	0.35m	5.5Kg	0.056	0.056	0.0056
palm	0.1m	2Kg	0.01	0.01	0.001
finger link	0.1m	1Kg	0.005	0.005	0.0005

TABLE I

DYNAMIC PARAMETERS OF SIMULATED ARM. I_{zz} IS THE MOMENT OF INERTIA ABOUT THE LINK AXIS. I_{xx} AND I_{yy} ARE THE REMAINING TWO ORTHOGONAL MOMENTS OF INERTIA. ALL FINGER LINKS HAVE THE SAME DYNAMIC PARAMETERS.

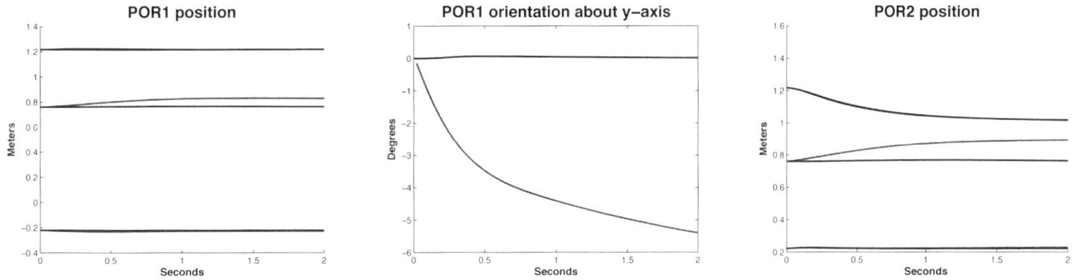

(a) (b) (c)

Fig. 3. Comparison between the dual-priority control law proposed in this paper (the solid black line) and a single-priority controller (the shaded blue line).

in the starting configuration, $p_2^* = p_2(0)$ and $r_2^* = r_2(0)$, the second-priority POR reference position is changed, $p_2^* = (0.96, 1.016, 0)^T$, and no external forces are applied to the manipulator. First, notice that the dual-priority control law (the solid black line in Figure 3) keeps POR 1 in approximately the same position and orientation during the interaction while the single-priority control law (the shaded blue line) allows POR 1 to move. This is not suprising because the single priority control law does not enforce the prioritization. Second, notice that Figure 3(c) shows that the single-priority control law gets closer to achieving the POR 2 objective than the dual-priority controller. Since the single-priority controller allowed POR 1 to move and rotate from the reference, the x-coordinate of the POR 2 position (the glue line in Figure 3(c)) moves toward the 0.96 reference (although it does not reach the reference because it is balancing the POR 1 objective as well). Nevertheless, since the POR 2 y-dimension (the top line in Figure 3(c)) is unconstrained with respect to the POR 1 objectives, both the single- and dual-priority control laws realize the y-dimension objective with approximately the same performance. In contrast, the dual-priority control law maintains constant POR 1 position and orientation. POR 2 position error is minimized in the directions that do not change POR 1 orientation and ignored in the directions that do.

VI. COMPARISON WITH SENTIS-KHATIB

Section III-E demonstrated that the proposed dual-priority control law has certain theoretical advantages relative to the Sentis-Khatib control law. These differences have an important effect on the overall performance of the control law. The two control laws are compared under the same conditions used above to compare with the single priority law. The simulation compared the Sentis-Khatib control law in Equation 22 with the proposed dual-priority control law in Equation 18. Both control laws are parameterized with the desired first- and second-priority impedances given in Equations 30 and 31.

Figure 4(a) and (b) demonstrates that both dual-priority control laws realize the first-priority impedance almost equally well (the proposed control law does a bit worse in keeping the orientation constraint.) This is expected since it is possible to substitute either control law into the manipulator dynamics (Equation 1), solve for the POR 1 accelerations, and arrive at Equation 2. However Figure 4(c) illustrates the differences regarding POR 2. The control law of this paper (the solid black line) converges to the position nearest (Euclidian norm) the POR 2 reference position while the Sentis-Khatib control law converges elsewhere (the shaded blue line). This result demonstrates that the theoretically suboptimal performance demonstrated in Equation 24 can have a significant practical impact.

VII. CONCLUSION

Multi-priority Cartesian impedance control is important because it enables the robot programmer to control a complex manipulator in terms of desired Cartesian impedances at multiple PORs. This is the natural extension of single end-effector control to branching manipulators. The prioritization is necessary to resolve conflicts between different objectives. Although multi-priority motion control can achieve prioritized motions that are similar to what can be achieved with multi-priority impedance control, these motions are not robust and predictable when the possibility of contact exists. Manipulation and assembly applications, in particular, require exactly the type of behavior that this paper has considered: motion

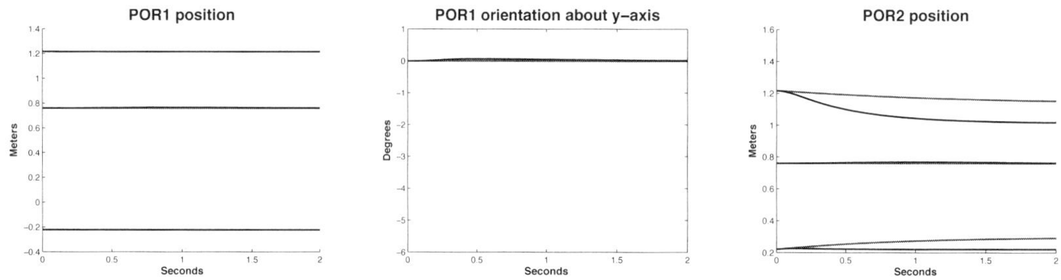

(a) (b) (c)

Fig. 4. Comparison between the dual-priority controller proposed in this paper (the solid black line) and the Sentis Khatib dual-priority controller (the shaded line). (a) and (b) show similar performance for both control laws with respect to the first priority impedance objective while (c) shows that the proposed controller has better performance with respect to the second priority impedance.

of a second-priority POR while the first-priority POR acts with a constant desired force or impedance. This paper makes two main contributions. First, we propose a locally optimal dual-priority impedance control law for the general case where end-effector load measurements may or may not be available. Second, we analyze the performance of the control law in theory and relative to the existing multi-priority impedance control literature.

References

[1] B. T. Inc., "Products - WAM Arm," *http://www.barrett.com/robot/products-arm.htm*.

[2] A. Albu-Schaffer, S. Haddadin, C. Ott, A. Stemmer, T. Wimbock, and G. Hirzinger, "The DLR lightweight robot lightweight design and soft robotics control concepts for robots in human environments," *Industrial Robot Journal*, vol. 34, pp. 376–385, 2007.

[3] A. Liegois, "Automatic supervisory control of the configuration and behavior of multibody mechanisms," *IEEE Trans. Syst. Man Cybernetics*, pp. 868–871, 1977.

[4] Y. Nakamura, *Advanced Robotics Redundancy and Optimization*. Addison-Wesley, 1991.

[5] L. Sciavicco and B. Siciliano, "A solution algorithm to the inverse kinematic problem for redundant manipulators," *IEEE Journal of Robotics and Automation*, vol. 4, no. 4, 1988.

[6] L. Sentis and O. Khatib, "Synthesis of whole-body behaviors through hierarchical control of behavioral primitives," *International Journal of Humanoid Robotics*, 2005.

[7] R. Platt, O. Brock, A. H. Fagg, D. Karupiah, M. Rosenstein, J. Coelho, M. Huber, J. Piater, D. Wheeler, and R. Grupen, "A framework for humanoid control and intelligence," in *Proceedings of the 2003 IEEE International Conference on Humanoid Robots*, October 2003.

[8] P. Chiacchio, S. Chiaverini, L. Sciavicco, and B. Siciliano, "Closed-loop inverse kinematics schemes for constrained redundant manipulators with task space augmentation and task priority strategy," *The International Journal of Robotics Research*, vol. 10, no. 4, pp. 410–425, 1991.

[9] S. Chiaverini, "Singularity-robust task-priority redundancy resolution for real-time kinematic control of robot manipulators," *IEEE Transactions on Robotics and Automation*, vol. 13, no. 3, pp. 398–410, 1997.

[10] G. Antonelli, "Stability analysis for prioritized closed-loop inverse kinematic algorithms for redundant robotic systems," *IEEE Transactions on Robotics*, vol. 25, no. 5, pp. 985–994, 2009.

[11] M. Huber, "A hybrid architecture for adaptive robot control," Ph.D. dissertation, U. Massachusetts, 2000.

[12] N. Mansard and F. Chaumette, "Task sequencing for sensor-based control," *IEEE Transactions on Robotics*, vol. 23, no. 1, pp. 60–72, 2007.

[13] N. Hogan, "Impedance control - an approach to manipulation: theory," *Journal of dynamic systems measurement and control*, vol. 107, pp. 1–7, 1985.

[14] C. Natale, B. Siciliano, and L. Villani, "Spatial impedance control of redundant manipulators," in *IEEE Int'l Conf. on Robotics and Automation*, 1999, pp. 1788–1793.

[15] A. Albu-Schffer and G. Hirzinger, "Cartesian impedance control techniques for torque controlled light-weight robots," in *IEEE Int'l Conf. on Robotics and Automation*, 2002, pp. 657–663.

[16] Y. Oh, W. Chung, and Y. Youm, "Extended impedance control of redundant manipulators based on weighted decomposition of joint space," *Journal of Robotic Systems*, vol. 15, no. 5, pp. 231–258, 1998.

[17] C. Ott, A. Kugi, and Y. Nakamura, "Resolving the problem of non-integrability of nullspace velocities for compliance control of redundant manipulators by using semi-definite lyapunov functions," in *IEEE Int'l Conf. on Robotics and Automation*, 2008, pp. 1999 - 2004.

[18] L. Sentis, "Synthesis and control of whole-body behaviors in humanoid systems," Ph.D. dissertation, Stanford University, July 2007.

[19] M. Spong, S. Hutchinson, and M. Vidyasagar, *Robot Modeling and Control*. John Wiley and Sons, Inc., 2005.

[20] R. Murray, Z. Li, and S. Sastry, *A Mathematical Introduction to Robotic Manipulation*. CRC Press, 1994.

Variable Impedance Control: A Reinforcement Learning Approach

Jonas Buchli, Evangelos Theodorou, Freek Stulp, Stefan Schaal
Computational Learning and Motor Control Lab
University of Southern California, Los Angeles, CA 90089
{buchli,etheodor,stulp,sschaal}@usc.edu

Abstract— **One of the hallmarks of the performance, versatility, and robustness of biological motor control is the ability to adapt the impedance of the overall biomechanical system to different task requirements and stochastic disturbances. A transfer of this principle to robotics is desirable, for instance to enable robots to work robustly and safely in everyday human environments. It is, however, not trivial to derive variable impedance controllers for practical high DOF robotic tasks. In this contribution, we accomplish such gain scheduling with a reinforcement learning approach algorithm, PI^2 (Policy Improvement with Path Integrals). PI^2 is a model-free, sampling based learning method derived from first principles of optimal control. The PI^2 algorithm requires no tuning of algorithmic parameters besides the exploration noise. The designer can thus fully focus on cost function design to specify the task. From the viewpoint of robotics, a particular useful property of PI^2 is that it can scale to problems of many DOFs, so that RL on real robotic systems becomes feasible. We sketch the PI^2 algorithm and its theoretical properties, and how it is applied to gain scheduling. We evaluate our approach by presenting results on two different simulated robotic systems, a 3-DOF Phantom Premium Robot and a 6-DOF Kuka Lightweight Robot. We investigate tasks where the optimal strategy requires both tuning of the impedance of the end-effector, and tuning of a reference trajectory. The results show that we can use path integral based RL not only for planning but also to derive variable gain feedback controllers in realistic scenarios. Thus, the power of variable impedance control is made available to a wide variety of robotic systems and practical applications.**

I. Introduction

Biological motor systems excel in terms of versatility, performance, and robustness in environments that are highly dynamic, often unpredictable, and partially stochastic. In contrast to classical robotics, mostly characterized by high gain negative error feedback control, biological systems derive some of their superiority from low gain compliant control with variable and task dependent impedance. If we adapt this concept of adaptive impedance for PD negative feedback control, this translates into time varying proportional and derivative gains, also known as gain scheduling. Finding the appropriate gain schedule for a given task is, however, a hard problem.

One possibility to overcome such problems is Reinforcement Learning (RL). The idea of RL is that, given only a reward function, the learning algorithm finds strategies that yield high reward through trial and error. As a special and important feature, RL accomplishes such optimal performance *without* knowledge of the models of the motor system and/or the environment. However, so far, RL does not scale well to high-dimensional continuous state-action control problems.

Closely related to RL is optimal control theory, where gain scheduling is a natural part of many optimal control algorithms. However, optimal control requires *model-based* derivations, such that it is frequently not applicable to complex robotic systems and environments, where models are unknown.

In this paper, we present a novel RL algorithm that does scale to complex robotic systems, and that accomplishes gain scheduling in combination with optimizing other performance criteria. Evaluations on two simulated robotic systems demonstrate the effectiveness of our approach. In the following section, we will first motivate variable impedance control. Then, we sketch our novel RL algorithms, called PI^2, and its applicability to learning gain scheduling. In the fourth section, we will present evaluation results on a 3 DOF and a 6 DOF robotic arm, where the task requires the robot to learn both a reference trajectory and the appropriate time varying impedance. We conclude with a review of related work and discussions of future directions.

II. Variable Impedance Control

The classical approach to robot control is negative feedback control with high proportional-derivative (PD) gains. This type of control is straightforward to implement, robust towards modeling uncertainties, and computationally cheap. Unfortunately, high gain control is not ideal for many tasks involving interaction with the environment, e.g. force control tasks or locomotion. In contrast, impedance control [5] seeks to realize a specific impedance of the robot, either in end-effector or joint space. The issue of specifying the target impedance, however, is not completely addressed as of yet. While for simple factory tasks, where the properties of the task and environment are know a priori, suitable impedance characteristics may be derivable, it is usually not easy to understand how impedance control is applied to more complex tasks such as a walking robot over difficult terrain or the manipulation of objects in daily life (e.g. pillows, hammers, cans, etc.). An additional benefit of variable impedance behavior in a robot comes from the added active safety due to soft "giving in", both for the robot and its environment.

In the following we consider robots with torque controlled joints. The motor commands $\mathbf{u}$ are calculated via a PD control law with feedforward control term $\mathbf{u}_{ff}$:

$$\mathbf{u} = -\mathbf{K}_P(\mathbf{q} - \mathbf{q}_d) - \mathbf{K}_D(\dot{\mathbf{q}} - \dot{\mathbf{q}}_d) + \mathbf{u}_{ff} \quad (1)$$

where $\mathbf{K}_P$, $\mathbf{K}_D$ are the positive definite position and velocity gain matrices, $\mathbf{q}$, $\dot{\mathbf{q}}$ are the joint positions and velocities, and $\mathbf{q}_d$, $\dot{\mathbf{q}}_d$ are the desired joint positions and velocities. The feedforward control term may come, for instance, from an inverse dynamics control component, or a computed torque control component [15]. Thus, the impedance of a joint is parameterized by the choice of the gains $\mathbf{K}_P$ ("stiffness") and $\mathbf{K}_D$ ("damping").

For many applications, the joint space impedance is, however, of secondary interest. Most often, regulating impedance matters the most at certain points that contact with the environment, e.g., the end-effectors of the robot. We therefore need to assess the impedance at these points of contacts rather than the joints. Joint space impedance is computed from the desired task space impedance $\mathbf{K}_{P,\mathbf{x}}, \mathbf{K}_{D,\mathbf{x}}$ by help of the Jacobian $\mathbf{J}$ of the forward kinematics of the robot as follows [15]:

$$\mathbf{K}_{P,\mathbf{q}} = \mathbf{J}^T\mathbf{K}_{P,\mathbf{x}}\ \mathbf{J} \quad \text{and} \quad \mathbf{K}_{D,\mathbf{q}} = \mathbf{J}^T\mathbf{K}_{D,\mathbf{x}}\ \mathbf{J} \quad (2)$$

Here we assume that the geometric stiffness due to the change of the Jacobian is negligible in comparison to the terms in Eq.(2). Regulating the task space impedance thus implies regulating the joint space impedance. Furthermore, this fundamental mathematical relationship between joint and task space also implies that a constant task stiffness in general means varying gains at the joint level.

In the next section we will sketch a reinforcement learning algorithm that is applied to learning the time dependent gain matrices.

III. Reinforcement Learning in High Dimensions – the PI^2 Algorithm

Reinforcement learning algorithms can be derived from different frameworks, e.g., dynamic programming, optimal control, policy gradients, or probabilistic approaches. Recently, an interesting connection between stochastic optimal control and Monte Carlo evaluations of path integrals was made [9]. In [18] this approach is generalized, and used in the context of model-free reinforcement learning with parameterized policies, which resulted in the PI^2 algorithm. In the following, we provide a short outline of the prerequisites and the development of the PI^2 algorithm as needed in this paper. For more details refer to [18].

The foundation of PI^2 comes from (model-based) stochastic optimal control for continuous time and continuous state-action systems. We assume that the dynamics of the control system is of the form

$$\dot{\mathbf{x}}_t = \mathbf{f}(\mathbf{x}_t, t) + \mathbf{G}(\mathbf{x}_t)(\mathbf{u}_t + \epsilon_t) = \mathbf{f}_t + \mathbf{G}_t(\mathbf{u}_t + \epsilon_t) \quad (3)$$

with $\mathbf{x}_t \in \Re^{n\times 1}$ denoting the state of the system, $\mathbf{G}_t = \mathbf{G}(\mathbf{x}_t) \in \Re^{n\times p}$ the control matrix, $\mathbf{f}_t = \mathbf{f}(\mathbf{x}_t) \in \Re^{n\times 1}$ the passive dynamics, $\mathbf{u}_t \in \Re^{p\times 1}$ the control vector and $\epsilon_t \in \Re^{p\times 1}$ Gaussian noise with variance $\mathbf{\Sigma}_\epsilon$. Many robotic systems fall into this class of control systems. For the finite horizon problem $t_i : t_N$, we want to find control inputs $\mathbf{u}_{t_i:t_N}$ which minimize the value function

$$V(\mathbf{x}_{t_i}) = V_{t_i} = \min_{\mathbf{u}_{t_i:t_N}} E_{\boldsymbol{\tau}_i}\left[R(\boldsymbol{\tau}_i)\right] \quad (4)$$

where R is the finite horizon cost over a trajectory starting at time t_i in state $\mathbf{x}_{t_i}$ and ending at time t_N

$$R(\boldsymbol{\tau}_i) = \phi_{t_N} + \int_{t_i}^{t_N} r_t\ dt \quad (5)$$

and where $\phi_{t_N} = \phi(x_{t_N})$ is a terminal reward at time t_N. r_t denotes the immediate reward at time t. $\boldsymbol{\tau}_i$ are trajectory pieces starting at $\mathbf{x}_{t_i}$ and ending at time t_N.

As immediate reward we consider

$$r_t = r(\mathbf{x}_t, \mathbf{u}_t, t) = q_t + \frac{1}{2}\mathbf{u}_t^T\mathbf{R}\mathbf{u}_t \quad (6)$$

where $q_t = q(\mathbf{x}_t, t)$ is an arbitrary state-dependent reward function, and $\mathbf{R}$ is the positive semi-definite weight matrix of the quadratic control cost. From stochastic optimal control [16], it is known that the associated Hamilton Jacobi Bellman (HJB) equation is

$$\partial_t V_t = q_t + (\nabla_{\mathbf{x}}V_t)^T\mathbf{f}_t - \frac{1}{2}(\nabla_{\mathbf{x}}V_t)^T\mathbf{G}_t\mathbf{R}^{-1}\mathbf{G}_t^T(\nabla_{\mathbf{x}}V_t) \quad (7)$$
$$+\frac{1}{2}trace\left((\nabla_{\mathbf{xx}}V_t)\mathbf{G}_t\mathbf{\Sigma}_\epsilon\mathbf{G}_t^T\right)$$

The corresponding optimal control is a function of the state and it is given by the equation:

$$\mathbf{u}(\mathbf{x}_{t_i}) = \mathbf{u}_{t_i} = -\mathbf{R}^{-1}\mathbf{G}_{t_i}^T(\nabla_{x_{t_i}}V_{t_i}) \quad (8)$$

We are leaving the standard development of this optimal control problem by transforming the HJB equations with the substitution $V_t = -\lambda\log\Psi_t$ and by introducing a simplification $\lambda\mathbf{R}^{-1} = \mathbf{\Sigma}_\epsilon$. In this way, the transformed HJB equation becomes a linear 2nd order partial differential equation. Due to the Feynman-Kac theorem [13, 25], the solution for the exponentially transformed value function becomes

$$\Psi_{t_i} = \lim_{dt\to 0}\int p\left(\boldsymbol{\tau}_i|\mathbf{x}_i\right)\exp\left[-\frac{1}{\lambda}\left(\phi_{t_N} + \sum_{j=0}^{N-1} q_{t_j}dt\right)\right]d\boldsymbol{\tau}_i \quad (9)$$

Thus, we have transformed our stochastic optimal control problem into an approximation problem of a path integral. As detailed in [18], it is not necessary to compute the value function explicitly, but rather it is possible to derive the optimal controls directly:

$$\mathbf{u}_{t_i} = \int P\left(\boldsymbol{\tau}_i\right)\mathbf{u}\left(\boldsymbol{\tau}_i\right)d\boldsymbol{\tau}_i \quad (10)$$
$$\mathbf{u}(\boldsymbol{\tau}_i) = \mathbf{R}^{-1}\mathbf{G}_{t_i}{}^T\left(\mathbf{G}_{t_i}\mathbf{R}^{-1}\mathbf{G}_{t_i}{}^T\right)^{-1}\left(\mathbf{G}_{t_i}\boldsymbol{\epsilon}_{t_i} - \mathbf{b}_{t_i}\right)$$

where $P\left(\boldsymbol{\tau}_i\right)$ is the probability of a trajectory $\boldsymbol{\tau}_i$, and $\mathbf{b}_{t_i}$ is a more complex expression, beyond the space constraints of this paper. The important conclusion is that it is possible to evaluate Eq. (10) from Monte Carlo roll-outs of the control system, i.e., our optimal control problem can be solved as an estimation problem.

A. The PI^2 Algorithm

The PI^2 algorithm is just a special case of the optimal control solution in Eq. (10), applied to control systems with parameterized control policy:

$$\mathbf{a}_t = \mathbf{g}_t^T(\boldsymbol{\theta} + \boldsymbol{\epsilon}_t) \tag{11}$$

i.e., the control command is generated from the inner product of a parameter vector $\boldsymbol{\theta}$ with a vector of basis function $\mathbf{g}_t$ – the noise $\boldsymbol{\epsilon}_t$ is interpreted as user controlled exploration noise.

A particular case of a control system with parameterized policy is the Dynamic Movement Primitives (DMP) approach introduced by [6]:

$$\begin{aligned}
\frac{1}{\tau}\dot{v}_t &= f_t + \mathbf{g}_t^T(\boldsymbol{\theta} + \boldsymbol{\epsilon}_t) && (12)\\
\frac{1}{\tau}\dot{q}_{d,t} &= v_t &&\\
f_t &= \alpha(\beta(g - q_{d,t}) - v_t) &&\\
\frac{1}{\tau}\dot{s}_t &= -\alpha s_t && (13)\\
[\mathbf{g}_t]_j &= \frac{w_j s_t}{\sum_{k=1}^{p} w_k}(g - q_0) && (14)\\
w_j &= \exp\left(-0.5 h_j (s_t - c_j)^2\right) && (15)
\end{aligned}$$

The intuition of this approach is to create desired trajectories $q_{d,t}, \dot{q}_{d,t}, \ddot{q}_{d,t} = \tau \dot{v}_t$ for a motor task out of the time evolution of a nonlinear attractor system, where the goal g is a point attractor and q_0 the start state. The parameters $\boldsymbol{\theta}$ determine the shape of the attractor landscape, which allows to represent almost arbitrary smooth trajectories, e.g., a tennis swing, a reaching movement, or a complex dance movement. While leaving the details of the DMP approach to [6], for this paper the important ingredients of DMPs are that i) the attractor system Eq. (12) has the same form as Eq. (3), and that ii) the p-dimensional parameter vector can be interpreted as motor commands as used in the path integral approach to optimal control. Learning the optimal values for $\boldsymbol{\theta}$ will thus create a optimal reference trajectory for a given motor task. The PI^2 learning algorithm applied to this scenario is summarized in Table I. As illustrated in [18, 19], PI^2 outperforms previous RL algorithms for parameterized policy learning by at least one order of magnitude in learning speed and also lower final cost performance. As an additional benefit, PI^2 has no open algorithmic parameters, except for the magnitude of the exploration noise $\boldsymbol{\epsilon}_t$ (the parameter λ is set automatically, cf. [18]). We would like to emphasize one more time that PI^2 *does not* require knowledge of the model of the control system or the environment.

Key Innovations in PI^2: In summary we list the key innovations in PI^2 that we believe lead to its superior performance. These innovations make applications like the the learning of gain schedules for high dimensional tasks possible.

- The basis of the derivation of the PI^2 algorithm is the transformation of the optimal control problem from a constrained minimization to a maximum likelihood formulation. This transformation is very critical since there is no need to calculate a gradient that is usually sensitive to noise and large derivatives in the value function.
- Paths with higher cost have lower probability. A clear intuition that has also rigorous mathematical representation through the exponentiation of the value function. This transformation is necessary for the linearization of HJB into the Chapman-Kolmogorov PDE.
- With PI^2 the optimal control problem is solved with the forward propagation of dynamics. Thus no backward propagation of approximations of the value function is required. This is a very important characteristic of PI^2 that allows for sampling (i.e. roll-out) based estimation of the path-integral.
- For high dimensional problems, it is not possible to sample the whole state space and that is the reason for applying path integral control in an iterative fashion to update the parameters of the DMPs.
- The derivation of an RL algorithm from first principles largely eliminates the need for open parameters in the final algorithm.

TABLE I

PSEUDOCODE OF THE $\mathbf{PI}^2$ ALGORITHM FOR A 1D PARAMETERIZED POLICY.

- **Given**:
  - An immediate cost function $r_t = q_t + \boldsymbol{\theta}_t^T \mathbf{R} \boldsymbol{\theta}_t$ (cf. Eq. (5))
  - A terminal cost term ϕ_{t_N} (cf. 5)
  - A stochastic parameterized policy $\mathbf{a}_t = \mathbf{g}_t^T(\boldsymbol{\theta} + \boldsymbol{\epsilon}_t)$ (cf. Eqs. (11) and (12))
  - The basis function $\mathbf{g}_{t_i}$ from the system dynamics (cf. 14)
  - The variance Σ_ϵ of the mean-zero noise $\boldsymbol{\epsilon}_t$
  - The initial parameter vector $\boldsymbol{\theta}$
- **Repeat** until convergence of the trajectory cost R:
  - Create K roll-outs of the system from the same start state $\mathbf{x}_0$ using stochastic parameters $\boldsymbol{\theta} + \boldsymbol{\epsilon}_t$ at every time step
  - **For** all K roll-outs, compute:
    * $P\left(\boldsymbol{\tau}_{i,k}\right) = \frac{e^{-\frac{1}{\lambda}S(\boldsymbol{\tau}_{i,k})}}{\sum_{k=1}^{K}[e^{-\frac{1}{\lambda}S(\boldsymbol{\tau}_{i,k})}]}$
    * $S(\boldsymbol{\tau}_{i,k}) = \phi_{t_N,k} + \sum_{j=i}^{N-1} q_{t_j,k} + \frac{1}{2}\sum_{j=i+1}^{N-1}(\boldsymbol{\theta} + \mathbf{M}_{t_j,k}\boldsymbol{\epsilon}_{t_j,k})^T \mathbf{R}(\boldsymbol{\theta} + \mathbf{M}_{t_j,k}\boldsymbol{\epsilon}_{t_j,k})$
    * $\mathbf{M}_{t_j,k} = \frac{\mathbf{R}^{-1}\mathbf{g}_{t_j,k}\,\mathbf{g}_{t_j,k}^T}{\mathbf{g}_{t_j,k}^T \mathbf{R}^{-1}\mathbf{g}_{t_j,k}}$
  - **For** all i time steps, compute:
    * $\delta\boldsymbol{\theta}_{t_i} = \sum_{k=1}^{K}\left[P\left(\boldsymbol{\tau}_{i,k}\right)\mathbf{M}_{t_i,k}\,\boldsymbol{\epsilon}_{t_i,k}\right]$
  - Compute $[\delta\boldsymbol{\theta}]_j = \frac{\sum_{i=0}^{N-1}(N-i)\, w_{j,t_i}\, [\delta\boldsymbol{\theta}_{t_i}]_j}{\sum_{i=0}^{N-1} w_{j,t_i}(N-i)}$
  - Update $\boldsymbol{\theta} \leftarrow \boldsymbol{\theta} + \delta\boldsymbol{\theta}$
  - Create one noiseless roll-out to check the trajectory cost $R = \phi_{t_N} + \sum_{i=0}^{N-1} r_{t_i}$. In case the noise cannot be turned off, i.e., a stochastic system, multiple roll-outs need to be averaged.

IV. VARIABLE IMPEDANCE CONTROL WITH PI^2

The PI^2 algorithm as introduced above seems to be solely suited for optimizing a trajectory plan, and not directly the controller. Here we will demonstrate that this is not the case, and how PI^2 can be used to optimize a gain schedule simultaneously to optimizing the reference trajectory. For this purpose, it is important to realize how Eq. (3) relates to a complete robotics system. We assume a d DOF robot that obeys rigid body dynamics. $\mathbf{q}^v$ denotes the joint velocities, and $\mathbf{q}^p$ the joint angle positions. Every DOF has its own reference trajectory from a DMP, which means that Eqs. (12) are duplicated for every DOF, while Eqs. (13), (14), and (15) are shared across all DOFs – see [6] for explanations on how to create multi-dimensional DMPs. Thus, Eq. (3) applied to this context, i.e. using rigid body dynamics equations, with $\mathbf{M}, \mathbf{C}, \mathbf{G}$ the Inertia matrix, Coriolis/centripedal and and gravity forces respectively, becomes:

$$\begin{aligned} \dot{\mathbf{q}}^v &= \mathbf{M}(\mathbf{q}^p)^{-1}\left(-\mathbf{C}(\mathbf{q}^p, \mathbf{q}^v) - \mathbf{G}(\mathbf{q}^p) + \mathbf{u}\right) \\ \dot{\mathbf{q}}^p &= \mathbf{q}^v \\ \frac{1}{\tau}\dot{s}_t &= -\alpha s_t \end{aligned} \quad (16)$$

where each element u_i of the control vector $\mathbf{u}$:

$$\begin{aligned} u_i &= -K_{P,i}\left(q_i^p - q_{d,i}^p\right) - \xi_i\sqrt{K_{P,i}}\left(q_i^v - q_{d,i}^v\right) \\ &+ \ u_{ff,i} \end{aligned} \quad (17)$$

The terms $q_{d,i}^v, q_{d,i}^p$ are the reference joint angle position and velocity of the ith DOF and they are given by the set of equations:

$$\begin{aligned} \frac{1}{\tau}\dot{q}_{d,i}^v &= \alpha(\beta(g_i - q_{d,i}^p) - q_{d,i}^v + \mathbf{g}_t^{i,T}(\boldsymbol{\theta}_{ref}^i + \boldsymbol{\epsilon}_t^i) \\ \frac{1}{\tau}\dot{q}_{d,i}^p &= q_{d,i}^v \end{aligned} \quad (18)$$

Note that in the control law in (17), we used Eq. (1) applied to every DOF individually using a time varying gain, and we inserted the common practice that the damping gain K_D^i is written as the square root of the proportional gain K_P^i with a user determined multiplier ξ^i. A critically important result of [18] is that for the application of PI^2 only those differential equations in Eq. (16) matter that have learnable parameter $\boldsymbol{\theta}^i$. Moreover, the optimization of these parameters is accomplished by optimizing the parameter vector of each differential equation independently (as shown in Table I), despite that the DOFs are coupled through the cost function. For this reason, PI^2 operates in a model free mode, as only one of the DMP differential equation per DOF is required, and all other equations, including the rigid body dynamics model, drop out.

For variable stiffness control, we exploit these insights and add one more differential equation per DOF in Eq. (16):

$$\dot{K}_{P,i} = \alpha_K\left(\mathbf{g}_{t,K}^{i,T}(\boldsymbol{\theta}_K^i + \boldsymbol{\epsilon}_{K,t}^i) - K_{P,i}\right) \quad (19)$$

$$[\mathbf{g}_{t,K}]_j = \frac{w_j}{\sum_{k=1}^p w_k} \quad (20)$$

This equation models the time course of the position gains, coupled to Eq. (15) of the DMP. Thus, $K_{P,i}$ is represented by a basis function representation linear with respect to the learning parameter $\boldsymbol{\theta}_K^i$, and these parameter are learned with the PI^2 algorithm following Table I. We will assume that the time constant $\frac{1}{\alpha_K}$ is so small, that for all practical purposes we can assume that $K_{P,i} = \mathbf{g}_{t,K}^{i,T}(\boldsymbol{\theta}_K^i + \boldsymbol{\epsilon}_{K,t}^i)$ holds at all times.

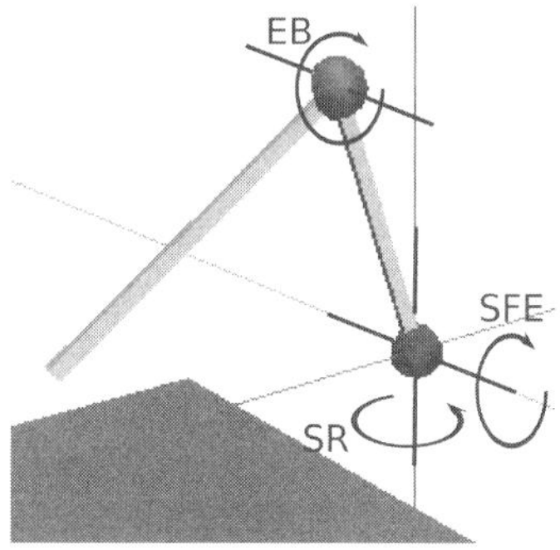

Fig. 1. 3-DOF Phantom simulation in SL.

Essentially equations 16,17,18 and 20 are incorporated in one stochastic dynamical system of the form of Eq. (3). In conclusion, we achieved a novel formulation of learning both the reference trajectory and the gain schedule for a multi-dimensional robotic system with model-free reinforcement learning, using the PI^2 algorithm and its theoretical properties as foundation of our derivations.

V. Results

We will now present results of applying the outlined algorithms to two simulated robot arms with 3 and 6 DOFs, respectively. For both robots, the immediate reward at time step t is given as:

$$r_t = W_{gain}\sum_i K_{P,t}^i + W_{acc}||\ddot{\mathbf{x}}|| + W_{subgoal}C(t) \quad (21)$$

Here, $\sum_i K_{P,t}^i$ is the sum over the proportional gains over all joints. The reasoning behind penalizing the gains is that low gains lead to several desirable properties of the system such as compliant behavior (safety and/or robustness [2]), lowered energy consumption, and less wear and tear. The term $||\ddot{\mathbf{x}}||$ is the magnitude of the accelerations of the end-effector. This quantity is penalized to avoid high-jerk end-effector motion. This penalty is low in comparison to the gain penalty.

The robot's primary task is to pass through an intermediate goal, either in joint space or end-effector space – such scenarios occur in tasks like playing tennis or table tennis. The component of the cost function $C(t)$ that represents this primary task will be described individually for each robot in the next sections. Gains and accelerations are penalized at each time step, but $C(t)$ only leads to a cost at specific time steps along the trajectory.

For both robots, the cost weights are $W_{subgoal} = 2000$, $W_{gain} = 1/N$, $W_{acc} = 1/N$. Dividing the weights by the number of time steps N is convenient, as it makes the weights independent of the duration of a movement.

A. Robot 1: 3-DOF Phantom

The Phantom Premium 1.5 Robot is a 3 DOF, two link arm. It has two rotational degrees of freedom at the base and one in

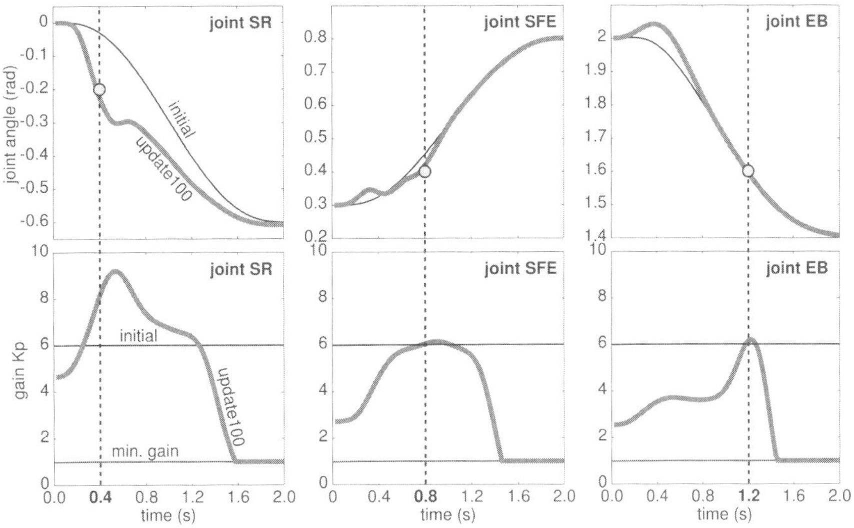

Fig. 2. Joint trajectories and gain schedules of the Phantom robot before learning (thin) and after 100 updates (thick). Round markers indicate intermediate subgoals.

the arm. We use a physically realistic simulation of this robot generated in SL [14], as depicted in Fig. 1.

The task for this robot is intentionally simple and aimed at demonstrating the ability to tune task relevant gains in joint space with straightforward and easy to interpret data.

The duration of the movement is $2.0s$, which corresponds to 1000 time steps at 500Hz servo rate. The intermediate goals for this robot are set as follows:

$$\begin{aligned} C(t) = \quad & \delta(t-0.4) \cdot \quad |\, q_{SR}(t)+0.2 \,| + \\ & \delta(t-0.8) \cdot \quad |\, q_{SFE}(t)-0.4 \,| + \\ & \delta(t-1.2) \cdot \quad |\, q_{EB}(t)-1.5 \,| \end{aligned} \tag{22}$$

This penalizes joint SR for not having an angle $q_{SR} = -0.2$ at time $t = 0.4s$. Joints SFE and EB are also required to go through (different) intermediate angles at times $0.8s$ and $1.2s$ respectively.

The initial parameters $\boldsymbol{\theta}^i$ for the reference trajectory are determined by training the DMPs with a minimum jerk trajectory [26] in joint space from $\mathbf{q}_{t=0.0} = [0.0\ 0.3\ 2.0]^T$ to $\mathbf{q}_{t=2.0} = [-0.6\ 0.8\ 1.4]^T$. The function approximator for the proportional gains of the 3 joints is initialized to return a constant gain of $6.0Nm/rad$. The initial trajectories are depicted as thin lines in Fig. 2, where the angles and gains of the three joints are plotted against time. Since the task of PI^2 is to optimize both trajectories and gains with respect to the cost function, this leads to a 6-D RL problem. The robot executes 100 parameter updates, with 4 noisy exploration trials per update. After each update, we perform one noise-less test trial for evaluation purposes.

Fig. 3 depicts the learning curve for the phantom robot (left), which is the overall cost of the noise-less test trial after each parameter update. The joint space trajectory and gain scheduling after 100 updates are depicted as thick solid lines in Fig. 2.

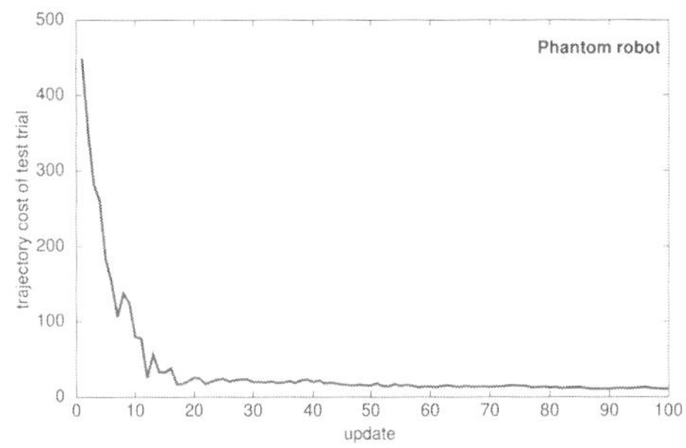

Fig. 3. Learning curve for the phantom robot.

From these graphs, we draw the following conclusions:

- PI^2 has adapted the initial minimum jerk trajectories such that they fulfill the task and pass through the desired joint angles at the specified times with only small error. These intermediate goals are represented by the circles on the graphs. The remaining error is a result of the trade-off between the different factors of the cost function (i.e. penalty for distance to goal vs. penalty for high gains).
- Because the magnitude of gains is penalized in general, they are low when the task allows it. After $t = 1.6s$, all gains drop to the minimum value[1], because accurate tracking is no longer required to fulfill the goal. Once the task is completed, the robot becomes maximally compliant, as one would wish it to be.
- When the robot is required to pass through the intermediate targets, it needs better tracking, and therefore higher gains. Therefore, the peaks of the gains correspond roughly to the times where the joint is required to pass through an intermediate point.

[1] We bound the gains between pre-specified maximum and minimum values. Too high gains would generate oscillations and can lead to instabilities of the robot, and too low gains lead to poor tracking such that the robot frequently runs into the joint limits.

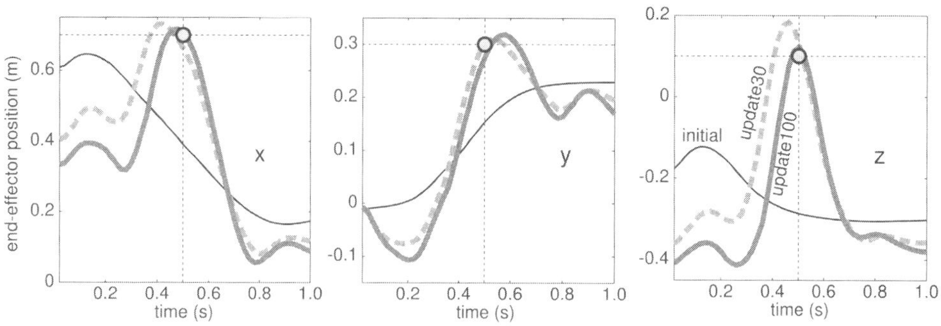

Fig. 4. End-effector trajectories of the Kuka robot before learning (thin), and after 30 (dashed) and 100 updates (thick).

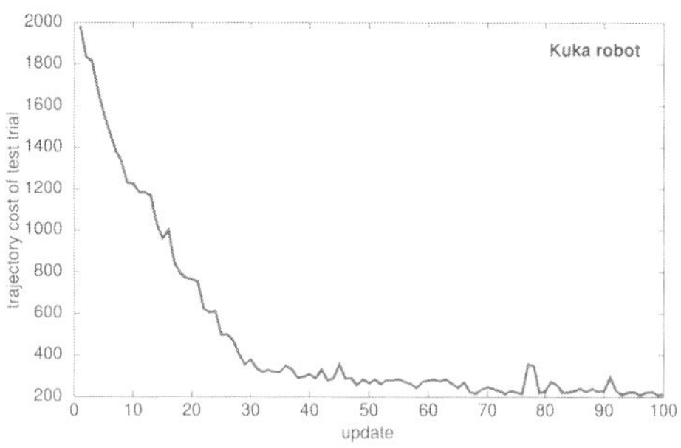

Fig. 5. Learning curve for the Kuka robot.

- Due to nonlinear effects, e.g., Coriolis and centripedal forces, the gain schedule shows more complex temporal behavior as one would initially assume from specifying three different joint space targets at three different times.

In summary, we achieved the objective of variable impedance control: the robot is compliant when possible, but has a higher impedance when the task demands it.

B. Robot 2: 6-DOF Kuka robot

Next we show a similar task on a 6 DOF arm, a Kuka Light-Weight Arm. This example illustrates that our approach scales well to higher-dimensional systems, and also that appropriate gains schedules are learned when intermediate targets are chosen in end-effector space instead of joint space.

The duration of the movement is $1.0s$, which corresponds to 500 time steps. This time, the intermediate goal is for the end-effector $\mathbf{x}$ to pass through $[\ 0.7\ 0.3\ 0.1]^T$ at time $t = 0.5s$:

$$C(t) = \quad \delta(t - 0.5)|\ \mathbf{x} - [\ 0.7\ 0.3\ 0.1]^T\ | \qquad (23)$$

The six joint trajectories are again initialized as minimum jerk trajectories. As before, the resulting initial trajectory is plotted as a thin line in Fig. 4. The initial gains are set to a constant $[60, 60, 60, 60, 25, 6]^T$. Given these initial conditions, finding the parameter vectors for DMPs and gains that minimizes the cost function leads to a 12-D RL problem. We again perform 100 parameter updates, with 4 exploration trials per update.

The learning curve for this problem is depicted in Fig. 5. The trajectory of the end-effector before learning and after 30 and 100 updates is depicted in Fig. 4. The intermediate goal at $t = 0.5$ is visualized by circles. Finally, Fig. 6 shows the gain schedules before learning and after 30 and 100 updates for the 6 joints of the Kuka robot.

From these graphs, we draw the following conclusions:

- PI^2 has adapted joint trajectories such that the end-effector passes through the intermediate subgoal at the right time. It learns to do so after only 30 updates (Figure 5).
- After 100 updates the peaks of most gains occur just before the end-effector passes through the intermediate goal (Figure 6), and in many cases decrease to the minimum gain directly afterwards. As with the phantom robot we observe high impedance when the task requires accuracy, and more compliance when the task is relatively unconstrained.
- The second joint (GA2) has the most work to perform, as it must support the weight of all the more distal links. Its gains are by far the highest, especially at the intermediate goal, as any error in this DOF will lead to a large end-effector error.
- The learning has two phases. In the first phase (plotted as dashed), the robot is learning to make the end-effector pass through the intermediate goal. At this point, the basic shape of the gain scheduling has been determined. In the second phase, PI^2 fine tunes the gains, and lowers them as much as the task permits.

VI. Related work

In optimal control and model based RL, Differential Dynamic Programming (DDP) [7] has been one of the most established and used frameworks for finite horizon optimal control problems. In DDP, both state space dynamics and cost function are approximated up to the second order. The assumption of stabilizability and detectability for the local approximation of the dynamics are necessary for the convergence of DDP. The resulting state space trajectory is locally optimal while the corresponding control policy consists of open loop feedforward command and closed loop gains relative to a nominal and optimal final trajectory. This characteristic allows the use of DDP for both planning and control gain scheduling problems. In [4, 24] DDP was extended to incorporate constrains in state and controls. In [10] the authors suggest computational

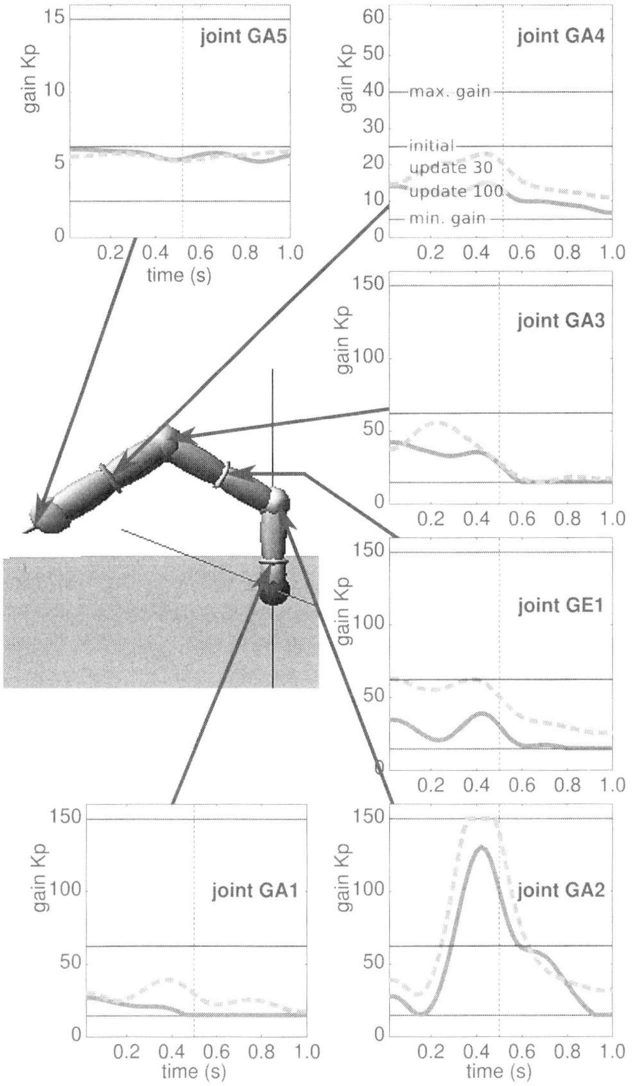

Fig. 6. Gain schedules of the six joints of the Kuka robot before learning (thin) and after 30 (dashed) and 100 updates (thick).

improvements to constrained DDP and apply the proposed algorithm to a low dimensional planning problem.

An example of a DDP application to robotics is in [12]. In this work, a min-max or Differential Game Theory approach to optimal control is proposed. There is a strong link between robust control frequency design analysis such as H_∞ control and the framework of Differential Game Theory [1]. Essentially the min-max DDP results in robust feedback control policies with respect to model uncertainty and unknown dynamics. Although, in theory, min-max DDP should resolve the issue of model uncertainty, it can lead to overly conservative control policies. The conservatism results from the need to guarantee that the game theoretic approach will be always stabilizable, i.e. making sure that the stabilizing controller wins. For linear and time invariant systems, such guarantee is feasible through γ-iteration [23]. However, for nonlinear systems providing this guarantee is not trivial.

The work on Receding Horizon DDP [17] provided an alternative and rather efficient way of computing local optimal feedback controls. Nevertheless, all the computations of optimal trajectories and control take place off-line and the model predictive component is only due to the fact that the final target state of the optimal control problem varies. Recent work on LQR-trees uses a simpler variation of DDP, the iterative Linear Quadratic Regulator (iLQR) [11], which is based on linear approximations of the state space dynamics, in combination with tools from Nonlinear Robust Control theory for region of attraction analysis. Given the local optimal feedback control policies, the sums of squares optimization scheme is used to quantify the size of of the basin of attraction, and provides so-called control funnels. These funnels improve sampling since they quantize the state space into attractor regions placed along the trajectories towards the target state. This is a model based approach and inherits all the problems of model based approaches to optimal control. In addition, even though sampling is improved, it is still an issue how LQR trees scale in high dimensional dynamical systems.

The path integral formalism for optimal control was introduced in [8, 9]. In this work, the role of noise in symmetry breaking phenomena was investigated in the context of stochastic optimal control. In [22], the path integral formalism is extended for stochastic optimal control of multi-agent systems, which is not unlike our multi DOF control systems.

Recent work on stochastic optimal control by [21, 20, 3] shows that for a class of discrete stochastic optimal control problems, the Bellman equation can be written as the KL divergence between the probability distribution of the controlled and uncontrolled dynamics. Furthermore, it is shown that the class of discrete KL divergence control problem is equivalent to the continuous stochastic optimal control formalism with quadratic cost control function and under the presence of Gaussian noise. In all this aforementioned work, both in the path integral formalism as well as in KL divergence control, the class of stochastic dynamical systems under consideration is rather restrictive since the control transition matrix is state independent. Moreover, the connection to direct policy learning in RL and model-free learning was not made in any of the previous projects. In [3], the stochastic optimal control problem is investigated for discrete state-action spaces, and therefore it is treated as Markov Decision Process (MDP). To apply our PI^2 algorithm, we do not discretize the state space and we do not treat the problem as an MDP. Instead we work in continuous state-action spaces which are suitable for performing RL in high dimensional robotic systems. To the best of our knowledge, our results present RL in one of the most high dimensional continuous state action spaces. In our derivations, the probabilistic interpretation of control comes directly from the Feynman-Kac Lemma. Thus we do not have to impose any artificial pseudo-probability treatment of the cost as in [3]. In addition, for continuous state-action spaces, we do not have to learn the value function as it is suggested in [3] via Z-learning. Instead we directly obtain the controls based on our generalization of optimal controls. In the previous work, the problem of how to sample trajectories is not addressed. Sampling is performed with the hope to cover all the relevant state space. We follow a rather different approach by incremental updating, which allows us to address robotic learning problems of the complexity and dimensionality of complete humanoid robots.

VII. Discussion

We presented a model-free reinforcement learning approach that can learn variable impedance control for robotic systems. Our approach is derived from stochastic optimal control with path integrals, a relatively new development that transforms optimal control problems into estimation problems. In particular, PI^2 goes beyond the original ideas of optimal control with path integrals by realizing the applicability to optimal control with parameterized policies and model-free scenarios.

The mathematical structure of the PI^2 algorithm makes it suitable to optimize simultaneously both reference trajectories and gain schedules. This is similar to classical differential dynamic programming (DDP) methods, but completely removes the requirements of DDP that the model of the controlled system must be known, that the cost function has to be twice differentiable in both state and command cost, and that the dynamics of the control system have to be twice differentiable. The latter constraints make it hard to apply DDP to tasks with discrete events, as is typical in force control and locomotion.

We evaluated our approach on two simulated robot systems, which posed up to 12 dimensional learning problems in continuous state-action spaces. The goal was to learn compliant control while fulfilling kinematic task constraints, like passing through an intermediate target. The evaluations demonstrated that the algorithm behaves as expected: it increases gains when needed, but tries to maintain low gain control otherwise. The optimal reference trajectory always fulfilled the task goal. Learning speed was rather fast, i.e., within at most a few hundred trials, the task objective was accomplished. From a machine learning point of view, this performance of a reinforcement learning algorithm is very fast.

The PI^2 algorithms inherits the properties of all trajectory-based learning algorithms in that it only finds locally optimal solutions. For high dimensional robotic system, this is unfortunately all one can hope for, as exploring the entire state-action space in search for a globally optimal solution is impossible.

Future work aims at applying these methods to actual robots for mobile manipulation and locomotion controllers. We believe that our methods are a major step towards realizing compliant autonomous robots that operate robustly in dynamic, stochastic environments, without harming other beings or themselves.

VIII. Acknowledgements

This research was supported in part by National Science Foundation grants ECS-0325383, IIS-0312802, IIS-0082995, IIS-9988642, ECS-0326095, ANI-0224419, the DARPA program on Learning Locomotion, the Multidisciplinary Research Program of the Department of Defense (MURI N00014-00-1-0637), and the ATR Computational Neuroscience Laboratories. J.B. was supported by a prospective researcher fellowship from the Swiss National Science Foundation. E.T. was supported by a Myronis Fellowship. F.S. was supported by a Research Fellowship from the German Research Foundation (DFG).

References

[1] T. Basar and P. Bernhard. *H-Infinity Optimal Control and Related Minimax Design Problems: A Dynamic Game Approach*. Birkhäuser, Boston, 1995.

[2] J. Buchli, M. Kalakrishnan, M. Mistry, P. Pastor, and S. Schaal. Compliant quadruped locomotion over rough terrain. In *IEEE/RSJ International Conference on Intelligent Robots and Systems*, pages 814–820, 2009.

[3] Todorov E. Efficient computation of optimal actions. *Proc Natl Acad Sci USA*, 106(28):11478–83, 2009.

[4] R. Fletcher. *Practical Methods of Optimization*, volume 2. Wiley, 1981.

[5] N. Hogan. Impedance control - an approach to manipulation. *ASME, Transactions, Journal of Dynamic Systems, Measurement, and Control*, 107:1–24, 1985.

[6] A.J. Ijspeert. Vertebrate locomotion. In M.A. Arbib, editor, *The handbook of brain theory and neural networks*, pages 649–654. MIT Press, 2003.

[7] D. H. Jacobson and D. Q. Mayne. *Differential Dynamic Programming*. Optimal Control. Elsevier Publishing Company, New York, 1970.

[8] H. J. Kappen. Linear theory for control of nonlinear stochastic systems. *Phys Rev Lett*, 95(20):200201, 2005.

[9] H.J. Kappen. Path integrals and symmetry breaking for optimal control theory. *Journal of Statistical Mechanics: Theory and Experiment*, 2005(11):P11011, 2005.

[10] G. Lantoine and R.P. Russell. A hybrid differential dynamic programming algorithm for robust low-thrust optimization. In *AAS/AIAA Astrodynamics Specialist Conference and Exhibit*, 2008.

[11] W. Li and E. Todorov. Iterative linear quadratic regulator design for nonlinear biological movement systems. In *Proceedings of the 1st International Conference on Informatics in Control, Automation and Robotics*, pages 222–229, 2004.

[12] J. Morimoto and C. Atkeson. Minimax differential dynamic programming: An application to robust biped walking. In *In Advances in Neural Information Processing Systems 15*, pages 1563–1570. MIT Press, Cambridge, MA, 2002.

[13] B. K. Oksendal. *Stochastic differential equations: an introduction with applications*. Springer, Berlin, New York, 6th edition, 2003.

[14] S. Schaal. The SL simulation and real-time control software package. Technical report, University of Southern California, 2007.

[15] L. Sciavicco and B. Siciliano. *Modelling and Control of Robot Manipulators*. Springer, London, New York, 2000.

[16] R.F. Stengel. *Optimal Control and Estimation*. Dover Publications, New York, 1994.

[17] Y. Tassa, T. Erez, and W. Smart. Receding horizon differential dynamic programming. In *Advances in Neural Information Processing Systems 20*, pages 1465–1472. MIT Press, Cambridge, MA, 2008.

[18] E. Theodorou, J. Buchli, and S. Schaal. Reinforcement learning in high dimensional state spaces: A path integral approach. *Journal of Machine Learning Research*, 2010. Accepted for publication.

[19] E. Theodorou, J. Buchli, and S. Schaal. Reinforcement learning of motor skills in high dimensions: A path integral approach. In *Proceedings of the IEEE International Conference on Robotics and Automation*, 2010.

[20] E. Todorov. Linearly-solvable markov decision problems. In *Advances in Neural Information Processing Systems 19*, pages 1369–1376. MIT Press, Cambridge, MA, 2007.

[21] E. Todorov. General duality between optimal control and estimation. In *Proc. of the 47th IEEE Conference on Decision and Control*, 2008.

[22] B. van den Broek, W. Wiegerinck, and B. Kappen. Graphical model inference in optimal control of stochastic multi-agent systems. *Journal of Artificial Intelligence Research*, 32:95–122, 2008.

[23] T. Vincent and W. Grantham. *Non Linear And Optimal Control Systems*. John Wiley & Sons Inc., 1997.

[24] S. Yakowitz. The stagewise Kuhn-Tucker condition and differential dynamic programming. *IEEE Transactions on Automatic Control*, 31(1):25–30, 1986.

[25] J. Yong. Relations among ODEs, PDEs, FSDEs, BSDEs, and FBSDEs. In *Proceedings of the 36th IEEE Conference on Decision and Control, 1997*, volume 3, pages 2779–2784, Dec 1997.

[26] M. Zefran, V. Kumar, and C.B. Croke. On the generation of smooth three-dimensional rigid body motions. *IEEE Transactions on Robotics and Automation*, 14(4):576–589, 1998.

A Fast Traversal Heuristic and Optimal Algorithm for Effective Environmental Coverage

Ling Xu and Tony Stentz
Robotics Institute, Carnegie Mellon University, Pittsburgh, PA, USA
{lingx, axs}@cs.cmu.edu

Abstract— **Tasks such as street mapping and security surveillance seek a route that traverses a given space to perform a function. These task functions may involve mapping the space for accurate modeling, sensing the space for unusual activity, or processing the space for object detection. With a prior map, an optimal path can be computed using a graph to represent the environment and generating the solution using known graph algorithms. However, the prior map may be inaccurate due to factors such as occlusion, outdatedness, dynamic objects, and resolution limitations. In this work, we address the NP-hard problem of optimal environmental coverage with incomplete prior map information.**

To utilize related algorithms in graph theory, we represent the environment as a graph. Using this representation, we present a graph coverage approach for optimal plan generation based on the Undirected Chinese Postman and Rural Postman problems. This approach produces a tractable solution through the use of low complexity algorithms in a branch-and-bound framework. Additionally, as the robot receives sensor updates during traversal of the environment, we update the graph to reflect those changes. The updated graph can be highly disconnected so computing an optimal solution can be NP-hard. To combat this, we introduce a heuristic for coverage path generation that helps maximize the connectivity of the updated graph. We evaluate our approach on a set of comparison tests in simulation.

I. Introduction

Many tasks, such as street sweeping, mail delivery, and robotic surveillance and patrol, require a robot to visit all points in an environment to accomplish a goal. These goals usually entail effectively mapping, sensing, or processing the space. In this work, we address this coverage problem where a robot is required to visit most locations in a given area. We assume a map of the environment is available. This is a reasonable assumption since information of most outdoor spaces can be obtained via satellite images. However prior maps can differ from the actual environment due to factors such as occlusions, datedness, and lower map resolutions. Even with an otherwise accurate map, dynamic conditions such as the presence of people or vehicles can diminish the effective accuracy of prior information. Therefore, an additional goal for the coverage problem we are addressing is to efficiently replan when changes occur in the environment.

In robotics, there is a number of related work in the area of continuous space coverage. In continuous space coverage, the robot must pass a detector over all points in the space in order to complete a task [1][2][3][4]. While these methods ensure completeness in terms of the area covered, they do not guarantee optimal path length. In graph theory and operations research, researchers have tackled this problem by representing the environment as a graph, and using routing algorithms such as the traveling salesman [5] [6] or postman problems [7] [8] to generate optimal solutions. Because we seek optimal paths for coverage efficiency, we choose to use the graph representation as the foundation of our solution approach.

In the graph representation, nodes in the graph denote locations in the environment and edges in the graph are the paths between the locations. For example, the map in Figure 1 is converted into a graph shown in Figure 2(a) by changing each street intersection into a node and each street into an edge. Each edge has a cost assigned to it where the cost can represent measurements such as Euclidean distance between locations, terrain traversability, travel time, or a combination of several metrics. Additionally, each edge is undirected meaning it can be traversed in both directions. Another example is a Voronoi diagram where the paths are edges in the graph and the path intersections are nodes. This is one way to generate optimal paths for some of the problems in continuous space coverage.

For our problem, we seek a coverage path that visits all the edges or a designated edge subset of the graph. This coverage problem can be modeled as an arc-routing problem. We focus on two types of arc-routing problems: the Chinese Postman Problem [9] and the Rural Postman Problem. The Chinese Postman Problem (CPP) seeks an optimal path that visits all the edges of the graph at least once, and the Rural Postman Problem (RPP) seeks an optimal path that visits a predefined subset of graph edges at least once. We define an optimal path as the lowest cost coverage path given the current graph information.

The CPP and RPP differ in two ways. The CPP has a polynomial time solution of $O(n^3)$ [7] [10]. It works well for applications where it is necessary to traverse every part of the space. For example, Sorensen uses the CPP to plan tours for farming machines in static known environments [11].

In many practical problems, it is not necessary to traverse all the edges in the graph. The RPP seeks a tour that traverses a required subset of the graph edges using the extra graph edges as travel links between the required edges. Unlike the CPP, the RPP is a NP-hard problem. Optimal solutions exist that formulate the RPP as an integer linear program and solve it using branch-and-bound [8]. One recent approach [12] introduced new dominance relations such as computing the minimum spanning tree on the connected components in a

Fig. 1. The map of an urban environment is shown where the box-like shapes represent buildings and the spaces between the buildings are roads.

(a)

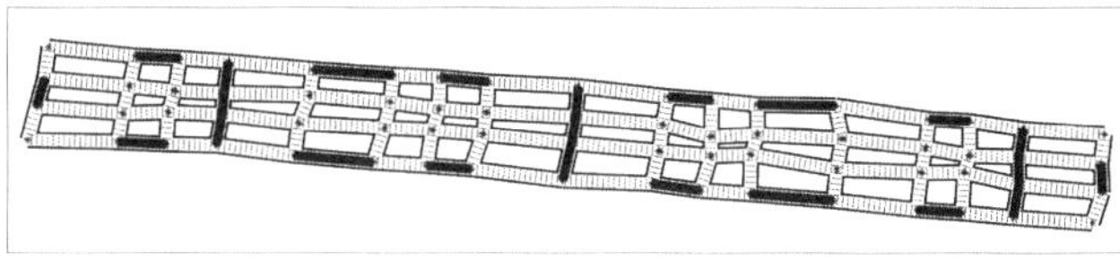

(b)

Fig. 2. (a) We represent the prior map as a graph where edges (white lines) denote the roads and vertices (stars) denote the road intersections. (b) Dotted and solid lines represent the CPP solution and are super-imposed on the above graph. Dotted lines denote edges that are traversed once and solid lines denote edges that are traversed more than once.

graph to solve large problem instances. Additionally, many TSP heuristics have been extended to the RPP [13]. For example, Christofides' approximation for the Euclidean TSP was modified for the undirected RPP and maintains its $\frac{3}{2}$ constant factor performance [14]. Improvements heuristics have been developed as postoptimality procedures for these approximation algorithms by modifying and shortening the suboptimal solution path to be as close to optimal as possible [15].

While most research on arc routing problems focus on the static environments, there has been work that address dynamic graphs. Moreira et al. present a heuristic-based approach for the Dynamic Rural Postman Problem (DRPP) [16]. They frame the problem as a machine cutting application where the graph changes as pieces of the cutting surface are cut and removed. They introduce and compare two heuristics for choosing the next edge to cut. While the run times they obtained were reasonable, the solution paths generated were not optimal.

In this paper, we present two contributions. First, we introduce a general approach for the online coverage problem with an incomplete prior map. As changes in the environment are detected, we present a method for updating the graph to reflect the new changes. Our approach uses the lower complexity Chinese Postman algorithm within a branch-and-bound framework to solve the harder Rural Postman problem. Second, we present a new heuristic for path generation that traverses the graph in such a way that when changes occur, the heuristic helps keep the new coverage problem polynomial.

The rest of the paper is organized as follows. In Section II, we introduce and describe each component of the coverage algorithm. In the next section, we explain the set of tests we conducted. The results are presented and discussed in Section IV. Finally, we conclude and list future directions for this work.

II. Coverage Algorithm

A. *Chinese Postman Problem*

We assume the environment is initially known in the form of a prior map. This prior map is converted into a graph structure with goal locations as nodes in the graph and paths between goals as edges in the graph. The first step in solving the coverage problem is to assume the prior map is accurate and generate a tour that covers all the edges in the graph. This problem can be represented as a Chinese Postman Problem (CPP).

Algorithm 1: Chinese Postman Problem Algorithm

```
Input: s, start vertex
       G, connected graph where each edge has a cost value
Output: P, tour found or empty if no tour found
1  if IsEven(G) then
2  |  P = FindEulerCycle(G, s);
3  else
4  |  O = FindOddVertices(G);
5  |  O' = FindAllPairsShortestPath(O);
6  |  Mate = FindMinMatching(O');
7  |  G' = (G, Mate);
8  |  P = FindEulerCycle(G', s);
9  end
10 return P;
```

The CPP optimal tour consists of traversing all the edges in the graph at least once and starting and ending at the same node. To solve this problem, we used the Edmonds and Johnson algorithm [17] [18], detailed in Algorithm 1.

The first step in the algorithm is to calculate the degree of each vertex. If all the vertices have even degree, then the algorithm finds an Euler tour using the End-Pairing technique (line 2). If any vertices have odd degree, a minimum weighted matching [19][20] among the odd degree vertices is found using an all pairs shortest path graph of the odd vertices (line 5). Because the matching algorithm requires a complete graph, the all pairs shortest pairs algorithm is a way to optimally connect all the odd vertices. The matching finds the minimum cost set of edges that connect the odd nodes (line 6). The edges in the matching are doubled in the graph making the degree of the odd nodes even (line 7). Finally, the algorithm finds a tour on the new Eulerian graph (line 8).

We ran the CPP algorithm on the example graph shown earlier. Since the graph is odd, a solution to the problem requires some of the edges to be traversed more than once. In Figure 2(b), the solution tour is depicted where the dotted and solid lines represent edges traversed once and edges traversed more than once, respectively.

B. Rural Postman Problem

In the Rural Postman Problem, there are two sets of graph edges: required and optional. We define the required subset of edge as *coverage edges*, and the optional subset as *travel edges*. Any solution would include all coverage edges and some combination of travel edges.

For combinatorial optimization problems such as the RPP, a common planning framework is branch-and-bound [21]. Branch-and-bound is a method of iterating through a set of solutions until an optimal solution is found. It consists of two steps, a branching step and a bounding step. In the branching step, the algorithm forms n branches of subproblems where each subproblem is a node in the branch-and-bound tree. The solution to any subproblem could lead to a solution to the original problem. In the bounding step, the algorithm computes a lower bound for the subproblem. These lower bounds enable branch-and-bound to guarantee a solution is optimal without having to search through all possible subproblems. Branch-and-bound is a general method for handling hard problems that are slight deviations from low complexity problems even though faster, ad hoc algorithms exist for each deviation. The process of partitioning the problem into subproblems works to our advantage since it enables the use of the applications of low complexity algorithms like the CPP as bounds on the subproblems.

In our approach, we use the branch-and-bound approach to iterate through all combinations of travel edges. We call the set of travel edges our partition set. At each branching step, we generate two subproblems by including and excluding a travel edge. Next, each subproblem in the branch-and-bound tree is solved using the CPP algorithm. The cost of the CPP solution is the lower bound on the cost of the RPP for the particular branch. If the travel set is large, this method can be computationally expensive.

To keep computation costs low, we reduce the partition set given to the branch and bound algorithm. First, we define a few terms used in the algorithm. A *coverage* or *travel vertex* is a vertex in the graph incident to only coverage or travel edges, respectively. A *border vertex* is a vertex in the graph incident to at least one coverage edge and at least one travel edge. A *travel path* is a a sequence of travel segments and travel vertices connecting a pair of border vertices. Finally, an *optimal travel path (OTP)* is a travel path connecting a pair of border vertices such that it is the lowest cost path (of any type) between the vertices, and the vertices are not in the same cluster (i.e., there is no path consisting of just coverage segments between them). In other words, OTPs are shortest paths between clusters of coverage segments that do not cut through any part of a coverage cluster. Additionally, we modify the CPP algorithm to permit the use of travel edges as part of the shortest path between the set of odd nodes. This allows the CPP to reason directly about the paths that cut through coverage clusters.

All the OTPs are computed by finding the lowest cost path p_{ij} (searching over the entire graph) between each pair of border vertices v_i and v_j in different clusters. If p_{ij} is a travel path, we save it as an OTP. If p_{ij} is not a travel path, then v_i and v_j do not have an OTP between them (i.e., p_{ij} = NULL). The OTPs become our partition set. The OTPs are unlabeled at the beginning of the algorithm. We iterate through the OTP set within the branch-and-bound framework. At each branch step, the algorithm generates a new subproblem by either including or excluding an OTP. At the beginning of the RPP algorithm shown in Algorithm 2, we assign cost 0 to the unlabeled OTPs and solve the problem with all the OTPs set as required edges using the CPP algorithm (lines 2,3). The problem and CPP cost are pushed onto a priority queue (line 4). The CPP cost is the lower bound on the problem since all the OTPs have zero cost. While the queue is not empty, the lowest cost subproblem is selected from the queue (lines 5,6).

Algorithm 2: Rural Postman Problem Algorithm

Input: s, start vertex
$G = (C, T)$, graph where each edge has a label and a cost value. C is the subset of coverage edges, and T is the subset of travel edges
OTP, subset of OTPs
Output: P, tour found or empty if no tour found

1 $pq = []$;
2 $G' = [G, OTP]$ where $\forall OTP$, cost(p_{ij}) = 0
3 $P = CPP(s, G')$;
4 $AddToPQ(pq, [G', P])$;
5 **while** $!isEmpty(pq)$ **do**
6     $[G', P] = PopLowestCost(pq)$;
7     $p_{ij} = FindMaxOTP(G')$;
8   **if** $p_{ij} == []$ **then**
9   return P;
10   **end**
11     $G'' = IncludeEdge(G', p_{ij})$;
12     $P1 = CPP(s, G'')$;
13     $AddToPQ(pq, [G'', P1])$;
14     $G'' = RemoveEdge(G', p_{ij})$;
15     $P2 = CPP(s, G'')$;
16     $AddToPQ(pq, [G'', P2])$;
17 **end**
18 return []

For the subproblem, the algorithm selects an unlabeled OTP p_{ij} with the highest path cost (line 7). By employing this strategy of choosing the OTP with the highest path cost, our aim is to increase the lower bound by the largest amount, which may help focus the search to the correct branch and prevent extraneous explorations. Once an OTP p_{ij} is selected, two branches are generated; the first branch includes p_{ij} in the solution (line 11), this time with the real path cost assigned, and the second branch omits p_{ij} from the solution (line 14). A solution to each branch is found using the CPP algorithm (lines 12,15). Because each solutions is generated with a cost of 0 assigned to the unlabeled OTPs in the subproblem, the costs of the inclusion and exclusion CPP solutions represent lower bounds on the cost of the RPP with and without using p_{ij} for travel, respectively. These new subproblems are added to the priority queue (lines 13, 16), and the algorithm iterates until the lowest cost problem in the queue contains no OTPs (line 8). The solution to this problem is the optimal solution to the RPP since it has either included or excluded every single OTP in the solution, and has a path cost that is equal to or lower than the lower bounds of the other branches. The branch-and-bound algorithm for the RPP is an exponential algorithm with a complexity of $O(|V|^3 2^t)$ where t is the number of OTPs and $|V|$ is the number of vertices in the graph.

C. Online Changes

Dynamic changes occur when the environment differs from the original map. There are two categories of planners that handle these differences. Contingency planners model the uncertainty in the environment and plan for all possible scenarios. Assumptive planners [22] presume the perceived world state is correct and plan based on this assumption. If disparities arise, the perceived state is corrected and replanning occurs. We choose to use the lower complexity assumptive planning in order to generate solutions quickly.

In our planner, an initial plan is found based on the graph of the environment. As the vehicle uncovers differences between the map and environment during traversal, the algorithm propagates them into the graph structure. This may require a simple graph modification such as adding, removing, or changing the cost of an edge. But it can also result in more significant graph restructuring. As mentioned earlier, these changes may convert the initial planning problem into an entirely different problem.

For the coverage problems we are addressing in this work, most changes to the environment are discovered when the robot is actively traversing the space. These online changes are typically detected when the robot is not at the starting location, but at a middle location along the coverage path. At this point, some of the edges have already been visited. Because it is not necessary to revisit the edges that are already traversed, the visited edges in the previous plan are converted to travel edges. As shown in Algorithm 3, if the unvisited edges in the new problem are connected, we run the CPP algorithm; otherwise, we run the RPP.

When environmental changes are found online, replanning is done on the updated graph with different starting and ending vertices. To remedy this disparity, we add an artificial edge from the current vehicle location c to the ending vertex s in the graph (line 3). This edge (c, s) is assigned a large cost value to prevent it from being doubled in the solution. Using this modified graph, a tour from s to s is found. The edge (c, s) is then deleted from the graph and from the tour (line 11). The algorithm adjusts the coverage path to start at the current location and travel to the end location (line 12). The table below shows the details for adjusting the path. The terms on the left hand side indicates the four possible initial path configurations where s→c indicates an edge. The terms in the middle are the procedures for adjusting the specific path, and the terms on the right are the final paths returned by the coverage algorithm.

s...s→(c...s)	⇒	(c...s)→s...s	⇒	c...s...s
(s...c)→s...s	⇒	Reverse(s...c)→s...s	⇒	c...s...s
s→(c...s)	⇒	(c...s)→s	⇒	c...s
(s...c)→s	⇒	Reverse(s...c)→s	⇒	c...s

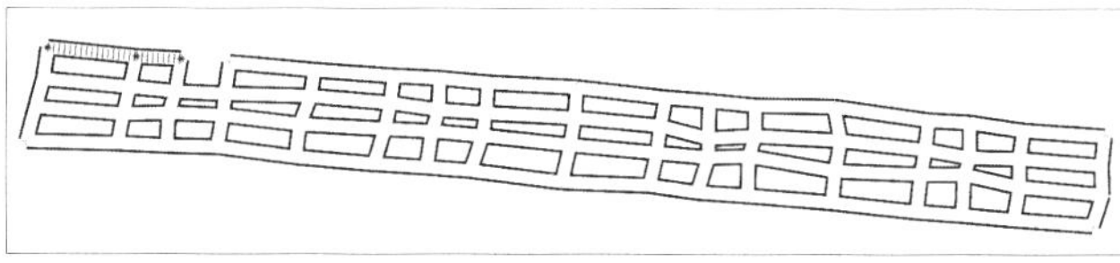

Fig. 3. During traversal of the CPP solution, the robot discovers that the third edge in the path is missing and ends the traversal. Dotted lines denote edges that are traversed once and solid lines denote edges that are traversed more than once.

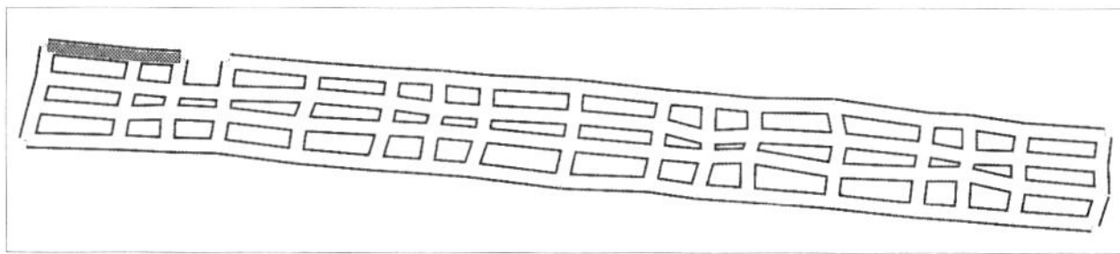

Fig. 4. To prevent visiting edges that have been already traversed, the graph is modified by setting the traversed edges to be travel edges which are represented by gray lines.

We now step through an example that illustrates the online coverage algorithm. In Figure 2(b), a CPP solution of the example graph is shown. In Figure 3, the vehicle traverses the initial CPP path until it discovers the third edge in its path is missing. The vehicle does not know about this change until it reaches the previous edge. The algorithm then sets the traversed edges in the path to be travel (Figure 4) and replans a new coverage tour shown in Figure 5. However, it encounters another missing edge as it travels along the new path. The traversed edges in the previous path are also converted to travel edges (Figure 6) and a final plan is found (Figure 7).

Algorithm 3: Online Coverage Algorithm

```
Input: s, start vertex,
       c, current vertex,
       G = (C, T), graph where each edge has a label and a
       cost value. C is the subset of coverage edges, and T is
       the subset of travel edges
       OTP, subset of OTPs
Output: P, tour found or empty if no tour found
1  G' = G;
2  if c ≠ s then
3  |  G' = [G, (c, s, INF)];
4  end
5  if IsConnected(C) then
6  |  P = CPP(s, G');
7  else
8  |  P = RPP(s, G', OTP);
9  end
10 if c ≠ s and P ≠ [] then
11 |  RemoveEdge(P, (c, s, INF));
12 |  AdjustPath(P, c, s);
13 end
14 return P;
```

D. Farthest Distance Heuristic

In our algorithm, when a robot encounters a change at a particular node, these changes are obstacles which prevent the robot from completing the current coverage solution. As a result, the problem needs to be re-solved with the previously

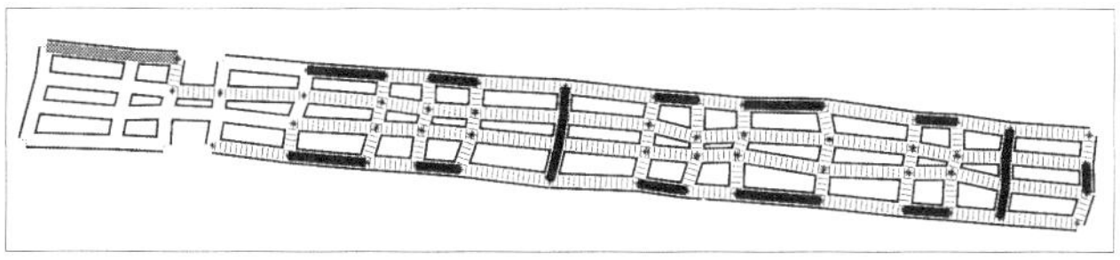

Fig. 5. The planner replans a new tour for the modified graph starting at the current location. During its new traversal, the robot discovers another edge in the path is missing and stops traversal.

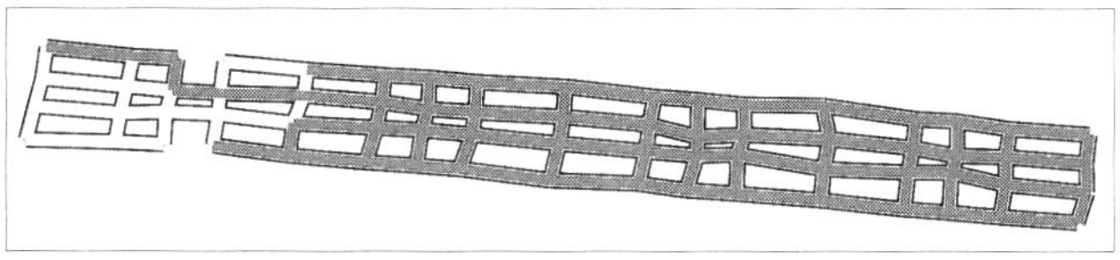

Fig. 6. For replanning, the traversed edges are again set to be travel edges, denoted by gray lines, in the graph.

traversed edges converted to travel edges. The number of optimal travel paths can be very large if the travel edges break the coverage subgraph into a large number of connected components. Therefore, to maintain efficiency, we want to keep the number of coverage clusters close to one. Ideally, we want the coverage subgraph and travel subgraph to be mutually independent. This would ensure that the required subgraph is connected, which is important for maintaining the coverage problem as a CPP, and maximize the likelihood that when the next change is detected, the travel edges do not disconnect the coverage components.

In the CPP algorithm, we use the End-Pairing technique to generate the Eulerian cycle from the graph. The algorithm consists of two steps. First, it builds cycles that intersect at at least one vertex. Next, the cycles are merged together two at a time by adding one cycle onto another at the intersecting vertex. The cycle building step is shown in Algorithm 4. During each step of the algorithm, edges are added to a path sequence and removed from the graph until the starting node of the path is encountered. In the original End-Pairing algorithm, the heuristic for choosing the next edge to add to the sequence consisted of picking a random edge incident to the current node.

To maintain a small coverage component set, we intuitively want to choose edges in such a way that the path travels away from the start and then travels back always visiting the farthest unvisited edges until it reaches the start. Essentially the coverage path should be always walking along the boundary of the coverage subgraph. This will allow the edges around the start to be as connected as possible while separating the coverage and travel subgraphs. This idea is translated into the Farthest Distance heuristic shown in equation 1 where s is the start node, i is the current node, $\{i, j\}$ is the set of edges incident to i, and D calculates the number of edges in the shortest path between s and j. To reduce computation time, we used D* Lite [23] to compute $D(s, j)$ since s is the same for each cycle generation step. In our CPP algorithm, we modified the End-Pairing algorithm to use the Farthest Distance heuristic to choose the next edge to add to the path.

$$e = argmax_{\forall\{i,j\}} D(s, j) \quad (1)$$

Algorithm 4: Cycle Building Algorithm

Input: s, start vertex,
G, graph
Output: C, cycle found

$C = [s]$;
$i = s$;
$e = NextEdgeHeuristic(G, s, i)$;
$i = OtherEndPoint(e, i)$;
while $i \neq s$ **do**
  $e = NextEdgeHeuristic(G, s, i)$;
  $i = OtherEndPoint(e, i)$;
  $C = [C; i]$;
  $RemoveEdge(G, e)$
end
return C;

III. Comparison Tests

The goal of our testing is to compare the Farthest Distance heuristic against the original heuristic of randomly selecting a neighboring edge as the next edge. For the tests, we vary four different parameters: traversal heuristic, graph, starting node s, and set of changes. At the beginning of the test, the graph is connected and has no travel edges. Using the CPP algorithm, a tour is computed. The test simulates a robot executing the tour starting at node s. If along the execution, the next edge to be traversed is in the change set, that edge is considered blocked. It is removed from the graph, and the graph is updated to reflect the visited edges and current vertex location. The updated graph is either a CPP or RPP, and the coverage algorithm calls the corresponding method to find a new solution. The execution is simulated again until another edge along the new path is found to be blocked or the traversal is completed.

We conducted two sets of tests. The first test set evaluated the two heuristics on rectilinear graphs, and the second set evaluated the heuristics on a real-world road network. We will first explain the procedure for each test set, and then present the metrics for comparing the two heuristics. The code ran on a machine with a 3.8GHz Intel Xeon processor with 5GB of RAM.

A. Rectilinear Graphs

Rectilinear graphs are graphs where the vertices were generated uniformly in the plane. Edges connect one vertex to its four closest neighbors, and are vertical and horizontal lines that

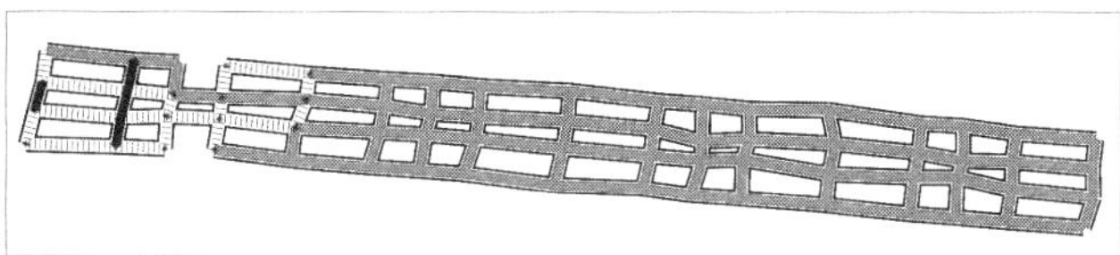

Fig. 7. The final path is replanned for the updated graph. This path starts at the node where the previous traversal ended.

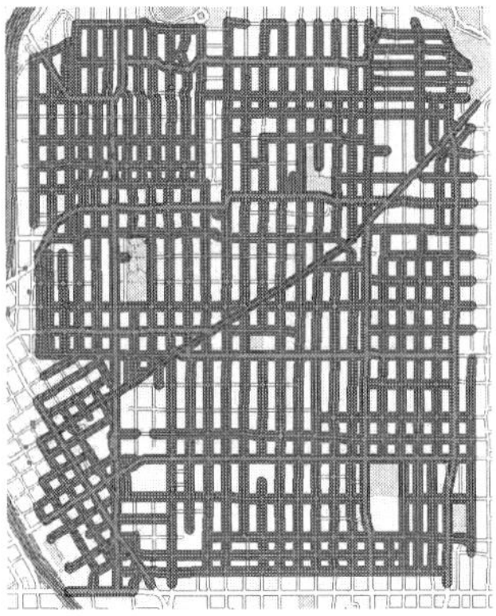

Fig. 8. This is the road network we used as our real-world example. This network covers an area of 1.5 miles by 2 miles. The edges and nodes included in our graph are highlighted in gray.

represent the Euclidean distance between the vertices. In other words, these graphs consist of a regular pattern of nodes and edges. We chose rectilinear graphs because they are similar to the networks used in many coverage applications. For example, in street coverage, the graph layout is similar to a grid where all streets (edges) meet at intersection points (nodes), and most intersections are four ways.

For our testing, three graph sizes were used; the sizes are $|V| = \{10 \times 10, 14 \times 14, 17 \times 17\}$. Five change sets were randomly generated for each of the three graphs. The change sets consisted of thirty different edges. We chose thirty because it is a significant number of changes (10% to 30% of the total edges in the graphs), but is a small enough number to keep computation low. For each of the graph-changeset combinations, we varied the starting node over all the nodes in the graph to avoid bias. Therefore, the coverage algorithm was called $3 \times 5 \times 30 \times |V|$ times for each of the two heuristics where $|V|$ is the number of nodes in the graph.

B. Real-world Example

The real-world example we used for testing is the road network of an urban neighborhood obtained from a dataset gathered by Newson and Krumm [24]. The network is shown in Figure 8. We converted the network into a graph with 764 vertices and 1130 edges. We generated five sets of changes for the data. Each change set consisted of five edges. Because the graph was so large, instead of varying the starting node over all nodes in the graph, we sampled a set of fifty distinct nodes for each change set. The samples were consistent for the two heuristics. In total, this test set yielded $5 \times 5 \times 50$ replans for the two heuristics.

C. Metrics

During each call to the coverage algorithm, the following items were measured: the number of connected coverage components in the graph, the number of optimal travel paths in the partition set, the number of branches in the search tree, the percentage of replanning calls that are CPP calls rather than RPP calls, and the computation time in seconds. Since the number of replanning calls was large, we placed a time limit on the branch-and-bound algorithm. The time limit was 70 seconds for rectilinear graphs and 100 seconds for the road network. If a replan was unable to produce a coverage path within the time limit, the particular test set failed.

IV. Results

Before presenting the results, we first highlight some aspects of the graphs to supplement the results. This information is shown in Table I. The first two columns display the number of nodes and edges in the graphs. The last two columns show the mean and standard deviation of the degrees of the vertices.

TABLE I

$\|V\|$	$\|E\|$	Mean Degree	Std Dev
10×10	180	3.6	0.57
14×14	364	3.7	0.50
17×17	544	3.8	0.47
764	1130	2.96	1.00

Results from the testing are presented in Tables II through V. The first table shows the average percentage of calls to the CPP algorithm over all trials, average computation time for successful trials, and the success percentage. The success percentage is the average number of successful replans over the total number of trials for each graph-heuristic combination. The second table shows the number of coverage components in the problem for each replan, number of OTPs when the number of components is greater than 1, and number of branches in the branch-and-bound tree for each RPP call. These values are computed only on the successful trials since the failed trials never return a solution path. Each column contains two numbers. The first number in the column represents the average for the metric and the second number represents the maximum.

A. Rectilinear Graphs

For the rectilinear graphs, the Farthest Distance heuristic performed a factor of two better than the random heuristic in percentage of CPP calls as shown in Table II. Using the heuristic, on average 96% of the replans resulted in graphs where the coverage subgraph was connected (number of connected components is one) and the lower complexity CPP was used. For the random heuristic, roughly half of the graphs were CPP graphs meaning half were the harder RPP problems. Furthermore, with our traversal heuristic, when the RPP algorithm did get called, the run time for all trials was never more than 70 seconds which results in 100% success. While the random heuristic did well on the 10×10 rectilinear trials with 84% of the replans successful, as the graph size got larger, the computation time for replans took longer and the success percentage went down to roughly 43%.

Looking at Table III, we can see that on average, the Farthest Distance heuristic keeps the number of connected coverage components smaller. Furthermore, when branch-and-bound is needed, the heuristic generates problems with a smaller set of OTPs. This smaller set translates to a shallower search tree. In terms of computation times, the Farthest Distance heuristic performs roughly 150 times better than the random heuristic.

B. Real-world Example

For the road network, the Farthest Distance heuristic performed more than a factor of two better than the random heuristic in maintaining connectivity among the coverage edges as shown in Table IV. When using the random heuristic, due to the high number of calls to the RPP algorithm, the run time on the majority of the trials exceeded the time limit. In contrast, the Farthest Distance heuristic had 98.4% success rate. Part of the reason for the drastic difference in successes was due to the large size of the graph (this graph is more than twice the size of the 17×17 graph). This resulted in the RPP algorithm needing more computation time. As shown in Table V, the number of connected components was similar for both heuristics. For the branch-and-bound data, we can see that the number of OTPs for the Farthest Distance heuristic was half the number of OTPs for the random heuristic which led to a smaller search tree.

C. Evaluation of the Coverage Algorithm

So far, we have shown comparison results regarding the Farthest Distance heuristic and the random heuristic. Next, we want to show some results of the coverage algorithm. From the road network data, we calculated the ratio of the travel edges over the total number of edges in the graphs for all the trials. The mean of the distribution was 0.44 with a standard deviation of 0.28. This denotes the travel set ranged from almost zero to almost the entire graph. Next, we calculated the size of the OTP sets that corresponded to each travel set (Figure 9). The ratio of the OTPs to travel edges has a mean of 0.08 with a standard deviation of 0.14. This indicates that our coverage algorithm dramatically reduced the partition set given to the branch-and-bound algorithm through the use of OTPs.

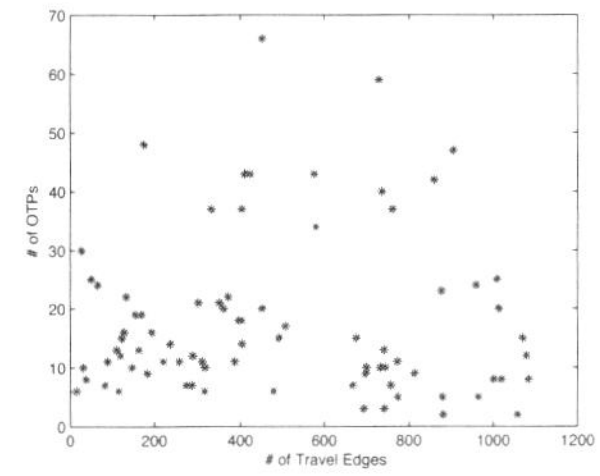

Fig. 9. This graph shows the correspondence between the number of travel edges, along the x-axis, and the number of OTPs, along the y-axis, for the road network data.

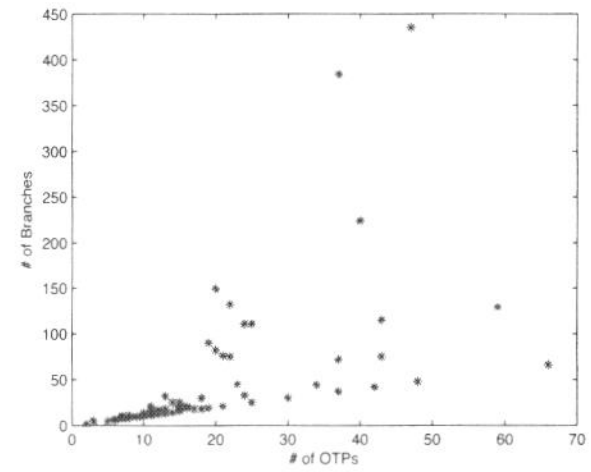

Fig. 10. This figure shows the relationship between the OTP set size, along the x-axis, and the number of branches in the search tree, along the y-axis, for the road network data.

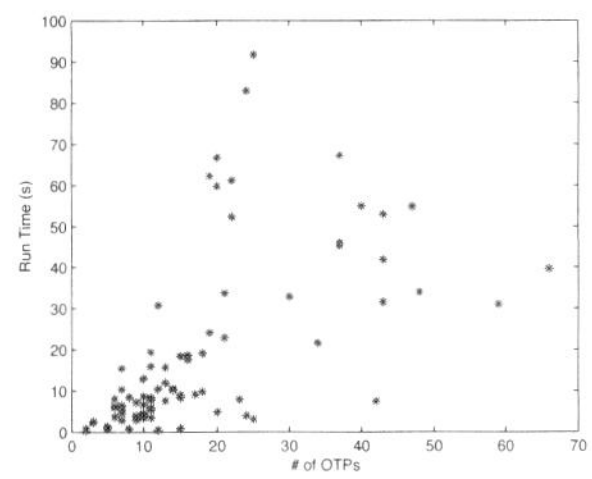

Fig. 11. This figure shows the relationship between the OTP set size, along the x-axis, and the computation time (shown in seconds), along the y-axis for the road network data.

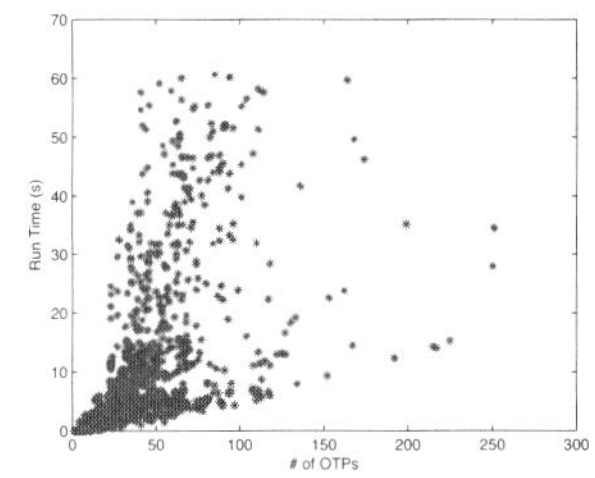

Fig. 12. This figure shows the relationship between the OTP set size, along the x-axis, and the computation time (shown in seconds), along the y-axis for the rectilinear graph data.

TABLE II

Computational results I for Rectilinear Graphs				
$\|V\|$	Heuristic	CPP Calls	Time(s)	Success Percentage
10×10	Random	46.67%	1.68	84.21%
	FarDist	92.19%	0.01	100.0%
14×14	Random	55.18%	3.02	51.08%
	FarDist	97.63%	0.02	100.0%
17×17	Random	59.59%	3.25	42.57%
	FarDist	98.66%	0.02	100.0%

TABLE III

Computational results II for Rectilinear Graph				
$\|V\|$	Heuristic	Components	OTPs	Branches
10×10	Random	2.11, 9.0	35.62, 153.2	352.45, 7823.8
	FarDist	1.08, 3.8	9.42, 38.8	15.86, 268.6
14×14	Random	1.67, 7.2	38.96, 204.4	276.03, 3773.6
	FarDist	1.03, 3.6	11.30, 36.6	24.42, 324.0
17×17	Random	1.52, 6.6	38.79, 244.2	182.52, 2135.2
	FarDist	1.01, 2.8	11.97, 30.4	18.66, 106.2

TABLE IV

Computational results I for Road Network graph				
$\|V\|$	Heuristic	CPP Calls	Time(s)	Success Percentage
764	Random	34.39%	12.63	30.40%
	FarDist	87.25%	1.14	98.40%

TABLE V

Computational results II for Road Network data				
$\|V\|$	Heuristic	Components	OTPs	Branches
764	Random	1.62, 4.0	23.59, 56.6	53.47, 326.4
	FarDist	1.14, 2.8	12.44, 40.0	18.65, 63.6

Next, we show how the smaller partition set affected the search tree and run time. From Figure 10, the maximum number of branches in the search tree was 435. Given that the largest number of travel edges was 1085, if we had used the travel edges as the partition set, the branch-and-bound problem corresponding to the maximum set would need to solve at least 1085 subproblems before finding the optimal solution. The smaller search tree translated into faster computation as evidenced in Figure 11. Similarly, the relationship between the number of OTPs and run time for the rectilinear graphs is shown in Figure 12.

D. Discussion

The results show that the Farthest Distance heuristic improves the percentage of CPP calls by a factor of two. The difference in the percentage between the rectilinear graphs and the road network is due to the lower degree of the vertices in the road network. As we can see from Figure 8, the road network is not a completely regular grid. There are intersections where more than four roads meet. Additionally, there are streets that dead end in one direction. One aspect of the graph that the heuristic relies on is that there is a path from the start node to another node j in the graph, and a path from j

back to the start. This assumption depends on path redundancy in the graph which may not hold for all graphs. For example, in a graph that contains bridges, the coverage problem will almost definitely become the more complex RPP when graph changes occur. This is due to the fact that converting a bridge edge to a travel edge will usually separate the start node and the remaining unvisited edges into different components. Hence, covering a graph with multiple bridges will limit the performance of not only the Farthest Distance heuristic, but also the random heuristic.

As shown in the results, the coverage algorithm greatly reduces the size of the partition set used to search for a solution. By limiting the branch-and-bound algorithm to only reason about the travel edges that make up the optimal path between boundary nodes, the faster CPP can be used to evaluate the remaining travel edges. For the RPP algorithm, the results show lower computation time on rectilinear graphs than for the road network graph. This is due to the smaller problem sizes of the rectilinear graphs. The larger problem size not only increases the run time of the CPP algorithm, but it also increases the time it takes to generate the OTP set.

V. Conclusion and Future Work

We present a general approach to the online coverage problem with an incomplete prior map. By representing the environment as a graph, our algorithm leverages the lower complexity Chinese Postman algorithm to produce optimal coverage solutions. This algorithm enables a faster run time by reducing the partition set to sequences of edges rather than single edges. Our algorithm is highly efficient when the number of OTP edges is small compared to the number of travel edges since it can exploit the faster CPP algorithm. This means a better connected graph can greatly decrease the computation time of our general coverage algorithm.

To improve the connectivity of the coverage graph when changes occur, we introduce a heuristic for generating the coverage path. Results from testing show that the Farthest Distance heuristic performs much better than the random heuristic in keeping the problem polynomial. One of the benefits of the Farthest Distance heuristic is that it is independent of our coverage algorithm. For example, it can be added to any coverage approach, either optimal or approximate, as a way to lower the complexity of the coverage problem when graph changes occur online.

For future work, we plan to test this method on a physical robot with real-time sensor updates to gather practical results of our approach. We also plan to extend this approach to directed and mixed graphs. Finally, we intend to investigate the potential of using the Farthest Distance heuristic on additional arc-routing problems.

Acknowledgment

This work was sponsored by the U.S. Army Research Laboratory, under contract Robotics Collaborative Technology Alliance (contract number DAAD19-01-2-0012). The views and conclusions contained in this document are those of the authors and should not be interpreted as representing the official policies, either expressed or implied, of the Army Research Laboratory or the U.S. Government.

References

[1] Howie Choset. Coverage for robotics - a survey of recent results. *Annals of Mathematics and Artificial Intelligence*, 31:113 – 126, 2001.

[2] Ercan U. Acar, Howie Choset, Alfred A. Rizzi, Prasad N. Atkar, and Douglas Hull. Morse decompositions for coverage tasks. *The International Journal of Robotics Research*, 21(4):331–344, 2002.

[3] Howie Choset, Sean Walker, Kunnayut Eiamsa-Ard, and Joel Burdick. Sensor-Based Exploration: Incremental Construction of the Hierarchical Generalized Voronoi Graph. *The International Journal of Robotics Research*, 19(2):126–148, 2000.

[4] E.U. Acar, H. Choset, and Ji Yeong Lee. Sensor-based coverage with extended range detectors. *IEEE Transactions on Robotics*, 22(1):189–198, Feb. 2006.

[5] S. Lin and B. W. Kernighan. An effective heuristic algorithm for the traveling-salesman problem. *Operations Research*, 21(2):498–516, 1973.

[6] David S. Johnson and Lyle A. Mcgeoch. The traveling salesman problem: A case study. In *Local Search in Combinatorial Optimization*. Wiley and Sons, 1997.

[7] H. A. Eiselt, Michel Gendreau, and Gilbert Laporte. Arc routing problems, part i: The chinese postman problem. *Operations Research*, 43(2):231–242, 1995.

[8] H. A. Eiselt, Michel Gendreau, and Gilbert Laporte. Arc routing problems, part ii: The rural postman problem. *Operations Research*, 43(3):399–414, 1995.

[9] M Guan. Graphic programming using odd and even points. *Chinese Mathematics*, 1:273–277, 1962.

[10] Christos H. Papadimitriou and Kenneth Steiglitz. *Combinatorial Optimization : Algorithms and Complexity*. Dover Publications, July 1998.

[11] C.G. Sorensen, T. Bak, and R.N. Jorgensen. Mission planner for agricultural robotics. *Agricultural Engineering International*, 2004.

[12] Gianpaolo Ghiana and Gilbert Laporte. A branch-and-bound algorithm for the Undirected Rural Postman Problem. *Mathematical Programming*, 87(3):467–481, 2000.

[13] Gilbert Laporte. Modeling and solving several classes of arc routing problems as traveling salesman problems. *Comput. Oper. Res.*, 24(11):1057–1061, 1997.

[14] G. N. Frederickson. Approximation algorithms for some routing problems. *Mathematical Programming*, 40:225–246, 1979.

[15] Alain Hertz, Gilbert Laporte, and Pierrette Nanchen Hugo. Improvement procedures for the undirected rural postman problem. *INFORMS J. on Computing*, 11(1):53–62, 1999.

[16] Luís M. Moreira, José F. Oliveira, A. Miguel Gomes, and J. Soeiro Ferreira. Heuristics for a dynamic rural postman problem. *Comput. Oper. Res.*, 34(11):3281–3294, 2007.

[17] Jack Edmonds and Ellis Johnson. Matching, Euler tours, and the Chinese Postman. *Mathematical Programming*, 5(1):88–124, 1973.

[18] Edward. Minieka. *Optimization algorithms for networks and graphs*. M. Dekker, New York :, 1978.

[19] Harold Neil Gabow. *Implementation of algorithms for maximum matching on nonbipartite graphs*. PhD thesis, Stanford, CA, USA, 1974.

[20] E. Rothberg. Implementation of H. Gabow's weighted matching algorithm (1992). ftp://dimacs.rutgers.edu/pub/netflow/matching/weighted/.

[21] E. L. Lawler and D. E. Wood. Branch-and-bound methods: A survey. *Operations Research*, 14(4):699–719, 1966.

[22] Illah Nourbakhsh and Michael Genesereth. Assumptive planning and execution: a simple, working robot architecture. *Autonomous Robots Journal*, 3(1):49–67, 1996.

[23] S. Koenig and M. Likhachev. Improved fast replanning for robot navigation in unknown terrain. In *Proceedings of the IEEE International Conference on Robotics and Automation*, volume 1, pages 968–975, 2002.

[24] Seattle road network data. http://research.microsoft.com/en-us/um/people/jckrumm/MapMatchingData/data.htm.

Stochastic Complementarity for Local Control of Discontinuous Dynamics

Yuval Tassa
Interdisciplinary Center for Neural Computation
Hebrew University
tassa@alice.nc.huji.ac.il

Emo Todorov
Computer Science and Engineering
University of Washington
todorov@cs.washington.edu

Abstract— **We present a method for smoothing discontinuous dynamics involving contact and friction, thereby facilitating the use of local optimization techniques for control. The method replaces the standard Linear Complementarity Problem with a Stochastic Linear Complementarity Problem. The resulting dynamics are continuously differentiable, and the resulting controllers are robust to disturbances. We demonstrate our method on a simulated 6-dimensional manipulation task, which involves a finger learning to spin an anchored object by repeated flicking.**

I. Introduction

Classic control methods focus on forcing a dynamical system to a reference trajectory. This approach is powerful but limited. For complex behaviors in underactuated domains, planning the desired trajectory cannot be easily separated from the control strategy. Optimal Control offers a comprehensive framework that solves both the planning and control problems by finding a policy which minimizes future costs. However, global methods for finding optimal policies scale exponentially with the state dimension, making them prohibitively expensive.

Local Optimal Control methods, or *trajectory optimizers*, play a key role in the search for nonlinear feedback controllers that are on par with biological motor systems. These methods find a solution in a small part of the state-space, but because their complexity scales polynomially, they may constitute the only feasible approach to tackling high-dimensional problems.

Locomotion and *hand-manipulation*, behaviors which encompass some of the most interesting control problems, are crucially dependent on contact and friction. These phenomena pose a problem for efficient variants of local methods, which require differentiability to some order. Hard contacts and joint limits are examples of dynamic discontinuities where velocities change instantaneously upon impact. Though these two phenomena can conceivably be modeled with stiff nonlinear spring-dampers, friction is inherently discontinuous and does not readily admit a smooth approximation.

If our goal is to produce an accurate simulation, it might be sensible to use a deterministic model of the dynamics, since macro-physical systems are often well-modeled as such. If however we wish to control the system, *noise* might qualitatively alter the optimal behavior. We argue that when the dynamics are discontinuous, an optimal controller must account for stochasticity in order to generate an acceptable policy. Because stochastic dynamics are inherently smooth (in the mean), differentiability issues automatically disappear.

A popular and established method for modeling hard unilateral constraints and friction, involves defining *complementarity conditions* on forces and velocities. Simulators based on this principle are called *time-stepping integrators* and solve a Linear Complementarity Problem at each time step. We propose to instead solve a Stochastic Linear Complementarity Problem, using the approach proposed in [1]. This method transforms the complementarity problem into a smooth nonlinear optimization which is readily solved.

The solution effectively describes new deterministic dynamics, that implicitly take into account noise around the contact. These modified dynamics can qualitatively be described as featuring a fuzzy "force-field" that extends out from surfaces, allowing both contact and friction to act at a distance. The size and shape of this layer are naturally determined by the noise distribution, without any free parameters.

We test our method on a simplified manipulation task. A two link planar finger must learn to spin an ellipse that is anchored to the wall by flicking at it. We use an off-line Model Predictive Control strategy to patch together a global, time-independent policy, which robustly performs the task for both the smoothed system and the original discontinuous one. We constructed an interactive simulation which allows the user to actively perturb the controlled system. The controller proved to be robust, withstanding all these disturbances.

II. A Qualitative Argument

Consider the simple and familiar task of holding an object. The force which prevents the object from falling is friction, related by Coulomb's law to the normal force exerted by the fingers. If we now slowly loosen our grip, there is no discernable change in the positions or velocities of the system until suddenly, when the the weight of the object penetrates the friction cone, sticking changes into slipping and the object drops from our hand.

In the Optimal Control context, the control signal realizes a tradeoff between a state-cost and a control-cost. Because the state-cost cannot change until the object begins to slip, a controller that is optimal with respect to deterministic dynamics will attempt to hold the object with the *minimum possible force*, i.e. on the very edge of the friction cone.

This delicate grip would be disastrously fragile – the smallest perturbation would cause the object to fall. If, however, there is uncertainty in our model, we can only maintain our grip *in probability*. By grasping with more force than is strictly necessary, we push probability-mass from the slip-regime into the stick-regime.

The moral is that optimal control of discontinuous dynamics cannot give acceptable results with a deterministic model. Our intuition of what constitutes a reasonable solution implicitly contains the notion of robustness, which requires the explicit modeling of noise near the discontinuities.

III. Background

A. Dynamics with unilateral contacts

The modeling and simulation of multibody systems with contacts and friction is a broad and active field of research [2]. One reasonable approach, not investigated here, is to model discontinuous phenomena with a continuous-time *hybrid* dynamical system which undergoes switching [3]. These methods require accurate resolution of collision and separation times, so fixed time-step integration with a fixed computational cost is impossible. Moreover, collision events can in principle occur infinitely often in a finite time, as when a rigid elastic ball bounces to rest. The appropriate control strategy would be to chop the trajectory into several first-exit problems, where contact surfaces serve as an exit manifold for one segment, while the post-collision state serves as an initial condition for the next one. It is not clear how a trajectory-optimizer for such a system could deal with unforeseen changes to the switch sequence, as when a foot makes grazing contact. Furthermore, such a scheme would mandate that every contact/stick-slip *configuration* be considered a hybrid system component, so their number would grow combinatorially.

Time-stepping avoids all of these problems. We give a brief exposition here and refer the interested reader to [4]. The equations of motion of a controlled mechanical system are

$$\mathbf{M}\dot{\mathbf{v}} = \mathbf{r} + \mathbf{u}$$
$$\dot{\mathbf{q}} = \mathbf{v}$$

where $\mathbf{q}$ and $\mathbf{v}$ are respectively the generalized coordinates and velocities, $\mathbf{M} = \mathbf{M}(\mathbf{q})$ is the mass matrix, $\mathbf{r} = \mathbf{r}(\mathbf{q}, \mathbf{v})$ the vector of total external forces (gravity, drag, centripetal, coriolis etc.) and $\mathbf{u}$ is the applied control signal (e.g. motor torques).

It is non-trivial to model discontinuous phenomena like rigid contact and Coulomb friction in continuous time. Doing so requires instantaneous momentum transfers via unbounded forces and the formalism of Measure Differential Inclusions to fully characterize solutions [5]. Time-stepping integrators rewrite the equations of motion in a discrete-time momentum-impulse formulation. Because the integral of force (the impulse) is always finite, unbounded quantities are avoided, and both *bounded* and *impulsive* forces can be properly addressed in one step.

For a timestep h, an Euler integration step for the momenta[1] $\mathbf{M}\mathbf{v}' = \mathbf{M}\mathbf{v}(t+h)$ and coordinates $\mathbf{q}' = \mathbf{q}(t+h)$ is:

$$\mathbf{M}\mathbf{v}' = h(\mathbf{r} + \mathbf{u}) + \mathbf{M}\mathbf{v}$$
$$\mathbf{q}' = \mathbf{q} + h\mathbf{v}'.$$

We want a unilateral constraint vector function $\mathbf{d}(\mathbf{q})$ to remain non-negative, for example a *signed distance* between objects, which is positive for separation, zero for contact and negative for penetration. We therefore demand that impulses λ be applied so that

$$\mathbf{d}(\mathbf{q}') \approx \mathbf{d}(\mathbf{q}) + h\mathbf{J}\mathbf{v}' \geq 0,$$

where $\mathbf{J} = \nabla\mathbf{d}(\mathbf{q})$. This leads to the following complementarity problem for $\mathbf{v}'$ and λ:

$$\mathbf{M}\mathbf{v}' = h(\mathbf{r} + \mathbf{u}) + \mathbf{M}\mathbf{v} + \mathbf{J}^{\mathsf{T}}\lambda \quad (1a)$$
$$\lambda \geq 0, \quad (1b)$$
$$\mathbf{d}(\mathbf{q}) + h\mathbf{J}\mathbf{v}' \geq 0, \quad (1c)$$
$$\lambda^{\mathsf{T}}(\mathbf{d}(\mathbf{q}) + h\mathbf{J}\mathbf{v}') = 0. \quad (1d)$$

Conditions (1b) and (1c) are read element-wise, and respectively constrain the contact impulse to be non-adhesive, and the distance to be non-penetrative. Condition (1d) asserts that $\mathbf{d}(\mathbf{q}') > 0$ (broken contact) and $\lambda > 0$ (collision impact), are mutually exclusive.

Since the mass matrix is always invertible, we can solve (1a) for $\mathbf{v}'$, and plug into (1c). Defining

$$\mathbf{A} = \mathbf{J}\mathbf{M}^{-1}\mathbf{J}^{\mathsf{T}} \quad (2a)$$
$$\mathbf{b} = \mathbf{d}(\mathbf{q})/h + \mathbf{J}\mathbf{v} + h\mathbf{J}\mathbf{M}^{-1}(\mathbf{r} + \mathbf{u}), \quad (2b)$$

we can now write (1) in standard LCP form:

$$\text{Find } \lambda \quad s.t. \quad 0 \leq \lambda \perp \mathbf{A}\lambda + \mathbf{b} \geq 0. \quad (3)$$

In order to solve for frictional impulses λ_f, the Coulomb friction law $\|\lambda_f\| \leq \mu\lambda$ must be incorporated. To maintain linearity, a polyhedral approximation to the friction cone is often used, though some new methods can handle a smooth one [7]. In the model of [8], the matrix $\mathbf{A}$ is no longer positive-definite, though a solution is guaranteed to exist. In the relaxed model of [9], $\mathbf{A}$ remains positive-definite. With either model the problem retains the LCP structure (3).

B. Smoothing methods for LCPs

The Linear Complementarity Problem (3) arises in many contexts and has stimulated considerable research [10][11]. One method of solving LCPs involves the use of a so-called NCP function $\phi(\cdot,\cdot)$ whose root satisfies the complementarity condition

$$\phi(a,b) = 0 \iff a \geq 0,\ b \geq 0,\ ab = 0.$$

Due to the element-wise nature of complementarity, is suffices to consider NCP functions with scalar arguments. Two popular examples are

$$\phi(a,b) = \min(a,b) \quad (4)$$

[1] In the derivation we assume $\mathbf{M}$ and $\mathbf{r}$ to be constant throughout the time-step, but for our simulations used a more accurate approximation [6].

and

$$\phi(a,b) = a + b - \sqrt{a^2+b^2}.$$

These functions reformulate the complementarity problem as a system of (possibly nonsmooth) nonlinear equations, whose residual can then be minimized. A particular method, closely related to the one used here, was presented by Chen and Mangasarian in [12] and proceeds as follows. Rewrite (4) as $\min(a,b) = a - \max(a-b, 0)$ and replace $\max(\cdot, 0)$ with a smooth approximation

$$s(x,\epsilon) \quad \rightarrow \quad \max(x,0) \quad \text{as} \quad \epsilon \downarrow 0.$$

One of several such functions proposed there is

$$s(x,\epsilon) = \epsilon \log(1 + e^{x/\epsilon}). \tag{5}$$

The authors then present an iterative algorithm whereby for a positive ϵ, the squared residual

$$r(a,b,\epsilon) = \big(a - s(a-b,\epsilon)\big)^2 \tag{6}$$

is minimized with a standard nonlinear minimization technique, ϵ is subsequently decreased, and the procedure repeated until a satisfactory solution is obtained.

C. Local optimization

Local Optimal Control methods, with roots in the venerable Maximum Principle [13], solve the control problem along a single *trajectory*, giving either an open-loop or a closed-loop policy that is valid in some limited volume around it. They are efficient enough to be used for real-time control of fast dynamics [14], and scale well enough to handle high-dimensional nonlinear mechanisms [15]. These methods can get stuck in local minima, though if neural controllers set the golden standard, it might be noted that biological suboptimal minima exist, somewhat anecdotally evidenced by the high-jump before Dick Fosbury. Their main detraction is that they solve the problem in only a small volume of space, namely around the trajectory. To remedy this, new methods [16] use Sum of Squares verification tools to measure the size of local basins-of-attraction and constructively patch together a global feedback controller. Finally, using the control-estimation duality, the powerful framework of estimation on graphical models is being brought to bear [17] on trajectory optimizers.

Though policy gradient methods like *shooting* can also be considered, we concentrate on Local Dynamic Programming, i.e. the construction of a local approximation to the Value function, and restrict ourselves to the finite-horizon case.

The control $\mathbf{u} \in \mathbb{R}^m$ affects the propagation of the state $\mathbf{x} \in \mathbb{R}^n$ through the general Markovian dynamics

$$\mathbf{x}' = \mathbf{f}(\mathbf{x},\mathbf{u}). \tag{7}$$

The cost-to-go starting from state $\mathbf{x}$ at time i with a control sequence $\mathbf{u}_{i:N-1} \equiv \{\mathbf{u}_i, \mathbf{u}_{i+1} \ldots, \mathbf{u}_{N-1}\}$, is the sum of running costs[2] $\ell(\mathbf{x},\mathbf{u})$ and final cost $\ell_f(\mathbf{x})$:

$$J_i(\mathbf{x}_i, \mathbf{u}_{i:N-1}) = \sum_{k=i}^{N-1} \ell(\mathbf{x}_k, \mathbf{u}_k) + \ell_f(\mathbf{x}_N),$$

[2]The possible dependence of ℓ on i is suppressed for compactness.

We define the optimal Value function at time i as the the cost-to-go given the minimizing control sequence

$$V_i^*(\mathbf{x}) \equiv \min_{\mathbf{u}_{i:N-1}} J_i(\mathbf{x}, \mathbf{u}_{i:N-1}).$$

Setting $V_N^*(\mathbf{x}) \equiv \ell_f(\mathbf{x}_N)$, the Dynamic Programming principle reduces the minimization over the entire sequence of controls to a sequence of minimizations over a single control, proceeding backwards in time:

$$V_i^*(\mathbf{x}) = \min_{\mathbf{u}}[\ell(\mathbf{x},\mathbf{u}) + V_{i+1}^*(\mathbf{f}(\mathbf{x},\mathbf{u}))] \tag{8}$$

1) First-Order Dynamic Programming: To derive a discrete-time equivalent of the Maximum Principle, we observe the following: Given a first-order approximation of the Value at $i+1$, if $\mathbf{f}$ is affine in $\mathbf{u}$ (which holds for mechanical systems) and ℓ is convex and smooth in $\mathbf{u}$ (so that $\nabla_{\mathbf{u}}\ell$ is invertible), then the minimizing $\mathbf{u}$ is given by:

$$\mathbf{u}_i^* = -\nabla_{\mathbf{u}}\ell^{-1}\left(\nabla_{\mathbf{u}}\mathbf{f}^{\mathsf{T}}\nabla_{\mathbf{x}}V_{i+1}\right) \tag{9}$$

with dependencies on $\mathbf{x}$ and $\mathbf{u}$ suppressed for readability. Once $\mathbf{u}_i^*$ is known, the approximation at time i is given by

$$\nabla_{\mathbf{x}}V_i(\mathbf{x}) = \nabla_{\mathbf{x}}\big(\ell(\mathbf{x},\mathbf{u}_i^*) + V_{i+1}(\mathbf{f}(\mathbf{x},\mathbf{u}_i^*))\big). \tag{10}$$

The first-order local dynamic programming algorithm proceeds by alternatingly propagating the dynamics forward with (7), and propagating $\mathbf{u}_i$ and $\nabla_{\mathbf{x}}V_i(\mathbf{x})$ backward with (9) and (10).

2) Second-Order Dynamic Programming: By propagating a quadratic model of $V_i(\mathbf{x})$, second-order methods can compute locally-linear policies. These provide both quadratic convergence rate and a more accurate, closed-loop controller. We define the *unminimized* Value function

$$Q_i(\delta\mathbf{x}, \delta\mathbf{u}) = \ell(\mathbf{x}+\delta\mathbf{x}, \mathbf{u}+\delta\mathbf{u}) + V_{i+1}(\mathbf{f}(\mathbf{x}+\delta\mathbf{x}, \mathbf{u}+\delta\mathbf{u}))$$

and expand to second order

$$\approx \frac{1}{2}\begin{bmatrix} 1 \\ \delta\mathbf{x} \\ \delta\mathbf{u} \end{bmatrix}^{\mathsf{T}} \begin{bmatrix} Q_0 & Q_x{}^{\mathsf{T}} & Q_u{}^{\mathsf{T}} \\ Q_x & Q_{xx} & Q_{xu} \\ Q_u & Q_{ux} & Q_{uu} \end{bmatrix} \begin{bmatrix} 1 \\ \delta\mathbf{x} \\ \delta\mathbf{u} \end{bmatrix}. \tag{11}$$

Solving for the minimizing $\delta\mathbf{u}$ we have

$$\delta\mathbf{u}^* = \operatorname*{argmin}_{\delta\mathbf{u}}[Q_i(\delta\mathbf{x},\delta\mathbf{u})] = -Q_{uu}^{-1}(Q_u + Q_{ux}\delta\mathbf{x}), \tag{12}$$

giving us both open-loop and linear-feedback control terms. This control law can then be plugged back into (11) to obtain a quadratic approximation of V_k. As in the first-order case, these methods proceed by alternating a forward pass which propagates the dynamics using the current policy, and a backward pass which re-estimates the Value and produces a new policy.

D. Optimal Control of discontinuous systems

To date, there has not been a profusion of research on Optimal Control of discontinuous dynamics. Stewart and Anitescu's recent contribution [18] includes a literature review. In that paper, the authors describe a method whereby a smooth approximation replaces the discontinuities. They show that in some limit, the solution of the smooth system converges to the solution of the nonsmooth one, and proceed to apply their method to the "Michael Schumacher" racing car problem. We would argue that though a solution is obtained, it might be a solution to the wrong problem. Qualitatively, it reaches the very limits of tire traction, and passes within a whisker of the walls. The smallest disturbance would send the car crashing. Their example is appropriate since this extreme driving well-describes the car racing profession, but the optimum is clearly non-robust. A driver who thinks that the road is slippery or the steering wheel is inaccurate, would most likely be considered a "better driver" by other standards.

The method presented in the next section describes a specific way of accounting for stochasticity when controlling dynamics with complementarity conditions, but in general, noise has a profound effect on optimal solutions. For deterministic continuous-time systems, even infinitely differentiable ones, the Value function, which satisfies the Hamilton Jacobi Bellman PDE, is often discontinuous. A solution always exists in the *Viscosity Solution* sense [19], but the fact that a special formalism is required is conspicuous. If, however, the dynamics are a stochastic diffusion with positive-definite noise covariance, the HJB equation gains a second-order Itô term, and the solution is always unique and smooth, regardless of discontinuities in the underlying dynamics.

IV. The Proposed Method

Instead of solving for contact impulses with an LCP, we propose instead to solve a Stochastic LCP. The incorporation of stochasticity makes the contact impulses a differentiable function of the state, facilitating the use of local control methods. Because a controller must always overcome noise (process, observation, modeling), the noise-induced smooth dynamics constitute a better model for control purposes, even if they are a worse approximation of the real physical mechanism.

A. New and old formulas for SLCPs

The study of Stochastic Linear Complementarity Problems is a fairly recent endeavor, see [20] for a survey. The general problem could be written

$$0 \leq \mathbf{x} \ \perp\ \mathbf{A}(\omega)\mathbf{x} + \mathbf{b}(\omega) \geq 0. \tag{13}$$

Where ω is random variable. In this form the problem is obviously not well-posed, since it is not clear in what sense $\mathbf{x}$ satisfies the constraints. The Expected Residual Minimization approach of [1], proposes that we minimize

$$r_{\text{ERM}}(\mathbf{x}) - \mathbf{E}\left[\|\Phi(\mathbf{x},\omega)\|^2\right] \tag{14}$$

where

$$\Phi(\mathbf{x},\omega)_i = \phi\Big(\mathbf{x}_i,\ \big(\mathbf{A}(\omega)\mathbf{x} + \mathbf{b}(\omega)\big)_i\Big)$$

is a vector of NCP residuals.

As detailed in section IV-B, in our case $\mathbf{A}$ can be considered constant to first order. This variant is discussed in section 3 of [1], where after a simple proof that $r_{\text{ERM}}(\mathbf{x})$ is continuously differentiable, it is explicitly computed for a uniformly distributed $\mathbf{b}$. Because we are ultimately trying to model a diffusion, the Normal distribution would be more appropriate. Considering scalar arguments a and $b_{\mathcal{N}}$, with

$$b_{\mathcal{N}} \sim p(b_{\mathcal{N}}) = \mathcal{N}(b_{\mathcal{N}}|b,\sigma) = \frac{1}{\sigma\sqrt{2\pi}}\exp\left(-\tfrac{1}{2}\left(\tfrac{b_{\mathcal{N}}-b}{\sigma}\right)^2\right)$$

and cumulative distribution

$$P(b_{\mathcal{N}}) = \int_{-\infty}^{b_{\mathcal{N}}} p(t)dt = \frac{1}{2}\left(1 + \operatorname{erf}\left(\tfrac{b_{\mathcal{N}}-b}{\sigma\sqrt{2}}\right)\right),$$

the expectation produces

$$\mathbf{E}\left[\min(a,b_{\mathcal{N}})^2\right] = \\ a^2 - \sigma^2(a+b)p(a) + \left(\sigma^2 + b^2 - a^2\right)P(a). \tag{15}$$

A possible variant of equation (14), would be to exchange squaring and expectation, giving what might be termed Residual Expectation Minimization:

$$r_{\text{REM}}(\mathbf{x}) = \|\,\mathbf{E}\left[\Phi(\mathbf{x},\omega)\right]\|^2 \tag{16}$$

Due to Jensen's inequality, $r_{\text{REM}}(\mathbf{x}) \leq r_{\text{ERM}}(\mathbf{x})$, and is thus a weaker bound, though their minima coincide as $\sigma \downarrow 0$. A benefit of this residual is that the expectation gives simpler formulae. In particular, if $b_{\mathcal{L}}$ has a Logistic distribution

$$b_{\mathcal{L}} \sim \mathcal{L}(b_{\mathcal{L}}|b,\sigma) = \frac{1}{\sigma}\Big(\exp\left(\tfrac{b_{\mathcal{L}}-b}{2\sigma}\right) + \exp\left(\tfrac{b-b_{\mathcal{L}}}{2\sigma}\right)\Big)^{-2}$$

then the expectation produces

$$\mathbf{E}[\min(a,b_{\mathcal{L}})] = a - \sigma\log(1 + e^{\frac{a-b}{\sigma}}). \tag{17}$$

It is clear that this residual is identical to (6) with the smoothing function (5), where σ takes the place of ϵ. This immediately provides us with a new interpretation to the method of [12], and gives us access to the literature which investigates it (e.g. [21]).

With these smooth approximations[3] to ϕ, the minimizing $\mathbf{x}$ is now a differentiable function of $\mathbf{A}$ and $\mathbf{b}$.

[3]Although we only used formulae (15) and (17) in our experiments, for completeness we also provide the expectation of the $\min(\cdot,\cdot)$ function with a Normally distributed argument:

$$\mathbf{E}[\min(a,b_{\mathcal{N}})] = a - \sigma^2 p(a) - (a-b)P(a),$$

and of $\min(\cdot,\cdot)^2$ with a Logistically distributed argument:

$$\mathbf{E}[\min(a,b_{\mathcal{L}})^2] = a^2 - 2a\sigma\log(1+e^{\frac{a-b}{\sigma}}) + 2\sigma^2\operatorname{Li}_2(-e^{\frac{a-b}{\sigma}}),$$

where $\operatorname{Li}_2(x)$ is the Dilogarithm function $\operatorname{Li}_2(x) = -\int_x^0 \frac{\log(1-t)}{t}dt$.

B. Determining the noise covariance

How should the noise covariance σ be determined? Assume that a gaussian state distribution is estimated from observations by some filter, and we are given

$$\Sigma_{\mathbf{q}} = \mathbf{E}[\mathbf{q}\mathbf{q}^\mathsf{T}] \quad \text{and} \quad \Sigma_{\mathbf{v}} = \mathbf{E}[\mathbf{v}\mathbf{v}^\mathsf{T}].$$

Examining (2a) we see that $\mathbf{A}$ is a function of the mass matrix $\mathbf{M}$ and the constraint Jacobian $\mathbf{J}$. Both of these depend on $\mathbf{q}$ but often smoothly and slowly. In contrast, since h must be small due to the Euler integration, its presence in the denominator of the first term of the RHS of (2b), suggests that the appropriate first-order approximation is

$$\Sigma_{\mathbf{b}} = \mathbf{E}[\mathbf{b}\mathbf{b}^\mathsf{T}] \approx \mathbf{J}\Sigma_{\mathbf{q}}\mathbf{J}^\mathsf{T} h^{-2}$$

To account for the second term as well, one would use

$$\Sigma_{\mathbf{b}} \approx \mathbf{J}(\Sigma_{\mathbf{q}} h^{-2} + \Sigma_{\mathbf{v}})\mathbf{J}^\mathsf{T}.$$

The element-wise nature of the complementarity conditions allows us to ignore the off-diagonal terms and use

$$\mathbf{b}(\omega)_i \sim \mathcal{N}(\omega|\mathbf{b}_i, \sqrt{(\Sigma_{\mathbf{b}})_{ii}}) \quad \text{or} \quad \mathcal{L}(\omega|\mathbf{b}_i, \sqrt{(\Sigma_{\mathbf{b}})_{ii}}), \tag{18}$$

to define the vector residual as

$$\Phi(\mathbf{x}, \omega)_i = \phi\Big(\mathbf{x}_i,\ \mathbf{A}_i\mathbf{x} + \mathbf{b}(\omega)_i\Big).$$

C. Making global controllers from trajectories

The output of the second-order trajectory optimizer of section III-C.2 is a time-dependent sequence of linear feedback policies $\mathbf{u}()_{1:N-1}$

$$\mathbf{u}(\mathbf{x})_i = \mathbf{u}_i - Q_{i\,uu}^{-1} Q_{i\,ux}(\mathbf{x} - \mathbf{x}_i).$$

Ideally, we would like to use an online Model Predictive Control strategy, whereby we iteratively use $\mathbf{u}()_1$ for one time-step and re-solve the problem. Our simulations were not fast enough for that (see below), so we resort to an off-line MPC strategy, as in [15]. Once a solution trajectory is obtained for some $\mathbf{x}_1$, we save $\mathbf{u}()_1$, and proceed to solve for a new trajectory starting at $\mathbf{x}_{1+d}$ for a small offset d. We can use the previous control sequence shifted by $-d$ and appended with d copies of the last control

$$\mathbf{u}^{\text{NEW}}()_{1:N-1} = [\mathbf{u}()_{1+d:N-1}\ \ \mathbf{u}()_{N-1}\ \ \mathbf{u}()_{N-1}\cdots], \tag{19}$$

as our initial guess for the policy of the shifted trajectory. Because we are usually not far from the optimum, we enjoy the quadratic convergence properties of second-order methods. The trajectory thus propagates forward, leaving behind it a trail of time-independent controllers $\mathbf{u}()_k$, all with horizon $T = hN$. We now use this collection of local linear controllers to construct a global controller that can be used online, by following a simple nearest neighbor rule:

$$\mathbf{u}(\mathbf{x}) = \mathbf{u}(\mathbf{x})_j \quad \text{with} \quad j = \operatorname*{argmin}_k \|\mathbf{x} - \mathbf{x}_k\|^2. \tag{20}$$

This is similar to the Trajectory Library concept [22]. If, as in the case below, the solution lies on a limit cycle, the time-independent trajectory will converge to it. Now we can perturb the initial state $\mathbf{x}_1$ every several time steps, and explore the state space around the limit cycle.

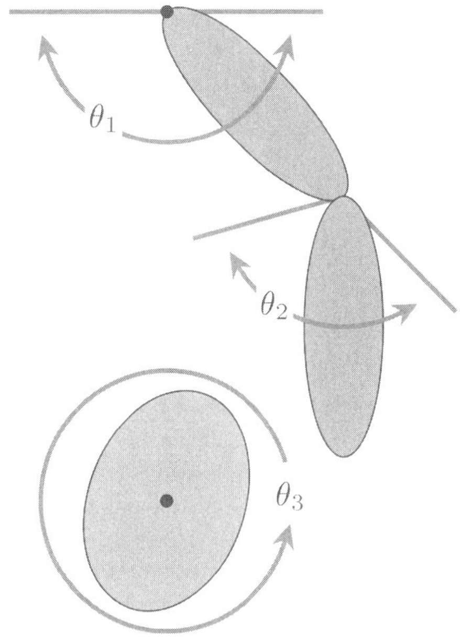

Fig. 1. The "flicking finger" dynamical system. The controller actuates θ_1 and θ_2 in order to spin the free ellipse around θ_3.

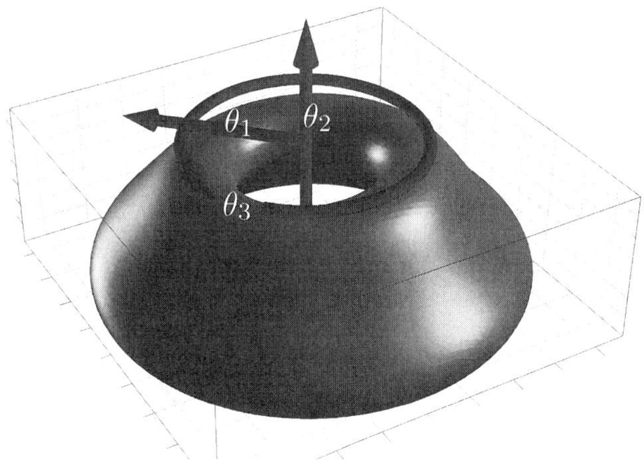

Fig. 2. The contact surface in the configuration space $[\theta_1\ \theta_2\ \theta_3]^\mathsf{T}$. The torus-like shape is the zero-valued isosurface of the distance function $\mathbf{d}(\mathbf{q})$. It corresponds to the set of points where the finger-tip makes contact with the free ellipse. Points inside and outside the surface correspond to penetration and broken contact respectively. A cylindrical coordinate system was chosen because θ_3 is a periodic variable (with period π rather than 2π due to symmetry).

V. Experiments

A. Setup

We performed our experiments with the 6-dimensional system of Figure 1. This planar system is composed of two articulated ellipses, with angles θ_1 and θ_2, and a free-spinning ellipse, whose angle is given by θ_3. The full state of the system is thus

$$\mathbf{x} = [\theta_1\ \theta_2\ \theta_3\ \dot\theta_1\ \dot\theta_2\ \dot\theta_3]^\mathsf{T}$$

The controller can apply two torques $\mathbf{u} = [u_1\ \ u_2]^\mathsf{T}$ to θ_1 and θ_2. We make it the controller's goal to spin the free ellipse in the positive direction, by defining a negative state-cost proportional to θ_3 with a quadratic control-cost:

$$\ell(\mathbf{x}, \mathbf{u}) = -c_x\theta_3 + c_u\|\mathbf{u}\|^2$$

Note that it would be very difficult to solve this task with control techniques that force the system to a pre-planned trajectory, since it is not clear how such a trajectory would be found.

In Figure 2, we depict the $\mathbf{d}(\mathbf{q}) = 0$ isosurface in the configuration space $[\theta_1\ \theta_2\ \theta_3]^\mathsf{T}$.

B. Parameters and methods

The following values can all be assumed to have the appropriate units of a self consistent unit system (e.g. MKS). The major and minor radii of the finger ellipses are .8 and .25 respectively. Those of the free ellipse are .7 and .5. The masses and moments of inertia correspond to a mass density of 1. The vertical distance between the two anchors (black dots in Figure 1) is 3. Angle limits are $-\pi \leq \theta_1 \leq 0$ and $-2\pi/3 \leq \theta_2 \leq 0$. The drag coefficients of the finger joints and of the free ellipse axis are .2, .2 and .7, respectively. Gravity in the vertical direction is -4. The control-cost coefficients were $c_u = 0.05$ and $c_x = 1$. The time step $h = 0.05$ and the number of time steps per trajectory $N = 75$, for a time horizon of $T = 3.5$.

Both angle constraints and the contact constraint were satisfied with impulses, as described in section III-A. We used a friction coefficient of $\mu = 0.5$, with the friction model of [9]. Since we did not estimate the dynamics, we used a constant $\sqrt{(\Sigma_\mathbf{b})_{ii}} = 0.1$, which was chosen as a trade-off between good convergence and small softening of the contact (described below). For the SLCP residual, both (15) and (17) gave qualitatively similar results, and we used (17) in the results below, because it is computationally cheaper.

We avoided the matrix inversions of (2) by directly solving the (stochastic) mixed complementarity problem of (1). This is easy to do with NCP functions, by setting $\phi(a, b) = b$ for those indexes where equality is desired.

We propagated the receding-horizon trajectory 400 times, with an offset $d = 4$, giving us a library of 400 locally linear feedback controllers. Every 15 time-steps, the velocities of initial state were perturbed by normal noise of variance 4. The reason only the velocities were perturbed is that random perturbations of the angles might lead to illegal (penetrative) configurations. When selecting the nearest neighbor of equation (20), we used a Euclidian norm scaled by the covariance of all the $\mathbf{x}_i$. In Figure 3, we plot the locations of the origins $\mathbf{x}_i$, projected onto the first 3 dimensions of the state-space.

Our simulator[4] was written in MATLAB for versatility and code readability, but is therefore not very efficient. One forward-backward pass of the local method took $\sim 5_s$ on a quad-core Core i7 machine. Starting from the zero policy, convergence was attained in ~ 60 passes. In receding-horizon mode, starting from the shifted previous policy (19), ~ 8 passes sufficed, due to quadratic convergence.

[4]Available online at http://alice.nc.huji.ac.il/~tassa/

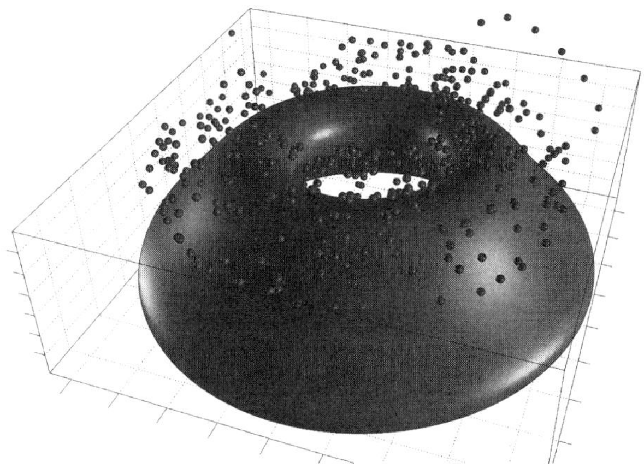

Fig. 3. The trajectory library. The dots correspond to the origins of 400 linear feedback controllers which we select with the nearest-neighbor rule (20). The global policy is effectively a Voronoi tessellation of these policies.

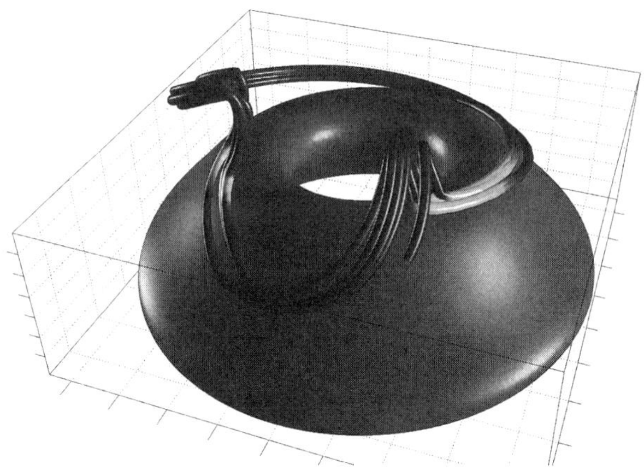

Fig. 4. The limit-cycle of a controlled trajectory. The color of the trajectory is proportional to the contact impulses, showing the points when flicking occurs.

C. Results

Our use of the SLCP when solving for velocities and impulses has the qualitative effect of creating a "force field" which extends from surfaces and lets contact and friction impulses act at a distance. If we could simulate the true stochastic system, some probability mass would experience contact before the mean would. The "fuzzy" force layer in our new deterministic system is an approximation to the mean force that would act on the mean of the distributed state.

The controller that was synthesized is very robust. We tested it on both the smoothed system and the original non-smooth one. For both systems, the controller dependably spun the ellipse, and recovered from disturbances. As an indicator of robustness, the controller worked well for time-steps different than the one which it was trained. We built a real-time simulator with a graphical front-end, which allows the user to perturb the system as it is being controlled, by interactively pulling on the ellipses with the mouse cursor. We were unable to perturb the system out of the controller's basin-of-attraction.

In order to assess the contribution of the smoothing to robustness, we solved again, with a smaller noise covariance

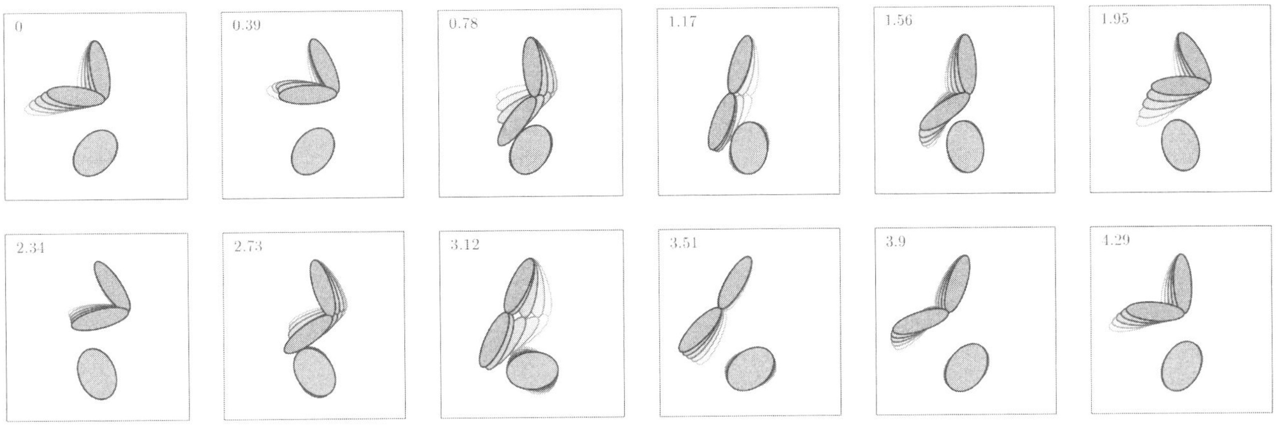

Fig. 5. Animation frames from the limit cycle. The third frame (time=0.78) and eighth frame (time=2.73) correspond to the contacts on the left and right side of Figure 4, respectively.

$\sqrt{(\Sigma_{\mathbf{b}})_{ii}} = 0.05$. The resulting controller was less robust in all of the senses described above. In particular, this controller sometimes entered into a non-productive limit-cycle where no contact was achieved. For smaller noise covariances the trajectory optimizer did not converge, so a direct comparison was not possible.

In Figure 4, we show the limit-cycle that a controlled trajectory makes in configuration space, with $h = 0.03$. The dynamics used were those of the original, nonsmooth system. The color of the trajectory is proportional to the contact impulses, showing the points when flicking occurs. The contact on the left of the figure corresponds to a weak tap that repositions the ellipse at a favorable angle, while the second one on the right side, exploits this position to deliver a stronger flick, that spins the ellipse. In Figure 5 we show animation frames from this sequence. The first and second contacts correspond to frame 3 (t=0.78) and frame 8 (t=2.73).

VI. Discussion

Our purpose here was to investigate the applicability of local methods to nonlinear control problems involving contact and friction. Our proposed method involves modifying the contact solver of the modeled dynamics, in a way which takes into account the controller's uncertainty regarding the state of the system. Though we present promising preliminary results that include a working robust controller, open issues remain.

Folding the noise into the dynamics, as we do here, is an approximation of what we would really like to do, namely simulate the full stochastic system, and measure costs WRT distributions rather than points. This option however has its own problems. The actual distributions which are propagated by the true dynamics become multimodal upon contact and require either a nontrivial parameterization, or a large number of "particles" for a non-parametric representation. In the case where distributions are represented as a mixture of samples, whether particles or σ-points, the dynamics would still be non-differentiable.

Local minima are still a significant problem for local methods like the ones used here. If we had not included gravity in the simulation, the initial trajectory with $\mathbf{u}_i = 0$ would have never achieved contact with the free ellipse, and the controller wouldn't have "known" about the possibility of contact. One option is to inject noise into the system during the initial stages of learning, for exploration purposes. Another option is to use the action-at-a-distance effect of the SLCP solution. By first solving for a system with very large postulated noise covariance, and then gradually reducing it, a scaffolding effect might be achieved.

The SLCP solution effectively replaces hard contact with a nonlinear spring-damper. However, unlike some arbitrary spring, it does not require setting the unknown parameters (spring and damper coefficients, form of the nonlinearity), and is derived in a principled way from the noise covariance. Additionally, unlike conventional springs which deform in proportion to the applied force, the forces which we compute scale with the effective inertia, so that a heavy and a light object experience the same smoothing.

A non-differentiable distance function $\mathbf{d}(\mathbf{q})$ would result in non-differentiable dynamics. For the ellipses used here, the signed distance is indeed differentiable, but this is not true for other shapes.

The complementarity conditions in (1) apply for quantities of different units: λ is an impulse while $\mathbf{d}(\mathbf{q}')$ is a distance. Because the smoothed NCP function $\phi(\cdot,\cdot)$ effectively mixes these quantities, a different choice of units would ostensibly lead to different results, which is clearly undesirable. This is a fundamental issue that requires further investigation. As famously observed by Stewart [4], "in many ways it is easier to write down a numerical method for rigid-body dynamics than it is to say exactly what the method is trying to compute".

Acknowledgment

This work was supported by the US National Science Foundation.

References

[1] X. Chen and M. Fukushima, "Expected residual minimization method for stochastic linear complementarity problems," *Math. Oper. Res.*, vol. 30, no. 4, pp. 1022–1038, 2005.

[2] F. Pfeiffer and C. Glocker, *Multibody dynamics with unilateral contacts*. Wiley-VCH, 1996.

[3] E. Westervelt, *Feedback control of dynamic bipedal robot locomotion*. Boca Raton: CRC Press, 2007.

[4] D. E. Stewart, "Rigid-Body dynamics with friction and impact," *SIAM Review*, vol. 42, no. 1, pp. 3–39, 2000.

[5] J. J. Moreau, "Unilateral contact and dry friction in finite freedom dynamics," *Nonsmooth mechanics and Applications*, p. 1–82, 1988.

[6] F. A. Potra, M. Anitescu, B. Gavrea, and J. Trinkle, "A linearly implicit trapezoidal method for integrating stiff multibody dynamics with contact, joints, and friction," *International Journal for Numerical Methods in Engineering*, vol. 66, no. 7, pp. 1079–1124, 2006.

[7] E. Todorov, "Implicit nonlinear complementarity: a new approach to contact dynamics," in *International Conference on Robotics and Automation*, 2010.

[8] M. Anitescu and F. A. Potra, "Formulating dynamic Multi-Rigid-Body contact problems with friction as solvable linear complementarity problems," *Nonlinear Dynamics*, vol. 14, no. 3, pp. 231–247, Nov. 1997.

[9] M. Anitescu, "Optimization-based simulation of nonsmooth rigid multibody dynamics," *Mathematical Programming*, vol. 105, no. 1, pp. 113–143, 2006.

[10] R. W. Cottle, J. Pang, and R. E. Stone, *The Linear Complementarity Problem*. SIAM, Oct. 2009.

[11] S. C. Billups and K. G. Murty, "Complementarity problems," *Journal of Computational and Applied Mathematics*, vol. 124, no. 1-2, pp. 303–318, Dec. 2000.

[12] C. Chen and O. L. Mangasarian, "A class of smoothing functions for nonlinear and mixed complementarity problems," *Computational Optimization and Applications*, vol. 5, no. 2, pp. 97–138, Mar. 1996.

[13] L. S. Pontryagin, V. G. Boltyanskii, R. V. Gamkrelidze, and E. F. Mishchenko, *The mathematical theory of optimal processes*. Interscience New York, 1962.

[14] P. Abbeel, A. Coates, M. Quigley, and A. Y. Ng, "An application of reinforcement learning to aerobatic helicopter flight," in *Advances in Neural Information Processing Systems 19: Proceedings of the 2006 Conference*, 2007, p. 1.

[15] Y. Tassa, T. Erez, and W. Smart, "Receding horizon differential dynamic programming," in *Advances in Neural Information Processing Systems 20*, J. Platt, D. Koller, Y. Singer, and S. Roweis, Eds. Cambridge, MA: MIT Press, 2008, p. 1465–1472.

[16] R. Tedrake, "LQR-Trees: feedback motion planning on sparse randomized trees," in *Proceedings of Robotics: Science and Systems (RSS)*, 2009.

[17] M. Toussaint, "Robot trajectory optimization using approximate inference," in *Proceedings of the 26th Annual International Conference on Machine Learning*, 2009.

[18] D. E. Stewart and M. Anitescu, "Optimal control of systems with discontinuous differential equations," *Numerische Mathematik*, 2009.

[19] W. H. Fleming and H. M. Soner, *Controlled Markov processes and viscosity solutions*. Springer Verlag, 2006.

[20] M. Fukushima and G. lin, "Stochastic equilibrium problems and stochastic mathematical programs with equilibrium constraints: A survey," *Pacific Journal of Optimization*, vol. to appear, 2010.

[21] X. Chen and P. Tseng, "Non-Interior continuation methods for solving semidefinite complementarity problems," *Mathematical Programming*, vol. 95, no. 3, pp. 431–474, Mar. 2003.

[22] M. Stolle and C. G. Atkeson, "Policies based on trajectory libraries," in *Proceedings of the International Conference on Robotics and Automation (ICRA 2006)*, 2006.

Distributed Optimization with Pairwise Constraints and Its Application to Multi-robot Path Planning

Subhrajit Bhattacharya
Department of Mechanical Engineering and Applied Mechanics
University of Pennsylvania
Philadelphia, PA 19104
Email: subhrabh@seas.upenn.edu

Vijay Kumar
Department of Mechanical Engineering and Applied Mechanics
University of Pennsylvania
Philadelphia, PA 19104
Email: kumar@seas.upenn.edu

Maxim Likhachev
Department of Computer and Information Science
University of Pennsylvania
Philadelphia, PA 19104
Email: maximl@seas.upenn.edu

Abstract— **Distributed approaches to constrained optimization problems have immense applications to multi-robot path planning, scheduling, task allocation and other problems requiring multiple robots to optimize a global objective function. The aim of these approaches is to solve a series of smaller optimization problems for each robot while sharing information among the robots, and in the process, solve the global optimization problem, which otherwise would have been intractable. Distributed approaches to separable convex optimization problems with linear constraints have been studied extensively in the past using techniques of dual and Lagrangian decomposition. In the present work, we investigate a distributed implementation of a general separable optimization problem with pair-wise non-linear constraints. On the theoretical side, we show the conditions under which the algorithm converges to an optimal solution. On the experimental side, we demonstrate the utility of the algorithm on the problem of multi-robot path planning with pair-wise distance constraints in large complex 2-D environments with obstacles.**

I. Introduction

Distributed implementation of optimization problems is an important field of research in distributed systems [15, 4]. In many optimization problems the joint state space of all the search variables is too big or too complex for the optimization problem to be solved centrally. At other times the complete information about all the state variables is not available to any central processor. Thus distributed implementation of such problems become indispensable. One example of particular interest to us is multi-robot planning problems. For instance, robots navigating towards their respective goals while staying within the communication range is one of the common planning problems in multi-robot robotics. Here the search variables are the robot trajectories, each of which theoretically lies in an infinite dimensional Hilbert space. Planning in the joint state-space of all the robots in such a case while satisfying complex constraints may be very expensive, if not practically impossible. Multi-robot path planning suffers from the inherent complexity resulting from the necessity of operating in Cartesian products of configuration and state spaces [7]. The continuous path planning problem is even more difficult to solve in a centralized setting [12] unless the problem is solved sequentially for each robot [16]. Open loop trajectory planning problems can be reduced to optimization problems. While completeness results are often possible [1] for simple problems with no constraints, it is difficult to respect more complex multi-robot constraints.

Another instance of distributed optimization problem is task allocation for multiple robots [10]. In these methods, one can impose rendezvous constraints at intermediate time points as tasks and reformulate the path planning problem as a task allocation problem. This then lends itself to auction-based solutions [8] for the team. However, these methods can produce highly sub-optimal solutions in the environments with obstacles without guarantees on convergence. There also exist an extensive research on distributed planning in a more general sense (a good survey can be found in [9]).

Separable optimization problems [4] (optimization problems that can be split up into simpler sub-problems involving only certain partitions of the variable set) with linear constraints have been studied extensively in the past and solved in a distributed fashion using techniques based on dual decomposition [15, 4]. Augmented Lagrangian type methods have been used for solving similar problems more efficiently [3, 14]. However such methods are limited to problems with linear constraints and rely on convexity of cost functions.

In the present work we investigate a distributed implementation of a separable optimization problem with *non-linear* constraints arising from coupling between pairs of robots. We do not make any assumption on the convexity of the cost or the constraint functions. Our theoretical analysis shows that the algorithm converges to an optimal solution under certain conditions. As a demonstration of the implementation of our algorithm we will mostly concentrate on solving the problem of path planning for teams of robots coupled with constraints on the distances between pairs of robots. Continuous motion planning is possible for such problems [2], but only practical in environments with moderate complexity. In this paper we explore discrete path planning algorithms for solving the individual simpler optimization problems in a large environment with obstacles and solve the global problem using our distributed optimization algorithm. We show that our approach is able to find efficiently optimal paths with complex cost functions, in arbitrarily complex environments, and with non-linear pair-wise constraints. We therefore demonstrate the utility and versatility of the proposed algorithm by solving large scale optimization problems in a distributed fashion, which otherwise would have been intractable.

II. Problem Definition

A. The Optimization Problem

In this paper we study the following problem: Find

$$\{\pi_1^*, \ldots, \pi_N^*\} = \text{argmin}_{\pi_1 \ldots \pi_N} \sum_{j=1 \ldots N} c_j(\pi_j) \qquad (1)$$

subject to the pairwise constraints

$$\Omega_{ij}(\pi_i^*, \pi_j^*) = 0, \qquad i, j = 1 \cdots N \qquad (2)$$

We call each π_i a *partition* of the set of search variables. In the context of multi-robot planning problem π_i will represent the path of robot i, while $c_j(\pi_j)$ will be the cost of path π_j. Ω_{ij} will represent the violation of the constraint between the trajectories of robot i and j.

B. Problem Assumptions

The following assumptions are made about the variables and functions appearing in the problem definition:

1. The optimization variables π_i lie in abstract vector spaces that are continuous, differentiable and simply connected. In the general case they may be considered as vectors in a finite dimensional Euclidean space, but they can lie (as we'll discuss later) in more complex spaces like infinite dimensional Hilbert spaces. We represent the space in which π_i lies as $\mathbb{H}$
2. The cost functions $c_i : \mathbb{H} \rightarrow \mathbb{R}_+$ are assumed to be continuous and smooth. In context of path planning for mobile robots, an example of such a cost function is the Euclidean length of the trajectories in an environment without obstacles. However, in our implementation we will later relax the condition. Note that we don't make any immediate assumption on the convexity of the cost functions.
3. The functions $\Omega_{ij} : \mathbb{H} \times \mathbb{H} \rightarrow \mathbb{R}$, defined for the unordered pair $\{i, j\}$, are continuous. Also, we require that the second derivatives of Ω_{ij} with respect to each of its parameters as well as the first mixed derivative are defined. That is, $\Omega_{ij}^{(0,2)}$, $\Omega_{ij}^{(2,0)}$ and $\Omega_{ij}^{(1,1)}$ are defined, where the superscripts denote the order of partial derivatives w.r.t. the respective parameters. A simple example of such a constraint function in context of multi-robot path planning is constraint on the distance between trajectories of two robots.
4. The functions Ω_{ij} are assumed to have the following properties
   i. Ω_{ij} is symmetric in its two parameters (i.e. $\Omega_{ij}(\pi_i, \pi_j) = \Omega_{ij}(\pi_j, \pi_i)$), and
   ii. $\Omega_{ij}^{(1,0)}(\pi_i, \pi_j) = -\Omega_{ij}^{(0,1)}(\pi_i, \pi_j)$.

   It is easy to note that these properties will be true if Ω_{ij} has the functional form $\Omega_{ij}(\pi_i, \pi_j) = G_{ij}(\pi_i - \pi_j)$, where $G_{ij} : \mathbb{H} \rightarrow \mathbb{R}$ is a continuous, smooth even function. Note that we don't make any immediate assumption on the convexity of Ω_{ij}.
5. We assume that besides c_r and Ω_{ij}, the quantities $c_r^{(2)}$, $\Omega_{ij}^{(1,0)}$, $\Omega_{ij}^{(2,0)}$ and $\Omega_{ij}^{(1,1)}$ are readily computable for a given set of input variables. This may be achieved either by knowing the expressions of the derivatives in closed form, or by computing them numerically.

If in a particular problem the cost and constraint functions are not smooth, they can always be approximated by smooth functions at the non-smooth regions using one of the many Mollification techniques [13]. We also note that inequality constraints can be converted to equality constraints (as required by 2) by taking $\max$ or $\min$ with zero, followed by Mollification treatments.

It is to be noted that in spite of the conditions imposed on Ω_{ij}, they can represent a wide variety of constraints, especially in robotics applications. The proposed functional form of the constraints can model constraints on communication, visibility, rendezvous, convoying, collision avoidance, and other more complex coordination between pairs of robots involving their trajectories as well as their derivatives (say, to incorporate dynamic and kinematic constraints).

III. The Algorithm

A pseudo-code of our algorithm is presented in Figure 1. The global optimization problem is decomposed into a series of lower-dimensional unconstrained optimization problems, each of which is solved in a single *partition*, π_r, of the search variables. In each iteration the penalty weights on the violation of constraints, W^k, are incremented along the direction given by V^k. The intuitive concept behind the algorithm is that we start off by solving the global unconstrained problem, which is completely decoupled (line 1). Then we gradually increase the penalty weights (line 6) for the constraints which are modeled as soft constraints. The directions in which we can change the weights to guarantee optimality and convergence are described later in the *Theoretical Analysis* section. With the new set of weights we solve a sub-problem, which is an unconstrained optimization problem on only a single partition, namely π_r (line 7). Thus we note that in each iteration only the variables in a single partition are changed, while the others remain unchanged.

An application of the algorithm in goal-directed navigation of multiple robots with rendezvous constraints at various points on the trajectory is illustrated in Figure 2. We note that the algorithm starts off with unconstrained trajectories at $k = 0$ and in each iteration only a single robot plans its trajectory. As the iterations progress, the robots gradually change their trajectories due to increase in penalty weights and try to satisfy the rendezvous constraints. Eventually they reach the global solution. It is to be noted that the gradual increment in the weights play a key role in ensuring that the solution is optimal. One-shot increase in the weights to large values to satisfy the constraints would have resulted in suboptimal trajectories since the robots are planning sequentially. This particular example is described in more details in the "Results" section of the paper.

procedure DistributedOptimization()

compute $\pi_i^0 = \text{argmin}_{\pi_i} c_i(\pi_i), \ \forall i \in \mathcal{N}^N$;
set $W_{ij}^0 = 0$ for all unordered $\{i,j\} \in \mathcal{P}^N$;
$r = r_0 \in \mathcal{S}$, $k = 0$;
while $\left(\Omega_{ij}(\pi_i^k, \pi_j^k) \neq 0 \ \forall \{i,j\} \in \mathcal{P}^N\right)$
    set $V^k = ComputeStepDirection(W^k, \{\pi\}^k, r)$
    set $W^{k+1} = W^k + \epsilon^k V^k$;
    compute $\pi_r^{k+1} = \text{argmin}_{\pi_r} \left[c_r(\pi_r) + \sum_{\{ir\} \in \mathcal{P}_r^N} W_{ir}^{k+1} \Omega_{ir}(\pi_i^k, \pi_r)\right]$;
    set $\pi_j^{k+1} = \pi_j^k$ for all other $j \neq r$;
    set $k = k+1$;
    set $r = r_k \in \mathcal{S}$;
end

Fig. 1. Distributed Optimization Algorithm

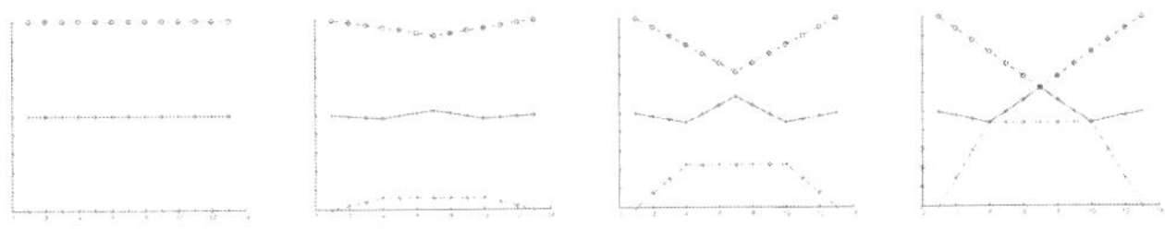

(a) Unconstrained plans, $k=0$ (b) $k=50$ (c) $k=150$ (d) Converged solution, $k=216$

Fig. 2. Demonstration of convergence towards global optimal solution with progress of iterations. In this example there are 3 robots with unconstrained trajectories 12 units long, parallel and separated by 5 units. Trajectories are defined by displacements along Y of 11 unit-spaced points on each.

The sets $\mathcal{N}^N$ is the set of all natural numbers from 1 to N, and $\mathcal{P}^N$ is the set of all unordered pairs of numbers in $\mathcal{N}^N$. The set S gives a sequence of partitions for which the individual sub-problems are solved in each iteration. $\mathcal{S} = \{r_0, r_1, r_2, \cdots\}$, where $r_i \in \{1, 2, \cdots, N\} \ \forall i \in \mathbb{N}$ can be constructed such that the partitions appear almost at equal frequency in the sequence. As we will see later, the sequence does not influence the convergence or optimality of the solution, but may influence the number of iterations required to converge.

Also, the step-sizes, ϵ^k, are small and determine the precision of the solution. The theoretical analysis in the paper are guaranteed to hold when the step-sizes are infinitesimal. But for practical purpose we choose finite and small step-sizes. The sequence $\mathcal{E} = \{\epsilon^0, \epsilon^1, \epsilon^2, \cdots\}$ may be predefined, set to a fixed constant step-size, or may be adaptive.

The procedure "*ComputeStepDirection*" in the Algorithm computes the direction in which to increase the penalty weights, W. We call this direction a *Step Direction*. In the following section (and especially Theorem 3) we will discuss the ways of computing such a direction.

IV. Theoretical Analysis

In this section we investigate the conditions under which the proposed algorithm will converge to an optimal solution. Theorem 2 along with Theorem 1 proves that under certain conditions the algorithm is guaranteed to converge to an optimal solution. Theorem 3 gives a prescription, using which will ensure that the said conditions hold.

A. Notations and Preliminaries

1) Unordered pair: We define the set of unordered pairs of natural numbers from 1 to N, and its subsets as follows:

$$\mathcal{P}^N = \{\{1,2\}, \{1,3\}, \cdots, \{1,N\}, \{2,3\}, \{2,4\}, \cdots, \{N-1,N\}\}$$

and, $\mathcal{P}_r^N = \{\{1,r\}, \cdots, \{r-1,r\}, \{r+1,r\}, \cdots, \{N,r\}\}$

In the following discussions, the subscripts "ij" of quantities (like W) or functions (like Ω) are elements from $\mathcal{P}^N$. Thus the order of the subscripts does not matter. When we write W (or W_1, or W_2), we mean the vector of length $\frac{1}{2}N(N-1)$ of all the W_{ij}'s. Also, often we will write $\{i,j\} \equiv \{k,r\} \in \mathcal{P}_r^N$ to indicate that $\{i,j\}$ is such that exactly one of i or j is equal to r, while the other one, which is not equal to r, is denoted by k.

2) Other notations:

i. For notational convenience we define the sets $\mathcal{N}^N = \{1, 2, \cdots, N\}$ and $\mathcal{N}_{-r}^N = \{1, 2, \cdots, r-1, r+1, \cdots N\}$

ii. We denote the set $\{\pi_1, \pi_2, \cdots, \pi_N\}$ as $\{\pi\}$. On similar lines, if there are arbitrary functions $\Upsilon_i : \mathbb{A} \rightarrow \mathbb{B}$, $i = \{1, 2, \cdots N\}$, we denote the collection of all these functions as $\{\Upsilon\} : \mathbb{A} \rightarrow \mathbb{B} \times \mathbb{B} \times \cdots \times \mathbb{B}$. Conversely, the jth element of $\{\Upsilon\}$ is denoted as $[\{\Upsilon\}]_j$ or Υ_j

iii. The subset of $\{\pi\}$ without the rth element is denoted by $\{\pi\}_{-r}$. Thus, $\{\pi\}_{-r} = \{\pi_1, \pi_2, \cdots, \pi_{r-1}, \pi_{r+1}, \cdots, \pi_N\}$

iv. If we have a smooth function $f : \mathcal{A}_1 \times \mathcal{A}_2 \times \cdots \mathcal{A}_n \rightarrow \mathbb{R}$, its derivatives are represented by $f^{(a_1, a_2, \cdots, a_n)}$. Thus, $f^{(1)} \equiv \nabla f$, $f^{(2)} \equiv \nabla^2 f$, $f^{(1,1)} \equiv \nabla_x \nabla_y f$, etc.

v. We define the Lagrangian of the global problem $\overline{U}$, and the Dual function $\overline{\Psi}$,

$$\overline{U}(\{\pi\}, W) := \sum_{k \in \mathcal{N}^N} c_k(\pi_k) + \sum_{\{kl\} \in \mathcal{P}^N} W_{kl} \Omega_{kl}(\pi_k, \pi_l)$$
$$\{\overline{\Pi}\}(W) := \arg\min_{\{\pi\}} \left[\overline{U}(\{\pi\}, W)\right]$$
$$\overline{\Psi}(W) := \min_{\{\pi\}} \left[\overline{U}(\{\pi\}, W)\right] = \overline{U}(\{\overline{\Pi}\}(W), W) \qquad (3)$$

In other words, $\{\overline{\Pi}\}(W)$ is the global optimum for the penalized objective function with W as penalty weights, and $\overline{\Psi}(W)$ is the optimal value.

vi. Similarly, we define for each *partition* π_r,

$$U_r(\pi_r, W_1, W_2) := c_r(\pi_r) + \sum_{\{kr\} \in \mathcal{P}_r^N} W_{1,kr} \Omega_{kr}(\overline{\Pi}_k(W_2), \pi_r)$$
$$\Pi_r(W_1, W_2) := \arg\min_{\pi_r} [U_r(\pi_r, W_1, W_2)]$$
$$\Psi_r(W_1, W_2) := \min_{\pi_r} [U_r(\pi_r, W_1, W_2)] = U_r(\Pi_r(W_1, W_2), W_1, W_2) \qquad (4)$$

That is, for a given value, $\{\overline{\Pi}\}_{-r}(W_2)$, of the partitions (except the rth one), $\Pi_r(W_1, W_2)$ gives the optimum for an individual sub-problem with W_1 as penalty weights.

vi. Finally, we define the following,

$$\mathbf{M}_r(W) = c_r^{(2)}(\overline{\Pi}_r(W)) + \sum_{\{lr\} \in \mathcal{P}_r^N} W_{lr} \Omega_{lr}^{(0,2)}(\overline{\Pi}_l(W), \overline{\Pi}_r(W))$$
$$\mathbf{N}_{lr}(W) = \Omega_{lr}^{(1,1)}(\overline{\Pi}_l(W), \overline{\Pi}_r(W)) \qquad (5)$$

Also, we note that the rth component of $\{\overline{\Pi}\}(W)$ is given by,

$$\begin{aligned} &\overline{\Pi}_r(W) \\ &= \arg\min_{\pi_r} \Big(c_r(\pi_r) + \textstyle\sum_{\{kr\} \in \mathcal{P}_r^N} W_{kr} \Omega_{kr}(\overline{\Pi}_k(W), \pi_r) + \sum_{k \in \mathcal{N}_{-r}^N} c_k(\overline{\Pi}_k(W)) \\ &\qquad + \textstyle\sum_{\{kl\} \in \mathcal{P}^N / \mathcal{P}_r^N} W_{kl} \Omega_{kl}(\overline{\Pi}_k(W), \overline{\Pi}_l(W))\Big) \\ &= \arg\min_{\pi_r} \Big(c_r(\pi_r) + \textstyle\sum_{\{kr\} \in \mathcal{P}_r^N} W_{kr} \Omega_{kr}(\overline{\Pi}_k(W), \pi_r)\Big) \\ &= \Pi_r(W, W) \end{aligned} \qquad (6)$$

Thus, $\overline{\Pi}_r^{(1)}(W) = \Pi_r^{(1,0)}(W, W) + \Pi_r^{(0,1)}(W, W)$ (7)

B. Theorems

Definition 1: [*Separable Optimal Flow*] Given the functions c_r, Ω_{ir} $\forall\{ir\} \in \mathcal{P}_r^N$ we call V a *Separable Optimal Flow Direction* and ϵ a *Separable Optimal Flow Step* at W for Ψ_r if and only if the following holds,

$$\begin{aligned}&\Psi_r(W+\epsilon V, W) - \Psi_r(W, W) \\ &\quad\le \Psi_r(W+\epsilon V, W+\epsilon V) - \Psi_r(W, W+\epsilon V) \\ &\text{and, } V_{ij}=0, \quad \forall\{i,j\} \text{ such that } r \notin \{i,j\}\end{aligned} \tag{8}$$

Together, V and ϵ are said to define a *Separable Optimal Flow* at W for Ψ_r.

Theorem 1: [*Optimality at each Iteration*] If the *Step Direction*, V^k, returned by procedure $ComputeStepDirection$ at the kth iteration in Line 5 of the Algorithm, along with the chosen *Step Size*, ϵ^k, define a *Separable Optimal Flow* at W^k for Ψ_{r_κ}, $\forall\ k$, then $\forall\ k$:

$$\begin{aligned}&\{\pi_1^k, \ldots, \pi_N^k\} \\ &\quad = \arg\min_{\{\pi\}} \left[\textstyle\sum_{i\in\mathcal{N}^N} c(\pi_i) + \sum_{\{ij\}\in\mathcal{P}^N} W_{ij}^k \cdot \Omega_{ij}(\pi_i, \pi_j)\right]\end{aligned}$$

Proof: Detailed proof can be found in Appendix ■

The result of the Theorem 1, in brief, can be stated as

$$\pi_i^k = \overline{\Pi}_i(W^k), \quad \forall i, k \tag{9}$$

The implication of the result is that there are specific directions (which we call *Separable Optimal Flow Directions*) in which we can increment the penalty weight vector W, such that the global optimum for the new set of penalty weights differs from the previous global optimum (i.e. optimum for the previous set of weights) in only one partition of the optimization variables, namely π_r. Thus, by moving along such a direction in k^{th} iteration, we only need to change π_{r_k}, and still remain at an optimum of the penalized net cost.

Definition 2: [*Ascent Direction*] We call V an *Ascent Direction* at W if and only if the following holds,

$$\sum_{\{ij\}\in\mathcal{P}^N} V_{ij}\Omega_{ij}(\overline{\Pi}_i(W), \overline{\Pi}_j(W)) > 0 \tag{10}$$

Theorem 2: [*Convergence of the Algorithm*] If the conditions in Theorem 1 hold, and the *Step Direction*, V^k, returned by the procedure $ComputeStepDirection$ at the kth iteration in Line 5 of the Algorithm is also an *Ascent Direction* at W^k for every k, then the Algorithm converges to an optimal solution if one exists.

Proof: Detailed proof can be found in Appendix ■

The result of Theorem 2 implies that if we always increment the penalty weights along directions that are both *Ascent Directions* and *Separable Optimal Flow Directions*, we will eventually converge to the global optimum, if it exists. Knowing the π_i^k's from previous iterations, it is easy to choose such a direction V^k for the current iteration as one that has positive inner product with the vector of all constraint violations at k^{th} iteration (i.e., the vector made of $\Omega_{ij}(\pi_i^k, \pi_j^k)$'s).

Theorem 3: [*Computation of Separable Optimal Flow Direction*] If the functions c_r and Ω_{ir} $\forall\{ir\} \in \mathcal{P}_r^N$ abide by the *Problem Assumptions*, we can find a *Separable Optimal Flow Direction* for Ψ_{r_k} at W^k defined by V^k, if it exists, along with a small enough *Step Size*, ϵ^k, at Line 5 of the Algorithm, using only the following quantities that are readily computable:

$$\begin{aligned}&W^k,\ c_i^{(2)}(\pi_i^k)\ \forall i \in \mathcal{N}^N, \\ &\quad\Omega_{ij}^{(0,2)}(\pi_i^k, \pi_j^k),\ \Omega_{ij}^{(1,1)}(\pi_i^k, \pi_j^k) \\ &\qquad\text{and } \Omega_{ij}^{(1,0)}(\pi_i^k, \pi_j^k) \quad \forall\{ij\} \in \mathcal{P}^N,\end{aligned}$$

In general we get to choose from a large set of possible *Separable Optimal Flow Directions*, thus giving us the opportunity to make it as an *Ascent Direction* as well.

Proof: Detailed proof can be found in Appendix ■

The theorem implies that we can compute *Separable Optimal Flow Directions*, if one exists, in terms of quantities that can be easily computed using the variables from the previous iteration. Within the limits of the error introduced by the finite step-size, we can compute a *Separable Optimal Flow*.

It is important to note that in general the freedom of choosing from a large set of possible *Separable Optimal Flow Direction* (set of linear combinations of the eigenvectors such that (30) is satisfied) gives us the opportunity to choose a vector V^k that is an *Ascent Direction* as well. In case the intersection between the set of *Separable Optimal Flow Directions* and the set of *Ascent Directions* is empty, we can choose to skip to the next element in the sequence $\{r_k\}$.

V. Results

A. An Exact Implementation

We demonstrate the algorithm using a MATLAB implementation of an idealized planning problem. The specific problem under consideration may be considered as a multi-robot goal-directed path planning in an environment without obstacles, where the trajectories of the robots are defined by displacements of unit-spaced points on the trajectories in vertical direction. Thus the trajectory, of a robot is given by $\pi_r = [start_r, y_{r2}, y_{r3}, \cdots, y_{rL}, goal_r]^T$, where $y_{ri}, i = 2\cdots L$ are the search variables and represent displacement of a point at x_i of the rth robot's trajectory. The constraints between robots a and b are rendezvous constraints defined by

$$\Omega_{ab}(\pi_a, \pi_b) = \left((y_{ac_1} - y_{bc_1})^2 + (y_{ac_2} - y_{bc_2})^2 + \cdots\right)^{1/2} = 0$$

where $c_1, c_2, \cdots$ are the points where a and b need to rendezvous. There are two components to the costs functions c_r - the length of the trajectory, and the integral of the square of accelerations over the trajectory. Assuming the robots have constant X-velocity, the net cost is thus given by a weighted sum of the two components,

$$\begin{aligned}c_r(\pi_r) = \alpha &\left((y_{r2} - start_r)^2 + (y_{r3} - y_{r2})^2 + \cdots\right)^{1/2} \\ &+ \beta\left(((y_{r4} - y_{r3}) - (y_{r3} - y_{r2}))^2 + \right. \\ &\qquad \left.((y_{r5} - y_{r4}) - (y_{r4} - y_{r3}))^2 + \cdots\right)\end{aligned}$$

The individual unconstrained optimization problems in Line 7 of the Algorithm were solved using MATLAB's *fminunc*.

Figure 2 demonstrates how the algorithm approaches the optimal feasible solution with progress of iterations. In this example we chose $\alpha = 1, \beta = 0$ and a constant step size of $\epsilon^k = 0.01$. The constraints are that the top robot needs to rendezvous with the middle one at the middle point (i.e. at x_7), while the bottom and middle ones rendezvous at two points (y_4 and y_{10}). Since it is a symmetric case with only trajectory length as cost, it's easy to compute the optimal solution separately by optimizing over just 2 variables. The optimal cost that way is found to be 43.946, while our algorithm terminates at a cost of

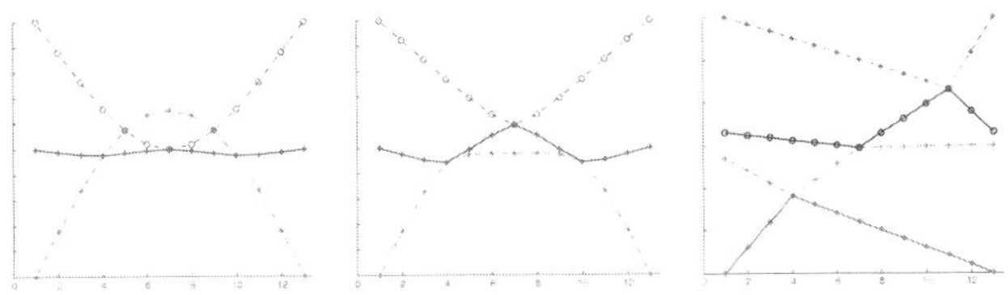

(a) Symmetric case: $\alpha = 0, \beta = 1$ (b) Symmetric case: $\alpha = 1, \beta = 0.2$ (c) An asymmetric case: $\alpha = 1, \beta = 0$

Fig. 3. Converged Solutions

43.962 for the said step size. Figures 3(a) and (b) illustrate the results for other values of α and β demonstrating the ability of the algorithm in dealing with complex cost functions. Figure 3(c) demonstrates an asymmetric case with 4 robots.

We attempted to solve these problems in a centralized fashion using MATLAB's Pseudo-Newton search method implemented in *fmincon*. Even when the gradient and Hessian functions were explicitly provided, it failed to find a solution satisfying the required tolerance with computation time limit of 20 minutes and 5000 as the limit on number of iterations. In contrast, our algorithm solved these problems in about a minute and a few hundred iterations.

B. A Discrete Implementation

In the following examples we perform a more realistic multi-robot path planning with pair-wise distance constraints in a complex environment with large number of obstacles. Thus, even the individual sub-problems that we need to solve at Line 7 of the Algorithm become significantly complex.

The three-dimensional state-space of each robot, $X - Y - Time$, was discretized into uniform cells, each cell representing nodes of the graph, and an A* graph search technique was employed to obtain an optimal path in the graph as an approximation of the trajectory of rth robot, π_r. The connectivity of the graph was such that a cell in the x, y, t space connected to its 8 neighboring cells and itself in x, y but with the time step incremented by one. This implies that any path in the graph will be discrete approximation of an element of $\mathbb{H}$. While an 8-connected grid is quick and efficient to perform search in, it confines the motion of the robot to 8 directions ($45°$ orientations). A consequence of this is that some seemingly sub-optimal solutions are actually optimal (minimum cost paths) in the 8-connected graph.

The cost function, c, is the Euclidean length of the trajectory. The constraints between the pairs of robots are defined by maximum distance, $\phi_{ij}(t)$, that the robots i and j can be apart at a given instant of time. Thus, the constraints are defined by,

$$\Omega_{ij}(\pi_i, \pi_j) = \int_0^T \varpi(\pi_i(t), \pi_j(t), \phi_{ij}(t)) = 0$$

where, $\varpi(\mathrm{s}, \mathrm{s}', p) = \max(0, d(\mathrm{s}, \mathrm{s}') - p)$ and d is a distance function.

1) Dealing with Obstacles: The algorithm and theoretical analysis described so far relies on the fact that the values of all the variables can be changed in small steps in each iterations. Unfortunately the presence of obstacles in the environment violates that condition. However the Optimality condition of Theorem 1 still holds in a single Homotopy class of all the trajectories. This means that as long as a trajectory changes in small steps and does not make big jump from one side of an obstacle to the other, Theorem 1 and the related results from Theorem 3 hold. Thus, whenever we get stuck in a particular homotopy class for all the trajectories, or whenever one of the trajectories makes a jump to a new homotopy class, we block this homotopy class as invalid and restart the planning. It is worth noting here that we use only *large* connected obstacles to characterize homotopy classes and avoid unnecessary creation of homotopy classes by small obstacles that do not effect optimality significantly. We primarily employed two different methods of blocking homotopy classes:

i. *Use of Blacklists* - We maintain a list of blocked regions (or balls) in the joint state-space of the robots violating one or more constraints. Every time we need to restart the planning as above, we update the "Blacklist" with the latest violation information in the homotopy class without a feasible solution. Thus, when we restart the planning we eventually will avoid the homotopy class, provided our choice of blocking ball was proper.

ii. *Systematic identification and blocking of Homotopy class* - Homotopy classes can be systematically identified using geometric methods [11] or complex analysis. We integrated such a method with our A* searches which enabled us to restrict searches in specific homotopy classes or block some homotopy classes.

In the following subsections we will present the performances of our algorithm in different environments.

C. Three Interconnected Rooms

The environment shown in Figure 4 consists of 3 interconnected rooms and 3 robots. The environment is made of discretized cells 89×94 in space and 200 time-steps. This results in the joint state-space of all the robots to have 10^{14} states. Our planner avoids planning in this huge state-space, but rather breaks up the problem into a series of plans in individual state-spaces with $< 2 \times 10^6$ states. The robots start from the left side of the environment and need to reach their goals on the right end. Figure 4(a) shows the unconstrained optimal plan. A constraint is defined between R_1 and R_3 such that they need to be within a distance of 2 discretization units at $t = 40$. And the constraint between R_1 and R_2 is such that they need to be within a distance of 2 discretization units at $t = 120$. The solution shown in Figure 4(b) was found in about one minute and 72 iterations on a computer running on 1.2GHz processor.

D. Extended Rendezvous

The environment shown in Figure 5 is similar to the previous example in size and discretization. In this example R_1 and R_2 need to traverse the environment from left to right, while R_3 needs to traverse it from top to bottom. Figure 5(a) shows the unconstrained optimal plan. Constraints between R_2 and R_3 is that they need to be within 2 discretization units distance from $t = 32$ to $t = 68$, and the constraints between R_1 and R_3 is that they need to be within 2 discretization units

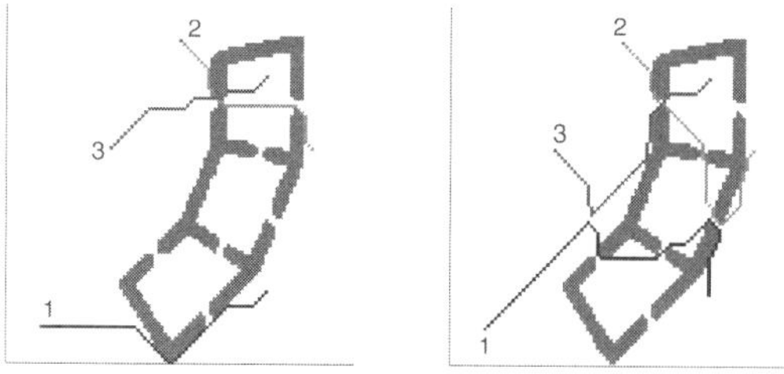

(a) Unconstrained plans (b) Converged feasible solution

Fig. 4. Planning in environment with three interconnected rooms

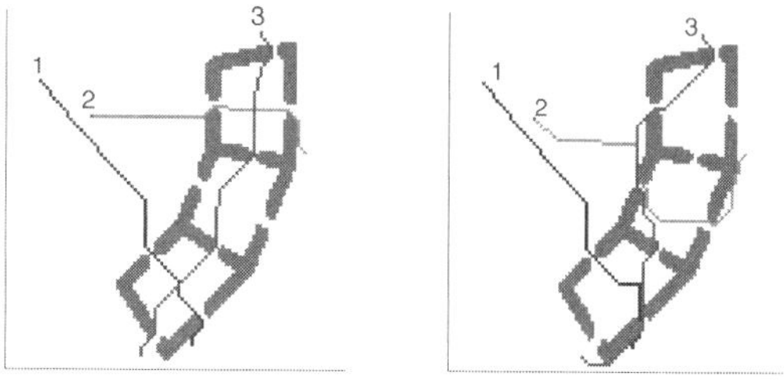

(a) Unconstrained plans (b) Converged feasible solution

Fig. 5. Extended rendezvous in environment with three interconnected rooms

distance from $t = 84$ to $t = 120$ The solution in Figure 5(b) was found after 75 iterations with $\epsilon = 0.1$ as the fixed step-size.

E. Extended Rendezvous in a Real Environment

Figure 6 shows a part of the 4th floor of Levine hall in University of Pennsylvania. The original map is 35 meters by 35 meters, discretized into 100x100 cells (each cell 35cmx35cm, almost the dimension of the Scarab robot, described later). There are 170 discretization steps in time. There are 3 robots and the unconstrained objectives, resulting in the solution in Figure 6(a), are that both Robot 1 (dashed trajectory) and Robot 2 (dash-dot trajectory) need to start at $t = 0$ inside the big room at the bottom and need to reach their respective goals by $t = 170$. Robot 3 (dotted trajectory) needs to start at $t = 45$ inside the small lower cubicle on the left side of the map and needs to reach the small storage space at the top right by $t = 120$. Figures 6(b) shows the converged feasible solution that satisfy the following constraints: A. Robot 2 needs to stay within 3 discretization units of robot 3 from $t = 60$ to $t = 80$; and B. Robot 1 then needs to stay within 3 discretization units of robot 3 from $t = 90$ to $t = 110$. Looking at the converged solution we observe that in order to satisfy constraint A, robot 2 loops it's trajectory around the central and lower cubicles, while to satisfy constraint B, robot 1 loops its trajectory around the central and upper cubicles.

F. Performance

We have successfully implemented the algorithm on teams of up to six robots and in 3 dimensions, that is planning in $X - Y - Z$. We do not present those results in the paper due to limitation on space, as well as to focus on the key features of the algorithm rather than the multitude of applications that it can have. More results and applications can be found in [5].

We tested the performance of our algorithm by running the previous example (*Extended rendezvous in a real environment*)

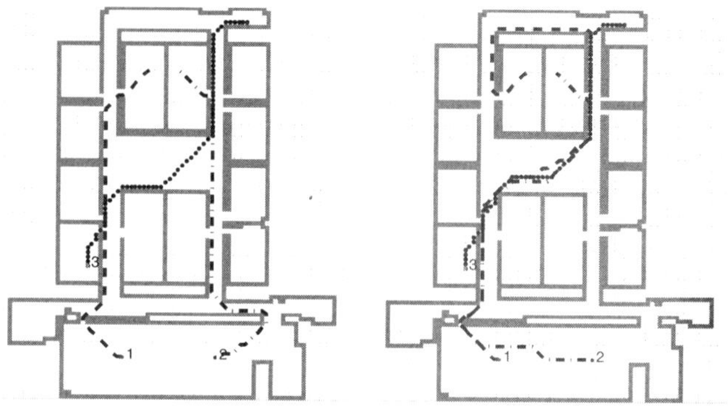

(a) Unconstrained plans (b) Converged feasible solution

Fig. 6. Extended rendezvous in a real environment

multiple times, but with randomized initial & goal states and randomized time spans for the constraints. The implementation of the problem was made in C++ and was run on a computer with Intel 1.2 GHz processor. The table below gives a summary of the run-time from 54 runs.

Min	**Max**	**Average**
28 s	169 s	59.7 s

The joint state-space of the three robots contains $\sim 170 \times (100 \times 100)^3 \simeq 1.7 \times 10^{14}$ states. Planning in such a huge graph is highly expensive, if not impossible. However our distributed planning technique was able to find the optimal solution up to the desired precision consistently and sufficiently fast, thus demonstrating the ability of the algorithm to solve large problems.

We also tested our algorithm in simulation using Gazebo, an open-source multi-robot simulator with accurate simulation of rigid-body physics. Experiment were also performed with *Scarab* mobile robot platforms along with an overhead LED tracking system to track the positions of the robots. In order to account for the non-zero radii of the robots, we performed a greedy collision avoidance during run-time. For controlling the non-holonomic robots, a feedback linearization technique was adopted. More details can be found in [5].

VI. Conclusion

We have developed an algorithm for solving large optimization problems with pair-wise nonlinear constraints and arbitrary objective functions in a distributed fashion. Our theoretical analysis gives the conditions under which the algorithm converges to an optimal solution and a prescription for making sure that those conditions are satisfied. We have successfully implemented the algorithm to solve a large multi-robot path planning problem in complex environment with arbitrary shaped obstacles and distance constraints using discrete graph searches for individual sub-problems.

Presently we are investigating the conditions under which *Separable Optimal Flow Direction* and *Ascent Direction* are guaranteed to exist. Generalizing each constrain to include more than two partitions is also under progress. In a nut-shell, besides solving a specific complex optimization problem in distributed fashion, our approach opens up a new direction for solving related classes of distributed optimization problems.

VII. Appendix

A. Proof of Theorem 1

The theorem clearly holds when for $k=0$ since $W_{ij}^0=0$ for all $\{ij\}\in\mathcal{P}^N$.

We prove the theorem by induction. Assume it holds for $k=0$ through $k=\kappa$. We will now prove that it continues to hold for the $k=\kappa+1$. By inductive assumption we have:

$$\{\pi_1^\kappa,\pi_2^\kappa,\cdots,\pi_N^\kappa\}= \arg\min_{\pi_1,\pi_2,\cdots,\pi_N}\Big(\sum_{i\in\mathcal{N}^N}c_i(\pi_i)+\sum_{\{ij\}\in\mathcal{P}^N}W_{ij}^\kappa\Omega_{ij}(\pi_i,\pi_j)\Big)$$

That is,
$$\pi_i^\kappa=\overline{\Pi}_i(W^\kappa),\quad \forall i\in\mathcal{N}^N \tag{11}$$

We also note that, by Line 8 of the Algorithm,
$$\pi_j^{\kappa+1}=\pi_j^\kappa=\overline{\Pi}_j(W^\kappa),\quad \forall\ j\in\mathcal{N}_{-r_\kappa}^N \tag{12}$$

We prove by contradiction. Let us assume that
$$\{\pi_1^{\kappa+1},\pi_2^{\kappa+1},\cdots,\pi_N^{\kappa+1}\}\neq \arg\min_{\pi_1,\pi_2,\cdots,\pi_N}\big(\sum_{i\in\mathcal{N}^N}c_i(\pi_i)+ \sum_{\{ij\}\in\mathcal{P}^N}W_{ij}^{\kappa+1}\Omega_{ij}(\pi_i,\pi_j)\big)$$

This implies that there exist $\pi_1',\pi_2',\cdots,\pi_N'$ given by
$$\{\pi_1',\pi_2',\cdots,\pi_N'\}= \arg\min_{\pi_1,\pi_2,\cdots,\pi_N}\big(\sum_{i\in\mathcal{N}^N}c_i(\pi_i)+ \sum_{\{ij\}\in\mathcal{P}^N}W_{ij}^{\kappa+1}\Omega_{ij}(\pi_i,\pi_j)\big)$$

such that,
$$\pi_i'=\overline{\Pi}_i(W^{\kappa+1}),\quad \forall i\in\mathcal{N}^N \tag{13}$$

and
$$\sum_{i\in\mathcal{N}^N}c_i(\pi_i^{\kappa+1})+\sum_{\{ij\}\in\mathcal{P}^N}W_{ij}^{\kappa+1}\cdot\Omega_{ij}(\pi_i^{\kappa+1},\pi_j^{\kappa+1}) > \sum_{i\in\mathcal{N}^N}c_i(\pi_i')+\sum_{\{ij\}\in\mathcal{P}^N}W_{ij}^{\kappa+1}\cdot\Omega_{ij}(\pi_i',\pi_j') \tag{14}$$

Again, by the algorithm, it holds that:
$$\pi_{r_\kappa}^{\kappa+1}=\arg\min_{\pi_{r_\kappa}}\big(c_{r_\kappa}(\pi_{r_\kappa}) + \sum_{\{ir_\kappa\}\in\mathcal{P}_{r_\kappa}^N}W_{ir_\kappa}^{\kappa+1}\cdot\Omega_{ir_\kappa}(\pi_i^\kappa,\pi_{r_\kappa})\big) \tag{15}$$

Noting that $\{ir_\kappa\}\in\mathcal{P}_{r_\kappa}^N\Rightarrow i\in\mathcal{N}_{-r_\kappa}^N$, and using equation (12), we get from (15),
$$\begin{aligned}&\pi_{r_\kappa}^{\kappa+1}=\Pi_{r_\kappa}(W^{\kappa+1},W^\kappa)\\ \Rightarrow\ &\Psi_{r_\kappa}(W^{\kappa+1},W^\kappa)\\ &=c_{r_\kappa}(\pi_{r_\kappa}^{\kappa+1})+\sum_{\{ir_\kappa\}\in\mathcal{P}_{r_\kappa}^N}W_{ir_\kappa}^{\kappa+1}\cdot\Omega_{ir_\kappa}(\pi_i^{\kappa+1},\pi_{r_\kappa}^{\kappa+1})\end{aligned} \tag{16}$$

Also, from (13) and (6),
$$\pi_{r_\kappa}'=\overline{\Pi}_{r_\kappa}(W^{\kappa+1})=\Pi_{r_\kappa}(W^{\kappa+1},W^{\kappa+1})$$

Thus,
$$\Psi_{r_\kappa}(W^{\kappa+1},W^{\kappa+1})=c_{r_\kappa}(\pi_{r_\kappa}')+\sum_{\{ir_\kappa\}\in\mathcal{P}_{r_\kappa}^N}W_{ir_\kappa}^{\kappa+1}\cdot\Omega_{ir_\kappa}(\pi_i',\pi_{r_\kappa}') \tag{17}$$

Thus from (14), (16) and (17), upon rearrangement,
$$\begin{aligned}&\sum_{i\in\mathcal{N}_{-r_\kappa}^N}\big(c_i(\pi_i^{\kappa+1})-c_i(\pi_i')\big)\\ &+\sum_{\{ij\}\in\mathcal{P}^N/\mathcal{P}_{r_\kappa}^N}W_{ij}^{\kappa+1}\cdot\big(\Omega_{ij}(\pi_i^{\kappa+1},\pi_j^{\kappa+1})-\Omega_{ij}(\pi_i',\pi_j')\big)\\ &>\Psi_{r_\kappa}(W^{\kappa+1},W^{\kappa+1})-\Psi_{r_\kappa}(W^{\kappa+1},W^\kappa)\end{aligned} \tag{18}$$

On the other hand, according to our inductive assumption, $\{\pi_1^\kappa,\pi_2^\kappa,\cdots,\pi_N^\kappa\}$ is a minimum for the global objective function with W^κ. Thus,
$$\begin{aligned}&\overline{U}(\{\pi_1^\kappa,\pi_2^\kappa,\cdots,\pi_N^\kappa\},W^k)\\ &\quad\le\overline{U}(\{\pi_1',\cdots,\pi_{r-1}',\Pi_r(W^\kappa,W^{\kappa+1}),\pi_{r+1}',\cdots,\pi_N'\},W^k)\end{aligned} \tag{19}$$

Now, since V^κ is a *Separable Flow Direction Direction* according to our hypothesis, it holds that $W_{ij}^{\kappa+1}-W_{ij}^\kappa=V_{ij}^\kappa=0\ \forall\{ij\}\in\mathcal{P}^N/\mathcal{P}_{r_\kappa}^N$. Again from (12) we have $\pi_j^{\kappa+1}=\pi_j^\kappa\ \forall j\in\mathcal{N}_{-r_\kappa}^N$. Using these, along with (11) and (13), we get from (19),
$$\begin{aligned}&\sum_{i\in\mathcal{N}_{-r_\kappa}^N}\big(c_i(\pi_i^{\kappa+1})-c_i(\pi_i')\big)\\ &+\sum_{\{ij\}\in\mathcal{P}^N/\mathcal{P}_{r_\kappa}^N}W_{ij}^{\kappa+1}\cdot\big(\Omega_{ij}(\pi_i^{\kappa+1},\pi_j^{\kappa+1})-\Omega_{ij}(\pi_i',\pi_j')\big)\\ &\le\Psi_{r_\kappa}(W^\kappa,W^{\kappa+1})-\Psi_{r_\kappa}(W^\kappa,W^\kappa)\end{aligned} \tag{20}$$

Thus from (18) and (20),
$$\begin{aligned}&\Psi_{r_\kappa}(W^\kappa+\epsilon^\kappa V^\kappa,W^\kappa+\epsilon^\kappa V^\kappa)-\Psi_{r_\kappa}(W^\kappa+\epsilon^\kappa V^\kappa,W^\kappa)\\ &\quad-\Psi_{r_\kappa}(W^\kappa,W^\kappa+\epsilon^\kappa V^\kappa)+\Psi_{r_\kappa}(W^\kappa,W^\kappa)<0\end{aligned} \tag{21}$$

However this is a contradiction to our assumption that the V^κ and ϵ^κ defines a *Separable Optimal Flow* at W^κ for Ψ_{r_κ}. Hence our assumption of the existence of $\{\pi_1',\pi_2',\cdots,\pi_N'\}$ was incorrect. This proves Theorem 1.

B. Proof of Theorem 2

Since $\overline{U}(\{\pi\},W)$ is linear in W, from [6] we know that $\overline{\Psi}(W)=\min_{\{\pi\}}\overline{U}(\{\pi\},W)$ is concave in W. Again by optimality condition, $\overline{U}^{(1,0)}(\{\overline{\Pi}\}(W),W)=0$

Thus,
$$\begin{aligned}&\overline{\Psi}(W)=\overline{U}(\{\overline{\Pi}\}(W),W)\\ \Rightarrow\ &\overline{\Psi}^{(1)}(W)=\overline{U}^{(0,1)}(\{\overline{\Pi}\}(W),W)\\ \Rightarrow\ &\Big[\overline{\Psi}^{(1)}(W)\Big]_{ij}=\Omega_{ij}(\overline{\Pi}_i(W),\overline{\Pi}_j(W))\end{aligned} \tag{22}$$

Thus, if V^k is an *Ascent Direction*, from (10) and (22) it will imply that in every iteration we move towards a direction in which $\overline{\Psi}$ increases. On the other hand, $\overline{\Psi}$ being concave, can have only an unique optimum, which is a maximum, when $\Big[\overline{\Psi}^{(1)}(W)\Big]_{ij}=0,\ \ \forall\{ij\}\in\mathcal{P}^N$. We thus note that the maximum, if it exists, is attained when all the constraints are satisfied. Hence taking steps towards the maximum means that we will eventually reach the feasible point, if it exists (within errors defined by the *Step Size*).

C. Proof of Theorem 3

From the definitions of U_r, Π_r and Ψ_r, by optimality condition, $U_r^{(1,0,0)}(\Pi_r(W_1,W_2),W_1,W_2)=0$,
$$\Rightarrow c_r^{(1)}(\Pi_r(W_1,W_2))+ \sum_{\{lr\}\in\mathcal{P}_r^N}W_{1,lr}\Omega_{lr}^{(0,1)}(\overline{\Pi}_l(W_2),\Pi_r(W_1,W_2))=0 \tag{23}$$

Differenciating (23) w.r.t. W_1, using *Problem Assumption* 4.ii., setting $W_1=W_2=W$, and using (6) & (7),
$$\mathbf{M}_r(W)\cdot\Big[\Pi_r^{(1,0)}(W,W)\Big]_{ij} =\begin{cases}\Omega_{kr}^{(1,0)}(\overline{\Pi}_k(W),\overline{\Pi}_r(W)), & \text{when }\{i,j\}\equiv\{k,r\}\in\mathcal{P}_r^N\\ 0, & \text{when }i\neq r,j\neq r\end{cases} \tag{24}$$

Similarly, differentiating (23) w.r.t. W_2, followed by setting $W_1 = W_2 = W$, and using (6) & (7), we get,

$$\mathbf{M}_r(W) \cdot \left[\Pi_r^{(0,1)}(W,W)\right]_{ij} + \sum_{\{lr\}\in\mathcal{P}_r^N} W_{lr}\Omega_{lr}^{(1,1)}(\overline{\Pi}_l(W),\overline{\Pi}_r(W)) \cdot \left[\overline{\Pi}_l^{(1)}(W)\right]_{ij} = 0 \tag{25}$$

Adding (25) and (24), and using (7),

$$\mathbf{M}_r(W) \cdot \left[\overline{\Pi}_r^{(1)}(W)\right]_{ij} + \sum_{\{lr\}\in\mathcal{P}_r^N} W_{lr}\mathbf{N}_{lr}(W) \cdot \left[\overline{\Pi}_l^{(1)}(W)\right]_{ij} = \begin{cases} \Omega_{kr}^{(1,0)}(\overline{\Pi}_k(W),\overline{\Pi}_r(W)), & \text{when } \{i,j\} \equiv \{k,r\} \in \mathcal{P}_r^N \\ 0, & \text{when } i \neq r, j \neq r \end{cases} \tag{26}$$

Equation (26) can be written as follows,

$$\begin{bmatrix} \mathbf{M}_1(W) & W_{12}\mathbf{N}_{12}(W) & W_{13}\mathbf{N}_{13}(W) & \cdots \\ W_{21}\mathbf{N}_{21}(W) & \mathbf{M}_2(W) & W_{23}\mathbf{N}_{23}(W) & \cdots \\ W_{31}\mathbf{N}_{31}(W) & W_{32}\mathbf{N}_{32}(W) & \mathbf{M}_3(W) & \cdots \\ \vdots & \vdots & \vdots & \ddots \end{bmatrix} \begin{bmatrix} \left[\overline{\Pi}_1^{(1)}(W)\right]_{mn} \\ \left[\overline{\Pi}_2^{(1)}(W)\right]_{mn} \\ \left[\overline{\Pi}_3^{(1)}(W)\right]_{mn} \\ \vdots \end{bmatrix} = \begin{bmatrix} \cdots \\ 0 \\ \Omega_{mn}^{(1,0)}(\overline{\Pi}_m(W),\overline{\Pi}_n(W)) \\ 0 \\ \cdots \\ 0 \\ \Omega_{mn}^{(1,0)}(\overline{\Pi}_n(W),\overline{\Pi}_m(W)) \\ 0 \\ \cdots \end{bmatrix} \begin{matrix} \\ \\ \rightarrow m^{th} \text{ row} \\ \\ \\ \\ \rightarrow n^{th} \text{ row} \\ \\ \\ \end{matrix} \tag{27}$$

For all $\{m,n\} \in \mathcal{P}^N$.

We note that equations (24) and (27) gives prescriptions for computing derivatives of Π_r and $\overline{\Pi}_r$ in terms of $\overline{\Pi}_r$ and the derivatives of c_r & Ω_{ij}. It is important to note that while the later are readily computable, the former aren't.

Again, differentiating (4) w.r.t. W_1 and using (23),

$$\left[\Psi_r^{(1,0)}(W_1,W_2)\right]_{ij} = \begin{cases} \Omega_{kr}(\overline{\Pi}_k(W_2),\Pi_r(W_1,W_2)), & \text{when } \{i,j\} \equiv \{k,r\} \in \mathcal{P}_r^N \\ 0, & \text{when } i \neq r, j \neq r \end{cases} \tag{28}$$

Differentiating (28) w.r.t. W_2, using *Problem Assumptions* 4.ii., followed by setting $W_1 = W_2 = W$, and using (6) & (7),

$$\left[\Psi_r^{(1,1)}(W,W)\right]_{ij,mn} = \begin{cases} \Omega_{kr}^{(1,0)}(\overline{\Pi}_k(W),\overline{\Pi}_r(W)) \cdot \left[\overline{\Pi}_k^{(1)}(W) - \overline{\Pi}_r^{(1)}(W)\right]_{mn} \\ \quad + \Omega_{kr}^{(1,0)}(\overline{\Pi}_k(W),\overline{\Pi}_r(W)) \cdot \left[\Pi_r^{(1,0)}(W,W)\right]_{mn}, \\ \qquad \text{when } \{i,j\} \equiv \{k,r\} \in \mathcal{P}_r^N \\ 0, \quad \text{when } i \neq r, j \neq r \end{cases} \tag{29}$$

Now using (24) and (27), we can compute $\left[\Psi_r^{(1,1)}(W,W)\right]_{ij,mn}$ in terms of the quantities mentioned in the statement of the theorem. It is important to note that the sole purpose of this dealing was to be able to express the derivatives of Π and $\overline{\Pi}$ in terms of computable quantities.

Once we obtain the complete matrix for $\Psi_r^{(1,1)}(W,W)$, we can choose a *Separable Optimal Flow Direction* as a linear combination of the right eigenvectors corresponding to the non-negative eigenvalues of $\Psi_r^{(1,1)}(W,W)$. Separable Optimal Flow Step, ϵ, is chosen according to a desired accuracy. Thus, if we choose $V = a_1\mathbf{e}_1 + a_2\mathbf{e}_2 + \cdots$, where $\mathbf{e}_1, \mathbf{e}_2, \cdots$ are the eigenvectors corresponding to non-negative eigenvalues $\lambda_1, \lambda_2, \cdots$, we get,

$$\begin{aligned} & \Psi_r(W+\epsilon V, W+\epsilon V) - \Psi_r(W, W+\epsilon V) \\ & \qquad - \Psi_r(W+\epsilon V, W) + \Psi_r(W,W) \\ \simeq\; & (\epsilon V)^T \left[\Psi_r^{(1,1)}(W,W)\right] (\epsilon V) \\ =\; & \epsilon^2 \;\; (a_1\mathbf{e}_1 + a_2\mathbf{e}_2 + \cdots)^T (a_1\lambda_1\mathbf{e}_1 + a_2\lambda_2\mathbf{e}_2 + \cdots) \\ =\; & \epsilon^2 \;\; (\lambda_1 a_1^2 + \lambda_2 a_2^2 + \cdots) \;\; \geq 0 \end{aligned} \tag{30}$$

We note that all the V_{ij} for which $i \neq r, j \neq r$ does not influnce the condition (30) since $\left[\Psi_r^{(1,1)}(W,W)\right]_{ij,mn} = 0$ when $i \neq r, j \neq r$. Thus we can always choose the V such that $V_{ij} = 0,\ i \neq r, j \neq r$. Thus, within the limits of the error introduced by the finite step-size, we can compute a *Separable Optimal Flow* if one exists. It is easy to see that if no positive real eigenvalue exists, then there does not exist a *Separable Optimal Flow* at W for that particular Ψ_r. Under such circumstances we can choose to skip to the next element in the sequence $\{r_k\}$ and retry.

Acknowledgment

We gratefully acknowledge support from ONR grant no. N00014-09-1-1052, NSF grant no. IIS-0427313, ARO grant no. W911NF-05-1-0219, ONR grants no. N00014-07-1-0829 and N00014-08-1-0696, and ARL grant no. W911NF-08-2-0004.

References

[1] Ali Ahmadzadeh, Gilad Buchman, Peng Cheng, Ali Jadbabaie, Jim Keller, Vijay Kumar, and George Pappas. Cooperative control of uavs for search and coverage. *Proceedings of the AUVSI Conference on Unmanned Systems*, 2006.

[2] Ali Ahmadzadeh, James Keller, George J. Pappas, Ali Jadbabaie, and Vijay Kumar. Critical cooperative surveillance and coverage with unmanned aerial vehicles. In Vijay Kumar Oussamma Khatib and Daniela Rus, editors, *International Symposium on Experimental Robotics*, STAR, Rio de Janeiro, July 2006. Springer-Verlag.

[3] Dimitri P. Bertsekas. *Dynamic Programming and Optimal Control*. Athena Scientific, 2007.

[4] Dimitri P. Bertsekas and John N. Tsitsiklis. *Parallel and Distributed Computation:Numerical Methods*. Prentice Hall, 1989.

[5] Subhrajit Bhattacharya, Maxim Likhachev, and Vijay Kumar. Distributed path consensus algorithm. Technical Report MS-CIS-10-07, University of Pennsylvania, 2010.

[6] Stephen Boyd and Lieven Vandenberghe. *Convex Optimization*. Cambridge University Press, 2009.

[7] J. Desai, J. Ostrowski, and V. Kumar. Modeling and control of formations of nonholonomic mobile robots. *IEEE Transactions on Robotics and Automation*, 17(6):905–908, December 2001.

[8] M Bernardine Dias, Robert Michael Zlot, Nidhi Kalra, and Anthony (Tony) Stentz. Market-based multirobot coordination: a survey and analysis. *Proc. of the IEEE*, 94(7):1257 – 1270, 2006.

[9] Edmund H. Durfee. Distributed problem solving and planning. *Lecture Notes in Computer Science*, 2086(2):118–149, 2001.

[10] Brian Gerkey and Maja Mataric. A formal analysis and taxonomy of task allocation in multi-robot systems. *Int'l. J. of Robotics Research*, 23(9):939–954, 2004.

[11] D. Grigoriev and A. Slissenko. Polytime algorithm for the shortest path in a homotopy class amidst semi-algebraic obstacles in the plane. In *ISSAC '98: Proceedings of the 1998 international symposium on Symbolic and algebraic computation*, pages 17–24, New York, NY, USA, 1998. ACM.

[12] Savvas G. Loizou and Kostas J. Kyriakopoulos. Closed loop navigation for multiple holonomic vehicles. In *Proc. of IEEE/RSJ Int. Conf. on Intelligent Robots and Systems*, pages 2861–2866, 2002.

[13] Diego A. Murio. *The Mollification Method and the Numerical Solution of Ill-Posed Problems*. Wiley-Interscience, 1993.

[14] I. Necoara and J.A.K. Suykens. Interior-point lagrangian decomposition method for separable convex optimization. *Journal of Optimization Theory and Applications*, 143:567–588, 2009.

[15] Sikandar Samar, Stephen Boyd, and Dimitry Gorinevsky. Distributed estimation via dual decomposition. 2007.

[16] H. Zhang, V. Kumar, and J. Ostrowski. Motion planning under uncertainty. In *IEEE International Conference on Robotics and Automation*, Leuven, Belgium, May 16-21 1998. IEEE.

PLISS: Detecting and Labeling Places Using Online Change-Point Detection

Ananth Ranganathan
Honda Research Institute, USA
aranganathan@honda-ri.com

Abstract—**We present PLISS (Place Labeling through Image Sequence Segmentation), a novel technique for place recognition and categorization from visual cues. PLISS operates on video or image streams and works by segmenting it into pieces corresponding to distinct places in the environment. An online Bayesian change-point detection framework that detects changes to model parameters is used to segment the image stream. Unlike current place recognition methods, in addition to using previously learned place models for labeling, PLISS can also detect and learn a previously unknown place or place category in an online manner. Moreover, since both the inferred boundaries of places (change-points) and the place labels are fully probabilistic, they can indicate when the inference is uncertain. New places and categories are detected using a systematic statistical hypothesis testing framework. We present extensive experiments on a large and difficult image dataset. We validate our claims by comparing results obtained using different types of features and by comparing results from PLISS against the state of the art.**

I. Introduction

Place recognition is the task of consistently labeling a particular place every time it is visited, while place categorization is the corresponding problem for a category of places. Such labels may range from "Place No. 1" and "Kitchen on 2nd floor with microwave and coffee machine", which are labels for particular places, to "Kitchen" and "corridor", which are category labels. Place recognition and categorization are essential in order for a robot or an intelligent agent to recognize places in a manner similar to that done by people. Place recognition facilitates human-robot communication [28] and is an integral part of semantic mapping procedures [35]. For instance, the label of a place strongly affects, among other things, the types of objects found there[29].

Most existing place recognition systems assume a finite set of place labels, which are learned offline from supervised training data. This is done using classifiers, which then categorize places based on input measurements during runtime. Classifier-based systems have the advantage of simplicity but also have many corresponding disadvantages

1) If a label category has large variation in measurements (for e.g., offices are vastly dissimilar in all aspects), a huge training set is necessary for adequate performance
2) The system is constrained to classify the input into the specified category set and cannot recognize the existence of a new label
3) Temporal consistency in the output labeling has to be externally enforced since the classifier simply labels each measurement individually

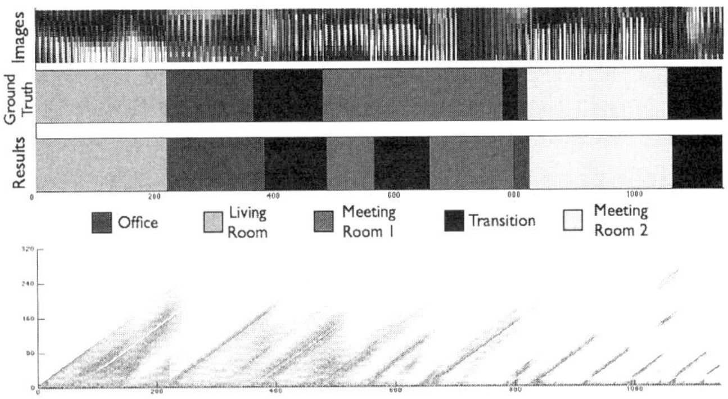

Figure 1. Place labeling using PLISS: Output is for an image sequence of 1043 frames containing 5 labels. Thumbnails of the images are shown on top, followed by groundtruth and maximum-likelihood output labeling. Many, but not all, changes in labels correspond to visible changes in appearance of the image thumbnails. The change-point posterior on segment lengths is shown at the bottom with darker regions denoting higher probability.

In this paper, we present a new place recognition method called PLISS, for Place Labeling through Image Sequence Segmentation, which tackles these problems. As its name suggests, PLISS works with video or image streams, thus intrinsically considering the temporal component of the problem. We take the novel approach of using change-point detection to segment the image streams into portions corresponding to places. Change-point detection is the problem of detecting abrupt changes to the parameters of a statistical model. The locations of these abrupt changes within the image stream are also the place boundaries, where a place is entered or exited.

We use an online Bayesian change-point detection algorithm that computes the probability of a change-point occurring at each timestep. The algorithm keeps track of all possibilities and does not make an irrevocable decision at any step. The probability of a change-point at any given timestep is obtained by combining a prior on the occurrence of change-points with the likelihood of the current measurement given all the possible scenarios in which change-points could have occurred in the past. We demonstrate that this computation can be done exactly for certain classes of prior and likelihood functions. Since the memory requirement for the exact algorithm increases linearly with the number of measurements processed thus far, we also provide an approximation using Rao-Blackwellized particle filters that is fast and requires only constant memory.

While change-point detection provides boundaries, the place

label is assigned probabilistically based on the measurement, the most recent label, and the change-point distribution. Statistical hypothesis testing is used to determine if the current measurement could have been generated by any of the pre-learned place models. If this is not the case, the measurement is declared to have come from a previously unknown place. In this manner, PLISS can systematically recognize a previously unknown place type and assign it a new label if required. Further, PLISS can learn and update place models online, which is helpful in cases where measurements from the same category show variation (albeit only if the variation is gradual). Hence, PLISS can operate with labeled data, which is used to learn place models at training time, or without it, when it recognizes new place categories as they arise and learns place models for them online.

We model places using the Multivariate Polya distribution, also known as the Dirichlet Compound Multinomial (DCM) model [19]. The DCM is a bag-of-words model and captures burstiness of the data, i.e. it models the observation that if a word occurs once in a document, it usually occurs repeatedly. We use the DCM model to model histogram measurements obtained by quantizing dense features computed on the images.

A sample result from PLISS on a sequence with 5 labels is shown in Figure 1. To test our claims, we present experiments on the large and difficult Visual Place Categorization (VPC) dataset [32]. We also provide comparisons using different types of features, as well as comparisons of the performance of PLISS against the VPC system [33].

II. Related Work

Work on place recognition in robotics has ranged over matching SIFT features across images [37] or other derived measures of distinctiveness for places such as Fourier signatures [18], subspace representations [11], and color histograms [31]. These methods have the disadvantage of not being able to generalize and also are invariant to perspective mainly through the use of omnidirectional images. Many approaches that generalize well, learn classifiers, such as SVMs, for each place from labeled data [25], [27]. A very recent system by Pronobis et al [24] focusses on merging cues from different sensing modalities, the output for each of which is obtained from individually trained SVMs. These, and similar classifier-based approaches cannot detect or learn previously unknown places and place categories, in contrast to PLISS. Place classification methods based on laser and sonar range scans also exist [13], [20], though we do not review them here.

In computer vision, place classification is also known as scene classification and image categorization. Two broad methods can be discerned - those that model local features [2] and distinctive parts of images[26], and those that extract global representations of images and learn from them[22][10][3]. The latter is closer to this work and includes the CENTRIST-based VPC system [33], and Spatial Pyramid Matching [15]. These methods also use classifiers to learn place categories and cannot detect new categories online. Further, they do not deal with image streams and videos except when filtering procedures are added on top to get temporal consistency [33]. Other approaches use keypoint matching for location recognition based on image retrieval [21][6] but cannot generalize.

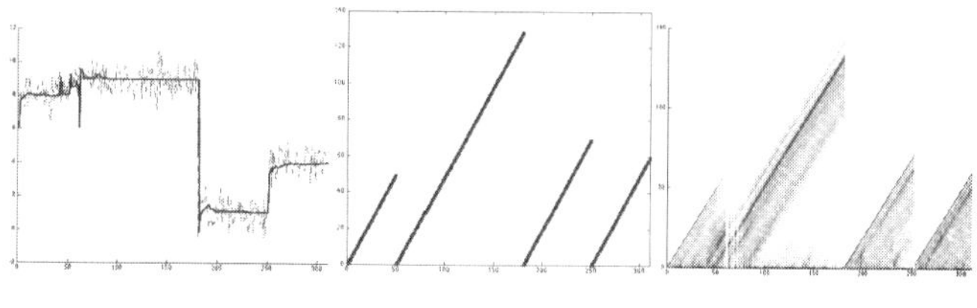

Figure 2. Change-point detection: (a) Univariate Gaussian input data (blue/dashed) with 4 segments and inferred segment means (in black) (b) Groundtruth for change-points shown as length of segments, which is the variable c_t used in our inference. (c) Output of change-point detection which is probability distribution on c_t.

Change-point detection has a long history in statistics and the best-known technique is the CUSUM detector [23] which involves piece-wise segments of Gaussian mean with noise. A more realistic method is segmented regression [7]. In our exposition, we closely follow the more general methods of [1], [9] which is applicable to conjugate-exponential models. Many applications in computer vision have used change-point detection previously[36][30][5]. However, to our knowledge, this is the first application of change-point detection to place recognition.

III. Place Recognition using Change-Points

We formulate the place recognition problem as follows. We are given a measurement stream that produces measurements at (possibly changing) intervals, which we will also call as timesteps. For each measurement, we are required to come up with a label corresponding to the type of the place. The place types are given in the form of L place models $M_1, M_2, \ldots, M_L$. We are also required to say if the measurement does not correspond to any of these labels.

We approach the problem by noting that the place label remains the same for the period of time when a robot is moving inside the particular type of place. It only changes sporadically when the robot travels into the next place. Thus, the measurement stream can be segmented into pieces corresponding to places, i.e. measurements in each segment are assumed to have been generated by the corresponding place model. The start and end of segments where the generating model changes, hence called change-points, provide a very strong indication regarding the place label.

We assume that a sequence of data $y_1, y_2, \ldots, y_t$ can be segmented into non-overlapping but adjacent segments. The boundaries between these segments are the change-points. We deal with the model-based change-point scenario here, wherein the form of the probability distribution in each segment remains the same and only the parameter value changes. We assume that the data are i.i.d within each segment and denote by c_t the length of the segment at time t. Note that c_t is also the time since the last change-point. If the current timestep is a change-point, then $c_t = 0$, and if no change-points have occurred yet, then $c_t = t$. A sample problem setup and output for change-point detection in the univariate Gaussian case in shown in Figure 2.

We denote the place label at time t as x_t^c. The place label is indexed by the current segment since the whole segment has a single label. However, this label is updated with each measurement, and hence the time index t is also used. The probability distribution over x_t^c is taken to be a discrete distribution of size L, one for each of the place models. The case where the place label takes none of these values is detected using statistical hypothesis testing.

We need to compute the joint posterior on c_t and x_t^c given the data, $p(c_t, x_t^c|y_{1:t})$, where $y_{1:t}$ denotes all the data from time 1 to time t. The posterior can be factored as

$$p(c_t, x_t^c|y_{1:t}) = p(c_t|y_{1:t})p(x_t^c|c_t, y_{1:t}) \quad (1)$$

The first term is the posterior over the segment length while the second term is the conditional posterior on the place label given the segment length. Note that the posterior over segment length is equivalent to inferring the change-points since $c_t = 0$ implies a change-point. Further, though the place label posterior is conditioned on the segment length, this does not imply that the place label can only change at a change-point. Since the place label is updated at every timestep, the algorithm can "change it's mind" regarding the place label at any point in time, given enough evidence.

We address change-point detection, i.e. computing $p(c_t|y_{1:t})$ in the next section. The place label posterior $p(x_t^c|c_t, y_{1:t})$ is addressed subsequently in Section V.

IV. Model-based Change-Point Detection

In the following exposition, we closely follow [1] and [9], both of which state essentially similar algorithms but with different state representations. We represent the likelihood of the data in segment c_t as $p(y_t|\xi_t^c)$ where ξ_t^c is a parameter set. The data inside each segment are assumed to be i.i.d and the parameters are assumed i.i.d according to a prior parameter distribution.

The change-point posterior from (1) can be expanded using Bayes law as

$$p(c_t|y_{1:t}) \propto p(y_t|c_t, y_{1:t-1})p(c_t|y_{1:t-1}) \quad (2)$$

The first term is the likelihood while the second term can be further expanded by marginalizing over the segment length at the previous timestep to yield a recursive formulation for c_t

$$p(c_t|y_{1:t-1}) = \sum_{c_{t-1}} p(c_t|c_{t-1})p(c_{t-1}|y_{1:t-1}) \quad (3)$$

where $p(c_t|c_{t-1})$ is the transition probability, $p(c_{t-1}|y_{1:t-1})$ is the posterior from the previous step, and we have made use of the fact that $c_1, c_2, \ldots, c_t$ form a Markov chain.

A. The Transition Probability

For characterizing the transition probability $p(c_t|c_{t-1})$ in (3), we note that the only two possible outcomes are $c_t = c_{t-1} + 1$ when there is no change-point at time t, and $c_t = 0$ otherwise. Hence, this is a prior probability on the "lifetime" of this particular segment where the segment ends if a change-point occurs. In survival analysis, such situations are routinely modeled using a hazard function, which represents the probability of failure in a unit time interval conditional on the fact that failure has not already occurred. If $H(.)$ is a hazard function, the transition probability can be modeled as

$$p(c_t|c_{t-1}) = \begin{cases} H(c_{t-1}+1) & \text{if } c_t = 0 \\ 1 - H(c_{t-1}+1) & \text{if } c_t = c_{t-1}+1 \end{cases} \quad (4)$$

In the simplest case where the probability of a change-point at every step is assumed constant, the length of a segment has to be modeled using an exponential distribution with time scale λ [9], so that $H(t) = 1/\lambda$.

B. The Data Likelihood

The data likelihood from (2) can be calculated only if we know the distribution parameter to use. Hence, we integrate over the parameter value using the parameter prior

$$p(y_t|c_t, y_{1:t-1}) = \int_{\xi^c} p(y_t|\xi^c)p(\xi^c|c_t, y_{t-1}^c) \quad (5)$$

where ξ^c is the model parameter for segment c_t, and y_{t-1}^c is the data from the current segment. The above integral can be computed in closed form if the two distributions inside the integral are in the conjugate-exponential family of distributions. Expensive numerical integrations have to be employed otherwise. In the following exposition, we assume the conjugate case.

Further, let the integrated function be denoted as $p(y_t|c_t, \eta_t^c)$ where η_t^c parametrizes the integrated data likelihood. Even though this function is usually not in the exponential family, it may be possible to update it directly using the sufficient statistics of the data corresponding to the current segment $\{y_{t-1}^c, y_t\}$, i.e. the integration need not be performed at every step. In the case where t is a change-point, the function is computed with prior values for $\eta_t^{(0)}$ (since $c_t = 0$) instead of any sufficient statistics.

C. Computational Cost

The algorithm described so far is exact. After n measurements, the possible segment lengths range from 0 to n, and the posterior contains the probability of all these cases. Further, if the optimal change-point locations are also needed, the posteriors from all timesteps have to be kept. Hence, the runtime and memory costs per timestep are $O(n)$ while the total memory cost is $O(n^2)$. These requirements are incompatible with long term online operation. Hence, we next provide a simple and intuitive particle filtering approximation that exhibits constant runtime and $O(n)$ total memory cost.

D. Online Operation Using Particle Filtering

We need the locations of change-points and the exact algorithm above accomplishes this by maintaining the posterior over segment lengths c_t for all t. We now approximate the posterior using N weighted particles, thereby obtaining a constant runtime algorithm.

Combining (2), (3), and (4) we get the segment length posterior as

$$p(c_t|y_{1:t}) \propto \begin{cases} w_t^{(0)} \sum_{c_{t-1}} H(c_{t-1}+1)\rho_{t-1} & \text{if } c_t = 0 \\ w_t^{(c)} \sum_{c_{t-1}} \{1 - H(c_{t-1}+1)\}\rho_{t-1} & \text{if } c_t = c_{t-1}+1 \end{cases} \tag{6}$$

where $w_t^{(c)} = p(y_t|c_t, y_{t-1}^c)$ and, for the case where t is a change-point and y_{t-1}^c is the empty set, $w_t^{(0)} = p(y_t|c_t)$. $\rho_{t-1} = p(c_{t-1}|y_{1:t-1})$ is the posterior from the previous timestep.

Clearly, the posterior (6) is amenable to straight-forward use in a particle filter with w_t as the particle weights. We use the optimal stratified resampling method of [8] that ensures runtime and memory usage proportional to the number of particles, i.e. $O(1)$ with regard to number of measurements seen so far. The particle weights are given by (5).

Note that since the likelihood parameters ξ^c in (5) are integrated out, this particle filter is Rao-Blackwellized [4] and has lower variance than a standard particle filter. This also makes the convergence of the algorithm more efficient.

V. Inferring the Place Label

The conditional posterior on the place label from (1) can be expanded using Bayes law as

$$p(x_t^c|c_t, y_{1:t}) \quad \propto \quad p(y_t^c|x_t^c, c_t)p(x_t^c|c_t) \tag{7}$$

where y_t^c is the data in the current segment, i.e. $y_t^c = \{y_{t-c_t}, \ldots, y_t\}$, since the place label only depends on these measurements. If we have L place models $M_1, M_2, \ldots, M_L$, the probabilities of each of these cases can be updated using (7). We simply use the label probability for the segment computed in the previous timestep as the prior, i.e. $p(x_t^c|c_t) = x_{t-1}^c$. For a new segment with $c_t = 0$, we set the prior to be a uniform distribution over the known labels.

Detection of an unknown place requires more involved calculation. In this case, we are required to show that the data does not arise from any of the models corresponding to known places. A systematic way of arriving at this conclusion involves statistical hypothesis testing. Hence, at each timestep, we perform L hypothesis tests to determine if the data arises from a known place. If these tests are expensive, they may also be performed once every T timesteps.

We now consider hypothesis testing for model M_i with parameter vector η so that the exact probability under this model is $p_0 = p(y_t^c|\eta)$. The significance of the observed data is the probability of all the data that is equally or less probable (more extreme) under η, $p_\sigma = \sum_{p(y|\eta)<p_0} p(y|\eta)$. This expression is exact but intractable for almost all but the most trivial models. We approximate using the likelihood ratio where the ratio of the model under the maximum likelihood parameter value and the true parameter value is computed

$$R \quad = \quad \frac{p_0}{p(y_t^c|\eta_{ml})} \tag{8}$$

where $\eta_{ml} = \text{argmax}_{\ \eta}\, p(y_t^c|\eta)$.

The statistic used in the hypothesis test is $-2\ln R$, where R is the likelihood ratio (8). This statistic can be shown to converge to the Chi-squared distribution with $k-1$ degrees of freedom [14] where k is the dimensionality of the parameter vector θ. The model M_i can be rejected if the Chi-squared probability is less than some threshold, usually set at 5% (0.05) or 1% (0.01). The test statistic converges to the Chi-squared distribution at the rate of $O(N^{-1})$, where N is the number of measurements used to compute the maximum likelihood parameter value η_{ml} [12]. In our case, each image feature is a measurement, and since the number of features per image is in the hundreds, convergence is not an issue.

We carry out the above test for each known place model and declare the place to be previously unknown if the tests reject all of them. Since the Chi-squared probabilities from the test do not say anything about the probability of the new label, the place distribution is set to a prior value for unknown places $p(x|\text{new label})$, which in our case is set so that the new label has probability 0.5 and all the other labels are equally probable. The new place label can be either stored for future reference along with the maximum likelihood model parameters η_{ml}, or be discarded if new places are of no interest. Thus, PLISS can detect new places and learn models for them online in contrast to most existing place categorization methods.

In terms of implementation, we augment the change-point algorithm of Section IV so that the discrete distribution on places is stored with each segment c_t. Similarly, in the particle filter, each particle maintains a place distribution. The place distribution becomes increasingly confident as the length of the segment c_t increases and is robust to noisy measurements and outliers. However, since it is also recomputed with each measurement, the algorithm does not make any irrevocable decision with regard to the place label and can "change its mind" given enough evidence. The cost of updating the place distribution is linear in the number of labels and hence, does not affect the runtime of the change-point algorithm.

VI. Measurements and Place Models

We model images using a "bag of words" model wherein a histogram is used to represent the image. These histograms are also used as input measurements to the PLISS algorithm. First, in an offline phase, SIFT features are computed on a dense grid on each of a set of images. The features are vector quantized using K-means to create a codebook of pre-specified size. Note that it is not necessary in this step for the image set used to be similar to the test images, though better results are obtained if this is true.

We compute a spatial pyramid [15] from the quantized SIFT features which is used as input to the PLISS system. The spatial pyramid is obtained by computing histograms, at various spatial resolutions, of feature frequencies in each of the codebook clusters. Following [15], we obtain successive spatial resolutions by dividing the image into a grid as shown in Figure 3. Note that only the histograms at the finest resolution need to be computed since the coarser resolution histograms can be obtained by simply adding the appropriate histograms at an immediately finer level. All the histograms

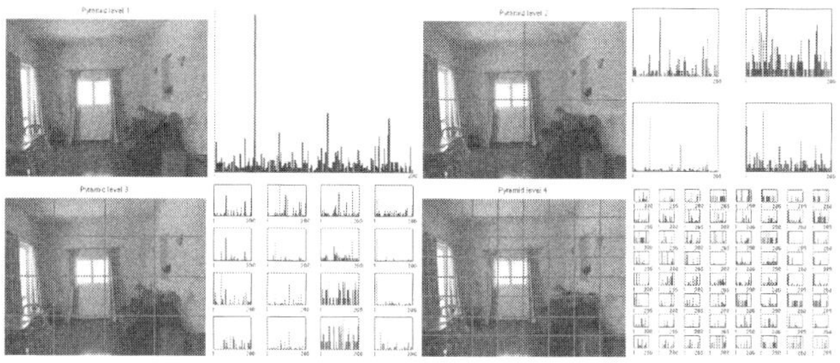

Figure 3. The Spatial Pyramid histogram: Histograms of clustered SIFT features, computed on image regions at different spatial resolutions, are concatenated to yield the representation of the image. The image regions are obtained by dividing it into successively finer grids.

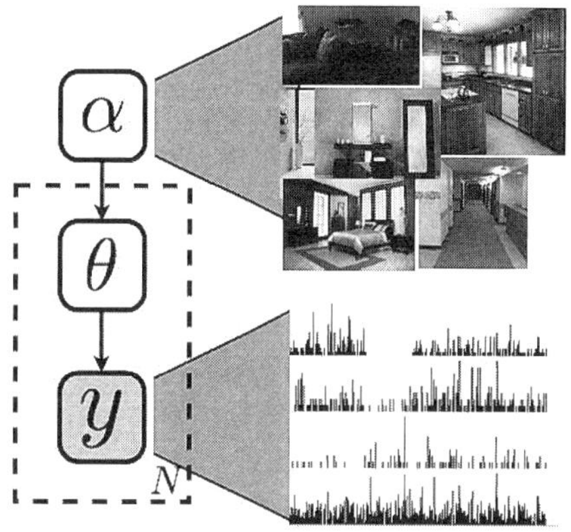

Figure 4. Graphical model illustrating the Multivariate Polya distribution. To obtain a measurement z, which is a quantized feature histogram, we first sample from a Dirichlet distribution with parameter α to obtain a Multinomial vector θ . This Multinomial distribution is, in turn, sampled to obtain the measurement histogram y. Note that a different θ has to be sampled for each y. Each α corresponds to a place or place category.

from the different grids are then concatenated to yield the spatial pyramid representation. The two parameters for computing the spatial pyramid are thus, the number of levels in the pyramid V and the number of clusters computed in SIFT space (size of the dictionary) K. SIFT features only have local information about an image patch while an image histogram only has global information. By combining both of these at different scales, the spatial pyramid obtains more fine-grained discriminative power.

In addition to SIFT features, we also compute spatial pyramids using two other features, CENTRIST [32] and texture, which we now describe. CENTRIST is based on the census transform [34], which is a local feature computed densely for every pixel of an image, and encodes the value of a pixel's intensity relative to that of its neighbors. It was originally introduced for identifying correspondence between local patches. The census transform is computed by considering a patch centered at every pixel. The transform value is a positive integer that takes a range of values depending on the size of the patch. For instance, a patch size of 3, where there are 8 pixels in the patch apart from the central pixel, yields transform values between 0 and 255 (2^8 values). This is equivalent to having a dictionary size of 256 without the need for the clustering step. Computing a histogram of these 256 values yields the CENTRIST (census transform histogram) descriptor. Additionally, the census transform itself is extremely efficient to compute. Hence, a spatial pyramid can be calculated much faster using CENTRIST than using SIFT features.

We also compute a spatial pyramid using texture. Texture is extracted using the 17-dimensional Leung-Malik filter bank [17]. Similar to the case of SIFT features, an image set is used on which texture is extracted and clustered using K-means to obtain a set of textons. Histograms containing the proportions of each of the textons are computed at different resolutions as before to obtain the texture-based spatial pyramid.

Place Models Using Spatial Pyramids

The place model is used to compute the measurement likelihood in (5). Since the measurements are histograms of word counts, we model them using a multinomial distribution having dimensions equal to the dictionary size. Further the prior over the multinomial parameter is the conjugate Dirichlet distribution to aid in ease of computation. Given a histogram measurement y, its likelihood according to (5) is

$$P(y|\alpha) \;=\; \int_{\theta} P(y|\theta)P(\theta|\alpha) \tag{9}$$

where $\theta = [\theta_1, \theta_2, \ldots, \theta_W]$ and $\alpha = [\alpha_1, \alpha_2, \ldots, \alpha_W]$ are the multinomial parameter and Dirichlet prior respectively, and W is the dictionary size. Assuming that the histogram y has bin counts given by $[n_1, n_2, \ldots, n_W]$, the distributions in the integrand above can be written as

$$p(y|\theta) \;=\; \frac{n!}{n_1!n_2!\ldots n_W!}\theta_1^{n_1}\theta_2^{n_2}\ldots\theta_W^{n_W} \tag{10}$$

$$p(\theta|\alpha) \;=\; \frac{\Gamma(\sum_{w=1}^{W}\alpha_w)}{\sum_{w=1}^{W}\Gamma(\alpha_w)}\theta_1^{\alpha_1-1}\theta_2^{\alpha_2-1}\ldots\theta_W^{\alpha_W-1} \tag{11}$$

The likelihood model in (9), where $P(y|\theta)$ is a multinomial distribution (10) and $P(\theta|\alpha)$ is a Dirichlet distribution (11), is called the Multivariate Polya model [19], or equivalently in document modeling, the Dirichlet Compound Multinomial (DCM) model [16]. The DCM distribution models burstiness in the data, i.e. given that a quantized feature appears once in a document, it is much more likely for it to occur multiple times rather than just once or twice. This is an intuitive observation in all realistic image data, particularly since we are dealing with densely computed features.

Performing the integration in (9), we get the final form of the likelihood, which is also our place model, as

$$P(y|\alpha) \;\propto\; \frac{n!}{\prod_{w=1}^{W} n_w}\frac{\Gamma(|\alpha|)}{\Gamma(n+|\alpha|)}\prod_{w=1}^{W}\frac{\Gamma(n_w+\alpha_w)}{\Gamma(\alpha_w)} \tag{12}$$

where $n = \sum_w n_w$, $|\alpha| = \sum_w \alpha_w$, and $\Gamma(.)$ denotes the Gamma function. Graphical intuition for the DCM model is provided by Figure 4. If the likelihood of a set of measurements is to be computed, then n is taken to be the total counts across all measurements, while n_w is the total count for a particular word.

Given a set of D images with features detected on them, the maximum likelihood value for α can be found by optimizing using gradient descent. It can be shown that this leads to the following fixed point update for the α parameter[19]

$$\alpha_w^{new} = \alpha_w \frac{\sum_{d=1}^{D} \psi(n_{dw} + \alpha_w) - \psi(\alpha_w)}{\sum_{d=1}^{D} \psi(n_{dw} + \alpha) - \psi(\alpha)} \quad (13)$$

where $\alpha = \sum_w \alpha_w$ as before, and $\psi(.)$ is the Digamma function, the derivative of the Gamma function. Faster but more complicated updates using Gauss-Newton iterations also exist [19].

We use DCM distributions as place models in PLISS. The α parameter for each place is learned from labeled images in an offline training phase, if training data is provided. During runtime, the distribution is used to compute the likelihood (5), and the α parameter is also updated after each measurement using the iterative rule (13) or the slightly faster Gauss-Newton updates. This facilitates online learning but if online learning is not required, the updated parameter can be discarded at the end of the segment.

Note that, since we are using spatial pyramids as input, it is only required to model the histograms at the finest level using the Multivariate Polya model since the coarser level histograms are simply summations of these. Thus, for a pyramid with V levels and level $V = 0$ denoting the whole image, the dimensionality of α is $4^V W$. However, a value of $V > 3$ has not been required in our experience. The expression for the hypothesis testing statistic, which is $-2 \ln R$ where R is the likelihood ratio, can also be easily obtained by substituting the distribution expression (12) into the likelihood ratio (8).

A recap of the overall PLISS algorithm using the DCM model is given in Algorithm 1.

Algorithm 1 Particle filtering algorithm for PLISS with online learning using the Multivariate Polya model

1) **Initialize:** Set prior parameter α_0. For all particles, $c_0 = 0$ and $x_0^0 = unif$ (unif. dist. over known labels).
2) **Update particle set:**
   For every timestep t do -
   For every particle $\{w_{t-1}, c_{t-1}, x_{t-1}^c, \alpha_{t-1}^c\}$ do
   - Create two new particles (1) No change-point case $l_1 = \{w_{t-1}, c_{t-1} + 1, x_{t-1}^c, \alpha_{t-1}^c\}$ (2) change-point $l_2 = \{w_{t-1}, 0, unif, \alpha_{t-1}^c\}$
   - Compute the prior for the (4) and update weights of l_1 and l_2
   - Using y_t, learn new parameter α_t^c for l_1 using (13) and set parameter α_t^0 for l_2 to α_0
   - Compute incremental weights for l_1 and l_2 from likelihood function (12) and multiply with particle weights
3) **Resample** from weights to get new set of particles
4) **Update place distributions:**
   For every particle $\{w_t, c_t, x_{t-1}^c, \alpha_t^c\}$ do
   - Perform a statistical hypothesis test using the likelihood ratio test (8) for each known place model
   - If a test indicates y_t^c to be arising from an existing place model, update x_{t-1}^c using (7) to get x_t^c
   - If all existing place models are rejected, create new place label and set x_t^c to the prior distribution $p(x|\text{new label})$

	Correct	False Positive	False Negative	% Correct
SIFT	62	6	14	81.6
CENTRIST (3x3)	55	11	21	72.3
CENTRIST (5x5)	59	6	17	77.6
Texture	51	3	25	67.1

(a)

	Labeling Accuracy (%)
SIFT	61.2
CENTRIST (3x3)	53.9
CENTRIST (5x5)	58.3
Texture	52.0

(b)

Figure 5. Results with no training data: (a)Change-point detection performance for PLISS using various feature types (b)Place labeling accuracy

VII. Experiments

We tested the PLISS system extensively on actual image data and describe the methodology and results here. The PLISS system was implemented in Matlab with no attempt made at any optimization. All parts of the algorithms, including feature detection, were performed in Matlab.

We use the Visual Place Categorization (VPC) dataset [32] for our experiments. The dataset consists of image sequences from six different homes, each containing multiple floors. The image set from each home consists of between 6000 and 10000 frames. In our experiments, we consider sequences from each floor to be a different image sequence. The dataset has been manually labeled into 5 categories to provide ground truth for the place categorization problem. In addition, a “transition” category is used to mark segments that do not correspond clearly to any class. The VPC dataset is significantly difficult since no effort has been made to keep all the images in the sequence informative. Thus, a number of images contain only a wall, which is something that could also be expected when a robot is moving around.

We computed SIFT features on a grid having width and height 8 pixels per cell. Features were computed on 16x16 image patches. The features were clustered to obtain a dictionary of size 256. For the CENTRIST features, we used patch sizes of 3x3 and 5x5. While the natural dictionary size of 256 was used for the 3x3 case, we clustered the census transform obtained from the 5x5 case to obtain 256 clusters. The number of textons generated for the texture features was also 256. We used a two-level pyramid to compute the spatial pyramid histogram in all cases. In experiments without training data, the dictionary was computed using K-means clustering of features extracted from a set of 500 images drawn randomly from the VPC data. Prior values for the place model parameter α to be used in the likelihood computation (5) at a change-point (when no data is available for the segment) were learned from this same 500 image set. The significance level for the hypothesis tests was set at 1%. The prior distribution for a new label $p(x|\text{new label})$ was set to 0.75 for the new label and equal probability for the others. All results were obtained using the particle filtering algorithm with 100 particles, as this gave similar results to the exact algorithm.

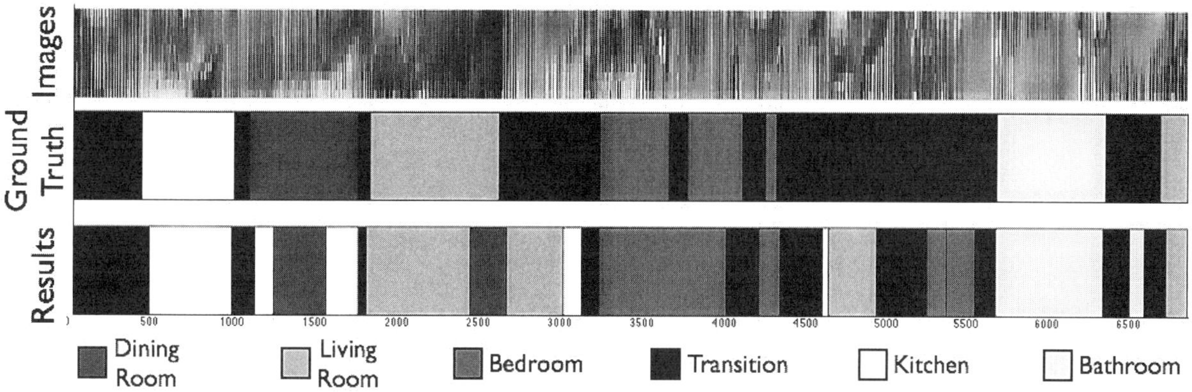

Figure 6. Maximum likelihood output labeling for the 1st floor image sequence from Home 3 of the VPC dataset, which is one of the more difficult ones and contains 6839 images. Thumbnails of original images, groundtruth labeling and the PLISS result are shown. Note that it is possible to spot many of the change-points from just the high-level image characteristics visible here but not all of them.

A. Experiments Without Offline Training

Almost all existing place recognition systems would be unusable if no training data were available. However, PLISS is still able to segment the data into pieces corresponding to different places and learn their models online. Hence, this experiment also tests the online learning capability of our system. Output was obtained for six image sequences from the VPC dataset containing a total of 14,346 frames. The total number of changepoints in these sequences, taken to be frames where the place label changes, was 76. The change-point detection performance using various features is shown in Figure 5(a). SIFT features perform best while texture features perform the worst. The system was declared to have found the change-point if it fired within 20 frames of the groundtruth. Note that many of the change-points, especially involving "transition" regions, are hard to recognize even for people.

Place labels for the sequences were also learnt online, and it was expected that the system would give the same place label when returning to a specific place but not when encountering another place with the same category label. The labeling was modified manually to reflect this for the six sequences and 39 distinct places were labeled. Results for place labeling are given in Figure 5(b). Considering that no training data has been used, the results are very promising when compared with the VPC system (Figure 8).

B. Place Recognition and Categorization

We next present experiments on specific place recognition which differs from the task of place categorization. The same six sequences were used, with 39 distinct labeled places, as in the previous experiment. We trained the system on every 3rd image and its corresponding label from these sequences to obtain the place models for each place. Note that k-means clustering and prior learning were done on the whole training set. Results obtained on the 9564 test image frames are given in Figure 7. The recognition rates are much higher since the training data and test data are similar, due to which all three features also yield similar results. While a more realistic test would also account for variation in lighting in the test data, the current results provide indication of the soundness of our approach even for a relatively large number of places.

We evaluated PLISS on the place categorization problem using the leave one out strategy followed by [32] to facilitate comparison. The system was trained on labeled data from 5 houses and tested on the 6th one. Average place categorization results from all 6 houses are reported. In practice, we only used every 3rd frame of the training set. We also omitted frames from the "transition" category during training and a model for this category was not learned. During runtime, any frame not belonging to one of 5 known labels was labeled as a "transition" frame. The maximum likelihood place labeling for a sequence is shown in Figure 6 along with groundtruth.

	Recognition Accuracy (%)
SIFT	84.2
CENTRIST (3x3)	81.6
CENTRIST (5x5)	83.0
Texture	81.2

Figure 7. Results for place recognition

Figure 8 shows the result of place categorization across all sequences for each of the six categories in the dataset. Results for the VPC system are taken from [32], which does not give performance for the "transition" class. PLISS using SIFT features performs similar to the VPC system on average and is better on a few individual classes. However, this is a very strong result considering that we have used a simple generative model while VPC uses SVM classifiers. Note that we also include results of the "transition" class, which PLISS can recognize reasonably well even without any training data provided in that category. The SVM-based VPC system would not be expected to perform well in this class since it is a catch-all "other" category with widely varying visual characteristics. PLISS with CENTRIST is a good compromise if a faster system is required.

VIII. Discussion

We have presented PLISS, a system for place recognition and categorization based on change-point detection. PLISS has significant advantages over existing methods of being able to detect and learn previously unknown places and place

	Bedroom	Kitchen	Living Room	Bath	Dining	Transition	Average
SIFT	61.72	51.11	29.08	69.61	14.94	42.82	44.88
CENTRIST (3x3)	60.26	43.36	17.49	63.31	13.32	36.38	39.02
CENTRIST (5x5)	61.81	48.27	24.07	65.68	16.18	37.43	42.24
Texture	56.17	39.41	18.92	60.15	8.84	31.91	35.9
VPC	*64.89*	*48.24*	*20.59*	*74.77*	*19.61*	-	*45.62*

Figure 8. Results for place categorization (% correctness): VPC average is over 5 categories while PLISS results include the "transition" category. VPC results have been taken from [32].

categories, of being able to learn online, and of being able to operate without any training data. Experiments on a difficult dataset show that, along with these advantages, PLISS also matches the state of the art in performance.

The basic assumption underlying PLISS is that places are sufficiently distinct to be identified visually. If this is not the case, i.e. the environment is severely perceptually aliased, performance will degrade as with any vision-based system. However, PLISS can be easily extended to incorporate multiple sensors to overcome such scenarios.

We envision significant performance improvements with more sophisticated place models. Such models may even incorporate object and context information from places, which cannot be done easily with SVM classifiers. Currently, PLISS does not distinguish between place labels and category labels and relies on the place models for making this distinction. It is future work to overcome this deficiency.

REFERENCES

[1] R.P. Adams and D.J.C. MacKay. Bayesian online changepoint detection. Technical report, University of Cambridge, Cambridge, UK, 2007. arXiv:0710.3742v1 [stat.ML].

[2] H. Andreasson, A. Treptow, and T. Duckett. Localization for mobile robots using panoramic vision, local features and particle filter. In *IEEE Intl. Conf. on Robotics and Automation (ICRA)*, 2005.

[3] A. Bosch, A. Zisserman, and X. Munoz. Image classification using random forests and ferns. In *Intl. Conf. on Computer Vision (ICCV)*, pages 1–8, 2007.

[4] G. Casella and C.P. Robert. Rao-Blackwellisation of sampling schemes. *Biometrika*, 83(1):81–94, March 1996.

[5] A. Taylan Cemgil, W. Zajdel, and B. Krose. A hybrid graphical model for robust feature extraction from video. In *IEEE Conf. on Computer Vision and Pattern Recognition (CVPR)*, 2005.

[6] M. Cummins and P. Newman. FAB-MAP: Probabilistic Localization and Mapping in the Space of Appearance. *The International Journal of Robotics Research*, 27(6):647–665, 2008.

[7] S. R. Esterby and A. H. El-Shaarawi. Inference about the point of change in a regression model. *Applied Statistics*, 30(3):277–285, 1981.

[8] P. Fearnhead and P. Clifford. Online inference for hidden markov models. *Journal of the Royal Statistical Society: Series B*, 65:887–899, 2003.

[9] P. Fearnhead and Z. Liu. On-line inference for multiple changepoint problems. *Journal of the Royal Statistical Society: Series B*, 69(4):589–605, 2007.

[10] L. Fei-Fei and P. Perona. A bayesian hierarchical model for learning natural scene categories. In *IEEE Conf. on Computer Vision and Pattern Recognition (CVPR)*, 2005.

[11] J. Gaspar, N. Winters, and J. Santos-Victor. Vision-based navigation and environmental representations with an omnidirectional camera. *IEEE Trans. Robot. Automat.*, 16(6):890–898, Dec 2000.

[12] W. Hoeffding. Asymptotically optimal tests for multinomial distributions. *Ann. Math. Statist.*, 36:369–408, 1965.

[13] B. Kuipers and P. Beeson. Bootstrap learning for place recognition. In *Proc.* 19th *AAAI National Conference on AI*, pages 174–180, Edmonton, Alberta, Canada, 2002.

[14] D. N. Lawley. A general method of approximating to the distribution of likelihood ratio criteria. *Biometrika*, 43:295–303, 1956.

[15] S. Lazebnik, C. Schmid, and J. Ponce. Beyond bags of features: spatial pyramid matching for recognizing natural scene categories. In *IEEE Conf. on Computer Vision and Pattern Recognition (CVPR)*, 2006.

[16] R.E. Madsen, D. Kauchak, and C. Elkan. Modeling word burstiness using the dirichlet distribution. In *Intl. Conf. on Machine Learning (ICML)*, pages 545–552, 2005.

[17] J. Malik, S. Belongie, T. Leung, and J. Shi. Contour and texture analysis for image segmentation. *Intl. J. of Computer Vision*, 43:7–27, June 2001.

[18] E. Menegatti, T. Maeda, and H. Ishiguro. Image-based memory for robot navigation using properties of the omnidirectional images. *Journal of Robotics and Autonomous Systems*, 47(4):251–267, 2004.

[19] T.P. Minka. Estimating a dirichlet distribution. 2003.

[20] O. Martínez Mozos, A. Rottmann, R. Triebel, P. Jensfelt, and W. Burgard. Semantic labeling of places using information extracted from laser and vision sensor data. In *In Proc. of the IEEE/RSJ IROS 2006 Workshop: From Sensors to Human Spatial Concepts*, 2006.

[21] D. Nistér and H. Stewénius. Scalable recognition with a vocabulary tree. In *IEEE Conf. on Computer Vision and Pattern Recognition (CVPR)*, 2006.

[22] A. Oliva and A. Torralba. Building the gist of a scene: The role of global image features in recognition. *Visual Perception, Progress in Brain Research*, 155, 2006.

[23] E. S. Page. Continuous inspection scheme. *Biometrika*, 41:100–115, 1954.

[24] A. Pronobis, O. M. Mozos, B. Caputo, and P. Jensfelt. Multi-modal semantic place classification. *Intl. J. of Robotics Research*, 2010. accepted.

[25] A. Rottmann, O. Martinez Mozos, C. Stachniss, and W. Burgard. Semantic place classification of indoor environments with mobile robots using boosting. In *Nat. Conf. on Artificial Intelligence (AAAI)*, 2005.

[26] C. Siagian and L. Itti. Biologically-inspired robotics vision monte-carlo localization in the outdoor environment. In *IEEE/RSJ Intl. Conf. on Intelligent Robots and Systems (IROS)*, 2007.

[27] A. Tapus, N. Tomatis, and R. Siegwart. Topological global localization and mapping with fingerprint and uncertainty. In *Proceedings of the International Symposium on Experimental Robotics*, 2004.

[28] E.A. Topp, H. Hüttenrauch, H.I. Christensen, and K.S. Eklundh. Bringing together human and robotic environment representations - a pilot study. In *IEEE/RSJ Intl. Conf. on Intelligent Robots and Systems (IROS)*, Beijing, China, October 2006.

[29] A. Torralba, K. P. Murphy, W. T. Freeman, and M. A. Rubin. Context-based vision system for place and object recognition. In *Intl. Conf. on Computer Vision (ICCV)*, volume 1, pages 273–280, 2003.

[30] G. Tsechpenakis, D. Metaxas, O. Hadjiliadis, and C. Neidle. Robust online change-point detection in video sequences. In *2nd IEEE Workshop on Vision for Human Computer Interaction (V4HCI), in conjunction with the IEEE Conference on Computer Vision and Pattern Recognition*, 2006.

[31] I. Ulrich and I. Nourbakhsh. Appearance-based place recognition for topological localization. In *IEEE Intl. Conf. on Robotics and Automation (ICRA)*, volume 2, pages 1023 – 1029, April 2000.

[32] J. Wu, H. Christensen, and J. M. Rehg. Visual place categorization: Problem, dataset, and algorithm. In *IEEE/RSJ Intl. Conf. on Intelligent Robots and Systems (IROS)*, 2009.

[33] J. Wu and J. M. Rehg. Where am i: Place instance and category recognition using spatial pact. In *IEEE Conf. on Computer Vision and Pattern Recognition (CVPR)*, 2008.

[34] R. Zabih and J. Woodfill. Non-parametric local transforms for computing visual correspondence. In *Eur. Conf. on Computer Vision (ECCV)*, volume 2, pages 151–158, 1994.

[35] H. Zender, P. Jensfelt, O. M. Mozos, G.-J. Kruijff, and W. Burgard. An integrated robotic system for spatial understanding and situated interaction in indoor environments. In *Nat. Conf. on Artificial Intelligence (AAAI)*, 2007.

[36] Y. Zhai and M. Shah. A general framework for temporal video scene segmentation. In *Intl. Conf. on Computer Vision (ICCV)*, volume 2, pages 1111–1116, 2005.

[37] Z. Zivkovic, O. Booij, and B. Kröse. From images to rooms. *Journal of Robotics and Autonomous Systems*, 55(5):411–418, 2007.

A Constant-Time Algorithm for Vector Field SLAM Using an Exactly Sparse Extended Information Filter

Jens-Steffen Gutmann, Ethan Eade, Philip Fong, and Mario Munich
Evolution Robotics, 1055 E. Colorado Blvd., Suite 410, Pasadena, CA 91106
{steffen,ethan,philip,mario}@evolution.com

***Abstract*— The constraints of a low-cost consumer product pose a major challenge for designing a localization system. In previous work, we introduced Vector Field SLAM [5], a system for simultaneously estimating robot pose and a vector field induced by stationary signal sources present in the environment. In this paper we show how this method can be realized on a low-cost embedded processing unit by applying the concepts of the Exactly Sparse Extended Information Filter [15]. By restricting the set of active features to the 4 nodes of the current cell, the size of the map becomes linear in the area explored by the robot while the time for updating the state can be held constant under certain approximations. We report results from running our method on an ARM 7 embedded board with 64 kByte RAM controlling a Roomba 510 vacuum cleaner in a standard test environment.**

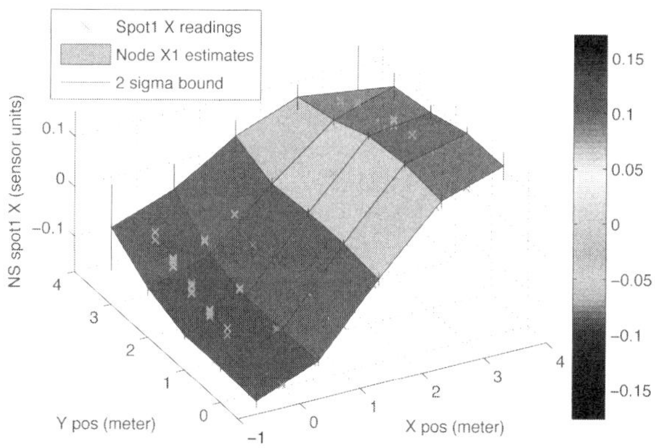

Fig. 1. Component *Spot1 X* of a vector field learned by EKF-SLAM.

I. Introduction

Deploying a localization system for the autonomous navigation of a low-cost consumer product is a non-trivial task. The majority of today's robotic floor cleaners employ relatively simple behaviors resulting in a more or less random movement in the environment. In order to achieve more systematic cleaning, the robot needs to estimate and track its position, a pre-requisite for mapping areas it has already cleaned and knowing where to go next.

Recently, a few notable systematic cleaners were introduced. Samsung's Hauzen vacuum cleaner, moving in straight lines while keeping its initial orientation, utilizes a SLAM system based on ceiling vision [9] for keeping track of its pose.

Neato Robotics uses a miniature laser range finder [11] on their XV-11 robot and employ standard range finder mapping techniques [10, 13] for estimating map and robot position.

Our solution to localization for a consumer product uses the information from time-stationary signals present in the environment. Examples of such signals are the signal strengths to WiFi stations or cellular networks, or the direction to unique beacons in the environment. These signals can often be well modelled when observed directly, i.e. sensor and signal source are on a direct path [4]. Signal reflections from walls and furniture, however, can disturb the measurements significantly.

Trying to model or discard signal reflections is a difficult task often resulting in unreliable and error-prone position estimates. Instead, recent research focused on building representations that map ground positions to expected signal values as they would appear at those locations. If learned off-line in a training phase, such a representation can practically be used for localization [6]. Ferris *at al.* showed that using Gaussian Process Latent Variable Models allows building of such signal maps from collected WiFi signal strength data [1].

In our work, we represent this signal map as a vector field over space approximated by a piece-wise linear function. Our model contains estimates of the signal values at so-called *node* positions of a regular grid laid onto the ground. Signal values at arbitrary positions are predicted by bilinear interpolation. Using this model, both vector field and robot pose are then learned through the application of simultaneous localization and mapping (SLAM). Previously, we introduced this approach as *Vector Field SLAM* and showed how an extended Kalman filter (EKF) as well as a non-linear optimization can accurately localize the robot and estimate signal maps in an environment equipped with a pair of active beacons [5]. As an example, Fig. 1 shows one component of a vector field learned by our EKF implementation of Vector Field SLAM.

Since the time and space complexity of EKF-SLAM is quadratic in the number of map features [13], an implementation on low-cost hardware with limited processing and memory resources is not practical. The contribution of this paper is the development of a constant-time algorithm with linear space requirements that allows running Vector Field SLAM on such hardware. We achieve this by a formulation of the exactly sparse extended information filter (ESEIF) [15] in which the set of *active* features is limited to the signal values of the four nodes of the cell occupied by the robot. In experiments, we verify that our application of ESEIF produces pose estimates comparable to our previous EKF solution.

This paper is organized as follows. After an introduction to Vector Field SLAM in the next section, we briefly describe the

ESEIF in Section III. Section IV presents our application of the ESEIF to Vector Field SLAM followed by an implementation using active beacons in Section V. Experimental results are described in Section VI. We draw conclusions in Section VII.

II. Vector Field SLAM

Vector Field SLAM learns the signal distribution of a time-invariant vector field over the environment while at the same time tracking the pose of the robot. A vector field of dimension M is defined as

$$\mathrm{VF} : \mathrm{SE}(2) \quad \to \quad \mathbb{R}^M \tag{1}$$

mapping a ground pose to a vector of signal values.

We assume that the dependency of signals on robot orientation can be fully characterized by some internal calibration parameters $\mathbf{c}$ of the sensor that allow for a *rotational variability* of measurements. For example, a WiFi receiver might show changes in signal strength on rotation caused by the directional sensitivity of the antenna. A sensor measuring bearing and elevation to beacons can show variations due to calibration errors of the sensor's vertical axis.

Under these assumptions we decompose the space of poses SE(2) into position and orientation. The vector field over position is then modeled as a piece-wise linear function by laying a regular grid of node positions $\mathbf{b}_i = (b_{i,x}, b_{i,y})^T$, $i = 1 \ldots N$ onto the ground. This creates cells with one node at each of the cell's four corners. Each node i holds a vector $\mathbf{m}_i \in \mathbb{R}^M$ describing the expected signal values when the robot is located at $\mathbf{b}_i$ and oriented in a preset direction θ_0.

For an arbitrary robot position with orientation θ_0, the signal values are computed by bilinear interpolation from the four nodes of the cell containing the robot. Let $\mathbf{x}_t = (x, y, \theta)^T$ be the robot pose and $\mathbf{b}_{i_0} \ldots \mathbf{b}_{i_3}$ the cell nodes enclosing the robot as shown in Fig. 2. The signal values at (x, y) with orientation θ_0 are then computed as:

$$h_0(x, y, \mathbf{m}_1 \ldots \mathbf{m}_N) \quad = \quad \sum_{j=0}^{3} w_j \mathbf{m}_{i_j} \tag{2}$$

where $\mathbf{m}_{i_0} \ldots \mathbf{m}_{i_3}$ are the signal values at the cell nodes and $w_0 \ldots w_3$ weights of the bilinear interpolation:

$$w_0 = \frac{(b_{i_1,x} - x)(b_{i_2,y} - y)}{(b_{i_1,x} - b_{i_0,x})(b_{i_2,y} - b_{i_0,y})} \tag{3}$$

$$w_1 = \frac{(x - b_{i_0,x})(b_{i_2,y} - y)}{(b_{i_1,x} - b_{i_0,x})(b_{i_2,y} - b_{i_0,y})} \tag{4}$$

$$w_2 = \frac{(b_{i_1,x} - x)(y - b_{i_0,y})}{(b_{i_1,x} - b_{i_0,x})(b_{i_2,y} - b_{i_0,y})} \tag{5}$$

$$w_3 = \frac{(x - b_{i_0,x})(y - b_{i_0,y})}{(b_{i_1,x} - b_{i_0,x})(b_{i_2,y} - b_{i_0,y})}. \tag{6}$$

The final signal values are computed by taking into account robot orientation θ and sensor calibration $\mathbf{c}$:

$$h(\mathbf{x}_t, \mathbf{c}, \mathbf{m}_1 \ldots \mathbf{m}_N) = h_R(h_0(x, y, \mathbf{m}_1 \ldots \mathbf{m}_N), \theta, \mathbf{c}). \tag{7}$$

Here h_R is a continuous function that transforms the interpolated signal values obtained through (2) by the robot orientation and calibration. This is usually a rotation by orientation

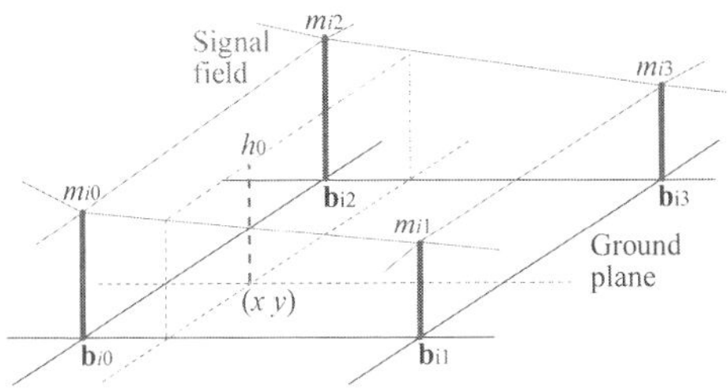

Fig. 2. Bilinear interpolation from cell nodes

θ followed by a correction with the rotational variability $\mathbf{c}$. We will provide a particular instance of h_R in an application using active beacons in Section V.

Robot path $\mathbf{x}_0 \ldots \mathbf{x}_T$, $\mathbf{x}_t \in \mathrm{SE}(2)$, rotational variability $\mathbf{c}$ and node values $\mathbf{m}_1 \ldots \mathbf{m}_N, \mathbf{m}_i \in \mathbb{R}^M$ are estimated through SLAM. Without loss of generality, $\mathbf{x}_0 = (0, 0, 0)^T$. At each time step $t = 1 \ldots T$ the robot receives a motion input $\mathbf{u}_t$ with covariance $\mathbf{R}_t$ and a measurement $\mathbf{z}_t$ with covariance $\mathbf{Q}_t$.

A motion model defined by a function g describes the motion of the robot since the previous time step:

$$\mathbf{x}_t \quad = \quad g(\mathbf{x}_{t-1}, \mathbf{u}_t) + \mathbf{e}_u \tag{8}$$

where $\mathbf{e}_u$ is a zero mean error with covariance $\mathbf{R}_t$.

Furthermore, our sensor model (7) predicts an observation given current robot pose, sensor calibration and features:

$$\mathbf{z}_t \quad = \quad h(\mathbf{x}_t, \mathbf{c}, \mathbf{m}_1 \ldots \mathbf{m}_N) + \mathbf{e}_z \tag{9}$$

where $\mathbf{e}_z$ is a zero mean error with covariance $\mathbf{Q}_t$.

In online SLAM [13], an algorithm estimates the state

$$\mathbf{y}_t \quad = \quad (\mathbf{x}_t, \mathbf{c}, \mathbf{m}_1 \ldots \mathbf{m}_N)^T \tag{10}$$

recursively over time. The EKF is one example of such an algorithm. Starting from an initial mean μ_0 and covariance Σ_0 the state is updated on motion and sensor observation using the well-known EKF equations [5, 12, 13].

In general, the initial state often contains a minimal set of map features. For Vector Field SLAM we include only the four nodes of the first cell in the initial state:

$$\mathbf{y}_0 \quad = \quad (\mathbf{x}_0, \mathbf{c}, \mathbf{m}_1 \ldots \mathbf{m}_4)^T. \tag{11}$$

An estimate for $\mathbf{y}_0$ can be obtained through non-linear optimization on a small set of collected measurements. In our approach we use RANSAC [2] for estimating a linear model of the vector field around the robot and then compute mean values $\hat{\mathbf{m}}_1 \ldots \hat{\mathbf{m}}_4$ of the four nodes. The initial mean and covariance are then found as

$$\mu_0 \quad = \quad (0, 0, 0, \hat{\mathbf{c}}, \hat{\mathbf{m}}_1 \ldots \hat{\mathbf{m}}_4)^T \tag{12}$$

$$\Sigma_0 \quad = \quad \mathrm{diag}(0, 0, 0, \infty \ldots \infty) \tag{13}$$

where $\hat{\mathbf{c}}$ is an initial guess of rotational variability. Although a tighter initial covariance can be found from the residual fitting error, in practice Eq. (13) is often sufficient since the node covariances decrease quickly after a few measurements.

During exploration new nodes need to be added to the state. In our approach these are created by a *structure copy*. Let $\mathbf{y}_t = (\mathbf{x}_t, \mathbf{c}, \mathbf{m}_1 \ldots \mathbf{m}_n)$ be the state vector at time t and

$\mathbf{b}_{n+1}$ the position of a new node. Let $\mathbf{b}_{j_1}$ and $\mathbf{b}_{j_2}$ be the positions of two nodes already contained in the state that lie on a straight line passing through $\mathbf{b}_{n+1}$. The new node $\mathbf{m}_{n+1}$ is then initialized by extrapolation from $\mathbf{m}_{j_1}$ and $\mathbf{m}_{j_2}$ as:

$$\mathbf{m}_{n+1} = \mathbf{A}_1\mathbf{m}_{j_1} + \mathbf{A}_2\mathbf{m}_{j_2} + \mathbf{e}_1 \qquad (14)$$

where $\mathbf{A}_1$ and $\mathbf{A}_2$ contain the extrapolation factors and $\mathbf{e}_1$ is a zero mean error term with a preset covariance $\mathbf{S}$ allowing the node values to change. An example of extrapolation factors is described in our implementation in Section V. The state of the EKF is then augmented with the new node and its covariance is expanded accordingly [5].

In our recent work we collected data on a robotic vacuum cleaner moving in an environment with active beacons. Our results demonstrate that both EKF-SLAM and an off-line GraphSLAM version of Vector Field SLAM significantly improve position estimates when compared to raw odometry or a direct triangulation of sensor data [5]. The limitations of EKF-SLAM are in the time and space complexity which is quadratic in the number of map features [13]. While this allows running the method on PC hardware, the requirements practically exceed the resources of low-cost embedded processors.

III. Exactly Sparse Extended Information Filter

In order to solve the problem of scalability in SLAM, representations of the state distribution different than the mean and covariance of an EKF have been researched. In a seminal paper, Thrun *et al.* [14] presented a sparse extended information filter (SEIF) with almost linear time complexity. Recently Walter *et al.* [15] further developed this approach to an exactly sparse extended information filter (ESEIF) producing estimates that are conservative relative to the EKF solution.

In general, the extended information filter (EIF) represents the state $\mathbf{y}_t$ at time t by an information vector η_t and information matrix Λ_t. The relation to the mean μ_t and covariance Σ_t of an EKF is given by:

$$\eta_t = \Sigma_t^{-1}\mu_t \qquad \Lambda_t = \Sigma_t^{-1}. \qquad (15)$$

On sensor observation the EIF state is updated as [14]:

$$\eta_t = \bar{\eta}_t + \mathbf{H}_y^T\mathbf{Q}_t^{-1}(\mathbf{z}_t - h(\bar{\mu}_t) + \mathbf{H}_y\bar{\mu}_t) \qquad (16)$$
$$\Lambda_t = \bar{\Lambda}_t + \mathbf{H}_y^T\mathbf{Q}_t^{-1}\mathbf{H}_y \qquad (17)$$

where $\bar{\eta}_t$ and $\bar{\Lambda}_t$ are the EIF state after a motion update (see below) and $\mathbf{H}_y$ is the Jacobian of h w.r.t. the state evaluated at an estimate $\bar{\mu}_t$ of the mean:

$$\mathbf{H}_y = \frac{\partial h}{\partial \mathbf{y}}(\bar{\mu}_t). \qquad (18)$$

The update time is quadratic in the number of observed features which is often small in practical applications.

The EIF update on motion is more complicated. It can be formulated by first augmenting the state with the new robot pose followed by a marginalization over the old pose [15]. Given a robot pose estimate μ_x and the EIF state at time $t-1$ separated into robot variables $\mathbf{x}$ and map variables $\mathbf{M}$:

$$\eta_{t-1} = \begin{pmatrix} \eta_x \\ \eta_M \end{pmatrix} \qquad \Lambda_{t-1} = \begin{pmatrix} \Lambda_{xx} & \Lambda_{xM} \\ \Lambda_{Mx} & \Lambda_{MM} \end{pmatrix} \qquad (19)$$

we compute information vector and matrix on motion as:

$$\bar{\eta}_t = \begin{pmatrix} \bar{\eta}_x \\ \bar{\eta}_M \end{pmatrix} \qquad \bar{\Lambda}_t = \begin{pmatrix} \bar{\Lambda}_{xx} & \bar{\Lambda}_{xM} \\ \bar{\Lambda}_{Mx} & \bar{\Lambda}_{MM} \end{pmatrix} \qquad (20)$$

where

$$\bar{\eta}_x = \Psi^T\eta_x + (\mathbf{R}_t + \mathbf{G}_x\Lambda_{xx}^{-1}\mathbf{G}_x^T)^{-1}\Delta_x \qquad (21)$$
$$\bar{\eta}_M = \eta_M - \Lambda_{Mx}(\Lambda_{xx}^{-1}\eta_x - \Omega\eta_x - \Psi\Delta_x) \qquad (22)$$
$$\bar{\Lambda}_{xx} = (\mathbf{R}_t + \mathbf{G}_x\Lambda_{xx}^{-1}\mathbf{G}_x^T)^{-1} \qquad (23)$$
$$\bar{\Lambda}_{xM} = \Psi^T\Lambda_{xM} \qquad \bar{\Lambda}_{Mx} = \Lambda_{Mx}\Psi \qquad (24)$$
$$\bar{\Lambda}_{MM} = \Lambda_{MM} - \Lambda_{Mx}(\Lambda_{xx}^{-1} - \Omega)\Lambda_{xM} \qquad (25)$$
$$\Delta_x = g(\mu_x, \mathbf{u}_t) - \mathbf{G}_x\mu_{\mathbf{x}} \qquad (26)$$
$$\Psi = \mathbf{G}_x^{-1} - \mathbf{G}_x^{-1}\mathbf{R_t}(\mathbf{R}_t + \mathbf{G}_x\Lambda_{xx}^{-1}\mathbf{G}_x^T)^{-1} \qquad (27)$$
$$\Omega = \Lambda_{xx}^{-1}\mathbf{G}_x^T(\mathbf{R}_t + \mathbf{G}_x\Lambda_{xx}^{-1}\mathbf{G}_x^T)^{-1}\mathbf{G}_x\Lambda_{xx}^{-1} \qquad (28)$$

and $\mathbf{G}_x$ is the Jacobian of motion model (8) w.r.t. robot pose:

$$\mathbf{G}_x = \frac{\partial g}{\partial \mathbf{x}}(\mu_x, \mathbf{u}_t). \qquad (29)$$

Our formulation through Eqs. (20-29) is equivalent to the one presented by Walter *et al.* [15] after utilizing the Woodbury matrix identity [3]. The advantage of Eqs. (20-29) is that the motion covariance $\mathbf{R}_t$ itself and not its inverse is used for updating the state. This allows for motions with arbitrarily small errors. The price of this formulation is that the motion Jacobian $\mathbf{G}_x$ must be invertible. In Section V, we show that this holds for a standard odometry motion model [13].

The most expensive operation of the motion update lies in Eq. (25) and is quadratic in the number of non-zero elements of the cross-information matrix Λ_{xM}, i.e. the number of map features that share information with the robot pose. In its plain form the EIF does not restrict this number and thus, does not provide a better solution than the EKF in terms of scalability [15]. The advancements of the SEIF and ESEIF are in controlling this number such that the space and time complexities are bounded.

The ESEIF divides map features into *active* and *passive* ones based upon whether or not they share information with the robot pose. The algorithm then poses an upper bound Γ_a on the number N_a of active features. As long as $N_a \leq \Gamma_a$ the state is updated on motion and sensor observation using the EIF Eqs. (16-29) as presented above.

When receiving an observation $\mathbf{z}_t$ that would cause an EIF update (16,17) to exceed the Γ_a threshold, a sparsification procedure takes place. For this $\mathbf{z}_t$ is partitioned into two sets:

$$\mathbf{z}_t = \{\mathbf{z}_\alpha, \mathbf{z}_\beta\} \qquad (30)$$

such that $|\mathbf{z}_\beta| \leq \Gamma_a$. The measurements $\mathbf{z}_\alpha$ are then used in a regular EIF update (16,17) followed by marginalizing over the robot pose and relocating it using the measurements $\mathbf{z}_\beta$.

Marginalization removes the robot variables from the state:

$$\check{\eta}_t = \bar{\eta}_M - \bar{\Lambda}_{Mx}\bar{\Lambda}_{xx}^{-1}\bar{\eta}_x \qquad (31)$$
$$\check{\Lambda}_t = \bar{\Lambda}_{MM} - \bar{\Lambda}_{Mx}\bar{\Lambda}_{xx}^{-1}\bar{\Lambda}_{xM} \qquad (32)$$

and can be computed in time quadratic in Γ_a.

For relocating the robot, the ESEIF computes a new robot pose $\mathbf{x}_t$ given the map features $\mathbf{M}_\beta$ and measurements $\mathbf{z}_\beta$:

$$\mathbf{x}_t = f(\mathbf{M}_\beta, \mathbf{z}_\beta) + \mathbf{e}_x \quad (33)$$

where $\mathbf{e}_x$ is a zero mean error with covariance $\mathbf{R}_x$. The final state is then computed by augmenting the new robot pose:

$$\breve{\eta}_t = \begin{pmatrix} \mathbf{R}_x^{-1}\left(f(\breve{\mu}_{M_\beta}, \mathbf{z}_\beta) - \mathbf{F}_M \breve{\mu}_t\right) \\ \breve{\eta}_t - \mathbf{F}_M^T \mathbf{R}_x^{-1}\left(f(\breve{\mu}_{M_\beta}, \mathbf{z}_\beta) - \mathbf{F}_M \breve{\mu}_t\right) \end{pmatrix} \quad (34)$$

$$\breve{\Lambda}_t = \begin{pmatrix} \mathbf{R}_x^{-1} & -\mathbf{R}_x^{-1}\mathbf{F}_M \\ -\mathbf{F}_M^T \mathbf{R}_x^{-1} & \breve{\Lambda}_t + \mathbf{F}_M^T \mathbf{R}_x^{-1} \mathbf{F}_M \end{pmatrix} \quad (35)$$

where $\mathbf{F}_M$ is the Jacobian of relocation model (33) w.r.t. map:

$$\mathbf{F}_M = \frac{\partial f}{\partial \mathbf{M}}(\breve{\mu}_{M_\beta}, \mathbf{z}_\beta). \quad (36)$$

This requires time quadratic in $|\mathbf{z}_\beta|$ and resets the set of active features to those in $\mathbf{z}_\beta$. Therefore, since $|\mathbf{z}_\beta| \leq \Gamma_a$ the computation time is fully controlled by parameter Γ_a.

In the ESEIF, an estimate of the mean μ_t is needed when computing Jacobians or updating the information vector. This mean can be found by solving a linear equation system:

$$\Lambda_t \mu_t = \eta_t. \quad (37)$$

A naïve approach of inverting Λ_t requires time cubic in the size of the state which destroys the scalability of the algorithm. As Walter *et al.* [15] pointed out we are often only interested in a subset of the mean. Therefore, (37) is partitioned into

$$\begin{pmatrix} \Lambda_{ll} & \Lambda_{lb} \\ \Lambda_{bl} & \Lambda_{bb} \end{pmatrix} \begin{pmatrix} \mu_l \\ \mu_b \end{pmatrix} = \begin{pmatrix} \eta_l \\ \eta_b \end{pmatrix} \quad (38)$$

where μ_l is a *local portion* to solve for and μ_b the *benign portion* for which we already have an estimate. This leads to:

$$\mu_l = \Lambda_{ll}^{-1}(\eta_l - \Lambda_{lb}\mu_b) \quad (39)$$

and requires only a subset of μ_b corresponding to the non-zero elements in Λ_{lb}, i.e. the Markov blanket of the local portion. The time for computing (39) is linear in the size of the Markov blanket and cubic in the size of the local portion.

The ESEIF has been shown to produce estimates close to those of an EKF with significant savings in memory usage and run-time improving the scalability of on-line SLAM [15].

IV. Vector Field SLAM using ESEIF

We now have the tools ready for using the ESEIF in Vector Field SLAM. For the upper bound Γ_a of active features, we seek the smallest possible number while still preserving a good level of information necessary for approximating the full Gaussian distribution of EKF-SLAM. By analyzing the sensor model (7), we note that robot pose $\mathbf{x}_t$, sensor calibration $\mathbf{c}$, and the four nodes $\mathbf{m}_{i_0} \ldots \mathbf{m}_{i_3}$ enclosing the robot are sharing information between each other. Our goal is to restrict the number of active features to this configuration, thus $\Gamma_a = 4$.

As long as the robot stays within the same cell, the ESEIF state is updated on sensor observation and motion through EIF Eqs. (16-29) where the set $\mathbf{M}$ in (19,20) include sensor calibration $\mathbf{c}$ and all nodes $\mathbf{m}_1 \ldots \mathbf{m}_n$ of the current state $\mathbf{y}_t$.

After moving into another cell, we reset the set of active features to the four nodes of the new cell when receiving a new sensor measurement. This involves performing the sparsification procedure of the ESEIF. In our sensor model (7) an observation $\mathbf{z}_t$ is associated with only the nodes of one cell. It is therefore not possible to partition $\mathbf{z}_t$ into two non-empty sets corresponding to different map features. We then choose the partitioning of

$$\mathbf{z}_\alpha = \emptyset \qquad \mathbf{z}_\beta = \mathbf{z}_t \quad (40)$$

i.e. all measurements are used for relocating the robot.

The ESEIF sparsification applies the marginalization in Eqs. (31,32). In Vector Field SLAM we not only marginalize out the robot pose $\mathbf{x}_t$, but also the rotational variability $\mathbf{c}$. This has certain advantages which we outline further below. Thus, for applying Eqs. (31,32) we set the robot variables to $\mathbf{x} = \{\mathbf{x}_t, \mathbf{c}\}$ and map variables to $\mathbf{M} = \{\mathbf{m}_1 \ldots \mathbf{m}_n\}$.

For re-inserting robot pose and rotational variability, we need to find a function f for Eq. (33) that generates these from the map $\mathbf{M}_\beta$ and observation $\mathbf{z}_\beta$. Unfortunately this turns out to be a difficult task. In Vector Field SLAM there can be areas where a single observation might not provide full information about the robot pose, e.g. when only a small subset of signals are observed, or when the measurement noise is large. Furthermore, for determining rotational variability, often measurements from different robot orientations are needed.

Instead of following Eqs. (34,35) we propose a two-step procedure for relocating robot pose and rotational variability. First, we augment the state with blank robot variables

$$\breve{\eta}_t = \begin{pmatrix} \breve{\Lambda}_{xx}\bar{\mu}_x \\ \breve{\eta}_t \end{pmatrix} \qquad \breve{\Lambda}_t = \begin{pmatrix} \breve{\Lambda}_{xx} & 0 \\ 0 & \breve{\Lambda}_t \end{pmatrix} \quad (41)$$

where $\bar{\mu}_x$ is the mean of robot variables before marginalization, and $\breve{\Lambda}_{xx}$ is a prior on robot pose and sensor calibration.

In the second step, the measurements $\mathbf{z}_t$ are used for updating this state using the regular EIF Eqs. (16,17). This creates new information between robot pose, sensor calibration and the nodes of the new cell, while all other cross information between robot variables and map are zero, thus $N_a \leq \Gamma_a$.

The prior $\breve{\Lambda}_{xx}$ plays an important role in our relocation procedure. For a truly conservative filter, the prior should be set to $\breve{\Lambda}_{xx} = 0$. The formulation then becomes equivalent to Eqs. (34,35) modulo different approximation errors in the involved Jacobians. In practical applications, however, it is often useful to restrict robot and sensor calibration to an area around the estimate before sparsification, since there was no physical cause that would have allowed them to change by a large amount. Since a non-zero prior can cause the filter to become over-confident, it has to be carefully chosen. In our application we maintain an estimate of the covariance Σ_{xx} of robot variables and compute $\breve{\Lambda}_{xx}$ as

$$\breve{\Lambda}_{xx} = (\Sigma_{xx} + \mathbf{R}_0)^{-1} \quad (42)$$

where $\mathbf{R}_0$ is a minimal covariance that needs to be tuned in order to ensure the filter produces non-optimistic estimates.

`Motion update:`

$(\eta_{t-1}, \Lambda_{t-1}) \xrightarrow[\mathbf{x}=\{\mathbf{x}_t\}, \mathbf{M}=\{\mathbf{c}, \mathbf{m}_1 \ldots \mathbf{m}_n\}]{(19\text{-}29)} (\bar{\eta}_t, \bar{\Lambda}_t)$

`Sensor observation update:`

if *robot in same cell* **then**

  Standard EIF update: $(\bar{\eta}_t, \bar{\Lambda}_t) \xrightarrow{(16\text{-}18)} (\eta_t, \Lambda_t)$

else

  Marginalize: $(\bar{\eta}_t, \bar{\Lambda}_t) \xrightarrow[\mathbf{x}=\{\mathbf{x}_t, \mathbf{c}\}, \mathbf{M}=\{\mathbf{m}_1 \ldots \mathbf{m}_n\}]{(31\text{-}32)} (\check{\eta}_t, \check{\Lambda}_t)$

  while *need new node* **do**

    Augment new node: $(\check{\eta}_t, \check{\Lambda}_t) \xrightarrow{(14,44)} (\check{\eta}_t, \check{\Lambda}_t)$

  end

  Augment blank: $(\check{\eta}_t, \check{\Lambda}_t) \xrightarrow[\mathbf{x}=\{\mathbf{x}_t, \mathbf{c}\}]{(41,42)} (\check{\eta}_t, \check{\Lambda}_t)$

  Standard EIF update: $(\check{\eta}_t, \check{\Lambda}_t) \xrightarrow{(16\text{-}18)} (\eta_t, \Lambda_t)$

end

Algorithm 1: ESEIF algorithm for Vector Field SLAM

For recovering an estimate of the mean μ_t we apply Eq. (39) where the local portion is $\{\mathbf{x}_t, \mathbf{c}, \mathbf{m}_{i_0} \ldots \mathbf{m}_{i_3}\}$. In the same way we obtain an estimate of covariance Σ_{xx} by replacing η_t in (39) with unit vectors corresponding to the robot variables.

The last missing piece in using the ESEIF for Vector Field SLAM is the initialization of state and new nodes. This follows the same steps as in EKF-SLAM. The initial state is found as

$$\eta_0 = \Sigma_0^{-1} \mu_0 \qquad \Lambda_0 = \Sigma_0^{-1} \tag{43}$$

where μ_0 and Σ_0 are computed by Eqs. (12,13). Inverting Σ_0 is trivial since the matrix is diagonal.

When extrapolating a new node through Eq. (14), the mean and covariance of the new node $\mathbf{m}_{n+1}$ are computed [5]. This requires knowledge of the mean and covariance of the participating nodes $\mathbf{m}_{j_1}$ and $\mathbf{m}_{j_2}$. While we maintain an estimate of the full mean vector, the covariances of $\mathbf{m}_{j_1}$ and $\mathbf{m}_{j_2}$ have to be recovered, an operation that is expensive in the information form. We approximate the covariance $\Sigma_{m_j m_j}$ of node $\mathbf{m}_j$ by conditioning it on all map features except those contained in the node's Markov blanket $\partial \mathbf{m}_j$, i.e. those features that share information with $\mathbf{m}_j$. Since each node is the corner of at most four cells, we have at most 8 nodes in $\partial \mathbf{m}_j$. The node covariance is then computed as:

$$\Sigma_{m_j m_j} = \left(\Lambda_{m_j m_j} - \Lambda_{m_j \partial m_j} \Lambda^{-1}_{\partial m_j \partial m_j} \Lambda_{\partial m_j m_j} \right)^{-1} \tag{44}$$

and is cubic in the size of the Markov blanket. It has been noted that this approximation can result in over-confident covariances [15]. We will revisit this issue in our conclusions.

Algorithm 1 summarizes our application of the ESEIF for Vector Field SLAM. The computation time is dominated by the covariance computation (44) for augmenting new nodes and the recovery of the mean vector (39). Fortunately new nodes are added rather infrequently and the time for computing (44) is constant since $|\partial \mathbf{m}_j| \leq 8$. The time for mean recovery (39) is also constant, since the size of the local portion is constant and its Markov blanket contains only nodes that share information with the four nodes of the local portion. Since each node has at most 8 neighbors, the size of the Markov blanket is constant. Thus, Vector Field SLAM using the ESEIF is an algorithm with constant time.

The memory requirements of the ESEIF are linear in the number of map nodes, due to the fact that each node shares information with at most 8 neighboring ones. Since the number of nodes is approximately proportional to the covered area of the environment, our algorithm requires memory linear in the size of the explored environment.

In the ESEIF marginalization (31,32) we chose to integrate out both robot pose $\mathbf{x}_t$ and sensor calibration $\mathbf{c}$. While it is possible to marginalize only over $\mathbf{x_t}$, the inclusion of $\mathbf{c}$ has two advantages. First, when augmenting the state with blank robot variables (41), process noise can be added to the sensor calibration through the covariance $\mathbf{R}_0$. This allows the sensor calibration to change over time or space, and ensures numerical stability. Without marginalizing over $\mathbf{c}$ any addition of process noise eventually fully populates Λ_t.

The second advantage of marginalizing out $\mathbf{c}$ is the constant size of the Markov blanket in (39). If $\mathbf{c}$ is not removed from the state, it creates shared information with all nodes and the time complexity of mean recovery becomes linear in the size of the explored area.

V. Implementation with Active Beacons

We have implemented Vector Field SLAM using ESEIF on a system using Northstar, a low-cost optical sensing system for indoor localization [16]. A beacon projects a pair of unique infrared patterns on the ceiling (see Fig. 3). An optical sensor on the robot detects these patterns and measures the direction to both spots. The sensor then reports the coordinates of both direction vectors projected onto the sensor plane.

Under ideal circumstances the reported spot coordinates change linearly with the robot position. However, infrared light reaches the sensor not only by direct line of sight but also through multiple paths by reflecting off walls and other objects, so the spot coordinates change in a non-linear way as the robot approaches an obstructed area. Fig. 4 shows the reported spot coordinate when moving the sensor along a straight line orthogonal to a wall located on the right. While the signal is quasi-linear on the left, reflections off walls distort the signal significantly until it bends over on the right. Fig. 4 also shows the piece-wise linear approximation of the signal

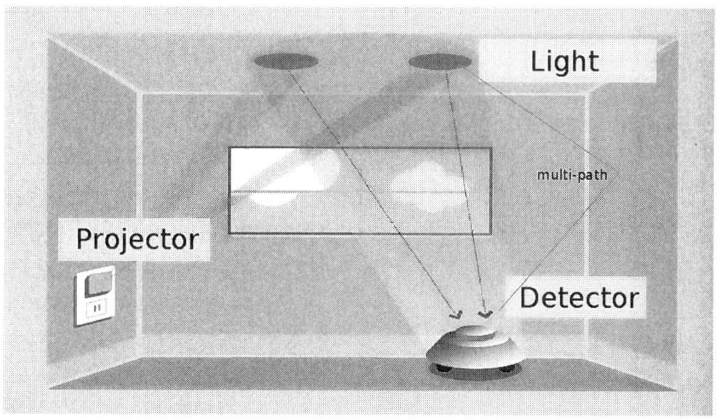

Fig. 3. The Northstar system. An optical sensor on the robot measures the bearing to two spots on the ceiling projected by an infrared beacon.

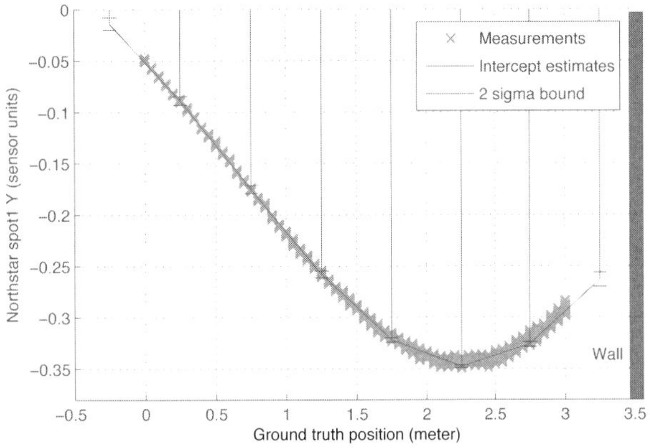

Fig. 4. Signal response of Northstar under multi-path conditions.

curve as computed by Vector Field SLAM together with 2σ intervals of the computed node covariances. Note that signal reflections can cause complex and unpredictable distortions relative to the ideal linear signal distribution.

The specific details for applying Vector Field SLAM to Northstar are as follows. The Northstar sensor provides a pair of spot coordinates, so $M = 4$ and

$$\mathbf{z}_t = (z_{x_1}, z_{y_1}, z_{x_2}, z_{y_2})^T. \quad (45)$$

The covariance $\mathbf{Q}_t$ can be derived from $\mathbf{z}_t$ along with two additional sensor outputs measuring the signal strength.

Due to tolerances in manufacturing and the mounting of the sensor on the robot, the sensor plane may not be perfectly horizontal. The result of such small angular errors is well-approximated by a coordinate offset for both spots. When rotating the sensor in place this offset becomes apparent as rotational variability. Thus, the calibration parameters are:

$$\mathbf{c} = (c_x, c_y)^T. \quad (46)$$

In the ideal case, the offset vanishes and we set $\hat{\mathbf{c}} = (0, 0)^T$.

When turning the sensor the spot coordinates change according to the rotation angle θ but in the opposite direction. The rotational component h_R of our model then becomes:

$$h_R(h_{x_1}, h_{y_1}, h_{x_2}, h_{y_2}, \theta, c_x, c_y) = \quad (47)$$
$$\begin{pmatrix} \cos\theta & \sin\theta & 0 & 0 \\ -\sin\theta & \cos\theta & 0 & 0 \\ 0 & 0 & \cos\theta & \sin\theta \\ 0 & 0 & -\sin\theta & \cos\theta \end{pmatrix} \begin{pmatrix} h_{x_1} \\ h_{y_1} \\ h_{x_2} \\ h_{y_2} \end{pmatrix} + \begin{pmatrix} c_x \\ c_y \\ c_x \\ c_y \end{pmatrix}$$

where $(h_{x_1}, h_{y_1}, h_{x_2}, h_{y_2})^T$ is the output vector of (2).

For the extrapolation of a node according to (14) we only consider node pairs which are equally spaced apart from the new node, and where the closer node is in the 8-neighborhood. The extrapolation factors in $\mathbf{A}_1$ and $\mathbf{A}_2$ are then set such that the direction between Northstar spots is copied from the closer node [5]:

$$\mathbf{A}_1 = -\frac{1}{2}\begin{pmatrix} 1 & 0 & 1 & 0 \\ 0 & 1 & 0 & 1 \\ 1 & 0 & 1 & 0 \\ 0 & 1 & 0 & 1 \end{pmatrix} \mathbf{A}_2 = \frac{1}{2}\begin{pmatrix} 3 & 0 & 1 & 0 \\ 0 & 3 & 0 & 1 \\ 1 & 0 & 3 & 0 \\ 0 & 1 & 0 & 3 \end{pmatrix}. \quad (48)$$

This completes the implementation for Northstar.

For the motion model (8) we employ a standard odometry model [13] where at each time step the robot translates along a direction and then rotates. Jacobian $\mathbf{G}_x$ and its inverse are:

$$\mathbf{G}_x = \begin{pmatrix} 1 & 0 & -\delta_y \\ 0 & 1 & \delta_x \\ 0 & 0 & 1 \end{pmatrix} \qquad \mathbf{G}_x^{-1} = \begin{pmatrix} 1 & 0 & \delta_y \\ 0 & 1 & -\delta_x \\ 0 & 0 & 1 \end{pmatrix} \quad (49)$$

where $\delta_x = x_t - x_{t-1}$ and $\delta_y = y_t - y_{t-1}$.

The motion covariance $\mathbf{R}_t$ is derived from input $\mathbf{u}_t$ [13].

VI. Results

For validating Vector Field SLAM using the ESEIF we collected sensor data of a robotic vacuum cleaner equipped with Northstar. Additionally, an optical motion capture system was used to obtain ground truth data [8]. We evaluated Vector Field SLAM on 9 different runs with increasing difficulty in distortion of the Northstar signal by multi-path. For all experiments we used a cell size of 1 meter in the vector field.

Fig. 5 shows the odometry data of the robot on run 3 as computed from its wheel encoders. In this and the following figures we superimpose the ground truth trajectory by finding scale and rigid transformation that best aligns the data with the one of the motion capture system.

The result of running ESEIF-SLAM on this data is shown in Fig. 6 where small ellipses drawn every 25 cm indicate the 1 σ pose uncertainty as computed by the filter. Compare this to the result obtained by EKF-SLAM in Fig. 7. Both methods compute the path of the robot close to the ground truth positions. The ESEIF results were obtained by setting $\mathbf{R}_0 = \mathrm{diag}(5^2\mathrm{cm}^2, 5^2\mathrm{cm}^2, 0.05^2\mathrm{rad}^2)$ when computing the prior in (42). Larger diagonals in $\mathbf{R}_0$ produce more conservative estimates while smaller ones cause over-confidence. The percentage of positions falling into a Mahanalobis distance of 4.61 of the filter estimates are 91 % and 92 % for ESEIF-SLAM and EKF-SLAM respectively. This indicates that both filters compute similar estimates and that the computed Gaussians have reasonable parameters.

Fig. 8 shows the mean position errors of odometry, pose directly computed from Northstar [16], EKF-SLAM, ESEIF-

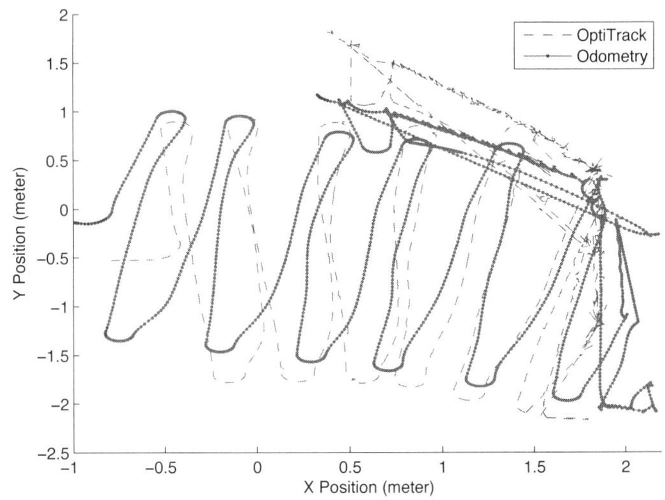

Fig. 5. Odometry information of vacuum cleaner on run 3.

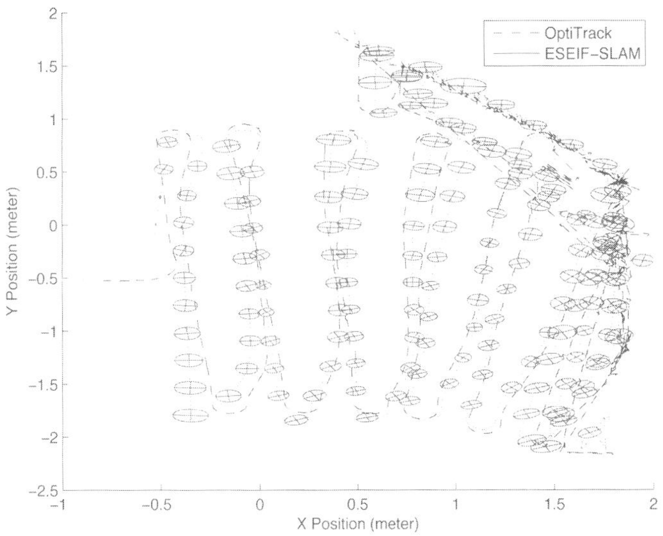

Fig. 6. Trajectory result of ESEIF-SLAM on run 3.

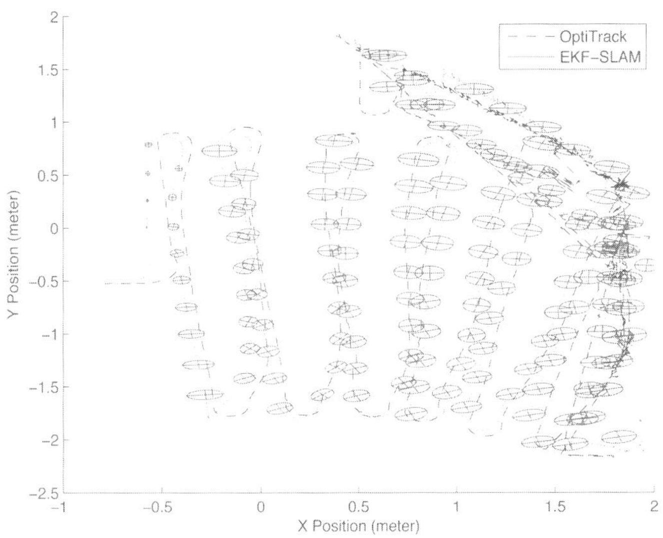

Fig. 7. Trajectory result of EKF-SLAM on run 3.

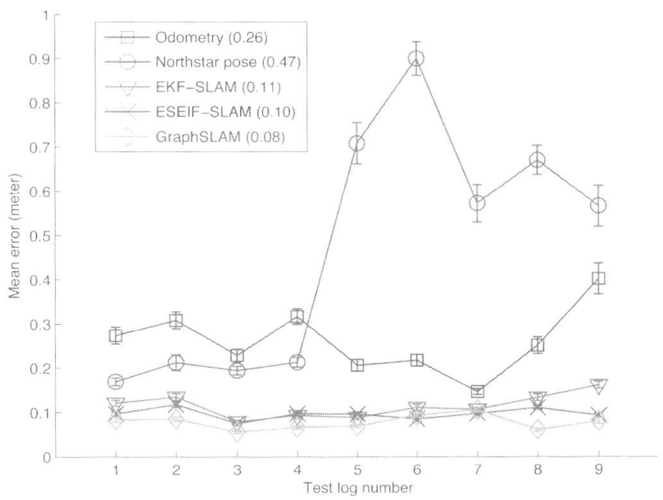

Fig. 8. Mean position errors over all 9 runs.

SLAM and an off-line GraphSLAM version [5] over all runs. Note how the error of Northstar increases from runs 1–4 to runs 5–9. The SLAM methods are able to cope with this as their mean error is not affected much. The overall mean error of ESEIF-SLAM is 10 cm which is slightly better than that of EKF-SLAM (11 cm) but worse than the full non-linear optimization of GraphSLAM (8 cm). All methods compare well to plain odometry (26 cm) or raw Northstar (47 cm).

We also implemented ESEIF-SLAM on an ARM 7 board with 64kByte memory connected to a Roomba 510 vacuum cleaner. A control program moves the robot around covering the environment in a systematic way based on the pose estimates provided by the ESEIF. We ran our system in a standard 5×4 meter test environment which is under discussion at the International Electrotechnical Commission (IEC) for evaluating the navigation capabilities of robotic floor cleaners. Fig. 9 shows the layout of this room and the trajectory of one robot run. The room is fully surrounded by walls. Solid blue (dark gray) areas mark obstacles like sofa, TV stand, and chair and table legs. The robot path starts in the upper right corner and is indicated as a white line. Areas covered by the vacuum cleaner are drawn in green (gray) while light red (light gray) marks unexplored terrain. The total run-time was 21 minutes.

One component of the vector field computed by the ESEIF in this run is shown in Fig. 10. While the sensor map contains large smooth areas, there are several non-linear regions. The big valley on the left corresponds to the sofa (largest obstacle block in Fig. 9). Furthermore, there are hills towards the front wall and the upper left corner.

We performed a total of 12 runs starting in each of the four corners of the IEC room, with the robot facing in three different directions. The position accuracy of all runs is shown in Fig. 11. The overall mean position error of the ESEIF is now 20 cm, which we attribute to the larger environment size.

For the computation of the ESEIF in these runs, a total of 42 nodes was necessary to cover the 6×5 meter environment. This leads to $3 + 2 + 42 \times 4 = 173$ variables in the ESEIF state. Thanks to the linear memory requirements, the size of the total data structure for ESEIF (including mean and robot covariance) fits into 12 kByte memory leaving room for covering larger environments and for other control structures.

The ARM 7 spends about 9 ms for a motion update and 39 ms for integrating a sensor observation. The most expensive operation is the recovery of mean and robot covariance, which involves solving a linear equation system in 21 variables through Cholesky decomposition. In summary, the total run-time of the ESEIF is about 48 ms and allows running localization with frame rates of up to 20 Hz.

VII. Conclusion

Vector Field SLAM using the ESEIF is a powerful tool for learning the distribution of time-stationary signals over the environment. By limiting the number of active features to the four nodes of the cell enclosing the robot, we obtain a constant-time SLAM algorithm with linear memory requirements. The cost of this achievement is a tradeoff in the

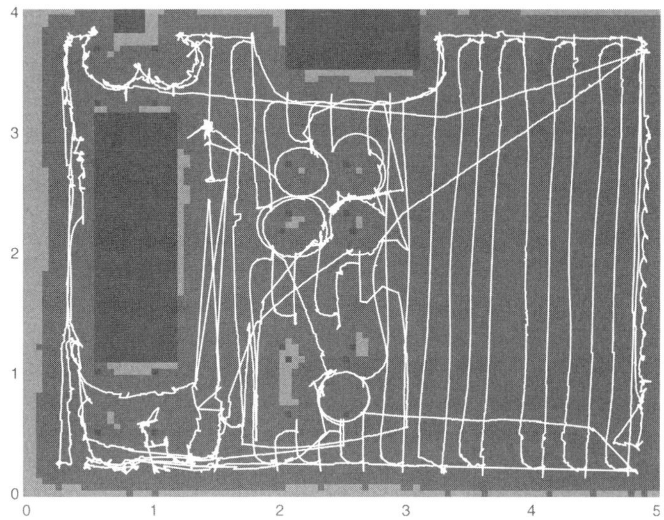

Fig. 9. Navigation run of Roomba 510 using the ESEIF in IEC room.

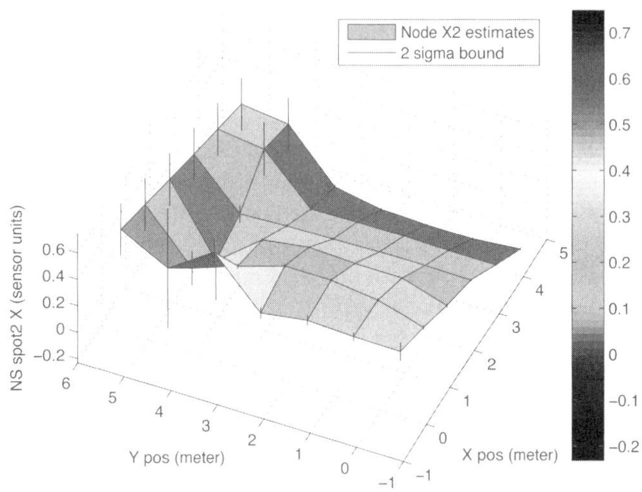

Fig. 10. Spot 2 X component estimated by ESEIF-SLAM in the IEC room.

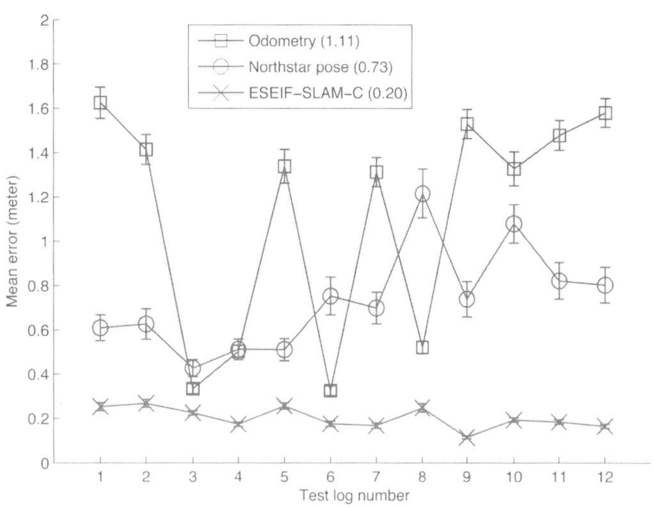

Fig. 11. Mean position errors over all 12 runs in the IEC room.

consistency of the filter. In order to arrive at our formulation, a number of approximations were made:

- Our procedure for relocating the robot requires the tuning of a prior and may result in over-confident estimates.
- The extrapolation of new nodes approximates covariances using Markov blankets which can result in over-confident covariances [15]. This can be compensated to some extent by using a larger preset covariance $\mathbf{S}$ in (14).
- The mean recovery computes only the values of robot pose, calibration and the nodes of the current cell. This may be insufficient in particular when closing loops.
- Like in other on-line SLAM methods, the errors introduced by evaluating motion and sensor Jacobians at the latest state estimate may cause inconsistent results [7].

Our experiments show that ESEIF-SLAM is capable of keeping a robotic vacuum cleaner localized in a home environment using low-cost sensors on embedded hardware. This opens the door for location-aware robotic consumer products such as autonomous and systematic floor cleaners.

REFERENCES

[1] B. Ferris, D. Fox, and N. Lawrence. WiFi-SLAM using Gaussian process latent variable models. In *Int. Joint Conference on Artificial Intelligence (IJCAI)*, 2007.

[2] M.A. Fischler and R.C. Bolles. Random sample consensus: A paradigm for model fitting with applications to image analysis and automated cartography. *Comm. of the ACM*, 24:381395, 1981.

[3] G.H. Golub and C.F. Van Loan. *Matrix Computations*. Johns Hopkins University Press, Baltimore, MD, 3rd edition, 1996.

[4] J. Graefenstein and M.E. Bouzouraa. Robust method for outdoor localization of a mobile robot using received signal strength in low power wireless networks. In *Int. Conference on Robotics and Automation (ICRA)*, 2008.

[5] J.-S. Gutmann, G. Brisson, E. Eade, P. Fong, and M. Munich. Vector Field SLAM. In *Int. Conference on Robotics and Automation (ICRA)*, Anchorage, 2010.

[6] A. Haeberlen, E. Flannery, A.M. Ladd, A. Rudys, D.S. Wallach, and L.E. Kavraki. Practical robust localization over large-scale 802.11 wireless networks. In *Proc. Int. Conference on Mobile Computing and Networking (MOBICOM)*, 2004.

[7] G.P. Huang, A.I. Mourikis, and S.I. Roumeliotis. Analysis and improvement of the consistency of Extended Kalman Filter based SLAM. In *Int. Conference on Robotics and Automation (ICRA)*, 2008.

[8] NaturalPoint Inc. Optitrack camera system, www.optitrack.com, 2009.

[9] Woo Yeon Jeong and Kyoung Mu Lee. CV-SLAM: A new ceiling vision-based SLAM technique. In *Int. Conference on Intelligent Robots and Systems (IROS)*, 2005.

[10] K. Konolige. Large-scale map-making. In *National Conference on Artificial Intelligence (AAAI)*, 2004.

[11] K. Konolige, J. Augenbraun, N. Donaldson, C. Fiebig, and P. Shah. A low-cost laser distance sensor. In *Int. Conference on Robotics and Automation (ICRA)*, 2008.

[12] R. Smith, M. Self, and P. Cheeseman. Estimating uncertain spatial relationships in robotics. In I.J. Cox and G.T. Wilfong, editors, *Autonomous Robot Vehicles*, pages 167–193. Springer-Verlag, 1990.

[13] S. Thrun, W. Burgard, and D. Fox. *Probabilistic Robotics*. MIT Press, Cambridge, MA, 2005.

[14] S. Thrun, Y. Liu, D. Koller, A.Y. Ng, Z. Ghahramani, and H. Durrant-Whyte. Simultaneous localization and mapping with sparse extended information filters. *Int. Journal of Robotics Research*, 23(7-8), 2004.

[15] Matthew R. Walter, Ryan M. Eustice, and John J. Leonard. Exactly sparse extended information filters for feature-based SLAM. *International Journal of Robotics Research*, 26(4):335–359, 2007.

[16] Y. Yamamoto, P. Prijanian, J. Brown, M. Munich, E. Di Bernardo, L. Goncalves, J. Ostrowski, and N. Karlsson. Optical sensing for robot perception and localization. In *Proc. Workshop on Advanced Robotics and its Social Impacts (ARSO)*, 2005.

Efficient Probabilistic Planar Robot Motion Estimation Given Pairs of Images

Olaf Booij and Ben Kröse
University of Amsterdam, The Netherlands
olaf.booij@xs4all.nl, b.j.a.krose@uva.nl

Zoran Zivkovic
NXP Research, the Netherlands
zoran.zivkovic@nxp.com

Abstract— **Estimating the relative pose between two camera positions given image point correspondences is a vital task in most view based SLAM and robot navigation approaches. In order to improve the robustness to noise and false point correspondences it is common to incorporate the constraint that the robot moves over a planar surface, as is the case for most indoor and outdoor mapping applications. We propose a novel estimation method that determines the full likelihood in the space of all possible planar relative poses. The likelihood function can be learned from existing data using standard Bayesian methods and is efficiently stored in a low dimensional look up table. Estimating the likelihood of a new pose given a set of correspondences boils down to a simple look up. As a result, the proposed method allows for very efficient creation of pose constraints for vision based SLAM applications, including a proper estimate of its uncertainty. It can handle ambiguous image data, such as acquired in long corridors, naturally. The method can be trained using either artificial or real data, and is applied on both controlled simulated data and challenging images taken in real home environments. By computing the maximum likelihood estimate we can compare our approach with state of the art estimators based on a combination of RANSAC and iterative reweighted least squares and show a significant increase in both the efficiency and accuracy.**

I. Introduction

Various vision based topological mapping [1, 2], view based geometrical mapping [3, 4] and robot navigation [5] approaches are based on the ability to compare pairs of images. A common way to do this is to automatically find similar looking image points [6], as done for two panoramic images in Figure 1. Because part of these point correspondences are the projections of the same 3D landmarks in the environment, they can be used to determine the relative camera pose up to an unknown scale [7]. A major challenge in determining the relative pose given point correspondences, is that a large percentage does not correspond to the same 3D landmark, but are so called *mismatches*. In addition the image point locations of correct matches are noisy, caused by for example noise of the imaging device and errors in the calibration.

To cope with this, so called *robust* algorithms are needed. There are three robust methods commonly used: RANSAC(RANdom SAmple Consensus) [8], M-Estimators (Maximum likelihood Estimators) [9] and the Hough Transform [10]. State of the art relative pose estimators first use RANSAC combined with the closed form Eight [11] or Five point algorithm [12] for an initial estimate and then apply robust iterative reweighted least squares (IRLS) techniques

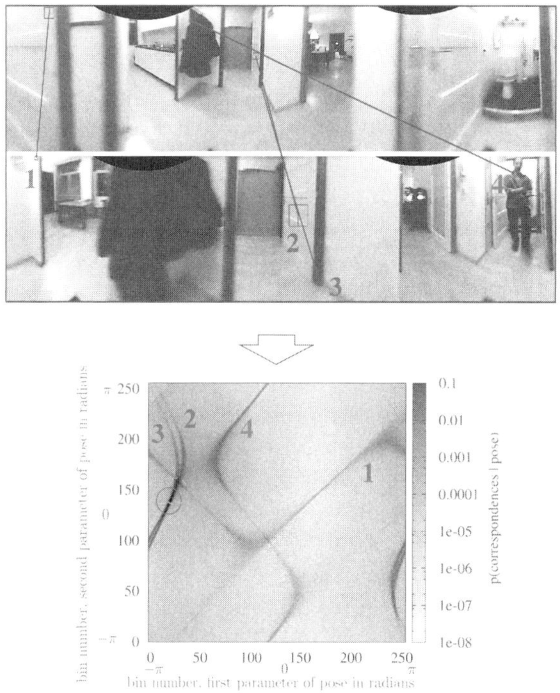

Fig. 1. An example. Applying standard SIFT matching on two panoramic images resulted in only four point correspondences including 2 mismatches (red numbers) and 2 almost degenerate correct matches (blue numbers). Still, the proposed method computes a full likelihood over the different possible relative robot poses, and the maximum likelihood (solid blue circle) is close to the ground truth (dashed green circle). The numbers relate the different correspondences in the image pair with the curves in the solution space.

such as M-Estimators in combination with the Eight point algorithm [11] to improve it. Both RANSAC and M-Estimators try to find a maximum likelihood solution by first rejecting mismatches and very noisy correspondences using an error threshold and base a least square solution on the remaining matches. Sophisticated techniques in Computer Vision try to determine this threshold from the image data itself [13, 14, 15]. In the field of Robotics, where characteristics of the camera are available, this threshold is usually determined through some calibration procedure. The Hough Transform on the other hand, if seen probabilistically, computes the full likelihood on a discrete grid of poses without making an explicit distinction between inliers and outliers. Because space requirements grow exponential with the number of parameters, it is in general not suited for pose estimation problems, although combinations of the Hough Transform with RANSAC [16] do exist, as well as methods that treat rotation and translation estimation separately [17, 18].

An approach to make pose estimation easier is to incorporate constraints on the possible relative poses. If the robot drives over a planar surface, then the camera can only rotate around a certain fixed axis which is perpendicular to the two dimensional translation direction. Given that the scale cannot be determined, the number of degrees of freedom for this planar relative pose problem reduces to two. This constraint can be used to improve indoor mapping application using wheeled robots [5], but also vision based outdoor mapping using vehicles driving over planar roads [1, 19, 20].

Commonly this constraint is imposed rather heuristically, for example by taking only the horizontal displacement of image points into account [4, 20]. A more proper solution is proposed by Brooks [21] which formulates a least square approach to the planar relative pose given noise free correspondences. This result was used in [22, 23, 2] for various robotics applications all in combination with RANSAC. In [22, 24] it was shown that two correspondences are enough to solve the problem and both suggest algorithms, which are briefly evaluated in combination with RANSAC. Makadia proposed a direct Hough based method to estimate planar relative poses [25]. However, it also used a Spherical Fourier Transform, which made it impractical due to its large computational costs.

The key concept of RANSAC based algorithms is to evaluate the solution space for randomly sampled poses. Discretizing and analyzing the whole solution space using a Hough-like approach potentially leads to much more robust results. In this paper we show that this is feasible for the planar motion estimation case without the additional approximations as in [17, 18]. Our results demonstrate the greatly increased robustness over the random sampling techniques.

Furthermore, we present a probabilistic representation where we learn the needed conditional distributions from real or simulated data without any parameters, except the size of the discretization grid. This allows fast tuning of the approach for different robots and cameras since most standard approaches have a set of parameters that need to be tuned [4, 3]. Additionally, the learned conditional distributions properly capture other sources of uncertainty that are difficult to explicitly account for, like the shaking of the robot and inaccurate camera calibration. Therefore, more accurate and robust results are obtained as demonstrated in the experimental section. Finally, testing the whole solution space potentially can be computationally expensive. We present an efficient implementation using a precomputed look up table (LUT). The geometry of the problem is analyzed and a parameterization is proposed to reduce the dimensionality of the needed LUT to only three dimensions. The experimental results show that the whole scheme can be even faster than the common sampling based approach while at the same time we get more robust and accurate results.

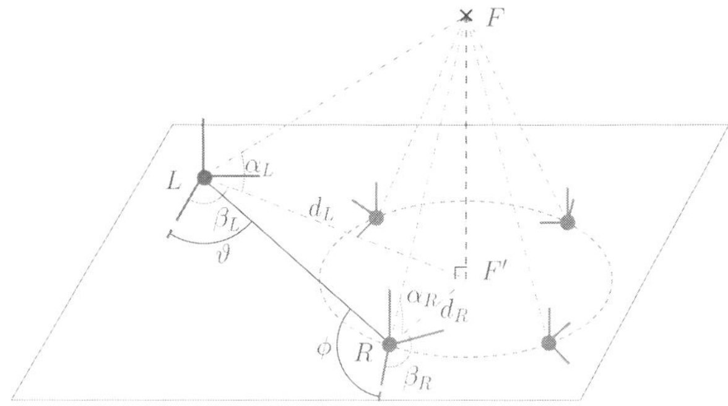

Fig. 2. 3D visualization of two cameras, L and R, positioned on the same plane both observing a landmark F. The dashed circle on the ground plane indicates the possible positions for robot R given the pose of L and observations of F.

The rest of the paper is organized as follows. First, in Section II we formalize the planar relative pose problem and describe how noise free correspondences relate to it. This relation is used in Section III to derive the novel estimator that can be trained using real image data. In Section IV we apply it on both simulated data and image data-sets taken in real home environments and compare the Maximum Likelihood estimates with the results from a planary constrained RANSAC combined with an M-Estimator. In Section V we discuss the qualitative advantage of having a full likelihood solution. Finally, in Section VI, we draw conclusions and propose some directions for improvement.

II. Relating Correspondences to Poses

The planar relative pose, which can be seen as a 2D translation and rotation, minus scale, can be parameterized in different ways. We choose to parameterize it using two angles ϑ and ϕ, see Figure 2. Angle ϑ denotes the direction of the translation, or heading, to robot R in the coordinate frame of robot L and angle ϕ denotes the heading to robot L in the coordinate frame of robot R. Another common parameterizations is using the heading and the rotation of robot R in the frame of robot L, such as in [21, 22]. However, our parameterization nicely reflects the symmetry of the problem.

It is, for now, assumed that image point correspondences are obtained by a noise free projection of landmarks, without mismatches. An image point is usually denoted by a 3D vector of unit length $\mathbf{x} = [x, y, z]'$, where the z-axis is pointing in the direction of the camera axis and the y-axis is in the planar case perpendicular to the ground plane. Here we denote an image point by its horizontal angle $\beta = \text{atan2}(z, x)$ and its vertical angle $\alpha = \arcsin(y)$. This is similar to the azimuth and elevation used in the inverse depth parameterization by landmark based SLAM methods [26]. If two robots L and R observe a landmark F, the point correspondence is thus denoted by α_L, β_L and α_R, β_R.

The 3D problem as shown in Figure 2 can be reduced to a simpler 2D problem, by projecting F on the plane getting F'. We define d_L and d_R as the distances of L to F' and R to F'

respectively. The length of F to F' can now be expressed by:

$$\overline{FF'} = \frac{\tan(\alpha_L)}{d_L} = \frac{\tan(\alpha_R)}{d_R}, \quad (1)$$

which results in the following ratio r of d_L and d_R:

$$r = \frac{d_L}{d_R} = \frac{\tan(\alpha_R)}{\tan(\alpha_L)}. \quad (2)$$

Angle $\angle F'RL$ can then be found using the Law of Sines:

$$\begin{aligned} \frac{\sin(\angle F'RL)}{d_L} &= \frac{\sin(\beta_L - \vartheta)}{d_R}, \\ \angle F'RL &= \arcsin\left(\frac{d_L}{d_R}\sin(\beta_L - \vartheta)\right). \quad (3) \end{aligned}$$

Angle ϕ can now be expressed as a function of angle ϑ and a single point correspondence by adding the horizontal observation angle β_R, and using Equation (2):

$$\phi = \beta_R + \arcsin\left(\frac{\tan(\alpha_R)}{\tan(\alpha_L)}\sin(\beta_L - \vartheta)\right). \quad (4)$$

We could also rewrite this formula into:

$$\vartheta = \beta_L + \arcsin\left(\frac{\tan(\alpha_L)}{\tan(\alpha_R)}\sin(\beta_R - \phi)\right), \quad (5)$$

which is evident given the symmetry between ϑ and ϕ.

We now obtained functions that map ϑ to ϕ and vice versa given a single point correspondence. These can be used to plot curves of possible relative robot poses as shown in Figure 3(a) for some randomly picked point correspondences. Looking closely at this figure one can see that some pairs of curves intersect each other twice. This shows that, although it was assumed that two point correspondences can be used to solve the relative pose problem [22, 24], in some cases two different relative robot poses result in the same two point correspondences.

Based on Equation 4 an algorithm was derived that computes both these solutions given two correspondences, which can be combined with hypothesize and test schemes such as RANSAC. The exposition of this "Planar Two point algorithm" is rather tedious and described in [27]. For completeness it is taken into account in the experiments.

III. Full likelihood estimator

We now use the relation between noise free correspondences and relative poses to develop an algorithm that can deal with noisy correspondences, including mismatches, see Figure 3(b). The problem can be formulated as determining the negative log likelihood of each pose (ϑ, ϕ) given n point correspondences $\{\xi_1, \ldots, \xi_n\}$, each parameterized by the angles $\xi_i = (\alpha_{Li}, \beta_{Li}, \alpha_{Ri}, \beta_{Ri})$ [28]:

$$\begin{aligned} \mathcal{L}(\vartheta, \phi) &= \sum_i \mathcal{L}_i(\vartheta, \phi) \quad (6) \\ &= \sum_i -\log p(\xi_i | \vartheta, \phi), \quad (7) \end{aligned}$$

where we use the common assumption that the correspondences are independent given the relative pose. We discretize

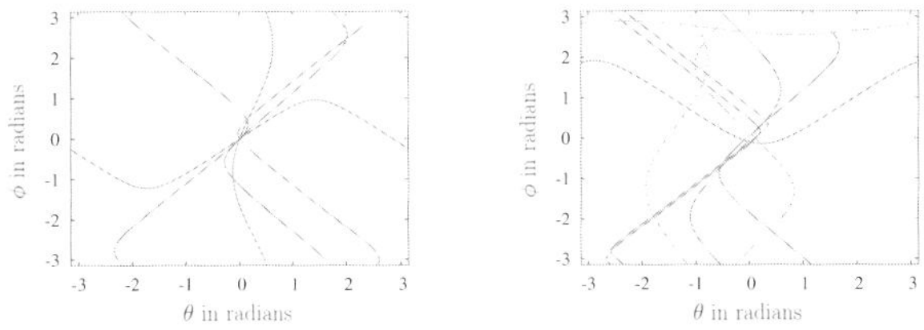

(a) (b)

Fig. 3. Visualization of the possible relative robot poses by plotting ϕ as a function of ϑ for multiple correspondences randomly picked using a relative pose with $\vartheta = \phi = 0$. (a) Generated using 5 noise free correspondences (red/solid curves). (b) Generated using 5 noisy correspondences and 3 mismatches (green/dashed curves). Note that, by coincidence, one of the mismatches corresponds to a curve close to the actual pose at $\vartheta = \phi = 0$.

the 2D likelihood into a 2D histogram. For each bin we need to sum the contribution of all point matches. Therefore for each point match we need to calculate the corresponding 2D histogram approximating its log likelihood contribution $\mathcal{L}_i(\vartheta, \phi)$, which is computational costly. In the next section, we show how this can be efficiently performed using a precomputed look-up table.

A. Look up table

The negative log likelihood of a single correspondence ξ is given by $-\log p(\alpha_L, \beta_L, \alpha_R, \beta_R | \vartheta, \phi)$. Thus, if we would naively construct a look up table, it would have 6 dimensions, 4 for the point correspondence angles and 2 for the relative pose. In order to keep the size of the LUT comprehensible, the 6D space should be discretized in large bins, resulting in a large discretization error. Fortunately, the dimensionality of the LUT can be reduced to only 3 dimensions by using the one point mapping function introduced in Section II and some common assumptions about the noise characteristics of image points.

The function given in Equation (4) can be written as

$$\phi - \beta_R + \arcsin\left(\frac{\tan(\alpha_R)}{\tan(\alpha_L)}\sin(\vartheta - \beta_L)\right) = 0. \quad (8)$$

As can be seen the terms α_L and α_R are only used in the combination $\frac{\tan(\alpha_R)}{\tan(\alpha_L)}$. Also, variables ϑ and β_L, and variables ϕ and β_R are only used in the combinations $\vartheta - \beta_L$ and $\phi - \beta_R$ respectively, which describe the horizontal angles to the landmark relative to the heading of the cameras. The joint probability of observing a correspondence under a certain relative pose can thus be represented as:

$$p(\alpha_L, \beta_L, \alpha_R, \beta_R, \vartheta, \phi) = p\left(\frac{\tan(\alpha_R)}{\tan(\alpha_L)}, \vartheta - \beta_L, \phi - \beta_R\right). \quad (9)$$

This holds if we assume that the noise of the horizontal and vertical view angles do not depend on their value, which is similar to the common assumption that the noise of the pixel locations is homogeneous.

Because of the symmetry of the representation of the relative planar pose, the two points of the point correspondences and the heading angles can be swapped giving the same result:

$$p\left(r, \vartheta - \beta_L, \phi - \beta_R\right) = p\left(\frac{1}{r}, \phi - \beta_R, \vartheta - \beta_L\right), \quad (10)$$

with $r = \frac{\tan(\alpha_R)}{\tan(\alpha_L)}$ for convenience. In practice this means that we only have to construct a LUT for $0 < r < 1$ and swap $\vartheta - \beta_L$ with $\phi - \beta_R$ and use $\frac{1}{r}$ instead of r if $r > 1$.

The likelihood can be determined from the joint probability by dividing by the probability of the pose $p(\vartheta, \phi)$:

$$\begin{aligned} p(\xi|\vartheta, \phi) &= p(\alpha_L, \beta_L, \alpha_R, \beta_R|\vartheta, \phi) && (11) \\ &= p(\alpha_L, \beta_L, \alpha_R, \beta_R, \vartheta, \phi)/p(\vartheta, \phi) && (12) \\ &= p\left(\frac{\tan(\alpha_{Ri})}{\tan(\alpha_{Li})}, \vartheta - \beta_{Li}, \phi - \beta_{Ri}\right)/p(\vartheta, \phi) && (13) \end{aligned}$$

During the construction the LUT, one can take care that the different relative poses are uniformly distributed as described below. This makes the likelihood proportional to the joint:

$$p(\xi|\vartheta, \phi) \propto p\left(\frac{\tan(\alpha_R)}{\tan(\alpha_L)}, \vartheta - \beta_L, \phi - \beta_R\right). \quad (14)$$

Efficiently determining a full likelihood over the relative poses given a set of correspondences is now straightforward. For each correspondence ξ_i we compute the value of $\frac{\tan(\alpha_{Ri})}{\tan(\alpha_{Li})}$ and pick the corresponding 2D slice of the look up table. Then we shift it in the direction of β_{Li} and β_{Ri}, wrapping the values at the borders. This results in the negative log likelihood of each pose given a correspondence, which can be summed for the different correspondences, resulting in a full likelihood. Algorithm 1 summarizes this procedure.

Algorithm 1 Full likelihood estimator using a LUT

```
HIST ← 2d array of zeros
for all correspondences {α_L, β_L, α_R, β_R} do
    compute r = tan(α_R) / tan(α_L)
    HIST += LUT(d(r)) shifted by d(β_L) and d(β_R)
             where d(·) denotes discretization
```

B. Constructing a look up table

The LUT representing the negative log likelihood can be constructed from existing data. This data can be generated by a simulator modeling the planar pose problem including a vision system. Better is to use a representative image set for which ground truth robot pose data is available. The main problem of real data is that the relative poses are in general not uniformly distributed. This invalidates the simplification proposed in Equation (14) and results in a bias of the LUT towards certain poses which are overrepresented in the dataset. However, we can easily compensate for this problem.

When constructing the LUT, we explicitly take the probability of the relative pose into account (see Equation (13)). In a first step, a 2D discretized probability $p(\vartheta, \phi)$ is constructed by making a histogram for all poses in the dataset and normalizing it. Then, in a second step, the dataset is used to build the 3D LUT like for the simulator, with the difference that for each pose correspondence it adds $\frac{1}{p(\vartheta,\phi)}$ to the 3D histogram. Again, each value of the histogram is replaced by its negative log, resulting in a proper LUT. Algorithm 2 summarizes the procedure to build a LUT. To keep the algorithm comprehensible, it ignores the symmetry between ϑ and ϕ.

A second problem, which can not be circumvented, is that the amount of data in an image dataset is limited. As a consequence we usually can not construct a LUT with a very high number of bins. In the next section we evaluate, among other things, the consequences of such a smaller LUT.

Algorithm 2 Constructing a LUT from real data

```
PP ← 2d array of zeros
for all train data {ϑ, φ, α_L, β_L, α_R, β_R} do
    PP(d(ϑ), d(φ)) += 1
LUT ← 3d array of zeros
for all train data {ϑ, φ, α_L, β_L, α_R, β_R} do
    compute r = tan(α_R) / tan(α_L)
    LUT(d(r), d(ϑ − β_L), d(φ − β_R)) += 1/ PP(d(ϑ), d(φ))
LUT ← - log each entry of LUT
```

IV. Experiments

By determining the Maximum Likelihood from the estimated full likelihood (LUT ML), we can compare our method for planar relative pose estimation with the state of the art robust methods using RANSAC combined with an M-Estimator as described briefly in Section I. We combined these with the Planar Two point, the Planar Three point and the general Eight point algorithm. We used simulated data and 4 datasets obtained from a robot with an omnidirectional vision system. The methods are compared on the basis of their robustness against mismatches and noise.

We evaluate the estimated heading and rotation angle as in [29] by taking the absolute difference with the ground truth values. Because these errors are not normally distributed we use the median to describe the error distribution. A robust way to describe the spread of these medians is to use the Median of Absolute Deviations (MAD). Another important evaluation criterion is the computational time used by the algorithms. For all experiments we report these times, as implemented in C++ and run on the single 2 Ghz CPU core of a Pentium PC.

A. Experiments on simulated data

Using simulated data allows us to control the projection noise and number of mismatches.

1) Data: Data was simulated by randomly picking a uniformly distributed point cloud of 3D landmarks inside a sphere of size 2 around the origin. Two random camera poses on a circle with radius 1 in the x-y plane around the origin are chosen. From these the ground truth values for ϑ and ϕ are determined. Note that the distribution of ϑ and ϕ is approximately uniform.

A set of point correspondences is constructed by projecting the landmarks on a spherical shaped image surface with a

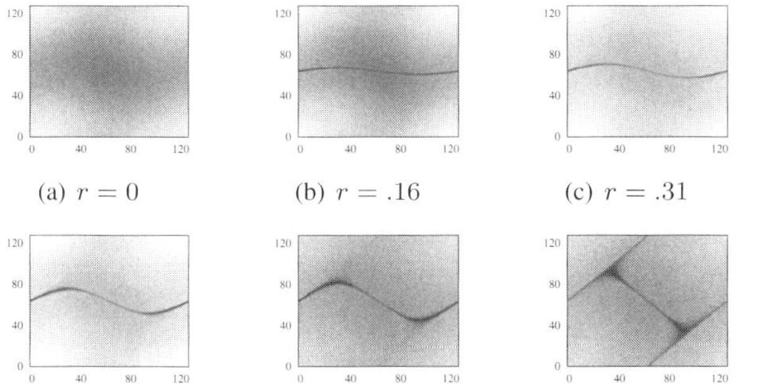

(a) $r = 0$ (b) $r = .16$ (c) $r = .31$

(d) $r = .55$ (e) $r = .78$ (f) $r = .99$

Fig. 4. Look up table obtained from simulated data as described in the experiment section.

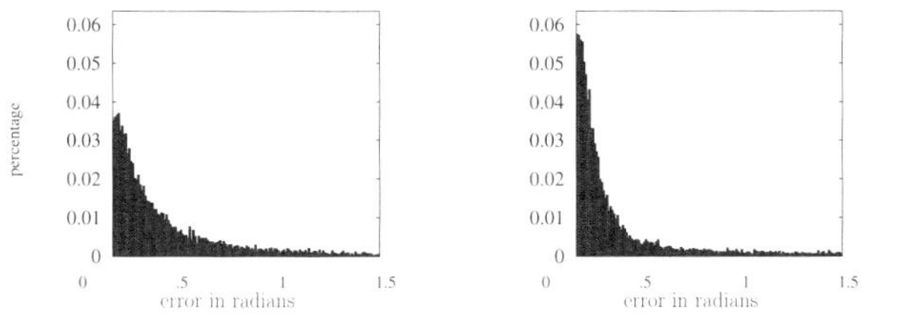

(a) Heading error 3pt. RANSAC + M-Estimator (b) Heading error LUT ML

Fig. 5. Comparison of the RANSAC+M-Estimator combined with the Three point algorithm and the proposed LUT ML method for 90% mismatches. The distribution of rotation errors shows a similar pattern.

radius of 1 around the camera pose. Thus, an ideal omnidirectional camera model is used with a full 360 degrees view angle in the horizontal and vertical direction. An amount of normally distributed noise with zero mean and standard deviation 0.01 is added to these projections. In addition, mismatches are added by creating false correspondences between projections of different landmarks. We use a mismatch rate of 90%.

2) Setup: A LUT was constructed using the same simulator as described above. The number of bins to represent ϕ, ϑ and r were all 128, which caused the computational time to be comparable with that of the RANSAC+M-Estimator. The number of point correspondences used was 10^{10}, which took 3 hours to build. The error threshold for both RANSAC and the M-Estimator were set according to the projection noise of the simulator. The number of RANSAC iterations was set to 100.

3) Resulting look up table: Figure 4 shows the resulting look up table. In 4(a) one can see that point correspondences with a r value close to zero, do not tell that much about the data. This is due to the fact that there is a high chance that it resulted from a mismatch. Note also that the histogram corresponding to a r value close to 1 in 4(f) is almost symmetric. Indeed if r is exactly 1 then $r = \frac{1}{r}$ and we could swap ϕ and θ.

4) Resulting distributions: The distribution of errors is given in Figure 5. The LUT ML method results is superior for 90% percent mismatches. Also, both distributions have long tails, indicating the need of robust statistics.

5) Sensitivity to mismatches: To test the robustness of the different methods to mismatches, we vary the number of mismatches, from 50% to 99%. In total 10^5 iterations were conducted. In Figures 6(a) and 6(e) the resulting estimation errors are plotted.

Both the heading and rotation results show the same trend. The error of the RANSAC+M-Estimator with the Eight point algorithm, which does not take planarity into account, increases fast if more than half of the correspondences are mismatches. The RANSAC+M-Estimator with the Two point algorithm is better, but is in turn outperformed by the Three point algorithm which starts diverging at 70% mismatches. The accuracy of the LUT ML estimator starts diverging at 85% and seems particularly robust against high mismatch rates. For mismatch rates under 60% the relative error is somewhat higher which can be explained by the discretization process used in the algorithm.

6) Sensitivity to violations of the planar assumption: In practice the motion of a robot is never strictly planar. Therefore, we test the behavior of the algorithms when small rotations around the x-axis, and around the z-axis, are added to the poses. We learn two separate LUTs for this experiment: one as described above using 10^7 simulated correspondences named LUT con, and one in which we additionally simulate small rotations around the x-axis, and around the z-axis, both up to .25 radians, named LUT rot. In Figures 6(b) and 6(f) the median errors are plot against the amount of pitch and roll, for a mismatch rate of 60%. As can be seen at an angle of .3 radians the RANSAC method and conventional LUT ML method have the same or worse error as the RANSAC with Eight point algorithm which is not influenced by non planar motion. Note that this corresponds to a percentage-grade of 31%. The LUT ML method with LUT con, learned to be more robust and is better than the RANSAC with Eight point algorithm up to .4 radians.

B. Experiments on real data

We compared the performance of our method with other methods on more than $3 * 10^6$ real image pairs.

1) Data: We used three distinct image datasets. The first two were obtained using our Nomad Super Scout II and the third by the 'Biron' robot from the University of Bielefeld. On both an omnidirectional vision system was mounted consisting of a conventional Firewire camera pointing upwards to a hyperbolic mirror. Note that the proposed methods do not require omnidirectional images. All sets are taken in real home environments. In the 'Almere 4' set there are some people walking in the rooms, the 'Spaan 1' set is taken during evening hours, and the 'Biron 1' set is taken in a feature poor home. The ground truth robot poses for the home sets were obtained by applying the SLAM algorithm described in [30] on laser scans and odometry. The home sets are available for research purposes, see [31] for details.

2) Setup: From every dataset we use every pair of images. We discard the images taken at the same position. Also, if images are taken at more than 5 meters apart for the Almere 4 set or more than 3 meters apart for the other sets, then the chance of finding point correspondences is small, so we also discard these pairs. Still, for each set there are around 10^6 image pairs left.

To extract point correspondences from the image pairs, the SIFT algorithm is used [6]. First omnidirectional images are mapped to panoramic images [32], from which the SIFT feature points are found. On average 238 features were found per image. These features are described by the standard SIFT descriptor of 128 dimensions. If two features in the same image have a small distance in descriptor space then they are removed. A set of point correspondences between two images is determined by applying the standard matching scheme as described in [6]. This resulted in on average 25 matches per image pair. The groundtruth relative pose was computed from the groundtruth robot positions.

3) Sensitivity to mismatches, trained with simulated data: We first use a LUT constructed using the simulator. In order to evaluate the performance of the methods we would like to vary the number of mismatches. This can not be controlled in real data, therefore we made subsets of the data on the basis of the distance between the poses. We assume that for larger distances it is more difficult to find matching features. In Figure 6(c) and 6(g) the heading and rotation error of the different methods is plot as a function of the distance between the images for dataset 'Almere 4'. It is clear that on a whole the errors are much larger than was the case for the simulation data. This is partly due to the fact that some of the views were obstructed by furniture, walls or people walking in the environment.

In the plot of the heading error (Figure 6(c)) one can see that the RANSAC+M-Estimator combined with the Two point algorithm is outperformed by the three point algorithm version, which in turn is clearly outperformed by the novel LUT method for distances larger than 1.5 meters. The accuracy of the RANSAC+M-Estimator combined with the Eight point algorithm is not much worse. This could have been caused by the fact that the robot was leaning over when accelerating, slightly violating the planarity constraint. For the rotation error (Figure 6(g)), the improvement of the ML estimator over the RANSAC+M-Estimator with Three Point is less clear.

4) Sensitivity to mismatches, trained with real data: Next we constructed a look up table using all the image pairs of the Almere 4 set that were within a 5 meter distance. We used two different binsizes, the first had 128 bins for all three dimensions and the other 16. Although about 10^6 image pairs were used, the LUT with 128^3 bins was not as smooth as the one learned with simulated data The proposed method was applied on the Spaan 1 set. The Maximum Likelihood solutions were compared to the RANSAC+M-Estimator combined with the Three point algorithm and solutions given two LUTs based on the simulator, also with 128^3 and 16^3.

In Figure 6(d) and 6(h) the results are shown. As can be seen the overall accuracy is less than for the Almere 4. A reason for this could be the motion blur, caused by the bad illumination of this dataset. The LUT ML with 128^3 bins based on the simulator performs best, followed by the 128^3 bins LUT based on the real images. Probably this somewhat worse performance is due to the limited number of training data. For the much smaller LUTs with 16^3 bins this seems to be less problematic, visible by the improvement of the LUT based on real images over the one based on simulated data.

5) Averages over the data sets: Application on other datasets and different bin sizes resulted in comparable errors. Figure 7 summarizes these results. Note that all three RANSAC based methods failed to robustly estimated poses for the difficult Biron 1 set as opposed to the proposed method.

6) Binsize vs CPU time: To evaluate the influence of different binsizes for the look up table, we tested the ML method for different numbers of bins. In Table I the average computational time in milliseconds is given per image pair for the 'Almere 4' set. Other datasets showed similar trends for both the LUTs constructed using the simulator and the real data. As can be seen small look up tables result in a large speed up, but this comes at the cost of more error (see Table 7). The RANSAC+M-Estimator combined with the Planar Three point algorithm is three times slower than the LUT method with 128^3 bins.

TABLE I

AVERAGE COMPUTATIONAL TIME USAGE PER RELATIVE POSE ESTIMATE IN MILLISECONDS FOR THE DIFFERENT METHODS.

ML				RANS+M-Est		
128	64	32	16	8pt	3pt	2pt
1.3	0.28	0.07	0.036	3.6	3.8	0.68

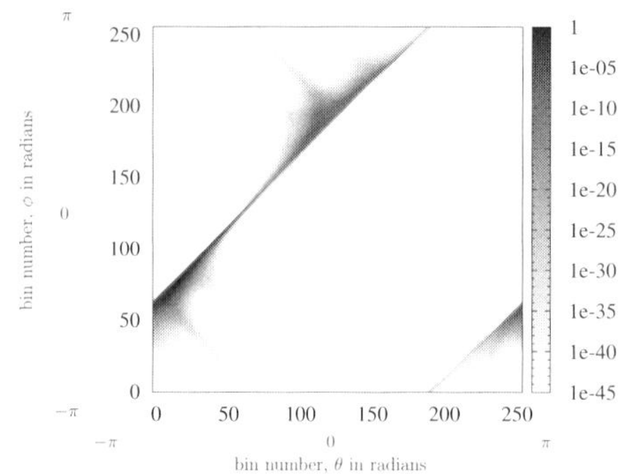

Fig. 8. The log posterior computed from the point correspondences between image 857 and 897 of Almere set 4. The distance between these camera positions was about 1 cm, while the rotation was 90 degrees.

V. DISCUSSION

An important advantage of the LUT based pose estimator is that it provides a full likelihood over the discretized space of possible relative poses. Thus, apart from finding a Maximum Likelihood solution, as shown in the Experiments section, this could make the method useful for a range of other applications.

A nice illustration of the usefulness of a full likelihood can be seen in Figure 8. It shows an example likelihood for a

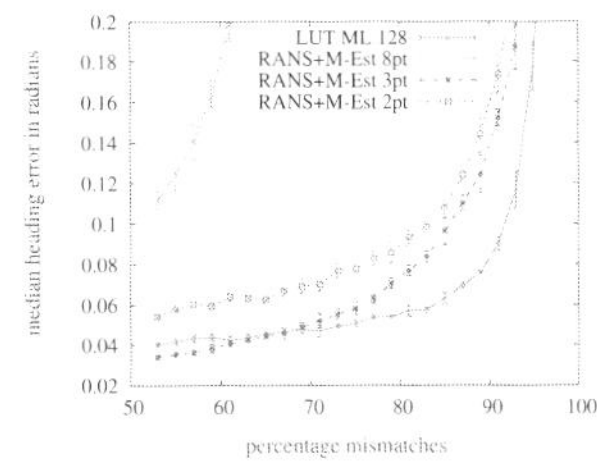

(a) Heading errors, simulation data

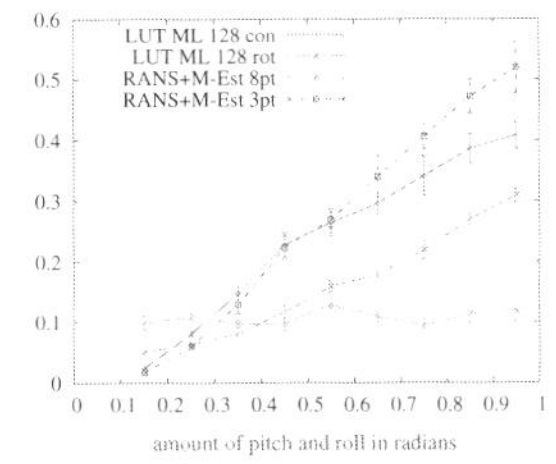

(b) Heading errors, non planar simulated data

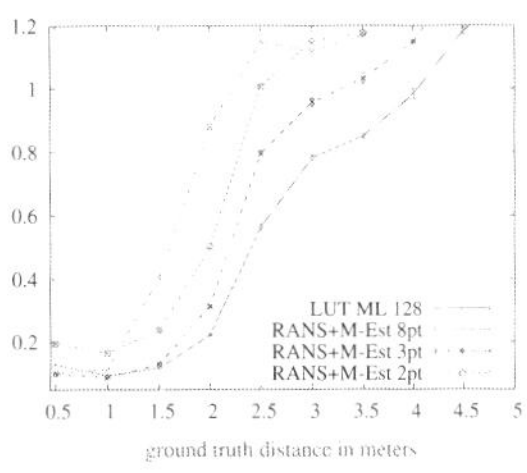

(c) Heading errors, Almere 4

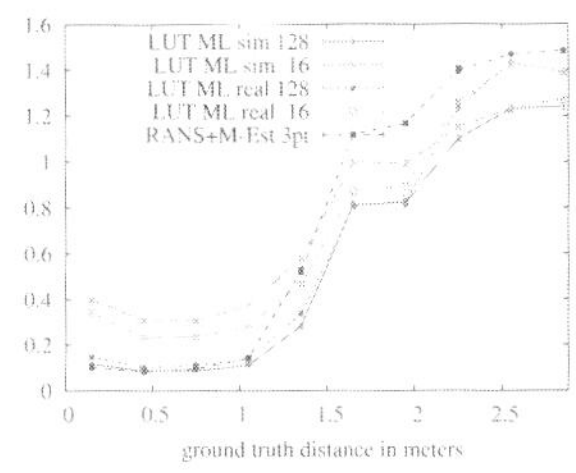

(d) Heading errors, Spaan 1

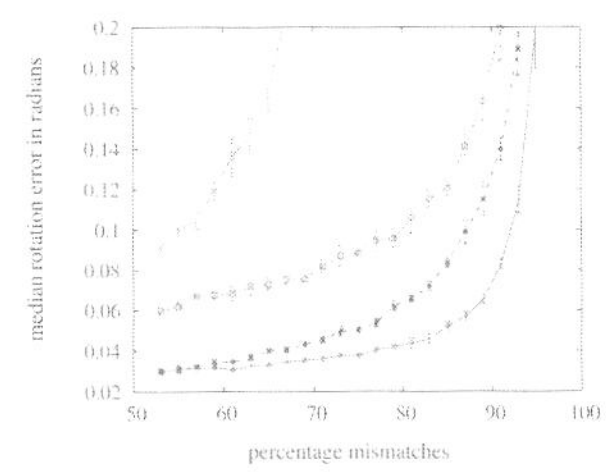

(e) Rotation errors, simulation data

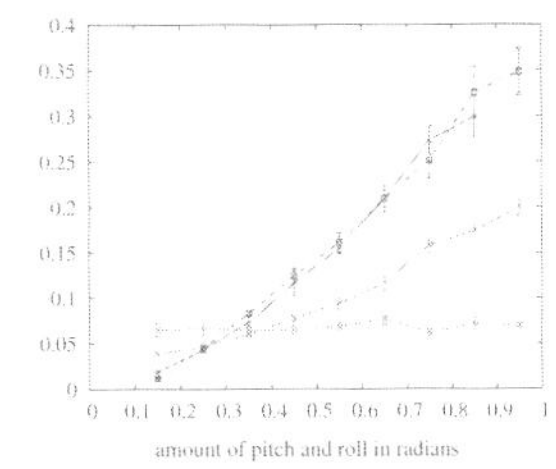

(f) Rotation errors, non planar simulated data

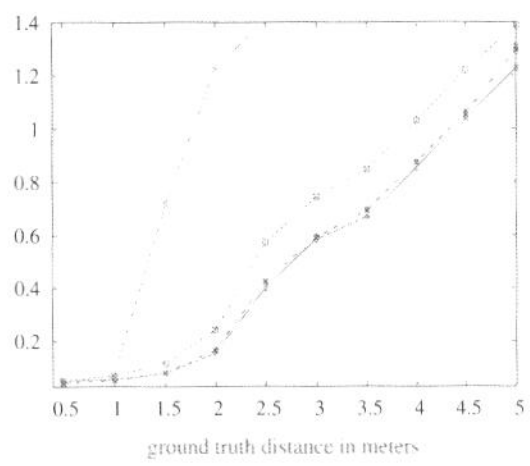

(g) Rotation errors, Almere 4

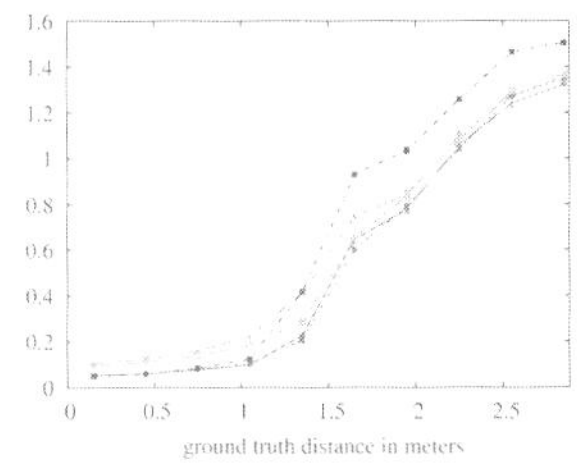

(h) Rotation errors, Spaan 1

Fig. 6. Comparison of LUT ML, trained using a simulator or real images and RANSAC+M-Estimator combined with different algorithms on the simulation dataset for different number of mismatches and amount of planar motion violation and Almere 4 and Spaan 1 for different distances between the image pairs. The MAD is used to draw confidence intervals of the medians.

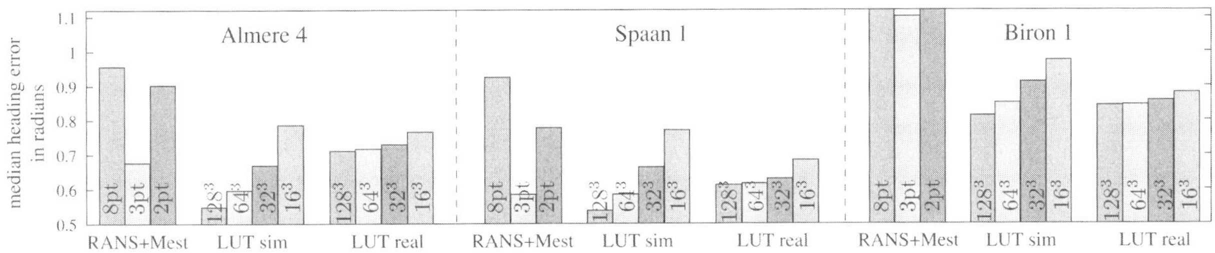

(a) Heading errors

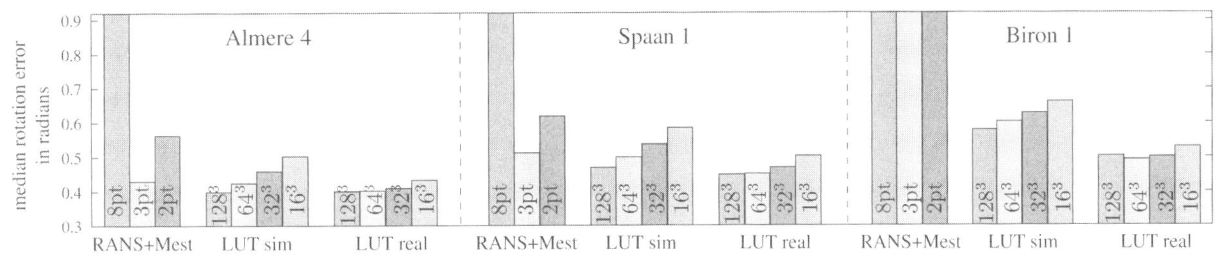

(b) Rotation errors

Fig. 7. (a) Heading errors and (b) rotation errors for the different estimators applied on all three datasets.

typical situation that occurred in the Almere 4 set. In this case the robot did not move forward but rotated on the spot. Thus, the heading of the robot, ϑ, can not be determined. This is correctly reflected by the estimated posterior. The rotation, on the other hand, can be determined. This can be seen by the diagonal relationship between ϑ and ϕ in the posterior.

The proposed method could be readily applied on particle filter based robot localization schemes [33] where each hypothesized robot pose can be weighed by the likelihood given newly acquired images. Also, geometric View based SLAM [3, 4] could benefit from the proposed method, because the uncertainty of the Maximum Likelihood can be estimated easily from the full likelihood. For example by fitting a Von Mises or mixture of Von Mises distributions on the discretized likelihood space [28].

Another task that is very much suited for the proposed example is that of topological mapping. Some state of the art topological mapping approaches use proper probabilistic data association techniques to compare pairs of images [1]. However, in addition they apply ad hoc rules to check whether the matched point correspondences fit in a certain local geometry, estimated using RANSAC. Because of the probabilistic nature of the proposed method, it is straightforward to combine it with these proper data association techniques, ending up in a fully probabilistic topological mapping method.

VI. Conclusion

In this paper we propose a novel approach to solve planar relative pose estimation from image point correspondences. We have shown the advantage of discretizing and analyzing the complete solution space, which is in the planar motion case 2 dimensional. Probabilistic methods were proposed that learn the likelihood over this space from a training set of representative images. Experiments on challenging image sets acquired in real homes showed a 20% increase in accuracy with respect to state of the art methods consisting of a planar constrained RANSAC and M-Estimators.

In addition an efficient technique was presented for building a concise look up table of the likelihood, reducing the estimation process to simple look ups. Computing a full likelihood given two images costs as little as 36 microsecond, as compared to the 3 milliseconds RANSAC uses. This could even be improved upon, for example, by using a multi-resolution approach as described in [34], in which a small look up table is used to isolate candidate areas for the ML solution, which are then investigated further using a bigger look up table. Another possibility is implementing the method on a GPU, which can much more quickly manipulate 2D histograms.

Continuing research will focus on taking advantage of the full likelihood that is estimated by the method. We foresee improvements in both geometric SLAM as in topological mapping, especially when it comes to uncertainty estimation.

Acknowledgments

We would like to thank Gwenn Englebienne and Leo Dorst for fruitful discussions and contributions to this paper.

References

[1] M. Cummins and P. Newman, "Highly scalable appearance-only SLAM - FAB-MAP 2.0," in *Proceedings of Robotics: Science and Systems (RSS)*, Seatle, USA, 2009.

[2] C. Valgren and A. J. Lilienthal, "Incremental spectral clustering and seasons: Appearance-based localization in outdoor environments," in *ICRA*. Pasadena, California, USA: IEEE, May 2008, pp. 1856–1861.

[3] R. Eustice, "Large-area visually augmented navigation for autonomous underwater vehicles," Ph.D. dissertation, MIT - Woods Hole Oceanographic Institute, June 2005.

[4] H. Andreasson, , T. Duckett, and A. J. Lilienthal, "A minimalistic approach to appearance based visual slam," *Transactions on Robotics*, vol. 24, no. 6, pp. 991–1001, October 2008.

[5] F. Fraundorfer, C. Engels, and D. Nister, "Topological mapping, localization and navigation using image collections," in *IROS*. San Diego, USA: IEEE/RSJ, November 2007, pp. 3872–3877.

[6] D. G. Lowe, "Distinctive image features from scale-invariant keypoints," *Int. Journal of Computer Vision*, vol. 60, no. 2, pp. 91–110, 2004.

[7] R. Hartley and A. Zisserman, *Multiple view geometry in computer vision, second edition*. Cambridge University Press, 2003.

[8] M. A. Fischler and R. C. Bolles, "Random sample consensus: A paradigm for model fitting with applications to image analysis and automated cartography," *Com. of the ACM*, vol. 24, no. 6, 1981.

[9] P. H. S. Torr and D. W. Murray, "The development and comparison of robust methods for estimating the fundamental matrix," *Int. Journal of Computer Vision*, vol. 24, no. 3, pp. 271–300, 1997.

[10] P. V. C. Hough, "Method and means for recognizing complex patterns," 1962, u.S. Patent 3,069,654.

[11] H. C. Longuet-Higgins, "A computer algorithm for reconstructing a scene from two projections," *Nature*, vol. 293, pp. 133–135, Sept. 1981.

[12] D. Nistér, "An efficient solution to the five-point relative pose problem," *IEEE Trans. Pattern Anal. Mach. Intell.*, vol. 26, no. 6, pp. 756–777, 2004.

[13] B. C. Matei, "Heteroscedastic errors-in-variables models in computer vision," Ph.D. dissertation, Rutgers University, 2001.

[14] H. Wang and D. Suter, "Robust adaptive-scale parametric model estimation for computer vision," *IEEE Trans. Pattern Anal. Mach. Intell.*, vol. 26, no. 11, pp. 1459–1474, 2004.

[15] P. H. S. Torr and A. Zisserman, "MLESAC: a new robust estimator with application to estimating image geometry," *Comput. Vis. Image Underst.*, vol. 78, no. 1, pp. 138–156, 2000.

[16] R. den Hollander and A. Hanjalic, "A combined RANSAC-Hough transform algorithm for fundamental matrix estimation," in *18th British Machine Vision Conference*. University of Warwick, UK, 2007.

[17] D. J. Heeger and A. D. Jepson, "Subspace methods for recovering rigid motion i: algorithm and implementation," *Int. J. Comput. Vision*, vol. 7, no. 2, pp. 95–117, 1992.

[18] A. Censi and S. Carpin, "HSM3D: Feature-less global 6DOF scan-matching in the Hough/Radon domain," in *ICRA*. Kobe, Japan: IEEE, May 2009.

[19] M. Pollefeys, D. Nistér, J. M. Frahm, A. Akbarzadeh, P. Mordohai, B. Clipp, C. Engels, D. Gallup, S. J. Kim, P. Merrell, C. Salmi, S. Sinha, B. Talton, L. Wang, Q. Yang, H. Stewénius, R. Yang, G. Welch, and H. Towles, "Detailed real-time urban 3d reconstruction from video," *Int. J. Comput. Vision*, vol. 78, no. 2-3, pp. 143–167, 2008.

[20] M. Milford and G. Wyeth, "Mapping a suburb with a single camera using a biologically inspired SLAM system," *IEEE Transactions on Robotics*, vol. 24, no. 5, pp. 1038–1053, 2008.

[21] M. Brooks, L. de Agapito, D. Huynh, and L. Baumela, "Towards robust metric reconstruction via a dynamic uncalibrated stereo head," *Image Vision Comput.*, vol. 16, no. 14, pp. 989–1002, 1998.

[22] D. Ortín and J. M. M. Montiel, "Indoor robot motion based on monocular images," *Robotica*, vol. 19, no. 3, pp. 331–342, 2001.

[23] J. Kosecká, F. Li, and X. Yang, "Global localization and relative positioning based on scale-invariant keypoints," *Robotics and Autonomous Systems*, vol. 52, no. 1, pp. 27–38, 2005.

[24] T. Goedemé, T. Tuytelaars, G. Vanacker, M. Nuttin, and L. V. Gool, "Feature based omnidirectional sparse visual path following," in *IROS*. Edmonton, Canada: IEEE/RSJ, August 2005, pp. 1003–1008.

[25] A. Makadia, D. Gupta, and K. Daniilidis, "Planar ego-motion without correspondences," in *WACV/Motion*. Los Alamitos, CA, USA: IEEE Computer Society, January 2005, pp. 160–165.

[26] J. Civera, A. J. Davison, and J. M. M. Montiel, "Inverse depth parametrization for monocular SLAM," *IEEE Transactions on Robotics*, vol. 24, no. 5, pp. 932–945, October 2008.

[27] O. Booij and Z. Zivkovic, "The planar two point algorithm," University of Amsterdam, Tech. Rep. IAS-UVA-09-05, 2009.

[28] C. M. Bishop, *Pattern Recognition and Machine Learning*, ser. Information Science and Statistics. Springer, 2006.

[29] D. H. Stewénius, C. Engels, and D. D. Nistér, "Recent developments on direct relative orientation," *ISPRS Journal of Photogrammetry and Remote Sensing*, vol. 60, no. 4, pp. 284–294, 2006.

[30] J. Folkesson, P. Jensfelt, and H. Christensen, "Vision SLAM in the measurement subspace," in *ICRA*. Barcelona, Spain: IEEE, April 2005, pp. 30–35.

[31] Z. Zivkovic, O. Booij, B. Kröse, E.Topp, and H.I.Christensen, "From sensors to human spatial concepts: an annotated dataset," *IEEE Transactions on Robotics*, vol. 24, no. 2, pp. 501–505, April 2008.

[32] R. Bunschoten, "Mapping and localization from a panoramic vision sensor," Ph.D. dissertation, University of Amsterdam, November 2003.

[33] H.-M. Gross and A. Koenig, "Robust omniview-based probabilistic self-localization for mobile robots in large maze-like environments," in *ICPR (3)*, 2004, pp. 266–269.

[34] E. B. Olson, "Real-time correlative scan matching," in *ICRA*. Kobe, Japan: IEEE, May 2009.

Efficient Non-parametric Surface Representations Using Active Sampling for Push Broom Laser Data

Mike Smith, Ingmar Posner, and Paul Newman
Oxford University, Mobile Robotics Group
{mike, hip, pnewman} @ robots.ox.ac.uk

Abstract—**This paper concerns the creation of efficient surface representations from laser point clouds. We produce a continuous, implicit, non-parametric representation with an update time that is constant. The algorithm places no restriction on the complexity of the underlying workspace surfaces and automatically prunes redundant data via an information theoretic criterion. This criterion makes the use of Gaussian Process regression a natural choice. We adopt a formulation which handles the typical non-functional relation between XY-location and elevation allowing us to map arbitrary environments. Results are presented that use real and synthetic data to analyse the trade-off between compression level and reconstruction error. We attain decimation factors in excess of two orders of magnitude without significant degradation in fidelity.**

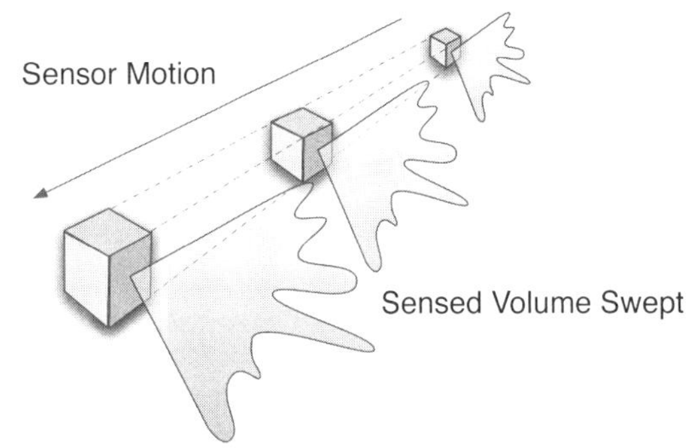

Figure 1. The geometry of a push broom laser. A single 2D sensor is pushed through the environment sweeping out a sensed volume

I. Introduction and Motivation

3D point clouds gathered from outdoor scenes are now ubiquitous. While visually appealing when rendered, they have limitations; two of which are addressed in this paper. Firstly, they are discrete samples – they do not readily admit access to the continuous nature of the workspace's surfaces. Secondly, they are inefficient in the sense of containing samples which, given other nearby samples and reasonable smoothness assumptions, are redundant and convey little information about the underlying scene geometry.

Our motivation lies in the value of obtaining richer workspace descriptions generated by a mobile robot. We restrict our attention to point clouds generated from the simplest and cheapest of 3D data acquisition systems – the push broom laser. Here, as shown in Figure 1, a single 2D laser scanner is pushed through the workspace and a point cloud is formed by aggregation. We describe a non-parametric method which simultaneously generates a continuous representation of the workspace from discrete laser samples and decimates the data set retaining only locally salient samples. We attain decimation factors in excess of two orders of magnitude without significant degradation in fidelity.

At the heart of our method is the application of a Gaussian Process (GP) to select support from a finite, time varying region. This GP is used to generate a predictive distribution of individual range measurements in constant time regardless of the quantity of the collected laser data. By evaluating the similarity between the distribution of the measurement (using a sensor model) and its GP-predicted distribution, we arrive at a criterion for keeping or dropping individual samples. Here we draw inspiration from work in sparse GPs [20], [2], [25]. We refer to this process as Active Sampling – the GP framework actively guides us in deciding which of many input points should be retained and later used for surface reconstructions.

Our approach admits the calculation of every (x, y, z) point that lies on a sampled workspace surface. A particular advantage of our formulation is the ability to represent common structures such as walls, which require multiple elevation values to be associated with the same position on the x-y-plane (Figure 2). Technically, this multivalued *relation* is *non-functional*, and we use this term to distinguish it from that of a single-valued, *well formed function*.

II. Related Work

Within the robotics community, 3D surface reconstruction from sparsely sampled environments is commonly motivated by the need to model an unknown terrain or to provide an accurate reconstruction of the workspace across scales. The latter is commonly achieved by meshing techniques (see, for example, [5], [7]) whereby every datum of a point cloud forms the vertex of a polygon. Often such techniques are augmented by a mesh decimation step which aims to obtain a good representation of the environment while at the same time reducing the amount of data to be stored. Commonly, mesh creation as well as the reduction in the number of vertices is based on sensor and environment dependent heuristics. Problems are encountered when the data arise from a non-homogeneous sampling of the workspace and/or coverage is incomplete – both are frequently the case in our problem domain.

An alternative approach considers a parametric representation based on geometric primitives [6], [28]. Planar representations are a popular choice since planes are pervasive in

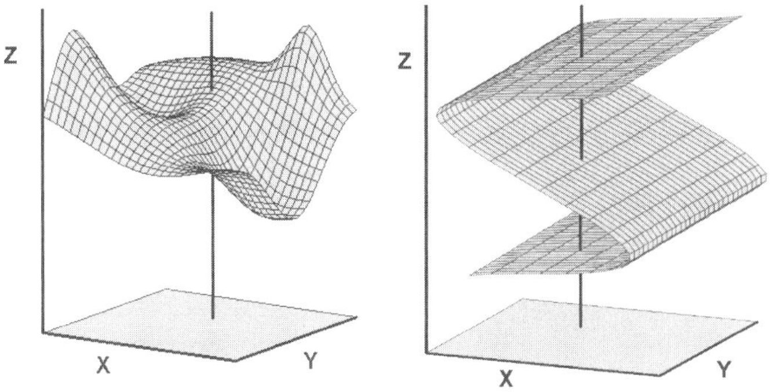

Figure 2. We wish to map scenes with multiple elevations for a given x, y point (right), rather than being restricted to a trivial case (left).

man-made environments. However, the reliance on geometric primitives places strong assumptions on the environment which are rarely justified in complex outdoor environments.

The requirement of robots to operate in ever larger environments provides ample incentive for research into suitable models of the terrain encountered. Consequently, much recent work in 3D reconstruction techniques trace their origin to this domain. Common approaches include discrete 2D, 2.5D and 3D methods such as [10], [1], [17], [29], which represent the world as a regular grid of cells or voxels. Each cell stores the probability of being occupied. As new sensor measurements arrive these probabilities are continuously updated. These representations are dependent on grid resolution and suffer from the necessity to strike a trade-off between resolution (i.e. memory requirements) and volume covered.

In recent years there has been a further trend in the terrain modelling community to employ non-parametric continuous models of surfaces. In this context, GP models are particularly favoured for their ability to handle incomplete data in a principled probabilistic fashion. Examples of such approaches include [9], [19], [30], although they were first introduced to the geostatistics field under the name *kriging* many years previously by [11] amongst others. While these approaches differ in their choice of (non-stationary) covariance function and GP sparsification method, they adopt the same parameterisation of the problem – that is they model a function $f : \mathbb{R}^2 \rightarrow \mathbb{R}$ which associates a single elevation value z with any given position (x, y) in 3D Euclidean space. While this well formed functional mapping is effective for terrain modelling it precludes application in the full 3D mapping case (Figure 2). Our work aims to overcome this limitation by introducing a more natural parameterisation of the surface reconstruction problem, as well as providing a principled method to sub-sample the data.

There has been much research into sparsification for GP regression, with an overview provided in [20]. Within this field, active subsampling strategies have been used to select information-rich data through use of information theoretic criteria [8], [3], [25]. Our work is particularly similar to [25], but we exploit the time-sequential nature of laser data (see Section IV-B) to form an exact and inexpensive predictive distribution for use in our decision criterion.

While we share the common goal of accurate 3D surface reconstruction with the literature discussed so far, this work bears closest relation to that of Gaussian Beam Processes [18]. The authors model laser data on a per-scan basis using a GP model to regress the range on bearing. However, while our model also regresses on range, we achieve an implicit model of the entire workspace through a sliding window approach and active data selection. This provides significant advantages above and beyond a mere extension of Gaussian Beam Processes to the 3D case.

III. Non-Functional Surface Representation via Sensor Configuration Space Parameterisation

Our aim is to form an implicit representation of workspace surfaces by processing data gathered from a laser sensor being moved through space along an arbitrary path. In particular, we pose queries of points in (x, y, z) on the workspace's surface as range queries along arbitrary rays emanating from the sensor at a point along the sensor's trajectory. Let a beam from the laser be parameterised as a point $\mathbf{q} \in \mathbb{S} \times \mathbb{R}$ where $\mathbf{q} = [\theta, t]^T$. θ denotes the 1D angular position of the laser beam (*not the whole sensor unit*) and t denotes the timestamp of the laser scan.

We describe $\mathbf{q}$ as a point in *sensor configuration space* $\mathcal{Q}$ as illustrated in Figure 3 and refer to it as a particular configuration of the laser sensor. Sensor configuration space provides a natural domain for range regression since it is closely related to the state of the laser and not that of the rest of the robotic system. The mapping from sensor configuration space to range is necessarily a well formed function – a laser sensor associates a single range r with any point in $\mathcal{Q}$ such that

$$G : \mathbb{S} \times \mathbb{R} \rightarrow \mathbb{R}, \mathbf{q} \mapsto r \tag{1}$$

Consider now a mapping from $\mathcal{Q}$ to the Euclidean workspace $\mathcal{W}$

$$E'(\mathbf{q}) = E(V(\mathbf{q}), G(\mathbf{q})) \tag{2}$$

where:

$$V : \mathbb{S} \times \mathbb{R} \rightarrow \mathbb{S}^3 \times \mathbb{R}^3, \mathbf{q} \mapsto \mathbf{p} \tag{3}$$

$$E : (\mathbb{S}^3 \times \mathbb{R}^3) \times \mathbb{R} \rightarrow \mathbb{R}^3, (\mathbf{p}, r) \mapsto \mathbf{x} \tag{4}$$

For every point $\mathbf{q}$ in $\mathcal{Q}$, $V(\mathbf{q})$ provides the six d.o.f. pose of the sensor's laser beam $\mathbf{p} \in \mathbb{S}^3 \times \mathbb{R}^3$ at the time a measurement is taken – represented as roll, pitch, yaw and position. $E(\mathbf{p}, r)$ maps a six d.o.f. laser beam pose $\mathbf{p}$ and a scalar range measurement r to a single point $\mathbf{x}$ in 3D Euclidean space. This parameterisation naturally eschews the non-functional relation between elevation z and (x, y) location that is commonly found in terrain mapping formulations (Figure 2). By keeping each operation distinct, we also decouple robot trajectory estimation $V(\mathbf{q})$, from that of the regression of the laser data $G(\mathbf{q})$[1]. This permits independent relaxation of the sensor trajectory (which, for example, could be in response to loop closure events).

To form an estimate of $G(\mathbf{q})$ at any arbitrary position in $\mathcal{Q}$ we turn to Gaussian Process Regression. Given a set of measurements $\mathcal{D} = \{(\mathbf{q}_i, r_i)\}_{i=1}^N$ at a query point $\mathbf{q}^*$

[1]Note that the regression will always be dependent on the ground-truth trajectory of the vehicle, but that no knowledge of this trajectory is required in our algorithm.

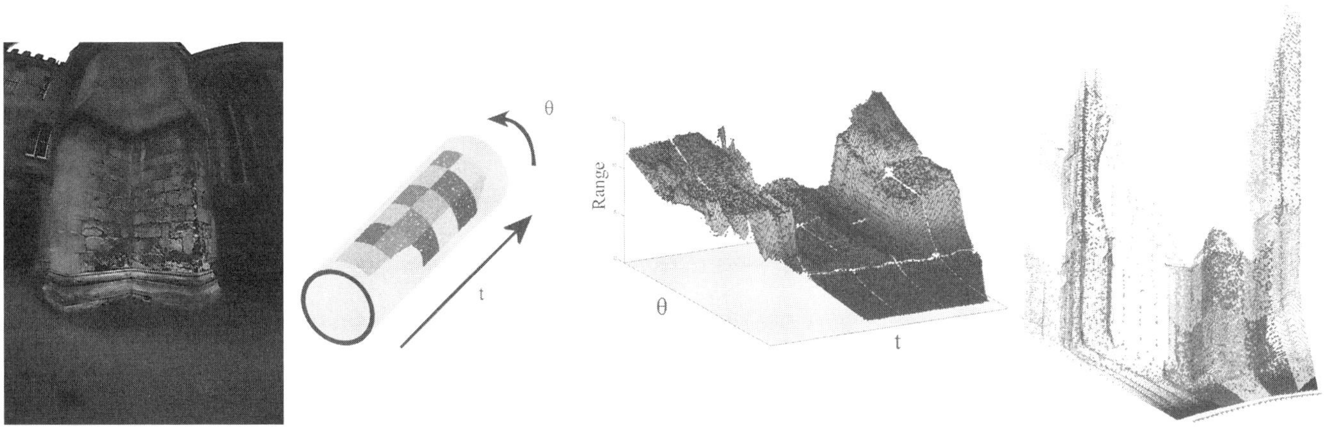

Figure 3. Parameterisation used in this work. (Left) A camera image used to provide the visual context for this example. The image is captured at the same time as the laser data plots in (left-mid) to (right), which forms the central part of exemplar d) in Figure 8 (see Section V). (Mid-left) Visualisation of sensor configuration space $\mathcal{Q}$. Every point $\mathbf{q}$ on this manifold is parameterised by time t and the angular position of the laser beam θ. (Mid-right) Laser range image. $\mathcal{Q}$ is unwrapped with t and θ of along the X and Y axes and output of the mapping $G(\mathbf{q})$, range predictions $\mathbf{r}$, along Z. (Right) 3D Euclidean space. Ranges $\mathbf{r}$ and vehicle poses $\mathbf{p}$ are mapped through $E(\mathbf{p}, \mathbf{r})$ to points in 3D. The robot trajectory is shown as coloured frames across the bottom. It is advised to view this figure in colour as a coloured tiling is used to the aid understanding of the mapping between sensor configuration space and euclidean space.

we can obtain a predictive distribution $p(r^*|\mathbf{q}^*, \mathcal{D})$. In the following section we present a brief summary of how this predictive distribution is obtained from a sliding window in sensor configuration space.

IV. Active Sliding Window using Gaussian Process Regression

Gaussian Processes (GPs) provide for non-parametric probabilistic regression over $\mathcal{Q}$. A GP consists of a collection of jointly Gaussian distributed random variables and describes a distribution over latent functions underlying observations. It is fully specified by mean $\mu(\mathbf{q})$ and covariance $k(\mathbf{q}, \mathbf{q}')$ functions. In our application we are concerned with estimating the mapping $G(\mathbf{q})$ corresponding to these latent functions. Given known ranges $\mathbf{r}$ from different configurations $Q = \{\mathbf{q}_i\}_{i=1}^N$ and a query point $\mathbf{q}^*$ with corresponding unknown target range r^* we can write:

$$\begin{bmatrix} \mathbf{r} \\ r^* \end{bmatrix} \sim \mathcal{N}\left(\begin{matrix} \mu(Q) \\ \mu(\mathbf{q}^*) \end{matrix}, \begin{bmatrix} K(Q,Q) + \sigma_m^2 I & \mathbf{k}(Q, \mathbf{q}^*) \\ \mathbf{k}(Q, \mathbf{q}^*)^T & k(\mathbf{q}^*, \mathbf{q}^*) + \sigma_m^2 \end{bmatrix} \right) \tag{5}$$

Each element of K is, in the stationary case, only on a suitably chosen distance metric $d = \|\mathbf{q} - \mathbf{q}'\|$ between two corresponding points $\mathbf{q}$ and $\mathbf{q}'$ in $\mathcal{Q}$. We explicitly account for noise in the training observations $\mathbf{r}$ through an additive white noise process of strength σ_m along the diagonal entries of K^2. The derivation of the mean $\mathbb{E}[r^*]$ and covariance $\mathbb{V}[r^*]$ of the predictive distribution $p(r^*|\mathbf{q}^*, \mathcal{D})$ for a deterministic $\mu(\mathbf{q}) = 0$ (as is commonly used [30]) are standard and can be found, for example, in [21]

$$\mathbb{E}[r^*] = \mathbf{k}(Q, \mathbf{q}^*)^T (K(Q,Q) + \sigma_m^2 I)^{-1} \mathbf{r} \tag{6}$$

$$\begin{aligned} \mathbb{V}[r^*] = \; & k(\mathbf{q}^*, \mathbf{q}^*) + \sigma_m^2 - \\ & \mathbf{k}(Q, \mathbf{q}^*)^T (K(Q,Q) + \sigma_m^2 I)^{-1} \mathbf{k}(Q, \mathbf{q}^*) \end{aligned} \tag{7}$$

[2]In this paper we have used $\sigma_m^2 = 0.01m^2$.

Throughout this work we use a member of the Matérn class of covariance functions as advocated in [27]. We note, however, that many others, including non-stationary covariance functions [9], could be adapted and substituted in its place (Section VI). The Matérn class is dependent on a shape parameter ν which regulates the smoothness of the interpolation. It equates to the more standard exponential covariance function as a special case when $\nu = \frac{1}{2}$, and the squared exponential as $\nu \to \infty$. As suggested in [21], we explored several common choices of $\nu = \{\frac{1}{2}, \frac{3}{2}, \frac{5}{2}, \infty\}$ over a number of workspaces varying in complexity. We found that $\nu = \frac{3}{2}$ consistently produced accurate surface reconstructions for a variety of support set sizes and length scales. Although the smoother covariance functions $\nu = \{\frac{5}{2}, \infty\}$ performed well for simple workspaces, they seemed over constrained in complex scenarios, and vice versa for the rough covariance function $\nu = \frac{1}{2}$. Thus

$$k(\mathbf{q}, \mathbf{q}')_{mat\acute{e}rn\,\nu=\frac{3}{2}} = \sigma_p^2 (1 + \frac{d\sqrt{3}}{l}) \exp(-\frac{d\sqrt{3}}{l}) \tag{8}$$

where σ_p^2 is the process noise, l is the length scale and $d = \|\mathbf{q} - \mathbf{q}'\|_{\mathcal{Q}}$ denotes the geodesic distance between $\mathbf{q}$ and $\mathbf{q}'$ in $\mathcal{Q}$, which for this sensor configuration is the ℓ_2 norm.

The quantities of data we consider render the application of a single monolithic GP infeasible – time complexity of a naïve implementation of GP regression is cubic in N, the size of $\mathcal{D}$. Instead, for each prediction we enforce a fixed predictive support window size n, formed from the closest (in terms of the ℓ_1 norm from t to t^*) points that have been actively accepted by our algorithm to guarantee a constant time algorithm. The time-sequential nature of laser data ensures that this support window slides across $\mathcal{Q}$ as the robot progresses along its trajectory.

We are interested in minimising the computational complexity of our algorithm for practical applications. In Section IV-B we devise an active sampling strategy to intelligently determine a salient subset of $\mathcal{D}$, $\mathcal{D}'$ to include in our active window. The advantages of this are two-fold: we can achieve a significant compression of $\mathcal{D}$, and we can re-use the inversion in Equations 6 and 7 across multiple predictions, thus decreasing computational cost. Where we must incorporate new points into the active window we update, rather than re-calculate the inversion (Section IV-C). This lowers the overall worst case prediction cost to $\mathcal{O}(n^2)$.

A. Stochastic Mean Function

We affect a stochastic non-zero mean function $\mu(\mathbf{q})$ through the use of explicit basis functions [21]. Explicit basis functions $\mathbf{h}(\mathbf{q})$ with weightings $\boldsymbol{\beta} \sim \mathcal{N}(\mathbf{b}, B)$, that are found in a traditional Bayesian weight space view of linear regression, are superimposed with the zero-mean GP $Z(\mathbf{q}) \sim \mathcal{GP}(0, k(\mathbf{q}, \mathbf{q}'))$ of Section IV

$$G(\mathbf{q}) = Z(\mathbf{q}) + \mathbf{h}(\mathbf{q})^T \beta \tag{9}$$

Crucially, in the absence of data, predictions using $G(\mathbf{q})$ tend to the surface estimates of the explicit basis functions, rather than that of the zero mean. One may view this formulation as the GP modelling the residuals between the implicit surface generated by the explicit basis function, and $\mathcal{D}$, and has been used successfully in robotics applications such as [14]. This is particularly beneficial when predicting at the boundary of our active window, and is the key to our compression rates (Section IV-B).

We take the limiting case as the prior on the weightings tends towards a uniform distribution, $(B^{-1} \rightarrow \mathbf{0})$. Following [15] we arrive at a predictive distribution that is independent of $\mathbf{b}$.

$$\mu_g(\mathbf{q}^*) = \mathbb{E}_Z[r^*] + \mathbf{u}^T \bar{\boldsymbol{\beta}} \tag{10}$$
$$\sigma_g^2(\mathbf{q}^*) = \mathbb{V}_Z[r^*] + \mathbf{u}^T (HK(Q,Q)^{-1}H^T)^{-1}\mathbf{u} \tag{11}$$

where $\mathbb{E}_Z[r^*]$ and $\mathbb{V}_Z[r^*]$ are given by Equation 6 and 7 respectively, $H = [\mathbf{h}(\mathbf{q}_1), \mathbf{h}(\mathbf{q}_2)...\mathbf{h}(\mathbf{q}_n)]$ and

$$\bar{\boldsymbol{\beta}} = (HK(Q,Q)^{-1}H^T)^{-1}HK(Q,Q)^{-1}\mathbf{r} \tag{12}$$
$$\mathbf{u} = \mathbf{h}(\mathbf{q}^*) - HK(Q,Q)^{-1}\mathbf{k}(Q, \mathbf{q}^*) \tag{13}$$

This formulation affords us the benefits of explicit basis functions without requiring additional parameters, and maintains the desirable predictive properties of GPs. Like Equation 6 and 7 an inverse must be calculated, but its dimension is the number of explicit basis functions chosen for the application. We choose the polynomial set of basis functions $\mathbf{h} = [1, \theta, \theta^2, \theta^3, t, t^2, t^3, \theta t, \theta t^2, \theta^2 t, \theta^2 t^2]^T$and have found cross terms above second order to have marginal overall impact on predictions.

B. Active Sampling through KL Divergence

In this Section we describe our method for active point selection. We decide if our prediction $r^* \sim \mathcal{N}(\mu_g, \sigma_g^2)$ (Equation 10 and 11) differs significantly from each measurement $r_m \sim \mathcal{N}(\mu_m, \sigma_m^2)$ at $\mathbf{q}^*$. We use the KL divergence between the two distributions to ascertain if the average additional information required to specify r_m as a result of using r^* (instead of the true r_m) is greater than a threshold κ. This threshold, naturally modulated by terrain complexity, sets decimation fraction $\frac{|\mathcal{D}|}{|\mathcal{D}'|}$. Thorough analysis on parameter choice can be found in Section V. The KL divergence between two 1D Gaussian distributions has a closed form solution:

$$D_{KL}(\mathcal{N}_m||\mathcal{N}_g) = \frac{1}{2}\left[log_e \frac{\sigma_g^2}{\sigma_m^2} + \frac{\mu_g^2 + \mu_m^2 - 2\mu_g\mu_m + \sigma_m^2}{\sigma_g^2} - 1 \right] \tag{14}$$

On adoption of any new point into the active support set we introspectively analyse the closest (ℓ_1 norm in $\mathcal{Q}$) r_m that had previously been decided redundant to determine if our belief has changed, given the new information. On adoption of this point introspection is then repeated for the next closest redundant r_m until there are no further adoptions. This approach is demonstrated in Figure 4. Introspection allows both sides of discontinuities to be analysed given the final sampling of the opposite side of the discontinuity. Figure 5 demonstrates the results of this reverse sweeping, and the typical $\mathcal{D}'$ that is stored and used for subsequent predictions instead of $\mathcal{D}$. In the worst case, where there are truly complex sections of the workspace, our algorithm performs as well as a naïve implementation by using all $\mathcal{D}$, while maintaining the ability to automatically subsample simple scenes. We note the resulting subsampling may also provide an information rich subset of $\mathcal{D}$. This may boost performance of other applications especially registration, where careful selection of points can increase accuracy and robustness [22].

C. Implementation Considerations

We seek to maximise the support set size for a given computational capacity. This provides more fully conditioned predictions, over a greater prediction window size, both while $\mathcal{D}'$ is constructed online and for subsequent surface querying. We use Cholesky decomposition and rank one updates of K to perform the expensive inversion in Equations 6 and 7, as is common in the GP literature [16], [24], [23]. Each time we accept a new point into our active set we first marginalise the most stale point (at the back of the active window), then expand the active set to include the new measurement.

Marginalisation: Given a positive definite matrix K, and its Cholesky factor C such that $C^T C = K$

$$K = \begin{bmatrix} K_{1,1} & \mathbf{k}_{1,2} & K_{1,3} \\ \mathbf{k}_{1,2}^T & k_{2,2} & \mathbf{k}_{2,3} \\ K_{1,3}^T & \mathbf{k}_{2,3}^T & K_{3,3} \end{bmatrix} C = \begin{bmatrix} C_{1,1} & \mathbf{c}_{1,2} & C_{1,3} \\ \mathbf{0} & c_{2,2} & \mathbf{c}_{2,3} \\ \mathbf{0} & \mathbf{0} & C_{3,3} \end{bmatrix}$$

we can remove the central row and column by marginalising C to:

$$\begin{bmatrix} C_{1,1} & C_{1,3} \\ \mathbf{0} & \Upsilon(C_{3,3}^T C_{3,3} + \mathbf{c}_{2,3}^T \mathbf{c}_{2,3}) \end{bmatrix} \tag{15}$$

Where Υ is the Cholesky update of [4] which exploits the special structure of $\mathbf{c}_{2,3}^T \mathbf{c}_{2,3}$, and is of complexity $O(n^2)$.

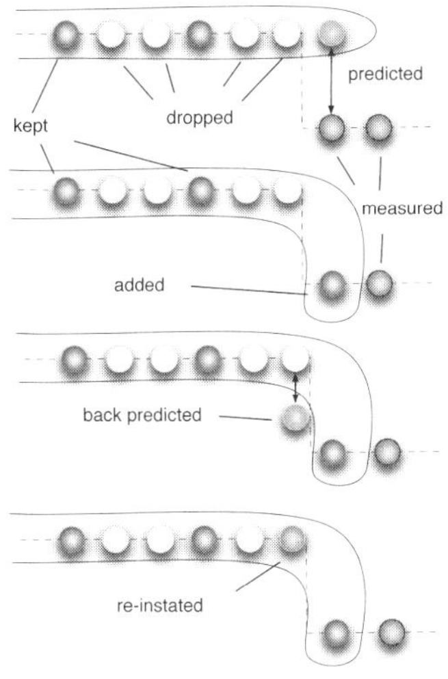

Figure 4. Active sampling of a 1D data series. Each node, connected by the chained lines, represents a measurement. Black indicates data accepted by the forward pass of the active sampling algorithm while the white have been rejected and the blue are yet to be observed. Green nodes are predictions made by our algorithm and red are points accepted through introspection. Progressing downwards, each subplot represents the algorithm at successive points in time. (top) The algorithm forms a prediction that has a KL divergence from the measurement that is greater than κ, hence it is adopted into the active support set (2nd row). (3rd row) The algorithm makes a back prediction which, where it was previously in agreement with the measurement, now has a significant error, that is greater than κ. (bottom) The measurement, which was previously rejected, is therefore accepted into the active region. It is recommended that this figure is viewed in colour.

An efficient implementation can be found in, for example, *cholupdate* of MATLAB®[12].

Expansion: Given $K = \begin{bmatrix} K_{1,1} & K_{1,3} \\ K_{1,3}^T & K_{3,3} \end{bmatrix}$ and corresponding Cholesky factor $R = \begin{bmatrix} R_{1,1} & R_{1,3} \\ \mathbf{0} & R_{3,3} \end{bmatrix}$, we wish to expand to C through insertion of a central row and column:

$$\begin{bmatrix} R_{1,1} & R_{1,1}^T \backslash \mathbf{k}_{1,2} & R_{1,3} \\ \mathbf{0} & \sqrt{k_{2,2} - \mathbf{c}_{1,2}^T \mathbf{c}_{1,2}} & \frac{\mathbf{k}_{2,3} - \mathbf{c}_{1,2}^T C_{1,3}}{c_{2,2}} \\ \mathbf{0} & \mathbf{0} & \Upsilon(R_{3,3}^T R_{3,3} - \mathbf{c}_{2,3}^T \mathbf{c}_{2,3}) \end{bmatrix} \quad (16)$$

where we have used $\backslash$ above to indicate the solution of $\mathbf{c}_{1,2}$ in the Equation $R_{1,1}^T \mathbf{c}_{1,2} = \mathbf{k}_{1,2}$, found by the use of backwards or forwards substitution for the triangular $R_{1,1}$.

V. Experimental Results

In this section we analyse how changing κ affects the decimation fraction of the original point cloud for data collected in a commonplace urban environment. Exemplars were obtained using the New College data set [26]. Data were gathered from a two wheeled Segway RMP~200 platform with two vertically aligned SICK LMS291-S14 laser sensors mounted on the side of the vehicle. Pose estimates were obtained using visual odometry from a forward facing stereoscopic camera [13]. Throughout we use the Matérn class of covariance functions, a length scale l of 8, a process variance σ_p^2 of $0.05m$ and a active window size of 200, all of which were found empirically to produce consistently accurate surface representations, for a given compression ratio, as discussed in Section IV. Error (in m) is defined as the ℓ_1 norm in range between a hold out set of measurements and corresponding predictions at the same point in $\mathcal{Q}$. Predictions are made as the centre of the active window passes the hold out set in $\mathcal{Q}$ and are conditioned on the current active set.

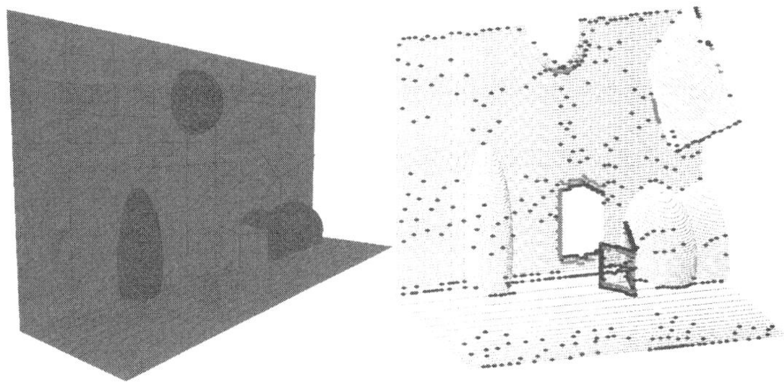

Figure 5. Result from using active sampling within our sensor configuration space formulation. (Left) CAD model used to generate synthetic data based on the wall section ($7 \times 3 \times 1.5m$) of exemplar a)(see Figure 8). (Right) Plot of synthetic data as a laser is dragged from left to right in front of the scene. Blue indicates all laser measurements, black denotes points selected on the forward pass of the algorithm, and red are points that have been chosen introspectively. We have used $\kappa = 0.8\,nats$ as in other examples (see Figures 8 and 10). Note that the floating cube in the centre of the scene generates a large ($1m$) discontinuity which is automatically highly sampled on both sides. For a plot indicating the effect of varying κ see Figure 7. It is recommended that this figure is viewed in colour.

A. Compression

Figure 6 depicts typical subsampling as κ is varied. For aggressive compression factors of 1000 we achieve a modest mean error of $0.3m$. For errors comparable with measurement precision of $0.015m$ we can achieve a compression factor of 6.

In Figure 7 results are collated across a range of κ for the four exemplars, a) through d), in Figure 8 and the two synthetic cases, e) and f) (Figure 5). Intuitively, as the threshold is increased an increasing compression ratio is produced, accompanied by an increase in error as depicted by the box plots in Figure 8. As one can discern from the images and CAD models, the relative positions of the curves correspond to the scene complexity: the more complex the scene, the lower the compression ratio. Scenes b) and d) are the most complex, with noisy foliage, measurements of ceilings behind window panes and discontinuities as great as $5m$, compared to that of the $1m$ discontinuity the person in c) presents. In all cases, these complex regions have been sampled most heavily – the outline of the person can be recognised in the Euclidean plot of c) in Figure 8.

B. Surface Error

Box plots indicate surface error distributions for each of the exemplars in Figure 8. As compression increases, error typically migrates from being that of a tightly bunched distribution around that of the measurement noise, to one with an

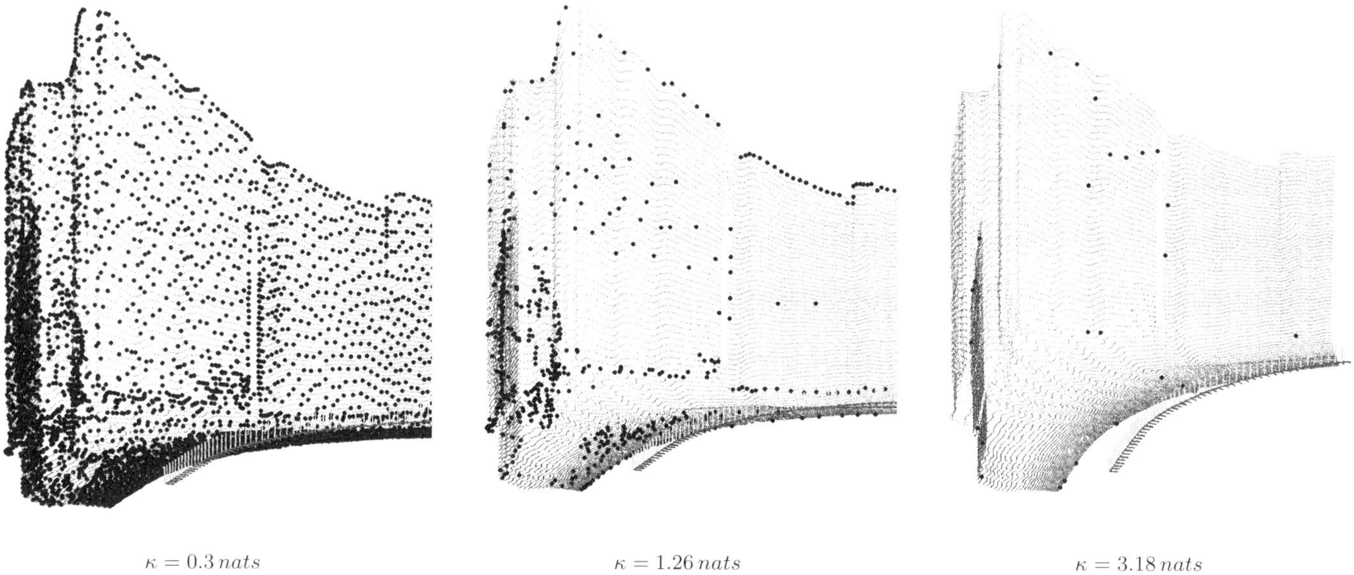

Figure 6. Point subsample plots for $\kappa = \{0.3, 1.26, 3.18\}\,nats$ for exemplar c) of Figure 8. The 28892 Original measurements are the smaller blue dots and points selected by our active sampling algorithm are in black. The vehicle trajectory is represented by the coloured frames across the bottom of each plot. (Left) 4543 laser measurements are used to produce predictions resulting in a mean error of $0.014m$ while in (middle) 323 points are used with a $0.177m$ average error. (Right) only 28 points have been selected, providing an average error of $0.32m$.

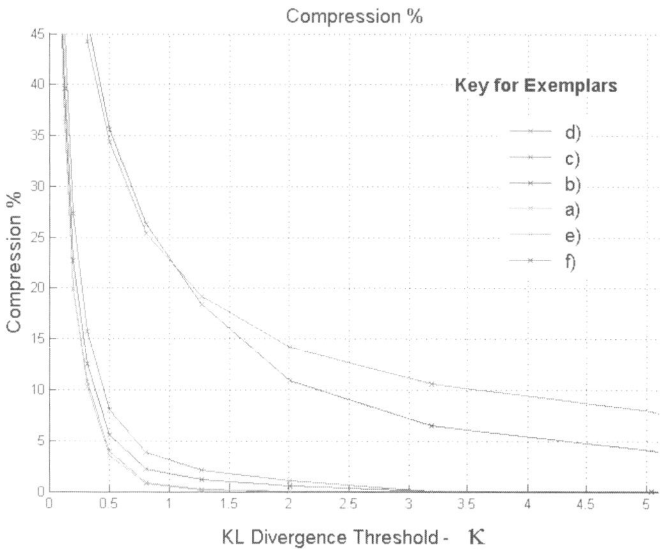

Figure 7. compression verses κ for a variety of urban environments. Images of the workspaces can be found in Figure 8 for exemplars a) through d) and a CAD model f) in Figure 5. Exemplar e) is the wall and ground plane of exemplar f), which is a simplified model of exemplar a). Compression, measured as the percentage of $\mathcal{D}$ selected by our algorithm, is plotted against κ at each of the configurations denoted by crosses. Greater compression results from higher thresholds and scenes that are less complex. It is recommended that this figure is viewed in colour.

increased mean and a significant tail – into the metre range in the worst case. A profile view of exemplar a) is used to highlight how errors change with κ in Figure 9.

To demonstrate the spatial distribution of error we provide the surface predictions of exemplar a) and b) coloured according to error in Figure 10. For exemplar b) we observe that the worst errors results at discontinuities and very noisy surfaces such as foliage. In these areas the performance is that of a naïve implementation that uses all $\mathcal{D}$ as discussed in Section IV-B.

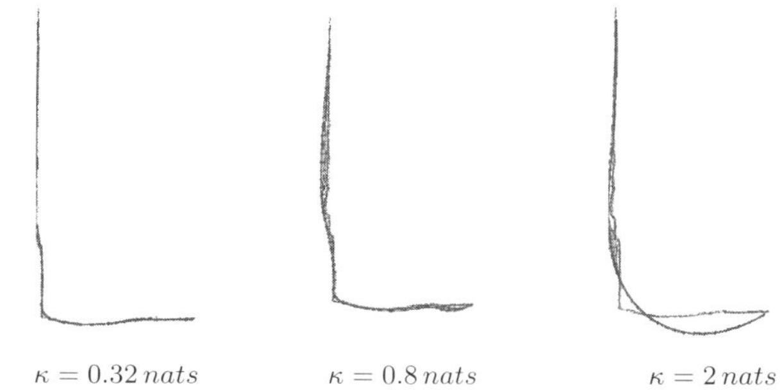

Figure 9. Profile view of a short section of wall and ground plane from exemplar a) (Figure 8 and 10), with $\kappa = \{0.32, 0.8, 2\}\,nats$ from left to right. $\mathcal{D}$ is shown in black, while predictions conditioned on $\mathcal{D}'$ are collated into the surfaces. Larger thresholds create larger point decimation, hence smoother surfaces that are in greater error.

VI. Conclusions

In this paper we introduced a novel parameterisation for push broom lasers. This formulation allows us to leverage techniques that require well-formed functions in what is in essence a non-functional problem domain. A Gaussian Process framework permits us a continuous surface representation using only a small subset of available points. Indeed, we can comfortably decimate point clouds of simple workspaces by factors in excess of two orders of magnitude, without significant degradation in fidelity. This is possible through use of a principled sampling strategy and a stochastic mean function. Further, we manage variables that are used throughout the active sampling process to ensure the algorithm is $\mathcal{O}(n^2)$ in

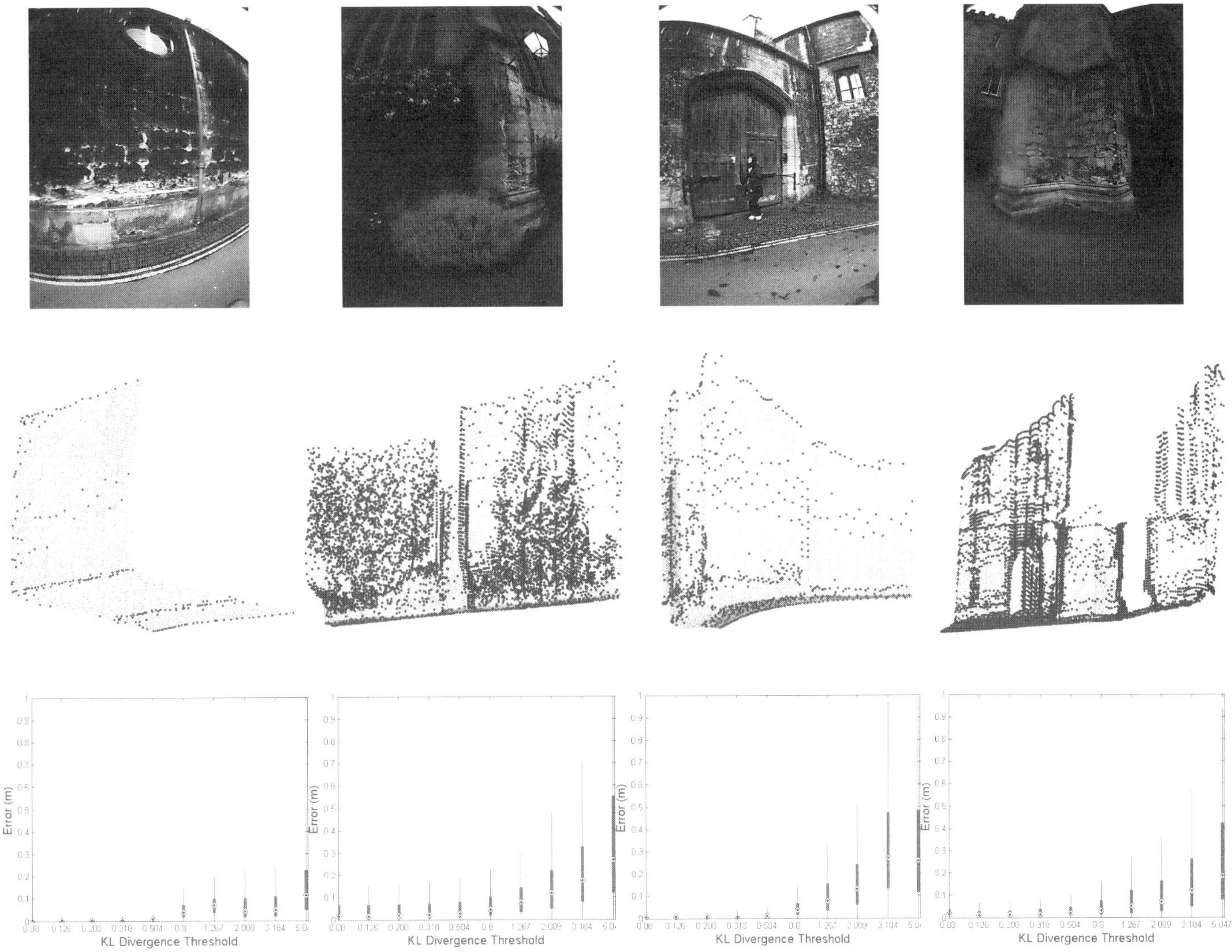

a) b) c) d)

Figure 8. real-world exemplars (column-wise). Images (top) were captured at the same time as the laser data (2nd row). Black points (2nd row) indicate actively sampled $\mathcal{D}'$ and the smaller blue points, raw $\mathcal{D}$, in Euclidean space for $\kappa = 0.8\,nats$. One can clearly observe that intersections of planes, a person, bushes and other complicated regions of $\mathcal{W}$ are more heavily sampled than simple regions. Box Plots (bottom) denote median and interquartile range for each of the logarithmically spaced thresholds in Figure 7. Whiskers are plotted that extend a maximum of one and a half times the interquartile range.

the size of the active window, which is constant, rather than $\mathcal{O}(N^3)$ in the size of the dataset.

Experimentation has provided insight into the compression ratio and surface reconstruction error that results from varying the KL divergence threshold used in our active sampling. To date we have considered the Matérn class of stationary covariance function, setting hyper parameters to best values found through search. We have experienced (Figure 10) that the accuracy of predictions in complex regions becomes that of the system with no subsampling. In order to enhance predictions in these complicated regions there is clear motivation to investigate possible alternative explicit basis functions, and non-stationary covariance functions such as [9], or [16]. The latter two approaches have the enticing benefit of automatically controlling hyper parameters, thereby allowing our predictions to be truly parameter free. We are also interested in online assessment of the active set size n to reduce computational load where a small n will suffice. In workspaces where the same regions are repeatedly mapped, one can imagine enormous compression ratios if we re-use the set $\mathcal{D}'$ that was created on the first iteration. In order to do this one would require a method to map between different sections of our sensor configuration space parameterisation. These developments are our next focus.

VII. Acknowledgments/Thanks

This work was funded through an Engineering and Physical Sciences Research Council CASE award with BAE Systems, and by the European Commission under grant agreement number FP7-231888-EUROPA. We would also like to thank Winston Churchill for provision of the CAD models used throughout this paper.

References

[1] J. Bares, M Hebert, T. Kanade, E. Krotkov, T. Mitchell, R. Simmons, and W. R. L. Whittaker. Ambler: An autonomous rover for planetary exploration. *IEEE Computer*, 22(1):18–26, 1989.

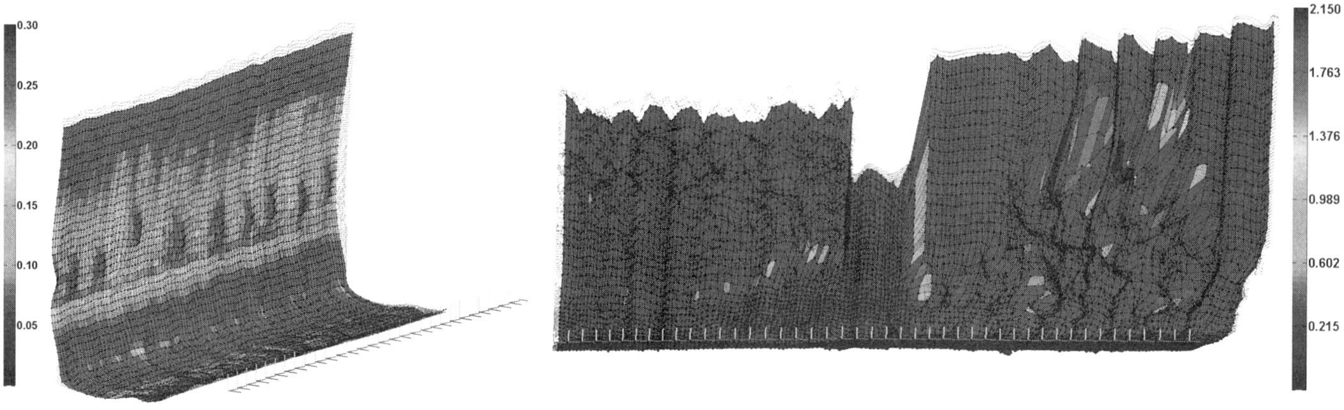

Max error: $0.26m$ Max error: $2.15m$

Figure 10. Surface predictions conditioned on $\mathcal{D}'$. Colouring is provided by the ℓ_1 norm in range from original points (in black) and the predicted surface, using $\kappa = 0.8\,nats$. Colour scaling is different for each plot, but both use a *jet* colour map with high errors in red and low errors in blue. (Left) and (right) illustrate simple and complex environments of exemplars a) and b) from Figure 8 respectively. A profile view of a short section of (Left) is also plotted in Figure 9. In (left), where there is an absence of large discontinuities, we notice the largest errors are away from $\mathcal{D}'$ (dark blue areas). In (right) largest errors occur where there are noisy surfaces and large discontinuities. In these regions there is automatically no data pruning and the accuracy of the surface is limited by that of a naïve implementation as discussed in Section IV-B. It is recommended that this figure is viewed in colour.

[2] L. Csató and M. Opper. Sparse on-line Gaussian processes. *Neural Computation*, 14(3):641–668, 2002.
[3] M.P. Deisenroth, C.E. Rasmussen, and J. Peters. Gaussian process dynamic programming. *Neurocomputing*, 72(7-9):1508–1524, 2009.
[4] J.J. Dongara, C.B. Moler, J.R. Bunch, and G.W. Stewart. *LINPACK users' Guide*. SIAM, 1979.
[5] C. Früh and A. Zakhor. Constructing 3d city models by merging aerial and ground views. *IEEE Comput. Graph. Appl.*, 23(6):52–61, 2003.
[6] D. Hähnel, W. Burgard, and S. Thrun. Learning compact 3D models of indoor and outdoor environment with a mobile robot. *Robotics and Autonomous Systems*, 44(1):15–27, 2003.
[7] H. Hoppe, T. DeRose, T. Duchamp, J. McDonald, and W. Stuetzle. Mesh optimization. In *Proceedings of the 20th annual conference on Computer graphics and interactive techniques*, pages 19–26. ACM New York, NY, USA, 1993.
[8] A. Krause, A. Singh, and C. Guestrin. Near-optimal sensor placements in Gaussian processes: Theory, efficient algorithms and empirical studies. *The Journal of Machine Learning Research*, 9:235–284, 2008.
[9] T. Lang, C. Plagemann, and W. Burgard. Adaptive non-stationary kernel regression for terrain modeling. In *Robotics: Science and Systems (RSS)*, 2007.
[10] M. C. Martin and H. Moravec. Robot evidence grids. Technical Report CMU-RI-TR-96-06, Robotics Institute, 1996.
[11] G. Matheron. The intrinsic random functions and their applications. *Advances in Applied Probability*, 5(3):439–468, 1973.
[12] The Mathworks. MATLAB 2008a. Natick, MA, 2008.
[13] P.M. Newman, G. Sibley, M. Smith, M. Cummins, A.R. Harrison, C. Mei, I. Posner, R. Shade, D. Schroeter, D.M. Cole, and I. Reid. Navigating, recognising and describing urban spaces with vision and laser. *IJRR*, 28(11-12):1406–1433, November 2009.
[14] D. Nguyen-Tuong and J. Peters. Using Model Knowledge for Learning Inverse Dynamics. In *IEEE International Conference on Robotics and Automation*, 2010.
[15] A. O'Hagan and J.F.C. Kingman. Curve fitting and optimal design for prediction. *Journal of the Royal Statistical Society. Series B (Methodological)*, 40:1–42, 1978.
[16] M.A. Osborne, S.J. Roberts, A. Rogers, S.D. Ramchurn, and NR Jennings. Towards real-time information processing of sensor network data using computationally efficient multi-output Gaussian processes. In *Proceedings of the 7th international conference on Information processing in sensor networks*, pages 109–120. IEEE Computer Society Washington, DC, USA, 2008.
[17] P. Pfaff and W. Burgard. An efficient extension of elevation maps for outdoor terrain mapping. In *Intl, Conf. on Field and Service Robotics (FSR)*, 2005.
[18] C. Plagemann, K. Kersting, P. Pfaff, and W. Burgard. Gaussian beam processes: A nonparametric bayesian measurement model for range finders. In *Robotics: Science and Systems (RSS)*, 2007.
[19] C. Plagemann, S. Mischke, S. Prentice, K. Kersting, N. Roy, and W. Burgard. Learning predictive terrain models for legged robot locomotion. In *Proc. of the IEEE/RSJ International Conference on Intelligent Robots and Systems (IROS)*, 2008.
[20] J. Quiñonero-Candela and C.E. Rasmussen. A unifying view of sparse approximate gaussian process regression. *J. Mach. Learn. Res.*, 6:1939–1959, 2005.
[21] C. E. Rasmussen and C. K. I. Williams. *Gaussian Processes for Machine Learning*. The MIT Press, 2006.
[22] S. Rusinkiewicz and M. Levoy. Efficient variants of the ICP algorithm. In *Proc. 3DIM*, pages 145–152, 2001.
[23] B. Scholkopf and A.J. Smola. *Learning with kernels*. the MIT Press, 2002.
[24] M. Seeger. Low rank updates for the Cholesky decomposition. Technical report, University of California at Berkeley, 2008.
[25] M. Seeger, C.K.I. Williams, and N.D. Lawrence. Fast forward selection to speed up sparse Gaussian process regression. In *Workshop on AI and Statistics*, volume 9, 2003.
[26] M. Smith, I. Baldwin, W. Churchill, R. Paul, and P.M. Newman. The new college vision and laser data set. *International Journal for Robotics Research (IJRR)*, 28(5):595–599, May 2009.
[27] M. L. Stein. *Interpolating Spatial Data*. Springer-Verlag, 1999.
[28] R. Triebel, W. Burgard, and F. Dellaert. Using hierarchical em to extract planes from 3d range scans. In *Int. Conf. Robotics and Automation (ICRA)*, 2005.
[29] R. Triebel, P. Pfaff, and W. Burgard. Multi-level surface maps for outdoor terrain mapping and loop closing. In *Intl. Conf. on Intelligent Robots and Systems (IROS)*, 2006.
[30] S. Vasudevan, F. Ramos, E. Nettleton, H. Durrant-Whyte, and A. Blair. Gaussian process modeling of large scale terrain. In *Int. Conf. Robotics and Automation (ICRA)*, 2009.

Sensor Placement for Improved Robotic Navigation

Michael P. Vitus
Department of Aeronautics and Astronautics
Stanford University
Email: vitus@stanford.edu

Claire J. Tomlin
Department of Electrical Engineering and Computer Sciences
University of California at Berkeley
Email: tomlin@eecs.berkeley.edu

Abstract—**Consider a set of fixed sensors used to estimate the state of a vehicle (e.g. position, orientation and velocity) while it attempts to follow a pre-planned trajectory. Since the sensor can only provide a measurement to the vehicle when it is within range, the deployment of the sensors will have a major impact on the ability of the vehicle to follow the trajectory. The problem addressed here is to optimally place the sensors in the environment such that the weighted function of the estimation error at each time-step is minimized. An optimization formulation is proposed that accounts for the uncertainty of the vehicle's state in determining whether it can receive a measurement from a sensor. A confidence level is introduced as a tuning parameter that controls the conservativeness of the solution. Consequently, the resulting solution increases the likelihood of the vehicle successfully following its intended trajectory. Finally, due to the interdependence between the sensors' positions, a novel incremental optimization algorithm is presented which significantly outperforms a standard nonlinear optimization procedure. Experimental and simulation results are shown which characterize the performance of the proposed algorithm.**

I. Introduction

Effective sensor deployment has most notably been studied with application to the Global Positioning System (GPS) [1]. In this system, the satellites effectively provide range measurements to the end user which are used to triangulate their position. The satellites' configuration heavily impacts the quality of the estimate of the user's position and has been extensively studied [2] [3].

Optimal sensor deployment strategies have also been studied for improving robotic localization. For example, Jourdan et al. [4] considered the case of deploying a network of static sensors that provide range measurements to the agent for localization. Most notably, they developed a locally optimal algorithm, significantly outperformed Simulated Annealing, to position the sensors on the boundaries of buildings to minimize the average position error bound over multiple agent locations. Zhang [5] examined the optimal orientation of sensors in 2D where the sensors can have different but constant measurement variances. A necessary condition was derived for the optimal orientation through minimizing the joint covariance matrix. This condition was then used to develop an $M - 3$ step algorithm that converged to the globally optimal solution, where M is the number of sensors. Zhang's formulation, however, does not consider the position of the sensors which could have a major impact on the quality of the solution.

A related problem to optimal sensor deployment is the generation of trajectories for mobile sensor platforms to improve target localization. In [6], the authors studied the case of an unmanned aerial vehicle (UAV) taking 3D-bearings-only measurements for the task of target localization. They setup an optimization problem that optimized the trajectory of the UAV to enhance the estimation performance characterized by the Fisher Information Matrix (FIM). Similarly, Sinha et al. [7] maximized the FIM as well as survivability and detection probability for a group of UAVs performing surveillance of several ground targets. Their solution technique used a gradient descent algorithm coupled with a genetic algorithm to search for the global minimum. Martinez et al. [8] investigated sensor placement and motion coordination strategies through the determinant of the FIM and characterized the global minima in the 2D case. They used the results to develop a motion coordination algorithm to dynamically control the mobile sensor network to an optimal deployment around the target.

Finally, the idea of using sensor optimization for simultaneous localization and mapping (SLAM) has also been considered. Strasdat et al. [9] explored the problem of SLAM for computational and/or memory limited systems, which can incorporate only a limited number of landmarks. Specifically, they proposed a landmark selection policy that identifies which landmarks are valuable for the robot to efficiently complete its navigation tasks. Through simulations and experimental demonstrations, they showed their algorithm outperformed handcrafted heuristics.

The sensor deployment problem considered in this paper is most similar to those considered in [4] and [5]. The problem involves a vehicle attempting to follow a pre-planned trajectory through the environment. Although [4] briefly touched on incorporating uncertainty, the proposed work presents a unifying framework for handling uncertainty of the vehicle's execution of the pre-planned trajectory. To increase the vehicle's likelihood of successfully following the trajectory and reaching its goal location, sensors are deployed in the environment to provide the vehicle with measurements when within range. A novel solution to this sensor deployment problem is presented which is applicable to any linear system with linear measurements.

There are many motivating examples for this work. One example is the placement of sensors to minimize the uncertainty of a vehicle following a pre-planned trajectory. There are several applications for which a pre-planned trajectory exists and is repeatedly used including automated supply chains, autonomous construction and disaster site cleanup. In these applications, the proposed work can provide better performance with fewer sensors than existing solutions. Another potential application is tracking building occupants' behavior for energy efficient control.

Another class of motivating problems is that of landmark placement in an environment, with the vehicle carrying a sensor to observe the landmarks for localization. For example, this work could be applied to feature selection for SLAM with computationally limited systems in which the exponential growth of requirements with number of landmarks is prohibitive. This problem is formulated in [9]. If the environment is dense in features, then the proposed algorithm could be applied without much change. With a small number of features, however, the expected distance to future landmarks plays a critical role and would need to be factored into the algorithm.

The work presented has several contributions. First, the sensor placement problem is formulated as an optimization program which minimizes the estimation error while the vehicle follows the pre-planned trajectory; this will maximize the likelihood of the vehicle successfully reaching its intended destination. In the original problem formulation, the discrete nature of the measurement region results in insufficient gradient information. Consequently, a continuous approximation is used that is more suitable for a numerical optimization solver. To ensure that the final solution is conservative, the uncertainty of the vehicle is taken into account when determining whether the vehicle can receive a measurement from the sensor. In this design, a confidence level is introduced as a tuning parameter that controls the degree of conservativeness of the resulting sensor deployment. Finally, an incremental optimization procedure is proposed that significantly outperforms both a simple, large nonlinear optimizer and Simulated Annealing.

The paper proceeds as follows. Section II describes the standard sensor placement problem formulation. In Section III, a continuous approximation of the measurement regions is presented. A solution that accounts for the uncertainty of the vehicle around the pre-planned trajectory is developed in Section IV. Simulation and experimental results are presented in Sections V and VI, respectively. The paper concludes with directions of future work and with an Appendix that analyzes the optimal measurement time for a simple system.

II. Problem Formulation

Consider the following linear stochastic system defined by,

$$x(k+1) = Ax(k) + Bu(k) + w(k), \forall k \in T_N \quad (1)$$

where $x(k) \in \mathbb{R}^n$ is the state of the system, $u(k) \in \mathbb{R}^p$ is the input of the system, $w(k) \in \mathbb{R}^n$ is the process noise and $T_N = \{0, \ldots, N-1\}$ is the horizon. The initial state, $x(0)$, is assumed to have a zero mean Gaussian distribution with covariance Σ_0 i.e., $x(0) \sim \mathcal{N}(0, \Sigma_0)$, and the process noise, $w(k)$, is assumed to have a zero mean Gaussian distribution, $w(k) \sim \mathcal{N}(0, \Sigma_w)$.

The system may estimate its own state using M sensors that can be placed in the environment. Let $\mathbb{M}$ be defined as $\{1, \ldots, M\}$. Each sensor, $i \in \mathbb{M}$, has position $s_i \in \mathbb{R}^d$, sensor measurements $C_i \in \mathbb{R}^{r\times n}$ and a maximum sensing radius $R_i \in \mathbb{R}$. The ith sensor's measurement is defined by $y_i(k) = C_i x(k) + v_i(k)$, where $y_i(k) \in \mathbb{R}^r$ and $v_i(k) \in \mathbb{R}^r$ are the measurement output and measurement noise of the ith sensor at time-step k, respectively. The measurement noise has a zero mean Gaussian distribution, $v_i(k) \sim \mathcal{N}(0, \Sigma_{v_i}), \forall i \in \mathbb{M}$.

By a standard result of linear estimation theory, the Kalman filter is the minimum mean square error estimator for the system considered. Let $\hat{\Sigma}_{k|k}$ be the covariance matrix of the optimal estimate of $x(k)$ given all of the measurements up to time-step k, let $\hat{\Sigma}_{k+1|k}$ be the covariance matrix of the predicted state, $x(k+1)$, given all of the measurements up to time-step k, and let $\hat{\Omega}_{k|k} = \hat{\Sigma}_{k|k}^{-1}$ be the information form of the standard Kalman filter. The information filter form has an important advantage over the standard Kalman filter recursion in that the measurement update is a simple sum over all the measurements, but this advantage could be negated by the increased computational cost in the motion update for large dimensional systems. The information form of the estimator recursion is as follows,

$$\hat{\Omega}_{k+1|k} = \left(A\Omega_{k|k}^{-1}A^T + \Sigma_w\right)^{-1} \quad (2)$$

$$\hat{\Omega}_{k+1|k+1} = \hat{\Omega}_{k+1|k} + \sum_{j=1}^{m} f_w\left(x(k+1), s_j, R_j, \hat{\Omega}_{k+1|k}\right) C_j^T \Sigma_{v_j}^{-1} C_j \quad (3)$$

with initial condition $\hat{\Omega}_{0|0} = \Sigma_0^{-1}$. The weighting function, $f_w(\cdot)$, indicates whether a sensor can provide a measurement or not. The general sensor placement problem is now stated.

$$\begin{array}{ll} \text{minimize} & V(s) = \sum_{k=1}^{N} \textbf{tr}\left(\hat{\Omega}_{k|k}^{-1}\right) \\ \text{subject to} & \\ & \text{Eqn. (2)} \ \forall k \in T_N \\ & \text{Eqn. (3)} \ \forall k \in T_N \end{array} \quad \text{(P2.1)}$$

There are several choices for the objective function, $V(s)$. Since the problem involves minimizing a matrix, it is desirable to find a scalar function for the objective to simplify the problem without significant loss of information. Three possible functions are the determinant, the maximum eigenvalue, or the trace. The main disadvantage of the determinant is that a small determinant can correspond to a very elongated ellipse. As opposed to the maximum of the eigenvalues, the trace represents the uncertainty in all directions equally and was chosen. To ensure a fair comparison between different algorithms, the summation over all time-steps is used since it balances the uncertainty of the vehicle throughout the trajectory.

The optimization problem P2.1 only has an analytical solution for very simple systems, such as the one presented in the Appendix; consequently a numerical optimization solver must be used. To aid the numerical solver in finding a locally optimal solution, the analytical gradient can be computed for this problem as stated in Algorithm 1, where $J(j)$ is the gradient of the j-th component of the concatenated sensor positions. A similar procedure can be used to calculate the analytical Hessian. Algorithm 1 is presented in a form that is easy to read but not necessarily the most efficient.

For the remainder of the paper, the problem is specialized for the case of a vehicle with state representing position, velocity, etc. Let $p(k) \in \mathbb{R}^d \subset x(k)$ be the position of the vehicle at the kth time-step. The true sensor measurement

Algorithm 1 Computation of the Gradient

for $j = 1, \ldots, m$ **do**

$\quad J(j) = 0,\ \hat{\Omega}_{0|0} = \Omega_0,\ \dfrac{\partial \hat{\Omega}^{-1}_{0|0}}{\partial s_j} = 0$

$\quad$ **for** $k = 1, \ldots, N$ **do**

$\quad\quad \bar{\Omega}_{k|k-1} = A\hat{\Omega}^{-1}_{k-1|k-1}A^T + \Sigma_w$

$\quad\quad \hat{\Omega}_{k|k} = \bar{\Omega}^{-1}_{k|k-1} + \sum_{i=1}^{m} f_w(x(k), s_i, R_i, \hat{\Omega}_{k|k-1}) C_i^T \Sigma_{v_i}^{-1} C_i$

$\quad\quad \dfrac{\partial \bar{\Omega}_{k|k-1}}{\partial s_j} = A \dfrac{\partial \hat{\Omega}^{-1}_{k-1|k-1}}{\partial s_j} A^T$

$\quad\quad \dfrac{\partial \hat{\Omega}_{k|k}}{\partial s_j} = -\bar{\Omega}^{-1}_{k|k-1} \dfrac{\partial \bar{\Omega}_{k|k-1}}{\partial s_j} \bar{\Omega}^{-1}_{k|k-1} + \dfrac{\partial f_w(x(k), s_j, R_j, \hat{\Omega}_{k|k-1})}{\partial s_j} C_j^T \Sigma_{v_j}^{-1} C_j$

$\quad\quad \dfrac{\partial \hat{\Omega}^{-1}_{k|k}}{\partial s_j} = -\hat{\Omega}^{-1}_{k|k} \dfrac{\partial \hat{\Omega}_{k|k}}{\partial s_j} \hat{\Omega}^{-1}_{k|k}$

$\quad\quad J(j) = J(j) + \mathbf{tr}\left(\dfrac{\partial \hat{\Omega}^{-1}_{k|k}}{\partial s_j}\right)$

$\quad$ **end for**

end for

weighting function in Eqn. (3) is defined as,

$$f_w(x, s, R, \hat{\Omega}) = \begin{cases} 1, & \text{if } \|p - s\| \leq R \\ 0, & \text{otherwise.} \end{cases} \quad (4)$$

There are two concerns with this formulation. First, the weighting function defined in Eqn. (4) produces an ill-formed optimization problem for numerical solvers because the gradient is 0 everywhere except when $\|p - s\| = R$, where it is undefined. Consequently, an approximate weighting function will have to be devised to provide more information to the numerical optimization solver. Second, only the pre-planned trajectory, x_{plan}, is known *a-priori* and the vehicle may stray away from it if the vehicle's process noise is large. Thus the vehicle's position, p, in the true sensor measurement weighting function cannot be readily evaluated.

III. Approximating the Weighting Function

To address the lack of gradient information in the true sensor weighting function, the function has been approximated by,

$$f_w(x, s, R, \hat{\Omega}) \approx g_w(d) = 1 - \frac{1}{1 + \exp(-(\alpha d + \beta))} \quad (5)$$

where $d = \|p - s\|$ is the distance between the sensor and the vehicle. The function is a transformed sigmoid decay function with two parameters that control the shape. The parameter α affects how quickly the function decays to zero beyond the maximum sensing radius R and is calculated through defining the value of the function at the maximum sensing radius. The parameter β controls the width of the shoulder. As β increases, the weighting function approaches the true sensing region. Figure 1 illustrates the weighting function for a maximum sensing radius of $R = 2$ meters, for an α parameter defined by $g_w(2) = 0.01$, and for values of $\beta = [5, 10, 15, 20, 25, 30]$. The weighting function is approximately 0 when the vehicle is outside the maximum sensing region and less than 1 when the vehicle is inside the sensing region. Therefore, this form of the weighting function is a conservative approximation of the true sensor weighting function which is a necessary condition to obtain a realistic solution. Since the function is continuous and smooth everywhere, it has a well defined derivative which is a major benefit for the numerical optimization solver.

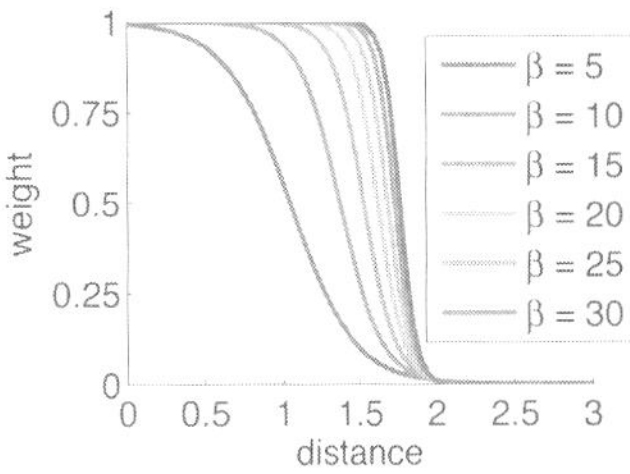

Fig. 1. The weighting function, $g_w(d)$, for various values of the parameter β with a maximum sensing radius of 2 meters and α is defined as $g_w(2) = 0.01$. As β increases, the weighting function approaches the true weighting function.

IV. Accounting for Uncertainty

For certain systems with large process noise, the pre-planned trajectory is not a good estimate of the vehicle's path if the system is not provided with enough measurements to accurately estimate its state. Consequently, Eqn. (5) cannot be used in its current form. Rather, the uncertainty in the vehicle's position, p, at each time-step will need to be taken into account. For the system considered, the estimate of the position of the vehicle at each time-step has a Gaussian distribution. Unfortunately the Gaussian distribution has infinite support, but luckily most of the probability distribution is centered closely around the mean. Therefore, only a region in the neighborhood of the pre-planned positions will have to be considered for conservatively estimating whether the vehicle will receive a measurement from a sensor. This region is represented by the ellipsoid $\mathcal{E}_\rho$. Consequently, d in Eqn. (5) can be replaced with $\Theta(p, s, \mathcal{E}_\rho) = \max_{o \in \mathcal{E}_\rho} \|o - s\|$, which is defined as the maximum distance from the sensor position, s, to any point contained within the ellipsoid, $\mathcal{E}_\rho$.

For a multivariate Gaussian random variable there is a natural choice for the ellipsoid that represents the uncertainty in the vehicle's position. Consider the d-D position of the vehicle, $p \sim \mathcal{N}(\mu, \Sigma)$, where $p \in \mathbb{R}^d$. A confidence level, δ, can be specified that results in a corresponding confidence ellipsoid, $\mathcal{E}_\rho$, such that, $\mathrm{p}(p \in \mathcal{E}_\rho) = \delta$. Given this definition, there are an infinite number of choices for the ellipsoid, but it will be restricted to be centered around the mean μ.

Define the random variable, $z \in \mathbb{R}$ such that $z = (p - \mu)^{\mathrm{T}} \Sigma^{-1} (p - \mu)$, which is a measure of the distance of p from μ. It is well known that z has a Chi-square distribution, $\mathcal{X}_d^2$. The confidence ellipsoid, parameterized by ρ, is then defined as, $\mathcal{E}_\rho = \{p \in \mathbb{R}^d | z \leq \rho\}$. Once the confidence level, δ, is specified, the confidence ellipsoid parameter is calculated via $\rho = F^{-1}_{\mathcal{X}_d^2}(\delta)$ where $F_{\mathcal{X}_d^2}$ is the chi-square cumulative distribution function. In this formulation, the confidence level is a tuning parameter that controls the conservativeness of the solution.

A. Example Objective Function

Several important properties of the objective function can now be demonstrated through a simple example. Consider the double integrator system with 1D position defined by,

$$A = \begin{bmatrix} 1 & \Delta t \\ 0 & 1 \end{bmatrix}, \quad B = \begin{bmatrix} 0.5\Delta t^2 \\ \Delta t \end{bmatrix}, \quad \Sigma_w = 1 \times 10^{-4} I,$$
$$C_i = \begin{bmatrix} 1 & 0 \end{bmatrix}, \quad \Sigma_{v_i} = 0.5, \quad R_i = 0.5, \forall i \in \{1, 2\}$$

where $\Delta t = 0.25$ seconds, $N = 200$, $\delta = 0.95$ and the pre-planned trajectory has equally spaced points between 0 and 10 meters. The logarithm of the objective function for this scenario is depicted in Figure 2, where s_1 and s_2 are the positions of the sensors. Note that the objective function is symmetric about the line $s_1 = s_2$ since the sensors are identical. The optimal sensor positions for the two sensors are $s_1 = 1.41$ and $s_2 = 4.54$ meters with an objective function value of 98.1.

Since the uncertainty of the vehicle position is included in determining when a measurement is allowed, the sensor positions are not very robust to positive perturbations to their positions. This is due to the fact that the process noise always increases the uncertainty of the vehicle. Consequently, if the sensor is pushed too far along the trajectory the vehicle's confidence ellipsoid will never be contained within the sensor footprint. This trend can be clearly seen in Figure 2 by the darker region located in the upper-right of the plot. If both sensors are perturbed by 0.05 meters into this region, the value of the objective function explodes to 2479 which is an increase of over 25 times. In contrast, if both sensors are perturbed by -0.1 the objective function value is 104.2 which is only a 1.06% increase. Therefore, one should be conservative in positioning the sensors since a small perturbation might result in a large detriment to the final solution.

This examples also illustrates that sensors closer to the start of the trajectory have a larger impact on the objective function. For example, if sensor s_1 is perturbed by $+0.05$ meters, then the objective function increases to 2479. However, if only sensor s_2 is perturbed by $+0.05$ meters, then the objective function only increases to 470. The reason behind this trend is due to the fact that if the first sensor is not allowed to perform any measurements then the subsequent sensors will be significantly less likely to provide a measurement. This property is dramatic for this example due to the small number of sensors. Small perturbations will have less of an impact if there are a large number of sensors over a short trajectory, since the interdependence is smaller.

B. Solution Methodology

The presented algorithm assumes the sensors are homogeneous to eliminate the impact on the solution due to the ordering of the sensors. Solving the optimization problem in P2.1 is difficult for two reasons. First, as stated in the previous section,

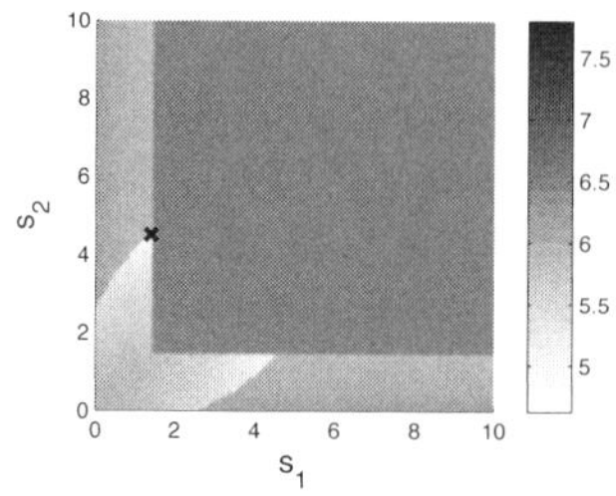

Fig. 2. The logarithm of the objective function for two identical sensors with the uncertainty of the position of the vehicle taken into account. The colormap ranges from light to dark for small to large values of the objective function. The optimal sensor positions are marked with an x. Note that since the sensors are identical the function is symmetric about the line $s_1 = s_2$.

the sensors' positions are highly dependent on each other and during the optimization procedure their position updates may have to be coordinated. Determining this level of coordination or interdependence strictly from the gradient information is difficult, if not infeasible, thus an intelligent algorithm must be designed. Second, the optimization program P2.1 is not convex; consequently, finding the globally optimal solution is difficult and the final solution is sensitive to the initial conditions since the optimizer will tend to find local minimas.

Due to the interdependence of the sensors, an incremental sensor placement algorithm is proposed; each additional sensor is placed in the environment while the previously placed sensors' positions are fixed. The motivation behind this type of algorithm lies within the observation in Section IV-A that sensors further along the trajectory are heavily dependent on the previous sensors' positions. For example, consider the objective function shown in Figure 2. If sensor s_1 is initially placed at 1 meter then the furthest that sensor s_2 can be placed is 4.1 meters. However, if s_1 is slightly perturbed to 1.1 meters then s_2 can be placed at 4.2 meters.

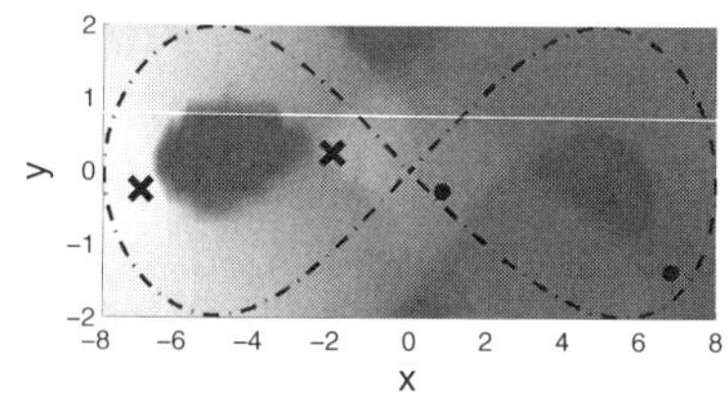

Fig. 3. This figure illustrates the objective function for the incremental addition of another sensor after two sensors have already been placed in the environment. The colormap ranges from light to dark for small to large values of the objective function. The pre-planned trajectory is shown as the dash-dotted line and the previous two sensor locations are indicated by the dots. There are two local minima for the additional sensor shown by the black x's.

While the incremental algorithm significantly reduces the the complexity of the optimization algorithm, the objective function is still plagued by many local minima. Consequently, a poor choice in the initial condition of the sensor position can lead to a poor solution. Unfortunately, there is no clear simple choice in choosing the initial condition for the optimization solver. Consider the objective function that is depicted in Figure 3 corresponding to the value of placing an additional sensor after two sensors have already been deployed in the environment. There are two local minima, highlighted by the x's, with objective value of 16.3 and 27.1 corresponding to a 66.7% difference. Since local minima are difficult for an optimizer to circumvent, the initial condition for each incremental sensor addition will be instrumental to the quality of the overall solution. In general, random sampling of the initial sensor position helps to prevent getting stuck in local minima but the state space to sample from could be quite large. Fortunately, the best sensor positions are typically located near the pre-planned trajectory as seen in Figure 3. Thus, the sampling space can be effectively limited to points along the pre-planned trajectory.

The overall algorithm is stated in Algorithm 2. The first step in the algorithm is to sample q positions from the pre-planned trajectory into S_{sample}. Then for each additional sensor, up to M sensors, the position from S_{sample} which minimizes the cost function $V(\{s, s_{\text{fixed}}\})$ is chosen. This position is then optimized through gradient descent while keeping all of the previous sensor positions fixed. This procedure is repeated until all of the sensors are placed. Since the incremental placement of sensors does not guarantee a locally optimal solution, a final optimization procedure is performed.

Algorithm 2 Solution Method

```
S_sample = sample q positions from x_plan
for i = 1, ..., M do
    s_i = arg min_{s ∈ S_sample} V({s, s_fixed})
    Optimize s_i through gradient descent
    s_fixed = {s_fixed, s_i}
end for
Optimize s_fixed through gradient descent
```

V. Simulation Results

Several simulation studies were performed to evaluate the performance of the presented sensor placement algorithm.

A. Scenarios

The following examples have double integrator dynamics for a 2D vehicle position with,

$$A = \begin{bmatrix} 1 & 0 & \Delta t & 0 \\ 0 & 1 & 0 & \Delta t \\ 0 & 0 & 1 & 0 \\ 0 & 0 & 0 & 1 \end{bmatrix}, \quad B = \begin{bmatrix} 0.5\Delta t^2 & 0 \\ 0 & 0.5\Delta t^2 \\ \Delta t & 0 \\ 0 & \Delta t \end{bmatrix}. \tag{6}$$

where $\Delta t = 0.25$ seconds. For all of the examples, the chosen parameters for the weighting function are $g_w(R) = 0.01$ and $\beta = 50$. In all the figures, the sensors are shown as numbered circles and the pre-planned trajectory is shown as the solid blue line with a subset of the uncertainty ellipsoids.

1) Position Sensing: The scenarios depicted in this section use position sensors with dynamics,

$$C_i = \begin{bmatrix} 1 & 0 & 0 & 0 \\ 0 & 1 & 0 & 0 \end{bmatrix}, \quad \Sigma_{v_i} = 0.1I$$

and a confidence level of $\delta = 0.85$. First, simple cases are considered with only a few sensors to aid understanding of the algorithm's choices. Figures 4 and 5 show the solution after the incremental sensor placement on the left and the final solution on the right.

Figure 4 shows an example with five sensors with a radius of sensing of 2 meters and process noise $\Sigma_w = 0.001I$. In this scenario, the final optimization results in moving sensor 1 further along the trajectory because sensor 2 was placed in the center. The objective value after the incremental placement step is 52.6 as compared to 36.9 after the final optimization.

For the scenario displayed in Figure 5, there are four sensors with a radius of sensing of 2 meters and process noise $\Sigma_w = 0.0001I$. After the final optimization, sensor 3 and 4 are pushed further back toward the start of the trajectory. Sensor 4 is positioned to sense more positions along the final bend of the pre-planned trajectory. In addition, sensor 1 is moved up to sense all of the positions in the first bend of the trajectory. The objective value after the incremental sensor placement step is 63.7 and after the final optimization it is 40.2.

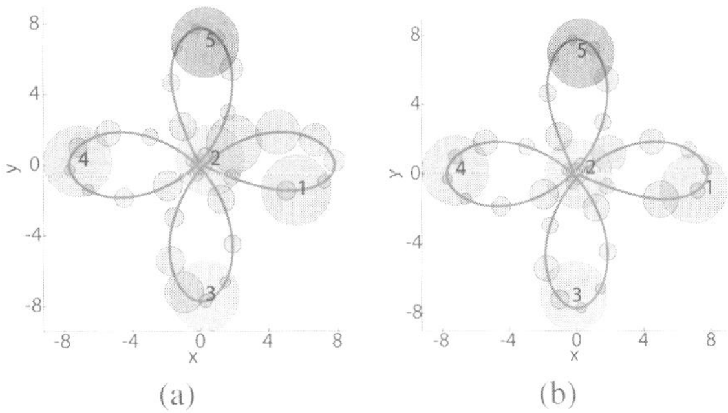

Fig. 4. The pre-planned trajectory is shown as the blue line, the uncertainty ellipsoids for a subset of the vehicle's pre-planned positions are shown as the blue ellipsoids and the sensor footprints are the numbered circles. (a) The solution after incrementally placing all of the sensors. (b) The solution after the last optimization step is performed. The main difference is that sensor 1 is allowed to move to the right since sensor 2 was placed in the center. The objective values for the two solutions are 52.6 and 36.9, respectively.

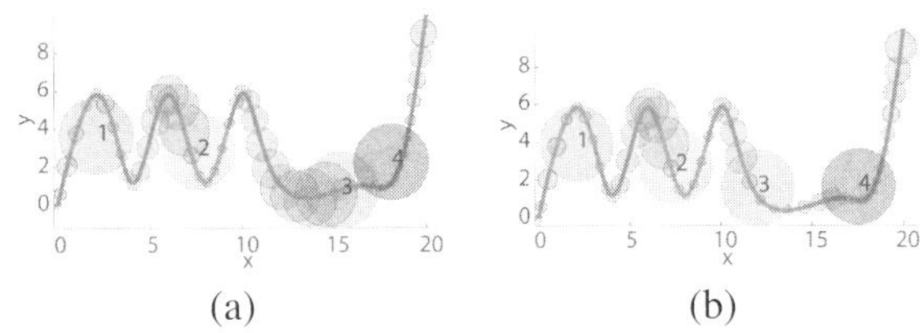

Fig. 5. (a) The solution after incrementally placing all of the sensors (b) The solution after the last optimization step is performed. The main difference between the two solutions is that sensor 3 and 4 are moved further back along the trajectory. The objective values for the two solutions are 63.7 and 40.2, respectively.

Figure 6 shows an example for a larger sized problem with 16 sensors and an extremely overlapping trajectory. The sensors have a radius of sensing of $R_i = 3$ meters. The process noise of the vehicle was set to $\Sigma_w = 0.01I$. In retrospect, the algorithm makes some fairly intelligent choices on where to place the sensors. In particular, sensors $2, 3, 6, 10, 11$ and 15 all take measurements at multiple parts of the trajectory. Also, sensors 5, 9 and 12 perform measurements for the entire time the vehicle is performing its loop.

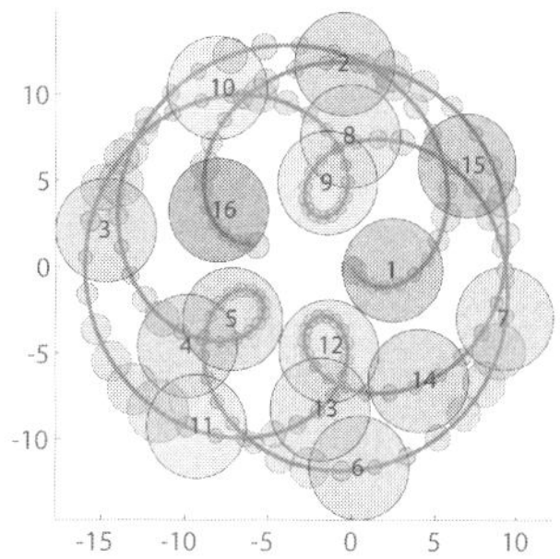

Fig. 6. The sensor placement solution for a sinuous trajectory for sixteen sensors with sensor range of 3 meters.

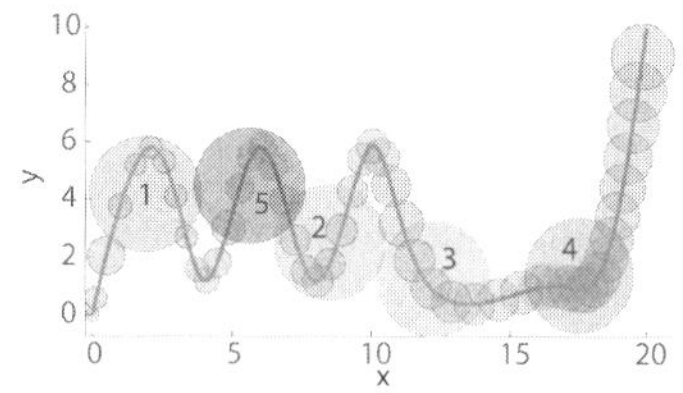

Fig. 7. The sensor placement solution for five velocity sensors.

2) Velocity Sensors: The last example shown in Figure 7 is for a system with velocity sensors with the same dynamics as the previous examples except,

$$\Sigma_w = 0.0001I, \quad C_i = \begin{bmatrix} 0 & 0 & 1 & 0 \\ 0 & 0 & 0 & 1 \end{bmatrix},$$
$$\Sigma_{v_i} = 0.001I, \quad R_i = 2, \ \forall i \in \{1, \ldots, 5\}$$

and a confidence level of $\delta = 0.95$. The sensor placements are similar to the previous example with position sensors but more sensors are required since the velocity sensors don't provide as much information.

B. Performance Analysis

To characterize the performance of the algorithm, the presented solution is compared with a nominal strategy which does not account for the position uncertainty. Three pre-planned trajectories were selected and the vehicle was simulated 500 times for each strategy. The first two use position sensors and the last one uses velocity sensors. Both sensor placement solutions for each trajectory are shown in Figure 8. The solution when accounting for the uncertainty of the vehicle is shown as the orange, solid circles with an x at the center and the solution without is displayed as the green, dashed circles with a dot at the center. The solution when not accounting for uncertainty was obtained through a gradient descent optimization algorithm because the interdependence of the sensors is drastically reduced in this formulation. In all cases, the vehicle starts at $(0, 0)$.

In all scenarios, the two different solutions are fairly similar at the beginning of the trajectory but differ drastically at the end of the trajectory. This difference arises primarily because the solution when planning with uncertainty has to shift the sensors closer to the beginning of the trajectory to account for the increasing uncertainty.

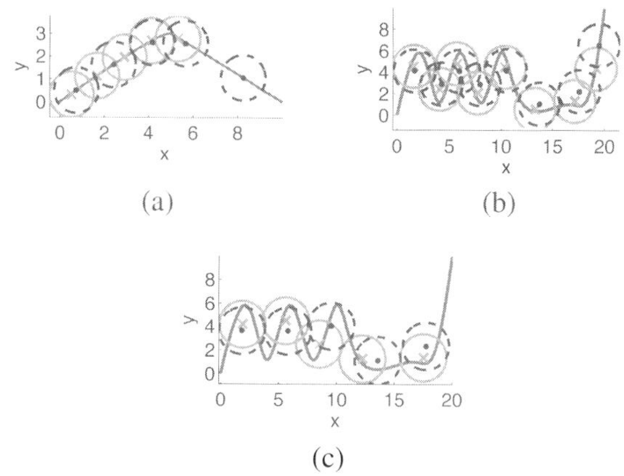

Fig. 8. Compares the sensor placement for three different scenarios with and without accounting for the uncertainty of the vehicle. The uncertainty included solution is depicted as the orange, solid circles with an x at the center and the solution not accounting for uncertainty is shown as the green, dashed circles with a dot at the center. For all scenarios, the major difference between the two solutions is the placement of the last sensors. (a) [Line] For a system with parameters: $\Sigma_w = 0.00005I$, $R_i = 1$ meter and $\Sigma_{v_i} = 0.05I$. (b) [Sinuous] The system parameters are: $\Sigma_w = 0.001I$, $R_i = 2$ meters and $\Sigma_{v_i} = 0.1I$. (c) [Vel. Sensors] The same parameters as previously used.

The performance results for the simulations are shown in Table I which presents several statistics for the positioning error at the final location. The statistics presented in the table are: the mean of the position error at the final time-step for all trajectories that didn't diverge, the maximum final positioning error and the percent of the trajectories that diverged. A divergent trajectory is classified as one in which the final position of the vehicle is outside of the uncertainty ellipsoid around the pre-planned trajectory. The uncertainty ellipsoid is calculated from the plan that accounts for the uncertainty of the vehicle since this is the most conservative prediction of the final position of the vehicle.

TABLE I

COMPARISON OF THE STATISTICS OF THE PERFORMANCE OF THE TWO DIFFERENT SOLUTIONS.

	With Uncertainty			Without Uncertainty		
	mean	max	% div.	mean	max	% div.
Line	0.95	2.73	0.0	0.27	25.10	6.8
Sinuous	0.70	2.19	0.2	0.37	22.92	3.4
Vel. Sensors	0.52	1.73	0.0	0.41	3.59	0.2

For all three scenarios, the mean performance of the vehicle is better when planning without uncertainty, but the max error and the percent diverged are significantly larger. The improved mean error performance when planning without uncertainty is expected since in this strategy, sensors can be placed further along the trajectory without penalty, thus reducing the final error for non-divergent trajectories. However, by not accounting for the uncertainty, the strategy is much more likely to result in divergent trajectories. It is also interesting to note that for the velocity sensors scenario, the performance statistics are very similar for both planning strategies. This is due to the fact that velocity sensors accumulate a large amount of drift in the position estimate of the vehicle. Thus, there is less benefit in incorporating the uncertainty in the planning stage.

C. Comparison to Simulated Annealing

For comparison, the proposed algorithm was evaluated against Simulated Annealing (SA). Since SA is a stochastic algorithm it has the potential to avoid local minima and approach the globally optimal solution. The proposed algorithm and SA were evaluated on three different scenarios. Figure 9 shows a histogram of the relative difference, $\frac{V_{SA}-V}{V}$, for 50 trials of each scenario.

In the first scenario, the pre-planned trajectory is a simple $45°$ line and there are 3 sensors with a sensing radius of 6 meters. Figure 9(a) shows a histogram of the relative performance between the proposed algorithm and SA. Even for this simple scenario, SA has difficultly consistently finding a reasonable solution. In 28% of the trials SA had an objective value larger than 50 times the proposed solution. In the next scenario, SA was run for the example shown in Figure 4 and the histogram of the relative performance is shown in Figure 9(b). For the final scenario, both algorithms were evaluated on the sinuous trajectory displayed in Figure 5 and the results are shown in Figure 9(c). In this example, SA was able to find a 20% better solution than the proposed algorithm 30% of the time, but 70% of the solutions were over 6 times worse than the proposed algorithm. Consequently, even though SA sometimes can find a better solution, the proposed algorithm consistently finds reasonable solutions. In addition, SA ran slower than the proposed algorithm by a factor of 22, 16 and 13.

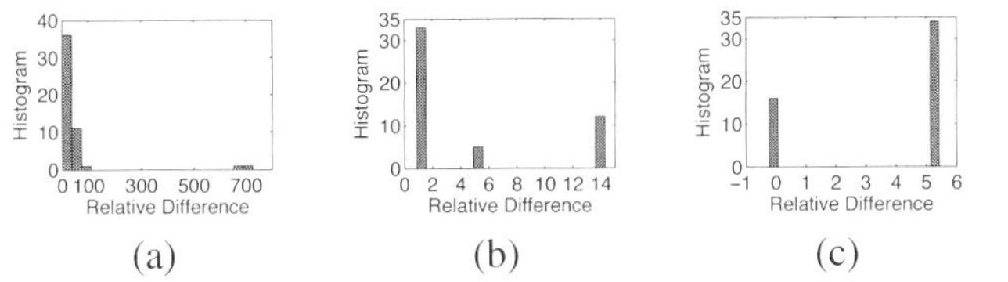

Fig. 9. Comparison of the relative performance between the proposed algorithm and Simulated Annealing for three different examples. A positive number means that Simulated Annealing found a worse solution than the proposed algorithm.

VI. Experimental Results

The sensor placement strategies were also evaluated on a quadrotor unmanned aerial vehicle. The vehicle has an onboard inertial measurement unit which provides three-axis attitude, attitude rate and acceleration measurements. An external Vicon [10] positioning system is also used to provide measurements of the vehicle's position with respect to a global coordinate frame.

With any aerial vehicle, the state (position, orientation and velocity) estimation is critical to a successful flight since small errors can easily lead to a disastrous crash. With this in mind, instead of withholding Vicon measurements when the vehicle is outside of the sensing regions, the vehicle was provided with degraded measurements to prevent any major catastrophes.

In the following experiments, the sensor placement strategies were evaluated on the estimation of the quadrotor's motion in the 2D x-y plane. For low speeds, the quadrotor's planar translational dynamics can be approximated by standard linear point mass dynamics as shown in Eqn. (1), (6). To follow the pre-planned trajectory, a controller with both feed-forward and feedback terms was utilized. The parameters used for the experimental tests were: $\Delta t = 0.05$ seconds, $\Sigma_w = 0.00001I$, $\Sigma_{v_i} = 0.001I$ and $R_i = 0.25$ meters.

The results of the experimental trials are depicted in Figure 10, with the planning with and without uncertainty cases shown on the left and right, respectively. The pre-planned trajectory is shown by the blue, dash-dotted line; the trajectory starts at $(-1, 0)$ meters and ends at $(-0.25, 0)$ meters and requires the vehicle to stop at each corner. Ten experimental trials were conducted for each sensor placement strategy. When accounting for uncertainty, nine out of the ten tests were successfully flown by the quadrotor vehicle. The test that failed required human intervention because the vehicle initially missed the second sensor's measurement region, causing the vehicle to oscillate around the trajectory. In contrast, the sensor placement strategy that didn't account for the uncertainty of the vehicle required human intervention for all ten trials to prevent a crash. The main cause of the drastic performance differences between the two plans is the placement of the first sensor. In particular, the solution without accounting for uncertainty does not place a sensor at the starting position, causing the vehicle to consistently overshoot the pre-planned trajectory at the first corner and subsequently oscillate out of control.

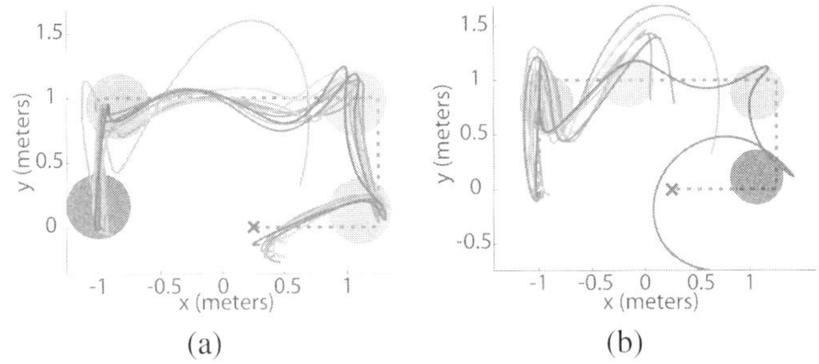

Fig. 10. Experimental data comparison for two different sensor placement strategies. The solid lines are the vehicle's trajectory in the 2D plane, the dash-dotted line is the pre-planned trajectory, the colored circles are the sensor footprints and the x mark is the final goal location. The experimental trials for planning with and without uncertainty are shown in (a) and (b), respectively. (a) Successfully completes nine out of ten trials. (b) Pilot intervention is required for all ten trials to prevent a crash.

VII. Conclusions

A generalized solution to the sensor deployment problem for a linear system with linear measurements was presented. To ensure the conservativeness of the solution, the uncertainty of the vehicle's position was accounted for when determining if the vehicle will receive a measurement. Finally, the sensor deployment strategy was demonstrated on a quadrotor platform, where the proposed algorithm showed significant performance increases over other strategies.

There are several interesting areas of future work that the authors wish to explore. First, the case of non-radial sensing regions due to obstacle occlusions should be explored which will increase the applicability of the algorithm. Another extension to the algorithm is to incorporate nonlinear system dynamics as well as nonlinear sensor measurements. Lastly, the authors wish to extend the algorithm to handle heterogeneous sensors; this presents a challenging addition since the ordering of the sensors heavily impacts the final solution.

ACKNOWLEDGMENT

The authors wish to thank Deborah Meduna for numerous discussions during the preparation of the paper.

APPENDIX
ANALYTICAL SOLUTION

For simple systems, an analytical solution to the optimization problem in P2.1 can be found. Consider a 1D system that has dynamics,

$$x_{k+1} = x_k + \Delta t u_k + w_k. \tag{7}$$

where $x_k \in \mathbb{R}$, $w_k \sim \mathcal{N}(0, \sigma_w)$ and $k \in \{1, \ldots, N\}$. In the current scenario, the system can only take a single measurement, $y_s = x_s + v$ with $v \sim \mathcal{N}(0, \sigma_v)$, at time-step, s. The optimal measurement time can be solved for by differentiating the objective function and setting it equal to 0 which results in:

$$s^* = \Bigg\{-\frac{\sigma_0}{\sigma_w}, \frac{1}{\sigma_w}\Bigg(-\frac{3}{4}(\sigma_0 + \sigma_v) + \frac{1}{4}(N+1)\sigma_w \pm \frac{1}{4}\sqrt{(\sigma_0 + 9\sigma_v + (N+1)\sigma_w)(\sigma_0 + \sigma_v + (N+1)\sigma_w)}\Bigg)\Bigg\}. \tag{8}$$

It is important to note that there is no guarantee that the value for s^* will be an integer. An integer solution, $\bar{s}^*$, can be determined by rounding s^*. The performance of this rounded solution is upper bounded away from the optimal solution by $V(\bar{s}^*) - V(s^*)$. Also, the optimal solution is not guaranteed to return a solution in the range $\{1, \ldots, N\}$, but these cases can still provide valuable insight into the optimal sensing time.

PERTURBATION ANALYSIS

An interesting analysis to perform on the optimal sensing time is how the solution varies due to perturbations in the system parameters.

Time Horizon N:

One parameter of interest is the time horizon. The derivative of the optimal sensing time with respect to the time horizon is,

$$\frac{\partial s^*}{\partial N} = \frac{1 + 5\sigma_v + \sigma_0 + (N+1)\sigma_w}{4\sqrt{(9\sigma_v + \sigma_0 + (N+1)\sigma_w)(\sigma_v + \sigma_0 + (N+1)\sigma_w)}}. \tag{9}$$

The steady state change as N approaches ∞ is $\lim_{N\to\infty} \frac{\partial s^*}{\partial N} = \frac{1}{2}$. Consequently, as N increases the optimal sensing time's dependence on the parameters of the system decreases; therefore, if the time horizon is increased by one time-step, the optimal sensing time will only change by 0.5.

Process Noise σ_w:

Characteristics of the optimal sensor measurement time with respect to changes in the process noise can also be determined analytically. Some of the important results from this analysis are as follows. As the process noise, σ_w, approaches 0 is $\lim_{\sigma_w \to 0^+} s^* = -\infty$, which is ill-formed because the sensing time is restricted to $s \in \{1, \ldots, N\}$, but a trend can still be extracted from the result. As the process noise decreases, the optimal sensing time also decreases toward the first possible measurement time. As σ_w approaches ∞ is $\lim_{\sigma_w \to \infty} s^* = \frac{1}{2}(N+1)$. Consequently, as the process noise grows toward infinity the optimal sensing time isn't dependent on any of the system parameters and approaches the middle of the time horizon.

Measurement Noise σ_v:

The properties of the optimal sensing time can also be analyzed with respect to the measurement noise. As the measurement noise, σ_v, tends to 0,

$$\lim_{\sigma_v \to 0^+} s^* = \frac{(N+1)\sigma_w - 2\sigma_0}{2\sigma_w} \le \frac{1}{2}(N+1) \tag{10}$$

which is bounded to the first half of the time horizon. As σ_v approaches ∞,

$$\lim_{\sigma_v \to \infty} s^* = \frac{2(N+1)\sigma_w - \sigma_0}{3\sigma_w} \le \frac{2}{3}(N+1) \tag{11}$$

which is dependent on the other system parameters but can be bounded to the first two-thirds of the time horizon.

REFERENCES

[1] P. Misra and P. Enge, *Global Positioning System: Signals, Measurements, and Performance*. Lincoln, MA, USA: Ganga-Jamuna Press, second ed., 2006.

[2] M. Phatak, "Recursive method for optimum GPS satellite selection," *Aerospace and Electronic Systems, IEEE Transactions on*, vol. 37, pp. 751–754, April 2001.

[3] M. Kihara and T. Okada, "A satellite selection method and accuracy for the global positioning system," *Navigation*, vol. 31, no. 1, 1984.

[4] D. B. Jourdan and N. Roy, "Optimal sensor placement for agent localization," in *Proceedings of the IEEE/ION Position Location and Navigation Symposium (PLANS 2006)*, (San Diego, CA), April 2006.

[5] H. Zhang, "Two-dimensional optimal sensor placement," *IEEE Transactions on Systems, Man, and Cybernetics*, vol. 25, May 1995.

[6] S. Ponda and E. Frazzoli, "Trajectory optimization for target localization using small unmanned aerial vehicles," in *AIAA Conf. on Guidance, Navigation, and Control*, (Chicago, IL), 2009.

[7] A. Sinha, T. Kirubarajan, and Y. Bar-Shalom, "Optimal cooperative placement of GMTI UAVs for ground target tracking," in *Proceedings of the IEEE Aerospace Conference*, (Big Sky, MT), March 2004.

[8] S. Martinez and F. Bullo, "Optimal sensor placement and motion coordination for target tracking," *Automatica*, vol. 42, pp. 661–668, April 2006.

[9] H. Strasdat, C. Stachniss, and W. Burgard, "Which landmark is useful? Learning selection policies for navigation in unknown environments," in *Proceedings of the IEEE International Conference on Robotics Automation (ICRA)*, 2009.

[10] Vicon, *Vicon MX Systems*. visited January 2010. http://www.vicon.com/products/viconmx.html.

Task-driven Tactile Exploration

Kaijen Hsiao
Willow Garage
Menlo Park, CA
Email: hsiao@willowgarage.com

Leslie Pack Kaelbling
Tomás Lozano-Pérez
MIT CSAIL
Cambridge, MA 02139
Email: lpk@mit.edu, tlp@mit.edu

Abstract— **This paper presents a decision-theoretic approach to problems that require accurate placement of a robot relative to an object in the world, including grasping and insertion. The decision process is applied to a robot hand with tactile sensors, to localize the object and ultimately achieve a target placement by selecting among grasping and information-gathering trajectories. The process is demonstrated in simulation and on a real robot.**

I. Introduction

Our goal is to develop a general-purpose strategy for task-driven manipulation of objects when there is uncertainty about the relative pose of the robot and the objects. This strategy applies to *relative placement* problems, which require the robot to achieve accurate placement with respect to a target object whose position is not accurately known. Placement problems include grasping (placing the robot relative to an object to be grasped), insertions (placing an object the robot is holding relative to another object), and other fine-motion tasks. In this paper, we focus on grasping.

Vision and range sensors can estimate the pose of an object, but there is still residual uncertainty. Tactile sensing, combined with proprioception, can give highly reliable information about object position. It is expensive to map out an entire object with tactile sensing, so we use the information requirements of the task to drive the sensing.

Decision theory frames problems of action selection when the true world state is unknown, providing a principled way to trade off the cost of performing information-gathering actions against the costs of performing inappropriate actions in the world. A decision-theoretic controller is constructed from two components: state estimation and action selection. The *state estimator* maintains a *belief state*, which is a probability distribution over the underlying, but not directly observable, states of the world. Each time an observation (such as a contact sensor reading) is made, the belief state is updated to incorporate the new information; each time an action is taken, the belief state is updated to reflect possible changes in the world state due to the action. The *action selection* component considers the current belief state and decides whether the state has been estimated sufficiently accurately to terminate and execute a final goal-achieving action, or whether additional observations should be made. If additional observations are to be made it chooses an action based on its potential for increasing the likelihood of success.

We consider several approximate decision procedures, based on *world relative trajectories* (WRTs) [11], which are robot arm trajectories parameterized with respect to a pose of an object in the world. In the simplest case, we have a single WRT, which would succeed as a terminal action if it were parameterized with the correct object pose. On every step, we execute that WRT, parameterized by the object pose that is most likely in the current belief state. The procedure is terminated when the estimated likelihood of success is high enough, and the WRT is executed one last time. A single WRT is not always enough to guarantee that the uncertainty will be reduced sufficiently, so we augment the set of WRTs with trajectories designed expressly with the goal of gathering information and/or re-orienting the object so it will be easier to interact with. Finally, we consider an extension to lookahead search, allowing the selection of an initial WRT because of its ability to gain information that will enable a subsequent WRT to be more effective.

II. Related work

Using tactile sensing to recognize and/or locate objects has a long history [16, 3, 7, 1], yet tactile sensing is used less often in robot manipulation than vision or range sensing. One possible reason is efficiency. Most work on tactile sensing has focused on recognizing/localizing objects in a task-independent manner and can be unnecessarily slow. Our goal is to integrate tactile sensing with the manipulation task, both in that the sensing arises from task-oriented motions and that the goal is to sense just enough to enable success on the task.

There are two paradigms for tactile recognition/localization. One obtains dense data, for example by surface scanning [2, 17]; the other uses sparse data directly via "contact probes" [9, 19]. Within the probe paradigm, there has been substantial work on "active" probing, choosing motions that best disambiguate among possible objects or poses [6, 19]. However, these probing motions have not typically been integrated into the goals of an overall manipulation task.

Our work fits within the general paradigm of motion planning under both sensing and control uncertainty. This problem has been addressed in non-probabilistic formulations (for example, [15, 20, 13]) and in probabilistic formulations (for example, [14]). Several previous approaches have used probabilistic state estimation to represent uncertainty and integrate observational information in manipulation problems [18, 8]. Hsiao et al. [10] frame the decision-making problem as a partially observable Markov decision process and solve it near-optimally, but can only address small problems. Cameron et

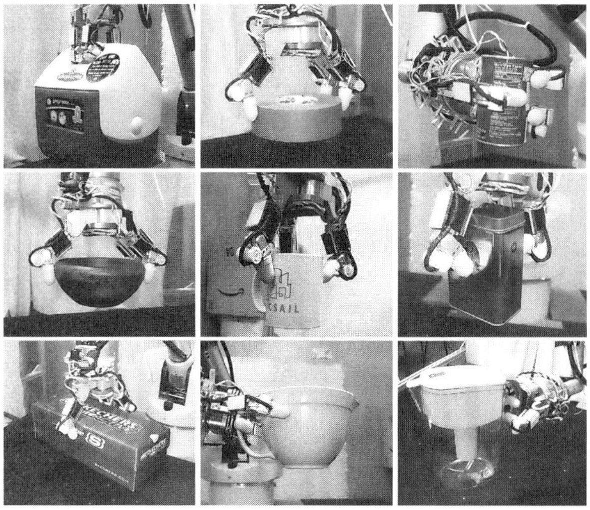

Fig. 1. Goal grasps for all objects except power drill.

al. [5] take a hypothesis-testing approach, applied to simple probes of a two-dimensional object.

This work builds on that of Hsiao et al. [11], who define WRTs and use them to limit the branching factor in lookahead search. That work showed, in simulation, that one-step lookahead, with respect to entropy reduction of the belief state, is an effective strategy for choosing grasping motions under uncertainty. However, reducing entropy in the belief state ignores the task requirements; we replace this criterion with estimated probability of task success. We also characterize the information conditions under which one can expect this type of strategy to produce correct results. We demonstrate the effectiveness of our methods both through simulation results and experiments with a real robot.

III. Action Selection

For concreteness, the rest of the paper discusses the problem of grasping an object using a pre-specified grasp (as shown in Figure 1), when there is uncertainty about the position of the object with respect to the robot, but it is important to keep in mind that the basic formulation is more broadly applicable.

A. Problem formulation

A grasp specification, G, consists of a set of relative poses for the hand and object, any of which is a successful grasp. The object is modeled as a 3D polygonal mesh; it is assumed to be positioned on a horizontal table with known z coordinate, resting on a known stable face. There are three degrees of pose uncertainty: x, y, θ. We call this space of object poses W.

A non-contact system (such as vision) generates an initial probability distribution over W; the sensing and estimation process will refine this distribution over time. The distribution will, in general, be multi-modal; it could be represented with various non-parametric or mixture distributions, but for simplicity, we use a multinomial distribution over a uniform discretization of W. This is our belief state, b; $b(s)$ is the probability of state s in distribution b.

The robot is a Barrett Arm and Hand; the space of possible actions is enormous, if viewed as a space of trajectories or velocity commands. We use an action space that consists of a small set of WRTs, each of which is a trajectory for the robot arm described relative to the pose of the object, typically ending with closing the fingers. Together, an object pose $w \in W$ and a WRT τ determine a robot trajectory $\tau(w)$. An action (a) consists of moving the arm to a common starting configuration, and executing $\tau(w)$ until it completes without making a contact or until a contact is detected, at which point the arm stops moving and each finger closes until it makes contact or is fully closed. Note that every action starts from the same starting configuration. A single observation (o) is composed of: joint-angle sensing, a 6-axis force-torque sensor on each of three fingers, and binary contact sensors on the palm and hand.

B. State estimation

After taking a new action a in belief state b, with underlying states s, and making observation o, the new belief state $b' = \mathrm{SE}(b, a, o)$ with underlying states s' is defined by

$$\mathrm{SE}(b, a, o)(s') = \frac{\Pr(o|s', a) \sum_s \Pr(s'|s, a) b(s)}{\Pr(o|b, a)} \ .$$

The state update depends on the observation model, $\Pr(o|s', a)$, and the state transition model, $\Pr(s'|s, a)$.

C. Decision procedures

An optimal behavior in this problem is a decision procedure that specifies which action to take in reaction to every possible belief state. We could assign a cost to taking actions, and then seek a policy that minimizes the expected cost of the robot's behavior, taking into account both the cost of executing actions for the purpose of gaining information and the cost of grasping the object incorrectly (including failing to grasp it at all). This problem is a POMDP [12], and can be computationally very difficult to solve in the general case.

We present two approximate, but efficient approaches to this problem. The policies that result are sub-optimal, in general. However, we seek to understand conditions that guarantee finite convergence to a desired grasp with high probability.

1) Single WRT: The simplest decision procedure assumes that we have a single WRT, τ^*, which, when executed with respect to the true object pose w^*, results in a grasp in G, the set of goal grasps. Letting b_t be the *belief state* at time t, the decision procedure is to: 1) Find the maximum *a posteriori* probability (MAP) pose $\hat{w}(b) = \arg\max_w b(w)$ and execute $\tau^*(\hat{w}(b))$. That is, to execute the grasping trajectory as if the object were at its most likely location. 2) Obtain observation o_{t+1} and update the belief state. 3) Terminate when a criterion on b is met, grasp the object using $\tau^*(\hat{w}(b))$ (if the hand is not already grasping at the new $\hat{w}(b)$), and pick it up.

The termination criterion depends on the expected loss of attempting to grasp based on the current belief state. Our loss function for executing a final grasp $\tau^*(w)$ and attempting to pick up the object is 0 if $\tau^*(w)$ results in a goal grasp and

1 otherwise. The expected loss, or *risk*, of executing $\tau^*(w)$, written $\rho(\tau^*(w), b)$, is an expectation of the loss taken with respect to belief distribution b.

We should select the w that minimizes $\rho(\tau^*(w), b)$ to parametrize the final grasp; that is the action that is optimal in the expected-loss sense. In practice, it can be expensive to evaluate ρ over the whole space W, so we commit to executing action $\tau^*(\hat{w}(b))$. Our termination condition is that the risk of this action be less than some risk threshold δ. In section III-D, we describe conditions under which this process requires a finite number of samples, in expectation, to terminate. It must be the case that each new grasping attempt yields information that ultimately decreases the risk.

2) Multiple WRTs: It may be that repeatedly executing τ^* will not give sufficient information to achieve the goal criterion. If the goal is to pick up a long object in the middle, repeated grasping will not give information about the object's displacement along its long axis. Thus, it is necessary to touch additional surfaces with the explicit purpose of gaining information. It may also be that the goal WRT is not executable: the object's handle, for example, may be out of reach of the robot, in which the robot must first grasp the object using a different grasp and re-orient it.

Given a set possible WRTs, τ_i, each of which has execution cost c_i, the decision procedure uses *finite-lookahead search* in an attempt to minimize the total cost of information gathering actions required to achieve the goal criterion.

1) We construct a search tree whose layers alternate between action choices of the robot and stochastic outcome "choices" of the environment. The root node is the current belief state, b. It branches on the choice of action a, and then on the possible observations o. The node reached from b via action a and observation o is a new belief state $b' = \mathrm{SE}(b, a, o)$. This process can be carried out to any depth; Figure 2 shows parts of a depth-2 tree.

For computational simplicity, the only actions considered in belief state b are $\tau_j(\hat{w}(b))$, for each $\tau_j \in \mathcal{T}$. We consider each possible WRT only with respect to the world state that is the most likely in b. The observation space is actually continuous; section IV-A.2 discusses the discretization process.

We are interested in $V(b)$, which is the expected cost of executing actions, starting in belief state b, until reaching a belief state for which $\rho_0(b)$, the probability of having a wrong grasp in belief state b, is less than δ. We use the tree to compute a finite-horizon approximation of $V(b)$ at a given b, based on backward induction. The value at the leaves is approximated as the risk of terminating and using the goal grasp in that belief state: $V_0(b) = \rho_0(b)$. Then, we compute

$$V_n(b) = \min_j \left[\sum_o \Pr(o \mid b, \tau_j(b)) U_n(\mathrm{SE}(b, \tau_j, o), \tau_j) \right] ,$$

where

$$U_n(b, \tau) = \begin{cases} \rho_0(b) & \text{if } \tau = \tau^* \text{ and } \rho_0(b) < \delta, \\ V_{n-1}(b) + c & \text{otherwise} \end{cases}$$

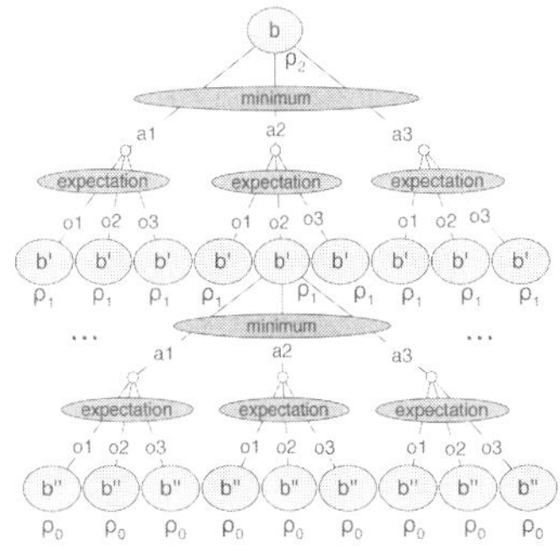

Fig. 2. Parts of a depth-2 search tree and associated value computation.

The function U captures the the fact that if the belief state meets the termination criteria, then the procedure terminates and there is no remaining action cost. If it does not, then the expected cost is as defined by V for the belief resulting from the action taken, plus the cost of the additional action.

2) Select the WRT τ_j that minimizes the expected value of V at the next step and execute it with respect to $\hat{w}(b)$.

3) Obtain observation o_{t+1}, and update the belief state.

4) Terminate when $\rho_0(b) < \delta$ after having grasped the object using τ^*, and pick it up. Otherwise, go to step 1.

In most situations, when re-positioning the object is not necessary, it is sufficient to set $k = 1$, that is, to select the WRT that will, in expectation, lead to a belief state that has the least risk with respect to executing the nominal grasp trajectory. However, as we illustrate experimentally in section V, looking deeper can cause the termination criterion to be reached faster.

D. Termination and correctness

We would like to understand how these decision procedures are likely to perform, depending on properties of the domain to which we apply them. There are two important questions: Will the procedure terminate in finite time? When it terminates, what is the likelihood that it will have selected a final action that meets the goal criterion? The answer depends on the informativeness of the observations and on the degree to which the actions change the pose of the target object.

1) Single WRT with object fixed: We begin by assuming that the object's pose does not change during the sensing process, that the loss is zero if and only if the goal WRT is executed with respect to the true underlying state, and that we have only a single WRT.

In the standard statistics problem of hypothesis testing with two hypotheses and a single observation action, the sequential decision process depends on the ratio of the probabilities of the two hypotheses in the posterior distribution. When the probability ratio goes outside of a fixed interval, then the sampling procedure is terminated and a hypothesis is selected. This procedure was shown by Wald [21] to terminate with a guaranteed risk after a finite number of trials, as long as the observation distribution is almost surely (with probability one) different, conditioned on the hypothesis.

This result is extended to the case of multiple discrete hypotheses in the M-ary sequential probability ratio test

(MSPRT), which is also guaranteed to terminate after finitely many trials whenever the observation distributions are almost surely different for each hypothesis [4]. These results hold whether the observation space is discrete or continuous. Our decision problem deviates from the MSPRT setting, in that different "experiments" are being chosen on each step, because each time the belief state is updated, the MAP world state is likely to change, and so the WRT is executed with respect to a different hypothesis about the object's pose.

Define the observation probability distribution for action $a = \tau^*(w_l)$ when the true world state is w_k to be

$$f_k^l(o) = \Pr(O = o \mid W = w_k, A = \tau^*(w_l))$$

and the *expected informativeness* of an action $\tau^*(w_l)$ to the distinction between states j and k to be:

$$I_{jk}^l = E_{f_k^l}\left[\sqrt{\frac{f_j^l(o)}{f_k^l(o)}}\right] = \int_o \sqrt{f_j^l(o)}\sqrt{f_k^l(o)}\,do$$

A clearly sufficient condition for termination is that, for all pairs of world states w_j, w_k, for all w_l, $I_{jk}^l \neq 1$; that is, that no matter how we parametrize our WRT, the observation distribution that it generates will be different across each pair of possible world states w_j, w_k. This satisfies the conditions for termination of the MSPRT. Of course, this won't be true in general. If w_l is a pose that is not spatially overlapping with either w_i or w_j, then no matter whether w_i or w_j is true, the probability of observing *no contact* is 1 (or high) when the robot attempts to grasp the object as if it were at w_l.

If w_k is the true state of the world, we can plausibly assume that, for all world states $j \neq k$:

$$I_{jk}^j < 1 \quad \text{and} \quad I_{jk}^k < 1 \tag{1}$$

What this means is that grasping as if the object were at pose j provides information that differentiates j from the true hypothesis k; and that grasping as if the object were at the true pose k provides information that differentiates pose k from all other possible poses. This is, of course, not true for many objects; if it is not, then the single-WRT process will not terminate and we need to use multiple WRTs.

It is important to see that taking uninformative actions is never destructive to the estimation process: it is simply a no-op. So, when the information ratios in (1) are less than 1, we can apply the MSPRT theorem as follows: In any sequence $a = a_0, \ldots, a_N$ where $n_j(a)$ of the trials in a are of action $\tau^*(w_j)$, then the probability that the decision procedure will not have terminated is a constant factor times $\max_j \left(I_{jk}^j\right)^{n_j(a)} \left(I_{jk}^k\right)^{n_k(a)}$. It will be sufficient to focus on the number of trials, $n_k(a)$, of $\tau^*(w_k)$ (which we will abbreviate as action k).

By a similar argument to the termination of the MSPRT, we can argue that any action other than action k will be exponentially unlikely to continue to be selected during an execution. So, very quickly, action k will predominate. The argument is as follows: any hypothesis j that is currently more likely than hypothesis k will be selected, but because $I_{jk}^j < 1$, it will drive down the likelihood of j with respect to k exponentially quickly, and therefore eventually not be selected. This will happen for each hypothesis $j \neq k$, until hypothesis k is selected. In case the likelihood of some j rises up above k, then action j will be selected, and it will drive the likelihood of j back down exponentially quickly.

The statistics literature provides arguments that probability ratio tests can be configured (by choosing termination criteria appropriately) to minimize total risk (when a cost is assessed for each sensing action). In our case, it is guaranteed that the risk of the final action is less than the maximum tolerable risk δ whenever the procedure terminates.

2) Goal sets: The goal set G may be such that executing $\tau^*(w)$ has 0 risk in a whole set of world states, not just the true state; for example, we may be indifferent about the orientation of a round can when it is grasped. In such a case, the requirements from formula 1 can be weakened. Let W_G be the set of $w \in W$ such that the grasp resulting from executing $\tau^*(w)$ is in G. Then, it is sufficient for termination that for all $j \in W \backslash W_G$ and for all $k \in W_G$, $I_{jk}^j < 1$ and $I_{jk}^k < 1$; that is, that the actions are discriminative between goal and non-goal w, but need not be discriminative within those sets.

3) Observations can move the object: It is more generally the case that the observation actions can change the state of the world, by moving the object as it is being sensed. If the information gained by each observation update compensates sufficiently for any additional entropy in the belief state introduced by the transition update, such that $\Pr(w_j)/\Pr(w_k)$ does not increase, in expectation, then the decision procedure will have a finite expected duration despite the object movement.

4) Multiple WRTs: We must increase the set of WRTs when the requirements from (1) are not satisfied for τ^*. Given a set of WRTs $\mathcal{T}$, it must be that, for any pair of world states $j \in W \backslash W_G$ and $k \in W_G$ that there exists some $\tau \in \mathcal{T}$ such that $I_{jk}^{j,\tau} < 1$ and some (not necessarily the same or different) $\tau \in \mathcal{T}$ such that $I_{jk}^{k,\tau} < 1$, where we have extended the definition of I to depend on τ in the obvious way.

We need to perform lookahead search when the execution of any single WRT is insufficient to yield an immediate decrease in risk. We can treat length-k sequences of the original actions as a new set of "macro" actions. If these macro actions, in combination, satisfy the requirements on informativeness, then the procedure will terminate and generate correct answers with high probability.

IV. Implementation

We implemented this method with a 7-DOF Barrett Arm and Hand, both in simulation (using ODE to simulate the physics of the world) and on an actual robot. The hand has ATI Nano17 6-axis force/torque sensors at the fingertips, and simple binary contact pads covering the inside (and some outside) surfaces of the fingers and the palm.

We used WRTs of three different types, for each target: 1) the goal WRT, τ^*, that grasps the object correctly if executed with respect to the correct world state, 2) information WRTs, that attempt to contact non-goal surfaces of the object and that sweep through the space to make an initial contact when uncertainty is high, and 3) re-orientation WRTs that use a grasp from above to rotate the object about its center of mass to make the goal WRT kinematically feasible.

We represent the belief state as a multinomial distribution over a three-dimensional grid of cells, with the x, y, and θ coordinates discretized into 31, 31, and 25 cells, respectively. A cell w comprises a set of actual poses. To handle this correctly in the transition and observation models, we should integrate over poses within w, which is computationally difficult. We instead treat cell w as if it were the pose in the center, which we call the *canonical pose* and write as $\bar{w}$.

A. Observations

In general, a WRT is executed in two phases: the arm is moved through the specified trajectory until a contact is felt anywhere on the hand or until the trajectory completes; then the three fingers are closed, each one terminating when it feels a contact or when it is fully closed. Each of these four motions (arm and three fingers) is treated as generating an observation tuple: $\langle \phi, c, \psi \rangle$, where ϕ is the observed pose of the robot at termination of the WRT based on the robot's proprioceptive sensors, c is a vector of readings from the contact sensors, and ψ is a representation of the 'swept path', that is, the volume of space through which the robot thinks it moved (based on proprioception) during the course of executing the WRT.

1) Observation model: For the purpose of belief-state update, we must specify an observation model, which is a probability distribution

$$\Pr(O = \langle \phi, c, \psi \rangle \mid W = w, A = \tau_i(w_j)) \ .$$

The size and complexity of the underlying state and observation spaces makes the modeling quite difficult. For tractability we make several assumptions: 1) There are no actual contacts that are not noticed or false triggering of the contact sensors. 2) The information gained from each aspect (arm, fingers) of the WRT is independent given the world state and action. 3) The swept path information is independent from the position and contact information. 4) The contact information at different contact points is independent.

In order to connect the observations to an underlying world state, it is necessary to reason about the (unobserved) true trajectory, ψ^* that the robot took. Letting τ_c stand for the commanded trajectory, we can write the observation probabilities $\Pr(\phi, c, \psi \mid w, \tau_c)$ as

$$\int_{\psi^*} \Pr(\psi^* \mid w, \tau_c) \Pr(\phi, c \mid \psi^*, w, \tau_c) \Pr(\psi \mid \psi^*, w, \tau_c, \phi) \ .$$

The robot is driven by a servo control loop that causes ψ, the observed trajectory, to track τ_c, the commanded trajectory, quite closely, so we can assume that, in the last term, $\psi = \tau_c$ (or, a prefix thereof, terminated at ϕ) and that it has probability 1. This integral is too difficult to evaluate, and so we approximate it by the maximum:

$$\Pr(\phi, c, \psi \mid w, \tau_c) \approx \max_{\psi^*} \Pr(\psi^* \mid w, \tau_c) \Pr(\phi, c \mid \psi^*, w, \tau_c) \ .$$

The scale of this value will be considerably different from the actual probability, but it can be normalized in the belief update.

No contact: When the robot observes no contact with the object, we assume that there was, in fact, no contact; so we consider only trajectories ψ^* in which no contact would occur. The value of ψ^* that is most likely, then, is the non-colliding trajectory that is as close to the observed ψ as possible.

Letting $d^*(\psi, w)$ be the depth of the deepest point of collision between ψ and the object at pose $\bar{w}$, assuming that the nearest collision-free trajectory is at distance $d^*(\psi, w)$ from ψ, and assuming that the likelihood of an observed trajectory is described by a Gaussian on its distance from the actual trajectory, we have

$$\Pr(\phi, None, \psi \mid w, \tau_c) \approx \mathcal{G}(d^*(\psi, w); 0, \sigma_p^2) = P_f(\psi, w) \ ,$$

where $\mathcal{G}$ is the Gaussian density function and σ_p^2 is a variance parameter. Although this approximation is efficient to compute, it can be inaccurate: there are situations in which the collision depth is small, but the distance between the sensed trajectory and the nearest non-colliding trajectory is quite large.

Contact: The observation probability, in the case of an observed contact, is the probability that as the robot executes commanded trajectory τ_c, that it will sense no contact up until ϕ, and then that it will sense the contacts c. We approximate the maximum of a product as a product of the maxima:

$$\begin{aligned} \Pr(\phi, c, \psi \mid w, \tau_c) &\approx \max_{\psi^*} \Pr(\psi^* \mid w, \tau_c) \Pr(\phi, c \mid \psi^*, w, \tau_c) \\ &\approx P_f(\lfloor \psi \rfloor, w) \max_{\phi^*} \Pr(\phi, c \mid \phi^*, w) \end{aligned}$$

where $\lfloor \psi \rfloor$ is the swept path, minus a short segment at the end, and ϕ^* is the final pose of ψ^*.

The fingertip sensors allow us to estimate the position and orientation of contacts, so each contact can be written as the pair $l_i(\phi, c), n_i(\phi, c)$, representing the location and normal of contact i. To do a careful job of estimating the probability of the contact, we would have to consider each pose ϕ^*, or possibly each face of the object, to find the most likely contact. Instead, as a fast approximation, we assume that the sensor reading was caused by contact with the closest object face, f^*, to $l_i(\phi)$, assuming the object is at pose $\bar{w}$. This choice maximizes the probability of the location, but not necessarily the normal. The final model is

$$\begin{aligned} \Pr(\phi, c, \psi \mid w, \tau_c) \approx\ & P_f(\lfloor \psi \rfloor, w) \mathcal{G}(d_l(l_i(\phi), f^*); 0, \sigma_l^2) \cdot \\ & \mathcal{G}(d_n(n_i(\phi), f^*); 0, \sigma_n^2) \end{aligned}$$

where $d_l(p, f)$ is the Euclidean distance from point p to face f and $d_n(n, f)$ is the angle between the vector n and the normal to face f, and σ_n^2 and σ_l^2 are variance parameters.

We model the fingertip position error with a standard deviation of 0.5 cm, and the fingertip normals with a standard deviation of 30 degrees. We model the location of a pad contact as the center of the contact pad, with a standard deviation of 1 cm, and we model the normal as the pad's surface normal, with a standard deviation of 90 degrees.

2) Reduced observation space: The space of possible observations is continuous and high-dimensional. In this section, we describe a very aggressive process for finding a small set of canonical observations to branch on during search. We always use the full observation, with the model described above, when doing belief-state estimation during execution.

The purpose of the lookahead search is to select actions that are most useful for gaining information. Reducing uncertainty in the orientation of the object, for example, is just as useful, no matter what the object's position is. Therefore, we can ignore the arm position at contact. For efficiency, but with a potential significant loss of effectiveness, we also ignore the information gained from ψ, the swept-volume aspect of the observation. So, we focus on reducing the space of possible contact observations c.

The observation distributions are all Gaussians centered at nominal observations, made assuming that the object is at one of the canonical poses $\bar{w}$ for $w \in W$. We take this discrete set of nominal observations (which in our implementation has about 24000 elements) as our starting set of possible observations.

Clustering: We start by clustering directly on contact observation vectors, using an agglomerative clustering method, with a distance metric that is a sum over the contacts of a weighted combination of the Euclidean distance between the contact locations and the angle between the contact normals; these distances are assumed to be infinite if one of the contact measurements is *None*. Each of the resulting clusters of observations is represented by its most likely observation.

Sub-sampling: Next, we prune observation clusters that are unlikely to occur throughout the state estimation process. In the early parts of the estimation process, we expect our belief distributions to be quite diffuse. Later in the estimation process, we expect the belief distributions to be concentrated around particular values of w, but, of course, we don't know which w. We do, however, know that later in the process, $\hat{w}$ will be near the true w, so many observations will be unlikely because they result from actions that are ill-matched to the true hypothesis.

We consider three different levels of uncertainty: the initial diffuse Gaussian, an intermediate one in which the standard deviations are reduced by $1/5$ in each dimension, and a focused one in which the deviations are quite small. We prune any observation clusters that have less than 0.01 probability of occurring under any of the three distributions. This reduces the observation cluster space in our examples to on the order of 50 elements.

Belief-dependent clustering: In the context of a particular search step, we have a current belief state b, and an action $\tau(\hat{w}(b))$. We can cluster observations based on their effect in this particular belief state. We do a further agglomerative clustering, grouping observations that lead to belief states that have similar variances in each dimension. In the current implementation, this process is only done for the initial belief state, where it gives the greatest leverage and cuts the number of clusters in half.

Belief-dependent sub-sampling: Finally, for any branch of the search tree, we sort all possible observation clusters by their probability in this belief state, and consider only the k most likely observations, whose summed probability is greater than 0.5. This is very aggressive, but it results in a manageable branching factor of between 1 and around 7.

B. Transition model

The transition model specifies $\Pr(W_{t+1} = w_j \mid W_t = w_i, A_t = \tau_c)$. We treat two cases separately: information and goal WRTs, which are not intended to change the object's pose (but which might do so inadvertently), and re-orientation WRTs, which explicitly attempt to move the object.

Before we incorporate information from the observation, the transition distribution is fairly diffuse: there is a chance the robot will miss the object entirely (and therefore leave it in the same pose), that the robot will graze it with one finger (and cause it to rotate), or that the robot will give it a solid shove. We compute a transition distribution that is already conditioned on some aspects of the observation, because by looking at all of the contact information jointly, we can estimate the force and torques that were applied to the object, and use that information to modulate the transition probabilities. This approach risks over-weighting the observation information, but seems to work well.

When no contacts are observed, we assume that the object was not contacted by the robot, and therefore was not moved. When contacts are observed, the transition distribution is a mixture of two Gaussians, one centered at the object's initial pose and one centered at a pose to which the object may have been "bumped" by the contacts:

$$\begin{aligned}\Pr(w_j' \mid w_i, \tau_c) \quad \propto \quad & (1 - p_b)\mathcal{G}(d_p(\bar{w}_i, \bar{w}_j); 0, \sigma_s^2) \\ & + p_b\mathcal{G}(d_p(bump(\bar{w}_i, \phi, c), \bar{w}_j); 0, \sigma_b^2)\ ,\end{aligned}$$

where p_b is the probability that the object will be bumped, d_p is a distance metric on poses that weights 1 cm of distance in position the same as 0.1 radians in rotation, and *bump* is a function that computes the most likely bump outcome from the old pose and the observation.

The *bump* pose is determined as follows. For each observed contact $c_i(\phi)$, we compute a unit force vector v_i applied at $l_i(\phi)$, in the direction $-n_i(\phi)$. We determine the center of mass of the object (assuming uniform density), compute the summed force and torque, and assume the object will translate a fixed distance per unit force in the direction of the net force, and rotate a fixed rotation per unit torque.

Because re-orientation has a moderately high probability of failure, we use a modified transition model, which is a mixture of three possible outcomes: the re-orientation fails entirely and the object stays in its initial pose, the re-orientation succeeds exactly, and the re-orientation leaves the object somewhere in between the start and goal poses. Each of the modes has a larger standard deviation than the corresponding standard deviation for other WRTs.

Fig. 3. Goal and information grasps for the power drill.

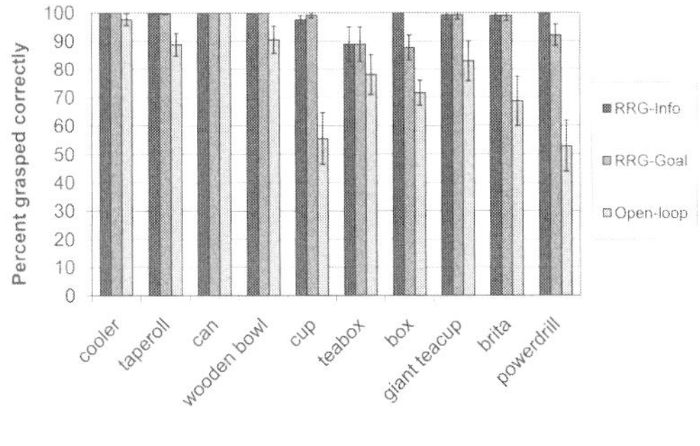

Fig. 4. Simulation results for all objects at low uncertainty.

V. Experiments

Our experiments used 10 different objects, shown with their goal grasps in Figures 1 and 3. The goal region for each object was hand-chosen to guarantee that being within the goal region ensures a stable grasp of the object. These regions are much larger for some objects than for others (for example, the goal region for the can is large, since the hand only has to envelop it), and the goal regions for the rotationally symmetric objects ignore the object orientation. In all experiments, the maximum number of actions allowed was 10; after the 9th action, if the goal criterion was not reached, the goal WRT was executed.

A. Simulation

In our simulation experiments, we compare three different strategies: 1) **Open-loop**: execute the goal WRT once and terminate; 2) **RRG-Goal**: execute the goal WRT repeatedly on the most likely state, terminating when the risk threshold is met; and 3) **RRG-Info**: choose among goal, re-orientation, and information WRTs, with a depth D lookahead-search decision procedure, terminating when the risk threshold is met. Each simulation experiments was carried out with at least 100 trials.

Figure 4 shows the results for experiments carried out in simulation with initial belief state a discretized Gaussian with standard deviations of 1 cm in x, 1 cm in y, and 3 degrees in θ, and lookahead-search depth $D = 2$. The chart shows the percentage of grasps that were executed successfully (with 90% confidence bounds), for each object placed at random positions drawn from the initial belief distribution, for the three algorithms. The risk threshold was chosen to correspond to a target success rate of 90%. Even at this low level of uncertainty, executing the goal WRT open-loop fails frequently for many of the objects. Using RRG-Goal allows us to succeed nearly all of the time, and using RRG-Info brings the success rate above 97% for all objects except the tea box, for which the decision procedure only selects the goal WRT, because it recognizes that it will still reach the target success rate.

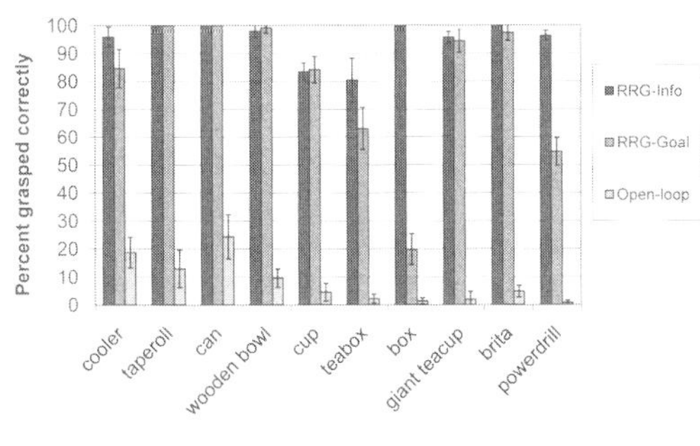

Fig. 5. Simulation results for all objects at high uncertainty.

Figure 5 shows the simulation results for higher levels of initial uncertainty (standard deviations of 5 cm in x, 5 cm in y, and 30 degrees in θ), again with $D = 2$. At this level of uncertainty, $\sim$14% of object positions are more than 10 cm away from the initial estimated position, and executing the goal WRT open-loop seldom succeeds. Using just RRG-Goal is sufficient for all of the objects except the box, as discussed earlier, and the power drill, for which the goal WRT grasps a nearly-cylindrical handle that gives it little information about the orientation for pressing the trigger. Using RRG-Info brings our success rate above 95% for all objects except the cup and tea box. The lower performance with these two objects is due to the relatively coarse state-space grid in our implementation, which is on the same order of size (1 cm) as the goal set for these relatively small objects. We expect that using an adaptive grid size would improve the performance.

Figure 6 shows the percentage of successful grasps (with 90% confidence bounds) for the power drill in simulation at the high level of uncertainty (5 cm/30 degree), where the target estimated level of success before termination was varied (from 10% to 90%) to generate data that shows the trade-off between the number of actions executed and the actual success rate. Note that in the left part of the graph we are using a low target level of success, which accounts for the low percentage of actual success. The five strategies used here are RRG-Goal, RRG-Info with lookahead-search depths of 3, 2, and 1, and RRG-Info-Entropy, which is like RRG-Info with lookahead-search of 1, but using entropy of the belief state at the leaves of the search tree (as in [11]) instead of risk. Each point on the graph represents the average number of actions taken before termination and the percent success over more than 100 simulated runs. Just executing the goal WRT repeatedly does not work well for this object, whereas searching with a depth of 1 works reasonably well. Note that using risk values at the leaves leads to a substantial improvement over using belief entropy. Recall that at lookahead-search depth 1, the decision procedure is considering plans with two actions, for example, an information WRT followed by the goal WRT. Increasing the depth to 2 causes the decision procedure to choose actions that may result in a lower probability of success after just 2 actions, but that pay off in terms of a higher probability of success later on. This is due to the fact that, although information WRT 1 (shown in Figure 3) provides information

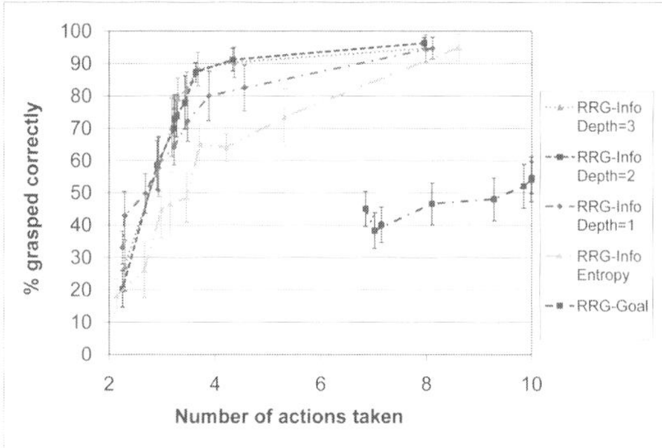

Fig. 6. Varying termination criterion trades off between grasp correctness and number of actions required; shown for power drill.

about all three dimensions at once and information WRT 2 only provides information about two dimensions, information WRT 2 for the power drill acts as a "funnel" for information WRT 1, enabling it to be effective more often. Increasing the lookahead-search depth to 3 yields no additional benefit. Although, for the drill, lookahead search with depth 1 has a success probability essentially identical to using a depth of 2 or 3 after 10 actions, searching deeper reduces the number of actions needed to succeed more than 90% of the time from an average of 7 actions to an average of 4 actions, which is a dramatic speedup. However, for most objects, using a depth of 1 works as well as using a depth of 2. This is significant since increasing the lookahead-search depth increases the action-selection time exponentially. In our (non-optimized Python) implementation, selecting the first action from among the 5 available power drill WRTs takes 3 seconds for a depth of 1; using a depth of 2 takes 10 times longer, and using a depth of 3 takes 60 times longer.

B. Real robot

On the real robot, we ran 10 experiments using RRG-Info with lookahead-search depth of 2, for the Brita pitcher and the power drill with high initial uncertainty. Both objects were grasped stably and lifted successfully 10 out of 10 times, with the trigger being pressed successfully on the power drill and the Brita pitcher being grasped properly by the handle. For the other 8 objects, we ran four experiments each: one at low uncertainty levels (1 cm/3 deg) and three at high uncertainty levels (5 cm/30 deg). 27 out of the 32 experiments succeeded. Two of the 5 failures (for the cooler and the can) were due to the robot contacting the object in a part of the hand with no sensors. One failure each (for the cup and tea box) were due to the coarseness of the state grid (as in the simulations). One failure with the giant tea cup was due to a flaw in our algorithm for computing collision depth for objects with thin features (like the tea cup handle).

Videos of our real robot experiments can be seen here: http://people.csail.mit.edu/kjhsiao/wrtpomdps

VI. Conclusion

We can draw several conclusions from this work. First, a small search depth is effective in our framework; a depth of 1 is usually sufficient, and a depth greater than 2 is generally not useful. This means that action selection does not have to be very expensive. Second, we can choose effective actions despite our aggressive observation clustering, designed to bring the observation branching factor to a manageable level. Third, the quality of the observation and transition models limit the effectiveness of our system; this presents substantial opportunities for further research. In particular, a more predictive transition model that could more accurately estimate how objects move when we bump into them could further improve our results. It could also enable us to add actions that purposely push objects in order to gain information, by, for instance, pushing them against walls.

We believe this work forms a step toward more general integration of tactile sensing and manipulation, ultimately supporting complex tasks such as multi-step assemblies.

Acknowledgement: This research was supported by the National Science Foundation under Grant No. 0712012.

References

[1] S. Akella and M. T. Mason. Using partial sensor information to orient parts. *IJRR*, 18(10):963–997, 1999.
[2] P. K. Allen and R. Bajcsy. Object recognition using vision and touch. In *IJCAI*, 1985.
[3] P. K. Allen and P. Michelman. Acquisition and interpretation of 3-D sensor data from touch. *IEEE Trans. on RA*, 6(4):397–404, 1990.
[4] C. W. Baum and V. V. Veeravalli. A sequential procedure for multihypothesis testing. *IEEE Trans. on Information Theory*, 40(6), 1994.
[5] A. Cameron and H. F. Durrant-Whyte. A Bayesian approach to optimal sensor placement. *IJRR*, 9(5):70–88, 1990.
[6] R. E. Ellis. Planning tactile recognition paths in two and three dimensions. *IJRR*, 11(2):87–111, 1992.
[7] M. Erdmann. Shape recovery from passive locally dense tactile data. In *WAFR*, 1998.
[8] K. Gadeyne, T. Lefebvre, and H. Bruyninckx. Bayesian hybrid model-state estimation applied to simultaneous contact formation recognition and geometrical parameter estimation. *IJRR*, 24:615, 2005.
[9] P. C. Gaston and T. Lozano-Perez. Tactile recognition and localization using object models: The case of polyhedra on a plane. *IEEE Trans. on PAMI*, 6:257–265, 1984.
[10] K. Hsiao, L. P. Kaelbling, and T. Lozano-Perez. Grasping POMDPs. *ICRA*, 2007.
[11] K. Hsiao, L. P. Kaelbling, and T. Lozano-Perez. Robust belief-based execution of manipulation programs. In *WAFR*, 2008.
[12] L. P. Kaelbling, M. L. Littman, and A. R. Cassandra. Planning and acting in partially observable stochastic domains. *Art. Intell.*, 101, 1998.
[13] J.C. Latombe. *Robot Motion Planning*. Kluwer Academic, 1991.
[14] Steven M. LaValle. *Planning Algorithms*. Cambridge U. Press, 2006.
[15] T. Lozano-Perez, M. T. Mason, and R. H. Taylor. Automatic synthesis of fine-motion strategies for robots. *IJRR*, 3(1), 1984.
[16] T. Okada and S. Tsuchiya. Object recognition by grasping. *Pattern Recognition*, 9(3):111–119, 1977.
[17] A. M. Okamura and M. R. Cutkosky. Feature detection for haptic exploration with robotic fingers. *IJRR*, 20(12):925–938, 2001.
[18] A. Petrovskaya and A. Y. Ng. Probabilistic mobile manipulation in dynamic environments, with application to opening doors. *IJCAI*, 2007.
[19] Steven Skiena. Problems in geometric probing. *Algorithmica*, 4(4):599–605, 1989.
[20] R. H. Taylor, M. T. Mason, and K. Y. Goldberg. Sensor-based manipulation planning as a game with nature. In *ISRR*, 1988.
[21] A. Wald. Sequential tests of statistical hypotheses. *The Annals of Mathematical Statistics*, 16(2), 1945.

On the Role of Hand Synergies in the Optimal Choice of Grasping Forces

Marco Gabiccini*† and Antonio Bicchi†‡

* Dept. of Mechanical, Nuclear and Production Eng.
University of Pisa, 56122 Pisa, Italy

† Interdept. Research Center "E. Piaggio"
University of Pisa, 56122 Pisa, Italy

‡ IIT - Istituto Italiano di Tecnologia,
16163 Genova, Italy

***Abstract*— Recent work on the analysis of natural and robotic hands has introduced the notion of *postural synergies* as a principled organization of their complexity, based on the physical characteristics of the hand itself. Such characteristics include the mechanical arrangements of joints and fingers, their couplings, and the low-level control reflexes, that determine the specific way the concept of "hand" is embodied in a human being or a robot. While the focus of work done so far with postural synergies has been on motion planning for grasp acquisition, in this paper we set out to investigate the role that different embodiments have on the choice of grasping forces, and on the ultimate quality of the grasp. Numerical results are presented showing quantitatively the role played by different synergies (from the most fundamental to those of higher-order) in making a number of different grasps possible. The effect of number and types of engaged synergies on the distribution of optimal grasp forces is considered. Moreover, robustness of results is investigated with respect to variation in uncertain parameters such as contact and joint stiffness.**

I. Introduction

Recent advances in neuroscience research have shown that the description of how the human hand moves during grasping is dominated by trajectories in a configuration space of much smaller dimension than the kinematic count would suggest. Such configuration space is sometimes referred to as the space of *postural synergies*, or the *eigengrasp space*.

One of the explanations for human efficiency in selecting appropriate grasps is that humans somehow unconsciously simplify the large search space through learning and experience. In a developmental perspective, it can be conjectured that learning is applied to a series of inner representations of the hand of increasing complexity, which varies with the experience and the degree of accuracy required. Santello et al. [1] investigated this hypothesis by collecting a large set of data containing grasping poses from subjects that were asked to shape their hands in order to mime grasps for a large set ($N = 57$) of familiar objects. Principal Components Analysis (PCA) of this data revealed that the first two principal components account for more than 80% of the variance, suggesting that a very good characterization of the recorded data can be obtained using a much lower-dimensional subspace of the hand DoF space. These and similar results seem to suggest that, out of the ca. 20 DoFs of a human hand, only two or three combinations can be used to shape the hand for basic grasps used in everyday life. It might also be speculated that higher order synergies can be recruited for executing more complex tasks, such as adaptive grasp force control, fine manipulation or haptic exploration.

One first explanation of the observed inter-digit coordination could be advanced in terms of mechanical constraints in the anatomy of the hand. More refined approaches recognize the role of peripheral and central nervous systems in establishing sensory-motor control synergies, as discussed, e.g., in [2] and [3]. Currently, investigations in the role and origin of synergies are being actively pursued by neuroscientists.

What the current knowledge about the neurophysiology of human hands already suggests at this point is that the brain uses the *hand* — meant as a cognitive entity for the organ of the sense of active touch — not as a mere collection of articular joints and muscles, but rather as an organized and ordered ensemble. The organization is dictated by principles that are embedded in the hand's embodiment - i.e. in its physical characteristics such as the mechanical arrangements of joints and fingers, their couplings, and the low-level control reflexes.

These ideas can be brought to use in robotics, as they suggest a new and principled way of simplifying the design and analysis of hands (as opposed to many empirical, often arbitrary design attempts), which has been the main roadblock for research in artificial hands in the past [4].

The application of synergy concepts has been pioneered in robotics by [5] and [6]. In [5], and later on in [7], the idea has been exploited in the dimensionality reduction of the search space in problems of automated grasp synthesis, and has been applied effectively to derive pre-grasp shapes for a number of complex robotic hands. In [6], authors designed a mechanical hand in which more or less accurate actuators are connected to different groups of mechanically interconnected joints, with a priority inspired by resemblance to postural synergies observed in human hands.

Much remains to be done to understand and exploit the implications of the synergy approach to analysis and design of artificial hands. For instance, the role of synergies in fine manipulation and haptic exploration are completely unexplored at present. Very little is known even about all grasping phases subsequent to grasp pre-shaping and contact acquisition, most notably on grasp force distribution and the fundamental problems of form and force closure.

In this paper, we study the effect of the number and type of engaged synergies on the distribution of optimal grasp forces and on the ultimate quality of the grasp.

To investigate grasp force distribution problems in basic whole-hand grasps executed by a hand with a limited number of (synergistic) DoFs, the analytical approach followed by the majority of grasp force studies, which abstract their analysis from the specific physical characteristics of the grasping hand, is unsuitable. We therefore introduce a novel analytical

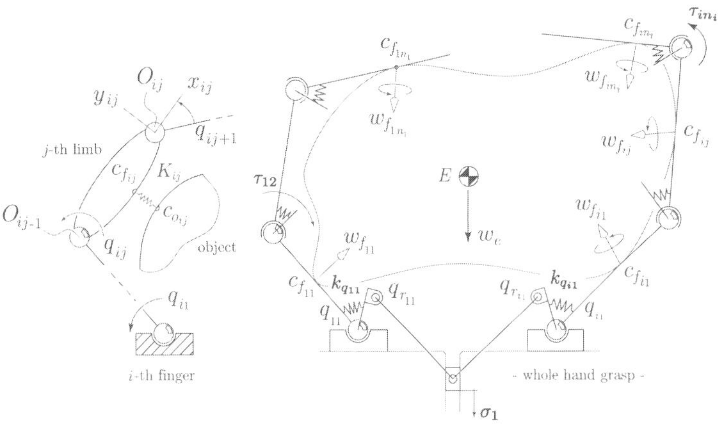

Fig. 1: Schematic of whole-limb manipulation with synergies.

Notation	Definition
P	palm (inertial) frame
S_{ij}	D.-H. limb frame, i^{th} finger, j^{th} limb
L_{ij}	local limb frame, i^{th} finger, j^{th} limb
C_{ij}	local contact frame, i^{th} finger, j^{th} limb
E	object frame
n	number of hand joints
$q \in \mathbb{R}^n$	actual joint angles
$q_r \in \mathbb{R}^n$	reference joint angles
$\tau \in \mathbb{R}^n$	joint torques
s	number of postural syergies
$\sigma \in \mathbb{R}^s$	synergistic displacements
$\eta \in \mathbb{R}^s$	synergistic generalized forces
c	dimension of the contact force/torque vector
f	contact force/torque vector
$\xi_f \in \mathbb{R}^c$	twists of the contact points on the fingers
$\xi_o \in \mathbb{R}^c$	twists of the contact points on the object
$u \in \mathbb{R}^6$	position and orientation of the object
$\xi_e \in \mathbb{R}^6$	object twist
$w_e \in \mathbb{R}^6$	object wrench
$J \in \mathbb{R}^{c\times n}$	hand Jacobian matrix
$S \in \mathbb{R}^{n\times s}$	synergy matrix
$G \in \mathbb{R}^{6\times c}$	grasp matrix

TABLE I: Notation for grasp analysis with postural synergies.

framework, which draws upon previous work on underactuated grasps. Numerical results are presented showing quantitatively the role played by different synergies (from the most fundamental to those of higher-order) in making possible a number of different grasps. As the analysis method we propose to solve force indeterminacies in the rigid-body system introduces a model of compliance in the system, an issue may arise of how strongly our results depend on a fundamentally uncertain and varying parameter such as compliance. To partially address this problem, we show that our numerical results are quite robust with respect to such uncertainties.

II. Preliminaries: Quasi-static Manipulation Model

A. Rigid Multibody Model

We model a cooperating manipulation system as a collection of an arbitrary number of robot "fingers" (i.e., simple chains of links connected through revolute or prismatic joints) attached to a common base "palm", and an object, which is in contact with all or some of the links. With reference to Fig. 1 and adopting the notation in Table I, let k be the number of fingers, n_i the number of degrees of freedom for the ith finger, and set $n = \sum_{i=1}^{k} n_i$ as the total number of hand degrees of freedom (DoFs). Moreover, let $q \in \mathbb{R}^n$ be the vector of joint angles for the whole hand, $q_i \in \mathbb{R}^{n_i}$ that for the ith finger, and $q_{ij} \in \mathbb{R}$ the angle of the jth limb on the ith finger.

According to standard conventions, we consider a fixed (palm) frame $\mathsf{P} = (O_p; x_p, y_p, z_p)$ and, for each finger in the hand, we attach a D.-H. frame $\mathsf{S}_{ij} = (O_{ij}; x_{ij}, y_{ij}, z_{ij})$ to its jth link, a local frame $\mathsf{L}_{ij} = (O_{l_{ij}}; x_{l_{ij}}, y_{l_{ij}}, z_{l_{ij}})$ to the center of the limb and a *normalized Gauss frame* $\mathsf{C}_{ij} = (C_{ij}; x_{c_{ij}}, y_{c_{ij}}, z_{c_{ij}})$ local to the surface of the fingertip, with its $z-$axis aligned with the outward normal. The frame attached to the object is $\mathsf{E} = (E; x_e, y_e, z_e)$, with origin coincident with the center of gravity G, i.e., $E \equiv G$.

Since only some of the fingers and/or some of the limbs in each finger may be in contact with the object at point $c_{f_{ij}} \in \mathbb{R}^3$, we define k sets of indices $\nu(i)$, $(i = 1, \ldots, k)$, each one describing the connectivity of the ith finger with the object. In view of further analysis, we distinguish between corresponding points $c_{f_{ij}}$ and $c_{o_{ij}}$ on the finger and on the object, respectively. On the contrary, under the hypothesis of small relative elastic rotations, we confuse local frames at the contact point on the object and the limbs with the unique C_{ij}.

In this analysis, the location of the contact points in space is assumed to be known, by either planning or sensing.

Let $f_{ij} \in \mathbb{R}^{c_{ij}}$ be the components in the contact frame C_{ij} of force/torque that can be transmitted through the contact at point $c_{f_{ij}}$. Its dimension depends on the contact type, e.g., $c_{ij} = 3$ for Point Contact With Friction (PCWF), $c_{ij} = 4$ for Soft Finger (SF), and its contribution to the actual wrench is characterized by the wrench basis H_{ij}^T [8].

Adopting the notation introduced in Table I, the *balance* and *congruence equations* for the object can be respectively written as

$$ {}^{P}w_e = Gf, \tag{1} $$

$$ \xi_o = G^T\, {}^{P}\xi_e, \tag{2} $$

where $G \in \mathbb{R}^{6\times c}$ is the grasp matrix.

Similarly, the *balance* and *congruence equations* for the hand are, respectively,

$$ \tau = J^T f, \tag{3} $$

$$ \xi_f = J\dot{q}, \tag{4} $$

where $J \in \mathbb{R}^{c\times n}$ is the hand Jacobian.

For the reader's convenience, a detailed derivation of eqs. (1)-(4) can be found in the Appendix.

B. Introducing Elasticity

Now, with reference to taxonomy in [9], in order to handle statically-indeterminate or *hyperstatic* grasps, which occur when $\mathcal{N}(G) \cap \mathcal{N}(J^T) \neq 0$, we follow [10], and introduce a set of virtual springs at the interface between corresponding contact points $c_{f_{ij}}$ and $c_{o_{ij}}$ on the fingers and the object. These result in a system of *linear* constitutive equations linking the components of relative displacements $\delta\xi_{of} := \delta\xi_o - \delta\xi_f$, that violate the constraints, to the corresponding contact force

$$ f = f_0 + \delta f, \quad \delta f = K\delta\xi_{of}, \tag{5} $$

where f_0 is the contact force in the reference configuration $\delta\xi_o = \delta\xi_f = 0$. According to [11], the stiffness matrix $K \in \mathbb{R}^{c\times c}$ can be computed as

$$ K = (C_s + JC_qJ^T)^{-1}, \tag{6} $$

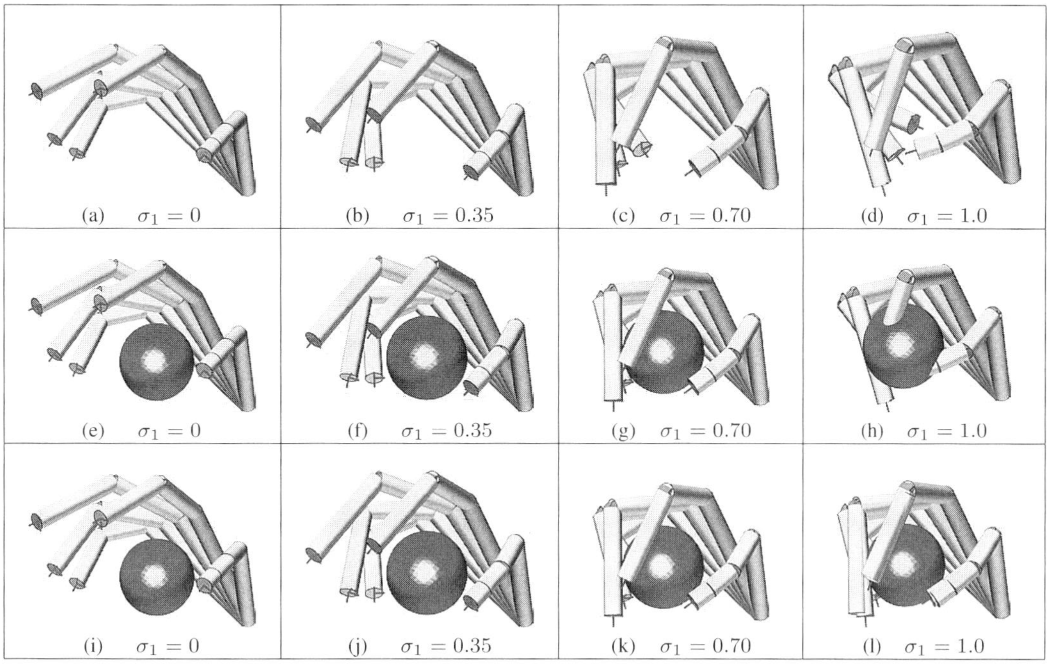

Fig. 2: Effect of the first synergy S_1 on the hand posture as a function of the normalized synergy coefficient $\sigma_1 \in [0, 1]$. Figs. (a)-(d): reference motion of the hand. Figs. (e)-(h): motion of the hand without contact interaction. Figs. (i)-(l): motion of the hand with contact interaction and joint elasticity.

where $C_s \in \mathbb{R}^{c \times c}$ is the structural compliance matrix (due to, e.g., the flexibility of limbs and fingerpads), and with $C_q \in \mathbb{R}^n$ the diagonal matrix whose element in position (k, k) is the compliance at the k-th joint. Joint compliance in animals is determined by the elastic properties of muscles and tendons and by modulation of the stretch reflex. Similar roles in robot hands are played by transmission and actuator compliance, and by the gain of the kth position servo. It should be noticed that in both cases joint compliance can be varied, both intentionally and not, although not necessarily in an independent way from joint to joint. The matrices C_s and C_q employed for the numerical tests reported in this paper are of the form

$$C_s = (1/k_{\text{stru}}) I_{c \times c}, \quad C_q = (1/k_{\text{ss}}) I_{n \times n}, \tag{7}$$

where k_{stru} (N/mm) is a value of structural stiffness, while k_{ss} (Nmm/rad) is an admissible value of joint stiffness.

III. Formulation of the Grasping Problem for The Hand Embodied

A. The "Softly Underactuated" Model

Consideration of synergies introduces a new vista on the grasp problem. A direct interpretation of the results described in [1] would imply that the joint configuration vector q could be represented as a function of fewer elements, collected in a *synergy vector* $\sigma \in \mathbb{R}^s$, as $q = q(\sigma)$, which effectively constraints the hand configuration in an $s-$dimensional sub-manifold of the joint configuration manifold. Synergistic hand velocities would belong to the tangent bundle to this manifold, and could be locally described by a linear map $\dot{q} = S(\sigma)\dot{\sigma}$. This situation is illustrated in Fig. 2, panels (a)-(d), where the reference posture of the hand is reported as a function of the synergy coefficient σ_1 scaling the first synergy vector S_1.

However, the sole kinematic model of the hand fails to describe the actual grasp of an object (Fig. 2, panels (e)-(h)). Therefore, contact forces must be brought into play if a realistic grasp analysis is in order.

Taking a step further, and in view of dealing with the most general case of statically-indeterminate grasps, both contact and joint compliances have to be included in the analysis.

Therefore, in the model we propose, the synergistic hand displacements $\delta\sigma \in \mathbb{R}^s$ do not command the joint displacements $\delta q \in \mathbb{R}^n$ *directly*, as assumed in the analysis in [7] and implemented in the design in [6]. Instead, the synergistic displacements input $\delta\sigma$ command the joint reference positions q_r, as described by the following linear equations (see also Fig. 1)

$$\delta q_r = S\,\delta\sigma, \quad S \in \mathbb{R}^{n \times s}\ (1 \leq s \leq n), \tag{8}$$

which, in turn, are related to the actual joint displacements by the constitutive equation

$$\delta q = \delta q_r - C_q \delta\tau, \tag{9}$$

where C_q is the joint compliance introduced in (7). It is worth mentioning that this model, whereby motion is controlled by a reference position and modulation of joint stiffness, has apparent similarities with the *equilibrium point hypothesis* in the motor control literature [12].

In the pre-grasp phase, forces are null $\delta\tau = 0$, hence $\delta q = \delta q_r$ and the reference and actual posture overlap perfectly (first three columns in Fig. 2). Hence, in this approach phase, the rigid synergy model $q = q(\sigma)$ is valid. When an actual grasp of an object occurs, however, the interference (contact) forces and hand compliance cause the actual hand to deviate from the reference hand (panel (l) in Fig. 2). Thus, in our hand model, the actual hand configuration is driven by synergies, but modifies its posture according to the object shape and compliance. We denote this as a *soft synergy* model of hands.

In the following of this paper, we employ data recorded in [1] and the definition of finger coordination patterns defined through PCA, to obtain numerical values for the *synergy matrix* S (aka "eigengrasp matrix", [7]). More specifically, different simplified synergy matrices S can be obtained extracting a number of columns from the orthogonal *full synergy matrix* $\bar{S} \in \mathbb{R}^{n \times n}$ obtaned from PCA data, and whose columns are ordered in an increasing order with respect to the relative contribution to the variance. Each of these extracted synergy matrices will represent a model of a specific (underactuated) hand.

For later use, let us further introduce the *balance* equation

$$\eta := S^T \tau = S^T J^T f, \tag{10}$$

where $\eta \in \mathbb{R}^s$ are the generalized *synergistic forces* corresponding to synergistic displacements.

We now shift our attention onto the effect of introducing motion coordination patterns for fingers on the ability of the hand to exert internal forces actively (full derivation of the analytical results is omitted for brevity).

B. General Solution of the Grasping Problem With Synergies

Extending the work [10], it can be shown that for a general grasping system with elastic contacts that applies a wrench w_e to the object, the general solution to the force distribution problem is given by

$$f = G_K^R w_e + \delta f_{hr_s} + \delta f_{ho_s}, \ \ \delta f_{hr_s} = E_s y, \ \delta f_{ho_s} = P_s z, \tag{11}$$

where $G_K^R = KG^T(GKG^T)^{-1} \in \mathbb{R}^{c \times 6}$ is the K-weighted pseudoinverse of G, providing the *particular solution* $G_K^R w_e$ that minimizes the potential energy $\frac{1}{2}\delta\xi_{of}^T K \delta\xi_{of}$, (see [13], [14]). In eq. (11), the columns of matrix $E_s \in \mathbb{R}^{c \times e_s}$ form a basis for the range space of the matrix $F_s \in \mathbb{R}^{c \times s}$, mapping the $\delta\sigma$'s into the *active internal forces* δf_{hr_s} that can be commanded by synergistic displacements

$$\delta f_{hr_s} = F_s \, \delta\sigma, \quad F_s := FS, \quad \text{rank}(F_s) = e_s. \tag{12}$$

The matrix $F \in \mathbb{R}^{c \times n}$ has expression

$$F := (I - G_K^R G) KJ, \quad \text{rank}(F) = e, \tag{13}$$

and maps independently controlled joint reference displacements δq_r's into *active internal forces*

$$\delta f_{hr} = F \, \delta q_r. \tag{14}$$

It is worth noting that if all the DoFs of the hand are independently controlled, $\delta q_r = \delta\sigma$, $S = I \in \mathbb{R}^{n \times n}$, and eq. (12) reduces to eq. (14).

The expression for δf_{hr_s} in eq. (11) is a parameterized version of the active homogeneous solution: optimal grasp force distributions can be found by minimizing a cost function with respect to $y \in \mathbb{R}^{e_s}$.

The third term δf_{ho_s} in eq. (11) is a fixed homogeneous solution representing *internal, passive (preload) contact forces*: this corresponds to contact forces that are preloaded at the beginning of the grasp operation (in the rest of the paper it will be assumed $z = 0 \in \mathbb{R}^{p_s}$). Matrix $P_s \in \mathbb{R}^{c \times p_s}$ represents a basis for this subspace. When all the hand joints are independently controlled, the basis matrix is denoted by $P \in \mathbb{R}^{c \times p}$, where p is the dimension of this subspace.

We summarize the above results by introducing the following subspaces

$$\mathcal{F}_h = \mathcal{R}(A) = \mathcal{N}(G) \subset \mathbb{R}^c, \tag{15}$$

$$\mathcal{F}_{hr_s} = \mathcal{R}(E_s) = \mathcal{N}(G) \cap \big(\mathcal{R}(KJS) + \mathcal{R}(KG^T)\big), \tag{16}$$

$$\mathcal{F}_{ho_s} = \mathcal{R}(P_s) = \mathcal{N}(G) \cap \mathcal{N}((JS)^T), \tag{17}$$

that yields the natural decomposition

$$\mathcal{F}_{h_s} = \mathcal{F}_{hr_s} \oplus \mathcal{F}_{ho_s}, \quad \mathbb{R}^c = \mathcal{F}_{h_s} \oplus \mathcal{R}(G_K^R) \tag{18}$$

In case of independent joint control ($S = I \in \mathbb{R}^{n \times n}$), active and passive subspaces in eqs. (15) reduce to $\mathcal{F}_{hr}$ and $\mathcal{F}_{ho}$, respectively, as in [10].

It is worth noting that among the free vectors $y \in \mathbb{R}^{e_s}$, $\hat{y}$ will denote a particular choice corresponding to an optimal grasp force distribution with respect to a chosen cost function.

From a computational point of view, the calculation of the desired basis matrix $E_s = \text{colbasis}(F_s)$ from eq. (12) is not optimal, since it entails the explicit calculation of G_K^R. A more efficient algorithm can be obtained by intersection of subspaces, observing that a consistent set of internal forces, parameterized by $x \in \mathbb{R}^h$, synergy displacements $\delta\sigma \in \mathbb{R}^s$, and object motions $\delta\xi_e \in \mathbb{R}^6$, must belong to the nullspace of $Q_s \in \mathbb{R}^{c \times (h+s+6)}$ (whose nullity is b_s), i.e.,

$$\underbrace{\begin{bmatrix} A & -KJS & KG^T \end{bmatrix}}_{Q_s} \begin{bmatrix} x \\ \delta\sigma \\ \delta\xi_e \end{bmatrix} = 0 \tag{19}$$

Defining $B_s \in \mathbb{R}^{(h+s+6) \times b_s}$, such that $\mathcal{R}(B_s) = \mathcal{N}(Q_s)$, and partitioning B_s as

$$B_s = \begin{bmatrix} B_{s_1}^T & B_{s_2}^T & B_{s_3}^T \end{bmatrix}^T, \tag{20}$$

where $B_{s_1} \in \mathbb{R}^{h \times b_s}$, $B_{s_2} \in \mathbb{R}^{s \times b_s}$, and $B_{s_3} \in \mathbb{R}^{6 \times b_s}$, such that

$$\begin{bmatrix} x^T & \delta\sigma^T & \delta\xi_e^T \end{bmatrix}^T = \begin{bmatrix} B_{s_1}^T & B_{s_2}^T & B_{s_3} \end{bmatrix}^T \gamma, \quad \gamma \in \mathbb{R}^{b_s}, \tag{21}$$

the subspace of active internal forces can be profitably obtained as

$$\mathcal{F}_{hr_s} = \mathcal{R}(AB_{s_1}), \quad E_s := \text{colbasis}(AB_{s_1}), \tag{22}$$

The synergy displacements $\delta\hat{\sigma}$ that must be commanded if a desired internal force $\delta\hat{f} = E_s\hat{y}$ is to be applied are given by

$$\delta\hat{\sigma} = B_{s_2}(AB_{s_1})^+ E_s \hat{y} \tag{23}$$

According to eq. (8) and eq. (23), the joint reference position is displaced by

$$\delta q_r = S B_{s_2}(AB_{s_1})^+ E_s \hat{y}, \tag{24}$$

and the object moves to a new equilibrium position defined by

$$\delta\xi_e = B_{s_3}(AB_{s_1})^+ E_s \hat{y} \tag{25}$$

Due to contact forces in the interaction with the object and the joint compliance, the hand joints moves differently with respect to the commanded reference, i.e., $\delta q \neq \delta q_r$. Their values can be explicitly calculated as

$$\delta q = \big[I - C_q J^T \big(I - G_K^R G\big) KJ\big] S \, \delta\hat{\sigma}, \tag{26}$$

where $\delta\hat{\sigma}$ can be recovered from (23). It is worth observing that in case of perfectly rigid joints, $C_q = 0 \in \mathbb{R}^{n \times n}$, and $\delta q = \delta q_r$. The other quantities are also modified due to the increased global stiffness, as evident from the definition of K in (6).

Finally, corresponding variation of the joint torques $\delta\tau$ can be obtained as

$$\delta\tau = J^T \big(I - G_K^R G\big) KJ \, S \, \delta\hat{\sigma}, \tag{27}$$

and the associated variation of the synergistic forces $\delta\eta$ is

$$\delta\eta = S^T \delta\tau \tag{28}$$

C. Qualitative Analysis of Hand Embodiment

With reference to eq. (12), the following relationships between ranks hold

$$\underbrace{\text{rank}(F_s)}_{e_s} = \underbrace{\text{rank}(S)}_{s} - \dim(\mathcal{N}(F) \cap \mathcal{R}(S)) \quad (29)$$

$$\text{rank}(F_s) \leq \min\{\text{rank}(F), \text{rank}(S)\} \quad (30)$$

$$\text{rank}(F_s) \geq \text{rank}(F) + \text{rank}(S) - n \quad (31)$$

Therefore, under the condition that each synergy S_i ($i \in \mathcal{S}$[1]) has a non-null projection onto $\mathcal{R}(F^T)$, we can assume that $e_s = s$. In this case, consider a fully actuated grasp with $\text{rank}(F) = e$. For the same grasp, consider underactuation and increase one by one the number of the engaged columns of S. As the number of synergies engaged increases, say $1 \leq s \leq e$, the dimension of the subspace $\mathcal{F}_{hr_s}$ of active internal forces also increases in the same manner. For $s > e$, according to eq. (30), $e_s = e$, and the dimension of the subspace $\mathcal{F}_{hr_s}$ reaches a plateau. This means that for fully actuated grasping systems characterized by an e-dimensional $\mathcal{F}_{hr}$, underactuation with a number of synergies $s \sim e$ does not endanger the ability to exert the *same* internal forces. If e is "small", say e.g. $s = 2$ or $s = 3$, application of an equally "small" number of synergies s results in a great control simplification without side effects on the grasping ability of the system, see results in sec. V-C.

A fundamental issue we want to investigate is the link between basic synergies, i.e., the first components obtained via PCA, accounting for much of the variance in geometric posture space, and the ability of the corresponding underactuated system to firmly grasp an object. However, the sole rank count in eq. (29) does not allow to compare *quantitatively* synergies with different shapes, i.e., different columns of S. Therefore, in this analysis the behaviour of a system underactuated by means of different number of synergies and, once fixed their number, with different synergies S_i, is presented. The performance parameter is the ability to attain *force-closure* conditions and, if this is the case, in obtaining lower values of a suitably defined cost function in determining optimal grasping forces. Then, we discuss robustness issues with respect to variations of the grasping stiffness: the range, going from values typical of robotic hands to those of the human hand, is elicited from [15], and [16].

IV. Force-Closure Problem with Synergies

At an intuitive level, the meaning of "force-closure" is that motions of the grasped object are completely restrained against arbitrary external disturbances, by virtue of the contact forces that the hand is capable to exert on the object. We emphasize here that this definition is very relevant to synergistically (under)actuated hands. Indeed (as opposite to the purely geometric nature of form-closure), force-closure involves consideration of which contact forces can be actively applied on the object by the specific hand under consideration. Under this regard, it clearly makes a difference if an object is grasped by a hand controlled by different numbers and types of synergies (corresponding to different active internal force subspaces, as discussed in the previous section).

Accordingly, we adopt here the definition of force-closure given in [17], which considers the case of underactuated hands:

[1] $\mathcal{S}$ is a set of indices used to select the corresponding columns in $\bar{S}$ to build S.

Definition 1 (Force-Closure): A grasp is defined Force-Closure if and only if the following conditions are satisfied:

1. forces in arbitrary directions are resistible, i.e. $\text{rank}(G) = 6$
2. the hand configuration is prehensile, i.e. $\exists y$ such that $f(y) \in \text{Int}(\mathcal{F})$ [2], with $f(y) = Ey$

In *Definition 1*, $\mathcal{F}$ is the composite friction cone defined as $\mathcal{F} = \mathcal{F}_{11} \times \cdots \times \mathcal{F}_{k,n_k}$. For brevity, we recall here only those types of friction cone implemented in our software, used to carry out numerical results in later sections. For a point contact with friction (PCWF), we have $f_{ij} \in \mathbb{R}^3$ and

$$\mathcal{F}_{ij} = \{f_{ij} \in \mathbb{R}^3 |\, f_{ij_3} \geq 0, \frac{1}{\mu_{ij}}(f_{ij_1}^2 + f_{ij_2}^2) \leq f_{ij_3}^2\}, \quad (32)$$

where f_{ij_3} is the normal component of the contact force at the point of contact c_{ij}, f_{ij_1}, f_{ij_2} the components in the tangential directions, and μ_{ij} the Coulomb friction coefficient.

For a soft-finger with elliptical friction limit approximation [18], we have $f_{ij} \in \mathbb{R}^4$ and

$$\mathcal{F}_{ij} = \{f_{ij} \in \mathbb{R}^4 |\, f_{ij_3} \geq 0, \frac{1}{\mu_{ij}}(f_{ij_1}^2 + f_{ij_2}^2) + \frac{1}{\mu_{ij_t}} f_{ij_4}^2 \leq f_{ij_3}^2\}, \quad (33)$$

where μ_{ij_t} is a proportionality constant between the torsion and shear limits. It is worth noting that the above models do not necessarily assume equal friction coefficients at all contact points: however, for brevity, in the numerical tests reported in sec. V-C and V-D, the friction coefficients will be assumed constant, i.e., $\mu_{ij} = \mu$ and $\mu_{ij_t} = 1$.

In view of formulating and solving the problem as a second order cone programming (SOCP) one, for which efficient algorithms and interfaces are today a mature technology, we recall that the satisfaction of the friction limit constraints in eq. (32) and (33) is equivalent to the positive definiteness of matrix $P = \text{Blockdiag}(P_{11}, \ldots, P_{k,n_k})$, $P \succ 0$, where the explicit expression for each P_{ij}, with $(i = 1, \ldots, k;\, j \in \nu(i))$, can be found in [18].

Then, consider the problem of finding the optimal distribution of contact forces f in the grasp of an object subject to the external load with regard to the minimization of a suitable cost function $\Psi(y)$. To formalize this problem, we give the following definition:

Definition 2 (Grasping Force Optimization): Given a grasp characterized by G_W^R, E_s, and P_s, and an object wrench $w_e \in \mathbb{R}^6$, find $\hat{y}$ in eq. (11), such that $f(\hat{y}) \in \text{Int}(\mathcal{F})$, and the cost function $\Psi(f(\hat{y}))$ is minimized.

In numerical tests, we assume: (i) zero preload at the beginning of the grasp ($z = 0$ in eq. (11)); (ii) zero net wrench applied to the object, $w_e = 0$; (iii) an auxiliary constraint on the minimum value for all the normal components f_n of the contact force. Under these hypotheses, the grasping force optimization problem is set up in the following way

$$\begin{aligned} &\hat{y} = \text{argmin}\, \Psi(y) \\ &\text{subject to } f = E_s\, y, \;\; P(f) \succ 0, \;\; f_n \succeq f_{min} \end{aligned} \quad (34)$$

Optimal contact force distribution is sought by employing, in turn, each of the following cost functions

$$\Psi_f(y) := \|f(y)\|_2, \;\; \Psi_\tau(y) := \|\tau(y)\|_2, \;\; \Psi_\eta(y) := \|\eta(y)\|_2,$$

[2] $\text{Int}(\mathcal{F})$ denote the internal part of the composite friction cone $\mathcal{F}$.

where $f \in \mathbb{R}^c$ are the contact forces, $\tau \in \mathbb{R}^{n_c}$ the joint torques for the contacting fingers in eq. (3), and $\eta \in \mathbb{R}^s$ the synergistic forces in eq. (10).

The problem is set up and solved as a semidefinite program (SDP) by employing the `CVX` modeling system for convex optimization based on MATLAB, see [19] and [20] for further details. The solver used is `SDPT3` [21], which implements an infeasible path-following algorithm for solving general SQLP – conic optimization problems involving semidefinite, second-order and linear cone constraints.

V. Numerical Tests

A. Paradigmatic Hand Model

The paradigmatic hand model is schematically represented in Fig. 3. This is the very same model for which detailed data were collected and presented in [1]. The fact that the large amount of high-quality data taken in [1] was kindly made available to us is a reason for this choice, along with the motivation that synergies can be defined for no other kinematic structure without a dose of arbitrarity.

The model has 15 DoFs corresponding to: 4 DoFs for the thumb: TR, TA, TM, TI (Thumb Rotation, Abduction, Metacarpal, Interphalangeal); 3 DoFs for the index: IA, IM, IP (Index Abduction, Metacarpal, Proximal interphalangeal); 2 DoFs for the middle: MM, MP (Middle Metacarpal, Proximal interphalangeal); 3 DoFs for the ring: RA, RM, RP (Ring Abduction, Metacarpal, Proximal interphalangeal); 3 DoFs for the little: LA, LM, LP (Little Abduction, Metacarpal, Proximal interphalangeal). It is worth noting that the middle finger has no abduction since it is considered the "reference finger" in the sagittal plane of the hand. Moreover, the Distal Interphalangeal (DI) angle is not present in none of the four fingers due to the limitation in the sensors embedded in the measuring glove employed, see [1] and references therein. With $\mathsf{P} = (O_p; x_p, y_p, z_p)$ with indicate the palm frame, and with $S_{ij} = (O_{ij}; x_{ij}, y_{ij}, z_{ij})$ and $\mathsf{C}_{ij} = (C_{ij}; x_{c_{ij}}, y_{c_{ij}}, z_{c_{ij}})$ the D.-H. and the normalized Gauss frame, respectively, for

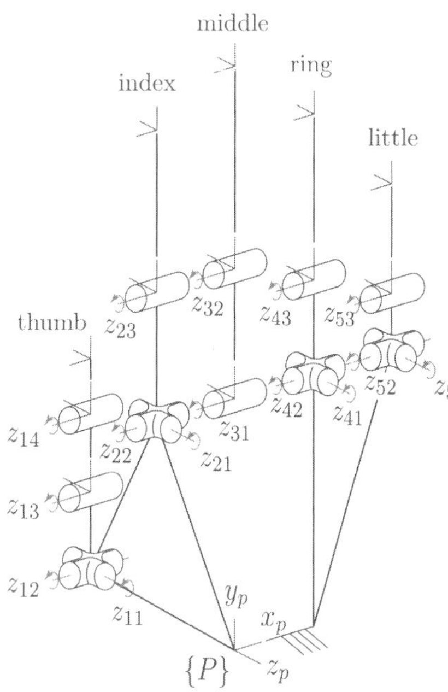

Fig. 3: 15 DoFs kinematic model of the paradigmatic hand.

Limb	a (mm)	α (rad)	d (mm)	q (rad)
l_{11}	0	$-\pi/2$	0	q_{11}
l_{12}	a_{12}	0	0	q_{12}
l_{13}	a_{13}	0	0	q_{13}
l_{14}	a_{14}	0	0	q_{14}

(a) D.-H. table for the Thumb finger.

Limb	a (mm)	α (rad)	d (mm)	q (rad)
l_{ij}	0	$-\pi/2$	0	q_{ij}
l_{ij}	a_{ij}	0	0	q_{ij}
l_{ij}	a_{ij}	0	0	q_{ij}

(b) D.-H. table for Index, Ring and Little fingers. Indices take the following values: $i = 2, 4, 5$ and $j = 1, 2, 3$.

Limb	a (mm)	α (rad)	d (mm)	q (rad)
l_{31}	a_{31}	0	0	q_{31}
l_{32}	a_{32}	0	0	q_{32}

(c) D.-H. table for the Middle finger.

TABLE II: D.-H. tables for the 15 DoFs paradigmatic hand.

the jth limb on the ith finger. The D.-H. tables for each finger/group of fingers are shown in Table IIa-IIc.

B. Assumptions in the Definition of Grasp Layouts

To define possible grasp configurations, we rely on data reported in [1]. In that work, no measurement of the object position, nor spatial location of the contact points, was registered (since the subjects were asked to shape their hands in order to mime grasps, with no physical object present). Therefore we made the following assumptions. Firstly, the hand configuration relative to the imagined grasp of an object, say a cherry, was defined as the mean joint configuration vector recorded in the grasp of that object, among a total of five trials. Therefore, the configuration of the hand is identified with the name of the object grasped. Secondly, a reasonable position of the center of gravity G for the grasped object was defined, taking into account both the hand configuration previously defined and a reasonable shape for that object. Then, the candidate contact point $c_{f_{ij}}$ on each limb l_{ij} was found as the nearest point on the same limb to G. Last step was to define the final grasp by selecting the limbs in contact, according to both the shape of the hand and the position of the object relative to the hand the same grasp suggests. The numerical data common to all tests are listed in Tab. III.

C. Precision grasp - The Cherry

As a first test case we consider the grasp of a cherry. The hand configuration, the estimated position of the center of gravity G of the object and the contact points are represented

Net Wrench w_e	Contact type	Frict. coeff. μ	$f_{min}(N)$
0	PCWF	1.5	0.1

TABLE III: Nominal data common to all tests.

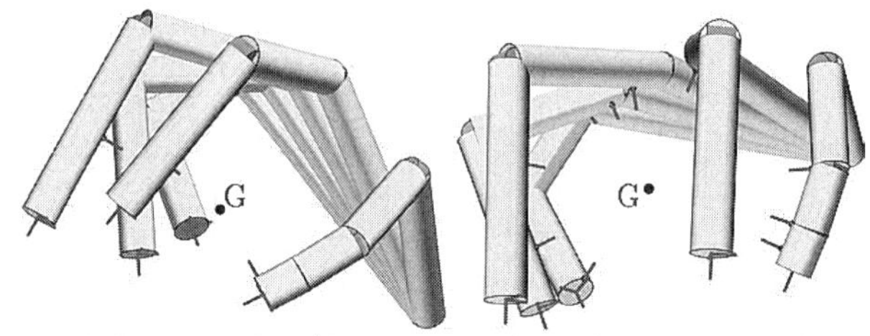

(a) Grasp of the Cherry (b) Grasp of the Ashtray

Fig. 4: Hand postures analyzed.

in Fig. 4 (a). The contact parameters employed are listed in Table III and, with reference to eqs. (6) and (7), the numerical value of the stiffness matrix K is specified by setting $k_{stru} = 1$ N/mm and $k_{\mathrm{ss}} = 100$ Nmm/rad. For this case, $\mathrm{rank}(G) = 6$, $c = 9$ and, with reference to eqs. (12), (13), and (15), $h = e = 3$, $p = 0$. The optimal contact force distribution is found with respect to y minimizing $\psi_f(y) = \|f(y)\|_2$.

Let us concentrate on the black curve in Fig. 5a. Interestingly enough, this grasp is force-closure even when engaging only the first synergy (S_1, first column of S), for which an optimal value of $\|f\|_2 = 0.349$ N, is found. When more synergies are engaged in the grasp, the norm of the contact force decreases as the dimension $e_s = s$ of $\mathcal{F}_{hr_s}$ increases, and there is a higher dimensional space where the optimal solution can be sought. Then, a plateau is obtained once the dimension e_s reaches e, as confirmed by rank considerations in eq. (29). Therefore, as far as controllable internal contact forces are concerned, no improvement can be obtained in the quality of the grasp by engaging a number of synergies $s > e$.

It is worth stressing that the trend in Fig. 5a is *highly dependent* on which synergies are considered. Had we engaged synergies in a backward fashion, i.e., from S_{15} to S_1, we would have obtained a completely different trend. Put another way, if we plug in *only one* synergy at a time, the grasp is force-closure only for synergies S_1, S_3 and S_4, meaning that maybe the first synergies are more fundamental to grasp objects than those of higher-order, but not in a strictly ordered fashion. However, we have no ambition of drawing general conclusions here, since fundamental roles are played by location of the contact points, surface normals, and types of contact constraints, which are here only reasonably estimated.

In order to test the robustness of the above trends, the previous analyses are repeated for different values of K obtained varying k_{ss} and k_{stru}, as shown in Figs. 5a-5b.

Interestingly, if we select $\Psi_\eta(y) = \|\eta\|_2$ as cost function, we obtain the trends depicted in Fig. 6a-6b. These show that, in terms of synergistic forces, adding more synergies than those strictly necessary to fulfill the dimension of $\mathcal{F}_{hr}$ worsens the cost, as the range of S (where we project the same τ's, see eq. (10)) increases. In other words, if a synergy is not actuated, synergistic forces in its direction are absorbed by the mechanical structure directly, and are not reflected in actuation costs.

Minimization of $\Psi_\tau(y) = \|\tau\|_2$ bears trends similar to those of Fig. 5, and are here omitted for brevity.

D. Power Grasp - The Ashtray

As second test case we consider the grasp of an ashtray. The hand configuration, the estimated position of the center of gravity G of the object and the contact points are represented in Fig. 4 (b). The contact parameters and the nominal stiffness values are the same employed in the previous case. This time, $\mathrm{rank}(G) = 6$, $c = 33$ and, with reference to eqs. (12), (13), and (15), $h = 27$, $e = 15$, $p = 12$. As intuitive, since $h > e = n$, increasing the number of synergies engaged results in a monotone decrease of the cost function, as elicited from Fig. 7, no matter the values of the grasping stiffness K. All curves present a rapid decrease for $1 \leq s \leq 3$, then a lighter decrease for $3 < s \leq 6$, after which no practical improvement is registered. Also interesting is that synergy S_1 represents the only "direction" along which we can obtain force-closure conditions by employing only a 1-dimensional subspace of $\mathbb{R}^{15}$, at least for nominal values of the parameters. No other synergy shows this property. To give an idea, to obtain force-closure without S_1, we need a basis composed of columns S_2

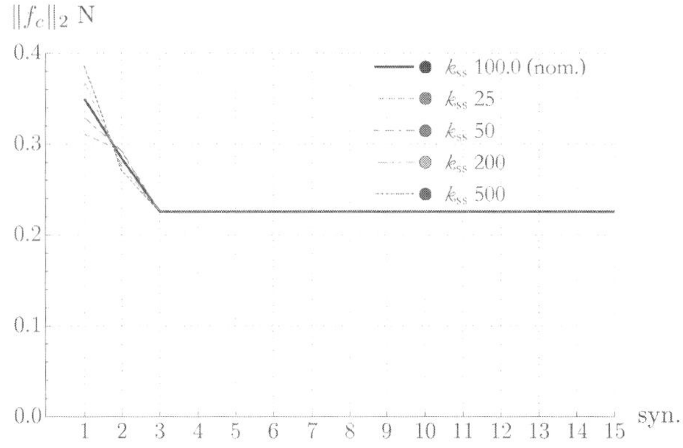

(a) Trends for variation of the steady-state gain k_{ss} (Nmm/rad).

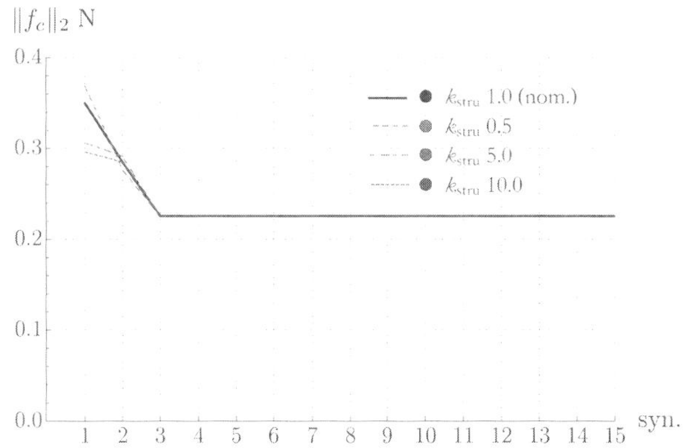

(b) Trends for variation of the structural stiffness k_{stru} (N/mm).

Fig. 5: The Cherry. Norm of the optimal contact force f w.r.t. the cost function $\|f\|_2$, with increasing number of synergies.

through S_9, that is a subspace seven-dimensional higher than before.

From Fig. 7, it is interesting to note that a wide variation of K can render infeasible the grasp with only S_1 engaged. However, adding just another column, e.g. S_2, solves already the problem.

Finally, if we select $\Psi_\eta(y) = \|\eta\|_2$ as cost function, we obtain the trends depicted in Fig. 8a-8b.

VI. CONCLUSION

The force decomposition and optimization problems in multiple whole-limb manipulation of hands with embodied synergies require an extension of existing analytical methods, and the consideration of compliance in the hand-object system.

We have presented two numerical case studies to characterize the role of different postural synergies in the ability of the hand to obtain force-closure grasps. The two case studies addressed a precision grasp and a power grasp, respectively, and are to be considered representative of a number of similar experiments, which could not be reported for space limitations.

The main results obtained from our investigations can be summarized as follows.

The force-closure property of grasps strongly depend on which synergies are used to control the hand. The first few synergies (the first one for the two case studies reported) are sufficient to establish force-closure. If the first few synergies are not actively controlled, force closure can only be obtained if many more DoFs (corresponding to higher-order synergies) are actuated.

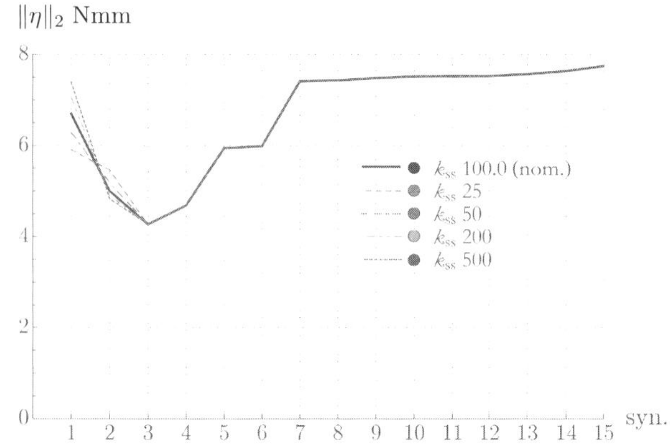

(a) Trends for variation of the steady-state gain k_{ss} (Nmm/rad).

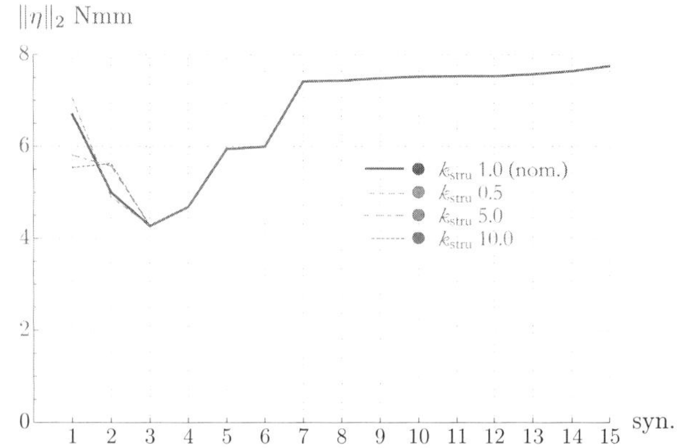

(b) Trends for variation of the structural stiffness k_{stru} (N/mm).

Fig. 6: The Cherry. Norm of the optimal synergistic force $\eta = S^T\tau$ w.r.t. the cost function $\|\eta\|_2$, with increasing number of synergies.

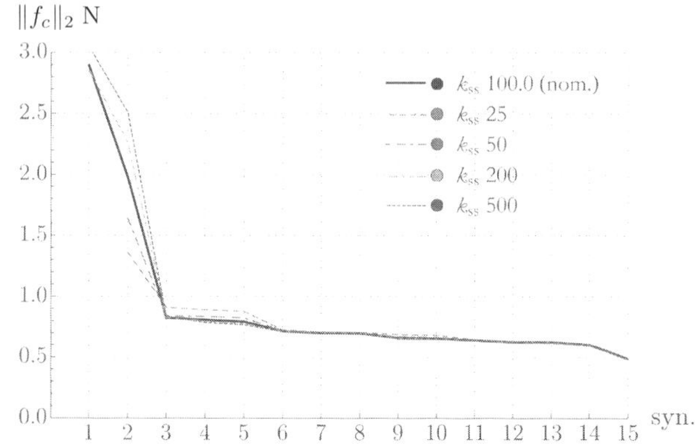

(a) Trends for variation of the steady-state gain k_{ss} (Nmm/rad).

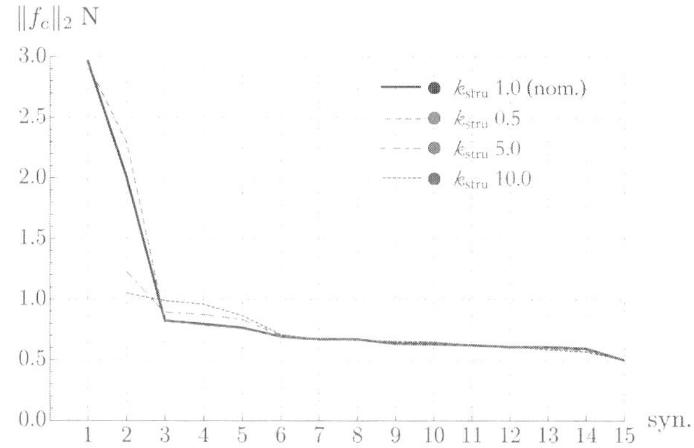

(b) Trends for variation of the structural stiffness k_{stru} (N/mm).

Fig. 7: The Ashtray. Norm of the optimal contact force f w.r.t. the cost function $\|f\|_2$, with increasing number of synergies.

A measure of the quality of the grasp (given in terms of the norm of contact forces needed to avoid slippage) is enhanced by increasing the number of actuated synergies, but only to a limited extent. No improvement is observed beyond the first three synergies in the precision grasp case, while continuous but small improvements are obtained in the whole-hand grasp case.

All the above results are consistently robust with respect to different values of stiffness parameters, which may reflect the uncertainty by which these parameters are known in human or robotic hand models, and/or the fact that grasp stiffness may be changed either voluntarily or not.

These results are consistent with the hypothesis that the first few postural synergies observed in grasp pre-shaping [1] are also crucial in grasping force optimization, when suitably translated from the kinematic configuraton space where they have been observed, to the force domain through the procedures illustrated in this paper.

Appendix

With reference to Fig. 1, and by employing the definition of the adjoint operator $\text{Ad}_{(R,d)}$ given in [22], the contribution of the ith finger, j^{th} limb to the components in P of the wrench $w_e \in \mathbb{R}^6$ exerted on the object is given by

$$^{P}w_{e_{ij}} = G_{ij} f_{ij}, \quad G_{ij} := \text{Ad}^T_{(^{c_{ij}}R_p,\, d^e_{c_{ij}})} H^T_{ij} \in \mathbb{R}^{6\times c_{ij}}, \tag{35}$$

where $^{c_{ij}}R_p$ is the rotation that aligns P onto C_{ij}, and $d^e_{c_{ij}}$ is the vector from C_{ij} to E. Let ϵ_{ij} be defined as follows

$$\epsilon_{ij} = \begin{cases} 1, & \text{if } j \in \nu(i) \\ 0, & \text{otherwise.} \end{cases} \tag{36}$$

Put $f_i = [\epsilon_{i_1} f^T_{i_1} \cdots \epsilon_{in_i} f^T_{in_i}]^T \in \mathbb{R}^{c_i}$, where $c_i = \sum_{j\in\nu(i)} c_{ij}$, and consider the contribution of the ith finger as

$$^{P}w_{e_i} = G_i f_i, \quad G_i := [\epsilon_{i1} G_{i1} \cdots \epsilon_{in_i} G_{in_i}] \in \mathbb{R}^{6\times c_i} \tag{37}$$

Stacking equation (37) for each finger, we can write the global grasp matrix as

$$^{P}w_e = Gf, \quad G := [G_1 \cdots G_k] \in \mathbb{R}^{6\times c}, \tag{38}$$

with $c = \sum_{i=1}^{k} c_i$.

Similarly, let $\xi_{c_{o_{ij}}} \in \mathbb{R}^{c_{ij}}$ be the local components in the constrained directions at point $c_{o_{ij}}$ due to a twist of the object $^{P}\xi_e$, with components in P. By setting $\xi_{o_i} = [\epsilon_{i1}\xi^T_{c_{o_{i1}}} \cdots \epsilon_{in_i}\xi^T_{c_{o_{in_i}}}]^T \in \mathbb{R}^{c_i}$, for the ith finger, and $\xi_o = [\xi^T_{o_1} \cdots \xi^T_{o_k}]^T \in \mathbb{R}^c$, for the whole hand, we can write by duality

$$\xi_o = G^T\, {}^{P}\xi_e, \quad G^T \in \mathbb{R}^{c\times 6} \tag{39}$$

We now consider the relationships for the statics and kinematics of the fingers. By employing the D.-H. convention, the

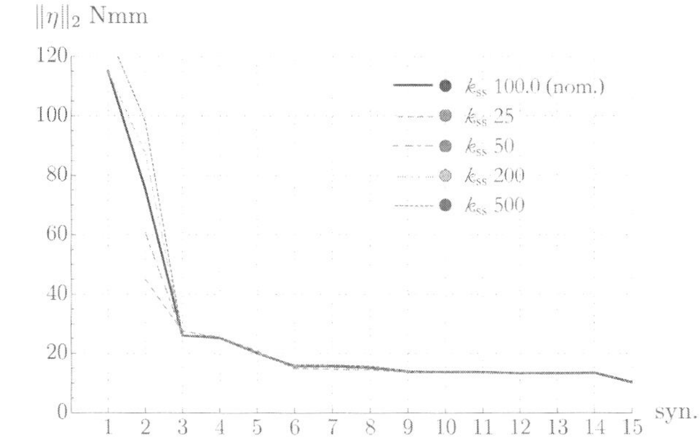

(a) Trends for variation of the steady-state gain k_{ss} (Nmm/rad).

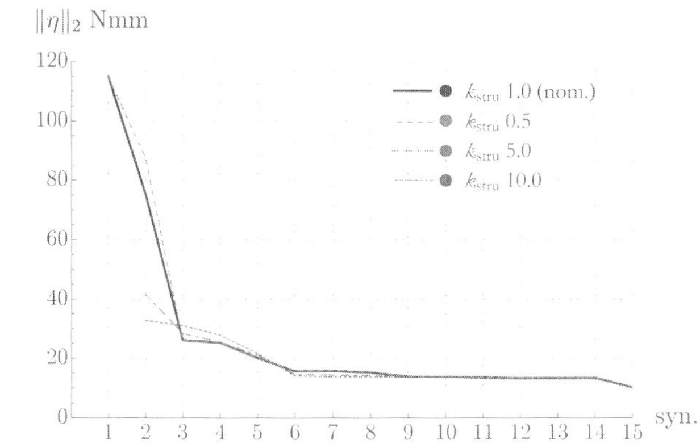

(b) Trends for variation of the structural stiffness k_{stru} (N/mm).

Fig. 8: The Ashtray. Norm of the optimal synergistic force $\eta = S^T \tau$ w.r.t. the cost function $\|\eta\|_2$, with increasing number of synergies.

local components in the constrained directions of the contact point $c_{f_{ij}}$ on the ith finger, j^{th} limb, can be written as

$$\xi_{c_{f_{ij}}} = J_{ij}\dot{q}_i, \quad J_{ij} := H_{ij}\,\mathrm{Ad}_{({}^{c_{ij}}R_p,\, d^{o_{ij}}_{c_{ij}})}\, J_{o_{ij}}(q_i), \qquad (40)$$

where $J_{o_{ij}} = [j_{i1} \ \cdots \ j_{ij} \ \ 0 \ \cdots \ 0] \in \mathbb{R}^{6\times n_i}$, with blocks defined as $(1 \le l \le j)$

$$j_{il} = \begin{cases} [z_{l-1}^T \quad 0^T]^T, & \text{for prismatic } l^{th} \text{ joint;} \\ [(d^{o_{f_i,l-1}}_{o_{f_{ij}}})^T \widehat{z}_{l-1} \quad z_{l-1}^T]^T, & \text{for revolute } l^{th} \text{ joint} \end{cases} \qquad (41)$$

By stacking (40) for all the limbs on the ith finger that are actually in contact, and setting $\xi_{f_i} = [\epsilon_{i1}\xi^T_{c_{f_{i1}}} \cdots \epsilon_{in_i}\xi^T_{c_{f_{in_i}}}]^T \in \mathbb{R}^{c_i}$, we obtain

$$\xi_{f_i} = J_i\dot{q}_i, \quad J_i := \left[\epsilon_{i1}J_{i1}^T \cdots \epsilon_{in_i}J_{in_i}^T\right]^T \in \mathbb{R}^{c_i\times n_i} \qquad (42)$$

Then, collecting the complete set velocities in twist $\xi_f = [\xi_{f_1}^T \cdots \xi_{f_k}^T]^T \in \mathbb{R}^c$, yields

$$\xi_f = J\dot{q}, \quad J := \mathrm{Blockdiag}(J_1, \ldots, J_k) \in \mathbb{R}^{c\times n} \qquad (43)$$

Finally, again by duality arguments, the map from the hand contact forces f to hand joint torques τ is given by

$$\tau = J^T f, \quad J = \mathrm{Blockdiag}(J_1^T, \ldots, J_k^T) \in \mathbb{R}^{n\times c} \qquad (44)$$

Acknowledgment

Authors wish to gratefully acknowledge Marco Santello for the inspiring discussions and for providing experimental data and parameters. This work is supported by the European Commission under CP grant no. 248587, "THE Hand Embodied", within the FP7-ICT-2009-4-2-1 program "Cognitive Systems and Robotics".

References

[1] M. Santello, M. Flanders, and J. F. Soechting, "Postural hand synergies for tool use," *J. Neurosci.*, vol. 18, no. 23, pp. 10 105–10 115, December 1998.

[2] C. R. Mason, J. E. Gomez, and T. J. Ebner, "Hand synergies during reach-to-grasp," *J Neurophysiol*, vol. 86, no. 6, pp. 2896–2910, December 2001.

[3] V. Cheung, A. d'Avella, M. Tresch, and E. Bizzi, "Central and sensory contributions to the activation and organization of muscle synergies during natural motor behaviors," *Journal of Neuroscience*, vol. 25, no. 27, pp. 6419–6434, 2005.

[4] A. Bicchi, "Hands for dextrous manipulation and robust grasping: a difficult road towards simplicity," *IEEE Trans. on Robotics and Automation*, vol. 16, no. 6, pp. 652–662, December 2000.

[5] M. Ciocarlie, C. Goldfeder, and P. Allen, "Dimensionality reduction for hand-independent dexterous robotic grasping," in *Proceedings of the IEEE/RSJ International Conference on Intelligent Robots and Systems*, 2007, pp. 3270–3275. [Online]. Available: http://dx.doi.org/10.1109/IROS.2007.4399227

[6] C. Brown and H. Asada, "Inter-finger coordination and postural synergies in robot hands via mechanical implementation of principal component analysis," in *IEEE-RAS International Conference on Intelligent Robots and Systems*, 2007, pp. 2877–2882.

[7] M. T. Ciocarlie and P. K. Allen, "Hand posture subspaces for dexterous robotic grasping," *The International Journal of Robotics Research*, vol. 28, no. 7, pp. 851–867, July 2009.

[8] R. Featherstone, *Rigid Body Dynamics Algorithms*, Springer, Ed. Springer, 2008.

[9] D. Prattichizzo and J. C. Trinkle, "Grasping," in *Springer Handbook of Robotics*. Springer, 2008.

[10] A. Bicchi, "On the problem of decomposing grasp and manipulation forces in multiple whole-limb manipulation," *Robotics and Autonomous Systems*, vol. 13, no. 2, pp. 127–147, July 1994.

[11] M. Cutkosky and I. Kao, "Computing and controlling the compliance of a robotic hand," *Robotics and Automation, IEEE Transactions on*, vol. 5, no. 2, pp. 151–165, 1989.

[12] A. G. Feldman and M. F. Levin, "The equilibrium-point hypothesis–past, present and future," *Adv. Exp. Med. Biol.*, no. 629, pp. 699–726, 2009.

[13] H. Hanafusa and H. Asada, "Stable prehension by a robot hand with elastic fingers," in 7th *ISIR*, Tokyo, 1997.

[14] J. Joh and H. Lipkin, "Lagrangian wrench distribution for cooperating robotic mechanisms," in *IEEE Conf. on Robotics and Automation*, 1991.

[15] I. Kao, M. Cutkosky, and R. Johansson, "Robotic stiffness control and calibration as applied to human grasping tasks," *Robotics and Automation, IEEE Transactions on*, vol. 13, no. 4, pp. 557–566, 1997.

[16] J. Friedman and T. Flash, "Task-dependent selection of grasp kinematics and stiffness in human object manipulation," *Cortex*, no. 43, pp. 444–460, 2007.

[17] A. Bicchi, "On the closure properties of robotic grasping," *The International Journal of Robotics Research*, vol. 14, no. 4, pp. 319–334, August 1995.

[18] M. Buss, H. Hashimoto, and J. B. Moore, "Dextrous hand grasping force optimization," *IEEE Trans. on Robotics and Automation*, vol. 12, no. 3, pp. 406–418, June 1996.

[19] M. Grant and S. Boyd, "Cvx: Matlab software for disciplined convex programming (web page and software)," http://stanford.edu/~boyd/cvx, 2004.

[20] ——, "Graph implementations for nonsmooth convex programs," in *Recent Advances in Learning and Control*, 2008, pp. 95–110.

[21] K. C. Toh, M. Todd, and R. Tutuncu, "Sdpt3 — a matlab software package for semidefinite programming," *Optimization Methods and Software*, vol. 1, pp. 545–581, 1999.

[22] R. M. Murray, Z. Li, and S. Sastry, *A Mathematical Introduction to Robotic Manipulation.* Boca Raton, FL: CRC Press, 1994.

Dynamic Constraint-based Optimal Shape Trajectory Planner for Shape-Accelerated Underactuated Balancing Systems

Umashankar Nagarajan
The Robotics Institute
Carnegie Mellon University
Pittsburgh, PA 15213
Email: umashankar@cmu.edu

Abstract— **This paper presents an optimal shape trajectory planner for *shape-accelerated underactuated balancing systems*, which are destabilized by gravitational forces. These systems have unactuated shape variables and fully actuated external variables. They also have the same number of actuated and unactuated degrees of freedom. Their equations of motion result in nonholonomic acceleration/dynamic constraints, which relate the acceleration of external variables to the position, velocity and acceleration of shape variables. This paper describes a procedure to use the dynamic constraints for planning shape trajectories, which when tracked will result in optimal tracking of desired external configuration trajectories. Examples of planned optimal shape trajectories for the 3D ballbot system, which is a 3D omnidirectional wheeled inverted pendulum, are also presented.**

I. Introduction

Underactuated mechanical systems are systems with fewer independent control inputs than the degrees of freedom [1]. Examples of underactuated systems can be found in robotic, aerospace, marine and locomotive applications. These systems are underactuated due to reasons like higher-order dynamics (*e.g.*, spacecraft, helicopter, underwater vehicles, flexible robots), actuator failures (*e.g.*, a failed manipulator joint, aircraft engine failure), and need for dynamic locomotion platforms (*e.g.*, Segway [2], ballbots [3]). A set of intentionally designed underactuated systems used in controls research include Acrobot, Pendubot, Cart-Pole, Beam-Ball and Rotating Pendulum [4].

In robotics, balancing (dynamically stable) mobile robots form a special class of underactuated systems. They include wheeled robots like Segway [2], ballbots [3] and legged robots like BigDog [5], MABEL [6]. Balancing robots will play a vital role in realizing the dream of placing robot workers in human environments. Unlike statically stable mobile robots, balancing (dynamically stable) mobile robots can: (*i*) be tall and skinny with high centers of gravity, (*ii*) have smaller footprints, and (*iii*) accelerate or decelerate quickly [3]. They can also be very effective mobile manipulators [7] with the ability to maintain postural stability, generate forces on external objects and withstand greater impact forces. These characteristics make them ideal for navigation and operation in cluttered human environments. Two-wheeled balancing robots, like Segway, have kinematic constraints restricting their direction of motion, whereas, single spherical-wheeled balancing robots, like ballbots, have omnidirectional motion that make them more suitable for operation in constrained spaces. In this paper, such omnidirectional balancing systems are referred to as *shape-accelerated underactuated balancing systems* (described in Sec. II-B).

An interesting and troubling factor in control and planning of such underactuated systems is the constraint on their dynamics by virtue of underactuation. These constraints are second-order nonholonomic [8] constraints, *i.e.*, non-integrable acceleration/dynamic constraints. These constraints restrict the family of trajectories that the configurations can follow. In control of balancing underactuated systems, which are destabilized by gravitational forces, it is important to maintain balance. Maintaining balance makes it difficult to accurately track any desired configuration trajectories.

In this paper, we propose a planning procedure to ensure good approximate tracking of desired trajectories for such underactuated balancing systems. For example, consider a 3D omnidirectional wheeled inverted pendulum robot like the ballbot [9]. We would like it to track some path on the floor, which could be a desired path from some global motion planning algorithm. The question to be answered here is: *how can the wheel/ball be moved along this path while balancing the pendulum?* The answer to this question is not trivial because of the dynamic coupling between the motion of the ball and the pendulum. Now, consider a balancing controller, as in [9], which ensures accurate tracking of desired lean trajectories for the pendulum, then the question, stated above, can be reformulated as: *how can the lean trajectories of the pendulum be chosen so that the ball moves along the desired path?* This is the question that is addressed in this paper.

In order to answer the above question, we should look into the nonlinear dynamics of the system. Given an accurate model, there have been a variety of controllers designed in nonlinear control literature, which use partial feedback linearization [10], to ensure approximate output tracking [11], [12]. But in real robots, we have to deal with model uncertainties, higher-order dynamics, nonlinear friction effects and disturbances. Controllers that cancel out nonlinear terms, assuming their accurate knowledge, can fail miserably on real robots. Though the model in-hand may not be accurate, it can often provide some useful information such as the dynamic constraints. In [13], an offline trajectory planning procedure that produces a class of parametric trajectories for the unactuated shape variables to reach static desired configurations using the dynamic constraint was presented. This planning procedure combined with 2-DOF control, an inner balancing control loop and an outer tracking control loop [9], proved to be very successful on the real robot. But, there are a few drawbacks with this approach: (i) it was not easily generalizable to arbitrary motions; (ii) it was computationally expensive; and (iii) it was an offline procedure.

In this paper, we present a generalized approach to generate shape trajectories ($e.g.$, lean angle trajectories in the ballbot), exploiting the structure of dynamic constraints for shape-accelerated underactuated balancing systems, which when tracked will result in optimal tracking of the desired motion. The advantages of this planning procedure are: (i) it can track any desired trajectories satisfying certain conditions (discussed in Sec. III-B); (ii) it is computationally less expensive; and (iii) it is fast and can be performed online on the real robot.

This paper is organized as follows: Sec. II presents the properties of shape-accelerated underactuated balancing systems; Sec. III discusses the special structure of the dynamic constraint for such systems and explains the shape trajectory planning procedure that ensures optimal tracking of desired external configuration trajectories; Sec. IV presents the results of the optimal shape trajectory planner for a 3D ballbot model; and finally, Sec. VI presents the conclusion.

II. Underactuated Mechanical Systems

The forced Euler-Lagrange equations of motion for a mechanical system are:

$$\frac{d}{dt}\frac{\partial \mathscr{L}}{\partial \dot{q}} - \frac{\partial \mathscr{L}}{\partial q} = F(q)\tau, \tag{1}$$

where, $q \in \mathbb{R}^n$ is the configuration vector, $\mathscr{L}(q,\dot{q}) = K(q,\dot{q}) - V(q)$ is the Lagrangian with kinetic energy K and potential energy V, $\tau \in \mathbb{R}^m$ is the control input and $F(q) \in \mathbb{R}^{n\times m}$ is the force matrix.

A mechanical system satisfying Eq. 1 is said to be an *underactuated system* [1] if $m < n$, $i.e.$, there are fewer independent control inputs than configuration variables.

Eq. 1 for an underactuated system can be written in matrix form as follows:

$$M(q)\ddot{q} + C(q,\dot{q})\dot{q} + G(q) = F(q)\tau, \tag{2}$$

where, $M(q) \in \mathbb{R}^{n\times n}$ is the inertia matrix, $C(q,\dot{q}) \in \mathbb{R}^{n\times n}$ is the matrix of Coriolis and centrifugal terms and $G(q) \in \mathbb{R}^{n\times 1}$ is the vector of gravitational forces.

A. External and Shape Variables

The configuration variables that appear in the inertia matrix are called *shape variables* (q_s), whereas, the configuration variables that do not appear in the inertia matrix are called *external variables* (q_x), $i.e.$, $\partial M(q)/\partial q_x = 0$.

Since the inertia matrix is independent of the external variables, the kinetic energy $K(q,\dot{q}) = \frac{1}{2}\dot{q}^T M(q)\dot{q}$ is also independent of the external variables, $i.e.$, $\partial K(q,\dot{q})/\partial q_x = 0$. In this case, the Lagrangian system is said to have kinetic symmetry [4].

Eq. 2 can be re-written as:

$$\begin{bmatrix} M_{xx}(q_s) & M_{xs}(q_s) \\ M_{sx}(q_s) & M_{ss}(q_s) \end{bmatrix} \begin{bmatrix} \ddot{q}_x \\ \ddot{q}_s \end{bmatrix} + \begin{bmatrix} h_x(q,\dot{q}) \\ h_s(q,\dot{q}) \end{bmatrix} = \begin{bmatrix} F_x(q) \\ F_s(q) \end{bmatrix}\tau, \tag{3}$$

where, $h(q,\dot{q}) = [h_x(q,\dot{q}), h_s(q,\dot{q})]^T$ is:

$$\begin{bmatrix} h_x(q,\dot{q}) \\ h_s(q,\dot{q}) \end{bmatrix} = \begin{bmatrix} C_{xx}(q,\dot{q}) & C_{xs}(q,\dot{q}) \\ C_{sx}(q,\dot{q}) & C_{ss}(q,\dot{q}) \end{bmatrix} \begin{bmatrix} \dot{q}_x \\ \dot{q}_s \end{bmatrix} + \begin{bmatrix} G_x(q) \\ G_s(q) \end{bmatrix}. \tag{4}$$

The underactuated systems can be classified based on whether the shape variables q_s are fully actuated, partially actuated or unactuated and based on the presence or lack of input couplings in the force matrix $F(q)$ [4].

B. Shape-Accelerated Underactuated Balancing Systems

In this paper, we are interested in shape-accelerated underactuated balancing systems, which form a special class of underactuated systems satisfying the following properties:

(i) The shape variables are unactuated and there is no input coupling, $e.g.$, $F_x(q) = I_m$ and $F_s(q) = 0$. A variety of underactuated systems can be transformed into *no input coupling* form with global change of coordinates [4].

(ii) There are equal number of actuated and unactuated variables, $i.e.$, $\dim(q_x) = \dim(q_s) = m$ and hence $n = 2m$.

In [4], underactuated systems satisfying properties (i)-(ii) are referred to as $Class-IIa$ underactuated systems.

(iii) The potential energy $V(q)$ is independent of q_x, $i.e.$, $\partial V(q)/\partial q_x = 0$. This implies that $G(q) = \partial V(q)/\partial q$ is also independent of q_x. Since both kinetic and potential energies are independent of q_x, the Lagrangian $\mathscr{L}$ is also independent of q_x, $i.e.$, $\mathscr{L}$ is symmetric w.r.t. q_x.

(iv) $M_{xx}(q_s)$ is constant, $i.e.$, $\partial M_{xx}(q_s)/\partial q_s = 0$.

(v) $M_{sx}(q_s)$ has differentially symmetric rows, $i.e.$, $\partial M_{sx}(q_s)/\partial q_s^i = \partial {M^i_{sx}}^T(q_s)/\partial q_s$, where $M^i_{sx}(q_s)$ refers to the ith row of $M_{sx}(q_s)$.

Properties (iii)-(v) make $h(q,\dot{q})$ independent of both q_x and $\dot{q}_x$.

(vi) The system has locally strong inertial coupling [1], $i.e.$, $\text{rank}(M_{sx}(q_s)) = n - m = m$ for all q in the neighborhood of the origin, where origin is the unstable equilibrium. A system satisfying the (local) strong inertial coupling is also known as *Internal/External Convertible System* [12].

(vii) $\partial G_s(q_s)/\partial q_s \neq 0$ at $q_s = 0$ and is invertible.

(viii) $\partial(M_{sx}(q_s)^{-1}G_s(q_s))/\partial q_s \neq 0$ at $q_s = 0$ and is invertible.

The significance of properties (vi)-(viii) will be explained in Sec. III, where they are used for the optimal shape trajectory planner design.

(ix) $\text{rank}(M_{ss}(q_s)) = m, i.e., M_{ss}(q_s)^{-1}$ exists.

(x) $\partial(M_{ss}(q_s)^{-1}G_s(q_s))/\partial q_s \neq 0$ at $q_s = 0$ and is invertible.

(xi) $M_{ss}(q_s)^{-1}M_{sx}(q_s) \neq 0$ at $q_s = 0$.

Properties (ix)-(xi) ensure that the Jacobian linearization (A, B) of the system (Eq. 3) at origin is controllable and its zero dynamics [10] is unstable at the origin. According to [12], an underactuated system that satisfies properties (ix)-(xi) is called a *balance system* and in this paper, we will refer to it as a *balancing system*.

The shape-accelerated underactuated balancing systems will have equations of motion of the form:

$$\begin{bmatrix} M_{xx} & M_{xs}(q_s) \\ M_{sx}(q_s) & M_{ss}(q_s) \end{bmatrix} \begin{bmatrix} \ddot{q}_x \\ \ddot{q}_s \end{bmatrix} + \begin{bmatrix} h_x(q_s, \dot{q}_s) \\ h_s(q_s, \dot{q}_s) \end{bmatrix} = \begin{bmatrix} \tau \\ 0 \end{bmatrix}, \quad (5)$$

where,

$$\begin{bmatrix} h_x(q_s, \dot{q}_s) \\ h_s(q_s, \dot{q}_s) \end{bmatrix} = \begin{bmatrix} 0 & C_{xs}(q_s, \dot{q}_s) \\ 0 & C_{ss}(q_s, \dot{q}_s) \end{bmatrix} \begin{bmatrix} \dot{q}_x \\ \dot{q}_s \end{bmatrix} + \begin{bmatrix} 0 \\ G_s(q_s) \end{bmatrix}. \quad (6)$$

We can see from Eq. 5 that the equations of motion of these systems are such that any non-zero shape configuration will result in acceleration of the external variables and in turn acceleration of the entire system, hence the name *shape-accelerated underactuated balancing systems*.

Some examples of shape-accelerated underactuated balancing systems are planar and 3D cart-pole system with unactuated lean angles, planar wheeled inverted pendulum (e.g. Segway [2] in a plane) and 3D omnidirectional wheeled inverted pendulum (e.g. the ballbot [13, 9]).

III. Dynamic Constraint-based Optimal Shape Trajectory Planner

In this paper, our objective is to plan shape trajectories, which when tracked will result in optimal tracking of desired external configuration trajectories. The planning procedure presented in this section exploits the special structure of dynamic constraints of shape-accelerated underactuated balancing systems described in Sec. II-B. So, let's first look at the dynamic constraint, its structure and the information that it provides.

A. Dynamic Constraint

The second set of m equations of motion associated with the unactuated shape variables in Eq. 5 given by

$$M_{sx}(q_s)\ddot{q}_x + M_{ss}(q_s)\ddot{q}_s + h_s(q_s, \dot{q}_s) = 0 \quad (7)$$

can be written as:

$$\Phi(q_s, \dot{q}_s, \ddot{q}_s, \ddot{q}_x) = 0. \quad (8)$$

Eq. 7 and Eq. 8 are called *second-order nonholonomic constraints*, *nonholonomic acceleration constraints*, or *dynamic constraints* because there exists no function Ψ such that $\ddot{\Psi} = \Phi(q_s, \dot{q}_s, \ddot{q}_s, \ddot{q}_x)$. The dynamic constraint equations are not even partially integrable, *i.e.*, they cannot be converted into first-order nonholonomic constraints. This is ensured by the fact that the gravitational vector $G(q_s)$ is not a constant and the inertia matrix $M(q_s)$ is dependent on the unactuated shape variables q_s. For a detailed discussion of these conditions, refer to [14].

Due to the properties listed in Sec. II-B, the dynamic constraint equations for shape-accelerated underactuated balancing systems are independent of the position and velocity of external variables. This special structure relates the acceleration of external variables to the position, velocity and acceleration of shape variables. In this section, we will attempt at understanding this relationship.

Let's consider the function

$$\begin{aligned} \Phi'(q_s, \ddot{q}_x) &= \Phi(q_s, 0, 0, \ddot{q}_x) \\ &= M_{sx}(q_s)\ddot{q}_x + h_s(q_s, 0) \\ &= M_{sx}(q_s)\ddot{q}_x + G_s(q_s). \end{aligned} \quad (9)$$

It follows from Eq. 8 that

$$\Phi'(q_s, \ddot{q}_x) = 0. \quad (10)$$

By implicit function theorem, if both $\partial\Phi'/\partial q_s$ and $\partial\Phi'/\partial\ddot{q}_x$ at $(q_s, \ddot{q}_x) = (0, 0)$ exist and are invertible, then there exists an invertible map $\Gamma : q_s \to \ddot{q}_x$ in the neighborhood of the origin such that $\Phi'(q_s, \Gamma(q_s)) = 0$ and $\Phi'(\Gamma^{-1}(\ddot{q}_x), \ddot{q}_x) = 0$. From Eq. 9, we get

$$\left.\frac{\partial\Phi'}{\partial q_s}\right|_{(q_s, \ddot{q}_x)=(0,0)} = \left.\frac{\partial G_s(q_s)}{\partial q_s}\right|_{q_s=0}, \quad (11)$$

$$\left.\frac{\partial\Phi'}{\partial\ddot{q}_x}\right|_{(q_s, \ddot{q}_x)=(0,0)} = \left.M_{sx}(q_s)\right|_{q_s=0}. \quad (12)$$

From properties (vi) and (vii) in Sec. II-B, we can see that the Jacobians in both Eq. 11 and Eq. 12 exist and are invertible and hence, there exists an invertible map $\Gamma : q_s \to \ddot{q}_x$ in the neighborhood of the origin such that $\Phi'(q_s, \Gamma(q_s)) = 0$ and $\Phi'(\Gamma^{-1}(\ddot{q}_x), \ddot{q}_x) = 0$. Γ can be derived directly from Eq. 9 and Eq. 10 as follows:

$$\begin{aligned} \ddot{q}_x &= -M_{sx}(q_s)^{-1}G_s(q_s) \\ &= \Gamma(q_s) \end{aligned} \quad (13)$$

in the neighborhood of the origin.

Jacobian linearization of Eq. 13 w.r.t. q_s at $q_s = 0$ gives

$$\begin{aligned} \left.\frac{\partial\ddot{q}_x}{\partial q_s}\right|_{q_s=0} &= -\left.\frac{\partial(M_{sx}(q_s)^{-1}G_s(q_s))}{\partial q_s}\right|_{q_s=0} \\ &= K^0_{q_s}. \end{aligned} \quad (14)$$

We know that $K^0_{q_s}$ is invertible by property (viii) in Sec. II-B. This implies, we have a linear map $K^0_{q_x} = (K^0_{q_s})^{-1}$ such that

$$q_s = K^0_{q_x}\ddot{q}_x, \quad (15)$$

and

$$\Phi'(K^0_{q_x}\ddot{q}_x, \ddot{q}_x) = 0 \quad (16)$$

in the neighborhood of the origin.

We can see that $\ddot{q}_x$ is a constant if and only if q_s is a constant. In order for the external configuration of the system in Eq. 5 to have a constant desired acceleration $\ddot{q}_x^d$, the system should stick to a constant shape configuration given by $q_s^d = K_{q_x}^0 \ddot{q}_x^d$.

When $\ddot{q}_x$ is not constant, $\dot{q}_s$ and $\ddot{q}_s$ are non-zero. Let's take $a = (q_s, \dot{q}_s, \ddot{q}_s)$ and $b = \ddot{q}_x$ and then Eq. 8 can be written as $\Phi(a, b) = 0$. Taking the Jacobian w.r.t. b at $(a, b) = (0, 0)$, we get

$$\left.\frac{\partial \Phi(a, b)}{\partial b}\right|_{(a,b)=(0,0)} = M_{sx}(q_s)\Big|_{q_s=0} \tag{17}$$

From property (vi) in Sec. II-B, we can see that Eq. 17 exists and is invertible. Hence, by implicit function theorem, there exists a map $\Gamma' : a \to b$ such that $\Phi(a, \Gamma'(a)) = 0$. The map Γ' is not invertible since $\partial\Phi(a, b)/\partial a$ at $(a, b) = (0, 0)$ exists but is not invertible.

We can see that in order to track a non-constant, time-varying $\ddot{q}_x^d(t)$, there is no function that maps $\ddot{q}_x^d(t)$ to $(q_s^d(t), \dot{q}_s^d(t), \ddot{q}_s^d(t))$ such that the dynamic constraints in Eq. 8 are satisfied. In the following subsection, we propose an optimal planner that ensures approximate tracking of $\ddot{q}_x^d(t)$.

B. Optimal Shape Trajectory Planner

In underactuated balancing systems, we are often interested in tracking desired trajectories for the external configuration variables without losing balance. Shape-accelerated underactuated balancing systems in Sec. II-B have constraints on the acceleration of these external variables w.r.t. the shape variables' position, velocity and acceleration as given below:

$$\begin{aligned} \ddot{q}_x &= -M_{sx}(q_s)^{-1}(M_{ss}(q_s)\ddot{q}_s + h_s(q_s, \dot{q}_s)) \\ &= \Gamma'(q_s, \dot{q}_s, \ddot{q}_s) \end{aligned} \tag{18}$$

So, for a desired acceleration trajectory for the external configuration $\ddot{q}_x^d(t)$, we would like to plan shape configuration trajectories $(q_s^p(t), \dot{q}_s^p(t), \ddot{q}_s^p(t))$, which when tracked will result in $\ddot{q}_x^p(t)$ (from Eq. 18) such that $\ddot{q}_x^p(t) = \ddot{q}_x^d(t)$. But, we have seen that $\Gamma'^{-1} : \ddot{q}_x(t) \to (q_s(t), \dot{q}_s(t), \ddot{q}_s(t))$ that satisfies $\Phi(\Gamma'^{-1}(\ddot{q}_x(t)), \ddot{q}_x(t)) = 0$ does not exist. Hence, we propose to find a map $\Omega : \ddot{q}_x^d(t) \to (q_s^p(t), \dot{q}_s^p(t), \ddot{q}_s^p(t))$ such that $\|\ddot{q}_x^p(t) - \ddot{q}_x^d(t)\|_2^2$ is minimized. Here, $\ddot{q}_x^p(t) = \Gamma'(q_s^p(t), \dot{q}_s^p(t), \ddot{q}_s^p(t))$.

Inspired from Eq. 15, we propose to use a linear map $K_{q_x} : \ddot{q}_x^d \to q_s^p$ such that

$$q_s^p(t) = K_{q_x} \ddot{q}_x^d(t), \tag{19}$$
$$\dot{q}_s^p(t) = K_{q_x} \dddot{q}_x^d(t), \tag{20}$$
$$\ddot{q}_s^p(t) = K_{q_x} \ddddot{q}_x^d(t). \tag{21}$$

The shape trajectory planning procedure can now be formulated as an optimization problem, where the elements of K_{q_x} are determined with the objective of minimizing the function

$$J = \|\ddot{q}_x^p(t) - \ddot{q}_x^d(t)\|_2^2, \tag{22}$$

where, $\ddot{q}_x^p(t) = \Gamma'(K_{q_x}\ddot{q}_x^d(t), K_{q_x}\dddot{q}_x^d(t), K_{q_x}\ddddot{q}_x^d(t))$. It is to be noted that the parameter space is m^2-dimensional and any optimization algorithm that solves a nonlinear least-squares problem can be used.

For a desired constant acceleration trajectory, $K_{q_x} = K_{q_x}^0$ ensures optimality. For any general $\ddot{q}_x^d(t)$, $K_{q_x} = K_{q_x}^0$ may not necessarily ensure optimality but will act as a good initial guess for the optimization process.

As one can see, the optimal shape trajectory planner described above ensures that $\ddot{q}_x^p(t)$ approximately tracks $\ddot{q}_x^d(t)$ but does not ensure that $q_x^p(t)$ approximately tracks $q_x^d(t)$ or $\dot{q}_x^p(t)$ approximately tracks $\dot{q}_x^d(t)$. This can be ensured only if the initial conditions for the external variables are met, *i.e.*, $q_x^p(0) = q_x^d(0)$ and $\dot{q}_x^p(0) = \dot{q}_x^d(0)$.

Given below are the conditions to be met in order to use the shape trajectory planner described above to track $q_x^d(t)$:

(i) $q_x^d(t)$ must at least be of *class* C^2, *i.e.*, $\dot{q}_x^d(t)$ and $\ddot{q}_x^d(t)$ exist and are continuous. If $\ddot{q}_x^d(t)$ does not exist then, the planned shaped trajectory $q_s^p(t)$ that is proportional to $\ddot{q}_x^d(t)$ will not exist as well.

(ii) $q_x^d(t)$ is preferred to be of *class* C^4, *i.e.*, first four derivatives exist and are continuous so as to ensure the existence of $q_s^p(t), \dot{q}_s^p(t), \ddot{q}_s^p(t)$ that depend on them. This condition also avoids discontinuities in the planned shape trajectories and its first two derivatives.

(iii) Initial conditions for the external variables are met, *i.e.*, $q_x^p(0) = q_x^d(0)$ and $\dot{q}_x^p(0) = \dot{q}_x^d(0)$. The desired position trajectory $q_x^d(t)$ can be tracked by approximate tracking of $\ddot{q}_x^d(t)$ only if the system starts at the correct initial position and velocity for the external variables.

IV. Example: 3D Ballbot Model

As an example, we present here the results of the optimal shape trajectory planner described in Sec. III-B for a 3D ballbot model, which is a 3D omnidirectional wheeled inverted pendulum. The ballbot (Fig. 1(a)) is modeled as a rigid cylindrical body on top of a rigid spherical wheel/ball with the following assumptions: (i) there is no slip between the ball and the floor, and (ii) there is no yaw/spinning motion for both the body and the ball, *i.e.*, they have two degrees-of-freedom each. It is to be noted that the results presented here are that of simulation and are not experimental results.

The forced Euler-Lagrange equations of motion are given by Eq. 1 with q_x corresponding to the configuration of the ball/wheel, and q_s corresponding to the configuration of the body (Euler angles). Here, $q_x = [\theta_x, \theta_y]^T$ and $q_s = [\phi_x, \phi_y]^T$. Planar versions of the ball and body configurations are shown in Fig. 1(b).

The ball configurations, θ_x and θ_y, are chosen such that the linear position of the ball/wheel (with radius r) on the XY ground plane is given by $x_w = r\theta_x$ and $y_w = r\theta_y$. This produces input coupling in $F(q)$:

$$F(q) = \begin{bmatrix} 1 & 0 \\ 0 & 1 \\ 0 & -1 \\ -1 & 0 \end{bmatrix}. \tag{23}$$

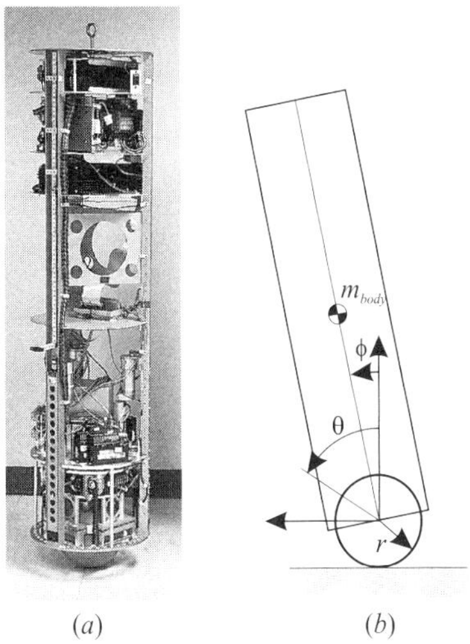

Fig. 1. (a) The ballbot balancing, (b) Planar ballbot model with ball and body configurations shown.

By global change of coordinates $\theta'_x = \theta_x - \phi_y$ and $\theta'_y = \theta_y + \phi_x$, the equations of motion with the new configuration vector $q' = [\theta'_x, \theta'_y, \phi_x, \phi_y]^T$ has no input coupling in $F'(q')$:

$$F'(q') = \begin{bmatrix} 1 & 0 \\ 0 & 1 \\ 0 & 0 \\ 0 & 0 \end{bmatrix}. \tag{24}$$

The new forced Euler-Lagrange equations are:

$$M'(q')\ddot{q}' + C'(q', \dot{q}')\dot{q}' + G'(q') = F'(q')\tau. \tag{25}$$

It is to be noted that the underactuated system in Eq. 25 satisfies all the properties of shape-accelerated underactuated systems and balancing systems listed in Sec. II-B. The expressions for M', C', G' are omitted here due to lack of space.

The last two equations of motion in Eq. 25 form the dynamic constraint equations of the system and these equations are used for the optimal shape trajectory planner in Sec. III-B. The optimization algorithm used here is Levenberg-Marquardt algorithm (LMA), which is a widely used tool for minimization problems in least-squares curve fitting and nonlinear programming. The dynamic equations were simulated in MATLAB and the optimization was implemented using MATLAB's *lsqnonlin* function. Some of the planning results are presented below.

A. Results of Optimal Shape Trajectory Planning

This section presents a variety of desired $x_w(t)$ and $y_w(t)$ that satisfy the conditions in Sec. III-B and the corresponding planned optimal shape trajectories, $\phi^p_x(t)$ and $\phi^p_y(t)$, which should be tracked to achieve them. It is important to note that the desired trajectories, $x_w(t)$ and $y_w(t)$, are trajectories of the center of the ball and not trajectories of the system's center of gravity. On a flat floor, these trajectories match the trajectories of the ball's contact point with the floor.

1) Straight Line Motion: Let's start with the simplest of trajectories that involve moving along a straight line between static configurations, *i.e.*, starting from rest at one point on the floor and coming to rest at another point on the floor. The desired $x_w(t)$ and $y_w(t)$ are chosen to be nonic (9th degree) polynomials in t so that their first four derivatives satisfy the boundary conditions.

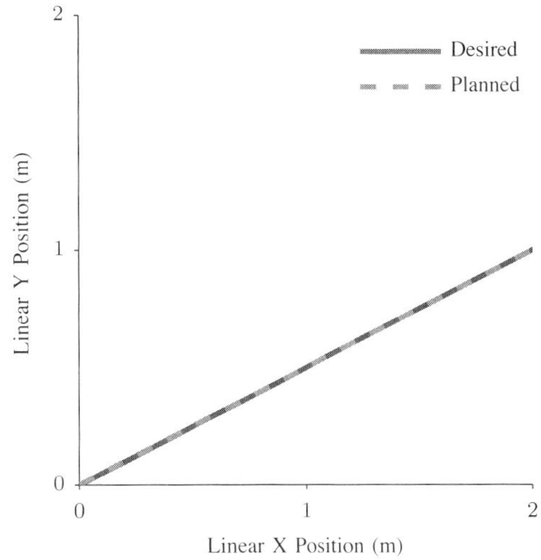

Fig. 2. Straight Line Motion - Linear XY (This figure is best viewed in color.)

$$x_w(t) = \sum_{i=0}^{9} a_i t^i,$$
$$y_w(t) = \sum_{i=0}^{9} b_i t^i, \tag{26}$$

where the coefficients $a_i s$ and $b_i s$ are determined based on the initial and final desired configurations.

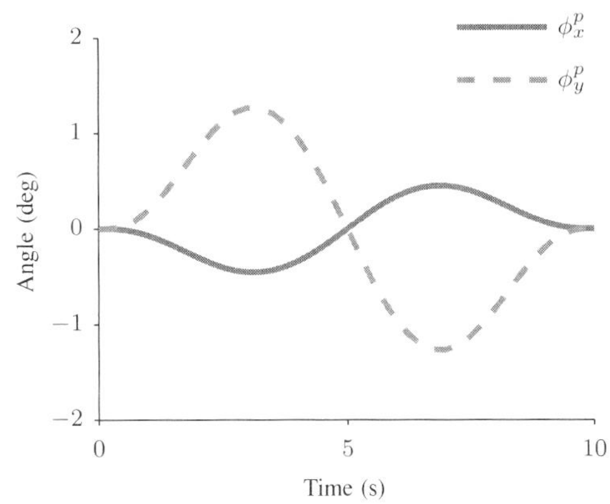

Fig. 3. Straight Line Motion - Planned Shape Trajectories

In Fig. 2, we can see the planned XY motion approximately tracking the desired XY motion. The planned shape trajectories that produce such a motion are shown in Fig. 3. The trajectories for the planned XY motion are obtained from Eq. 18. The tracking error statistics are: RMSE = 0.0147 m and Maximum Error = 0.0238 m.

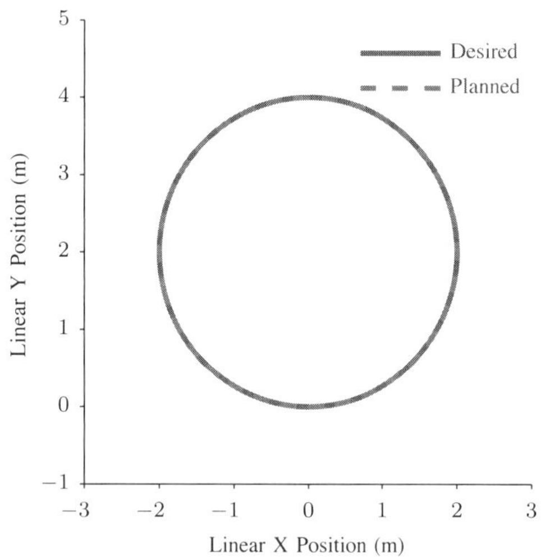

Fig. 4. Circular Motion - Linear XY (This figure is best viewed in color.)

2) Circular Motion: Here, we would like the robot to move along a circular path on the floor. The desired $x_w(t)$ and $y_w(t)$ are:

$$x_w(t) = R\sin(\omega t),$$
$$y_w(t) = R(1 - \cos(\omega t)), \tag{27}$$

where, $R = 2$ m and $\omega = \pi/10$ rad/s.

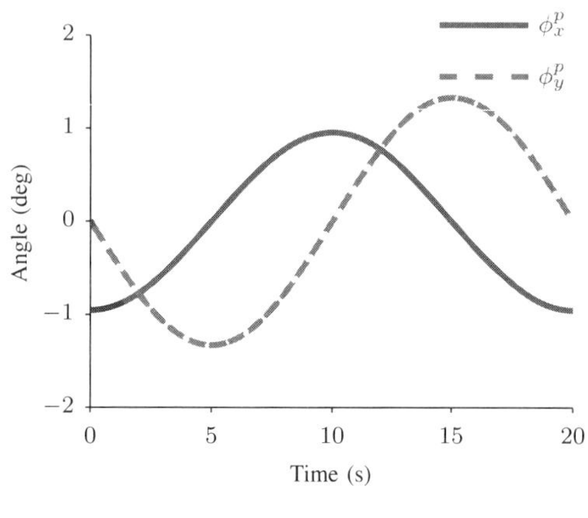

Fig. 5. Circular Motion - Planned Shape Trajectories

In Fig. 4, we can see the planned XY motion approximately tracking the desired XY motion. The planned shape trajectories that produce such a motion are shown in Fig. 5. The trajectories for the planned XY motion are obtained from Eq. 18. It is to be noted that the results presented here are observed only when the appropriate initial conditions are met. The tracking error statistics are: RMSE = 4.3x10^{-4} m and Maximum Error = 5.6x10^{-4} m.

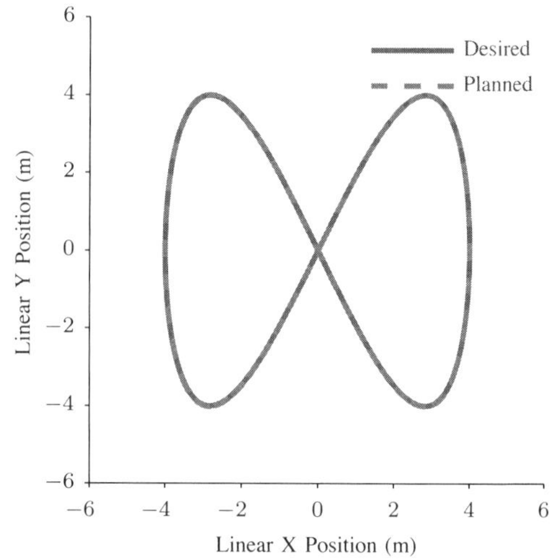

Fig. 6. Figure-8 Motion - Linear XY (This figure is best viewed in color.)

3) Figure-8 Path: Here, we would like the robot to move along a figure-8 path on the floor. The desired $x_w(t)$ and $y_w(t)$ are:

$$x_w(t) = A_x\sin(\omega_x t),$$
$$y_w(t) = A_y\sin(\omega_y t), \tag{28}$$

where, $A_x = 4$ m, $\omega_x = \pi/20$ rad/s, $A_y = 4$ m, and $\omega_y = \pi/10$ rad/s.

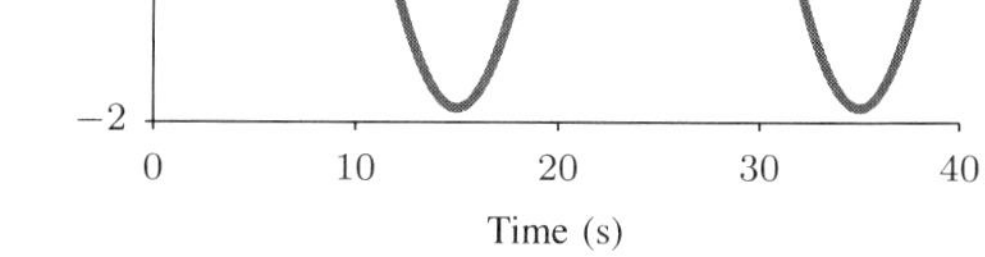

Fig. 7. Figure-8 Motion - Planned Shape Trajectories

In Fig. 6, we can see the planned XY motion approximately tracking the desired XY motion. The planned shape trajectories that produce such a motion are shown in Fig. 7. It is to be noted that the results presented here are observed only when the appropriate initial conditions are met. The trajectories for the planned XY motion are obtained from Eq. 18. The tracking error statistics are: RMSE = 0.0105 m and Maximum Error = 0.0177 m.

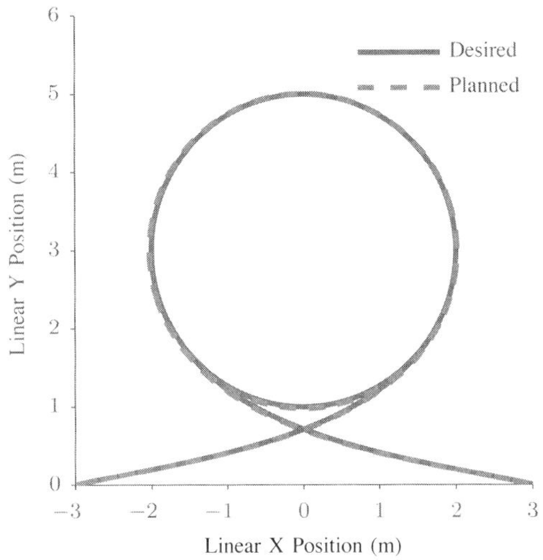

Fig. 8. Join-Circle-Leave Motion - Linear XY (This figure is best viewed in color.)

4) Join-Circle-Leave Motion: The tracking of a circle, sine wave and figure-8 paths require that the appropriate initial conditions are met. But in reality, we would like to start at any random initial configuration and track a particular path on the floor. Here, we present the result of starting from rest, joining a circular path at a desired configuration, making 1.5 revolutions, leaving the circular path at a diametrically opposite point and coming to rest. This motion consists of 3 different trajectories fused together. The *joining* and *leaving* trajectories are nonic (9th degree) polynomials that satisfy the initial and final configurations required to join and leave the circular path respectively. The nonic polynomials ensure that the first four derivatives of the fused trajectories are continuous and not just piecewise continuous.

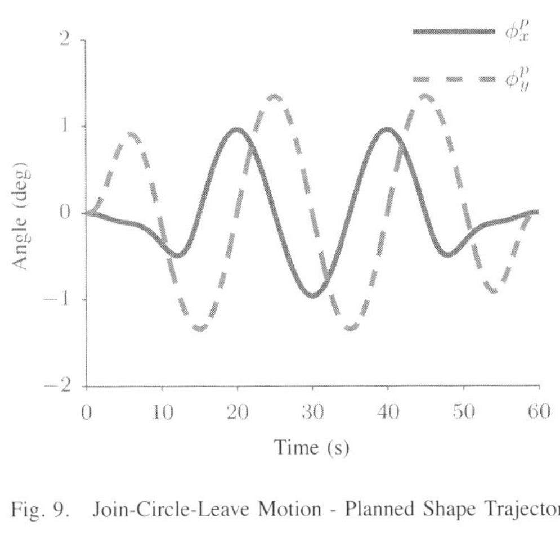

Fig. 9. Join-Circle-Leave Motion - Planned Shape Trajectories

In Fig. 8, we can see the planned XY motion approximately tracking the desired XY motion. The planned shape trajectories that produce such a motion are shown in Fig. 9. The trajectories for the planned XY motion are obtained from Eq. 18. The tracking error statistics are: RMSE = 0.0167 m and Maximum Error = 0.0286 m.

B. Real-Time Planning

For all the results presented in Sec. IV-A, the optimization tolerance values for both the residual norm and parameter values were set to $< 10^{-4}$. On a standard Intel Core-2 Duo processor, the optimization implementation in MATLAB converges in < 1 second. This ensures that the optimal shape trajectory planner presented in this paper can be used for real-time planning on the robot.

V. Balancing and Tracking Control

The entire planning procedure presented in this paper assumes that there exists a balancing controller, similar to the one in [9], which has good shape trajectory tracking performance. Given the balancing controller and the optimal shape trajectory planner, we can achieve good approximate tracking of the desired external configuration trajectories. But it is to be noted that this tracking is open-loop and with wrong initial conditions, we have no way to ensure approximate tracking of the desired external configuration trajectories. Moreover, while testing on the real robot, modeling uncertainties, unmodeled dynamics, nonlinear friction effects and noise may prevent good approximate tracking. Hence, we should have an external trajectory tracking controller, similar to the one in [13], to ensure better tracking.

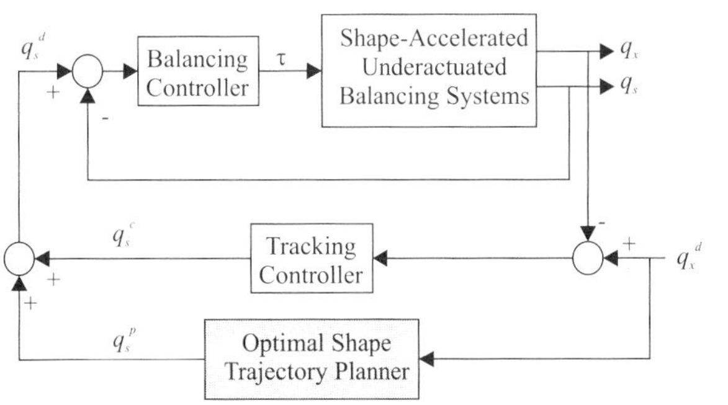

Fig. 10. Control Architecture

We propose to follow [13] in using the control architecture shown in Fig. 10 for good approximate tracking of desired external configuration trajectories (q_x^d) on the real robot. The balancing controller tracks the desired shape trajectories (q_s^d), which are a sum of planned (q_s^p) and compensation (q_s^c) shape trajectories. The planned shape trajectories are given by the optimal shape trajectory planner, whereas, the compensation shape trajectories are provided by the tracking controller, which tries to compensate for the deviation of the external configuration trajectories from the desired ones.

In Fig. 11, we show the result of the control architecture in Fig. 10 tracking the external configuration trajectories corresponding to a circular motion starting with zero initial conditions, *i.e.*, there is no initial linear velocity in the x direction, as in Fig. 4. We can see that the motion asymptotically converges to the circular motion, which shows the effectiveness of the tracking controller in compensating for the errors due to wrong initial conditions.

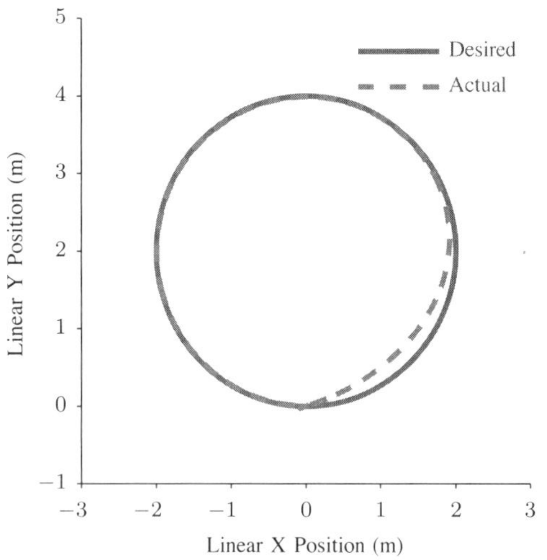

Fig. 11. Tracking Circular Motion with Zero Initial Conditions (This figure is best viewed in color.)

The contribution of this paper is the optimal shape trajectory planner, whereas, the balancing and tracking controller designs are borrowed from our previous work [13]. In [13], we have shown that the control architecture in Fig. 10 is capable of successfully tracking straight line motions on the real robot. Based on these previous successful experimental results, we strongly believe that this control architecture will enable the real robot to track the motions presented in Sec. IV-A. It is to be noted that the planner in [13] was designed based on our intuition of how the robot should lean in order to achieve a straight line motion, whereas, in this paper, we have presented a generalized shape trajectory planner that will enable the ballbot to approximately track any desired external configuration trajectory.

VI. Conclusion

An algorithm to plan optimal shape trajectories for shape-accelerated underactuated balancing systems was presented. These shape trajectories, which when tracked will result in approximate tracking of the desired external configuration trajectories subject to conditions on differentiability of the desired trajectory and correct initial conditions. The planning procedure presented assumes that there exists a balancing controller that can accurately track desired shape trajectories. A feedback trajectory tracking controller used in combination with the optimal shape planner can ensure better tracking when the initial conditions are not met.

In the 3D ballbot example, we have seen that given $(x_w^d(t), y_w^d(t)) \in C^4$, we can use the optimal shape trajectory planner to obtain the desired shape trajectories, which when tracked will result in the desired motion. In a general motion planning setting, any standard path planning algorithm like A^* can be used to obtain a desired $(x_w(s), y_w(s))$ path. In order to use the optimal shape trajectory planner discussed in this paper as a tool for overall motion planning, we have to parametrize this path in time. We propose that using *nonic splines*, which ensure that the first four derivatives for the overall trajectory will be continuous throughout its domain, will help us achieve this goal. Future work will include experimental testing of the planned shape trajectories presented here and design of a global motion planner that provides desired external configuration trajectories for the optimal shape trajectory planner.

VII. Acknowledgements

This work was supported in part by NSF grants IIS-0308067 and IIS-0535183. The author thanks Ralph Hollis, George Kantor, Howie Choset and RSS reviewers for their valuable comments that helped in improving this paper.

References

[1] M. W. Spong, "The control of underactuated mechanical systems," in *First International Conference on Mechatronics*, Mexico City, 1994.

[2] H. G. Nguyen, J. Morrell, K. Mullens, A. Burmeister, S. Miles, N. Farrington, K. Thomas, and D. Gage, "Segway robotic mobility platform," in *SPIE Proc. 5609: Mobile Robots XVII*, Philadelphia, PA, October 2004.

[3] R. Hollis, "Ballbots," *Scientific American*, pp. 72–78, October 2006.

[4] R. Olfati-Saber, "Nonlinear control of underactuated mechanical systems with application to robotics and aerospace vehicles," Ph.D. dissertation, Massachusetts Institute of Technology, February 2001.

[5] M. Raibert, K. Blankespoor, G. Nelson, R. Playter, and the BigDog Team, "Bigdog, the rough-terrain quadraped robot," in *In Proc. 17th World Congress of the International Federation of Automatic Control*, 2008, pp. 10 822–10 825.

[6] J. Grizzle, J. Hurst, B. Morris, H.-W. Park, and K. Sreenath, "Mabel, a new robotic bipedal walker and runner," in *American Control Conference*, St. Louis, MO, June 2009.

[7] P. Deegan, B. Thibodeau, and R. Grupen, "Designing a self-stabilizing robot for dynamic mobile manipulation," in *Robotics: Science and Systems - Workshop on Manipulation for Human Environments*, August 2006.

[8] J. R. Ray, "Nonholonomic constraints," *American Journal of Physics*, vol. 34, pp. 406–408, 1966.

[9] U. Nagarajan, A. Mampetta, G. Kantor, and R. Hollis, "State transition, balancing, station keeping, and yaw control for a dynamically stable single spherical wheel mobile robot," *Proc. IEEE Int'l. Conf. on Robotics and Automation*, May 2009.

[10] A. Isidori, *Nonlinear Control Systems*. Berlin: Springer-Verlag, 1989.

[11] A. Isidori and C. I. Byrnes, "Output regulation of nonlinear systems," *IEEE Transactions on Automatic Control*, vol. 35, no. 2, pp. 131–140, 1990.

[12] N. H. Getz, "Dynamic inversion of nonlinear maps with applications to nonlinear control and robotics," Ph.D. dissertation, University of California at Berkeley, December 1995.

[13] U. Nagarajan, G. Kantor, and R. Hollis, "Trajectory planning and control of an underactuated dynamically stable single spherical wheeled mobile robot," *Proc. IEEE Int'l. Conf. on Robotics and Automation*, May 2009.

[14] G. Oriolo and Y. Nakamura, "Control of mechanical systems with second-order nonholonomic constraints: Underactuated manipulators," in *In Proc. 30th IEEE Conference on Decision and Control*, vol. 3, 1991, pp. 2398–2403.

Design and Optimization Strategies for Muscle-like Direct Drive Linear Permanent Magnet Motors

Bryan P. Ruddy and Ian W. Hunter
BioInstrumentation Laboratory
Depatment of Mechanical Engineering
Massachusetts Institute of Technology
Cambridge, Massachusetts 02139–4301
Email: {ruddy, ihunter}@mit.edu

Abstract—**We report a new approach to the design of direct-drive linear permanent magnet motors for use in general-purpose robotic actuation, with particular attention to applications in bird-scale flapping-wing robots. We show a simple, quantitative analytical modeling framework for this class of actuators, and demonstrate inherent scaling properties that allow the production of motors with force densities and efficiencies comparable to those of biological muscles. We will show how this model leads to a set of practical design specifications for muscle-like motors, and examine the resulting trade-off between thermal management and motor fabrication complexity.**

I. Introduction

Actuator performance is one of the main technological shortcomings preventing the development of successful general-purpose robots. While mainstream electric, pneumatic, and hydraulic actuators are capable of power densities, force densities, and efficiencies far in excess of those of natural actuators (i.e. muscles), they rarely can achieve all of these properties simultaneously or during transients [1], [2]. Furthermore, traditional actuators work best at size and force scales well in excess of those associated with the human body, rendering most industrial robots unsafe and too large for human interaction [3]. As a result, bioinspired robots lag well behind in performance compared to the organisms inspiring them, and haptic interfaces tend to be weak compared with their human operators.

Biological muscle has several properties that make it particularly difficult to mimic with conventional actuators [4]. Muscle has a variable stiffness, allowing it to passively stabilize a joint, and is generally associated with compliant tendons capable of efficient mechanical energy storage. It has very low inertia, yet it can generate a very high force, sustained across a displacement of up to 40% of the muscle's length. Muscle does this while remaining relatively efficient and powerful, with efficiency during cyclic motion estimated between 10% and 40% [5] and power densities near 100 W/kg.

A common approach to the problem of making robotic actuators behave more like biological muscle has been the development of artificial muscle materials, capable of undergoing shape change in response to electrical, thermal, or chemical stimulus [6]. While some of these materials have exciting long-term prospects, at present they are only suitable for niche applications, and fundamental progress in materials science will be required to improve their performance to compete with natural muscle or even with conventional actuators.

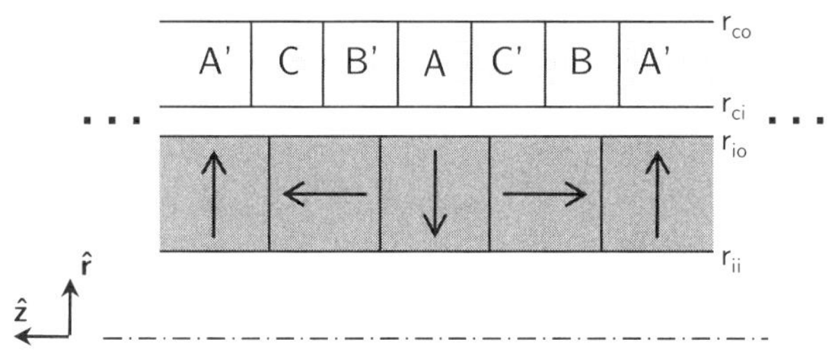

Fig. 1. The tubular motor configuration considered in this work is shown here, denoted the "HG" configuration. Nearest the center (indicated by a dot-dashed line) is a quasi-Halbach magnet array, in which blocks with arrows are magnets, with the arrows indicating the direction of magnetization. The axial magnets in the array serve to provide closed flux paths, enhancing the field produced. Outside the magnets lies a coil, here shown arranged in a typical three-phase winding (A, B, and C denote phases, with negative current in the primed windings). This magnet and coil structure is repeated periodically along the z-axis to form the motor. The radial dimensions in the model (r_{ii}, r_{io}, r_{ci}, and r_{co}) are as labeled.

Since artificial muscles are not yet able to improve upon conventional actuators, how can conventional actuators be improved to better compete with muscle? For the purposes of this work, we will focus on the potential development of electromagnetic (EM) actuators. In particular, the simplicity of direct-drive EM actuators is appealing as a canvas for improvements, and we will specifically discuss linear permanent-magnet (PM) direct drive actuators. Permanent magnet motors have favorable scaling properties [7], as do direct-drive linear actuators. While direct-drive motors are generally known for having low force densities and low efficiencies [2], their performance envelope is determined by their electromagnetic and thermal design, which can be modeled from basic physics, rather than by the tribological properties of gears, which are both difficult to model and difficult to improve.

At present, general models for the maximum theoretical performance of linear PM motors in terms of fundamental physics are unavailable [7], [8]; our objective will therefore be to develop such a general model, and to use it to find ways to improve the force capability and efficiency of these motors to a level competitive with biological muscle. Using our model, we will establish that direct-drive linear PM motors can be constructed with power densities, force densities, and efficiencies all simultaneously exceeding those of biological muscle. These improved properties can be achieved through a combination of active cooling, parts miniaturization, and careful electromagnetic design. We will show that the primary physical mechanism allowing this improved performance is the enhanced current density allowed by miniaturization and cooling, well in excess of that allowed by conventional motor design.

We will apply this model to determine the feasibility of direct-drive linear PM motors in a specific new application, as the wing actuators in bird-scale flapping-wing flying robots. Specifically, we will show the potential of such motors to sinusoidally drive a load of 1000 N/kg at a peak velocity of 1 m/s while retaining efficiencies above 40%, as compared to the efficiency of just 10% to 20% for bird muscles under similar loading conditions [9], [10] .

II. Fundamental Motor Physics

In essence, we can consider a PM motor as an active material actuator in which a body force is produced on a conductor (the "active material") proportional to the surrounding magnetic flux density, B, and to the applied current density, J. For this basic model, we will assume that the magnetic field is produced "for free", consider it as a material property, and neglect the magnets necessary to produce it in a real motor. We will likewise ignore any practical restrictions on the provision of current, and assume that the conductor can be excited by a spatially uniform current density. Thus, the 0-D motor model consists of a bare conductor carrying a uniform current density immersed in a uniform magnetic field that is exactly perpendicular to the current. While this model can describe rotary and linear motors equally well, we will restrict our analysis to that of linear motors.

The force production and power consumption of a PM actuator, its basic input-output relations, are given by the Lorentz force law and by Ohm's law,

$$F = BJV \text{ and} \tag{1}$$

$$P = \frac{J^2 V}{\sigma}, \tag{2}$$

where F is the force produced by the motor, V is the volume of active material, P is the electrical power converted by the motor into heat, and σ is the conductivity of the active material. Here, we do not consider superconducting motors, which have zero Ohmic losses but instead have other power loss mechanisms and require energy-hungry cryogenics. Both of these quantities are directly proportional to the amount of active material—depending on the application, it may be preferable to describe this amount in terms of volume or in terms of mass. For this analysis, we will consider mass-sensitive applications; an alternative formulation in terms of volume can be straightforwardly derived.

We would like to describe motor performance in terms of parameters that do not couple motor size and drive current. One such parameter is the ratio of the force generated to the square root of the power consumed ϵ, commonly known as the motor constant and listed on motor datasheets as K_m. The force generated per unit mass F', which we will call the force density, is another such parameter:

$$F' = \frac{BJ}{\rho_c}, \tag{3}$$

$$\epsilon \equiv \frac{F}{\sqrt{P}} = B\sqrt{\frac{\sigma M}{\rho_c}}. \tag{4}$$

Here, ρ_c is the mass density of the active material and M is the total mass of active material. Note that each performance parameter is independently related to a single motor design parameter, in addition to the material properties (B, σ, and ρ_c): the force density achievable depends on the maximum allowable current density, while the motor constant depends on the size of the motor. The maximum current density, in turn, is restricted either by the maximum amount of heat that can be removed from and tolerated by the motor or by non-thermal damage mechanisms (demagnetization and electromigration).

We can, in turn, relate these performance parameters to the energy efficiency of a linear PM motor operating on a given work loop. Consider perhaps the simplest situation, in which an actuator of total stroke ℓ_s sinusoidally drives an ideal dashpot at a frequency ω, with a resulting peak force amplitude of F_0. In this case, the instantaneous mechanical power P_m and electrical power P_e are given by

$$P_m = \frac{\omega \ell_s F_0}{2} \cos^2 \omega t, \tag{5a}$$

$$P_e = P_m + \frac{F_0^2}{\epsilon^2} \cos^2 \omega t. \tag{5b}$$

We can then integrate over one period in time to find the mechanical and electrical work per cycle W_m and W_e, respectively, calculate the efficiency $\eta = W_m/W_e$, and define a dimensionless efficiency parameter $\hat{\eta}$

$$\eta = \frac{1}{1+\hat{\eta}}, \tag{6a}$$

$$\hat{\eta} \equiv \frac{2F_0}{\omega \ell_s \epsilon^2} = \underbrace{\frac{\rho_c}{\sigma B^2}}_{\text{materials}} \times \underbrace{\frac{2}{\omega \ell_s}}_{\text{velocity}} \times \underbrace{F_0'}_{\substack{\text{force}\\ \text{density}}} \tag{6b}$$

Here, we have broken out the components of ϵ to explain the scaling of $\hat{\eta}$, and F_0' denotes the peak force amplitude F_0 divided by the actuator mass M. Three factors are seen to contribute to the efficiency parameter: the first is related to the material properties of the actuator, the second is equal to the inverse of the peak actuator velocity, and the third is equal to the peak force density required by the work loop.

At first glance, the expression for $\hat{\eta}$ in Equation 6b is disappointing, as it links an increase in force density directly to a decrease in efficiency. However, the efficiency remains high for even relatively large values of force density if copper and the best available magnets (1.47 T remanence NdFeB) are used. For reference, it is estimated [10] that bird flight is less than 20% efficient. Using work loop parameters similar to those seen by pigeon flight muscles [9] ($2/\omega L = 1$ m/s, $F_0' = 1000$ N/kg) we find $\hat{\eta} = 7.1 \times 10^{-2}$, or that the efficiency is 93%. Thus, the actuator properties of copper wire in a magnetic field are sufficient to achieve very high efficiency even when operating under the relatively slow, heavily-loaded conditions of flapping-wing flight. While we anticipate that the capability of a real motor will be greatly reduced due to the need for magnets and spatially-varying fields, there is a considerable performance margin available.

III. Actuator Analytical Modeling and Optimization

A. Tubular Motor Electromagnetic Model

In order to model a real motor, we need to account for the magnetic field that can be produced by a given set of magnets, rather than assume a field provided by the environment. Fortunately for this process, modern rare-earth magnet materials behave linearly within their performance envelope [11], and so their fields can be computed from Poisson's equation. For maximum flexibility and ease of computation, we will consider linear motors with periodic magnet and coil structures, referred to as permanent magnet synchronous motors in the literature; the behavior of aperiodic motors such as voice coils can be derived from that of periodic motors by accounting for end effects [12], [13].

The magnets and coils in a linear motor can be arranged in many different ways [14], each amenable to a different set of analytical techniques. Here, we will consider a tubular motor configuration due to its efficient copper usage and intrinsic symmetry. Computation of the magnetic fields and resulting performance of tubular motors has been performed in many different ways [15]–[17], but the solutions reported in the literature tend to have numerical stability problems, and are presented in a manner that can make direct comparison between different configurations and different scales difficult.

In this work, we chose to directly solve Poisson's equation in cylindrical coordinates for several different motor topologies. Here we will present the analysis for a motor comprising a quasi-Halbach magnet array arranged to project a magnetic field radially outwards, surrounded by a layer of coils, as shown in Figure 1 and referred to henceforth as the "HG" configuration (Halbach-Gap). A Halbach array [11], [18] is chosen to produce closed magnetic flux paths, and thus strong fields, without using iron. An ideal Halbach array has a continuously varying direction of magnetization, which is very difficult to fabricate, but a segmented array (quasi-Halbach) reaches 90% of the ideal performance with only four uniform magnets per period [18]. While the HG configuration is relatively ineffective, we consider it here due to the relative brevity of its field model, and easy extension to a more-efficient structure with Halbach arrays on both sides of the air gap. Results for other magnet configurations, including configurations with iron components and completely closed flux paths, will be presented in a future publication.

The magnetic field can be determined by first breaking down the magnetization into a Fourier series:

$$\boldsymbol{M} = \sum_{n \text{ odd}} \left[M_{rn} \cos(nkz)\, \hat{\boldsymbol{r}} + M_{zn} \sin(nkz)\, \hat{\boldsymbol{z}} \right], \tag{7a}$$

$$M_{rn} = \frac{4B_r}{n\pi\mu_0} \sin\left(\frac{n\pi\delta}{2}\right), \tag{7b}$$

$$M_{zn} = -\frac{4B_r}{n\pi\mu_0} \cos\left(\frac{n\pi\delta}{2}\right). \tag{7c}$$

In these equations, $\boldsymbol{M}$ is the magnetization vector, M_{rn} and M_{zn} are the radial and axial components of the magnetization, respectively, n is the harmonic order, k is the wavenumber of the magnet array (equal to 2π divided by the pole pitch p), B_r is the remanence of the permanent magnet material, and δ is the fractional width of the radial magnets in the array. Note that the magnetization, and thus the overall magnetic field, has only odd harmonics. The current density in the coil region can similarly be described in terms of a (scalar) Fourier series.

With the magnetization and current density described thusly, we can proceed to solve the appropriate form of Poisson's equation in the magnet, coil, and air regions of the motor:

$$\nabla^2 \boldsymbol{A}_{\text{magnet}} = -\mu_0 \nabla \times \boldsymbol{M}, \tag{8a}$$

$$\nabla^2 \boldsymbol{A}_{\text{coil}} = -\mu_0 J_n \cos(nkz)\, \hat{\boldsymbol{\theta}}, \tag{8b}$$

$$\nabla^2 \boldsymbol{A}_{\text{air}} = \boldsymbol{0}, \tag{8c}$$

where $\boldsymbol{A}$ is the magnetic vector potential and J_n is the Fourier series coefficient of the current density in the circumferential direction. Here, we have assumed that both the magnet and the conductor have unity relative permeability. The solution in the air regions is simply a linear combination of modified Bessel functions, but solution of the equations in the magnet and coil regions leads to inhomogeneous Bessel equations,

$$\frac{\partial^2 A}{\partial r} + \frac{1}{r}\frac{\partial A}{\partial r} - \left(c^2 + \frac{1}{r^2}\right) A(r) = d, \tag{9}$$

where A is a function of the radial coordinate only and c and d are constants. These equations are often solved in the literature using Green's functions [15] or by approximating the radial magnets' magnetization as being nonuniform in order to give a solution in terms of elementary functions [16]. However, there exists a set of named functions that solve this equation, the modified Struve functions [17], [19]. So long as care is taken in the formulation of the solutions to avoid numerical stability problems, the modified Struve functions can be efficiently calculated using the power series and asymptotic series presented in [19], [20].

Using the modified Struve functions L_0 and L_1, the magnetic field in each region can be found, as shown in Equations 10. Here, I_0 and I_1 refer to modified Bessel functions of the first kind, K_0 and K_1 refer to modified Bessel functions of the second kind, a and b are arbitrary constants, and $\boldsymbol{H}$ is the magnetic field vector. These solutions must then be patched together using appropriate boundary conditions for the magnetic field and the magnetic flux density to find values for the constants a and b in each region, and to thus find the magnetic field throughout the motor structure.

To calculate the force produced by the motor, we can first compute the Maxwell stress tensor at the inner and outer surfaces of the coil region. We can then multiply the stress at the inner and outer coil surfaces by the appropriate circumferences, add them, and average over the length of a pole pitch to find the force per unit length generated by the motor. We can approximate the force production as being done by only the first spatial harmonics of the current and magnetic field distributions [18], greatly simplifying the calculation process. By gathering all of the dimensional parameters together, we find an expression for the total force generated by the motor,

$$F = \frac{B_r J_1 \ell}{k^2} \hat{f}, \tag{11a}$$

$$\hat{f} \equiv \pi \left(\hat{a}_c \hat{b}_m + \hat{a}_m \hat{b}_c \right), \tag{11b}$$

where $\hat{f}$ is given by Equation 11b. Here, ℓ is the overall length of the motor, J_1 is the first harmonic of the current density, and $\hat{f}$ is a dimensionless parameter that depends only on ratios between the pole pitch and the radii of the motor components.

The dimensionless force $\hat{f}$ depends on four dimensionless parameters related to the magnetic fields: $\hat{a}_c$ and $\hat{b}_c$, describing the coil, and $\hat{a}_m$ and $\hat{b}_m$, describing the magnets. $\hat{a}_c$ and $\hat{b}_c$, given below in Equations 12a and 12b, do not depend upon the particular arrangement of magnets, but the dimensionless magnet parameters $\hat{a}_m$ and $\hat{b}_m$ must be found and tabulated for each magnet configuration by solving the magnetic field boundary conditions. In Equations 12a and 12b, r_{ci} is the coil inner radius, and r_{co} is the coil outer radius.

$$\begin{aligned}\hat{a}_c = kr_{ci}\Big[&\Big(\mathrm{I}_1(kr_{ci}) - \mathrm{L}_1(kr_{ci})\Big)\mathrm{K}_0(kr_{ci}) + \\ &\Big(\mathrm{I}_0(kr_{ci}) - \mathrm{L}_0(kr_{ci})\Big)\mathrm{K}_1(kr_{ci})\Big] - \\ kr_{co}\Big[&\Big(\mathrm{I}_1(kr_{co}) - \mathrm{L}_1(kr_{co})\Big)\mathrm{K}_0(kr_{co}) + \\ &\Big(\mathrm{I}_0(kr_{co}) - \mathrm{L}_0(kr_{co})\Big)\mathrm{K}_1(kr_{co})\Big],\end{aligned} \tag{12a}$$

$$\begin{aligned}\hat{b}_c = kr_{ci}\Big[&\Big(\mathrm{I}_0(kr_{ci}) - \mathrm{L}_0(kr_{ci})\Big)\mathrm{I}_1(kr_{ci}) - \\ &\Big(\mathrm{I}_1(kr_{ci}) - \mathrm{L}_1(kr_{ci})\Big)\mathrm{I}_0(kr_{ci})\Big] - \\ kr_{co}\Big[&\Big(\mathrm{I}_0(kr_{co}) - \mathrm{L}_0(kr_{co})\Big)\mathrm{I}_1(kr_{co}) - \\ &\Big(\mathrm{I}_1(kr_{co}) - \mathrm{L}_1(kr_{co})\Big)\mathrm{I}_0(kr_{co})\Big].\end{aligned} \tag{12b}$$

For the HG configuration, $\hat{a}_m$ and $\hat{b}_m$ are given by Equations 13a and 13b. The dimensionless magnet parameters, in this case, depend upon r_{ii}, the inner radius of the magnets inside the coil, and r_{io}, the outer radius of the magnets inside the coil. Note that all radii are non-dimensionalized by the wavenumber k.

The remaining basic performance measures for the motor can also now be written:

$$P = \frac{\ell J_1^2}{\sigma k^2} \hat{P}, \tag{14}$$

$$M = \frac{\rho_c \ell}{k^2} \hat{m}. \tag{15}$$

$$\begin{aligned}\boldsymbol{H}_{\text{magnet}} = \sum_{n\ \text{odd}} &\Big[\ a\,\mathrm{I}_1(nkr) + b\,\mathrm{K}_1(nkr) + \frac{\pi}{2} M_{rn}\left(\mathrm{I}_1(nkr) - \mathrm{L}_1(nkr)\right) - M_{rn}\Big]\,\hat{\boldsymbol{r}}\cos(nkz) + \\ &\Big[-a\,\mathrm{I}_0(nkr) + b\,\mathrm{K}_0(nkr) - \frac{\pi}{2} M_{rn}\left(\mathrm{I}_0(nkr) - \mathrm{L}_0(nkr)\right) - M_{zn}\Big]\,\hat{\boldsymbol{z}}\sin(nkz),\end{aligned} \tag{10a}$$

$$\begin{aligned}\boldsymbol{H}_{\text{coil}} = \sum_{n\ \text{odd}} &\Big[\ a\,\mathrm{I}_1(nkr) + b\,\mathrm{K}_1(nkr) - \frac{\pi J_n}{2nk}\left(\mathrm{I}_1(nkr) - \mathrm{L}_1(nkr)\right)\Big]\,\hat{\boldsymbol{r}}\sin(nkz) + \\ &\Big[-a\,\mathrm{I}_0(nkr) + b\,\mathrm{K}_0(nkr) + \frac{\pi J_n}{2nk}\left(\mathrm{I}_0(nkr) - \mathrm{L}_0(nkr)\right)\Big]\,\hat{\boldsymbol{z}}\cos(nkz),\end{aligned} \tag{10b}$$

$$\boldsymbol{H}_{\text{air}} = \sum_{n\ \text{odd}} \Big[\ a\,\mathrm{I}_1(nkr) + b\,\mathrm{K}_1(nkr)\Big]\,\hat{\boldsymbol{r}}\cos(nkz) + \Big[-a\,\mathrm{I}_0(nkr) + b\,\mathrm{K}_0(nkr)\Big]\,\hat{\boldsymbol{z}}\sin(nkz). \tag{10c}$$

$$\hat{a}_m = 0 \tag{13a}$$

$$\begin{aligned}\hat{b}_m = \frac{\pi}{2}\sin\left(\frac{\pi\delta}{2}\right)\Big[&kr_{ii}\Big(\left(\mathrm{I}_0(kr_{ii}) - \mathrm{L}_0(kr_{ii})\right)\mathrm{I}_1(kr_{ii}) - \left(\mathrm{I}_1(kr_{ii}) - \mathrm{L}_1(kr_{ii})\right)\mathrm{I}_0(kr_{ii})\Big) - \\ &kr_{io}\Big(\left(\mathrm{I}_0(kr_{io}) - \mathrm{L}_0(kr_{io})\right)\mathrm{I}_1(kr_{io}) - \left(\mathrm{I}_1(kr_{io}) - \mathrm{L}_1(kr_{io})\right)\mathrm{I}_0(kr_{io})\Big)\Big] + \\ \cos\left(\frac{\pi\delta}{2}\right)\Big[&kr_{io}\,\mathrm{I}_1(kr_{io}) - kr_{ii}\,\mathrm{I}_1(kr_{ii})\Big]\end{aligned} \tag{13b}$$

where $\hat{P}$ is the dimensionless power parameter and $\hat{m}$ is the dimensionless mass parameter. The dimensionless power parameter is independent of motor configuration, while the dimensionless mass parameter changes with both the motor configuration and the motor materials. For example, in Equation 17 below, the dimensionless mass parameter is given for the HG configuration:

$$\hat{P} \equiv \frac{\pi}{2}\left((kr_{co})^2 - (kr_{ci})^2\right), \tag{16}$$

$$\hat{m} = \pi\left[\frac{\rho_m}{\rho_c}\left((kr_{io})^2 - (kr_{ii})^2\right) + (kr_{co})^2 - (kr_{ci})^2\right], \tag{17}$$

where ρ_m is the mass density of the magnet. We can now also compute the motor constant ϵ,

$$F' = \frac{B_r J_1}{\rho_c}\frac{\hat{f}}{\hat{m}}, \tag{18}$$

$$\epsilon = B_r\hat{\epsilon}\sqrt{\frac{\sigma M}{\rho_c}}, \quad \hat{\epsilon} \equiv \frac{\hat{f}}{\sqrt{\hat{P}\hat{m}}}. \tag{19}$$

Note that the expressions for the motor constant and the force density are almost identical to those presented in Section II, with the addition of dimensionless correction factors $\hat{\epsilon}$ and $\hat{f}/\hat{m}$ representing the effects of the spatially varying magnetic fields and the mass of the magnets. The functional form of Equations 19 and 18 do not depend on the details of the magnet and conductor configuration; the motor configuration and geometry affect only the value of the dimensionless correction factors.

The dimensionless motor constant $\hat{\epsilon}$ depends only on the ratios of the layer radii to the pole pitch, and can be shown to be convex in these dimensionless lengths. There is a broad optimum range of geometry for the HG magnet configuration, near dimensionless geometry $kr_{ii} = 0.86$, $kr_{io} = 3.0$, $kr_{ci} = 3.03$, $kr_{co} = 3.87$, and $\delta = 0.41$. (An air gap between the magnets and coils equal to 1% of the radius is included as a thermal expansion and manufacturing tolerance allowance.) With this geometry, the maximum value of the dimensionless motor constant is $\hat{\epsilon} = 0.21$. Other configurations yield better optimum values for the dimensionless motor constant; for instance the results of Trumper *et al.* for a planar Halbach array motor configuration [18] can be transformed into this framework to give a value of $\hat{\epsilon} = 0.37$.

B. Finite Element Validation of Electromagnetic Model

In order to check the accuracy of our analytical solution for the motor magnetic fields, we performed a finite element (FEA) simulation of the optimum HG motor configuration given in the previous section and compared the results to those derived analytically for an identical motor. The FEA was performed using a remanence of 1.47 T and a magnet relative permeability of 1.05 in ANSYS. Special attention to numerical instabilities that arise in the analytical model when the value of nkr exceeds 30 at the outside of the motor is needed to analytically compute the magnetic field to high harmonic order. Contour plots of the radial magnetic flux density are shown in Figure 2, for both the analytical model of Equations 10 and the FEA model. These plots show good qualitative agreement, primarily differing due to the inherent limitations of a Fourier series representation for the discontinuous magnetization.

FEA was also used to make quantitative assessments of the analytical solution's veracity. Figure 3 shows a comparison of the radial flux density in the middle of the air gap between the analytical and FEA models, with approximately 8% discrepancy. This difference can be explained primarily by the non-unity relative permeability of the magnets in the FEA model. For a slightly different motor geometry, the linear force density under a square wave current density excitation was computed using both analytical and FEA techniques; the analytical method predicted a force between 1% and 3% lower than that from the FEA.

C. Thermal Model

While Equation 19 describes the efficiency of the motor while operating under any conditions and at any size scale, we need to specify the maximum current density J_1 in order to determine the achievable force density. The dominant physical process limiting the current density is Joule heating of the wire and magnets beyond their maximum operating temperatures. In order to specify the maximum current density, then, we can use a basic heat transfer model [21] to relate it to the maximum temperature rise ΔT.

For brevity, consider a simple, planar heat transfer model for the HG motor configuration in which it is cooled solely by convection on its exterior, as characterized by a convection coefficient $\bar{h}$. From elementary heat transfer theory, we can calculate a thermal resistance, R_{th} from the inner surface of the coil to the ambient according to

$$R_{th} = \frac{r_{co} - r_{ci} + 2\kappa_c\bar{h}^{-1}}{4\pi\kappa_c r_{co}}, \tag{20}$$

where κ_c is the thermal conductivity of the coil. We can then proceed to calculate the motor temperature as a function of the time-average (RMS) current density $\bar{J}_1$,

$$\Delta T = \frac{\pi\left(r_{c,o}^2 - r_{c,i}^2\right)R_{th}\bar{J}_1^{\,2}}{2\sigma} = \frac{\bar{J}_1^{\,2}}{\sigma\kappa_c k^2}\hat{P}\hat{R}_{th}, \tag{21a}$$

$$\hat{R}_{th} \equiv \pi\kappa_c R_{th} = \frac{kr_{co} - kr_{ci} + 4\pi \mathrm{Bi}^{-1}}{4kr_{co}}, \tag{21b}$$

$$\mathrm{Bi} \equiv \frac{p\bar{h}}{\kappa_c}, \tag{21c}$$

where we have defined a Biot number Bi in terms of the pole pitch p. Combining this expression (Equation 21a) with Equation 18 and the definition of $\hat{\epsilon}$, we find the perhaps surprising result that the thermal constraint is equivalent to a constraint on the pole pitch:

$$p \le \frac{2\pi B_r}{\rho_c \bar{F}'}\sqrt{\pi\sigma\kappa_c\Delta T} \times \frac{\hat{\epsilon}}{\sqrt{\hat{m}\hat{R}_{th}}}, \tag{22}$$

where $\bar{F}'$ denotes the RMS force density. This relationship suggests that it should be possible to achieve an arbitrarily

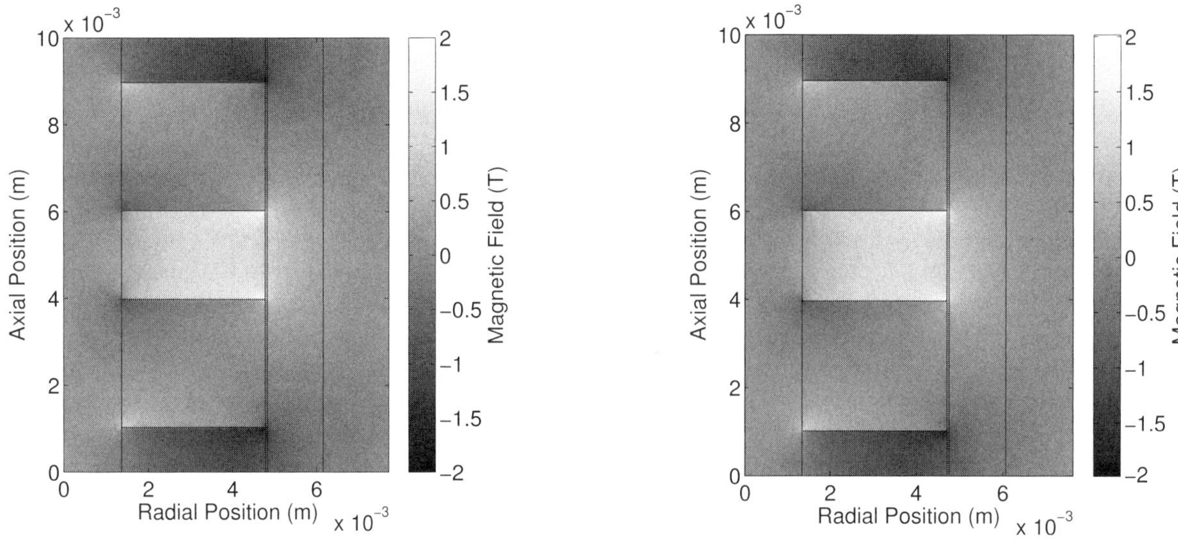

Fig. 2. A plot of the radial magnetic flux density for an optimized HG configuration motor ($kr_{ii} = 0.86$, $kr_{io} = 3.0$, $kr_{ci} = 3.03$, $kr_{co} = 3.87$, $\delta = 0.41$) as computed using ANSYS finite element software (right) and using the analytical model of Equations 10 with 75 harmonic terms (left). Qualitative agreement between the analytical and numerical flux plots is excellent, primarily differing in the sharpness of the boundaries between radial and axial magnets.

high force density so long as the pole pitch is made sufficiently small. While this is contrary to standard motor design practice, which is done by simply assuming a maximum current density (typically 10^7 $\mathrm{A/m^2}$), it is in agreement with the scaling described by Cugat *et al.* [7] and can be intuitively explained with reference to microelectronics, where current densities over 10^9 $\mathrm{A/m^2}$ are routinely used in small bond wires and interconnects.

As a numerical example consider once again the work loop parameters associated with pigeon flight muscles [9], which produce an RMS force density $\bar{F}' \approx 700$ N/kg, the optimum geometry for the HG configuration, a coil thermal conductivity of $\kappa_c \approx 1$ W/m · K, and a maximum temperature rise of $100°$ C. If we wish to have a pole pitch of greater than 10 mm, we will require a heat transfer coefficient $\bar{h} \geq 400$ $\mathrm{W/m^2K}$. This heat transfer coefficient can be achieved through vigorous air cooling or by simple liquid cooling [21].

D. Model Summary and Application

Just as in the analysis of the ideal motor in Section II, certain basic scaling properties of tubular linear synchronous motors are readily apparent. The motor constant ϵ in Equation 19 scales according to the same parameters as it did in Equation 4, with an additional dimensionless parameter describing the deviations from the ideal motor. Thus, even with the permanent magnet structure included, the motor constant still scales with the square root of the motor mass.

We can also examine the impact of the dimensionless motor constant correction factor on the motor efficiency,

$$\hat{\eta} = \underbrace{\frac{\rho_c}{\sigma B_r^2}}_{\text{materials}} \times \underbrace{\frac{2}{\omega \ell_s}}_{\text{velocity}} \times \underbrace{F_0'}_{\substack{\text{force}\\\text{density}}} \times \underbrace{\hat{\epsilon}^{-2}}_{\substack{\text{motor}\\\text{geometry}}} \tag{23}$$

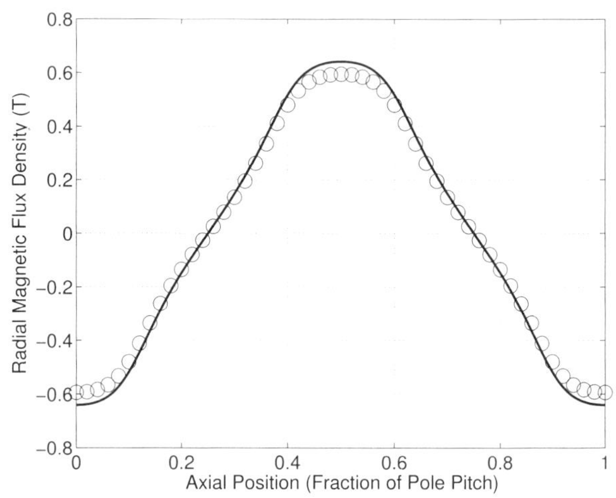

Fig. 3. A comparison of the radial magnetic flux density in the center of the coil of the HG motor depicted in Figure 2 between the 75-harmonic analytical model (solid line) and the FEA model (open circles) The maximum deviation occurs near the peak of the flux waveform, where the analytical solution is approximately 8% greater in magnitude than the FEA result.

Here, we can see that $\hat{\epsilon}$ is a dimensionless efficiency parameter that can be used to compare the efficiencies of different motor configurations and geometric ratios at any size scale. We have shown that, for the tubular HG configuration, the optimum $\hat{\epsilon} = 0.21$, and we can use the results of other magnetic field calculations to find the optimum $\hat{\epsilon}$ for additional configurations. In some literature [22] a similar parameter to $\hat{\epsilon}$, the "motor constant square density" is used to characterize motor performance, but this parameter does not separate the influence of material selection from that of geometric design.

The overall size of the motor components scales with the

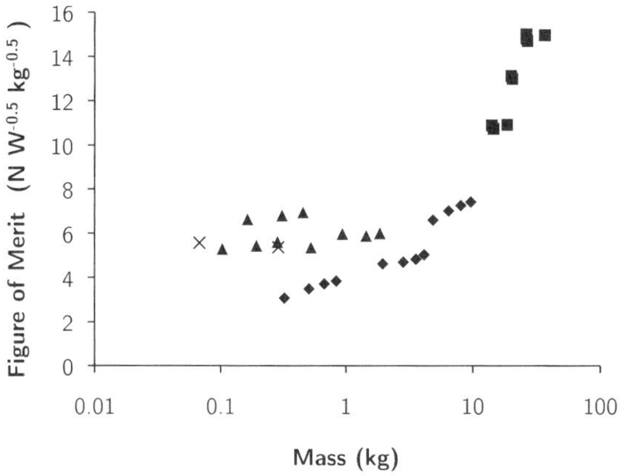

Fig. 4. The ratio of the motor constant to the square root of the mass is plotted for 30 different linear synchronous motors from four different manufacturers: Copley (diamonds), Moog (squares), Anorad (triangles), and Faulhaber (crosses). With the exception of the Moog motors, this figure of merit is independent of size. Compare with the theoretical figure of merit for an HG motor of 26 $\mathrm{N} \cdot \mathrm{W}^{-0.5} \cdot \mathrm{kg}^{-0.5}$.

Fig. 5. A photograph of the prototype wing-flapping test-bed, showing the polyimide film wing, the brass wing spar, and the actuator attachment point. The wing has a span of 250 mm, and is connected to the actuator with a 10 mm moment arm. The springs connected to the wing spar have a very low stiffness, and are used solely to keep the wing centered in its range of motion. Not shown are the actuator under test or the LVDT position transducer.

pole pitch, and thus with the required force density:

$$p \leq \underbrace{2\pi \frac{B_r\sqrt{\pi\sigma\kappa_c}}{\rho_c}}_{\text{materials}} \times \underbrace{\sqrt{\Delta T}}_{\text{thermal limit}} \times \underbrace{\frac{1}{\bar{F}'}}_{\text{force density}} \times \underbrace{\sqrt{\frac{\hat{\epsilon}}{\hat{m}\hat{R}_{th}}}}_{\text{motor geometry}}. \tag{24}$$

If internal thermal conduction limits the heat transfer out of the motor, the pole pitch scales inversely with the required force density. If convective thermal resistance dominates heat transfer out of the motor, the required pole pitch scales as the inverse square of the specified force density–high force densities call for highly miniaturized coils and magnetic structures, which may be difficult to fabricate. Short pole pitches also require long motors, with many repeating units and thus a high part and/or winding count.

Since the only motor dimensions that appear outside the dimensionless parameters are the pole pitch and overall length, the effect of the radial geometry of any motor configuration is universal and independent of the desired performance. Thus, the radial dimensions can be optimized for each configuration to yield the highest possible value of the dimensionless motor constant $\hat{\epsilon}$. The use of a closed-form analytical model renders geometric optimization very computationally efficient, and a modest computer proves capable of examining thousands of geometries per second. Furthermore, $\hat{\epsilon}$ is convex in most geometric parameters, although some configurations perform best as the overall radial scale tends towards infinity (i.e. a flat plate rather than a tube.)

E. Comparison with Commercial Motors

While motor manufacturers do not release enough data for a full evaluation of Equations 23 and 24, we can nonetheless verify the general scaling of the model against existing motors. The ratio $\epsilon/\sqrt{M}$, which should depend only on motor internal geometry and on materials choice, is plotted for a variety of commercially-available linear synchronous motors across three orders of magnitude in size in Figure 4. With the exception of the largest motors, which incorporate more mechanical components than the others, this ratio is approximately constant for each motor type. There is considerable variability in motor performance, however, with this figure of merit varying by a factor of three between different motor types. By comparison, with 1.47 T magnets and an optimum HG geometry, the model of this work suggests a figure of merit of 26 $\mathrm{N} \cdot \mathrm{W}^{-0.5} \cdot \mathrm{kg}^{-0.5}$, about twice as high as the large Moog motors can offer. The difference can be ascribed to the mechanical components of the Moog motor (bearings, sealed housing, etc.) and certain non-idealities of motor design not covered by this model (copper packing factor, end effects, etc.).

A similar examination of the pole pitch and force density of commercial motors verifies the scaling of Equation 24. For example, Anorad's passively-cooled 15 mm pole pitch motors average 80 N/kg while their 30 mm motors average 55 N/kg, in close agreement with the expected scaling for convection-dominated thermal resistance. Active cooling in these motors results in very modest improvements in force density, perhaps due to poor coupling between the cooling fluid and the coils.

IV. Applications to Flapping Flight

Our motor model gives us a way to determine the size, complexity, and cooling requirements of a linear PM motor for any given performance requirements, including requirements traditionally [1] thought outside the capabilities of such motors. As we have considered previously in this work, flapping-wing flight is a problem in robotics that is particularly constrained by actuator performance. The limited performance of conventional actuators is usually overcome by driving the wings symmetrically with a single high-speed rotary motor or engine and a gear reduction system, but this design does not allow the asymmetric flapping that is essential to low-speed maneuverability [23]. To achieve full control authority over the robot's trajectory, each wing needs to be actuated independently with variable stroke. The simplest way to achieve this is to use direct-drive linear actuators to move each wing.

Using Equation 23, we can determine the efficiency achievable with motors in such an actuation scheme. If we assume that motors of an HG configuration will be used, and continue to match the properties of the pigeon pectoralis muscle, we find that the efficiency of our hypothetical two-actuator direct-drive bird robot will be 40%, well in excess of the efficiency of birds in nature.

In order to further explore this application, we have begun construction of a flapping wing actuator test-bed, shown in Figure 5. A wing with a passive feathering axis is connected to a conventional linear EM actuator (BEI Kimco LA25-42-000A) with similar dimensions and geometry to a pigeon wing. This apparatus will be used to test motors designed and built according to the principles described in this work, and to obtain suitable comparison data using off-the-shelf actuators.

V. Discussion and Future Work

In conclusion, by developing a non-dimensionalized model for linear PM motor performance, we have established that linear motors are capable of extremely high, to date unrealized performance. With suitable provision of active cooling and miniaturized components, linear PM motors can achieve force densities more than an order of magnitude higher than those of commercially-available motors while retaining acceptable efficiency. This capability opens up a large range of new applications in robotics, from locomotion and manipulation to haptic interfaces, at the relatively small cost of requiring custom motor design for each application.

To obtain a more accurate estimate of motor performance, our analysis framework can be modified to include non-idealities such as the effects of less than 100% copper fill in the coils, imperfect coil winding factors, and overhung and underhung coil configurations. It can also be adapted to handle saturable iron flux return paths, to accurately describe magnets that are simpler in fabrication than Halbach arrays. We will also use the lessons taught by our model to construct a prototype motor to illustrate the theoretically achievable performance.

References

[1] J. D. Madden, "Mobile robots: Motor challenges and materials solutions," *Science*, vol. 318, no. 5853, pp. 1094–1097, 2007.

[2] J. M. Hollerbach, I. W. Hunter, and J. Ballantyne, "A comparative analysis of actuator technologies for robotics," in *The Robotics Review 2* (O. Khatib, J. J. Craig, and T. Lozano-Perez, eds.), (Cambridge, MA), pp. 299–342, MIT Press, 1992.

[3] A. Bicchi and G. Tonietti, "Fast and 'soft-arm' tactics," *IEEE Robotics & Automation Magazine*, vol. 11, pp. 22–33, June 2004.

[4] F. E. Zajac, "Muscle and tendon: Properties, models, scaling, and application to biomechanics and motor control," *Critical Reviews in Biomedical Engineering*, vol. 17, no. 4, pp. 359–411, 1989.

[5] N. P. Smith, C. J. Barclay, and D. S. Loiselle, "The efficiency of muscle contraction," *Progress in Biophysics and Molecular Biology*, vol. 88, no. 1, pp. 1 – 58, 2005.

[6] J. D. W. Madden, N. A. Vandesteeg, P. A. Anquetil, P. G. A. Madden, A. Takshi, R. Z. Pytel, S. R. Lafontaine, P. A. Wieringa, and I. W. Hunter, "Artificial muscle technology: physical principles and naval prospects," *IEEE Journal of Oceanic Engineering*, vol. 29, pp. 706–728, Jul. 2004.

[7] O. Cugat, J. Delamare, and G. Reyne, "Magnetic micro-actuators and systems (MAGMAS)," *IEEE Transactions on Magnetics*, vol. 39, pp. 3607–3612, Nov. 2003.

[8] E. R. Laithwaite, "The goodness of a machine," *Proceedings of the IEE*, vol. 112, pp. 538–542, Mar. 1965.

[9] A. A. Biewener, W. R. Corning, and B. W. Tobalske, "*In vivo* pectoralis muscle force-length behavior during level flight in pigeons (*Columba livia*)," *Journal of Experimental Biology*, vol. 201, pp. 3293–3307, 1998.

[10] A. Kvist, Å. Lindström, M. Green, T. Piersma, and G. H. Visser, "Carrying large fuel loads during sustained bird flight is cheaper than expected," *Nature*, vol. 413, pp. 730–732, Oct. 2001.

[11] K. Halbach, "Design of permanent multipole magnets with oriented rare earth cobalt material," *Nuclear Instruments and Methods*, vol. 169, pp. 1–10, 1980.

[12] M. G. Lee, S. Q. Lee, and D.-G. Gweon, "Analysis of Halbach magnet array and its application to linear motor," *Mechatronics*, vol. 14, pp. 115–128, 2004.

[13] D. Golda and M. L. Culpepper, "Modeling 3D magnetic fields for precision magnetic actuators that use non-periodic magnet arrays," *Precision Engineering*, vol. 32, pp. 134–142, Apr. 2008.

[14] E. R. Laithwaite, "Linear electric machines—a personal view," *Proceedings of the IEEE*, vol. 63, pp. 250–290, Feb. 1975.

[15] J. Wang, G. W. Jewell, and D. Howe, "A general framework for the analysis and design of tubular linear permanent magnet machines," *IEEE Transactions on Magnetics*, vol. 35, pp. 1986–2000, May 1999.

[16] N. Bianchi, "Analytical field computation of a tubular permanent-magnet linear motor," *IEEE Transactions on Magnetics*, vol. 36, pp. 3798–3801, Sep. 2000.

[17] Wijono and H. Arof, "Open circuit field distribution and induced voltage of a cylindrical permanent magnet linear generator," *American Journal of Applied Sciences*, vol. 4, no. 11, pp. 912–917, 2007.

[18] D. L. Trumper, M. E. Williams, and T. H. Nguyen, "Magnet arrays for synchronous machines," in *Conference Record of the 1993 IEEE Industry Applications Society Annual Meeting*, pp. 9–18, Oct. 1993.

[19] M. Abramowitz and I. A. Stegun, eds., *Handbook of Mathematical Functions with Formulas, Graphs, and Mathematical Tables*. New York: Dover, tenth ed., 1964.

[20] S. Zhang and J. Jin, *Computation of Special Functions*. New York: Wiley, 1996.

[21] J. H. Lienhard, IV and J. H. Lienhard, V, *A Heat Transfer Textbook*. Cambridge, MA: Phlogiston Press, 3rd ed., 2008.

[22] T. Mizuno, M. Kawai, F. Tsuchiya, M. Kosugi, and H. Yamada, "An examination for increasing the motor constant of a cylindrical moving magnet-type linear actuator," *IEEE Transactions on Magnetics*, vol. 41, pp. 3976–3978, Oct. 2005.

[23] B. W. Tobalske, "Biomechanics of bird flight," *Journal of Experimental Biology*, vol. 210, pp. 3135–3146, 2007.

Study of Group Food Retrieval by Ants as a Model for Multi-robot Collective Transport Strategies

Spring Berman*[1], Quentin Lindsey[†], Mahmut Selman Sakar[‡1], Vijay Kumar[†], and Stephen Pratt[§]
*School of Engineering and Applied Sciences, Harvard University, Cambridge, MA 02138 USA
[†]GRASP Laboratory, University of Pennsylvania, Philadelphia, PA 19104 USA
[‡]Department of Mechanical Engineering, Massachusetts Institute of Technology, Cambridge, MA 02139 USA
[§]School of Life Sciences, Arizona State University, Tempe, AZ 85287 USA

Abstract— **Group food retrieval in some ant species serves as a useful paradigm for multi-robot collective transport strategies that are decentralized, scalable, and do not require *a priori* information about the payload. We investigate this phenomenon in *Aphaenogaster cockerelli* in order to extract the ants' roles during transport, the rules that govern their actions, and the individual forces that they apply to guide a food item to their nest. To measure these forces, we designed and fabricated elastic structures with calibrated stiffness properties, induced ants to retrieve the structures, and tracked the resulting deformations with a camera. We then developed a hybrid system model of the ant behaviors that were observed in the experiments. We conducted simulations of the behavioral model that incorporate a quasi-static model of planar manipulation with compliant attachment points. Our simulations qualitatively replicate individual ant activity as well as certain macroscopic features of the transport.**

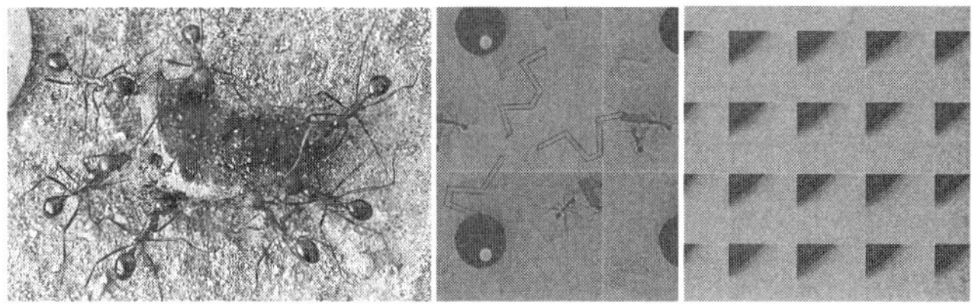

Fig. 1. From left to right: *A. cockerelli* ants retrieving a piece of fig and an elastic structure (springs are outlined for clarity), described in Section III-A, and a SCARAB differential-drive robot [15] equipped with a single-actuator gripper with passive compliance.

I. INTRODUCTION

While there are various approaches to cooperative robotic manipulation, there are few decentralized approaches that are applicable to large groups of mobile robots. Such approaches can provide a greater degree of flexibility and robustness in construction and manufacturing applications. Cooperative food retrieval in ants is a striking example of a fully decentralized manipulation strategy that is scalable in the number of transporters and successful for a wide range of loads and environments. This biological phenomenon offers inspiration for the analogous problem in robotics: a group of robots is tasked to manipulate an arbitrarily-shaped payload, which is too heavy for a single robot to move, to a target destination without *a priori* knowledge about the payload or obstacles in the environment. The robots must rely on local sensing and no explicit communication in order for the strategy to be scalable.

To extract the rules that govern ant transport behavior and better understand the mechanisms of successful collective transport, we study this phenomenon in *Aphaenogaster cockerelli* (Fig. 1), a monomorphic ant species that is common in the deserts of the southwestern United States. To the authors' knowledge, this is the first work that investigates the mechanics of cooperative transport in ants. We fabricate elastic two-dimensional structures that we use as vision-based ant force sensors and videotape ant retrieval of the structures. Using the video data, we quantitatively illustrate salient features of ant transport, including the emergence of consensus. We then develop a dynamic and behavioral model of ant transport in a step toward implementation on robotic platforms such as the one in Fig. 1. We validate the model by reproducing the observed characteristics of transport through simulations.

[1]This work was done when Spring Berman (email: sberman@eecs.harvard.edu) and Mahmut Selman Sakar were affiliated with the GRASP Laboratory at the University of Pennsylvania.

II. LITERATURE REVIEW

A. Collective transport in ants

Group retrieval has evolved independently many times in the ants, but it is developed to much higher levels in some species than in others [6]. Groups can lift more weight per worker than solitary carriers, enhancing the colony's efficiency at food collection. The small colonies of *A. cockerelli* are skillful at group retrieval [5]. When a solitary forager finds an item too large for her to move, she summons a team of nestmates within a few minutes. They quickly make off with the food, pre-empting competitors whose mass recruitment abilities and greater aggression would otherwise allow them to displace the relatively timid *A. cockerelli*.

Collective transport is a stigmergic process in the sense that the medium of communication between ants is the progress of the transport itself [22]. The tractive resistance of the prey stimulates orientational and positional changes by the ants and recruitment behavior [21], [23]. Transport has been observed to consist of an "uncoordinated" phase in which ants arrive at the prey, either by chance or via recruitment, and pull on it in all directions, followed by a "coordinated" phase during which a preferred direction emerges from the ants' exertions and transport occurs [12]. The phase transition has been attributed to spatial rearrangements [16], a decrease in number of ants, realignments and rotation of the prey [23], and tensions acting as positive feedback when associated with success [12].

There has been some debate over the existence of cooperation among ants during transport. On one hand, ants often appear to behave as though they were isolated, at times pulling on the load in opposite directions. Indeed, they frequently exhibit the same behaviors in solitary transport and group transport [21], [23]. However, the concurrent behaviors of multiple ants during group transport produces some cooperative features: groups can move prey significantly faster [2], develop higher mean power [23], and exert greater forces than single ants [21]. The consensus is that this apparent collaboration does not arise from conscious deliberation, learning, or communication, but is rather a juxtaposition of uncoordinated individual forces on the prey with a strong component toward the nest, the common destination of the ants [12].

B. Multi-robot transport systems

Many approaches to cooperative robotic manipulation rely on centralized or leader/follower schemes that require knowledge of the load geometry and possibly the contact forces [17]. Several decentralized manipulation schemes have been developed for motion and force control of a payload by a fixed group of robots in an obstacle-free environment [8], [13], [17], [20], [24]. Groups of two robots only were used for experimental verification [8], [17], [24], and [8] requires communication among platforms. Other decentralized multi-robot transport approaches include object closure and caging, which compose the robot behaviors of approach, surround, and push in order to geometrically enclose and move an object in a predictable manner [4], [19]. An ant-inspired decentralized approach to multi-robot box pushing, in which robots switch between simple behaviors in response to locally sensed cues, is presented in [9].

III. Elastic Vision-Based Force Sensor

A. Design

Preliminary trials were conducted with plastic disks, which were coated with fig paste to induce retrieval. A group of about 15 ants was able to transport a disk of mass $m = 1.6$ g and diameter $d = 4.6$ cm. The ants were better at transport when the perimeter of the disk had small tabs, measuring 2 mm long and 0.5 mm wide, for their mandibles to grasp. More than 20 ants participated in carrying a larger disk, with $m = 3.4$ g and $d = 6.9$ cm, although they were not as effective as the group carrying the smaller disk.

Using these dimensions as guidelines, we developed and tested several planar elastic structures. Each consisted of a circular load ringed with several springs that were tipped with thin tabs (see the labels on the structure in Fig. 2a). In all of the structures, some ants gripped the springs themselves rather than the tabs; however, their contributions to the spring deflections did not seem significant since most ants gripped the tabs and bars, which were smeared with fig paste.

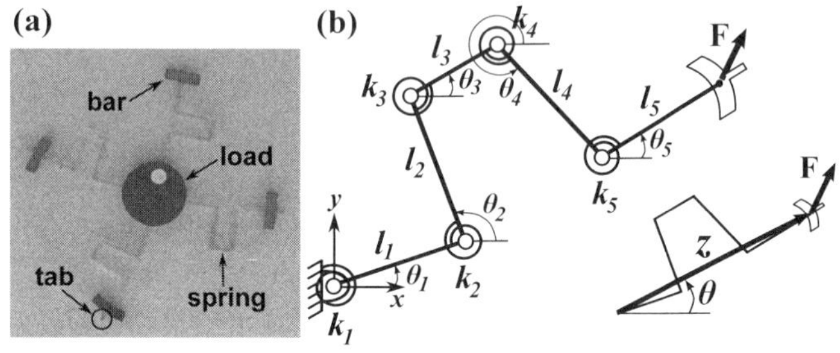

Fig. 2. (a) Elastic structure 2 with labels. (b) General pseudo-rigid-body model of a spring on the structure (see Section III-D).

The first prototype ($m = 1.3$ g, $d = 8.5$ cm) was laser cut from 0.75-mm-thick Lexan material. A group of 12-15 ants carried the structure effectively, but they produced negligible spring deformations, even when several ants pulled on one spring. To obtain measurable deformations, we switched to fabricating 3.2-mm-thick structures out of the soft material polydimethylsiloxane, or PDMS (see Section III-B). We chose PDMS for its mechanical compliance, manufacturability, and biocompatibility [26]. When pulled by ants, a second prototype ($m = 2.6$ g, $d = 4.5$ cm) and a structure we will call structure 1 ($m = 2.3$ g, $d = 6.1$ cm) showed significant spring deformations in the radial and tangential directions, respectively, but not in both directions. For structure 2 ($m = 1.6$ g, $d = 7.7$ cm), we modified the spring geometry, shown in Fig. 2b, to have more similar radial and tangential stiffnesses. The lengths of the spring members in mm are $l_1 = 9.8$, $l_2 = l_4 = 11.9$, $l_3 = 5.6$, $l_5 = 10.9$, and the spring width is $h = 0.8$ mm. Since the μ for PDMS is relatively large regardless of surface, we chose to reduce this quantity in structures 1 and 2 by applying DuPont™ Teflon® tape to the bottom of the loads.

B. Fabrication

The elastic structures were fabricated using the transfer molding procedure shown in Fig. 3. The process started with the fabrication of molds of the structure shape. The substrate used for the mold was acrylic (Lucite International) because of its strength, elasticity, and ease of manufacturing. A flatbed laser cutter (Universal Laser Systems X-660) was used to cut the negative structure features (Fig. 3a). Then the mold and a 5 cm × 5 cm glass plate were silanized by vapor deposition of (tridecafluoro-1,1,2,2-tetrahydrooctyl)-1-trichlorosilane for 1h at room temperature in a vacuum chamber, and the mold was fixed on the glass plate.

The PDMS elastomer base (Dow Corning, MI) was polymerized by mixing 10:1 (w/w) ratio with curing agent. The solution was stirred and degassed under vacuum for 20 min, and the mold was filled with the viscous solution. Excess solution was scraped off the surface of the mold with a clean razor blade (Fig. 3b). The polymer was cured for 2h at 80°C on a hot plate (Fig 3c). Once cured, the structure (Fig. 2a) was carefully removed with a pair of tweezers.

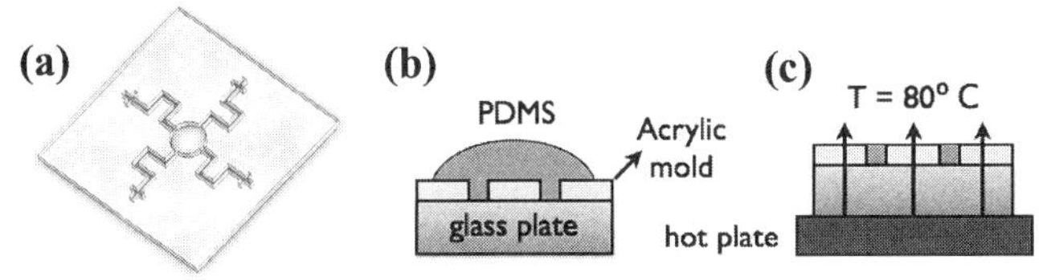

Fig. 3. Elastic structure fabrication process.

C. Calibration

The calibration setup uses a micro/macro manipulation apparatus to create a lookup table of applied force vs. spring tip displacement. The load of the structure is clamped to an acrylic mount such that one spring is in the appropriate configuration. This mount is attached to a motorized rotary stage (New Focus™), which is mounted on a H107 ProScan™II xy stage with 1 micron repeatability, step size of 5 microns, and travel of 10.16 cm × 7.62 cm. A needle, which is attached to a load cell (Transducer Techniques®) in compression configuration and allowed to rotate freely, is inserted into the bar at the tip of the spring. Using an automated routine, the stage moves to predetermined locations on a rectangular grid, and a force measurement is taken using the load cell at each location. To determine the radial spring force, the longitudinal axis of the spring is aligned parallel to the axis of the load cell. After the routine has swept the entire grid, the spring is returned to its rest configuration and rotated with the rotary stage 90° clockwise so that its latitudinal axis is aligned with the load cell in order to measure the tangential forces. Half of the grid is swept in that configuration so that the load cell is in compression. The spring is then rotated 180° counterclockwise so that the other half of the grid is swept.

D. Model

We develop a simple model of the springs that can be used to quickly estimate their radial and tangential stiffnesses and to design a spring geometry with desired stiffness properties. Since the springs consist of flexible members that can undergo large deflections, we apply a *pseudo-rigid-body model*, which describes the behavior of compliant mechanisms [7].

Fig. 2b shows the most general case of the spring model for structure 2. Each member i is represented as a rigid beam attached at a pin joint, where a torsional spring with spring constant k_i models the beam's resistance to deflection. Since the members are fixed-fixed segments, we define $k_i = 2\gamma K_\theta EI/l_j$ [7], where $j = 1$ for $i = 1$ and $j = 2$ otherwise. Here $\gamma = 0.85$, $K_\theta = 2.68$, E is Young's modulus, and I is the moment of inertia of the beam. Let $\Delta\theta_i$ be the angular displacement of beam i from its undeflected configuration. The moment at each pin joint is $\mathbf{T_i} = -k_i \Psi_i \hat{\mathbf{k}}$, where $\Psi_1 = \Delta\theta_1, \Psi_2 = \Delta\theta_2 - \Delta\theta_1, \Psi_3 = \Delta\theta_2 - \Delta\theta_3, \Psi_4 = \Delta\theta_4 - \Delta\theta_3, \Psi_5 = \Delta\theta_4 - \Delta\theta_5$.

We define three cases of the applied force $\mathbf{F}$ in polar coordinates, given in Table I. We use the principle of virtual work to solve for the force-displacement relations. In each case, we model the spring as a 1-DOF system with generalized coordinate θ_q and include a subset $\mathcal{I}$ of the torsional springs, corresponding to members that display the most noticeable compliance over the calibration displacement range. We compare the force-displacement models to data from seven calibration trials on the same spring. The calibrated forces are obtained as x and y components and must be transformed into F_r, F_θ beforehand. For cases 2 and 3, the F_θ measurements are interpolated along the circles $(x - l_1)^2 + y^2 = (l_3 + l_5)^2$ and $x^2 + y^2 = (l_1 + l_3 + l_5)^2$, respectively.

TABLE I

SPRING MODEL PARAMETERS FOR DIFFERENT LOADING CONDITIONS

	$\mathbf{F}$	θ_1	θ_2	θ_3	θ_4	θ_5	$\mathcal{I}$
1	$F_r\mathbf{e}_r,\ \theta = 0$	0	θ_q	0	$2\pi - \theta_q$	0	{2,...,5}
2	$F_\theta\mathbf{e}_\theta,\ \theta \geq 0$	0	θ_q	$\theta_q - \frac{\pi}{2}$	$\theta_q + \pi$	$\theta_q - \frac{\pi}{2}$	{2}
3	$F_\theta\mathbf{e}_\theta,\ \theta < 0$	θ_q	$\theta_q + \frac{\pi}{2}$	θ_q	$\theta_q + \frac{3\pi}{2}$	θ_q	{1}

1) Case 1: Let r denote the spring displacement from its undeformed state. We solve for F_r in terms of θ_2, which is related to r geometrically:

$$F_r = \frac{\pi/2 - \theta_2}{2l_2 \sin(\theta_2)} \sum_{i=2}^{5} k_i, \qquad \theta_2 = \cos^{-1}\left(\frac{r}{2l_2}\right). \tag{1}$$

Using calibration data, the least squares estimate of E is 1.66 MPa, which is within the typical range for PDMS (360 kPa to ~3 MPa) [1]. Fig. 4a shows that equation (1) closely matches the calibration curve.

2) Case 2: We solve for F_θ in terms of θ_2. Then from $l_1 + (l_3 + l_5)\sin\theta_2 = z\cos\theta$, where $z = ||\mathbf{z}||$:

$$F_\theta = \frac{k_2}{z - l_1\cos\theta} \cos^{-1}\left(\frac{z\cos\theta - l_1}{l_3 + l_5}\right), \quad \theta \geq 0. \tag{2}$$

As Fig. 4b shows, equation (2) matches the averaged calibration data fairly well. Note that the measured F_θ has much greater variability than F_r.

3) Case 3: We solve for F_θ in terms of θ_1, which due to the constraints on the other members is equal to θ:

$$F_\theta = k_1\theta/z, \quad \theta < 0. \tag{3}$$

Fig. 4b shows that the model underestimates $|F_\theta|$ in this case. A closer fit to the data can be achieved by multiplying k_1 by 5, as shown by the light dashed plot.

IV. EXPERIMENTS

A. Video recording of group retrieval

Experiments were performed in *A. cockerelli*'s natural Sonoran desert environment in South Mountain Park, Phoenix, Arizona. Ants were induced to retrieve an elastic structure placed approximately one meter from an active nest entrance. The tabs at the ends of the springs were made attractive to the ants by smearing them with fig paste. The structure was placed on a flat wooden board (50 cm × 40 cm) lined with ivory copy paper. This provided a smooth surface across which the ants could drag the structure, as well as a featureless background to aid in automated video tracking. A high definition video camera (Panasonic HVX200 or Canon HG20) was fixed above

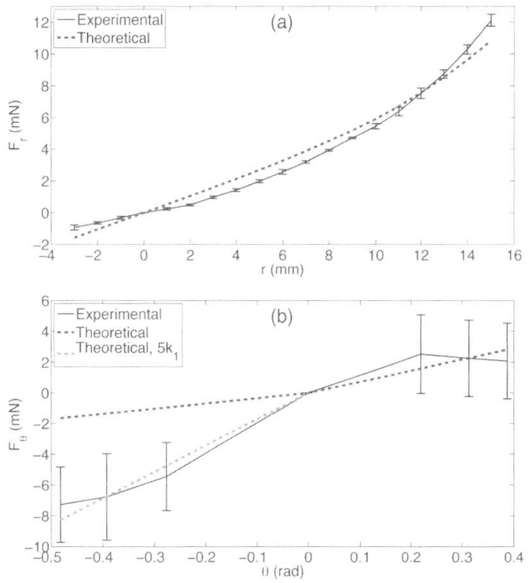

Fig. 4. Experimental and theoretical force-displacement curves: (a) case 1, (b) cases 2 and 3. Measurements are averaged over seven calibration trials. Error bars show standard deviations.

the structure on the vertical post of a copy stand. Its field of view encompassed a region of about 35 cm × 20 cm. A screen blocked direct sunlight from this region, to avoid shadows that could hinder video tracking. Fig baits in the vicinity of the structure were used to attract foragers. Once an ant discovered the structure, she soon initiated recruitment after unsuccessful attempts at moving it herself. A group of four to six ants then cooperated to move the structure toward the nest.

B. Video data image processing

In order to extract the structure configuration and the spring deformations from the video recordings of the transport, we tracked certain features of the structure, including the load and the tips of the springs. We covered the load with a black circular label marked with a small white circle offset from the center. Using Swistrack [14], a general purpose vision processing software, we used thresholding and blob detection to label and track the load center and the dot. The load orientation was obtained by comparing the relative position of the centers of the dot and the load. The bar on each spring was marked with a red, blue, or green label and was tracked using similar methods. We calculated the deformation of a spring from the position of its label, the load pose, and other geometric quantities.

C. Results

1) Mechanics of transport: We found that typical structure inertial forces were on the order of 10^{-4} mN, which was negligible compared to the friction force on the structures, $\mu mg = 22$ mN, where $\mu = 1.4$ was measured from inclined plane tests with the structures. This indicates that the structures undergo quasi-static motion. The structures were often observed to move by stick-slip, which resulted in noisy data on load speed, v_l. The average ant force $||\mathbf{F}_i||$ was calculated to be 10.5 ± 5.0 mN (sample size $n = 10906$); 99.1% of the samples did not exceed 30.0 mN.

2) Cooperative features of transport: Fig. 6 and 8 show snapshots of recorded transports with structures 1 and 2, respectively, as well as the evolution of the load configuration over both trials. Fig. 5 shows the linear and angular speeds of the load over time for the trial in Fig. 6. Fig. 7 displays the sum of the ant *interaction forces* [10] for the trial in Fig. 8, each defined as the projection of the difference in forces applied by two ants onto their relative position vector:

$$||\mathbf{F}_{ij}^{int}|| = (\mathbf{F}_i - \mathbf{F}_j) \cdot (\mathbf{r}_i - \mathbf{r}_j)/||\mathbf{r}_i - \mathbf{r}_j|| \ . \qquad (4)$$

This quantity is a measure of the degree of cooperation between two ants; it is zero when ants are perfectly cooperating.

We see evidence of an initial transport phase of low coordination among the ants followed by a more highly coordinated phase, as has been documented in the literature on group transport in ants (see Section II-A). Fig. 5 shows that structure 1 initially moves slowly on average, and then its speed sharply increases to significantly higher values at $t = 68$ s. As Fig. 6 reveals, the transition between these phases of speed is initiated by the ants' reorientation of the structure a few degrees counterclockwise. The trial with structure 2 indicates that the phases can also be characterized in terms of the sum of the interaction forces, with the transition occurring as a sizable reduction in this quantity. Fig. 7 shows that this sum decreases to about half its initial value on average after $t = 32$ s. From the snapshots in Fig. 8, this increase in cooperation is apparent in the reorientation of the ants to pull with larger force components toward the nest during the second phase.

3) Load speed saturation with increased group size: Fig. 9 shows that three ants are needed to move structure 2, and that when more than five ants participate in transport, there is on average no increase in the load speed. This is due to increased interference between ants during transport with higher populations.

V. Simulation of Cooperative Manipulation

In order to simulate realistic motions of the elastic structures that are caused by the ants, we use a planar manipulation model that predicts the motion of an object similar in construction to the structures. In developing the model, we assume quasi-static motion, which was justified experimentally in Section IV-C.1. In addition, we require that friction forces on the springs exceed the spring restorative forces so that a deformed spring cannot, in the absence of ants, cause the entire structure to move.

The simulated object and its associated notation is shown in Fig. 10. The position and orientation of a reference frame O_b that is fixed to the load is (x, y, θ) in an inertial frame O_w. For each spring $j \in \{1, ..., J\}$, we define a frame R_j at (x_{R_j}, y_{R_j}), a point along the arc where the spring is attached to the load, and a frame P_j at (x_{P_j}, y_{P_j}), a point at the tip of the spring. The orientation of R_j in frame O_w is denoted by θ_{R_j}, and

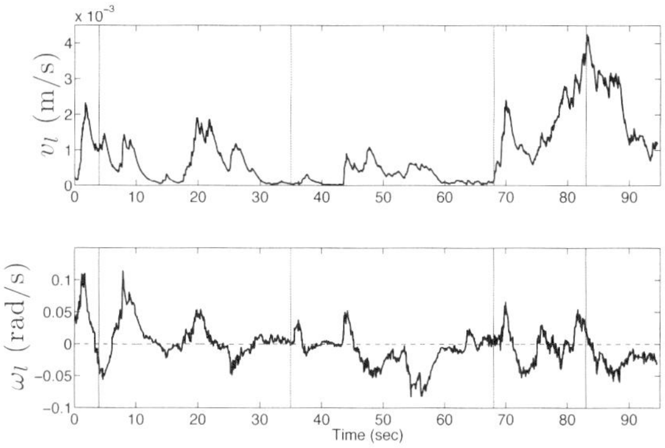

Fig. 5. Linear and angular load speeds over time for the trial in Fig. 6. Vertical lines indicate the times of the snapshots.

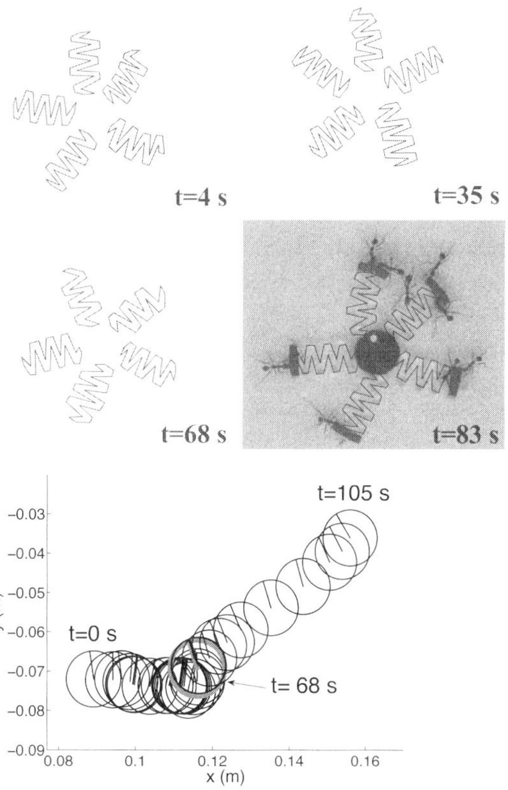

Fig. 6. Snapshots of ant transport with structure 1 (top); load position and orientation (bottom). The structure springs are outlined for clarity. The thick gray circle in the bottom plot signifies the load at $t = 68$ s.

the orientation of P_j is θ_{P_j}, which is set to the orientation of frame O_w. We define a homogeneous transformation matrix

$$\mathbf{A}(x, y, \theta) = \begin{bmatrix} \cos\theta & -\sin\theta & x \\ \sin\theta & \cos\theta & y \\ 0 & 0 & 1 \end{bmatrix}. \tag{5}$$

Then we can represent the configurations just described as ${}^{O_w}\mathbf{A}_{O_b} \equiv \mathbf{A}(x, y, \theta)$, ${}^{O_w}\mathbf{A}_{R_j} \equiv \mathbf{A}(x_{R_j}, y_{R_j}, \theta_{R_j})$, and ${}^{O_w}\mathbf{A}_{P_j} \equiv \mathbf{A}(x_{P_j}, y_{P_j}, \theta_{P_j})$.

Using the notation in [18], we define the load twist in the

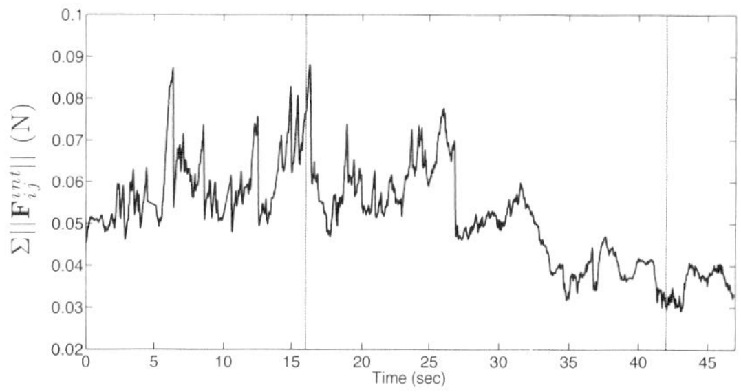

Fig. 7. Sum of interaction forces over time for the trial in Fig. 8. Vertical lines indicate the times of the snapshots.

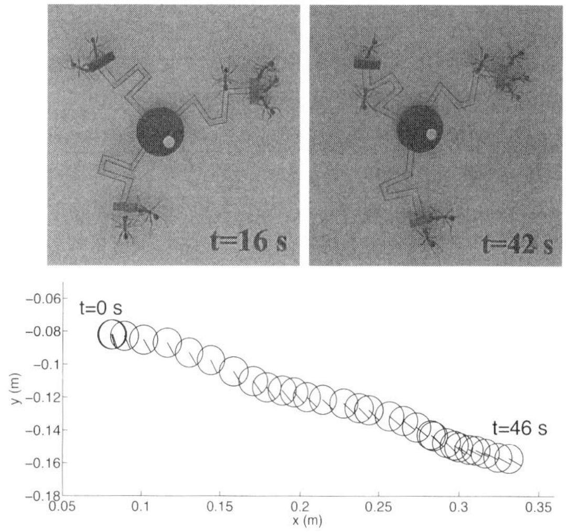

Fig. 8. Snapshots of ant transport with structure 2 (top); load position and orientation (bottom). The structure springs are outlined for clarity.

inertial frame as

$$\hat{\xi}_{load} = {}^{O_w}\mathbf{A}_{O_b}^{-1}\, {}^{O_w}\dot{\mathbf{A}}_{O_b} \tag{6}$$

or in terms of the vector

$$\xi_{load} = \begin{bmatrix} \dot{x} & \dot{y} & \dot{\theta} \end{bmatrix}^T. \tag{7}$$

Similarly, we can define the displacement of frame P_j from its undeformed configuration in the frame R_j:

$${}^{R_j}\mathbf{A}_{P_j} = {}^{O_w}\mathbf{A}_{R_j}^{-1}\, {}^{O_w}\mathbf{A}_{P_j}, \tag{8}$$

$$\hat{\xi}_{sp,j} = {}^{R_j}\mathbf{A}_{P_j}^{-1}\, {}^{R_j}\dot{\mathbf{A}}_{P_j}. \tag{9}$$

We model each spring j as a composition of an extension spring and a rotary spring, resulting in a wrench $\mathbf{w}_j$ in the frame R_j with the following form:

$$\mathbf{F}_j = -\frac{\kappa_\tau \theta_{sp,j}}{l}\begin{bmatrix} -\sin(\theta_{P_j} + \theta) \\ \cos(\theta_{P_j} + \theta) \end{bmatrix} - \kappa_\rho (l - l_o)\begin{bmatrix} \cos(\theta_{P_j} + \theta) \\ \sin(\theta_{P_j} + \theta) \end{bmatrix}$$
$$\mathbf{w}_j = \begin{bmatrix} \mathbf{F}_j \\ x_{sp,j} F_{j,y} - y_{sp,j} F_{j,x} \end{bmatrix}, \tag{10}$$

where $l = (x_{sp,j}^2 + y_{sp,j}^2)^{1/2}$ is the deformed length of the spring, l_0 is the free length of the spring, κ_τ is the rotational torque constant, and κ_ρ is the extension spring constant.

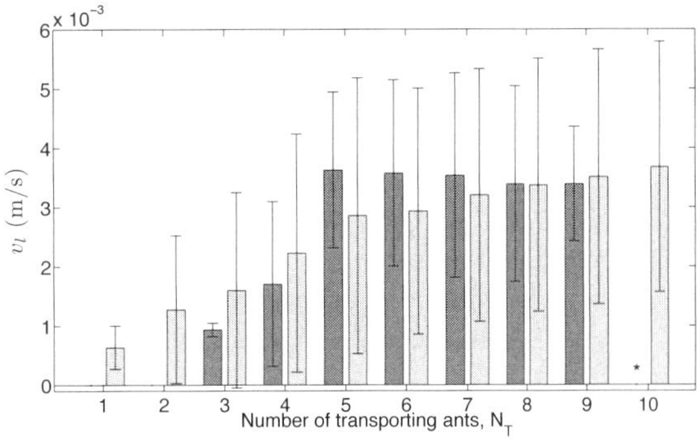

Fig. 9. Average load speed vs. transport group size in the experimental trials with structure 2 (dark gray bars) and simulation trials (light gray bars), which are described in Section VII. Error bars show standard deviations. Experiment sample sizes for $N_T = 4-9$ are $n =$ 561, 453, 4154, 2448, 1866, 182; data for $N_T = 1-3$ were obtained manually, and $*$ indicates that no data were available. Simulation sample sizes for $N_T = 1-10$ are $n =$ 1186594, 19971, 50522, 36362, 35459, 4956, 17700, 38777, 46289, 19234.

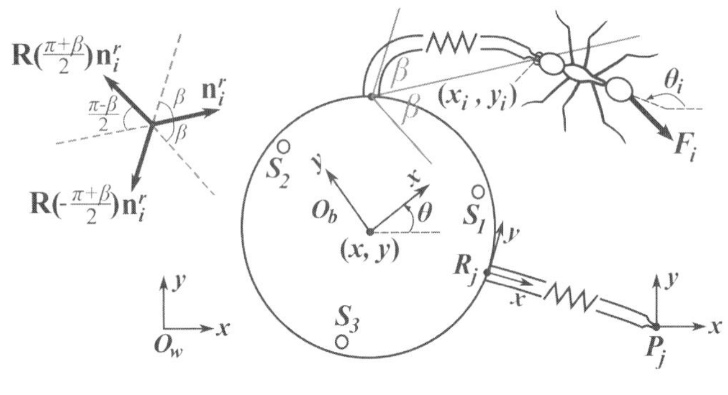

Fig. 10. Notation used in Sections V and VI for a simulated elastic structure with a transporting ant (two springs are shown).

To model friction, we use a finite support point approach similar to the one in [3]. We assume that the object is supported by three support points S_i, $i = 1,2,3$, whose coordinates are $(x_{S,i}, y_{S,i})$ in frame O_b. The friction forces λ_t comprise an unknown 6×1 vector:

$$\lambda_t = [\lambda_{t,1}\ \lambda_{t,2}\ \lambda_{t,3}]^T, \quad \lambda_{t,i} = [\lambda_{t,i,x}\ \lambda_{t,i,y}]. \tag{11}$$

The friction cone at the ith support point with a *known* normal force $\lambda_{n,i}$ is denoted by $\mathcal{FC}_i$ and is defined by Coulomb friction: $0 \leq \|\lambda_{t,i}\| \leq \mu\lambda_{n,i}$. We can write the 2×1 velocity vector of the support point for a load twist as:

$$v_{t,i} = \begin{bmatrix} 1 & 0 & -y_{S,i} \\ 0 & 1 & x_{S,i} \end{bmatrix} \xi_{load} \equiv \mathbf{B}_i\, \xi_{load}. \tag{12}$$

From Coulomb's law, the friction forces are equal to $\mu\lambda_{n,i}$ and are opposite to the direction of slip, except if the slip is zero, in which case the magnitude is indeterminate. This can be written explicitly as:

$$\lambda_{t,i}(\xi_{load}) \in \underset{\lambda_i \in \mathcal{FC}_i}{\arg\min}\ v_{t,i}(\xi_{load})^T \lambda_i \tag{13}$$

or in aggregate form,

$$\lambda_t(\xi_{load}) \in \underset{\lambda_i \in \mathcal{FC}_i}{\arg\min}\ \xi_{load}^T \mathbf{B}^T \lambda, \tag{14}$$

where $\mathbf{B}^T = [\mathbf{B}_1^T\ \mathbf{B}_2^T\ \mathbf{B}_3^T]$.

We can determine the motion of the load by solving the following max-min problem,

$$\underset{\xi_{load} \in \Sigma}{\text{maximize}} \quad \underset{\lambda_{t,i} \in \mathcal{FC}_i}{\text{minimize}} \quad L(\xi, \lambda_t) \tag{15}$$

where L is the instantaneous power from the friction forces and the wrenches caused by the spring deformations:

$$L(\xi, \lambda_t) = \xi_{load}^T \mathbf{B}^T \lambda_t + \sum_{j=1}^{J} \xi_{sp,j}^T \mathbf{w}_j. \tag{16}$$

VI. Behavioral Model

We developed a model of the ant transport behavior from qualitative observations of the videos of transport trials. The transport strategy is represented as a hybrid system with probabilistic transitions between task modes. Fig. 10 illustrates the notation that we use in describing the model.

We consider a population of N ants, each represented as a point-mass agent governed by a kinematic model,

$$\dot{\mathbf{q}}_i = \mathbf{u}_i \tag{17}$$

where $\mathbf{q}_i = [\mathbf{r}_i\ \theta_i]^T = [x_i\ y_i\ \theta_i]^T$ denotes the position of a point associated with ant i and the ant's orientation, both in the inertial frame, and $\mathbf{u}_i = [u_i^x\ u_i^y\ u_i^\theta]^T$ is the vector of control inputs.

We model the ants as switching between two behavioral modes, each corresponding to a different $\mathbf{u}_i$. In the behavior *Search for grasp point*, an ant i moves toward the structure in search of a protrusion to latch onto with its mandibles. If the outermost tip of a spring enters the sensing range of the ant and the spring is not already occupied by another ant, the ant latches onto the spring with probability p_1. Once attached, the ant is in the *Transport mode* and pulls on the spring in a manner that directs the structure toward its estimate of the nest. The ant reverts to mode *Search for grasp point* with a probability per time step $p_{2,i}(t)$, which is defined as a sigmoid function of the magnitude of the force $\mathbf{F}_i(t)$ that the ant applies to the spring at time t:

$$p_{2,i}(t) = \left(1 + e^{-\zeta(\|\mathbf{F}_i(t)\| - 0.5F_m)}\right)^{-1}, \tag{18}$$

where F_m is a force close to the maximum force that the ant can apply and $\zeta > 1$. We chose a sigmoid function to reflect the ants' tendencies to release the spring more often at large deflections and to occasionally lose their grip or lose interest even at low deflections. The ants stop moving when $\|\mathbf{r} - \mathbf{r}_n\| \leq \rho$, where $\mathbf{r} = [x\ y]^T$ and $\mathbf{r}_n = [x_n\ y_n]^T$, the position of the nest, for some small constant ρ.

A. Search for grasp point

We define v_i and ω_i as the forward linear and angular speeds, respectively, of ant i. An ant's estimates of the structure position (x, y), given by $(\hat{x}, \hat{y})$, are defined as values of the random variables $\hat{X} \sim \mathcal{N}(x, \sigma^2)$ and $\hat{Y} \sim \mathcal{N}(y, \sigma^2)$, where σ^2 is a specified variance. These values are updated at each time step. The ant's desired orientation θ_i^d is defined as the angle of the vector from (x_i, y_i) to $(\hat{x}, \hat{y})$. A proportional controller is used to steer the ant's orientation to θ_i^d:

$$\omega_i = k e_i^\theta \ , \quad e_i^\theta = (\theta_i^d - \theta_i) \in [-\pi, \pi] \ . \tag{19}$$

A feedback linearization scheme is used to relate v_i and ω_i to the linear velocities of (x_i, y_i). Hence the control inputs are

$$\begin{bmatrix} u_i^x \\ u_i^y \\ u_i^\theta \end{bmatrix} = \begin{bmatrix} \cos\theta_i & -r\sin\theta_i \\ \sin\theta_i & r\cos\theta_i \\ 0 & 1 \end{bmatrix} \begin{bmatrix} v_i \\ \omega_i \end{bmatrix}, \tag{20}$$

where r is a small offset distance along the longitudinal axis of the ant.

B. Transport

An ant i in this mode is grasping the tip of spring j. The ant pulls on its attachment point, whose coordinates are given by $(x_i, y_i) = (x_{P_j}, y_{P_j})$, with a force $\mathbf{F}_i$. Let $\mathbf{n}_i^f$ be the unit vector in the direction of this force. Also define $\mathbf{n}_i^n$ as the unit vector from (x_i, y_i) toward the nest coordinates, (x_n, y_n), and $\mathbf{n}_i^r$ as the unit vector from (x_{R_j}, y_{R_j}), to (x_i, y_i). The angles of these vectors in the inertial frame are $\theta_i + \pi$, θ_i^n, and θ_i^r, respectively. To navigate to their nest (i.e., determine θ_i^n), ants use visual landmarks and path integration with celestial cues as a compass [25].

We observed that an ant grasping a spring will attempt to rotate the load into an orientation in which the ant can pull it while walking backward toward the nest. To emulate this behavior, which has also been noted in [23], we specify that at each time t, the ant can begin to reorient itself to any angle between $\theta_i^{min}(t) = \theta_i^r(t) - \beta$ and $\theta_i^{max}(t) = \theta_i^r(t) + \beta$, where $\beta \leq \pi/2$. Similarly to equation (19), a proportional controller steers θ_i:

$$\omega_i = k e_i^\theta \ , \quad e_i^\theta = (\theta_i^d - (\theta_i + \pi)) \in [-\pi, \pi] \ . \tag{21}$$

Fig. 10 illustrates a division of the plane into three regions that are bounded by light dashed lines; in the diagram, $\mathbf{R}(\theta)$ is the 2D rotation matrix. The desired orientation θ_i^d at a given time depends on which of the regions contains $\mathbf{n}_i^n$:

$$\theta_i^d = \begin{cases} \theta_i^n & \text{if} \quad \mathbf{n}_i^r \cdot \mathbf{n}_i^n \geq \cos\beta \\ \theta_i^{max} & \text{if} \quad \mathbf{R}(\frac{\pi+\beta}{2})\mathbf{n}_i^r \cdot \mathbf{n}_i^n \geq \cos\frac{\pi-\beta}{2} \\ \theta_i^{min} & \text{if} \quad \mathbf{R}(-\frac{\pi+\beta}{2})\mathbf{n}_i^r \cdot \mathbf{n}_i^n \geq \cos\frac{\pi-\beta}{2} \end{cases} \tag{22}$$

The control inputs are defined as follows:

$$\mathbf{u}_i = [-v_i \cos\theta_i \ \ -v_i \sin\theta_i \ \ \omega_i]^T \ . \tag{23}$$

VII. Collective Transport Simulation

We simulated transport scenarios with different numbers of ants and springs on the elastic structure. The configurations of the structure and the ants are updated at each time step using the dynamic manipulation model in Section V and the behavioral model in Section VI. The simulated structure has the dimensions and weight of structure 2, with spring constants $\kappa_\tau = 4.52 \times 10^{-4}$ $\frac{\text{Nm}}{\text{rad}}$ and $\kappa_\rho = 0.664$ $\frac{\text{N}}{\text{m}}$ obtained from the spring model in Section III-D. We selected $p_1 = 0.2$ and $\beta = 60°$ to emulate the observed ant behavior and $F_m = 20$ mN to achieve maximum ant forces near the experimental value of ~ 30 mN (see Section IV-C.1). We also set $k = 10$ s^{-1}, $r = 2$ mm, $\zeta = 100$, and $\sigma^2 = 0.01$. The ants initially have uniformly randomly distributed positions and orientations. All ants start in *Search for grasp point* mode, in which they move at $v_i = 0.08$ m/s, which was estimated from video data. The nest is located at (0.2 m, -0.17 m).

Fig. 12 shows snapshots of a simulation with four ants and four springs and the resulting load trajectory and orientation, and Fig. 11 gives the corresponding load speeds and sum of interaction forces over time. We also conducted trials with one ant/one spring, five ants/ten springs, and ten ants/ten springs. In the simulations, the average ant force $||\mathbf{F}_i||$ was 5.0 ± 5.1 mN and the maximum force was 36.5 mN (sample size $n = 520057$), which are comparable to the experimental values. The interaction force sum in Fig. 11 varies over a similar range to the sum in Fig. 7. Fig. 9 shows that the simulations produced average v_l that were fairly close to the observed v_l for $N_T = 3 - 9$. In addition, the figure illustrates that the simulations reproduce the effect of diminishing increases in average v_l with larger transport populations, although the speed may not have saturated yet at $N_T = 10$.

The simulations replicate not only the successful transport of the structure to the nest, as shown by the trajectory in Fig. 12, but also the distinctive property of a phase transition to a higher degree of coordination among individuals. The speed v_l in Fig. 11 displays a similar trend to the experimental v_l in Fig. 5: a phase of low average speed with relatively small peaks is followed by a phase of increased speed. In the simulation, this transition is also marked by increased angular rotation of the load, as shown by the plot of ω_l. While the sum of the interaction forces in Fig. 11 does not obviously signal the phase change as it does in Fig. 7, it displays a sharp reduction around $t = 83$ s that coincides with the jump in v_l. The snapshots in Fig. 12 show that the the transporting ants are arranged in a less cooperative pose in the first phase than in the second, as occurs in the snapshots of the experiments.

VIII. Conclusion

We have presented a study of the mechanics of cooperative prey retrieval in *A. cockerelli* ants based on experimental trials with vision-based force sensors. The data show that this distributed transport system is characterized by an initial disordered phase that transitions to a coordinated phase of increased load speed and a higher degree of cooperation among the transporters, as well as a plateau in the progress of the task after a threshold population is reached. We developed a quasi-static dynamic manipulation model and an ant behavioral model based on observations and incorporated these models into a simulation that replicates the experimental phenomena.

In future work, we would like to conduct further experiments to extract the ants' changing roles, examine the mechanics of lifting during transport, and investigate the role of sustained interaction forces in helping to avoid obstacles and overcome deadlocks. Another objective is the implementation of the ant-like cooperative transport strategies on the robots in

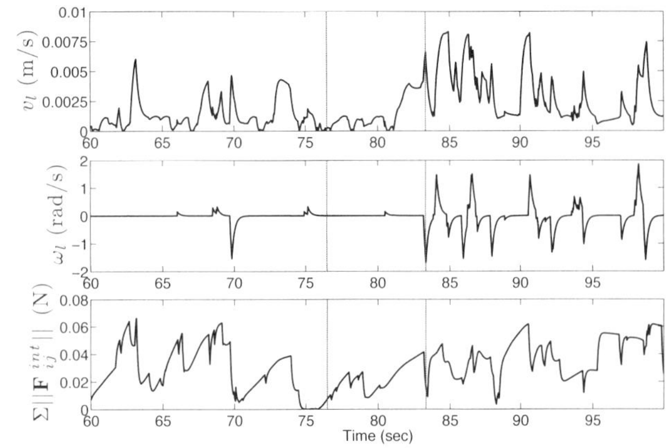

Fig. 11. Linear and angular load speeds and sum of interaction forces over time for the simulated trial in Fig. 12. Vertical lines indicate the times of the snapshots.

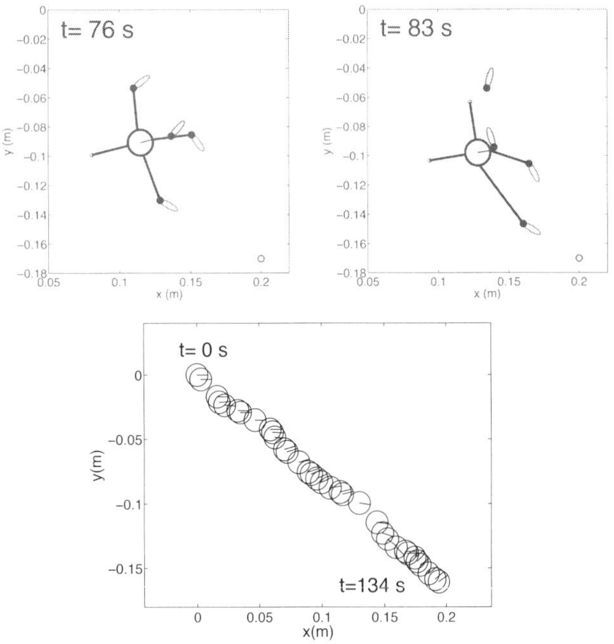

Fig. 12. Snapshots of simulated ant transport (top two plots); load position and orientation (bottom). In the top two plots, ants that are attached to the tips of the springs are in *Transport* mode, the other ants are in *Search for grasp point* mode, and the nest is indicated by the circle at (0.2 m, -0.17 m).

Fig. 1. The dynamic manipulation model has been adapted to these robots and validated in [11]. The implementation will require decentralized approaches for locating attachment points on an object, such as through a vision-based method; grasping these points firmly; agreeing on the target destination through a consensus algorithm if it is not preprogrammed; and navigating toward this location while avoiding obstacles and other robots. While ants can sidestep, the robots are nonholonomic and so they cannot imitate these maneuvers, which must be taken into account in their motion control.

Acknowledgments

We gratefully acknowledge support from NSF IIS-0427313, ARO Grant W911NF-05-1-0219, and ONR Grants N00014-07-1-0829 and N00014-08-1-0696.

References

[1] D. Cappelleri, G. Piazza, and V. Kumar. Two-dimensional, vision-based μN force sensor for microrobotics. In *Proc. Int'l. Conf. on Robotics and Automation (ICRA)*, Kobe, Japan, 2009.

[2] R. Chauvin. Le transport de proies chez les fourmis. Y-a-t-il entr'aide? *Behaviour*, 2(4):249–256, 1950.

[3] P. Cheng, J. Fink, V. Kumar, and J. S. Pang. Cooperative towing with multiple robots. *ASME J. Mechanisms and Robotics*, 1(1), 2009.

[4] J. Fink, N. Michael, and V. Kumar. Composition of vector fields for multi-robot manipulation via caging. In *Robotics: Science and Systems (RSS)*, 2007.

[5] B. Holldobler, R. C. Stanton, and H. Markl. Recruitment and food-retrieving behavior in Novomessor (Formicidae, Hymenoptera: I. Chemical signals). *Behav. Ecol. Sociobiol.*, 4:163–181, 1978.

[6] B. Holldobler and E. O. Wilson. *The Ants*. Harvard University Press, Cambridge, MA, 1990.

[7] L. Howell. *Compliant Mechanisms*. John Wiley & Sons, Inc., New York, 2001.

[8] O. Khatib, K. Yokoi, K. Chang, D. Ruspini, R. Holmberg, and A. Casal. Coordination and decentralized cooperation of multiple mobile manipulators. *J. of Robotic Systems*, 13(11):755–764, 1996.

[9] C. R. Kube and E. Bonabeau. Cooperative transport by ants and robots. *Robotics and Autonomous Systems*, 30:85–101, 2000.

[10] V. Kumar and K. Waldron. Force distribution in closed kinematic chains. *IEEE J. Robotics and Automation*, 4(6):657–664, 1988.

[11] Q. Lindsey, M. Shomin, and V. Kumar. Cooperative quasi-static planar manipulation with multiple robots. In *ASME IDETC/CIE*, 2010.

[12] Arnaud Lioni. *Ch. VII: Le transport collectif chez Oecophylla longinoda. From: Auto-assemblage et transport collectif chez Oecophylla. Unpublished.* PhD thesis, Univ. Libre de Bruxelles, Belgium, 2000.

[13] Y. Liu, S. Arimoto, V. Parra-Vega, and K. Kitagaki. Decentralized adaptive control of multiple manipulators in co-operations. *Int'l. J. of Control*, 67(5):649–673, 1997.

[14] T. Lochmatter, P. Roduit, C. Cianci, N. Correll, J. Jacot, and A. Martinoli. SwisTrack - A Flexible Open Source Tracking Software for Multi-Agent Systems. In *Proc. IEEE/RSJ Int'l. Conf. on Intelligent Robots and Systems (IROS)*, pages 4004–4010, 2008.

[15] N. Michael, J. Fink, and V. Kumar. Experimental testbed for large multirobot teams: verification and validation. *IEEE Robotics and Automation Magazine*, 15(1):53–61, 2008.

[16] M. W. Moffett. Cooperative food transport by an Asiatic ant. *National Geographic Research*, 4(3):386–394, 1988.

[17] G. Montemayor and J. Wen. Decentralized collaborative load transport by multiple robots. In *Proc. Int'l. Conf. on Robotics and Automation (ICRA)*, pages 372–377, 2005.

[18] R. Murray, Z. Li, and S. Sastry. *A Mathematical Introduction to Robotic Manipulation*. CRC Press, Florida, 1994.

[19] G. A. S. Pereira, V. Kumar, and M. F. M. Campos. Decentralized algorithms for multi-robot manipulation via caging. *Int'l. Journal of Robotics Research*, 23:783–795, 2002.

[20] H. Su and V. Krovi. Decentralized dynamic control of a nonholonomic mobile manipulator collective: a simulation study. In *Proc. ASME Dynamic Systems and Control Conf. (DSCC)*, Ann Arbor, MI, 2008.

[21] J. H. Sudd. The transport of prey by an ant, Pheidole crassinoda. *Behaviour*, 16:295–308, 1960.

[22] J. H. Sudd. How insects work in groups. *Discovery*, 24(6):15–19, 1963.

[23] J. H. Sudd. The transport of prey by ants. *Behaviour*, 25:234–271, 1965.

[24] D. Sun and J. K. Mills. Manipulating rigid payloads with multiple robots using compliant grippers. *IEEE/ASME Trans. on Mechatronics*, 7(1):23–34, 2002.

[25] R. Wehner. The architecture of the desert ant's navigational toolkit (Hymenoptera: Formicidae). *Myrmecological News*, 12:85–96, 2009.

[26] Y. Zhao, C. C. Lim, D. B. Sawyer, L. Ronglih, and X. Zhang. Cellular force measurements using single-spaced polymeric microstructures: isolating cells from base substrate. *J. Micromech. Microeng.*, 15:1649–1656, 2005.

Incremental Sampling-based Algorithms for Optimal Motion Planning

Sertac Karaman Emilio Frazzoli

Abstract— **During the last decade, incremental sampling-based motion planning algorithms, such as the Rapidly-exploring Random Trees (RRTs), have been shown to work well in practice and to possess theoretical guarantees such as probabilistic completeness. However, no theoretical bounds on the quality of the solution obtained by these algorithms, e.g., in terms of a given cost function, have been established so far. The purpose of this paper is to fill this gap, by designing efficient incremental sampling-based algorithms with provable optimality properties. The first contribution of this paper is a negative result: it is proven that, under mild technical conditions, the cost of the best path returned by RRT converges almost surely to a non-optimal value, as the number of samples increases. Second, a new algorithm is considered, called the Rapidly-exploring Random Graph (RRG), and it is shown that the cost of the best path returned by RRG converges to the optimum almost surely. Third, a tree version of RRG is introduced, called RRT*, which preserves the asymptotic optimality of RRG while maintaining a tree structure like RRT. The analysis of the new algorithms hinges on novel connections between sampling-based motion planning algorithms and the theory of random geometric graphs. In terms of computational complexity, it is shown that the number of simple operations required by both the RRG and RRT* algorithms is asymptotically within a constant factor of that required by RRT.**

I. Introduction

Even though modern robots may posses significant differences in sensing, actuation, size, application, or workspace, the motion planning problem, i.e., the problem of planning a dynamically feasible trajectory through a complex environment cluttered with obstacles, is embedded and essential in almost all robotics applications. Moreover, this problem has several applications in other disciplines such as verification, computational biology, and computer animation [1]–[5].

Motion planning has been a highly active area of research since the late 1970s. Early approaches to the problem has mainly focused on the development of *complete* planners (see, e.g., [6]), which find a solution if one exists and return failure otherwise. However, it was established as early as 1979 that even a most basic version of the motion planning problem, called the piano mover's problem, is known to be PSPACE-hard [7], which strongly suggests that complete planners are doomed to suffer from computational complexity.

Tractable algorithms approach the motion planning problem by relaxing the completeness requirement to, for instance, *resolution completeness*, which amounts to finding a solution, if one exists, when the resolution parameter of the algorithm is set fine enough. Most motion planning methods that are based on gridding or cell decomposition fall into this category. A more recent line of research that has achieved a great success has focused on the construction of paths connecting randomly-sampled points. Algorithms such as Probabilistic RoadMaps (PRM) [8] have been shown to be *probabilistically complete*, i.e., such that the probability of finding a solution, if one exists, approaches one as the number of samples approaches infinity.

The authors are with the Laboratory for Information and Decision Systems, Massachusetts Institute of Technology, Cambridge, MA.

The PRM algorithm constructs a graph of feasible paths off-line, and is primarily aimed at multiple-query applications, in which several motion-planning problems need to be solved in the same environment. Incremental sampling-based algorithms have been developed for single-query, real-time applications; among the most influential of these, one can mention Rapidly-exploring Random Trees (RRT) [9], and the algorithm in [10]. These algorithms have been shown to be probabilistically complete, with an exponential decay of the probability of failure. Moreover, RRTs were demonstrated in various robotic platforms in major robotics events (see, e.g., [11]).

A class of incremental sampling-based motion planning algorithms that is worth mentioning at this point is the Rapidly-exploring Random Graphs (RRGs), which were proposed in [12] to find feasible trajectories that satisfy specifications other than the usual "avoid all the obstacles and reach the goal region". More generally, RRGs can handle specifications given in the form of deterministic μ-calculus, which includes the widely-used Linear Temporal Logic (LTL). RRGs incrementally build a graph of trajectories, since specifications given in μ-calculus, in general, require infinite-horizon looping trajectories, which are not included in trees.

To address the challenges posed by real-time applications, state-of-the-art motion planning algorithms, such as RRTs, are tailored to return a feasible solution quickly, paying almost no attention to the "quality" of the solution. On the other hand, in typical implementations [11], the algorithm is not terminated as soon as the first feasible solution is found; rather, all the available computation time is used to search for an improved solution, with respect to a performance metric such as time, path length, fuel consumption, etc. A shortcoming of this approach is that there is no guarantee that the computation will eventually converge to optimal trajectories. In fact, despite the clear practical need, there has been little progress in characterizing optimality properties of sampling-based motion planning algorithms, even though the importance of these problems was emphasized in early seminal papers such as [9].

Yet, the importance of the quality of the solution returned by the planners has been noticed, in particular, from the point of view of incremental sampling-based motion planning. In [13], Urmson and Simmons have proposed heuristics to bias the tree growth in the RRT towards those regions that result in low-cost solutions. They have also shown experimental results evaluating the performance of different heuristics in terms of the quality of the solution returned. In [14], Ferguson and Stentz have considered running the RRT algorithm multiple times in order to progressively improve the quality of the solution. They showed that each run of the algorithm results in a path with smaller cost, even though the procedure is not guaranteed to converge to an optimal solution.

To the best of the authors' knowledge, this paper provides the first thorough analysis of optimality properties of incremental sampling-based motion planning algorithms. In particular, it is shown that the probability that the RRT converges to an optimal solution, as the number of samples approaches infinity, is zero under some reasonable technical assumptions. In fact, the RRT algorithm almost always converges to a non-optimal solution. Second, it is shown that the probability of the same event for the RRG algorithm is one. That is, the RRG algorithm is asymptotically optimal, in the sense that it converges to an optimal solution almost surely as the number of samples approaches infinity. Third, a novel variant of the RRG algorithm is introduced, called RRT*, which inherits the asymptotic optimality of the RRG algorithm while maintaining a tree structure. To do so, the RRT* algorithm essentially "rewires" the tree as it discovers new lower-cost paths reaching the nodes that are already in the tree. Finally, it is shown that the asymptotic computational complexity of the RRG and RRT* algorithms is essentially the same as that of RRTs.

To the authors' knowledge, the algorithms considered in this paper are the first computationally efficient incremental sampling-based motion planning algorithms with asymptotic optimality guarantees. Indeed, the results in this paper imply that these algorithms are optimal also from an asymptotic computational complexity point of view, since they closely match lower bounds for computing nearest neighbors. The key insight is that connections between vertices in the graph should be sought within balls whose radius vanishes with a certain rate as the size of the graph increases, and is based on new connections between motion planning and the theory of random geometric graphs [15], [16].

The paper is organized as follows. Section II lays the ground in terms of notation and problem formulation. Section III is devoted to the introduction of the RRT and RRG algorithms. In Section IV, these algorithms are analyzed in terms of probabilistic completeness, asymptotic optimality, and computational complexity. The RRT* algorithm is presented in Section V, where it is shown that RRT* inherits the theoretical guarantees of the RRG algorithm. Experimental results are presented and discussed in Section VI. Concluding remarks and directions for future work are given in Section VII.

Due to space limitations, results are stated without formal proofs. An extended version of this paper, including proofs of the major results, technical discussions, and extensive experimental results, is available [17]. An implementation of the RRT* algorithm in the C language is available at `http://ares.lids.mit.edu/software`.

The focus of this paper is on the basic problem of navigating through a connected bounded subset of a d-dimensional Euclidean space. However, the proposed algorithms also extend to systems with differential constraints, as shown in [18].

II. Preliminary Material

A. Notation

A sequence on a set A, denoted as $\{a_i\}_{i \in \mathbb{N}}$, is a mapping from $\mathbb{N}$ to A with $i \mapsto a_i$. Given $a, b \in \mathbb{R}$, the closed interval between a and b is denoted by $[a, b]$. The Euclidean norm is denoted by $\| \cdot \|$. Given a set $X \subset \mathbb{R}^d$, the closure of X is denoted by $\mathrm{cl}(X)$, the Lebesgue measure of X, i.e., its volume, is denoted by $\mu(X)$. The closed ball of radius $r > 0$ centered at $x \in \mathbb{R}^d$ is defined as $\mathcal{B}_{x,r} := \{y \in \mathbb{R}^d \mid \|y - x\| \leq r\}$. The volume of the unit ball in $\mathbb{R}^d$ is denoted by ζ_d.

Given a set $X \subset \mathbb{R}^d$, and a scalar $s \geq 0$, a path in X is a continuous function $\sigma : [0, s] \to X$, where s is the length of the path defined in the usual way. Given two paths in X, $\sigma_1 : [0, s_1] \to X$, and $\sigma_2 : [0, s_2] \to X$, with $\sigma_1(s_1) = \sigma_2(0)$, their concatenation is denoted by $\sigma_1|\sigma_2$, i.e., $\sigma = \sigma_1|\sigma_2 : [0, s_1 + s_2] \to X$ with $\sigma(s) = \sigma_1(s)$ for all $s \in [0, s_1]$, and $\sigma(s) = \sigma_2(s - s_1)$ for all $s \in [s_1, s_1 + s_2]$. The set of all paths in X with nonzero length is denoted by Σ_X. The straight continuous path between $x_1, x_2 \in \mathbb{R}^d$ is denoted by $\mathtt{Line}(x_1, x_2)$.

Let $(\Omega, \mathcal{F}, \mathbb{P})$ be a probability space. A random variable is a measurable function from Ω to $\mathbb{R}$; an extended random variable can also take the values $\pm\infty$. A sequence $\{\mathcal{Y}_i\}_{i \in \mathbb{N}}$ of random variables is said to converge surely to a random variable $\mathcal{Y}$ if $\lim_{i \to \infty} \mathcal{Y}_i(\omega) = \mathcal{Y}(\omega)$ for all $\omega \in \Omega$; the sequence is said to converge almost-surely if $\mathbb{P}(\{\lim_{i \to \infty} \mathcal{Y}_i = \mathcal{Y}\}) = 1$.

B. Problem Formulation

In this section, two variants of the path planning problem are presented. First, the feasibility problem in path planning is formalized, then the optimality problem is introduced.

Let X be a bounded connected open subset of $\mathbb{R}^d$, where $d \in \mathbb{N}$, $d \geq 2$. Let X_{obs} and X_{goal}, called the *obstacle region* and the *goal region*, respectively, be open subsets of X. Let us denote the *obstacle-free space*, i.e., $X \setminus X_{\mathrm{obs}}$, as X_{free}. Let the *initial state*, x_{init}, be an element of X_{free}. In the sequel, a path in X_{free} is said to be a *collision-free path*. A collision-free path that starts at x_{init} and ends in the goal region is said to be a *feasible path*, i.e., a collision-free path $\sigma : [0, s] \to X_{\mathrm{free}}$ is feasible if and only if $\sigma(0) = x_{\mathrm{init}}$ and $\sigma(s) \in X_{\mathrm{goal}}$.

The *feasibility problem* of path planning is to find a feasible path, if one exists, and report failure otherwise.

Problem 1 (Feasible planning) *Given a bounded connected open set $X \subset \mathbb{R}^d$, an obstacle space $X_{\mathrm{obs}} \subset X$, an initial state $x_{\mathrm{init}} \in X_{\mathrm{free}}$, and a goal region $X_{\mathrm{goal}} \subset X_{\mathrm{free}}$, find a path $\sigma : [0, s] \to X_{\mathrm{free}}$ such that $\sigma(0) = x_{\mathrm{init}}$ and $\sigma(s) \in X_{\mathrm{goal}}$, if one exists. If no such path exists, then report failure.*

Let $c : \Sigma_{X_{\mathrm{free}}} \to \mathrm{R}_{>0}$ be a function, called the *cost function*, which assigns a non-negative cost to all nontrivial collision-free paths. The *optimality problem* of path planning asks for finding a feasible path with minimal cost.

Problem 2 (Optimal planning) *Given a bounded connected open set X, an obstacle space X_{obs}, an initial state x_{init}, and a goal region X_{goal}, find a path $\sigma^* : [0, s] \to \mathrm{cl}(X_{\mathrm{free}})$ such that (i) $\sigma^*(0) = x_{\mathrm{init}}$ and $\sigma^*(s) \in X_{\mathrm{goal}}$, and (ii) $c(\sigma^*) = \min_{\sigma \in \Sigma_{\mathrm{cl}(X_{\mathrm{free}})}} c(\sigma)$. If no such path exists, then report failure.*

III. ALGORITHMS

In this section, two incremental sampling-based motion planning algorithms, namely the RRT and the RRG algorithms, are introduced. Before formalizing the algorithms, let us note the primitive procedures that they rely on.

Sampling: The function `Sample` : $\mathbb{N} \rightarrow X_{\rm free}$ returns independent identically distributed (i.i.d.) samples from $X_{\rm free}$.

Steering: Given two points $x, y \in X$, the function `Steer` : $(x, y) \mapsto z$ returns a point $z \in \mathbb{R}^d$ such that z is "closer" to y than x is. Throughout the paper, the point z returned by the function `Steer` will be such that z minimizes $\|z - y\|$ while at the same time maintaining $\|z - x\| \leq \eta$, for a prespecified $\eta > 0$, i.e., $\texttt{Steer}(x, y) = \mathrm{argmin}_{z \in \mathbb{R}^d, \|z-x\| \leq \eta} \|z - y\|$.

Nearest Neighbor: Given a graph $G = (V, E)$ and a point $x \in X_{\rm free}$, the function `Nearest` : $(G, x) \mapsto v$ returns a vertex $v \in V$ that is "closest" to x in terms of a given distance function. In this paper, we will use Euclidean distance (see, e.g., [9] for alternative choices), i.e., $\texttt{Nearest}(G = (V, E), x) = \mathrm{argmin}_{v \in V} \|x - v\|$.

Near Vertices: Given a graph $G = (V, E)$, a point $x \in X_{\rm free}$, and a number $n \in \mathbb{N}$, the function `Near` : $(G, x, n) \mapsto V'$ returns a set V' of vertices such that $V' \subseteq V$. The `Near` procedure can be thought of as a generalization of the nearest neighbor procedure in the sense that the former returns a collection of vertices that are close to x, whereas the latter returns only one such vertex that is the closest. Just like the `Nearest` procedure, there are many ways to define the `Near` procedure, each of which leads to different algorithmic properties. For technical reasons to become clear later, we define $\texttt{Near}(G, x, n)$ to be the set of all vertices within the closed ball of radius r_n centered at x, where $r_n = \min\left\{ \left(\frac{\gamma}{\zeta_d} \frac{\log n}{n} \right)^{1/d}, \eta \right\}$, and γ is a constant. Hence, the volume of this ball is $\min\{\gamma \frac{\log n}{n}, \zeta_d \eta^d\}$.

Collision Test: Given two points $x, x' \in X_{\rm free}$, the Boolean function $\texttt{ObstacleFree}(x, x')$ returns `True` iff the line segment between x and x' lies in $X_{\rm free}$, i.e., $[x, x'] \subset X_{\rm free}$.

Both the RRT and the RRG algorithms are similar to most other incremental sampling-based planning algorithms (see Algorithm 1). Initially, the algorithms start with the graph that includes the initial state as its single vertex and no edges; then, they incrementally grow a graph on $X_{\rm free}$ by sampling a state $x_{\rm rand}$ from $X_{\rm free}$ at random and extending the graph towards $x_{\rm rand}$. In the sequel, every such step of sampling followed by extensions (Lines 2-5 of Algorithm 1) is called a single *iteration* of the incremental sampling-based algorithm.

Hence, the body of both algorithms, given in Algorithm 1, is the same. However, RRGs and RRTs differ in the choice of the vertices to be extended. The `Extend` procedures for the RRT and the RRG algorithms are provided in Algorithms 2 and 3, respectively. Informally speaking, the RRT algorithm extends the nearest vertex towards the sample. The RRG algorithm first extends the nearest vertex, and if such extension is successful, it also extends all the vertices returned by the `Near` procedure, producing a graph in general. In both cases, all the extensions resulting in collision-free trajectories are added to the graph as edges, and their terminal points as new vertices.

Algorithm 1: Body of RRT and RRG Algorithms

1 $V \leftarrow \{x_{\rm init}\}$; $E \leftarrow \emptyset$; $i \leftarrow 0$;
2 **while** $i < N$ **do**
3     $G \leftarrow (V, E)$;
4     $x_{\rm rand} \leftarrow \texttt{Sample}(i)$; $i \leftarrow i + 1$;
5     $(V, E) \leftarrow \texttt{Extend}(G, x_{\rm rand})$;

Algorithm 2: $\texttt{Extend}_{RRT}(G, x)$

$V' \leftarrow V$; $E' \leftarrow E$;
$x_{\rm nearest} \leftarrow \texttt{Nearest}(G, x)$;
$x_{\rm new} \leftarrow \texttt{Steer}(x_{\rm nearest}, x)$;
if $\texttt{ObstacleFree}(x_{\rm nearest}, x_{\rm new})$ **then**
    $V' \leftarrow V' \cup \{x_{\rm new}\}$;
    $E' \leftarrow E' \cup \{(x_{\rm nearest}, x_{\rm new})\}$;
return $G' = (V', E')$

IV. ANALYSIS

A. Convergence to a Feasible Solution

In this section, the feasibility problem is considered. It is proven that the RRG algorithm inherits the probabilistic completeness as well as the exponential decay of the probability of failure (as the number of samples increase) from the RRT. These results imply that the RRT and RRG algorithms have the same performance in producing a solution to the feasibility problem as the number of samples increase.

Sets of vertices and edges of the graphs maintained by the RRT and the RRG algorithms can be defined as functions from the sample space Ω to appropriate sets. More precisely, let $\{\mathcal{V}_i^{\rm RRT}\}_{i \in \mathbb{N}}$ and $\{\mathcal{V}_i^{\rm RRG}\}_{i \in \mathbb{N}}$, sequences of functions defined from Ω into finite subsets of $X_{\rm free}$, be the sets of vertices in the RRT and the RRG, respectively, at the end of iteration i. By convention, we define $\mathcal{V}_0^{\rm RRT} = \mathcal{V}_0^{\rm RRG} = \{x_{\rm init}\}$. Similarly, let $\mathcal{E}_i^{\rm RRT}$ and $\mathcal{E}_i^{\rm RRG}$, defined for all $i \in \mathbb{N}$, denote the set of edges in the RRT and the RRG, respectively, at the end of iteration i. Clearly, $\mathcal{E}_0^{\rm RRT} = \mathcal{E}_0^{\rm RRG} = \emptyset$.

An important lemma used for proving the equivalency between the RRT and the RRG algorithms is the following.

Lemma 3 *For all $i \in \mathbb{N}$ and all $\omega \in \Omega$, $\mathcal{V}_i^{\rm RRT}(\omega) = \mathcal{V}_i^{\rm RRG}(\omega)$ and $\mathcal{E}_i^{\rm RRT}(\omega) \subseteq \mathcal{E}_i^{\rm RRG}(\omega)$.*

Lemma 3 implies that the paths discovered by the RRT

Algorithm 3: $\texttt{Extend}_{RRG}(G, x)$

$V' \leftarrow V$; $E' \leftarrow E$;
$x_{\rm nearest} \leftarrow \texttt{Nearest}(G, x)$;
$x_{\rm new} \leftarrow \texttt{Steer}(x_{\rm nearest}, x)$;
if $\texttt{ObstacleFree}(x_{\rm nearest}, x_{\rm new})$ **then**
    $V' \leftarrow V' \cup \{x_{\rm new}\}$;
    $E' \leftarrow E' \cup \{(x_{\rm nearest}, x_{\rm new}), (x_{\rm new}, x_{\rm nearest})\}$;
    $X_{\rm near} \leftarrow \texttt{Near}(G, x_{\rm new}, |V|)$;
    for *all* $x_{\rm near} \in X_{\rm near}$ **do**
        if $\texttt{ObstacleFree}(x_{\rm new}, x_{\rm near})$ **then**
            $E' \leftarrow E' \cup \{(x_{\rm near}, x_{\rm new}), (x_{\rm new}, x_{\rm near})\}$;
return $G' = (V', E')$

algorithm by the end of iteration i is, essentially, a subset of those discovered by the RRG by the end of the same iteration.

An algorithm addressing Problem 1 is said to be *probabilistically complete* if it finds a feasible path with probability approaching one as the number of iterations approaches infinity. Note that there exists a collision-free path starting from x_{init} to any vertex in the tree maintained by the RRT, since the RRT maintains a connected graph on X_{free} that necessarily includes x_{init}. Using this fact, the probabilistic completeness property of the RRT is stated alternatively as follows.

Theorem 4 (see [9]) *If there exists a feasible solution to Problem 1, then* $\lim_{i\to\infty} \mathbb{P}\left(\{\mathcal{V}_i^{\text{RRT}} \cap X_{\text{goal}} \neq \emptyset\}\right) = 1$.

An *attraction sequence* [9] is defined as a finite sequence $\mathcal{A} = \{A_1, A_2, \dots, A_k\}$ of sets as follows: (i) $A_0 = \{x_{\text{init}}\}$, and (ii) for each set A_i, there exists a set B_i, called the *basin* such that for any $x \in A_{i-1}$, $y \in A_i$, and $z \in X \setminus B_i$, there holds $\|x - y\| \leq \|x - z\|$ Given an attraction sequence $\mathcal{A}$ of length k, let p_k denote $\min_{i\in\{1,2,\dots,k\}} \left(\frac{\mu(A_i)}{\mu(X_{\text{free}})}\right)$.

The following theorem states that the probability that the RRT algorithm fails to return a solution, when one exists, decays to zero exponentially fast.

Theorem 5 (see [9]) *If there exists an attraction sequence $\mathcal{A}$ of length k, then* $\mathbb{P}\left(\{\mathcal{V}_i^{\text{RRT}} \cap X_{\text{goal}} = \emptyset\}\right) \leq e^{-\frac{1}{2}(i\,p_k - 2k)}$.

With Lemma 3 and Theorems 4 and 5, the following theorem is immediate.

Theorem 6 *If there exists a feasible solution to Problem 1, then* $\lim_{i\to\infty} \mathbb{P}\left(\{\mathcal{V}_i^{\text{RRG}} \cap X_{\text{goal}} \neq \emptyset\}\right) = 1$. *Moreover, if an attraction sequence $\mathcal{A}$ of length k exists, then* $\mathbb{P}\left(\{\mathcal{V}_i^{\text{RRG}} \cap X_{\text{goal}} = \emptyset\}\right) \leq e^{-\frac{1}{2}(i\,p_k - 2\,k)}$.

B. Asymptotic Optimality

This section is devoted to the investigation of optimality properties of the RRT and the RRG algorithms. First, under some mild technical assumptions, it is shown that the probability that the RRT converges to an optimal solution is zero. However, the convergence of this random variable is guaranteed, which implies that the RRT converges to a *non*-optimal solution with probability one. On the contrary, it is subsequently shown that the RRG algorithm converges to an optimal solution almost-surely.

Let $\{\mathcal{Y}_i^{\text{RRT}}\}_{i\in\mathbb{N}}$ be a sequence of extended random variables that denote the cost of a minimum-cost path contained within the tree maintained by the RRT algorithm at the end of iteration i. The extended random variable $\mathcal{Y}_i^{\text{RRG}}$ is defined similarly. Let c^* denote the cost of a minimum-cost path in $\text{cl}(X_{\text{free}})$, i.e., the cost of a path that solves Problem 2.

Let us note that the limits of these two extended random variable sequences as i approaches infinity exist. More formally, notice that $\mathcal{Y}_{i+1}^{\text{RRT}}(\omega) \leq \mathcal{Y}_i^{\text{RRT}}(\omega)$ holds for all $i \in \mathbb{N}$ and all $\omega \in \Omega$. Moreover, $\mathcal{Y}_i^{\text{RRT}}(\omega) \geq c^*$ for all $i \in \mathbb{N}$ and all $\omega \in \Omega$, by optimality of c^*. Hence, $\{\mathcal{Y}_i^{\text{RRT}}\}_{i\in\mathbb{N}}$ is a surely non-increasing sequence of random variables that is surely lower-bounded by c^*. Thus, for all $\omega \in \Omega$, the limit $\lim_{i\to\infty} \mathcal{Y}_i^{\text{RRT}}(\omega)$ exists. The same argument also holds for the sequence $\{\mathcal{Y}_i^{\text{RRG}}\}_{i\in\mathbb{N}}$.

1) Almost Sure Nonoptimality of the RRT: Let Σ^* denote the set of all optimal paths, i.e., the set of all paths that solve Problem 2, and X_{opt} denote the set of states that an optimal path in Σ^* passes through, i.e., $X_{\text{opt}} = \cup_{\sigma^*\in\Sigma^*} \cup_{\tau\in[0,s^*]} \{\sigma^*(\tau)\}$. Consider the following assumptions.

Assumption 7 (Zero-measure Optimal Paths) *The set of all points in the state-space that an optimal trajectory passes through has measure zero, i.e.,* $\mu(X_{\text{opt}}) = 0$.

Assumption 8 (Sampling Procedure) *The sampling procedure is such that the samples* $\{\texttt{Sample}(i)\}_{i\in\mathbb{N}}$ *are drawn from an absolutely continuous distribution with a continuous density function $f(x)$ bounded away from zero on* X_{free}.

Assumption 9 (Monotonicity of the Cost Function) *For all $\sigma_1, \sigma_2 \in \Sigma_{X_{\text{free}}}$, the cost function c satisfies the following:* $c(\sigma_1) \leq c(\sigma_1|\sigma_2)$.

Assumption 7 rules out trivial cases, in which the RRT algorithm can sample exactly an optimal path with non-zero probability. Assumption 8 also ensures that the sampling procedure can not be tuned to construct the optimal path exactly. Finally, Assumption 9 merely states that extending a path to produce a longer path can not decrease its cost.

Recall that d denotes the dimensionality of the state space. The negative result of this section is formalized as follows.

Theorem 10 *Let Assumptions 7, 8, and 9 hold. Then, the probability that the cost of the minimum-cost path in the RRT converges to the optimal cost is zero, i.e.,*

$$\mathbb{P}\left(\left\{\lim_{i\to\infty} \mathcal{Y}_i^{\text{RRT}} = c^*\right\}\right) = 0,$$

whenever $d \geq 2$.

The key idea in proving this result is that the probability of extending a node on an optimal path (e.g., the root node) goes to zero very quickly, in such a way that any such node will only have a finite number of children, almost surely. Because of Assumptions 7 and 8, this implies the result.

As noted before, the limit $\lim_{i\to\infty} \mathcal{Y}_i^{\text{RRT}}(\omega)$ exists and is a random variable. However, Theorem 10 directly implies that this limit is strictly greater than c^* with probability one, i.e., $\mathbb{P}\left(\{\lim_{i\to\infty} \mathcal{Y}_i^{\text{RRT}} > c^*\}\right) = 1$. In other words, it is established, as a corollary, that the RRT algorithm converges to a nonoptimal solution with probability one.

It is interesting to note that, since the cost of the best path returned by the RRT algorithm converges to a random variable, say $\mathcal{Y}_\infty^{\text{RRT}}$, Theorem 10 provides new insight explaining the effectiveness of approaches as in [14]. In fact, running multiple instances of the RRT algorithm amounts to drawing multiple samples of $\mathcal{Y}_\infty^{RRT}$.

2) Almost Sure Optimality of the RRG: Consider the following set of assumptions, which will be required to show the asymptotic optimality of the RRG.

Assumption 11 (Additivity of the Cost Function) *For all $\sigma_1, \sigma_2 \in \Sigma_{X_{\text{free}}}$, the cost function c satisfies the following:* $c(\sigma_1|\sigma_2) = c(\sigma_1) + c(\sigma_2)$.

Assumption 12 (Continuity of the Cost Function) *The cost function c is Lipschitz continuous in the following sense: there exists some constant κ such that for any two paths $\sigma_1 : [0, s_1] \to X_{\text{free}}$ and $\sigma_2 : [0, s_2] \to X_{\text{free}}$, $|c(\sigma_1) - c(\sigma_2)| \leq \kappa \sup_{\tau \in [0,1]} \|\sigma_1(\tau\, s_1) - \sigma_2(\tau\, s_2)\|$.*

Assumption 13 (Obstacle Spacing) *There exists a constant $\delta \in \mathbb{R}_+$ such that for any point $x \in X_{\text{free}}$, there exists $x' \in X_{\text{free}}$, such that (i) the δ-ball centered at x' lies inside X_{free}, i.e., $\mathcal{B}_{x',\delta} \subset X_{\text{free}}$, and (ii) x lies inside the δ-ball centered at x', i.e., $x \in \mathcal{B}_{x',\delta}$.*

Assumption 12 ensures that two paths that are very close to each other have similar costs. Let us note that several cost functions of practical interest satisfy Assumptions 11 and 12. Assumption 13 is a rather technical assumption, which ensures existence of some free space around the optimal trajectories to allow convergence. For simplicity, it is assumed that the sampling is uniform, although the results can be directly extended to more general sampling procedures.

Recall that d is the dimensionality of the state-space X, and γ is the constant defined in the `Near` procedure. The positive result that states the asymptotic optimality of the RRG algorithm can be formalized as follows.

Theorem 14 *Let Assumptions 11, 12, and 13 hold, and assume that Problem 1 admits a feasible solution. Then, the cost of the minimum-cost path in the RRG converges to the optimal cost almost-surely, i.e.,*

$$\mathbb{P}\left(\left\{\lim_{i \to \infty} \mathcal{Y}_i^{\text{RRG}} = c^*\right\}\right) = 1,$$

whenever $d \geq 2$ and $\gamma > \gamma_L := 2^d(1 + 1/d)\mu(X_{\text{free}})$.

This result relies on the fact that a random geometric graph with n vertices formed by connecting each vertex with vertices within a distance of $d_n = \gamma' (\log n \,/\, n)^{1/d}$ will result in a connected graph almost surely as $n \to \infty$, whenever γ' is larger than a certain lower bound γ_1 [19]. In fact, the bound on γ' is a *tight threshold* in the sense that there exists an upper bound $\gamma_2 < \gamma_1$ such that, if $\gamma' < \gamma_2$, then the resulting graph will be disconnected almost surely [19]. This result strongly suggests that shrinking the ball in the `Near` procedure faster than the rate proposed will not yield an asymptotically optimal algorithm. The authors have experienced this fact in simulation studies: setting γ to around one third of γ_L does not seem to provide the asymptotic optimality property. On the other hand, as it will be shown in the next section, if the size of the same ball is reduced slower than the proposed rate, then the asymptotic complexity of the resulting algorithm will not be the same as the RRT. Hence, scaling r_n with $(\log n \,/\, n)^{1/d}$ in the `Near` procedure, surprisingly, achieves the perfect balance between asymptotic optimality and low computational complexity, since relevant results in the theory of random geometric graphs and lower bounds on nearest neighbor computation strongly suggest that a different rate will lose either the former or the latter while failing to provide an extra benefit in any of the two.

C. Computational Complexity

The objective of this section is to compare the computational complexity of RRTs and RRGs. It is shown that these algorithms share essentially the same asymptotic computational complexity in terms of the number of calls to *simple operations* such as comparisons, additions, and multiplications.

Consider first the computational complexity of the RRT and the RRG algorithms in terms of the number of calls to the primitive procedures introduced in Section III. Notice that, in every iteration, the number of calls to `Sample`, `Steer`, and `Nearest` procedures are the same in both algorithms. However, number of calls to `Near` and `ObstacleFree` procedures differ: the former is never called by the RRT and is called at most once by the RRG, whereas the latter is called exactly once by the RRT and at least once by the RRG.

Let $\mathcal{O}_i^{\text{RRG}}$ be a random variable that denotes the number of calls to the `ObstacleFree` procedure by the RRG algorithm in iteration i. Notice that, as an immediate corollary of Lemma 3, the number of vertices in the RRT and RRG algorithms is the same at any given iteration. Let $\mathcal{N}_i$ be the number of vertices in these algorithms at the end of iteration i. The following theorem establishes that the expected number of calls to the `ObstacleFree` procedure in iteration i by the RRG algorithm scales logarithmically with the number of vertices in the graph as i approaches infinity.

Lemma 15 *In the limit as i approaches infinity, the random variable $\mathcal{O}_i^{\text{RRG}} / \log(\mathcal{N}_i)$ is no more than a constant in expectation, i.e., $\limsup_{i \to \infty} \mathbb{E}\left[\frac{\mathcal{O}_i^{\text{RRG}}}{\log(\mathcal{N}_i)}\right] \leq \phi$, where $\phi \in \mathbb{R}_{>0}$ is a constant that depends only on the problem instance.*

However, some primitive procedures clearly take more computation time to execute than others. For a meaningful comparison, one should also evaluate the time required to execute each primitive procedure in terms of the number of simple operations (also called steps) that they perform. This analysis shows that the expected number of simple operations performed by the RRG is asymptotically within a constant factor of that performed by the RRT, which establishes that the RRT and the RRG algorithms have the same asymptotic computational complexity in terms of the number of steps that they perform.

First, notice that `Sample`, `Steer`, and `ObstacleFree` procedures can be performed in a constant number of steps, i.e., independent of the number of vertices in the graph.

Second, consider the computational complexity of the `Nearest` procedure. The problem of finding a nearest neighbor is widely studied, e.g., in the computer graphics literature. Even though algorithms that achieve sub-linear time complexity are known [20], lower bounds suggest that nearest neighbor computation requires at least logarithmic time [21]. In fact, assuming that the `Nearest` procedure computes an approximate nearest neighbor (see, e.g., [21] for a formal definition) using the algorithm given in [21], which is optimal in fixed dimensions from a computational complexity point of view closely matching a lower bound for tree-based algorithms, the `Nearest` algorithm has to run in $\Omega(\log n)$ time as formalized in the following lemma.

Let $\mathcal{M}_i^{\mathrm{RRT}}$ be the random variable that denotes the number of steps executed by the RRT algorithm in iteration i.

Lemma 16 *Assuming that* `Nearest` *is implemented using the algorithm given in [21], which is computationally optimal in fixed dimensions, the number of steps executed by the RRT algorithm at each iteration is at least order* $\log(\mathcal{N}_i)$ *in expectation in the limit, i.e., there exists a constant* $\phi_{RRT} \in \mathbb{R}_{>0}$ *such that* $\liminf_{i \to \infty} \mathbb{E}\left[\frac{\mathcal{M}_i^{\mathrm{RRT}}}{\log(\mathcal{N}_i)}\right] \geq \phi_{RRT}$.

Likewise, problems similar to that solved by the `Near` procedure are also widely studied in the literature, generally under the name of *range search problems*, as they have many applications in, for instance, computer graphics [20].

Similar to the nearest neighbor search, computing approximate solutions to the range search problem is computationally easier. A range search algorithm is said to be ε-approximate if it returns all vertices that reside in the ball of size r_n and no vertices outside a ball of radius $(1+\varepsilon)\, r_n$, but may or may not return the vertices that lie outside the former ball and inside the latter ball. In fixed dimensions, computing ε-approximate solutions can, in fact, be done in logarithmic time using polynomial space, in the worst case [22].

Note that the `Near` procedure can be implemented as an approximate range search while maintaining the asymptotic optimality guarantee. Notice that the expected number of vertices returned by the `Near` procedure also does not change, except by a constant factor. Hence, the `Near` procedure can be implemented to run in order $\log n$ expected time in the limit and linear space in fixed dimensions.

Let $\mathcal{M}_i^{\mathrm{RRG}}$ denote the number of steps performed by the RRG algorithm in iteration i. Then, together with Lemma 15, the discussion above implies the following lemma.

Lemma 17 *Assuming that the* `Near` *procedure is implemented using the algorithm given in [22], the number of steps executed by the RRG algorithm at each iteration is at most order* $\log(\mathcal{N}_i)$ *in expectation in the limit, i.e., there exists a constant* $\phi_{RRG} \in \mathbb{R}_{>0}$ *such that* $\limsup_{i \to \infty} \mathbb{E}\left[\frac{\mathcal{M}_i^{\mathrm{RRG}}}{\log(\mathcal{N}_i)}\right] \leq \phi_{RRG}$.

Finally, by Lemmas 16 and 17, we conclude that the RRT and the RRG algorithms have the same asymptotic computational complexity as stated in the following theorem.

Theorem 18 *Under the assumptions of Lemmas 16 and 17, there exists a constant* $\phi \in \mathbb{R}_{>0}$ *such that*

$$\limsup_{i \to \infty} \mathbb{E}\left[\frac{\mathcal{M}_i^{\mathrm{RRG}}}{\mathcal{M}_i^{\mathrm{RRT}}}\right] \leq \phi.$$

V. A Tree Version of the RRG Algorithm

Maintaining a tree structure rather than a graph may be advantageous in some applications, due to, for instance, relatively easy extensions to motion planning problems with differential constraints, or to cope with modeling errors. The RRG algorithm can also be slightly modified to maintain a tree structure, while preserving the asymptotic optimality properties as well the computational efficiency. In this section a tree version of the RRG algorithm, called RRT*, is introduced and analyzed.

A. The RRT* Algorithm

Given two points $x, x' \in X_{\mathrm{free}}$, recall that $\mathtt{Line}(x, x') : [0, s] \to X_{\mathrm{free}}$ denotes the path defined by $\sigma(\tau) = \tau x + (s - \tau)x'$ for all $\tau \in [0, s]$, where $s = \|x' - x\|$. Given a tree $G = (V, E)$ and a vertex $v \in V$, let `Parent` be a function that maps v to the unique vertex $v' \in V$ such that $(v', v) \in E$.

The RRT* algorithm differs from the RRT and the RRG algorithms only in the way that it handles the `Extend` procedure. The body of the RRT* algorithm is presented in Algorithm 1 and the `Extend` procedure for the RRT* is given in Algorithm 4. In the description of the RRT* algorithm, the cost of the unique path from x_{init} to a vertex $v \in V$ is denoted by $\mathtt{Cost}(v)$. Initially, $\mathtt{Cost}(x_{\mathrm{init}})$ is set to zero.

Algorithm 4: $\mathtt{Extend}_{RRT^*}(G, x)$

1 $V' \leftarrow V$; $E' \leftarrow E$;
2 $x_{\mathrm{nearest}} \leftarrow \mathtt{Nearest}(G, x)$;
3 $x_{\mathrm{new}} \leftarrow \mathtt{Steer}(x_{\mathrm{nearest}}, x)$;
4 **if** $\mathtt{ObstacleFree}(x_{\mathrm{nearest}}, x_{\mathrm{new}})$ **then**
5     $V' \leftarrow V' \cup \{x_{\mathrm{new}}\}$;
6     $x_{\min} \leftarrow x_{\mathrm{nearest}}$;
7     $X_{\mathrm{near}} \leftarrow \mathtt{Near}(G, x_{\mathrm{new}}, |V|)$;
8   **for** *all* $x_{\mathrm{near}} \in X_{\mathrm{near}}$ **do**
9   **if** $\mathtt{ObstacleFree}(x_{\mathrm{near}}, x_{\mathrm{new}})$ **then**
10             $c' \leftarrow \mathtt{Cost}(x_{\mathrm{near}}) + c(\mathtt{Line}(x_{\mathrm{near}}, x_{\mathrm{new}}))$;
11   **if** $c' < \mathtt{Cost}(x_{\mathrm{new}})$ **then**
12                 $x_{\min} \leftarrow x_{\mathrm{near}}$;
13     $E' \leftarrow E' \cup \{(x_{\min}, x_{\mathrm{new}})\}$;
14   **for** *all* $x_{near} \in X_{\mathrm{near}} \setminus \{x_{\min}\}$ **do**
15   **if** $\mathtt{ObstacleFree}(x_{\mathrm{new}}, x_{\mathrm{near}})$ *and* $\mathtt{Cost}(x_{\mathrm{near}}) > \mathtt{Cost}(x_{\mathrm{new}}) + c(\mathtt{Line}(x_{\mathrm{new}}, x_{\mathrm{near}}))$ **then**
16             $x_{\mathrm{parent}} \leftarrow \mathtt{Parent}(x_{\mathrm{near}})$;
17             $E' \leftarrow E' \setminus \{(x_{\mathrm{parent}}, x_{\mathrm{near}})\}$;
              $E' \leftarrow E' \cup \{(x_{\mathrm{new}}, x_{\mathrm{near}})\}$;
18 **return** $G' = (V', E')$

Similar to the RRT and RRG, the RRT* algorithm first extends the nearest neighbor towards the sample (Lines 2-3). However, it connects the new vertex, x_{new}, to the vertex that incurs the minimum accumulated cost up until x_{new} and lies within the set X_{near} of vertices returned by the `Near` procedure (Lines 6-13). RRT* also extends the new vertex to the vertices in X_{near} in order to "rewire" the vertices that can be accessed through x_{new} with smaller cost (Lines 14-17).

B. Convergence to a Feasible Solution

For all $i \in \mathbb{N}$, let $\mathcal{V}_i^{\mathrm{RRT}^*}$ and $\mathcal{E}_i^{\mathrm{RRT}^*}$ denote the set of vertices and the set of edges of the graph maintained by the RRT* algorithm, at the end of iteration i. The following lemma is the equivalent of Lemma 3.

Lemma 19 *For all* $i \in \mathbb{N}$ *and all* $\omega \in \Omega$, $\mathcal{V}_i^{\mathrm{RRT}^*}(\omega) = \mathcal{V}_i^{\mathrm{RRG}}(\omega)$, *and* $\mathcal{E}_i^{\mathrm{RRT}^*}(\omega) \subseteq \mathcal{E}_i^{\mathrm{RRG}}(\omega)$.

From Lemma 19 and Theorem 6, the following theorem, which asserts the probabilistic completeness and the exponential decay of failure probability of the RRT* algorithm, is immediate.

Theorem 20 *If there exists a feasible solution to Problem 1, then* $\lim_{i\to\infty} \mathbb{P}\left(\left\{\mathcal{V}_i^{\mathrm{RRT}^*} \cap X_{\mathrm{goal}} \neq \emptyset\right\}\right) = 1$. *Moreover, if an attraction sequence* $\mathcal{A}$ *of length* k *exists, then* $\mathbb{P}\left(\left\{\mathcal{V}_i^{\mathrm{RRT}^*} \cap X_{\mathrm{goal}} = \emptyset\right\}\right) \leq e^{-\frac{1}{2}(i\,p_k - 2\,k)}$.

C. Asymptotic Optimality

Let $\mathcal{Y}_i^{\mathrm{RRT}^*}$ be a random variable that denotes the cost of a minimum cost path in the tree maintained by the RRT* algorithm, at the end of iteration i. The following theorem ensures the asymptotic optimality of the RRT* algorithm.

Theorem 21 *Let Assumptions 11, 12, and 13 hold. Then, the cost of the minimum cost path in the RRT* converges to* c^* *almost surely, i.e.,* $\mathbb{P}\left(\left\{\lim_{i\to\infty} \mathcal{Y}_i^{\mathrm{RRT}^*} = c^*\right\}\right) = 1$.

D. Computational Complexity

Let $\mathcal{M}_i^{\mathrm{RRT}^*}$ be the number of steps performed by the RRT* algorithm in iteration i. The following theorem follows from Lemma 19 and Theorem 18.

Theorem 22 *Under the assumptions of Theorem 18, there exists a constant* ϕ *such that* $\limsup_{i\to\infty} \mathbb{E}\left[\frac{\mathcal{M}_i^{\mathrm{RRT}^*}}{\mathcal{M}_i^{\mathrm{RRT}}}\right] \leq \phi$.

VI. Simulations

This section presents simulation examples. A thorough simulation study of the algorithms can be found in [17].

The RRT and the RRT* algorithms are run in a square environment with obstacles and the cost function is set to the Euclidean path length. The trees maintained by the algorithms at different stages are shown in Figure 1. The figure illustrates that the RRT algorithm does not considerably improve the solution, whereas the RRT* algorithm converges towards an optimal solution by finding a feasible solution of the homotopy class that the optimal path lies in. An important difference between the RRT and the RRT* algorithms is the ability of the latter to efficiently consider different homotopy classes. Thus, in an environment cluttered with obstacles, the cost of first feasible solution found by the RRT or the RRT* algorithms can be drastically higher than the optimal cost. Although the RRT* algorithm efficiently improves the solution over time, the RRT algorithm tends to get stuck with the first solution found. In fact, Monte-Carlo runs of both algorithms, as shown in Figure 2.(a)-(b), illustrate that generally the RRT does not improve the first solution found, whereas the RRT* algorithm improves the solution significantly within the first few thousand iterations, for this particular scenario. Moreover, the cost of the best path in RRT seems to have high variance, while after a few thousand iterations the costs of the best path in the RRT* is almost the same in all runs, as expected from the theoretical results presented in the previous sections. Finally, the relative complexity of the two algorithms is demonstrated in Monte-Carlo runs in Figure 2.(c). Notice that the ratio of the running time of the algorithms up until a certain iteration converges to a constant as the number of iterations increases. Note that the convergence to this constant is achieved when the free space is "fully explored", i.e., almost uniformly filled with the nodes of the trees. However, before then the complexity of the RRT* is much lower than the complexity in the limit value. In fact, the average amount of time that the RRT* algorithm takes for finding a feasible solution was found to be no more than five times that of the RRT, in this particular scenario. Moreover, the first solution found by the RRT* generally costs considerably less than that found by the RRT.

VII. Conclusions and Future Work

This paper presented the results of a thorough analysis of the RRT and RRG algorithms for optimal motion planning. It is shown that, as the number of samples increases, the RRT algorithm converges to a sub-optimal solution almost surely. On the other hand, it is proven that the RRG algorithm has the asymptotic optimality property, i.e., almost sure convergence to an optimal solution, which the RRT algorithm lacked. The paper also proposed a novel algorithm called the RRT*, which inherits the asymptotic optimality property of the RRG, while maintaining a tree structure rather than a graph. The RRG and the RRT* were shown to have no significant overhead when compared to the RRT algorithm in terms of asymptotic computational complexity. Experimental evidence demonstrating the effectiveness of the proposed algorithms and supporting the theoretical claims was also provided.

The results reported in this paper can be extended in a number of directions, and applied to other sampling-based algorithms other than RRT. First of all, the proposed approach, building on the theory of random graphs to adjust the length of new connections can enhance the computational efficiency of PRM-based algorithms. Second, the algorithms and the analysis should be modified to address motion planning problems in the presence of differential constraints, also known as kino-dynamic planning problems. A third direction is the optimal planning problem in the presence of temporal/logic constraints on the trajectories, e.g., expressed using formal specification languages such as Linear Temporal Logic, or the μ-calculus. Such constraints correspond to, e.g., rules of the road constraints for autonomous ground vehicles, mission specifications for autonomous robots, and rules of engagement in military applications. Ultimately, incremental sampling-based algorithms with asymptotic optimality properties may provide the basic elements for the on-line solution of differential games, as those arising when planning in the presence of dynamic obstacles.

Finally, it is noted that the proposed algorithms may have applications outside of the robotic motion planning domain. In fact, the class of incremental sampling algorithm described in this paper can be readily extended to deal with problems described by partial differential equations, such as the eikonal equation and the Hamilton-Jacobi-Bellman equation.

Acknowledgments

The authors are grateful to Professors M.S. Branicky and G.J. Gordon for their insightful comments on a draft version of this paper. This research was supported in part by the Michigan/AFRL Collaborative Center on Control Sciences, AFOSR grant no. FA 8650-07-2-3744.

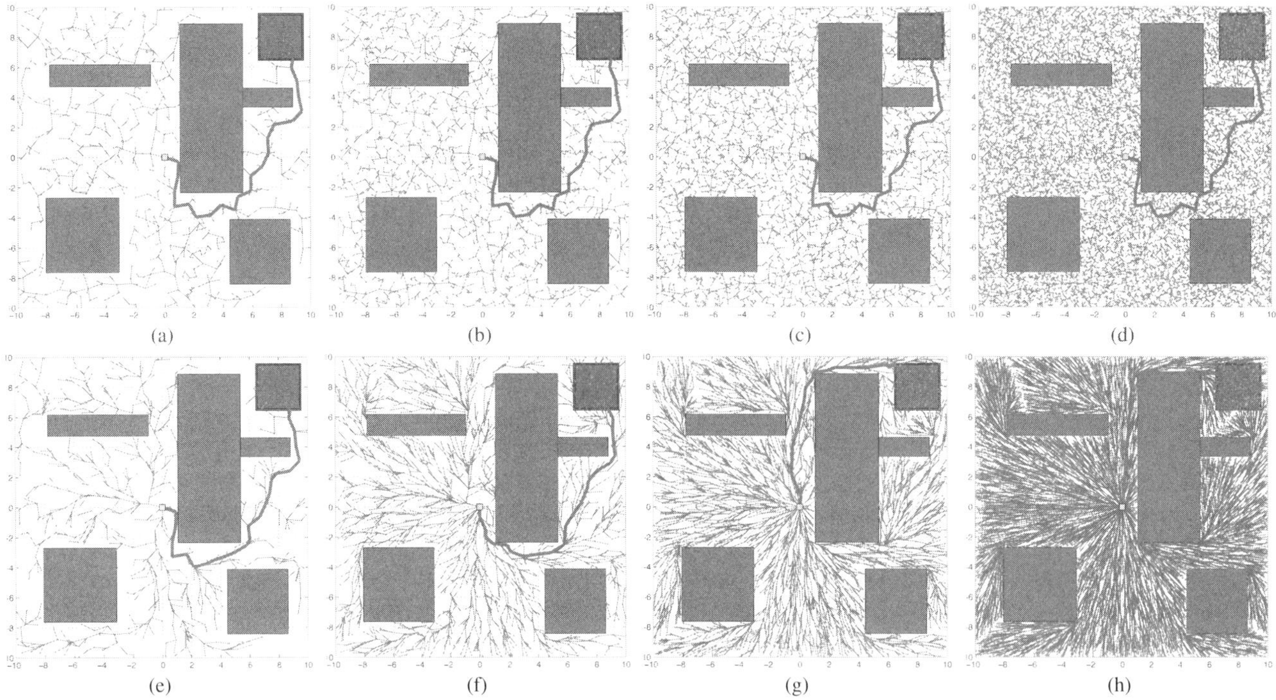

Fig. 1. A Comparison of the RRT* and RRT algorithms on a simulation example. The tree maintained by the RRT algorithm is shown in (a)-(d) in different stages, whereas that maintained by the RRT* algorithm is shown in (e)-(h). The tree snapshots (a), (e) are at 1000 iterations , (b), (f) at 2500 iterations, (c), (g) at 5000 iterations, and (d), (h) at 15,000 iterations. The goal regions are shown in magenta. The best paths that reach the target are highlighted with red.

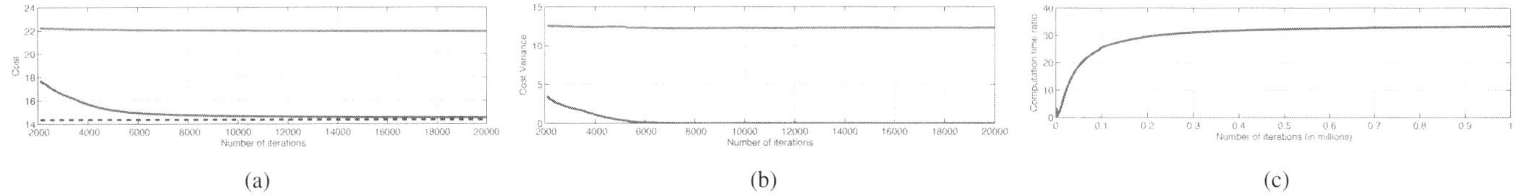

Fig. 2. The cost of the best paths in the RRT (shown in red) and the RRT* (shown in blue) plotted against iterations averaged over 500 trials in (a). The optimal cost is shown in black. The variance of the trials is shown in (b). A comparison of the running time of the RRT* and the RRT algorithms averaged over 50 trials is shown in (c); the ratio of the running time of the RRT* over that of the RRT up until each iteration is plotted versus the number of iterations.

REFERENCES

[1] J. Latombe. Motion planning: A journey of robots, molecules, digital actors, and other artifacts. *International Journal of Robotics Research*, 18(11):1119–1128, 1999.

[2] A. Bhatia and E. Frazzoli. Incremental search methods for reachability analysis of continuous and hybrid systems. In R. Alur and G.J. Pappas, editors, *Hybrid Systems: Computation and Control*, number 2993 in Lecture Notes in Computer Science, pages 142–156. Springer-Verlag, Philadelphia, PA, March 2004.

[3] M. S. Branicky, M. M. Curtis, J. Levine, and S. Morgan. Sampling-based planning, control, and verification of hybrid systems. *IEEE Proc. Control Theory and Applications*, 153(5):575–590, Sept. 2006.

[4] J. Cortes, L. Jailet, and T. Simeon. Molecular disassembly with RRT-like algorithms. In *IEEE International Conference on Robotics and Automation (ICRA)*, 2007.

[5] Y. Liu and N.I. Badler. Real-time reach planning for animated characters using hardware acceleration. In *IEEE International Conference on Computer Animation and Social Characters*, pages 86–93, 2003.

[6] J. T. Schwartz and M. Sharir. On the 'piano movers' problem: II. general techniques for computing topological properties of real algebraic manifolds. *Advances in Applied Mathematics*, 4:298–351, 1983.

[7] J.H. Reif. Complexity of the mover's problem and generalizations. In *Proceedings of the IEEE Symposium on Foundations of Computer Science*, 1979.

[8] L.E. Kavraki, P. Svestka, J Latombe, and M.H. Overmars. Probabilistic roadmaps for path planning in high-dimensional configuration spaces. *IEEE Transactions on Robotics and Automation*, 12(4):566–580, 1996.

[9] S. M. LaValle and J. J. Kuffner. Randomized kinodynamic planning. *International Journal of Robotics Research*, 20(5):378–400, May 2001.

[10] D. Hsu, R. Kindel, J. Latombe, and S. Rock. Randomized kinodynamic motion planning with moving obstacles. *International Journal of Robotics Research*, 21(3):233–255, 2002.

[11] Y. Kuwata, J. Teo, G. Fiore, S. Karaman, E. Frazzoli, and J.P. How. Real-time motion planning with applications to autonomous urban driving. *IEEE Transactions on Control Systems*, 17(5):1105–1118, 2009.

[12] S. Karaman and E. Frazzoli. Sampling-based motion planning with deterministic μ-calculus specifications. In *IEEE Conference on Decision and Control (CDC)*, 2009.

[13] C. Urmson and R. Simmons. Approaches for heuristically biasing RRT growth. In *Proceedings of the IEEE/RSJ International Conference on Robotics and Systems (IROS)*, 2003.

[14] D. Ferguson and A. Stentz. Anytime RRTs. In *Proceedings of the IEEE/RSJ International Conference on Intelligent Robots and Systems (IROS)*, 2006.

[15] M. Penrose. *Random Geometric Graphs*. Oxford University Press, 2003.

[16] J. Dall and M. Christensen. Random geometric graphs. *Physical Review E*, 66(1):016121, Jul 2002.

[17] S. Karaman and E. Frazzoli. Incremental sampling-based algorithms for optimal motion planning. http://arxiv.org/abs/1005.0416.

[18] S. Karaman and E. Frazzoli. Optimal kinodynamic motion planning using incremental sampling-based methods. In *Proceedings of the IEEE Conference on Decision and Control (CDC)*, 2010. Submitted.

[19] S. Muthukrishnan and G. Pandurangan. The bin-covering technique for thresholding random geometric graph properties. In *Proceedings of the sixteenth annual ACM-SIAM symposium on discrete algorithms*, 2005.

[20] H. Samet. *Design and Analysis of Spatial Data Structures*. Addison-Wesley, 1989.

[21] S. Arya, D. M. Mount, R. Silverman, and A. Y. Wu. An optimal algorithm for approximate nearest neighbor search in fixed dimensions. *Journal of the ACM*, 45(6):891–923, November 1999.

[22] S. Arya and D. M. Mount. Approximate range searching. *Computational Geometry: Theory and Applications*, 17:135–163, 2000.

Stochastic Modeling of the Expected Time to Search for an Intermittent Signal Source Under a Limited Sensing Range

Dezhen Song, Chang-Young Kim, and Jingang Yi

***Abstract*— A mobile robot is deployed to search for a stationary target that intermittently emits short duration signals. The searching mission is accomplished as soon as the robot receives a signal from the target. However, the robot cannot perceive the signal unless the target is within its limited sensing range. Therefore, the time to search the target is inherently random and hence unknown despite its importance in many searching and rescue applications. Here we propose the expected searching time (EST) as a metric to evaluate different robot motion plans under different robot configurations. We derive a closed form solution for computing the EST. To illustrate the EST model, we present two case studies. In the first case, we analyze two common motion plans: a slap method and a random walk. The EST analysis shows that the slap method is asymptotically faster than the random walk when the searching space size increases. In the second case, we compare a team of n low-cost equally-configured robots with a super robot that has the sensing range equal to that of the summation of the n robots. The EST analysis shows that the low-cost robot team takes $\Theta(1/n)$ time and the super robot takes $\Theta(1/\sqrt{n})$ time as $n \to \infty$. In both cases, our EST model successfully demonstrates its ability in assessing the searching performance. The analytical results are also confirmed in simulation.**

I. Introduction

Mobile robots are often employed to perform searching tasks such as finding a black box in a remote area after an airplane crash, searching victims after an earthquake or a mine collapse disaster, or locating artifacts on the ocean floor. In many cases, the target can intermittently emit short duration signals to assist searching. For example, an airplane black box transmits radio signals periodically. An earthquake victim may knock the rubble from time to time. The searching task is accomplished once the robot detects the signal emitted by the target. However, the robot usually has a limited sensing range and cannot detect the target that is out of the range. It seems straightforward that we can use the traditional coverage-based motion plans to guide the robot to cyclically scan the searching space to locate the target. However, the time to search the target is inherently random and hence remains unknown despite its importance in many searching and rescue applications.

This work was supported in part by the National Science Foundation under CAREER grant IIS-0643298 and MRI-0923203.

D. Song and C. Kim are with CSE Department, Texas A&M University, College Station, TX 77843, USA, (emails: dzsong@cse.tamu.edu and kcyoung@cse.tamu.edu).

J. Yi is with MAE Department, Rutgers University, Piscataway, NJ 08854 USA, (email: jgyi@rutgers.edu).

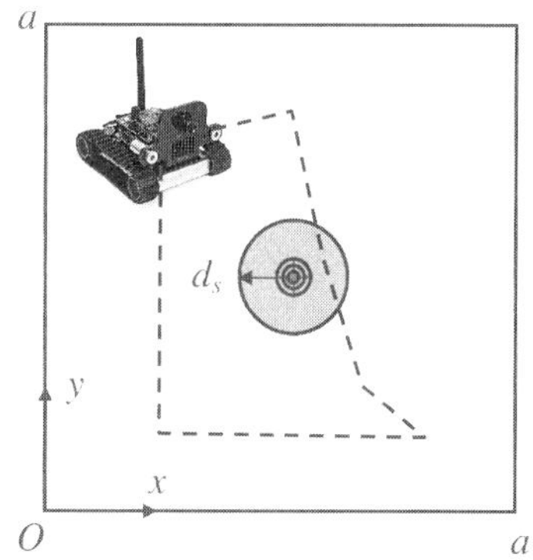

Fig. 1. A robot attemps to search for a target (the red dot) that intermittently emits short duration signals in a square. The gray circle is the region that the robot can sense the signal from the target. The dashed line is a robot trajectory.

To address this new problem, we propose the expected searching time (EST) as a metric for the searching ability. We model the searching process as a delayed alternating renewal process and derive the EST as a function of the searching space size, the signal transmission rate, and the robot sensing range. The resulting closed-form solution of the EST can be used to analyze the searching efficiency for different robot configurations and searching plans. Since the model components can be obtained from online measurements and known robot parameters, a great benefit of the resulting model is that it is capable of predicting the EST for an ongoing searching process. This characteristic is important for time-critical searching and rescue applications.

The contributions of the paper are trying to bring analytical results to interpret well-known searching strategies. The EST analysis not only can reveal our common believes about existing searching methods but also predict how the effectiveness of those methods changes as trajectory selection, sensing range, searching space size, and robot distribution change. Building on the latest development in random walk in constrained space from stochastic modeling community, our analysis for the first time show that traditional slap method ($\Theta(a^2)$) actually is asymptotically faster than the random walk ($O(a^2 \log a)$) for a squared searching space of the size length a, which is never known before. In the second case, we compare a team of n identically-configured low-cost robots with a super robot that has the sensing coverage equal to the summation of the n low-cost robots. The EST analysis shows that the low-cost robot team outperforms the super robot because its EST is $\Theta(1/n)$ while the EST for the super robot is $\Theta(1/\sqrt{n})$ as $n \to \infty$.

Again, this new analytical result has not been seen before and is important for developing new search strategies.

The analytical results are confirmed in simulation for both cases. The EST model successfully demonstrates its ability in assessing the searching performance under different robot configurations and motion plans.

The rest of the paper is organized as follows. We begin with the related work in Section II. We define the problem in Section III. We derive the EST in a closed form in Section IV. The two case studies are presented in Section V. The analytical results are validated in simulation in Section VI before we conclude the paper.

II. Related Work

Searching an object in physical space is one of the most important tasks for robots or humans. When prior information such as the spatial distribution of the target is known, this is comparable to the foraging behavior of animal [1]. However prior target information is often not available. If the target is continuously emitting signals, just simply scanning the entire searching space once enables the robot to find the target. Since the worst case for the searching time is the time to cover the entire searching space, the searching problem becomes a coverage problem [2]–[4]. For a known environment, a coverage problem for a single robot often employs different approaches to decompose the searching space and output a continuous path that allows the robot to cover the entire searching space. If the searching space can be modeled as a set of w-disjoint discrete choices, searching a target with a limited sensing range and w-choice is known as a w-lane Cow-Path problem [5].

While the running time is well understood for the coverage problems [6], [7], this is not the case when the searching process depends on the signal emitted by the target because the collocation of the robot and the target does not necessarily mean that the target is found. When the target is not emitting a signal, the robot cannot find the target. The robot has to keep scanning the searching space. The deterministic coverage algorithm becomes a Las Vegas algorithm [8] where the target will eventually be found but the searching time is random. However, the searching time can be crucial for many searching tasks. For example, victims of an earthquake often have limited survival time. Although searching itself is a very old problem, few models exist for analyzing the effectiveness of a searching strategy when the source is intermittently emitting signals.

Another set of related work is robot exploration and mapping problems where the environment is not previously known [9]. The task is not only to cover the entire space but also to output the true representations of the environment. Recent advances in using a multi-robot team to perform exploration and mapping tasks mainly focuses on the coordination of the robot/sensor team [10]–[15] under various dynamics, communication, sensing, and energy constraints. Although not directly applicable to our problem, researchers have accumulated interesting empirical results that are using a team of low-cost robots usually performs faster and more fault-tolerant [14] than a single expensive robot. This really inspires our problem because we want to see if our analytical model can show similar results under similar constraints/conditions.

Our group has built experience in searching for targets that intermittently transmit signals by developing algorithms and systems to detect an unknown wireless sensor network [16]–[18]. In these problems, the robot can accumulate the information about the target location over time through the signal strength readings and antenna models. The searching problem is less difficult because the robot can utilize the information in the planning process. However, such information is often not available in many searching tasks, which is the focus of this paper.

III. Problem Definition

As illustrated in Fig. 1, a single robot searches for a single target in a squared 2D Euclidean space with a side length of a. Define d_s as the maximum sensing distance of the sensor on the robot. The robot travels at the constant speed of v. To formulate the problem and focus on the most relevant issues, we make the following assumptions,

1) There is no prior information about possible locations of the target. Therefore, the target is assumed to be uniformly distributed in the searching space. This is actually the most difficult searching case.
2) The target transmits short duration signals periodically according to a Poisson process with a known rate λ. The signal duration is short due to energy concerns. A Poisson process is a good approximation to a general random arrival process in stochastic modeling [19]. In some cases, the target may be a continuous beacon; but it is very difficult to be detected due to environment conditions or unreliable sensing, which can also be modeled as a target with intermittent signals.
3) During the searching process, either the target is static or its movements are negligible in comparison to d_s. The searching space is much larger than the sensing distance: $a \gg d_s$.

Condition 1 (Sensing Condition): The robot cannot sense the signal unless an actively-transmitting target is within distance d_s due to the sensing range limit.

As illustrated in Fig. 1, this defines a circle centered at the target with the radius of d_s, which is the region that robot has a chance to sense the target. We refer to the region as "the circle" in the rest of the paper. Due to the fact that the robot does not know the location of the target, the actual position of the circle in the searching space is also unknown.

Condition 2 (Termination Condition): The searching task is accomplished as soon as the robot senses a signal.
Condition 2 implies that the robot cannot find an inactive target even it is collocated with the target. For example, an airplane is not be able to notice the survivor on an island if the person does not send a signal (e.g. fire or smoke). On the other hand, only one signal reception is needed in the searching process. Conditions 1 and 2 establish a new type of searching problem as oppose to a regular coverage problem. Let us define T_s as the searching time for the robot to find the target. Therefore, our problem is defined as follows,

Problem 1 (The EST Computation): Given λ, d_s, and a, calculate the EST $E(T_s)$, where $E(\cdot)$ denotes the expected value function.

IV. MODELING

One immediate question about Problem 1 is whether we can obtain the EST without referring to or being limited to a particular motion plan. To address this dependency, we first characterize the motion plans based on their outcomes before modeling the EST.

A. Characterizing Planners

Periodically, the robot planner outputs a motion plan and the system is naturally a repetitive scanning process. We name its trajectory in each period as a *tour*. The dashed line in Fig. 1 illustrates a tour.

Definition 1: A *tour* starts at the moment when the robot enters the target circle and ends at next moment when the robot enters the circle again.
Tours may be quite different based on the planner. For example, tour length varies each time if the robot follows a random walk. As another example, a deterministic planner usually has a fixed tour trajectory.

Based on Condition 2, we know that the robot does not accumulate the knowledge regarding the target location from tours to tours because no signal has been perceived before the moment the searching mission is accomplished. Hence we can treat each tour as independently and identically distributed (*i.i.d.*). This allows us to model the searching process as a renewal process.

When a tour begins, the robot first spends some time inside the circle, which is defined as τ_{IN}. After that, the robot leaves the circle and spends some time before entering the circle again, where the next tour starts. This yields an alternating renewal process. The duration outside the circle is defined as τ_{OUT}. Hence $\tau_{\text{IN}} + \tau_{\text{OUT}}$ is the overall duration for the tour.

B. Modeling the EST

Without loss of generality, we assume the robot starts the searching process from the origin which is on the boundaries of the searching space. It takes some time to reach the circle where the first tour starts. Define the time as delay D. From Conditions 1 and 2, we know that the robot cannot find the target in D. The searching process is a delayed alternating renewal process. Define T_s^c as the time to find the target after the robot enters the repetitive tours. Hence, the EST is

$$E(T_s) = E(D) + E(T_s^c). \tag{1}$$

Define N as the number of signal transmissions during τ_{IN} in a tour. Since the arrival process of the signal transmission is Poisson, N conforms to a Poisson distribution,

$$P(N=k) = \frac{e^{-\lambda\tau_{\text{IN}}}(\lambda\tau_{\text{IN}})^k}{k!}, \quad k = 0, 1, 2, ..., \infty. \tag{2}$$

We know that event $N > 0$ means that at least one signal transmission happens during τ_{IN}. This means the target is found. Therefore, we can compute $E(T_s^c)$ by conditioning on N,

$$E(T_s^c) = E(T_s^c|N>0)P(N>0) + E(T_s^c|N=0)P(N=0), \tag{3}$$

where $P(N > 0) = 1 - e^{-\lambda\tau_{\text{IN}}}$ and $P(N = 0) = e^{-\lambda\tau_{\text{IN}}}$ according to (2).

Now let us compute $E(T_s^c|N > 0)$. Since event $N > 0$ is equivalent to event $T_s^c \le \tau_{\text{IN}}$, we have

$$\begin{aligned} E(T_s^c|N>0) &= E(T_s^c|T_s^c \le \tau_{\text{IN}}) \\ &= \frac{1}{\lambda} - \frac{\tau_{\text{IN}}e^{-\lambda\tau_{\text{IN}}}}{1-e^{-\lambda\tau_{\text{IN}}}} \end{aligned} \tag{4}$$

because the conditional distribution $T_s^c|T_s^c \le \tau_{\text{IN}}$ is a truncated exponential distribution. It is worth noting that (4) is valid only if $\tau_{\text{IN}} > 0$. This is guaranteed according to Definition 1. On the other hand, we know

$$E(T_s^c|N=0) = \tau_{\text{IN}} + \tau_{\text{OUT}} + E(T_s^c) \tag{5}$$

because the robot cannot find the inactive target in the current tour and has to start all over again in next tour.

Plugging (4) and (5) into (3) and (1), we have the following Theorem.

Theorem 1: Given the expected time $E(D)$ for the robot to reach the circle, the Poisson arrival rate of signal λ of the target, the traveling time τ_{IN} inside the circle, and the traveling time τ_{OUT} outside the circle, the EST of the target is

$$E(T_s) = E(D) + \frac{1}{\lambda} + \tau_{\text{OUT}}\frac{e^{-\lambda\tau_{\text{IN}}}}{1-e^{-\lambda\tau_{\text{IN}}}}. \tag{6}$$

Theorem 1 has a surprisingly succinct format revealing the relationship between the EST and the corresponding variables. To further explain (6), let us consider the following extreme cases:

Case 1: When $\lambda \to \infty$, it means that the target continuously transmits signals. An example is that a lost hiker keeps fire burning. Hence the light and the smoke of the fire become the continuous signal. Now the searching time becomes the time that it takes for the robot to enter the circle. The problem degenerates to the traditional coverage problem where $E(T_s) = E(D)$.

Case 2: When $\tau_{\text{OUT}} = 0$, it means that the signal emitted by the target is so powerful that the circle defined by d_s can cover the entire searching space. In this case, $E(D) = 0$. Hence $E(T_s) = 1/\lambda$. This is sensible because the result means the robot can find the target as soon as it emits a signal.

Case 3: When $\tau_{\text{IN}} \to 0$, which happens when d_s is infinitesimally small, we have $E(T_s) \to \infty$. This result conforms to our expectation.

Remark 1: It is worth noting that (6) does not depend on a particular motion plan or the shape/dimension of the searching space, which makes it widely applicable in practice. Actually, the EST can be also applied to analyze searching tasks carried by humans. In many cases, the signal transmission rate λ is known; $E(D)$ can be estimated based on observations; τ_{IN} can be estimated based on d_s and v; and τ_{OUT} can be measured based on observations that how often a robot would revisit a region with the same size of the circle. Based on the known information and online measurements, we can even predict the EST for an ongoing searching process regardless its motion plan, which is of great importance in applications where the searching time literally means life or death.

V. Analysis of Common Searching Strategies

Theorem 1 can be used to analyze the searching performance under different robot motion plans and configurations. We begin with demonstrating how Theorem 1 can reveal the difference between two motion plans from common coverage methods, namely, the slap method and the random walk.

A. The Slap Method

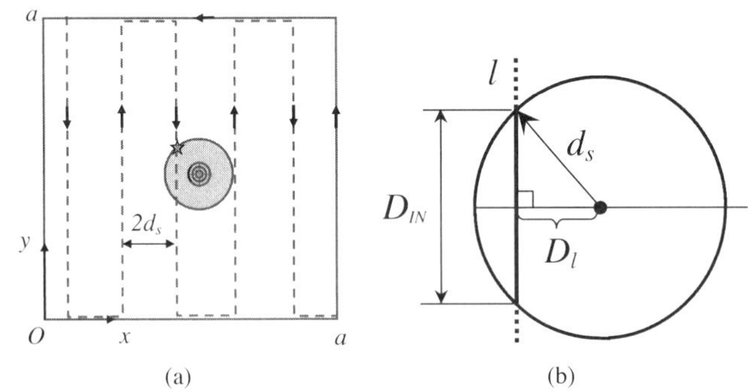

Fig. 2. (a) A sample motion plan for the slap method. (b) An illustration of how a tour (line l) intersects the circle of the target.

The slap method [20], also known as the trapezoidal decomposition [21] in robotics, sequentially scans the entire searching space back and forth. Fig. 2(a) gives the robot motion plan for the square case. The plan is a set of y-axis parallel lines (appears to be vertical lines in Fig. 2(a)) that cover the entire searching space. The vertical lines are inter-connected using the boundaries of the searching space to formulate a complete tour. To guarantee an intersection between the circle and the tour, the distance between adjacent vertical lines is set to be $2d_s$.

The red "$\star$" in Fig. 2(a) is the starting point of the tour. Since tours are exactly the same in the slap method, the subsequent tours start exactly at the same location. The overall tour length is approximately $a^2/(2d_s)$. Given the robot speed v, we know it takes

$$\tau_{\text{IN}} + \tau_{\text{OUT}} \approx \frac{a^2}{2vd_s} \tag{7}$$

time for the robot to finish the tour. Since the target could be anywhere in the searching space with equal probabilities, we know that

$$E(D) \approx (\tau_{\text{IN}} + \tau_{\text{OUT}})/2 = \frac{a^2}{4vd_s}. \tag{8}$$

The remaining undetermined variable is τ_{IN}. Let us define D_{IN} as the distance traveled inside the circle. D_{IN} is the length of intersection when the line intersects the circle as illustrated in Fig. 2(b). Here we ignore the boundary effect where the circle is not a full circle because $a \gg d_s$. Line l in Fig. 2(b) is a part of the tour. When l intersects the circle, we define D_l as the distance between the center of the circle and the line. Since the target is uniformly distributed in the 2D space, $D_l \sim U(0, d_s)$ is uniformly distributed. From Fig. 2(b), we know

$$\tau_{\text{IN}} = \frac{D_{\text{IN}}}{v} = \frac{2\sqrt{d_s^2 - D_l^2}}{v}. \tag{9}$$

Plugging (7), (8) and (9) into (6) and conditioning on D_l, we have,

$$E(T_s|D_l) \approx \frac{a^2}{4vd_s} + \frac{1}{\lambda} + \left(\frac{a^2}{2vd_s} - \frac{2\sqrt{d_s^2 - D_l^2}}{v}\right)\phi(\lambda, D_l), \tag{10}$$

where

$$\phi(\lambda, D_l) = \frac{1}{e^{\frac{2\lambda\sqrt{d_s^2 - D_l^2}}{v}} - 1}. \tag{11}$$

Since $a \gg d_s$, $\tau_{\text{OUT}} \gg \tau_{\text{IN}}$, and $\frac{2\sqrt{d_s^2 - D_l^2}}{v}$ is negligible if compared with $\frac{a^2}{2vd_s}$, we have,

$$E(T_s|D_l) \approx \frac{a^2}{4vd_s} + \frac{1}{\lambda} + \frac{a^2}{2vd_s}\phi(\lambda, D_l). \tag{12}$$

Hence we have the EST for the slap method,

$$\begin{aligned} E(T_s) &= \int_{\delta=0}^{d_s} E(T_s|D_l = \delta)\frac{1}{d_s}d\delta \\ &\approx \frac{a^2}{4vd_s} + \frac{1}{\lambda} + \frac{a^2}{2vd_s}g(d_s, \lambda) \end{aligned} \tag{13}$$

where

$$g(d_s, \lambda) = E(\phi(\lambda, D_l)) = \int_{\delta=0}^{d_s} \frac{1}{d_s}\phi(\lambda, D_l)d\delta. \tag{14}$$

Let $\delta = d_s \cos\theta$, we can transform (14) into

$$g(d_s, \lambda) = \int_{\theta=0}^{\pi/2} \frac{1}{e^{\frac{2\lambda d_s \sin\theta}{v}} - 1}\sin\theta d\theta. \tag{15}$$

When λ and d_s/v are very small, (14) can be further simplified,

$$g(d_s, \lambda) \approx \frac{\pi v}{4\lambda d_s} - 1. \tag{16}$$

Remark 2: Eq. (13) also suggests that a fast robot (large v) with great sensing distance d_s reduces the EST. This conclusion agrees with our intuition that mobility and sensing are the key elements in searching. However, it also takes a target's cooperation to further reduce the EST. When the robot reaches its speed and sensing limit, the only way to reduce EST is to increase λ. Of course, the target usually has energy constraints and cannot arbitrarily increase λ.

The analysis assumes the distance between vertical lines is $2d_s$, which ensures there is only one intersection between the circle and the tour. When a smaller spacing is used, the overall tour length increases and so does τ_{IN}. The analysis is slightly more complicated because it needs to be conditioned on the number of intersections between the tour and the circle. The results actually share a similar format with (15) and the same asymptotical properties with respect to a, v, and λ. Since our focus is to compare the asymptotic behavior of the slap method with that of the random walk, we omit the analysis here.

Another reasonable concern is that whether ignoring the

boundary effect impacts the final result. When the circle is located at the boundary of the square, distance D_{IN} cannot be computed using (9). Since the target has to be located within d_s distance of the boundary to create the scenario, the probability that such event happens is less than $\frac{4d_s a}{a^2} = \frac{4d_s}{a} \ll 1$, since $d_s \ll a$. Hence its impact to the final EST is ignorable because D_{IN} for such case is not significantly different from that of the non-boundary case. Therefore, we will ignore boundary effect in the rest of the paper.

B. Random Walk

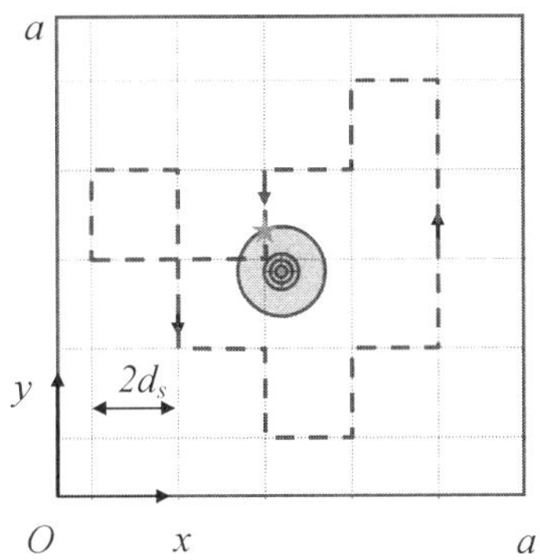

Fig. 3. An illustration of robot motion plan based on a 2D lattice-based random walk.

Another popular motion plan is to employ a 2D random walk to traverse the searching space. As illustrated in Fig. 3, we partition the entire searching space using a 2D finite lattice with a spacing of $2d_s$ in each dimension. Denoting N_s as the number of lattice nodes, we have $N_s = \frac{a^2}{4d_s^2}$ nodes. Finer lattice is possible but usually associated with higher energy cost because the robot has to make a lot more turns.

The robot always moves from one lattice node to its neighboring node with equal probabilities. The robot does not cross the boundaries. According to [22], this is a finite 2D lattice with reflective boundaries. Recall that a tour starts at the moment the robot enters the circle. Since the robot might not enter the circle at the exactly same location in different tours, each tour is not necessarily a completely closed curve as that in the slap method case. The closed curve tour in Fig. 3 only happens with a probability of $1/4$.

To compute the EST in (6), we need to compute $E(D)$. Recall that the robot always starts at origin. Given the location of target (X_t, Y_t), computing the mean time that it takes the robot to follow the random walk to reach a particular location (X_t, Y_t) is the mean first passage time (MFPT) [23], [24] problem in stochastic modeling. The exact solution to this problem is expressed in the format of pseudo Green functions and cannot be explicitly analyzed. Since $a \gg d_s$, there are a large number of nodes $\frac{a^2}{4d_s^2}$ in the 2D lattice and each robot move takes $\frac{2d_s}{v}$ time. Hence we can apply the recent results of MFPT using its asymptotic format in [25],

$$E(D|X_t, Y_t) \approx \frac{a^2}{2vd_s}\left(\alpha_0 + \alpha_1 \ln\sqrt{X_t^2 + Y_t^2}\right), \quad (17)$$

where α_0 and α_1 are constants and can be determined by Monte Carlo methods. According to [25], α_0 and α_1 strikingly do not depend on lattice size but local transitional properties. Hence,

$$E(D) = \int_0^a \int_0^a E(D|X_t = x, Y_t = y)\frac{1}{a^2}dxdy, \quad (18)$$

$$\approx \frac{\alpha_0 a^2}{2vd_s} + \frac{\alpha_1}{2vd_s}\int_0^a \int_0^a \ln\sqrt{x^2 + y^2}dxdy. \quad (19)$$

Since

$$\int_0^a \int_0^a \ln\sqrt{x^2 + y^2}dxdy = a^2 \ln a + \frac{\pi + 2\ln 2 - 6}{4}a^2$$

we have

$$E(D) \approx \frac{a^2}{2vd_s}\left(\alpha_0 + \alpha_1 \ln a + \alpha_1 \frac{\pi + 2\ln 2 - 6}{4}\right). \quad (20)$$

The remaining unknown term in (6) is $E(\tau_{\text{OUT}} \frac{e^{-\lambda\tau_{\text{IN}}}}{1-e^{-\lambda\tau_{\text{IN}}}})$. Given the robot speed v, τ_{IN} is uniquely determined by the distance in the circle D_{IN}, which is independent of the overall trajectory. Also $E(\tau_{\text{OUT}}) \approx E(\tau_{\text{OUT}} + \tau_{\text{IN}})$ given that $a \gg d_s$. Hence,

$$E(\tau_{\text{OUT}} \frac{e^{-\lambda\tau_{\text{IN}}}}{1 - e^{-\lambda\tau_{\text{IN}}}}) \approx E(\tau_{\text{OUT}} + \tau_{\text{IN}})E(\frac{e^{-\lambda\tau_{\text{IN}}}}{1 - e^{-\lambda\tau_{\text{IN}}}}). \quad (21)$$

Since the 2D lattice-based random walk is undirected and symmetric in transitional probability, we know that the stationary probability of staying inside the circle is $p_c = \frac{\pi d_s^2}{a^2}$. Therefore, we know the following is true according to Renewal Reward theorem,

$$\frac{E(\tau_{\text{IN}})}{E(\tau_{\text{OUT}} + \tau_{\text{IN}})} = p_c = \frac{\pi d_s^2}{a^2}. \quad (22)$$

Plugging (22) into (21), we have

$$E\left(\tau_{\text{OUT}} \frac{e^{-\lambda\tau_{\text{IN}}}}{1 - e^{-\lambda\tau_{\text{IN}}}}\right) \approx \frac{a^2}{\pi d_s^2}E(\tau_{\text{IN}})E\left(\frac{e^{-\lambda\tau_{\text{IN}}}}{1 - e^{-\lambda\tau_{\text{IN}}}}\right). \quad (23)$$

Now, we focus on the computation of τ_{IN}. Since the lattice has a spacing of $2d_s$, two scenarios exist when the tour on the lattice intersects the circle: i) the nearest lattice point on the tour is inside the circle and ii) the nearest lattice point on the tour is outside the circle as illustrated in Fig. 4. Let us define events that scenarios i) and ii) happen as events E_i and E_o, respectively. Since the circle center is uniformly located in the searching space,

$$P(E_i) = \frac{\pi d_s^2}{4d_s^2} = \frac{\pi}{4} = 1 - P(E_o). \quad (24)$$

When event E_o happens, we know that the robot trajectory intersects the circle as a straight line as shown in Fig. 4(a). Hence we have

$$\tau_{\text{IN}}|E_o = \frac{D_{\text{IN}}}{v}, \quad (25)$$

where D_{IN} is defined in (9) and the right side of $|$ is the condition for the equality to be true. This is a notation convention widely used in stochastic modeling [19]. Hence

$$E(\tau_{\text{IN}}|E_o) = \frac{\pi d_s}{2v}, \text{ and} \quad (26)$$

$$E\left(\frac{e^{-\lambda\tau_{\text{IN}}}}{1 - e^{-\lambda\tau_{\text{IN}}}}|E_o\right) = g(d_s, \lambda). \quad (27)$$

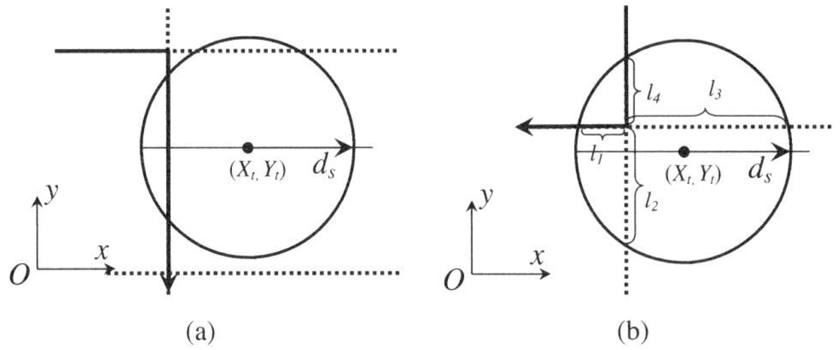

Fig. 4. An illustration of how the robot trajectory in solid line intersects the circle. (a) Scenario i): when the nearest lattice point on tour is located outside the circle. (b) Scenario ii): when the nearest lattice point on tour is located inside the circle. The dashed line in the figure is part of the lattice.

When event E_i happens, one lattice point is inside the circle. As illustrated in Fig. 4(b), the lattice point inside the circle partitions the lattice edges inside the circle into four parts: l_1, l_2, l_3, and l_4. When a robot trajectory intersects the circle, the part of the trajectory inside the circle can be divided into two segments, which are defined as L' and L'', respectively. L' refers to the segment that the robot takes to arrive at the lattice node and L'' refers to the segment that the robot takes to leave the circle. Hence

$$\tau_{\text{IN}}|E_i = \frac{L' + L''}{v}.$$

Since L' and L'' have equal probabilities to take l_1, l_2, l_3, and l_4, there is a total of 16 combinations. Conditioning on the 16 (L', L'') combinations and the circle center location (X_t, Y_t), we get the same results as shown in (26) and (27). Combining those results for the E_i and E_o events by conditioning on them, we have the unconditional expected values,

$$E(\tau_{\text{IN}}) = \frac{\pi d_s}{2v}, \text{ and} \tag{28}$$

$$E\left(\frac{e^{-\lambda\tau_{\text{IN}}}}{1 - e^{-\lambda\tau_{\text{IN}}}}\right) = g(d_s, \lambda), \tag{29}$$

where $g(\cdot)$ is defined in (15). Plugging (20), (23), (28), and (29) into (6), we can obtain the EST for the random walk case,

$$\begin{aligned} E(T_s) \approx& \frac{a^2}{2vd_s}\left(\alpha_0 + \alpha_1 \ln a + \alpha_1 \frac{\pi + 2\ln 2 - 6}{4}\right) \\ &+ \frac{1}{\lambda} + \frac{a^2}{2d_s v} g(d_s, \lambda). \end{aligned} \tag{30}$$

Comparing (30) to (13) , we have the follow conclusion,

Corollary 1: With the same field side length a, the sensing range d_s, and the signal transmission rate λ, the $E(T_s)$ value of the slap method is asymptotically smaller than that of the random walk when $a \to \infty$.

Proof: It is straightforward because $E(T_s) = \Theta(a^2)$ for the slap method from (13) while $E(T_s) = \Theta(a^2 \ln a)$ for the random walk according to (30). ■

C. Analysis of Different Robot Configurations

Theorem 1 can also be used to analyze cases under different robot configurations. Here we compare two configurations.

A low-cost robot team (LCRT) case: We have n identically-configured low-cost robots. To coordinate the searching, we partition the searching space into n sub square fields with an area of a^2/n each and allocate one robot for each sub square field.

A single expensive robot (ASER) case: We have an expensive robot equipped with a very capable sensor that has a sensing area equal to the combination of those of the n low-cost robots. If each of the low-cost robot has a sensing range of d_s, then the area of the combined sensing region for n robots is $n\pi d_s^2$. Therefore, the sensing distance for the expensive robot is set to $d_s' = \sqrt{n} d_s$ to ensure the same-sized sensing coverage at any given time.

We are now ready to compare these two robot configurations. Since the slap method is asymptotically faster than the random walk, we build on the slap method results in (30). For the LCRT, only one robot actually has the target in its sub field. Hence, the rest of $n-1$ robots are irrelevant in the searching process. Comparing with the original EST in (30), we just need to replace a with $\frac{a}{\sqrt{n}}$. Defining the searching time for the LCRT as T_s', we have

$$E(T_s') \approx \frac{a^2}{4vnd_s} + \frac{1}{\lambda} + \frac{a^2}{2vnd_s} g(d_s, \lambda). \tag{31}$$

Defining the searching time for the ASER as T_s'', we have

$$E(T_s'') \approx \frac{a^2}{4v\sqrt{n}d_s} + \frac{1}{\lambda} + \frac{a^2}{2v\sqrt{n}d_s} g(\sqrt{n}d_s, \lambda). \tag{32}$$

From (15), it is not difficult to see that

$$g(\sqrt{n}d_s, \lambda) \to 0 \quad \text{as} \quad n \to \infty. \tag{33}$$

Therefore, we have the following conclusion,

Corollary 2: When traveling at the same velocity v, the low-cost robot team can find the target asymptotically faster than the single expensive robot does when n increases, if $1/\lambda$ is not the dominating factor in the EST.

Proof: From (32) and (33), we know $E(T_s'') = \Theta(\frac{1}{\sqrt{n}} + \frac{1}{\lambda})$. From (31), we know $E(T_s') = \Theta(\frac{1}{n} + \frac{1}{\lambda})$. Hence the conclusion follows. ■

It is actually rather surprising to see the result in Corollary 2 at the first sight. We have not expected such a significant difference in the comparison. This conclusion is rather interesting because it shows that an expensive robot with superior sensing capability is not as good as a large number of low-cost robots with less capable sensors when searching for targets that intermittently transmits short duration signals.

Remark 3: This analysis also shows that if there are cost functions associated with the number of robots, different sensor options, or different velocity options available, we can use the EST results as an objective function to optimize the robot configuration for the task.

VI. Experiments

We test our results using Monte Carlo simulation. The simulation program is written in Microsoft Visual C++.Net 2005 on a Desktop PC with an 32-bit Windows XP Professional Edition OS. The Desktop PC has an Intel 2.13 GHz Core2Duo CPU with 2GB RAM and a 250 GB Hard disk.

The experimental results are illustrated in Figs. 5 and 6. Each data point in both figures is an average of 10,000

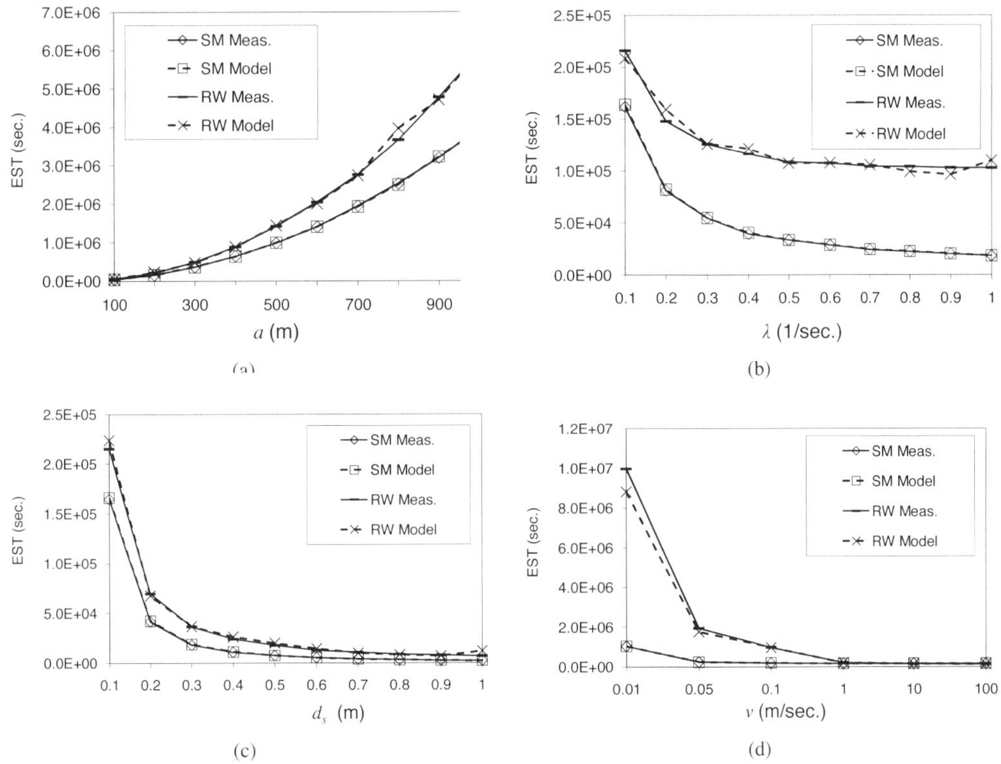

Fig. 5. Simulation results in (a), (b), (c), and (d) for validating Theorem 1 with respect to a, λ, d_s, and v, respectively. SM stands for the slap method. RW stands for the random walk. Model means the model prediction of the EST. Meas. means the measured mean searching time.

independent trials. At the beginning of each trial, we reset the robot position to be at $(0, 0)$ and generate the target location according to a 2D uniform distribution. We then run the robot according to the selected motion plan and finish the trial as soon as the target is found.

A. Validating Theorem 1 and Corollary 1

We test Theorem 1 using both the slap method and the random walk because Theorem 1 is supposed to be independent of motion plans. The simulation is set up with different a, λ, d_s, and v settings in Table I. In each setting, we collect both the model predicted EST and the measured mean searching time. The measured mean searching time is the average of the searching time over the 10k trials (the "Meas." values in Fig. 5). The model predicted ESTs, which are the "Model" values in Fig. 5, refer to the predicted ESTs according to the measured D, λ, τ_{IN}, and τ_{OUT} values in the experiment. In other words, we record their values and average them over the 10k trials and to obtain the estimation of $1/\lambda$, $E(D)$, and $E(\tau_{\text{OUT}}\frac{e^{-\lambda\tau_{\text{IN}}}}{1-e^{-\lambda\tau_{\text{IN}}}})$. We then feed them into (6) to obtain the model prediction of the EST.

As illustrated in Fig. 5, the model prediction is fairly consistent with the measured mean searching time under all settings. There are more fluctuations between the model prediction and the measured mean searching time in random walk-based results than that of the slap method. This is expected because of more random factors associated with the random walk. Under the same trial number, the results from the random walk should contain more randomness.

Figure	a (m)	λ (1/sec.)	d_s (m)	v (m/s)
Fig. 5(a)	100-1000	0.1	1.0	1.0
Fig. 5(b)	200	0.1-1.0	1.0	1.0
Fig. 5(c)	200	0.1	1-10	1.0
Fig. 5(d)	200	0.1	1.0	0.01-100
Fig. 6	200	0.1	1.0	1.0

TABLE I
PARAMETER SETTINGS FOR RESULTS IN FIG. 5 AND FIG. 6.

The curve trends with respect to a, λ, d_s and v in Fig. 5 are also consistent with our analysis in (13) and (30). The EST increases as the field side length a increases. The EST decreases as λ, d_s, and v increase. All figures show that the random walk is slower than the slap method. In particular, Fig. 5 (a) is consistent with the asymptotical difference in Corollary 1.

B. Validating Corollary 2

We have also implemented both LCRT and ASER robot configurations. Again, the parameter settings are in the last row of Table I. The measured ESTs for both the configurations are shown in Fig. 6. It is clear that the EST for the LCRT is always much smaller than that of the ASER. This is consistent with Corollary 2. Curves in the figure also show the trend that the EST decreases as the n increases. This is consistent with our analysis. Also, as n gets very big, the curve levels at a non-zero value. This indicates that the signal transmission rate dominates the searching time. On the other hand, it is not desirable to arbitrarily increase n because the marginal benefit would decrease.

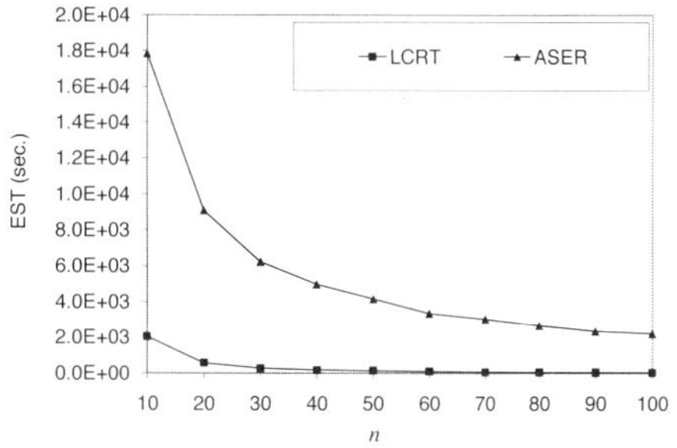

Fig. 6. Simulation results for comparing two robot configurations

VII. Conclusions and Future Work

We analytically modeled the expected searching time for a robot with a limited sensing range to search for a target that intermittently emits short duration signals. We presented the closed-form model for the EST. The EST model is motion-plan independent and can be used to analyze different motion plans or robot configurations in two case studies. In the first case, we analyzed the slap method and the random walk and found that the slap method is asymptotically faster than the random walk. In the second case, the EST model revealed the interesting result that a low-cost robot team is always asymptotically faster than an expensive robot when the sensory coverage is the same. In both cases, the results demonstrated the usefulness and the capability of our EST analysis. Our theoretically results were extensively tested using simulation. The simulation results were consistent with the model.

This work will lead to a rich set of exciting future work. As an extension, we can analyze cases where multiple targets are needed to be searched. We can also develop the EST metrics for the searching of a moving target, an un-cooperating target, or multiple sensor combinations. Different sensor models can also be considered. Also, the searching space may exist obstacles. Applications of these extended results will be wide-ranging.

Acknowledgement

Thanks for E. Frew, J. Xiao, and R. Volz for their insightful discussions. Thanks for Y. Xu, J. Zhang, A. Aghamohammadi, and W. Li for their inputs and contributions to the Networked Robots Laboratory in Texas A&M University.

References

[1] E. Gelenbe, N. Schmajuk, J. Staddon, and J. Reif, "Autonomous search by robots and animals: A survey," *Robotics and Autonomous Systems*, vol. 22, no. 1, pp. 23–34, 1997.

[2] H. Choset, "Coverage for robotics a survey of recent results," *Annals of Mathematics and Artificial Intelligence*, vol. 31, no. 1-4, pp. 113–126, Oct. 2001.

[3] S. Martinez, J. Cortes, and F. Bullo, "Motion coordination with distributed information," *IEEE Control Systems Magazine*, vol. 27, no. 4, pp. 75–88, August 2007.

[4] M. Schwager, D. Rus, and J. Slotine, "Decentralized, adaptive control for coverage with networked robots," *International Journal of Robotics Research*, vol. 28, no. 3, pp. 357–375, March 2009.

[5] M. Kao, J. Reif, and S. Tate, "Searching in an unknown environment: An optimal randomized algorithm for the cow-path problem," *Information and Computation*, vol. 131, no. 1, pp. 63–79, Nov. 1996.

[6] A. López-Ortiz and S. Schuierer, "Online parallel heuristics and robot searching under the competitive framework," in *SWAT '02: Proceedings of the 8th Scandinavian Workshop on Algorithm Theory*, 2002, pp. 260–269.

[7] R. Baeza-Yates, J. Culberson, and J. Rawlins, "Searching in the plane," *Information and Computation*, vol. 106, no. 2, pp. 234–252, Oct. 1993.

[8] R. Motwani and R. Raghavan, Eds., *Randomized Algorithms*. Cambridge Univeristy Press, 1995.

[9] S. Thrun, W. Burgard, and D. Fox, *Probabilistic Robotics*. MIT Press, 2005.

[10] K. Laventall and J. Cortes, "Coverage control by multi-robot networks with limited-range anisotropic sensory," *International Journal of Control*, vol. 82, no. 6, pp. 1113 – 1121, June 2009.

[11] P. Brass, "Bounds on coverage and target detection capabilities for models of networks of mobile sensors," *ACM Transactions on Sensor Networks (TOSN)*, vol. 3, no. 2, p. 9, June 2007.

[12] W. Burgard, M. Moors, D. Fox, R. Simmons, and S. Thrun, "Collaborative multi-robot exploration," in *IEEE International Conference on Robotics and Automation, San Francisco, CA*, Apr. 2000, pp. 476–481.

[13] M. Batalin and G. Sukhatme, "The design and analysis of an efficient local algorithm for coverage and exploration based on sensor network deployment," *IEEE Transactions on Robotics*, vol. 23, no. 4, pp. 661–675, August 2007.

[14] W. Burgard, M. Moors, C. Stachniss, and F. Schneider, "Collaborative multi-robot exploration," *IEEE Transactions on Robotics*, vol. 21, no. 3, pp. 376–386, June 2005.

[15] D. Fox, J. Ko, K. Konolige, B. Limketkai, D. Schulz, and B. Stewart, "Distributed multirobot exploration and mapping," *Proceedings of The IEEE*, vol. 94, no. 7, pp. 1325–1339, July 2006.

[16] D. Song, J. Yi, and Z. Goodwin, "Localization of unknown networked radio sources using a mobile robot with a directional antenna," in *the American Control Conference (ACC), New York City, USA, July*, 2007, pp. 5952–5957.

[17] D. Song, C. Kim, and J. Yi, "Simultaneous localization of multiple unknown CSMA-based wireless sensor network nodes using a mobile robot with a directional antenna," *Journal of Intelligent Service Robots*, vol. 2, no. 4, pp. 219–233, Oct. 2009.

[18] ——, "Monte carlo simultaneous localization of multiple unknown transient radio sources using a mobile robot with a directional antenna," in *IEEE International Conference on Robotics and Automation (ICRA), Kobe, Japan*, May 2009.

[19] S. Ross, *Introduction to Probability Models, Ninth Edition*. Academic Press, 2007.

[20] F. Preparata and M. Shamos, *Computational Geometry: An Introduction*. Springer-Verlag, Berlin, 1985.

[21] J. Latombe, *Robot Motion Planning*. Kluwer Acadmic, Boston, MA, 1991.

[22] M. Brummelhuis and H. J. Hilhorst, "How a random walk covers a finite lattice," *Physica A: Statistical Mechanics and its Applications*, vol. 185, no. 1-4, pp. 35–44, June 1992.

[23] S. Condamin, O. Bénichou, and M. Moreau, "First-passage times for random walks in bounded domains," *Physical Review Letters*, vol. 95, no. 26, p. 260601, Dec 2005.

[24] J. D. Noh and H. Rieger, "Random walks on complex networks," *Physical Review Letters*, vol. 92, no. 11, p. 118701, Mar. 2004.

[25] S. Condamin, O. Benichou, V. Tejedor, R. Voituriez, and J. Klafter, "First-passage times in complex scale-invarant media," *Nature*, vol. 450, pp. 77–80, Nov. 2007.

Closing the Learning-Planning Loop with Predictive State Representations

Byron Boots
Machine Learning Dept.
Carnegie Mellon University
Pittsburgh, PA, USA
Email: beb@cs.cmu.edu

Sajid M. Siddiqi
Robotics Institute
Carnegie Mellon University
Pittsburgh, PA, USA
Email: siddiqi@google.com

Geoffrey J. Gordon
Machine Learning Dept.
Carnegie Mellon University
Pittsburgh, PA, USA
Email: ggordon@cs.cmu.edu

***Abstract*— A central problem in artificial intelligence is to choose actions to maximize reward in a partially observable, uncertain environment. To do so, we must *learn* an accurate model of our environment, and then *plan* to maximize reward. Unfortunately, learning algorithms often recover a model which is too inaccurate to support planning or too large and complex for planning to be feasible; or, they require large amounts of prior domain knowledge or fail to provide important guarantees such as statistical consistency. To begin to fill this gap, we propose a novel algorithm which provably learns a compact, accurate model directly from sequences of action-observation pairs. To evaluate the learner, we then *close the loop* from observations to actions: we plan in the learned model and recover a policy which is near-optimal in the *original* environment (not the model). In more detail, we present a spectral algorithm for learning a Predictive State Representation (PSR). We demonstrate the algorithm by learning a model of a simulated high-dimensional, vision-based mobile robot planning task, and then performing approximate point-based planning in the learned model. This experiment shows that the learned PSR captures the essential features of the environment, allows accurate prediction with a small number of parameters, and enables successful and efficient planning. Our algorithm has several benefits which have not appeared together in any previous PSR learner: it is computationally efficient and statistically consistent; it handles high-dimensional observations and long time horizons by working from real-valued features of observation sequences; and finally, our close-the-loop experiments provide an end-to-end practical test.**

I. Introduction

Planning a sequence of actions or a policy to maximize reward has long been considered a fundamental problem for autonomous agents. In the hardest version of the problem, an agent must form a plan based solely on its own experience, without the aid of a human engineer who can design problem-specific features or heuristics; it is this version of the problem which we must solve to build a truly autonomous agent.

Partially Observable Markov Decision Processes (POMDPs) [23, 3] are a general framework for single-agent planning. POMDPs model the state of the world as a latent variable and explicitly reason about uncertainty in both action effects and state observability. Plans in POMDPs are expressed as *policies*, which specify the action to take given any possible probability distribution over states. Unfortunately, exact planning algorithms such as *value iteration* [23] are computationally intractable for most realistic POMDP planning problems. Furthermore, researchers have had only limited success learning POMDP models from data.

Predictive State Representations (PSRs) [11] and the closely related *Observable Operator Models (OOMs)* [8] are generalizations of POMDPs that have attracted interest because they both have greater representational capacity than POMDPs and yield representations that are *at least* as compact [20, 4]. In contrast to the latent-variable representations of POMDPs, PSRs and OOMs represent the state of a dynamical system by tracking occurrence probabilities of a set of future events (called *tests* or *characteristic events*) conditioned on past events (called *histories* or *indicative events*). Because tests and histories are observable quantities, it has been suggested that learning PSRs and OOMs should be easier than learning POMDPs. And, many successful approximate planning techniques for POMDPs can be used to plan in PSRs and OOMs with minimal adjustment.

The quality of an optimized policy for a POMDP, PSR, or OOM depends strongly on the accuracy of the model: inaccurate models typically lead to useless plans. We can specify a model manually or learn one from data, but due to the difficulty of learning, it is far more common to see planning algorithms applied to hand-specified models, and therefore to small systems where there is extensive and goal-relevant domain knowledge. For example, recent extensions of approximate planning techniques for PSRs have only been applied to hand-constructed models [9, 7].

Learning models for planning in partially observable environments has met with only limited success. As a result, there have been few successful attempts at closing the loop by learning a model from an environment, planning in that model, and testing the plan in the environment. For example, Expectation-Maximization (EM) [1] does not avoid local minima or scale to large state spaces; and, although many learning algorithms have been proposed for PSRs [21, 30, 13, 26, 2] and OOMs [8, 5, 31], none have been shown to learn models that are accurate enough for planning.

Several researchers have, however, made progress in the problem of planning using a learned model. In one instance [17], researchers obtained a POMDP heuristically from the output of a model-free algorithm [12] and demonstrated planning on a small toy maze. In another instance [16], researchers used Markov Chain Monte Carlo (MCMC) inference both to learn a factored Dynamic Bayesian Network (DBN) representation of a POMDP in a small synthetic network administration domain, as well as to perform online planning. Due to the cost of the MCMC sampler used, this approach is still impractical for larger models. In a final example, researchers learned Linear-Linear Exponential Family PSRs from an agent traversing a simulated environment, and found a policy using a policy gradient technique with a parameterized function of the learned PSR state as input [29, 27]. In this case both the learning and the planning algorithm were subject to local optima. In addition, the authors determined that the learned model was too inaccurate to support value-function-based planning methods [27]. Perhaps the closest prior work is that of Rosencrantz et al. [15]: like our method, their algorithm uses a straightforward sequence of algebraic operations to derive parameter estimates from matrices of observable statistics. However, Rosencrantz et al. do not attempt to provide a detailed proof of consistency such as the one in Equations 4(a–c) and Section III below. And, their method does not easily generalize to allow real-valued observations, "indicative features," and "characteristic features," as we derive in Sections III-A and III-B below; in our experience, using these features (instead of discrete, mutually-exclusive events) greatly reduces the variance of our estimated model parameters. Finally, the experiments of Rosencrantz et al. focus on the observation-only case, rather than on estimating the effects of actions and using the learned model for planning. (We describe several more-minor differences between the algorithms below in Section III.)

The current paper differs from these and other previous examples of planning in learned models: it both uses a principled and provably statistically consistent model-learning algorithm, and demonstrates positive results on a challenging high-dimensional problem with continuous observations. In particular, we propose a novel, consistent spectral algorithm for learning a variant of PSRs called *Transformed PSRs* [15] directly from execution traces. The algorithm is closely related to subspace identification for learning Linear Dynamical Systems (LDSs) [22, 25] and spectral algorithms for learning Hidden Markov Models (HMMs) [6] and Reduced-Rank HMMs [18]. We then demonstrate that this algorithm is able to learn compact models of a difficult, realistic dynamical system without any prior domain knowledge built into the model or algorithm. Finally, we perform approximate point-based value iteration (PBVI) in the learned compact models, and demonstrate that the greedy policy for the resulting value function works well in the original (not the learned) system. To our knowledge this is the first research that combines all of these achievements, closing the loop from observations to actions in an unknown nonlinear, non-Gaussian system with no human intervention beyond collecting the raw transition data and specifying features.

II. Predictive State Representations

A predictive state representation (PSR) [11] is a compact and complete description of a dynamical system. PSRs represent state as a set of predictions of observable experiments or *tests* that one could perform in the system. Specifically, a test of length k is an ordered sequence of action-observation pairs $\tau = a_1^\tau o_1^\tau \dots a_k^\tau o_k^\tau$ that can be executed and observed at a given time. Likewise, a *history* is an ordered sequence of action-observation pairs $h = a_1^h o_1^h \dots a_t^h o_t^h$ that has been executed and observed prior to a given time. The *prediction* for a test τ is the probability of the sequence of observations $\tau^O = o_1^\tau, \dots, o_k^\tau$ being generated, given that we intervene to take the sequence of actions $\tau^A = a_1^\tau, \dots, a_k^\tau$. If the observations produced by the dynamical system match those specified by the test, the test is said to have *succeeded*. The key idea behind a PSR is that, if we know the expected outcomes of executing all possible tests, then we also know everything there is to know about the state of a dynamical system.

In PSRs, actions in tests are *interventions*, not observations. We use a single vertical bar to indicate conditioning and a double vertical bar to indicate intervening: e.g., $\Pr[\tau^O \mid h \,||\, \tau^A]$ is the probability of the observations in test τ, given an observed history h, and given that we intervene to execute the actions in τ. We write $Q(h)$ for the *prediction vector* of success probabilities for a set of tests $Q = \{q_i\}$:

$$Q(h) = [\Pr[q_1^O \mid h \,||\, q_1^A], ..., \Pr[q_{|Q|}^O \mid h \,||\, q_{|Q|}^A]]^\mathsf{T}$$

Knowing the probabilities of some tests may allow us to compute the probabilities of other tests. That is, given a test τ and a set of tests Q, there may exist a *prediction function* f_τ such that $\Pr[\tau^O \mid h \,||\, \tau^A] = f_\tau(Q(h))$ for all histories h. In this case, we say $Q(h)$ is a *sufficient statistic* for τ.

Formally, a PSR is a tuple $\langle A, O, Q, F, m_1 \rangle$. A is the set of possible actions, and O is the set of possible observations. Q is a *core* set of tests, i.e., a set whose prediction vector $Q(h)$ is a sufficient statistic for *all* tests. F is the set of prediction functions f_τ for all tests τ (which must exist since Q is a core set), and $m_1 = Q(\epsilon)$ is the initial prediction vector after seeing the empty history ϵ. In this work we will restrict ourselves to *linear* PSRs, in which all prediction functions are linear: $f_\tau(Q(h)) = r_\tau^\mathsf{T} Q(h)$ for some vector $r_\tau \in \mathbb{R}^{|Q|}$.

Since $Q(h)$ is a sufficient statistic for all tests, it is a *state* for our PSR: i.e., we can remember just $Q(h)$ instead of h itself. After taking action a and seeing observation o, we can update $Q(h)$ recursively: if we write M_{ao} for the matrix with rows $r_{ao\tau}^\mathsf{T}$ for $\tau \in Q$, then we can use Bayes' Rule to show:

$$Q(hao) = \frac{M_{ao}Q(h)}{\Pr[o \mid h \,||\, a]} = \frac{M_{ao}Q(h)}{m_\infty^\mathsf{T} M_{ao}Q(h)} \tag{1}$$

where m_∞ is a normalizer, defined by $m_\infty^\mathsf{T} Q(h) = 1\ (\forall h)$.

Specifying a PSR involves first finding a core set of tests Q, called the *discovery problem*, and then finding the parameters M_{ao}, m_∞, and m_1 for these tests, called the *learning problem*. A core set Q for a linear PSR is said to be *minimal* if the tests in Q are linearly independent [8, 20], i.e., no one test's prediction is a linear function of the other tests' predictions. The discovery problem is usually solved by searching for linearly independent tests by repeatedly performing Singular Value Decompositions (SVDs) on collections of tests [30]. The learning problem is then solved by regression.

Transformed PSRs (TPSRs) [15] are a generalization of PSRs: TPSRs maintain a small number of sufficient statistics which are *linear combinations* of a (potentially very large and non-minimal) set of test probabilities. Accordingly, TPSRs can be thought of as *linear transformations* of regular PSRs. Therefore, TPSRs include PSRs as a special case, since this transformation can be the identity. The main benefit of TPSRs is that, given a core set of tests, the parameter learning problem can be solved and a large step toward solving the discovery problem can be achieved in closed form, as we will see below. In this respect, TPSRs are closely related to the transformed representations of LDSs and HMMs found by *subspace identification* [25, 22, 6].

A. Observable Representations

Our learning algorithm is based on an *observable representation* of a PSR, that is, one where each parameter corresponds directly to observable quantities. This representation depends on a core set of tests $\mathcal{T}$ (not necessarily minimal). It also depends on a set $\mathcal{H}$ of *indicative events*, that is, a mutually exclusive and exhaustive partition of the set of all possible histories. We will assume $\mathcal{H}$ is *sufficient* (defined below).

For purposes of gathering data, we assume that we can sample from some distribution ω over histories; our observable representation depends on ω as well. E.g., ω might be the steady-state distribution of some exploration policy. Note that this assumption means that we *cannot* estimate m_1, since we don't have samples of trajectories starting from m_1. So, instead, we will estimate m_*, an arbitrary feasible state, which is enough information to enable prediction. If we make the stronger assumption that we can repeatedly reset our PSR to its starting distribution, a straightforward modification of our algorithm will allow us to estimate m_1 as well.

We define several observable matrices in terms of $\mathcal{T}$, $\mathcal{H}$, and ω. After each definition we show how these matrices relate to the parameters of the underlying PSR. These relationships will allow us to define an equivalent TPSR, and will also be key tools in designing our learning algorithm and showing its consistency. The first observable matrix is $P_\mathcal{H} \in \mathbb{R}^{|\mathcal{H}|}$, containing the probabilities of every event $H_i \in \mathcal{H}$ when we sample a history h according to ω:

$$[P_\mathcal{H}]_i \equiv \Pr[H_i] \implies P_\mathcal{H} = \Pr[\mathcal{H}] \tag{2a}$$

Here we have defined $\Pr[H_i]$ to mean $\Pr[h \in H_i]$, and $\Pr[\mathcal{H}]$ to mean the vector whose elements are $\Pr[H_i]$ for $H_i \in \mathcal{H}$.

The next observable matrix is $P_{\mathcal{T},\mathcal{H}} \in \mathbb{R}^{|\mathcal{T}|\times|\mathcal{H}|}$, whose entries are *joint* probabilities of tests $\tau_i \in \mathcal{T}$ and indicative events $H_j \in \mathcal{H}$ when we sample $h \sim \omega$ and take actions τ_i^A:

$$\begin{aligned}
[P_{\mathcal{T},\mathcal{H}}]_{i,j} &\equiv \Pr[\tau_i^O, H_j \,||\, \tau_i^A] \equiv \mathbb{E}[r_{\tau_i}^\mathsf{T} Q(h) \mid H_j] \Pr[H_j] \\
&\equiv r_{\tau_i}^\mathsf{T} s_{H_j} \Pr[H_j] \\
\implies P_{\mathcal{T},\mathcal{H}} &= RSD
\end{aligned} \tag{2b}$$

Here we define $s_{H_j} = \mathbb{E}[Q(h) \mid H_j]$ to be the expected state given indicative event H_j; and as above, the vector r_{τ_i} lets us compute the probability of test τ_i given the state. Finally, we let $R \in \mathbb{R}^{|\mathcal{T}|\times|Q|}$ be the matrix with rows $r_{\tau_i}^\mathsf{T}$, $S \in \mathbb{R}^{|Q|\times|\mathcal{H}|}$ be the matrix with columns s_{H_j}, and $D = \operatorname{diag}(\Pr[\mathcal{H}])$.

Eq. (2b) tells us that the rank of $P_{\mathcal{T},\mathcal{H}}$ is no more than $|Q|$, since its factors R and S each have rank at most $|Q|$. At this point we can define a *sufficient* set of indicative events as promised: it is a set of indicative events for which the rank of $P_{\mathcal{T},\mathcal{H}}$ is *equal* to $|Q|$.

The final observable matrices are $P_{\mathcal{T},ao,\mathcal{H}} \in \mathbb{R}^{|\mathcal{T}|\times|\mathcal{H}|}$, one matrix for each action-observation pair. Entries of $P_{\mathcal{T},ao,\mathcal{H}}$ are probabilities of *triples* of an indicative event, the next action-observation pair, and a subsequent test, if we intervene to execute a and τ_i^A:

$$\begin{aligned}
[P_{\mathcal{T},ao,\mathcal{H}}]_{i,j} &\equiv \Pr[\tau_i^O, o, H_j \,||\, a, \tau_i^A] \\
&= \mathbb{E}[\Pr[\tau_i^O, o \mid h \,||\, a, \tau_i^A] \mid H_j] \Pr[H_j] \\
&= \mathbb{E}[\Pr[\tau_i^O \mid h, o \,||\, a, \tau_i^A] \Pr[o \mid h \,||\, a] \mid H_j] \Pr[H_j] \\
&= \mathbb{E}[r_{\tau_i}^\mathsf{T} Q(hao) \Pr[o \mid h \,||\, a] \mid H_j] \Pr[H_j] \\
&= \mathbb{E}[r_{\tau_i}^\mathsf{T} M_{ao} Q(h) \mid H_j] \Pr[H_j] \quad \text{by Eq. (1)} \\
&= r_{\tau_i}^\mathsf{T} M_{ao} s_{H_j} \Pr[H_j] \\
\implies P_{\mathcal{T},ao,\mathcal{H}} &= R M_{ao} S D
\end{aligned} \tag{2c}$$

Just like $P_{\mathcal{T},\mathcal{H}}$, the matrices $P_{\mathcal{T},ao,\mathcal{H}}$ have rank at most $|Q|$ due to their factors R and S.

We can also relate the factor S to the parameters m_∞ and m_*: for m_∞, we start from the fact $m_\infty^\mathsf{T} S = 1^\mathsf{T}$ (since each column of S is a vector of core-test predictions), repeatedly multiply by S and $S^\dagger$, and use the fact $SS^\dagger = I$.

$$m_\infty^\mathsf{T} S = 1_k^\mathsf{T} \implies m_\infty^\mathsf{T} S S^\dagger = 1_k^\mathsf{T} S^\dagger \implies m_\infty^\mathsf{T} = 1_k^\mathsf{T} S^\dagger \tag{3a}$$

$$\implies m_\infty^\mathsf{T} S = 1_k^\mathsf{T} S^\dagger S \implies 1_k^\mathsf{T} = 1_k^\mathsf{T} S^\dagger S \tag{3b}$$

Here, $k = |\mathcal{H}|$ and 1_k denotes the ones-vector of length k. For m_*, we note that each column of S is a feasible state, and so we can take m_* to be any convex combination of the columns of S. In particular, we will take $m_* = S \Pr[\mathcal{H}]$.

We now define a TPSR in terms of the matrices $P_\mathcal{H}$, $P_{\mathcal{T},\mathcal{H}}$, $P_{\mathcal{T},ao,\mathcal{H}}$ and an additional matrix $U \in \mathbb{R}^{|\mathcal{T}|\times|Q|}$ such that $U^\mathsf{T} R$ is invertible. A natural choice for U is the left singular vectors of $P_{\mathcal{T},\mathcal{H}}$, although a randomly-generated U will work with probability 1. We can think of the columns of U as specifying a core set Q' of *linear combinations* of tests which define a state vector for our TPSR. Finally we simplify the expressions using Equations 2(a–c) to show that our parameters are only a similarity transform away from the original PSR parameters:

$$b_* \equiv U^\mathsf{T} P_{\mathcal{T},\mathcal{H}} 1_k = U^\mathsf{T} RSD 1_k = (U^\mathsf{T} R) m_* \tag{4a}$$

$$\begin{aligned}
b_\infty^\mathsf{T} &\equiv P_\mathcal{H}^\mathsf{T} (U^\mathsf{T} P_{\mathcal{T},\mathcal{H}})^\dagger = 1_n^\mathsf{T} S^\dagger S D (U^\mathsf{T} P_{\mathcal{T},\mathcal{H}})^\dagger \\
&= 1_n^\mathsf{T} S^\dagger (U^\mathsf{T} R)^{-1} (U^\mathsf{T} R) S D (U^\mathsf{T} P_{\mathcal{T},\mathcal{H}})^\dagger \\
&= 1_n^\mathsf{T} S^\dagger (U^\mathsf{T} R)^{-1} = m_\infty^\mathsf{T} (U^\mathsf{T} R)^{-1}
\end{aligned} \tag{4b}$$

$$\begin{aligned} B_{ao} &\equiv U^\top P_{\mathcal{T},ao,\mathcal{H}}(U^\top P_{\mathcal{T},\mathcal{H}})^\dagger \\ &= U^\top R M_{ao} S D (U^\top P_{\mathcal{T},\mathcal{H}})^\dagger \\ &= U^\top R M_{ao}(U^\top R)^{-1}(U^\top R) S D (U^\top P_{\mathcal{T},\mathcal{H}})^\dagger \\ &= (U^\top R) M_{ao} (U^\top R)^{-1} \end{aligned} \tag{4c}$$

The derivation of Eq. 4b makes use of Eqs. 3a and 3b. To get $b_1 = (U^\top R)m_1$ instead of b_* in Eq. 4a, replace $P_{\mathcal{T},\mathcal{H}}1_k$, the vector of expected probabilities of tests for $h \sim \omega$, with the vector of probabilities of tests for $h = \epsilon$.

Given these parameters, we can calculate the probability of observations $o_{1:t}$, given that we start from state m_1 and intervene with actions $a_{1:t}$. Here we write the product of a sequence of transitions as $M_{ao_{1:t}} = M_{a_1o_1}M_{a_2o_2}\ldots M_{a_to_t}$.

$$\begin{aligned} \Pr[o_{1:t} \| a_{1:t}] &= m_\infty^\top (U^\top R)^{-1}(U^\top R) M_{ao_{1:t}} (U^\top R)^{-1}(U^\top R) m_1 \\ &= b_\infty^\top B_{ao_{1:t}} b_1 \end{aligned} \tag{5}$$

In addition to the initial TPSR state b_1, we define normalized conditional *internal states* b_t by a recursive update:

$$b_{t+1} = \frac{B_{ao_t} b_t}{b_\infty^\top B_{ao_t} b_t} \tag{6}$$

It is straightforward to show recursively that b_t can be used to compute conditional observation probabilities. (If we only have b_* instead of b_1, then initial probability estimates will be approximate, but the difference in predictions will disappear over time as our process mixes.) The linear transform from a TPSR internal state b_t to a PSR internal state $q_t = Q(h_t)$ is given by $q_t = (U^\top R)^{-1} b_t$. If we chose U by SVD, then the prediction of tests $\Pr[\mathcal{T}^O \mid h \,\|\, \mathcal{T}^A]$ at time t is given by $Ub_t = UU^\top R q_t = R q_t$ (since by the definition of SVD, U is orthonormal and the row spaces of U and R are the same).

III. Learning TPSRs

Our learning algorithm works by building empirical estimates $\widehat{P}_{\mathcal{H}}$, $\widehat{P}_{\mathcal{T},\mathcal{H}}$, and $\widehat{P}_{\mathcal{T},ao,\mathcal{H}}$ of the matrices $P_{\mathcal{H}}$, $P_{\mathcal{T},\mathcal{H}}$, and $P_{\mathcal{T},ao,\mathcal{H}}$ defined above. To build these estimates, we repeatedly sample a history h from the distribution ω, execute a sequence of actions, and record the resulting observations.

Once we have computed $\widehat{P}_{\mathcal{H}}$, $\widehat{P}_{\mathcal{T},\mathcal{H}}$, and $\widehat{P}_{\mathcal{T},ao,\mathcal{H}}$, we can generate $\widehat{U}$ by singular value decomposition of $\widehat{P}_{\mathcal{T},\mathcal{H}}$. We can then learn the TPSR parameters by plugging $\widehat{U}$, $\widehat{P}_{\mathcal{H}}$, $\widehat{P}_{\mathcal{T},\mathcal{H}}$, and $\widehat{P}_{\mathcal{T},ao,\mathcal{H}}$ into Equation 4[1]:

$$\begin{aligned} \widehat{b}_* &= \widehat{U}^\top \widehat{P}_{\mathcal{H}} 1_k \\ \widehat{b}_\infty &= (\widehat{P}_{\mathcal{T},\mathcal{H}}^\top \widehat{U})^\dagger \widehat{P}_{\mathcal{H}} \\ \widehat{B}_{ao} &= \widehat{U}^\top \widehat{P}_{\mathcal{T},ao,\mathcal{H}} (\widehat{U}^\top \widehat{P}_{\mathcal{T},\mathcal{H}})^\dagger \end{aligned}$$

As we include more data in our averages, the law of large numbers guarantees that our estimates $\widehat{P}_{\mathcal{H}}$, $\widehat{P}_{\mathcal{T},\mathcal{H}}$, and $\widehat{P}_{\mathcal{T},ao,\mathcal{H}}$ converge to the true matrices $P_{\mathcal{H}}$, $P_{\mathcal{T},\mathcal{H}}$, and $P_{\mathcal{T},ao,\mathcal{H}}$ (defined in Equation 2). So by continuity of the formulas above, if our system is truly a PSR of finite rank, our estimates $\widehat{b}_*$, $\widehat{b}_\infty$, and $\widehat{B}_{ao}$ converge to the true parameters up to a linear transform—that is, our learning algorithm is *consistent*. Although parameters estimated with finite data can sometimes lead to negative probability estimates when filtering or predicting, this problem can be avoided in practice by thresholding the predicted probabilities by some small positive probability.

Note that the learning algorithm presented here is distinct from the TPSR learning algorithm presented in Rosencrantz et al. [15]. In addition to the differences mentioned above, a key difference between the two algorithms is that here we estimate the *joint* probability of a past event, a current observation, and a future event in the matrix $\widehat{P}_{\mathcal{T},ao,\mathcal{H}}$, whereas in [15], the authors instead estimate the probability of a future event, *conditioned* on a past event and a current observation. To compensate, Rosencrantz et al. later multiply this estimate by an approximation of the probability of the current observation, conditioned on the past event. Rosencrantz et al. also derive the approximate probability of the current observation differently: as the result of a regression instead of directly from empirical counts. Finally, Rosencrantz et al. do not make any attempt to multiply by the marginal probability of the past event, although this term cancels in the current work. In the absence of estimation errors, both algorithms would arrive at the same answer, but taking errors into account, they will typically make different predictions. The difficulty of extending the Rosencrantz et al. algorithm to handle real-valued features stems from the difference between joint and conditional probabilities: the observable matrices in Rosencrantz et al. are conditional expectations, so their algorithm depends on being able to condition on discrete indicative events or observations. In contrast, the next section shows how to extend our algorithm to use real-valued features.

A. Learning TPSRs with Features

In data gathered from complex real-world dynamical systems, it may not be possible to find a reasonably-sized core set of discrete tests $\mathcal{T}$ or sufficient set of indicative events $\mathcal{H}$. When this is the case, we can generalize the TPSR learning algorithm and work with *features* of tests and histories, which we call *characteristic features* and *indicative features* respectively. In particular let $\mathcal{T}$ and $\mathcal{H}$ be large sets of tests and indicative events (possibly too large to work with directly) and let $\phi^{\mathcal{T}}$ and $\phi^{\mathcal{H}}$ be shorter vectors of characteristic and indicative features. The matrices $P_{\mathcal{H}}$, $P_{\mathcal{T},\mathcal{H}}$, and $P_{\mathcal{T},ao,\mathcal{H}}$ will no longer contain probabilities but rather *expected values* of features or products of features. For the special case of features that are *indicator functions* of tests and histories, we recover the probability matrices from Section II-A.

Here we prove the *consistency* of our estimation algorithm using these more general matrices as inputs. In the following equations $\Phi^{\mathcal{T}}$ and $\Phi^{\mathcal{H}}$ are matrices of characteristic and indicative features respectively, with first dimension equal to the number of characteristic or indicative features and second dimension equal to $|\mathcal{T}|$ and $|\mathcal{H}|$ respectively. An entry of $\Phi^{\mathcal{H}}$ is the expectation of one of the indicative features given the occurrence of one of the indicative events. An entry of $\Phi^{\mathcal{T}}$

[1]The learning strategy employed here may be seen as a generalization of Hsu et al.'s spectral algorithm for learning HMMs [6] to PSRs. Note that since HMMs and POMDPs are a proper subset of PSRs, we can use the algorithm in this paper to learn back both HMMs and POMDPs in PSR form.

is the weight of one of our tests in calculating one of our characteristic features. With these features we generalize the matrices $P_{\mathcal{H}}$, $P_{\mathcal{T},\mathcal{H}}$, and $P_{\mathcal{T},ao,\mathcal{H}}$:

$$
\begin{aligned}
&[P_{\mathcal{H}}]_i \equiv \mathbb{E}(\phi_i^{\mathcal{H}}(h)) = \textstyle\sum_{H\in\mathcal{H}} \Pr[H]\Phi_{iH}^{\mathcal{H}} \\
&\implies P_{\mathcal{H}} = \Phi^{\mathcal{H}} \Pr[\mathcal{H}] && \text{(7a)} \\
&[P_{\mathcal{T},\mathcal{H}}]_{i,j} \equiv \mathbb{E}(\phi_i^{\mathcal{T}}(\tau^O)\cdot\phi_j^{\mathcal{H}}(h)\,||\,\tau^A) \\
&= \textstyle\sum_{\tau\in\mathcal{T}}\sum_{H\in\mathcal{H}} \Pr[\tau^O, H\,||\,\tau^A]\Phi_{i\tau}^{\mathcal{T}}\Phi_{jH}^{\mathcal{H}} \\
&= \textstyle\sum_{\tau\in\mathcal{T}} r_\tau^{\mathsf{T}}\Phi_{i\tau}^{\mathcal{T}}\sum_{H\in\mathcal{H}} s_H \Pr[H]\Phi_{jH}^{\mathcal{H}} && \text{by Eq. (2b)} \\
&\implies P_{\mathcal{T},\mathcal{H}} = \Phi^{\mathcal{T}} RSD\Phi^{\mathcal{H}^{\mathsf{T}}} && \text{(7b)} \\
&[P_{\mathcal{T},ao,\mathcal{H}}]_{i,j} \equiv \mathbb{E}(\phi_i^{\mathcal{T}}(\tau^O)\cdot\phi_j^{\mathcal{H}}(h)\cdot\delta(o)\,||\,a,\tau^A) \\
&= \textstyle\sum_{\tau\in\mathcal{T}}\sum_{H\in\mathcal{H}} \Pr[\tau^O, o, H\,||\,a,\tau^A]\Phi_{i\tau}^{\mathcal{T}}\Phi_{jH}^{\mathcal{H}} \\
&= \left(\textstyle\sum_{\tau\in\mathcal{T}} r_\tau^{\mathsf{T}}\Phi_{i\tau}^{\mathcal{T}}\right) M_{ao}\left(\textstyle\sum_{H\in\mathcal{H}} s_H \Pr[H]\Phi_{jH}^{\mathcal{H}}\right) && \text{by Eq. (2c)} \\
&\implies P_{\mathcal{T},ao,\mathcal{H}} = \Phi^{\mathcal{T}} RM_{ao}SD\Phi^{\mathcal{H}^{\mathsf{T}}} && \text{(7c)}
\end{aligned}
$$

where $\delta(o)$ is an indicator for observation o. The parameters of the TPSR (b_*, b_∞, and B_{ao}) are now defined in terms of a matrix U such that $U^{\mathsf{T}}\Phi^{\mathcal{T}}R$ is invertible (we can take U to be the left singular values of $P_{\mathcal{T},\mathcal{H}}$), and in terms of the matrices $P_{\mathcal{H}}$, $P_{\mathcal{T},\mathcal{H}}$, and $P_{\mathcal{T},ao,\mathcal{H}}$. We also define a new vector e s.t. $\Phi^{\mathcal{H}^{\mathsf{T}}}e = 1_k$; this means that the ones vector 1_k^{T} must be in the row space of $\Phi^{\mathcal{H}}$. Since $\Phi^{\mathcal{H}}$ is a matrix of features, we can always ensure that this is the case by requiring one of our features to be a constant. Finally we simplify the expressions using Equations 7(a–c) to show that our parameters are only a similarity transform away from the original PSR parameters:

$$
\begin{aligned}
b_* &\equiv U^{\mathsf{T}}P_{\mathcal{T},\mathcal{H}}e = U^{\mathsf{T}}\Phi^{\mathcal{T}}RSD\Phi^{\mathcal{H}^{\mathsf{T}}}e \\
&= U^{\mathsf{T}}\Phi^{\mathcal{T}}RSD1_k = (U^{\mathsf{T}}\Phi^{\mathcal{T}}R)m_* && \text{(8a)} \\
b_\infty^{\mathsf{T}} &\equiv P_{\mathcal{H}}^{\mathsf{T}}(U^{\mathsf{T}}P_{\mathcal{T},\mathcal{H}})^\dagger = 1_n^{\mathsf{T}}S^\dagger SD\Phi^{\mathcal{H}^{\mathsf{T}}}(U^{\mathsf{T}}P_{\mathcal{T},\mathcal{H}})^\dagger \\
&= 1_n^{\mathsf{T}}S^\dagger(U^{\mathsf{T}}\Phi^{\mathcal{T}}R)^{-1}(U^{\mathsf{T}}\Phi^{\mathcal{T}}R)SD\Phi^{\mathcal{H}^{\mathsf{T}}}(U^{\mathsf{T}}P_{\mathcal{T},\mathcal{H}})^\dagger \\
&= 1_n^{\mathsf{T}}S^\dagger(U^{\mathsf{T}}\Phi^{\mathcal{T}}R)^{-1} = m_\infty^{\mathsf{T}}(U^{\mathsf{T}}\Phi^{\mathcal{T}}R)^{-1} && \text{(8b)} \\
B_{ao} &\equiv U^{\mathsf{T}}P_{\mathcal{T},ao,\mathcal{H}}(U^{\mathsf{T}}P_{\mathcal{T},\mathcal{H}})^\dagger \\
&= U^{\mathsf{T}}\Phi^{\mathcal{T}}RM_{ao}(U^{\mathsf{T}}\Phi^{\mathcal{T}}R)^{-1}(U^{\mathsf{T}}\Phi^{\mathcal{T}}R)SD\Phi^{\mathcal{H}^{\mathsf{T}}}(U^{\mathsf{T}}P_{\mathcal{T},\mathcal{H}})^\dagger \\
&= (U^{\mathsf{T}}\Phi^{\mathcal{T}}R)M_{ao}(U^{\mathsf{T}}\Phi^{\mathcal{T}}R)^{-1} && \text{(8c)}
\end{aligned}
$$

Just as in the beginning of Section III, we can estimate $\widehat{P}_{\mathcal{H}}$, $\widehat{P}_{\mathcal{T},\mathcal{H}}$, and $\widehat{P}_{\mathcal{T},ao,\mathcal{H}}$, and then plug the matrices into Equations 8(a–c). Thus we see that if we work with characteristic and indicative features, and if our system is truly a TPSR of finite rank, our estimates $\widehat{b}_*$, $\widehat{b}_\infty$, and $\widehat{B}_{ao}$ again converge to the true PSR parameters up to a linear transform.

B. Kernel Density Estimation for Continuous Observations

For continuous observations, we use Kernel Density Estimation (KDE) [19] to model the observation probability density function (PDF). Although use of kernel density estimation for continuous observations in PSRs is not new [28], extending our algorithm to use KDE results, for the first time, in a statistically consistent learning algorithm for PSRs with continuous observations. We use a fraction of the training data points as kernel centers, placing one multivariate Gaussian kernel at each point.[2] The KDE of the observation PDF is a convex combination of these kernels; since each kernel integrates to 1, this estimator also integrates to 1. KDE theory [19] tells us that, with the correct kernel weights, as the number of kernel centers and the number of samples go to infinity and the kernel bandwidth goes to zero (at appropriate rates), the KDE estimator converges to the observation PDF in L_1 norm. The kernel density estimator is completely determined by the normalized vector of kernel weights; therefore, if we can estimate this vector accurately, our estimate of the observation PDF will converge to the observation PDF as well. Hence our goal is to predict the correct expected value of this normalized kernel vector given all past observations. In the continuous-observation case, we can still write our latent-state update in the same form, using a matrix B_{ao}; however, rather than learning each of the uncountably-many B_{ao} matrices separately, we learn one base operator per kernel center, and use convex combinations of these base operators to compute observable operators as needed. For details on practical aspects of learning with continuous observations, see Section IV-B.

C. Practical Computation of the TPSR Parameters

As defined above, each element of $\widehat{P}_{\mathcal{H}}$ is the empirical expectation (over histories sampled from ω) of the corresponding element of the indicative feature vector—that is, element i is $\frac{1}{w}\sum_{t=1}^{w}\phi_{it}^{\mathcal{H}}$, where $\phi_{it}^{\mathcal{H}}$ is the ith indicative feature evaluated at the tth sampled history. Similarly, each element of $\widehat{P}_{\mathcal{T},\mathcal{H}}$ is an empirical expectation of the *product* of one indicative feature and one characteristic feature, if we sample a history from ω and then follow an appropriate sequence of actions. We can compute all elements of $\widehat{P}_{\mathcal{T},\mathcal{H}}$ from a single sample of trajectories if we sample histories from ω, follow an appropriate exploratory policy, and then importance-weight each sample [2]: $[\widehat{P}_{\mathcal{T},\mathcal{H}}]_{ij}$ is $\sum_{t=1}^{w}\alpha_t\phi_{it}^{\mathcal{T}}\phi_{jt}^{\mathcal{H}}$, where α_t is an importance weight. (In our experiments below, the importance weights are constant by design, and therefore cancel out.)

Once we have constructed $\widehat{P}_{\mathcal{T},\mathcal{H}}$, we can compute $\widehat{U}$ as the matrix of left singular vectors of $\widehat{P}_{\mathcal{T},\mathcal{H}}$. One of the advantages of subspace identification is that the complexity of the model can be tuned by selecting the number of singular vectors in $\widehat{U}$, at the risk of losing prediction quality.

Finally, since there may be many large matrices $\widehat{P}_{\mathcal{T},ao,\mathcal{H}}$, rather than computing them directly, we instead compute $\widehat{U}^{\mathsf{T}}\widehat{P}_{\mathcal{T},ao,\mathcal{H}}$ for each pair a, o. The latter matrices are much smaller, and in our experiments, we saved substantially on both memory and runtime by avoiding construction of the larger matrices. To construct $\widehat{U}^{\mathsf{T}}\widehat{P}_{\mathcal{T},ao,\mathcal{H}}$, we restrict to those training trajectories in which the action at the middle time step is a. Then, each element of $\widehat{P}_{\mathcal{T},ao,\mathcal{H}}$ is a weighted empirical expectation (among the restricted set of trajectories) of the product of one indicative feature, one characteristic feature, and element o of the observation kernel vector. So,

[2]We use a general elliptical covariance matrix, chosen by PCA: that is, we use a spherical covariance after projecting onto the eigenvectors of the covariance matrix of the observations, and scaling by the square roots of the eigenvalues.

$$\widehat{U}^\top \widehat{P}_{\mathcal{T},ao,\mathcal{H}} = \sum_{t=1}^{w_a} \alpha_t^a (\widehat{U}^\top \phi_t^{\mathcal{T}})(\phi_t^{\mathcal{H}})^\top \frac{1}{Z_t} K(o_t - o) \quad (9)$$

where $K(\cdot)$ is the kernel function and Z_t is the kernel normalization constant computed by summing over the observation kernels for each o_t. As above, in our experiments, the importance weights α_t^a are uniform.

IV. Experimental Results

We have introduced a novel algorithm for learning TPSRs directly from data, as well as a kernel-based extension for modeling continuous observations. We judge the quality of our TPSR learning algorithm by first learning a model of a challenging non-linear, partially observable, controlled domain directly from sensor inputs and then "closing the loop" by *planning* in the learned model. Successful planning is a much stronger result than standard dynamical system evaluations such as one-step squared error or prediction log-likelihood. Unlike previous attempts to learn PSRs, which either lack planning results [15, 28], or which compare policies within the learned system [29], we compare our resulting policy to a bound on the best possible solution in the original system and demonstrate that the policy is close to optimal.

A. The Autonomous Robot Domain

Our simulated autonomous robot domain consisted of a simple 45×45 unit square arena with a central obstacle and brightly colored walls (Figure 1(A-B)), containing a robot of radius 2 units. The robot could move around the floor of the arena and rotate to face in any direction. The robot had a simulated 16×16 pixel color camera, with a focal plane one unit in front of the robot's center of rotation, and with a visual field of $45°$ in both azimuth and elevation, corresponding to an angular resolution of $\sim 2.8°$ per pixel. The resulting 768-element pattern of unprocessed RGB values was the only input to the robot (images were *not* preprocessed to extract features), and each action produced a new set of pixel values. The robot was able to move forward 1 or 0 units, and simultaneously rotate $15°$, $-15°$, or $0°$, resulting in 6 unique actions. In the real world, friction, uneven surfaces, and other factors confound precisely predictable movements. To simulate this uncertainty, a small amount of Gaussian noise was added to the translation (mean 0, standard deviation .1 units) and rotation (mean 0, standard deviation $5°$) components of the actions. The robot was allowed to occupy any real-valued (x, y, θ) pose that didn't intersect a wall; in case of an attempt to drive through a wall, we interrupted the commanded motion just before contact, simulating an inelastic collision.

The autonomous robot domain was designed to be a difficult domain comparable to the most complex domains that previous PSR algorithms have attempted to model. In particular, the domain in this paper was modeled after the autonomous robot domains found in recent PSR work [28, 29]. The proposed problem, learning a model of this domain and then planning in the learned model, is quite difficult. The autonomous robot has *no* knowledge of any of the underlying properties of the domain, e.g. the geometry of the environment or the robot motion model; it only has access to samples of the 256 pixel features, and how these features change as actions are executed. Writing a correct policy for a specific task in this domain by hand would be at best tedious—and in any case, as mentioned above, it is often impractical to hand-design a policy for an autonomous agent, since doing so requires guessing the particular planning problems that the agent may face in the future. Furthermore, the continuous and non-linear nature of this domain makes learning models difficult. For example, a POMDP model of this domain would require a prohibitively large number of hidden states, making learning and planning next to impossible. PSRs are able to overcome this problem by *compactly* representing state in a low-dimensional real-valued space, and the algorithm presented in this work allows us to efficiently learn the parameters of the PSR in closed form.

B. Learning a Model

We learn our model from a sample of 10000 short trajectories, each containing 7 action-observation pairs. We generate each trajectory by starting from a uniformly randomly sampled position in the environment and executing a uniform random sequence of actions. We used the first $l = 2000$ trajectories to generate kernel centers, and the remaining $w = 8000$ to estimate the matrices $P_{\mathcal{H}}$, $P_{\mathcal{T},\mathcal{H}}$, and $P_{\mathcal{T},ao,\mathcal{H}}$.

To define these matrices, we need to specify a set of indicative features, a set of observation kernel centers, and a set of characteristic features. We use Gaussian kernels to define our indicative and characteristic features, in a similar manner to the Gaussian kernels described above for observations; our analysis allows us to use arbitrary indicative and characteristic features, but we found Gaussian kernels to be convenient and effective. Note that the resulting features over tests and histories are just *features*; unlike the kernel centers defined over observations, there is no need to let the kernel width approach zero, since we are not attempting to learn accurate PDFs over the histories and tests in $\mathcal{H}$ and $\mathcal{T}$.

In more detail, we define a set of 2000 *indicative kernels*, each one centered at a sequence of 3 observations from the initial segment of one of our trajectories. We choose the kernel covariance using PCA on these sequences of observations, just as described for single observations in Section III-B. We then generate our indicative features for a new sequence of three observations by evaluating each indicative kernel at the new sequence, and normalizing so that the vector of features sums to one. (The initial distribution ω is, therefore, the distribution obtained by initializing uniformly and taking 3 random actions.) Similarly, we define 2000 *characteristic kernels*, each one centered at a sequence of 3 observations from the end of one of our sample trajectories, choose a kernel covariance, and define our characteristic feature vector by evaluating each kernel at a new observation sequence and normalizing. Finally, we define 500 *observation kernels*, each one centered at a single observation from the middle of one of our sample trajectories, and replace each observation by its corresponding vector of normalized kernel weights. Next, we construct the matrices $\widehat{P}_{\mathcal{H}}$, $\widehat{P}_{\mathcal{T},\mathcal{H}}$, and $\widehat{P}_{\mathcal{T},ao,\mathcal{H}}$ as the empirical

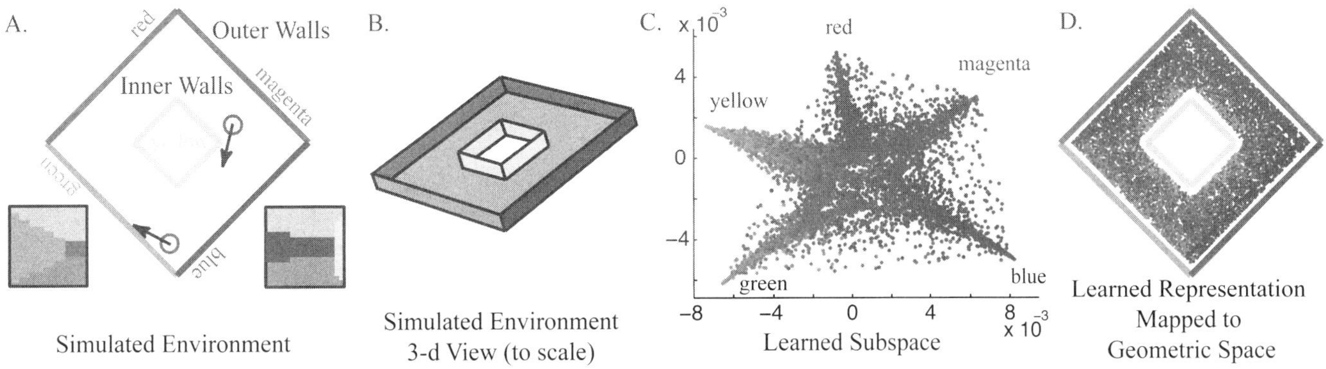

Fig. 1. Learning the Autonomous Robot Domain. Full-color figures are available in the online proceedings: http://www.roboticsproceedings.org/rss06/p36.html (A) The robot uses visual sensing to traverse a square domain with multi-colored walls and a central obstacle. Examples of images recorded by the robot occupying two different positions in the environment are shown at the bottom of the figure. (B) A to-scale 3-dimensional view of the environment. (C) The 2nd and 3rd dimension of the learned subspace (the first dimension primarily contained normalization information). Each point is the embedding of a single history, displayed with color equal to the average RGB color in the first image in the highest probability test. The star-shaped manifold captures the visual space of the robot with each "point" of the star containing concentrations of embeddings that predict images predominantly composed of a particular color. (D) The same points in (C) projected onto the environment's geometric space, demonstrating that the manifold sensibly captures features of geometric space.

expectations over our 8000 training trajectories according to the equations in Section III. Finally we chose $|Q'| = 5$ as the dimension of our TPSR, the smallest dimension that was able to produce high quality policies (see Section IV-D below).

C. Qualitative Evaluation

Having learned the parameters of the TPSR according to the algorithm in Section III, we can use the model for prediction, filtering, and planning in the autonomous robot domain. We first evaluated the model *qualitatively* by projecting the sets of histories in the training data onto the learned TPSR state space: $\widehat{U}^\top \widehat{P}_{\mathcal{H}}$. We colored each datapoint according to the average of the red, green, and blue components of the highest probability observation following the projected history. The features of the low dimensional embedding clearly capture the topology of the major features of the robot's visual environment (Figure 1(C-D)), and continuous paths in the environment translate into continuous paths in the latent space (Figure 2(B)).

D. Planning in the Learned Model

To test the quality of the learned model, we set up a navigation problem where the robot was required to plan to reach a goal image (looking directly at the blue wall). We specified a large reward (1000) for this observation, a reward of -1 for colliding with a wall, and 0 for every other observation. We learned a reward function by linear regression from the embedded histories $\widehat{U}^\top \widehat{P}_{\mathcal{H}}$ to the observed immediate rewards. We used the learned reward function to compute an approximate state-action value function via the PSR extension of the Perseus variant of PBVI [14, 24, 9, 7] with discount factor $\gamma = .8$, a prediction horizon of 10 steps, and with the 8000 embedded histories as the set of belief points. The learned value function is displayed in Figure 2(A). We then computed an initial belief by starting with b_* and tracking for 3 action-observation pairs, and executed the greedy policy for our learned value function. Examples of paths planned in the learned model are presented in Figure 2(B); the same paths are shown in geometric space in Figure 2(C). (Recall that the robot only has access to images, *never* its own position.)

The reward function encouraged the robot to navigate to a specific set of points in the environment, so the planning problem can be viewed as a shortest path problem. Even though we don't encode this intuition into our algorithm, we can use it to quantitatively evaluate the performance of the policy in the original system. First we randomly sampled 100 initial histories in the environment and asked the robot to plan paths for each based on its learned policy. We compared the number of actions taken both to a random policy and to the *optimistic path*, calculated by A* search in the robot's configuration space given the *true underlying position*. Note that this comparison is somewhat unfair: in order to achieve the same cost as the optimistic path, the robot would have to know its true underlying position, the dynamics would have to be deterministic, all rotations would have to be instantaneous, and the algorithm would need an unlimited amount of training data. Nonetheless, the results (Figure 2(D)) indicate that the performance of the TPSR policy is close to this optimistic bound. We think that this result is remarkable, especially given that previous approaches have encountered significant difficulty modeling continuous domains [10] and domains with similarly high levels of complexity [29].

V. Conclusions

We have presented a novel *consistent* subspace identification algorithm that simultaneously solves the *discovery* and *learning* problems for TPSRs. In addition, we provided two extensions to the learning algorithm that are useful in practice, while maintaining consistency: characteristic and indicative features only require one to know relevant features of tests and histories, rather than core sets of tests and sufficient sets of histories, while kernel density estimation can be used to find observable operators when observations are real-valued. We also showed how point-based approximate planning techniques can be used to solve the *planning* problem in the learned model. We demonstrated the representational capacity of our model and the effectiveness of our learning algorithm by learning a compact model from simulated autonomous robot vision data. Finally, we closed the loop by successfully planning with the learned models. To our knowledge this is the first instance

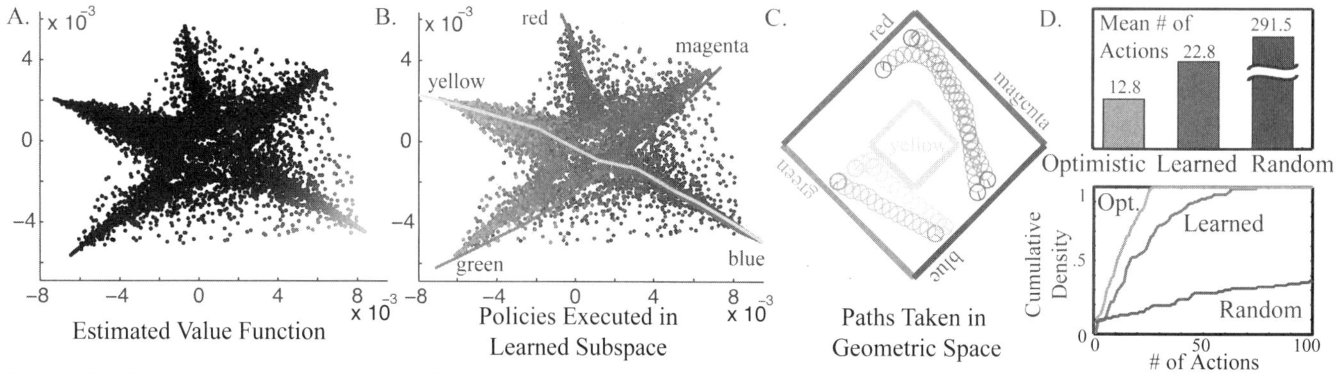

Fig. 2. Planning in the learned state space. (A) The value function computed for each embedded point; lighter indicates higher value. (B) Policies executed in the learned subspace. The red, green, magenta, and yellow paths correspond to the policy executed by a robot with starting positions facing the red, green, magenta, and yellow walls respectively. (C) The paths taken by the robot in geometric space while executing the policy. Each of the paths corresponds to the path of the same color in (B). The darker circles indicate the starting and ending positions, and the tick-mark indicates the robot's orientation. (D) Analysis of planning from 100 randomly sampled start positions to the target image (facing blue wall). In the bar graph: the mean number of actions taken by the optimistic solution found by A* search in configuration space (left); taken by executing the policy found by Perseus in the learned model (center); and taken by executing a random policy (right). Line graph illustrates the cumulative density of the number of actions given the optimal, learned, and random policies.

of learning a model for a simulated robot in a nonlinear, non-Gaussian, partially observable environment of this complexity using a consistent algorithm and successfully planning in the learned model. We compare the policy generated by our model to a bound on the best possible value, and determine that our policy is close to optimal.

We believe the spectral PSR learning algorithm presented here, and subspace identification procedures for learning PSRs in general, can increase the scope of planning under uncertainty for autonomous agents in previously intractable scenarios. We believe that this improvement is partly due to the greater representational power of PSRs as compared to POMDPs, and partly due to the efficient and statistically consistent nature of the learning method.

Acknowledgements

SMS (now at Google) was supported by the NSF under grant number 0000164, by the USAF under grant number FA8650-05-C-7264, by the USDA under grant number 4400161514, and by MobileFusion/TTC. BB was supported by the NSF under grant number EEEC-0540865. BB and GJG were supported by ONR MURI grant number N00014-09-1-1052.

References

[1] J. Bilmes. A gentle tutorial on the EM algorithm and its application to parameter estimation for Gaussian mixture and hidden Markov models. Technical Report ICSI-TR-97-021, 1997.
[2] M. Bowling, P. McCracken, M. James, J. Neufeld, and D. Wilkinson. Learning predictive state representations using non-blind policies. In *Proc. ICML*, 2006.
[3] A. R. Cassandra, L. P. Kaelbling, and M. R. Littman. Acting optimally in partially observable stochastic domains. In *Proc. AAAI*, 1994.
[4] E. Even-Dar, S. M. Kakade, and Y. Mansour. Planning in POMDPs using multiplicity automata. In *UAI*, 2005.
[5] A. K. H. Jaeger, M. Zhao. Efficient training of OOMs. In *NIPS*, 2005.
[6] D. Hsu, S. Kakade, and T. Zhang. A spectral algorithm for learning hidden Markov models. In *COLT*, 2009.
[7] M. T. Izadi and D. Precup. Point-based planning for predictive state representations. In *Proc. Canadian AI*, 2008.
[8] H. Jaeger. Observable operator models for discrete stochastic time series. *Neural Computation*, 12:1371–1398, 2000.
[9] M. R. James, T. Wessling, and N. A. Vlassis. Improving approximate value iteration using memories and predictive state representations. In *AAAI*, 2006.
[10] N. K. Jong and P. Stone. Towards employing psrs in a continuous domain. Technical Report UT-AI-TR-04-309, University of Texas at Austin, 2004.
[11] M. Littman, R. Sutton, and S. Singh. Predictive representations of state. In *Advances in Neural Information Processing Systems (NIPS)*, 2002.
[12] A. McCallum. *Reinforcement Learning with Selective Perception and Hidden State*. PhD thesis, University of Rochester, 1995.
[13] P. McCracken and M. Bowling. Online discovery and learning of predictive state representations. In *Proc. NIPS*, 2005.
[14] J. Pineau, G. Gordon, and S. Thrun. Point-based value iteration: An anytime algorithm for POMDPs. In *Proc. IJCAI*, 2003.
[15] M. Rosencrantz, G. J. Gordon, and S. Thrun. Learning low dimensional predictive representations. In *Proc. ICML*, 2004.
[16] S. Ross and J. Pineau. Model-based Bayesian reinforcement learning in large structured domains. In *Proc. UAI*, 2008.
[17] G. Shani, R. I. Brafman, and S. E. Shimony. Model-based online learning of POMDPs. In *Proc. ECML*, 2005.
[18] S. Siddiqi, B. Boots, and G. J. Gordon. Reduced-rank hidden Markov models. In *Proceedings of the Thirteenth International Conference on Artificial Intelligence and Statistics (AISTATS-2010)*, 2010.
[19] B. W. Silverman. *Density Estimation for Statistics and Data Analysis*. Chapman & Hall, 1986.
[20] S. Singh, M. James, and M. Rudary. Predictive state representations: A new theory for modeling dynamical systems. In *Proc. UAI*, 2004.
[21] S. Singh, M. L. Littman, N. K. Jong, D. Pardoe, and P. Stone. Learning predictive state representations. In *Proc. ICML*, 2003.
[22] S. Soatto and A. Chiuso. Dynamic data factorization. Technical report, UCLA, 2001.
[23] E. J. Sondik. *The Optimal Control of Partially Observable Markov Processes*. PhD thesis, Stanford University, 1971.
[24] M. T. J. Spaan and N. Vlassis. Perseus: Randomized point-based value iteration for POMDPs. *Journal of Artificial Intelligence Research*, 24:195–220, 2005.
[25] P. Van Overschee and B. De Moor. *Subspace Identification for Linear Systems: Theory, Implementation, Applications*. Kluwer, 1996.
[26] E. Wiewiora. Learning predictive representations from a history. In *Proc. ICML*, 2005.
[27] D. Wingate. *Exponential Family Predictive Representations of State*. PhD thesis, University of Michigan, 2008.
[28] D. Wingate and S. Singh. On discovery and learning of models with predictive representations of state for agents with continuous actions and observations. In *Proc. AAMAS*, 2007.
[29] D. Wingate and S. Singh. Efficiently learning linear-linear exponential family predictive representations of state. In *Proc. ICML*, 2008.
[30] B. Wolfe, M. James, and S. Singh. Learning predictive state representations in dynamical systems without reset. In *Proc. ICML*, 2005.
[31] M. Zhao, H. Jaeger, and M. Thon. A bound on modeling error in observable operator models and an associated learning algorithm. *Neural Computation*, 2009.

Belief Space Planning Assuming Maximum Likelihood Observations

Robert Platt Jr., Russ Tedrake, Leslie Kaelbling, Tomás Lozano-Pérez
Computer Science and Artificial Intelligence Laboratory
Massachusetts Institute of Technology
{rplatt,russt,lpk,tlp}@csail.mit.edu

Abstract— **We cast the partially observable control problem as a fully observable underactuated stochastic control problem in belief space and apply standard planning and control techniques. One of the difficulties of belief space planning is modeling the stochastic dynamics resulting from unknown future observations. The core of our proposal is to define deterministic belief-system dynamics based on an assumption that the maximum likelihood observation (calculated just prior to the observation) is always obtained. The stochastic effects of future observations are modeled as Gaussian noise. Given this model of the dynamics, two planning and control methods are applied. In the first, linear quadratic regulation (LQR) is applied to generate policies in the belief space. This approach is shown to be optimal for linear-Gaussian systems. In the second, a planner is used to find locally optimal plans in the belief space. We propose a replanning approach that is shown to converge to the belief space goal in a finite number of replanning steps. These approaches are characterized in the context of a simple nonlinear manipulation problem where a planar robot simultaneously locates and grasps an object.**

I. Introduction

Control problems in partially observable environments are important to robotics because all robots ultimately perceive the world through limited and imperfect sensors. In the context of robotic manipulation, tactile and range sensors mounted on the manipulator near the contact locations can provide a tremendous advantage in precision over remote sensing [1, 2]. However, co-locating the sensors with the contacts this way complicates planning and control because it forces the system to trade off sensing and acting. It essentially requires the system to solve a difficult instance of the partially observable control problem, often modeled as a partially observable Markov decision process (POMDP). Unfortunately this problem has been shown to be PSPACE complete, even for a finite planning horizon, discrete states, actions, and observations [3].

One solution to the partially observable control problem is to form plans in the "belief space" of the manipulator - the space of all possible distributions over the state space. The controller then selects actions based not only on the current most-likely state of the robot, but more generically on the information available to the robot. A hallmark of belief-space planning is the ability to generate information-gathering actions. However, planning in the belief space is challenging for a number of reasons. Even coarse finite-dimensional approximations of the belief state distributions require planning in dimensions that are much larger than the original state space. Furthermore, the resulting belief state dynamics are nonlinear, underactuated (number of control inputs is smaller than the dimension of the belief space), and stochastic (transitions depend on observations which have not yet been made).

A number of powerful tools exist for planning and control of high-dimensional non-linear underactuated systems. In order to apply these tools, this paper defines nominal belief space dynamics based on an assumption that all future observations will obtain their maximum likelihood values (this assumption is also made in [4, 5]). During execution, the system tracks the true belief based on the observations actually obtained. Departures from the nominal belief dynamics caused by these unexpected observations are treated as Gaussian process noise. As a result, it is possible to apply standard control and planning techniques. In particular, we use linear quadratic regulation (LQR) to calculate belief space policies based on a local linearization of the belief space dynamics. In spite of this linearization, the resulting belief space policy is shown to be optimal for underlying linear-Gaussian systems. For non-linear systems, it produces reasonable policies within a local region about the linearization point. When large observations cause system belief to leave the locally stabilized region, we propose replanning from the new belief state. We analyze this replanning approach and demonstrate that, under certain conditions, it is guaranteed to ultimately reach a goal region in belief space in a finite number of replanning steps.

A. Related work

Finding an exact optimal solution to a POMDP is an intractable problem [3]. As a result, research has focused on various approaches to approximating the solution. One approach is the 'most-likely state' approximation. This method assumes that the true state of the MDP that underlies the POMDP is in fact the mode of the current belief state. Actions are taken according to the optimal policy in the underlying MDP. The approach considers stochasticity in the underlying process dynamics, but assumes no state uncertainty exists in the future. More sophisticated versions of this approximation include Q-MDP [6] and FIB [7]. A fundamental failing of these approaches is that the system never takes actions for the explicit purpose of reducing uncertainty because the planner assumes that no uncertainty exists.

Another approach that is applicable to some kinds of

POMDPs with continuous state, action, and observation spaces is the belief roadmap approach [8]. This method ranks paths through a probabilistic roadmap defined in the underlying state space in terms of the change in covariance over the path. Van der Berg *et. al.* propose a related approach where a set of potential trajectories are evaluated by tracking the belief state along each trajectory and selecting the one that minimizes the likelihood of failure [9]. In contrast to this class of work, the current paper proposes planning directly in the belief space. This enables our planner to utilize knowledge of the belief dynamics during the planning process rather than evaluating a set of paths through the underlying space.

Our approach is also related to the 'determinize-and-replan' approximation to MDPs that assumes world dynamics are deterministic for the purposes of planning. It takes the first action, observes the actual resulting state, and then replans. This approach, as embodied in FF-Replan, has been very successful (won the ICAPS06 planning competition) in a variety of contexts [10]. Our approach can be viewed as 'determinize-and-replan', applied to POMDPs. It has the significant advantage over the most-likely state approach that it can and does explicitly plan to gain information. And, by replanning when surprising observations are obtained, it remains robust to unlikely outcomes.

Two approaches closely related to our current proposal are the nominal belief-state observation (NBO) of Miller *et. al.* [4] and the work of Erez and Smart [5]. Both approaches plan in belief space using extended Kalman filter dynamics that incorporate an assumption that observations will be consistent with a maximum likelihood state. In relation to these works, this paper has a greater focus on methods for control and replanning. We analyze the conditions under which LQR in belief space is optimal and show that our replanning framework must eventually converge.

II. Problem Specification

We reformulate the underlying partially observable problem as a fully observable belief space problem with an associated cost function.

A. Underlying system

Consider the following partially observable control problem. Let $x_t \in X$ be the unobserved d-dimensional combined state of the robot and the environment at time t. Although the state is not observed directly, noisy observations, $z_t \in Z$, are available as a non-linear stochastic function of x_t:

$$z_t = g(x_t) + \omega, \tag{1}$$

where g is the deterministic component of the measurement function and ω is zero-mean Gaussian noise with possibly state-dependent covariance, W_t. This paper only considers the case where the underlying process dynamics are deterministic:

$$x_{t+1} = f(x_t, u_t). \tag{2}$$

Both g and f are required to be a differentiable functions of x_t and u_t. Although we expect our technique to extend to systems with noisy process dynamics, consideration is limited to deterministic systems to simplify the analysis.

B. Belief system

Although the underlying state of the system is not directly observed, we assume that the controller tracks a fixed parameterization of a probability distribution over the state, $P(x)$. The parameters of the probability density function (pdf) of this distribution will be referred to as the "belief state" and can be tracked using Bayesian filtering as a function of control actions and observations:

$$P(x_{t+1}) = \eta P(z_{t+1}|x_{t+1}) \int_x P(x_{t+1}|x, u_t)P(x),$$

where η is a normalization constant. Notice that since the belief update is a function of the measured observations accrued during execution, it is impossible in general to predict ahead of time exactly how system belief will change until the observations are made.

For the rest of this paper, we will focus on Gaussian belief state dynamics where the extended Kalman filter belief update is used. On each update, the extended Kalman filter linearizes the process and observation dynamics (Equations 2 and 1) about the current mean of the belief distribution:

$$x_{t+1} \approx A_t(x_t - m_t) + f(m_t, u_t), \tag{3}$$

and

$$z_t \approx C_t(x_t - m_t) + g(m_t) + \omega, \tag{4}$$

where m_t is the belief state mean, and $A_t = \frac{\partial f}{\partial x}(m_t, u_t)$, and $C_t = \frac{\partial g}{\partial x}(m_t)$, are Jacobian matrices linearized about the mean and action.

For the Gaussian belief system, the distribution is a Gaussian with mean, m_t, and covariance, Σ_t: $P(x) = \mathcal{N}(x|m_t, \Sigma_t)$. This belief state will be denoted by a vector, $b_t = (m_t^T, s_t^T)^T$, where $s = (s_1^T, \ldots, s_d^T)^T$ is a vector composed of the d columns of $\Sigma = (s_1, \ldots, s_d)$. If the system takes action, u_t, and makes observation, z_t, then the EKF belief state update is:

$$\Sigma_{t+1} = \Gamma_t - \Gamma_t C_t^T (C_t \Gamma_t C_t^T + W_t)^{-1} C_t \Gamma_t, \tag{5}$$

$$m_{t+1} = f_t + \Gamma_t C_t^T (C_t \Gamma_t C_t^T + W_t)^{-1}(z_{t+1} - g(f_t)), \tag{6}$$

where

$$\Gamma_t = A_t \Sigma_t A_t^T$$

and f_t denotes $f(m_t, u_t)$.

C. Cost function

In general, we are concerned with the problem of reaching a given region of state space with high certainty. For a Gaussian belief space, this corresponds to a cost function that is minimized at zero covariance. However, it may be more important to reduce covariance in some directions over others. Let $\{\hat{n}_1, \ldots, \hat{n}_k\}$ be a set of unit vectors pointing in k directions in which it is desired to minimize covariance and let the relative importance of these directions be described by the weights, $w_1, \ldots, w_k$. For a given state-action trajectory,

$b_{\tau:T}, u_{\tau:T}$, we define a finite horizon quadratic cost-to-go function:

$$J(b_{\tau:T}, u_{\tau:T}) = \sum_{i=1}^{k} w_i \left(\hat{n}_i^T \Sigma_T \hat{n}_i\right)^2 + \sum_{t=\tau}^{T-1} \tilde{m}_t^T Q \tilde{m}_t + \tilde{u}_t^T R \tilde{u}_t, \tag{7}$$

where $\tilde{m}_t = m_t - \bar{m}_t$ and $\tilde{u}_t = u_t - \bar{u}_t$ are the mean and action relative to a desired state-action point or trajectory, $\bar{m}_t$ and $\bar{u}_t$, and Σ_T is the covariance matrix at the end of the planning horizon. The first summation penalizes departures from zero covariance along the specified directions on the final timestep. The second summation over the planning horizon penalizes departures from a desired mean and action with positive definite cost matrices Q and R. While Equation 7 is quadratic, the first summation is not expressed in the standard form. However, its terms can be re-written in terms of s, the vector of the columns of Σ:

$$(\hat{n}_i^T \Sigma \hat{n}_i)^2 = s^T L_i s,$$

where the cost matrix, L_i, is a function of $\hat{n}_i$:

$$L_i = \begin{pmatrix} \hat{n}_i n_{i,1} \\ \vdots \\ \hat{n}_i n_{i,d} \end{pmatrix} \begin{pmatrix} \hat{n}_i n_{i,1} \\ \vdots \\ \hat{n}_i n_{i,d} \end{pmatrix}^T,$$

where $n_{i,j}$ is the jth element of $\hat{n}_i$. As a result, we re-write the cost-to-go function as:

$$J(b_{\tau:T}, u_{\tau:T}) = s_T^T \Lambda s_T + \sum_{t=\tau}^{T-1} \tilde{m}_t^T Q \tilde{m}_t + \tilde{u}_t^T R \tilde{u}_t, \tag{8}$$

where

$$\Lambda = \sum_{i=1}^{K} w_i L_i.$$

Although not included in Equation 8, all plans in the belief space are required to satisfy certain final value constraints. First, a constraint on the final value mean of the belief system is specified: $m_T = \bar{m}_T$. B-LQR (Section IV) incorporates this constraint by augmenting the cost-to-go function (Equation 14). Direct transcription (Section V-A) incorporates this constraint directly.

For a policy defined over the belief space,

$$u_t = \pi(b_t),$$

the expected cost-to-go from a belief state, b_τ, at time τ with a planning horizon $T-\tau$ (the planning horizon could be infinite) is:

$$J^\pi(b_\tau) = \mathcal{E}_{z_{\tau:T-1}} \left\{ J(b_{\tau:T}, \pi(b_{\tau:T-1})) \right\}, \tag{9}$$

where the expectation is taken over future observations. The optimal policy minimizes expected cost: $\pi^*(b_\tau) = \arg\min_\pi J^\pi(b_\tau)$.

III. Simplified belief space dynamics

In order to apply standard control and planning techniques to the belief space problem, it is necessary to define the dynamics of the belief system. Given our choice to use the EKF belief update, Equations 5 and 6 should be used. Notice that Equation 6 depends on the observation, z_t. Since these observations are unknown ahead of time, it should be necessary to evaluate the expected value of seeing each observation and take the expectation over all observations. However, since it is difficult to evaluate this marginalization, we make a key simplifying assumption for the purposes of planning and control: we assume that future observations are normally distributed about a maximum likelihood. If action u_t is taken from belief state b_t, then the maximum likelihood observation is:

$$z_{ml} = \arg\max_z P(z|b_t, u_t). \tag{10}$$

Evaluating the maximum likelihood observation for the EKF using Equation 10, we have:

$$\begin{aligned} z_{ml} &= \arg\max_z \int P(z|x_{t+1}) P(x_{t+1}|b_t, u_t) dx_{t+1} \\ &\approx \arg\max_z \int \mathcal{N}(z|C_t(x_{t+1} - f_t) + g(f_t), W_t) \\ &\qquad \mathcal{N}(x_{t+1}|f_t, \Gamma_t) dx_{t+1} \\ &= \arg\max_z \mathcal{N}(z|g(f_t), C_t \Gamma_t C_t^T + W_t) \\ &= g(f_t). \end{aligned}$$

Substituting into Equation 6, and restating Equation 5, the simplified dynamics are:

$$b_{t+1} = F(b_t, u_t), \tag{11}$$

where F evaluates to the b_{t+1} corresponding to m_{t+1} and Σ_{t+1},

$$m_{t+1} = f_t + v, \tag{12}$$

$$\Sigma_{t+1} = \Gamma_t - \Gamma_t C_t^T (C_t \Gamma_t C_t^T + W_t)^{-1} C_t \Gamma_t, \tag{13}$$

and v is Gaussian noise.

IV. LQR in belief space

Linear quadratic regulation (LQR) is an important and simple approach to optimal control [11]. While LQR is optimal only for linear-Gaussian systems, it is also used to stabilize non-linear systems in a local neighborhood about a nominal point or trajectory in state space. LQR in belief space (B-LQR) is the application of LQR to belief space control using the simplified belief system dynamics (Equations 12 and 13). Since the belief system dynamics are always non-linear, policies found by B-LQR can be expected to be only locally stable. However, as we show below, it turns out that B-LQR is optimal and equivalent to linear quadratic Gaussian (LQG) control for systems with linear-Gaussian process and observation dynamics. Moreover, B-LQR produces qualitatively interesting policies for systems with non-linear dynamics.

We are interested in the finite horizon problem with a final state constraint on the mean of the belief system. The final state constraint is accommodated by adding an additional term to Equation 8 that assigns a large cost to departures from the desired final mean value. Also, the LQR cost function used in the current work is linear about the planned trajectory:

$$\begin{aligned} J(b_{\tau:T}, u_{\tau:T}) &= \tilde{m}_T^T Q_{large} \tilde{m}_T + \tilde{s}_T^T \Lambda \tilde{s}_T \\ &+ \sum_{t=\tau}^{T-1} \tilde{m}_t^T Q \tilde{m}_t + \tilde{u}_t^T R \tilde{u}_t. \qquad (14) \end{aligned}$$

Notice that the second term in the above does not measure covariance cost about zero (that would be $s_T^T \Lambda s_T$). Instead, we keep the system linear by measuring cost relative to the belief trajectory, $\bar{s}_T$. Nevertheless, note that it should be possible measure covariance cost relative to zero by using an affine rather than linear version of the LQR controller.

Since B-LQR is operating in belief space, is necessary to linearize Equation 11 about a nominal belief and action, denoted $\bar{b}_t$ and $\bar{u}_t$:

$$\mathbb{A}_t = \frac{\partial F}{\partial b}(\bar{b}_t, \bar{u}_t)$$

and

$$\mathbb{B}_t = \frac{\partial F}{\partial u}(\bar{b}_t, \bar{u}_t).$$

Notice that the mean of the Gaussian belief system has the same nominal dynamics as the underlying system. As a result, $\mathbb{A}_t$ always has the form:

$$\mathbb{A}_t = \begin{pmatrix} A_t & \mathbf{0} \\ \frac{\partial s_{t+1}}{\partial m_t} & \frac{\partial s_{t+1}}{\partial s_t} \end{pmatrix},$$

where $A_t = \frac{\partial f}{\partial x}(\bar{m}_t, \bar{u}_t)$ (see Equation 3). Also, note that since the control input never directly affects covariance, $\mathbb{B}$ always has the form

$$\mathbb{B}_t = \begin{pmatrix} B_t \\ 0 \end{pmatrix},$$

where $B_t = \frac{\partial f}{\partial u}(\bar{m}_t, \bar{u}_t)$. Finally, since Equation 14 only assigns departures from the mean a recurring cost, the recurring cost matrix for belief state is

$$\mathbb{Q} = \begin{pmatrix} Q & \mathbf{0} \\ \mathbf{0} & \mathbf{0} \end{pmatrix}.$$

Using the above, the Riccati equation for the discrete time finite horizon problem is:

$$\begin{aligned} \mathbb{S}_t &= \mathbb{Q} + \mathbb{A}_t^T \mathbb{S}_{t+1} \mathbb{A}_t \\ &- \mathbb{A}_t^T \mathbb{S}_{t+1} \mathbb{B}_t (\mathbb{B}_t^T \mathbb{S}_{t+1} \mathbb{B}_t + R_t)^{-1} \mathbb{B}_t^T \mathbb{S}_{t+1} \mathbb{A}_t, (15) \end{aligned}$$

where $\mathbb{S}_t$ is the expected cost-to-go matrix for linear-Gaussian systems, $J^{\pi^*}(b_t) = b_t^T \mathbb{S}_t b_t$. The optimal action for linear-Gaussian systems is:

$$u^* = -(\mathbb{B}_t^T \mathbb{S}_{t+1} \mathbb{B}_t + R_t)^{-1} \mathbb{B}_t^T \mathbb{S}_{t+1} \mathbb{A}_t b_t. \qquad (16)$$

The following theorem demonstrates that B-LQR is equivalent to LQG control for linear-Gaussian systems.

Theorem 1: If the cost function is of the form of Equation 14, and the underlying process and observation dynamics (Equations 1 and 2) are linear, then B-LQR is optimal.

Proof: We show that under the conditions of the theorem, B-LQR is equivalent to LQG and is optimal as a result.

First, if the underlying observation dynamics are linear, then $\frac{\partial s_{t+1}}{\partial m_t} = 0$ and $\mathbb{A}_t$ is block diagonal. Also, note that if the cost is of the form in Equation 14, then the state cost, $\mathbb{Q}$, is also block diagonal with respect to mean and covariance (m and s). As a result of these two facts, the solution to the belief space Riccati equation (Equation 15) at time t, $\mathbb{S}_t$, always has the form,

$$\mathbb{S}_t = \begin{pmatrix} S_t & \mathbf{0} \\ \mathbf{0} & P_t \end{pmatrix},$$

where S_t is the solution of the Riccati equation in the underlying space,

$$\begin{aligned} S_t &= Q + A_t^T S_{t+1} A_t \\ &- A_t^T S_{t+1} B_t (B_t^T S_{t+1} B_t + R_t)^{-1} B_t^T S_{t+1} A_t, \end{aligned}$$

and P_t is arbitrary. Substituting into Equation 16, we have:

$$u^* = -(B_t^T S_{t+1} B_t + R_t)^{-1} B_t^T S_{t+1} A_t m_t.$$

Since this is exactly the solution of the LQG formulation, and LQG is optimal for linear-Gaussian process and observation dynamics, we conclude the B-LQR is optimal. ∎

V. PLANNING

Since B-LQR is based on a local linearization of the simplified belief dynamics, it is clear that planning methods that work directly with the non-linear dynamics can have better performance. A number of planning methods that are applied to underactuated systems are relevant to the belief space planning problem: rapidly expanding random trees (RRTs) [12], LQR-trees [13], and nonlinear optimization methods [14]. Presently, we use an approach based on a nonlinear optimization technique known as direct transcription. The resulting trajectory is stabilized using time varying B-LQR.

A. Direct transcription

Direction transcription is an approach to transcribing the optimal control problem to a nonlinear optimization problem. Suppose we want to find a path from time 1 to T starting at b_1 that optimizes Equation 8. Direction transcription parameterizes the space of possible trajectories by a series of k segments. Let δ be a user-defined integer that defines the length of each segment in time steps. Then the number of segments is $k = \frac{T}{\delta}$. Let $b'_{1:k}$ and $u'_{1:k-1}$ be sets of belief state and action variables that parameterize the trajectory in terms of the segments. Segment i begins at time $i\delta$ and ends at time $i\delta + \delta - 1$. The cost function, Equation 8, is approximated in terms of these segments:

$$\begin{aligned} J(b_{1:T}, u_{1:T}) &\approx \hat{J}(b'_{1:k}, u'_{1:k}) \\ &= s^T \Lambda s + \sum_{j=1}^{k} \tilde{m}_i'^T Q \tilde{m}_i' + \tilde{u}_i'^T R \tilde{u}_i', (17) \end{aligned}$$

where, for the purposes of planning, $\tilde{m}'_i$ is measured with respect to the final value constraint, $\bar{m}'_i = \bar{m}_T$. The belief state on the last time step of this segment can be found by integrating F over δ:

$$\phi(b'_i) = F(b'_i, u_i) + \sum_{t=i\delta}^{i\delta+\delta-1} F(b_{t+1}, u_i) - F(b_t, u_i). \quad (18)$$

It is now possible to define the nonlinear optimization problem that approximates the optimal control problem. We want assignments to the variables, $b'_{1:k}$ and $u'_{1:k}$, that minimize the approximate cost, $\hat{J}(b'_{1:k}, u'_{1:k})$ subject to constraints that require each segment to be dynamically consistent with its neighboring segments:

$$\begin{aligned} b'_2 &= \phi(b'_1, u'_1) \\ &\vdots \\ b'_k &= \phi(b'_{k-1}, u'_{k-1}). \end{aligned} \quad (19)$$

Since we have a final value constraint on the mean component of belief, an additional constraint is added:

$$m'_k = \bar{m}_T. \quad (20)$$

By approximating the optimal control problem using Equations 18 through 20, it is possible to apply any optimization method available. In the current work, we use sequential quadratic programming (SQP) to minimize the Lagrangian constructed from the costs and constraints in Equations 18- 20. At each iteration, this method takes a Newton step found by solving the Karush-Kuhn-Tucker (KKT) system of equations. Ultimately SQP converges to a local minimum. For more information, the reader is recommended to [14].

B. Replanning strategy

```
Input  : initial belief state, b.
Output: vector of control actions, u_{1:T}.
while b not at goal do
    (ū_{1:T}, b̄_{1:T}) = create_plan(b);
    for t ← 1 to T − 1 do
        u_t = lqr_control(b_t, ū_t, b̄_t);
        b_{t+1} = EKF(b_t, u_t, z_t);
        if m̃_{t+1} > θ then break;
    end
    b = b_{t+1}
end
```

Algorithm 1: Belief space planning algorithm.

Since the actual belief transition dynamics are stochastic, a mechanism must exist for handling divergences from the planned trajectory. One approach is to use time varying B-LQR about the planned path to stabilize the trajectory. While this works for small departures from the planned trajectory, replanning is needed to handle larger divergences. We propose the basic replanning strategy outlined in Algorithm 1 (above). In the *create_plan* step, the algorithm solves for a belief space trajectory that satisfies the final value constraints on the mean using direct transcription. Next, *lqr_control* solves for and executes a locally stable policy about the trajectory. The EKF step tracks belief state based on actual observations. When the mean component of belief departs from the planned trajectory by more than a given threshold, θ,

$$m_t - \bar{m}_t = \tilde{m}_t > \theta,$$

the for loop breaks and replanning occurs. It is assumed that the threshold, θ, is chosen such that B-LQR is stable within the threshold.

C. Analysis of belief space covariance during replanning

Under Algorithm 1, the mean component of belief can be shown to reach the final value constraint in a finite number of replanning steps. Essentially, we show that each time the belief system deviates from the planned trajectory by more than θ, covariance decreases by a finite amount. Each time this occurs, it becomes more difficult for the belief state mean to exceed the threshold again. After a finite number of replanning steps, this becomes impossible and the system converges to the final value constraint. In the following, we shall denote the spectral norm of a matrix, A, by $\|A\|_2$. The spectral norm of a matrix evaluates to its largest singular value.

We require the following conditions to be met:

1) The observation covariance matrix is positive definite and there exists a strictly smaller matrix, W_{min}, such that $W > W_{min}$,
2) the underlying process dynamics are deterministic,
3) A_t has eigenvalues no greater than one,
4) the observations are bounded, $\|z_t\| < z_{max}$,

Lemma 1: Suppose that Algorithm 1 executes under the conditions above. Then, the change in covariance between time t_a when $\tilde{m}_a = 0$ and time t_b when $\tilde{m}_b > \theta$ (the replanning threshold) is lower bounded by

$$\|\Sigma_a - \Sigma_b\|_2 \geq \left(\frac{\theta}{\tau \|W_{min}^{-\frac{1}{2}}\|_2 z_{max}} \right)^2,$$

where $\tau = t_b - t_a$.

Proof: Since m changes by at least θ over τ timesteps, there is at least one timestep between t_a and t_b (time t_c) where $\Delta m_c = \|\tilde{m}_{c+1} - \tilde{m}_c\| > \frac{\theta}{\tau}$. This implies:

$$\begin{aligned} \frac{\theta}{\tau} &\leq |\Gamma_c C_c^T (C_c \Gamma_c C_c^T + W_c)^{-1} (z_{c+1} - \bar{z}_{c+1})| \\ &\leq \|\Gamma_c C_c^T (C_c \Gamma_c C_c^T + W_c)^{-1}\|_2 z_{max}. \end{aligned}$$

Since

$$\|(C_c \Gamma_c C_c^T + W_c)^{-\frac{1}{2}}\|_2 \leq \|W_{min}^{-\frac{1}{2}}\|_2,$$

and using the properties of the spectral norm, we have that:

$$\frac{\theta}{\tau} \leq \|\Gamma_c C_c^T (C_c \Gamma_c C_c^T + W_c)^{-\frac{1}{2}}\|_2 \|W_{min}^{-\frac{1}{2}}\|_2 z_{max}.$$

Dividing through by $\|W_{min}^{-\frac{1}{2}}\|_2 z_{max}$ and squaring the result, we have:

$$\left(\frac{\theta}{\tau \|W_{min}^{-\frac{1}{2}}\|_2 z_{max}} \right)^2 \leq \|\Gamma_c C_c^T (C_c \Gamma_c C_c^T + W_c)^{-1} C_c \Gamma_c\|_2.$$

Considering the covariance update equation (Equation 13), and considering that $\|A_c\|_2 \leq 1$, it must be that

$$\|\Sigma_a - \Sigma_b\|_2 \geq \left(\frac{\theta}{\tau \|W_{min}^{-\frac{1}{2}}\|_2 z_{max}} \right)^2.$$

■

Since Lemma 1 establishes that Σ decreases by a constant amount each time algorithm 1 replans, the following Theorem is able to conclude that after a finite number of replans, the belief state mean converges exponentially toward the final value constraint. We require the following additional assumptions:

1) The planner always finds a trajectory such that the belief space mean satisfies the final value constraint on the final timestep, T.
2) B-LQR stabilizes the trajectory within the replanning threshold, θ.

Theorem 2: Under the conditions of Lemma 1 and the conditions above, Algorithm 1 causes the mean of belief state to be exponentially convergent to the final value condition after a finite number of replanning steps.

Proof: Under Algorithm 1, the system does one of two things: 1) the mean of the belief system never exceeds the replan threshold and B-LQR gives us exponential convergence of the mean to the final value condition, or 2) the mean exceeds the replan threshold. In the second case, we have by Lemma 1 that Σ decreases by a fixed minimum amount in one or more directions. For an initial Σ with finite variance, this is only possible a finite number of times. After that, condition 1 above becomes true and the mean of the belief system is exponentially convergent. ■

VI. Experiments

We explored the capabilities of our approach to belief space planning in two experimental domains: the light-dark domain and the planar grasping domain.

A. The light-dark domain

In the light-dark domain, a robot must localize its position in the plane before approaching the goal. The robot's ability to localize itself depends upon the amount of light present at its actual position. Light varies as a quadratic function of the horizontal coordinate. Depending upon the goal position, the initial robot position, and the configuration of the light, the robot may need to move away from its ultimate goal in order to localize itself. Figure 1 illustrates the configuration of the light-dark domain used in our experiments. The goal position is at the origin, marked by an X in the figure. The intensity in the figure illustrates the magnitude of the light over the plane. The robot's initial position is unknown.

The underlying state space is the plane, $x \in \mathbb{R}^2$. The robot is modeled as a first-order system such that robot velocity is determined by the control actions, $u \in \mathbb{R}^2$. The underlying system dynamics are linear with zero process noise, $f(x_t, u_t) = x_t + u$. The observation function is identity,

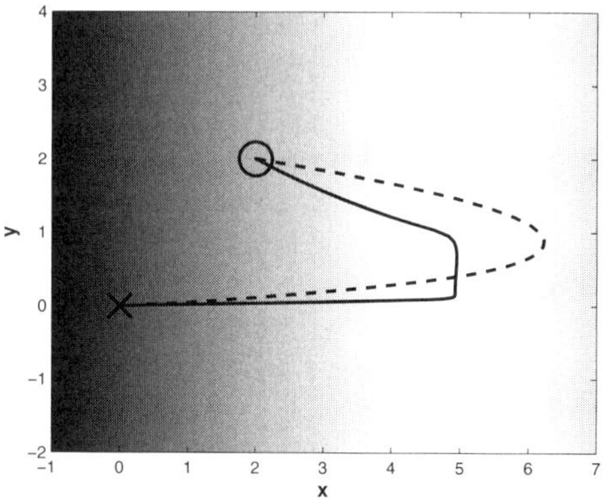

Fig. 1. Comparison of the mean of the planned belief state trajectory found by B-LQR (the dashed line) with the locally optimal trajectory found by direct transcription (the solid line).

$g(x_t) = x_t + \omega$, with zero-mean Gaussian observation noise a function of state, $\omega \sim \mathcal{N}(\cdot|0, w(x))$, where

$$w(x) = \frac{1}{2}(5 - x_x)^2 + \text{const.}$$

has a minimum when $x_x = 5$, where x_x is the first element of x. Belief state was modeled as an isotropic Gaussian pdf over the state space: $b = (m, s) \in \mathbb{R}^2 \times \mathbb{R}^+$. The cost function (Equation 8) used recurring state and action costs of $R = diag(0.5, 0.5)$ and $Q = diag(0.5, 0.5)$, and a final cost on covariance, $\Lambda = 200$. B-LQR had an additional large final cost on mean. Direct transcription used a final value constraint on mean, $m = (0, 0)$, instead. The true initial state was $x_1 = (2.5, 0)$ and the prior belief (initial belief state) was $b_1 = (2, 2, 5)$. The replanning threshold was $\theta = 0.1$. At each replanning step, direct transcription was initialized with a random trajectory. B-LQR linearized the belief system dynamics about $(0, 0, 0.5)$.

1) Results and discussion: Figure 1 shows solutions to the light-dark problem domain found by B-LQR (the dotted line) and by direct transcription (the solid line). The B-LQR trajectory shows the integrated policy assuming that the assumed observation dynamics were always obtained. The direct transcription trajectory shows the initial plan found before replanning. Most importantly, notice that even though the B-LQR trajectory is based on a poor linearization of the belief dynamics, it performs surprisingly well (compare with the locally optimal direct transcription trajectory). However, it is clear that B-LQR is sub-optimal because whereas the locally optimal trajectory lingers in the minimum-noise region at $x_x = 5$, the B-LQR trajectory overshoots past the minimum noise point to $x_x = 6$.

Figure 2 illustrates the behavior of the replanning performed by Algorithm 1. The dotted line shows the mean of the belief space trajectory that was found on the first planning step. The solid line shows the actual trajectory. Whereas the system expected that it began execution at $x = (2, 2)$, it

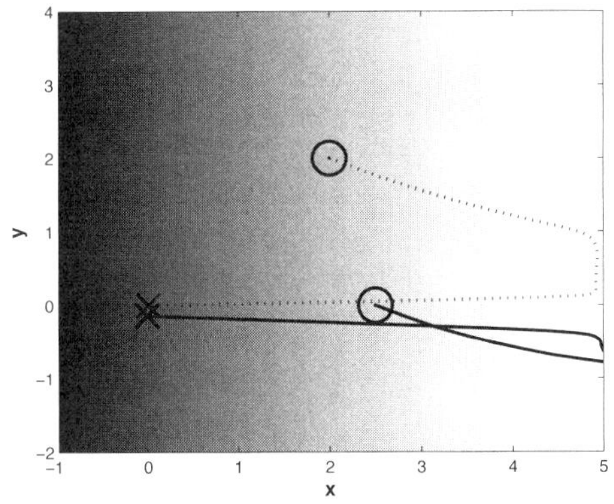

Fig. 2. Comparison of the true robot trajectory (solid line) and the mean of the belief trajectory that was initially planned by direct transcription (dotted line).

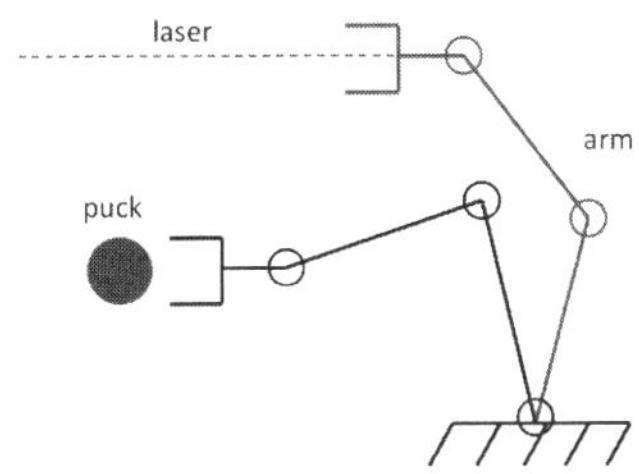

Fig. 3. Laser-grasp domain. A range-finding laser (the dashed line) points out from the robot end effector. The objective is to move the end-effector to a point just in front of the puck on the left (the manipulator approach configuration).

actually began execution at $x = (2.5, 0)$. As a result of this confusion, it initially took actions consistent with its initial belief. However, as it moved into the light, it quickly corrected its misperception. After reaching the point of maximum light intensity, the system subsequently followed a nearly straight line toward the goal.

B. Laser-grasp domain

In the laser-grasp domain, a planar robot manipulator must locate and approach a round puck as illustrated in Figure 3. The robot manipulator position is known, but the puck position is unknown. The robot locates the puck using a range-bound laser range finder that points out from the end-effector along a line. The end-effector always points horizontally as indicated. In order to solve the task, the manipulator must move back and forth in front of the puck so that the laser (the laser is always switched on) detects the puck location and then move to the grasp approach point. The robot is controlled by specifying Cartesian end-effector velocities.

1) Setup: The underlying state space, $x \in \mathbb{R}^2$, denotes the position of the manipulator relative to an "approach point" defined directly in front of the object. Although the end-effector position is assumed to be known completely, state is

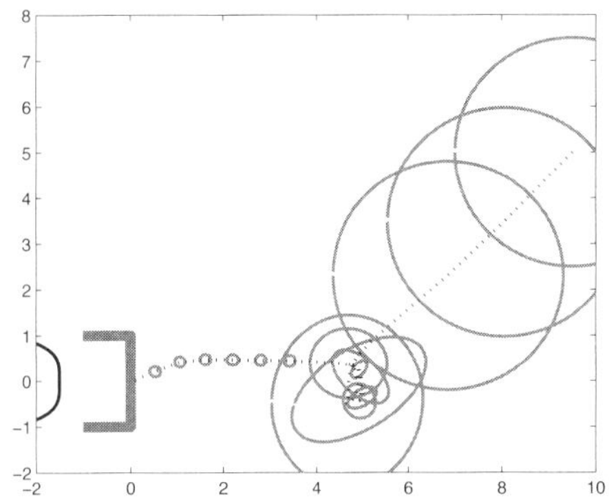

Fig. 4. Trajectory found using direct transcription for the laser-grasp domain. The dotted line denotes the mean of the belief state trajectory. The ellipses sample the covariance matrix at various points along the trajectory. The half circle on the left represents the puck. Just to the right of the puck, the end-effector is illustrated at the approach point.

not observed directly because the object position is unknown. The control action, $u \in \mathbb{R}^2$, specifies the end-effector velocity:

$$f(x_t, u_t) = x_t + u_t.$$

In order to get a smooth measurement function that is not discontinuous at the puck edges, the puck was modeled using a symmetric squashing function about the origin. The result was roughly circular with a "radius" of approximately 0.65. The measurement gradient, C, was zero outside of that radius. State dependent noise was defined to be large when the laser scan line was outside the radius of the puck (modeling an unknown and noisy background). The noise function also incorporated a low-amplitude quadratic about $x_x = 5$ modeling a sensor with maximum measurement accuracy at 5 units away from the target. The belief space was modeled as a non-isotropic Gaussian in the plane: $b = (m, s) \in \mathbb{R}^2 \times \mathbb{R}^3$. The parameterization of covariance in $\mathbb{R}^3$ rather than $\mathbb{R}^4$ is a result of incorporating the symmetry of the covariance matrix into the representation. The cost function was of the form of Equation 8 with $Q = diag(10, 10, 10, 10, 10)$, $R = diag(10, 10)$, and $\Lambda = diag(10000, 0, 0, 10000)$ (denoting no preference for covariance in one direction over another).

2) Results and discussion: Figure 4 shows a representative plan found by direct transcription that moves the end-effector along the dotted line from its starting location in the upper right to the goal position in front of the puck (The geometry of the experiment roughly follows Figure 3). The initial belief state was $b_1 = (9, 5, 5, 0, 5)$. The ellipses in Figure 4 illustrate the planned trajectory of the Gaussian. First, notice that the system plans to move the laser in front of the puck so that it may begin to localize it. Covariance does not change until the end-effector is actually in front of the puck. Also, notice that the plan lingers in front of the puck near the optimal sensor range. During this time, the trajectory makes short jumps up and down, apparently "scanning" the puck. Finally, as time

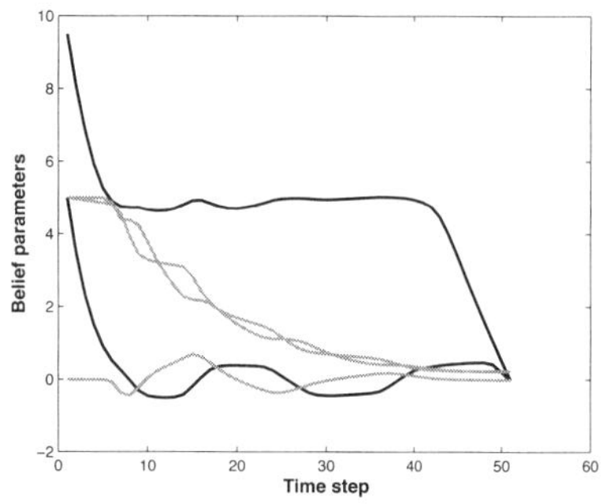

Fig. 5. Planned belief trajectory as a function of time step. The two black lines denote the mean of the belief. The three gray lines denote the elements of covariance. Notice that as the plan is "scanning" the puck, different elements of covariance change in alternation.

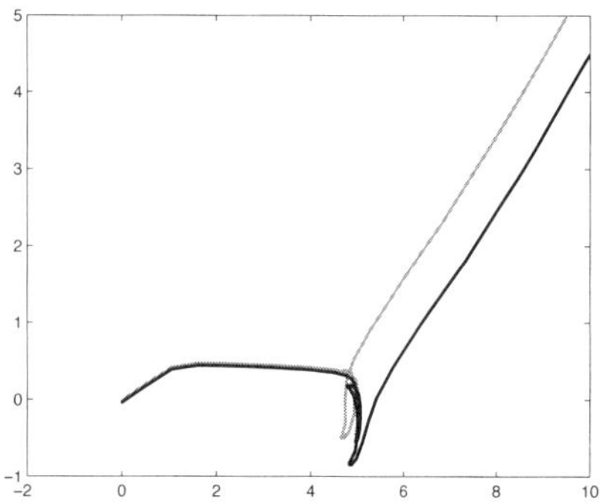

Fig. 6. Comparison between the initially planned trajectory (the gray line) and the actual trajectory (the black line).

approaches the planning horizon, the end-effector moves off to the approach point. Figure 5 provides a more in-depth look at what is going on while the plan scans the puck. First, notice that the plan spends almost all its time in the "sweet spot" scanning the puck. Second, notice that different elements of covariance change during different phases of the scanning motion. This suggests that during scanning, the plan actively alternates between reducing covariance in different directions. The effect results from the fact that the dimension of the observation (a scalar scan depth) is one while the Gaussian belief is over a two-dimensional space. At a given point in time, it is only possible to minimize one dimension of covariance and the plan alternates between minimizing various different dimensions, resulting in the scanning behavior. Finally, Figure 6 illustrates the behavior the replanning strategy where time varying B-LQR stabilization was used. Initially, the mean of the system prior is at $m = (9, 5)$ but the true state is at $x = (10, 4.5)$. The incorrect belief persists until the system reaches a point in front of the puck. At this point, the incorrect prior is gradually corrected during the scanning process until the true state of the system finally reaches the origin.

VII. Conclusion

This paper explores the application of underactuated planning and control approaches to the belief space planning problem. Belief state dynamics are underactuated because the number of controlled dimensions (the parameters of a probability distribution) exceeds the number of independent control inputs. As a result, the dynamics are constrained in a way that can make planning difficult. Our contribution is to recast the belief space planning problem in such a way that conventional planning and control techniques are applicable. As a result, we are able to find belief space policies using linear quadratic regulation (LQR) and locally optimal belief space trajectories using direct transcription. We provide theoretical results characterizing the effectiveness of a plan-and-replan strategy. Finally, we show that the approach produces interesting and relevant behaviors on a simple grasp problem where it is necessary to acquire information before acting.

Acknowledgement

This work was sponsored by the Office of Secretary of Defense under Air Force Contract FA8721-05-C-0002.

References

[1] C. Corcoran and R. Platt, "Tracking object pose and shape during robot manipulation based on tactile information," in *IEEE Int'l Conf. on Robotics and Automation*, vol. 2, 2010.
[2] A. Petrovskaya, O. Khatib, S. Thrun, and A. Ng, "Bayesian estimation for autonomous object manipulation based on tactile sensors," in *IEEE Int'l Conf. on Robotics and Automation*, 2006, pp. 707–714.
[3] C. Papadimitriou and J. Tsitsiklis, "The complexity of markov decision processes," *Mathematics of Operations Research*, vol. 12, no. 3, pp. 441–450, 1987.
[4] S. Miller, A. Harris, and E. Chong, "Coordinated guidance of autonomous uavs via nominal belief-state optimization," in *American Control Conference*, 2009, pp. 2811–2818.
[5] T. Erez and W. Smart, "A scalable method for solving high-dimensional continuous pomdps using local approximation," in *Proceedings of the International Conference on Uncertainty in Artificial Intelligence*, 2010.
[6] M. Littman, A. Cassandra, and L. Kaelbling, "Learning policies for partially observable environments: Scaling up," in *Proceedings of the Twelfth International Conference on Machine Learning*, 1995.
[7] M. Hauskrecht, "Value-function approximations for partially observable markov decision processes," *Journal of Artificial Intelligence Research*, vol. 13, pp. 33–94, 2000.
[8] S. Prentice and N. Roy, "The belief roadmap: Efficient planning in linear pomdps by factoring the covariance," in *12th International Symposium of Robotics Research*, 2008.
[9] J. Van der Berg, P. Abbeel, and K. Goldberg, "Lqg-mp: Optimized path planning for robots with motion uncertainty and imperfect state information," in *Proceedings of Robotics: Science and Systems (RSS)*, 2010.
[10] S. Yoon and R. Fern, A. Givan, "FF-replan: A baseline for probabilistic planning," in *Proceedings of the International Conference on Automated Planning and Scheduling*, 2007.
[11] D. Bertsekas, *Dynamic Programming and Optimal Control: 3rd Edition*. Athena Scientific, 2007.
[12] S. LaValle and J. Kuffner, "Randomized kinodynamic planning," *International Journal of Robotics Research*, vol. 20, no. 5, pp. 378–400, 2001.
[13] R. Tedrake, "LQR-trees: Feedback motion planning on sparse randomized trees," in *Proceedings of Robotics: Science and Systems (RSS)*, 2009.
[14] J. Betts, *Practical methods for optimal control using nonlinear programming*. Siam, 2001.

Motion Planning under Bounded Uncertainty Using Ensemble Control

Aaron Becker
Department of Electrical and Computer Engineering
University of Illinois at Urbana-Champaign
abecker5@illinois.edu

Timothy Bretl
Department of Aerospace Engineering
University of Illinois at Urbana-Champaign
tbretl@illinois.edu

***Abstract*—This paper considers the problem of motion planning for a nonholonomic unicycle despite uncertainty that scales both the forward speed and the turning rate by an unknown but bounded constant. We model the unicycle as an ensemble control system, show that the position of this ensemble is controllable, and derive motion planning algorithms to steer this position between a given start and goal. We apply our work to a differential-drive robot with unknown but bounded wheel radius, and validate our approach with hardware experiments.**

I. Introduction

In this paper we consider the problem of motion planning for a nonholonomic unicycle in the presence of uncertainty that scales both the forward speed and the turning rate by an unknown but bounded constant. We focus on the unicycle because its kinematic model can be used to represent many different car-like vehicles of practical interest. We focus on bounded uncertainty of this special form—scaling the control inputs by an unknown constant—both because it serves as a useful pedagogical example for the development of new planning algorithms and because it is encountered in a surprising variety of real robotic systems.

In particular, consider a single unicycle that rolls without slipping. We describe its configuration by $q = (x, y, \theta)$ and its configuration space by $\mathcal{Q} = \mathbb{R}^2 \times \mathbb{S}^1$. The control inputs are the forward speed $u_1 \in \mathbb{R}$ and the turning rate $u_2 \in \mathbb{R}$. Corresponding to these inputs, we define vector fields $g_1, g_2 \colon \mathcal{Q} \to T_q\mathcal{Q}$ by

$$g_1(q) = \begin{bmatrix} \cos q_3 \\ \sin q_3 \\ 0 \end{bmatrix} \qquad g_2(q) = \begin{bmatrix} 0 \\ 0 \\ 1 \end{bmatrix}$$

and write the kinematics of the unicycle in the standard form

$$\dot{q}(t) = g_1(q(t))u_1(t) + g_2(q(t))u_2(t). \tag{1}$$

Given start and goal configurations q_{start} and q_{goal}, the motion planning problem for a single unicycle is to find inputs

$$u_1(t) \colon [0, T] \to \mathbb{R}$$
$$u_2(t) \colon [0, T] \to \mathbb{R},$$

that result in $q(0) = q_{\text{start}}$ and $q(T) = q_{\text{goal}}$ for free final time T. Depending on the nature of the unicycle (e.g., car-like, differential-drive, etc.) these inputs might be subject to constraints like $|u_2(t)| \leq |u_1(t)| \leq 1$ that enforce a maximum speed and a minimum turning radius. Similarly, we might require that $q(t) \in \mathcal{Q}_{\text{free}}$ to consider collision-avoidance.

We will solve this same motion planning problem, but under uncertainty that can be captured by scaling both the forward speed u_1 and the turning rate u_2 by some unknown, bounded constant. The resulting kinematic model has the form

$$\dot{q}(t) = \epsilon \left(g_1(q(t))u_1(t) + g_2(q(t))u_2(t) \right), \tag{2}$$

where $\epsilon \in [1 - \delta, 1 + \delta]$ for some $0 \leq \delta < 1$. However, rather than try to steer a single robot governed by the uncertain kinematic model (2), our approach is to steer an uncountably infinite collection of robots parameterized by ϵ, each one governed by the exact kinematic model

$$\dot{q}(t, \epsilon) = \epsilon \big(g_1 \left(q(t, \epsilon) \right) u_1(t) + g_2 \left(q(t, \epsilon) \right) u_2(t) \big). \tag{3}$$

Following the terminology introduced by recent work in control theory [1]–[7], we call this fictitious collection of robots an *ensemble* and call the model (3) an *ensemble control system*. The idea is that if we can find a single set of inputs $u_1(t)$ and $u_2(t)$ that result in $q(0, \epsilon) = q_{\text{start}}$ and $q(T, \epsilon) = q_{\text{goal}}$ for all $\epsilon \in [1 - \delta, 1 + \delta]$, then we can certainly guarantee that the actual robot, which corresponds to one particular value of ϵ, will move from start to goal.

We will begin in Section II with a brief review of ensemble control theory and other related work. In Section III, we will proceed to show that although the system (3) is not fully controllable, it is nonetheless controllable with respect to position—in other words, we can achieve any arbitrary position $x(T, \epsilon)$ and $y(T, \epsilon)$, even if we cannot achieve any arbitrary heading $\theta(T, \epsilon)$. Given that one can interpret what we are doing as steering an uncountably infinite number of unicycles under the constraint that every one of them receives the same control input, this controllability result may come as a bit of a surprise. Based on this result, we will derive a practical motion planning algorithm in Section IV that steers the system (2) from start to goal, regardless of ϵ. Finally, in Section V we will apply our work to a differential-drive robot with unknown but bounded wheel radius, showing that (2) is an appropriate model and validating our approach to motion planning with hardware experiments. We do not consider obstacle-avoidance within the scope of this paper—this and several other possible extensions will be discussed in our concluding remarks (Section VI).

II. Related Work

A. Ensemble Control

Ensemble control, as presented in [1]–[7], extends the theory of nonlinear controllability from finite-dimensional systems, for example of the form (1) in Section I, to a particular class of infinite-dimensional systems characterized by a dispersion parameter, for example of the form (3) where this parameter is ϵ. Such an extension is necessary because infinite-dimensional systems typically evolve on configuration spaces that are not compact, a condition that is used in the proof of standard controllability theorems. Chow's theorem, for instance, implies that the drift-free system (1) is locally controllable because the vector fields g_1, g_2, and

$$[g_1, g_2] = \frac{\partial g_2}{\partial q} g_1 - \frac{\partial g_1}{\partial q} g_2$$

are linearly independent and span the tangent space $T_q\mathcal{Q}$ at every configuration $q \in \mathcal{Q}$ (e.g., see [8]), but this theorem says nothing about the nearly identical system (3). To get around this problem, we take the same basic approach as in [7], using repeated bracketing to get higher-order powers of ϵ and then using polynomial approximation to construct arbitrary vector flows. However, systems like ours are ignored by [7] after noting that the drift-free control system

$$\dot{q}(t,\epsilon) = \epsilon \sum_{i=1}^{m} g_i\left(q(t,\epsilon)\right) u_i(t)$$

is not controllable if $g_1, \ldots, g_m$ generate a nilpotent Lie algebra. We will show explicitly that (3) is not fully controllable in Section III, but we will then proceed to show that this system is nonetheless controllable with respect to position.

The origins of this approach are within the physics community. In this context, an "ensemble" is a very large collection of identical or nearly identical molecules, atoms, or elementary particles, and the goal of "ensemble control" is to manipulate the average properties of such an ensemble. Early work in this area was done, for example, by Simon van der Meer, who won the 1984 Nobel prize in physics for controlling the density at which circulating protons are packed in an accelerator using applied magnetic fields [9]. The more recent work of Brockett, Khaneja, and Li has found primary application so far to quantum systems, for example manipulating nuclear spins in Nuclear Magnetic Resonance (NMR) spectroscopy [1]–[7]. Robotics researchers are also beginning to adopt the term "ensemble," for example in the context of multi-robot formations [10], but the formal methodology of ensemble control has yet to be applied. We should also note that other approaches to dealing with infinite-dimensional systems (such as taking advantage of differential flatness) have been developed in parallel, as in [11].

B. Motion Planning Under Uncertainty

There is a vast literature on motion planning under uncertainty in robotics, excellent reviews of which may be found in texts such as [12]–[14] and examples of which range from early work on preimage backchaining [15] to very recent work on needle-steering using the stochastic motion roadmap [16]. As one example, we have drawn particular inspiration from work on sensorless manipulation [17]. In this work, like our own, the basic idea is to explicitly maintain the set of all possible robot configurations and to select a sequence of actions that reduces the size of this set and drives it toward some goal configuration. Carefully selected primitive operations can make this easier. For example, sensorless manipulation strategies often use a sequential composition of primitive operations, "squeezing" a part either virtually with a programmable force field or simply between two flat, parallel plates [18]. Sensorless manipulation strategies also may take advantage of limit cycle behavior, for example engineering fixed points and basins of attraction so that parts only exit a feeder when they reach the correct orientation [19], [20]. These two strategies have been applied to a much wider array of mechanisms such as vibratory bowls and tables [21], [22] or assembly lines [18], [23], [24], and have also been extended to situations with stochastic uncertainty [25], [26] and closed-loop feedback [27], [28]. Our interest in this particular collection of work also stems from our belief that ensemble control theory may provide new insight into sensorless manipulation of large numbers of objects at once.

III. Analysis of Controllability

A. Method of Approach

In this section we will establish controllability results for the systems (1) and (3), which we described in Section I but reprint here for convenience:

$$\dot{q}(t) = g_1(q(t))u_1(t) + g_2(q(t))u_2(t) \tag{1}$$
$$\dot{q}(t,\epsilon) = \epsilon\big(g_1\left(q(t,\epsilon)\right)u_1(t) + g_2\left(q(t,\epsilon)\right)u_2(t)\big). \tag{3}$$

In both of these expressions, the control vector fields g_1, g_2 are given by

$$g_1(q) = \begin{bmatrix} \cos q_3 \\ \sin q_3 \\ 0 \end{bmatrix} \quad \text{and} \quad g_2(q) = \begin{bmatrix} 0 \\ 0 \\ 1 \end{bmatrix}.$$

In our presentation, we follow the general framework outlined by [8] (also see [14], [29], [30]). To begin, recall that any drift-free control system of the form

$$\dot{q}(t) = \sum_{i=1}^{m} g_i(q(t))u_i(t) \tag{4}$$

is *controllable* if for every $q_{\text{start}}, q_{\text{goal}} \in \mathcal{Q}$ there exist inputs $u_1(t), \ldots, u_m(t)$ such that if $q(0) = q_{\text{start}}$ then $q(T) = q_{\text{goal}}$ for some $T > 0$. We assume here that the configuration space $\mathcal{Q}$ is a smooth n-dimensional manifold and that $(u_1, \ldots, u_m) \in U$ for a constraint set $U \subset \mathbb{R}^m$ containing the origin in the interior of its convex hull.

In many drift-free control systems of interest, the number of inputs m is strictly less than the dimension n of the tangent space $T_q\mathcal{Q}$ at each configuration $q \in \mathcal{Q}$. For example, in our system (1) the number of inputs is $m = 2$ and the number of dimensions is $n = 3$, reflecting the nonholonomic constraint

$$\dot{q}_1 \sin q_3 - \dot{q}_2 \cos q_3 = 0$$

that says a unicycle—rolling without slipping—cannot instantaneously move sideways. The situation is worse for the ensemble control system (3), where again $m = 2$ but $n = \infty$. The key question is if these differential constraints are integrable—in other words, if we are "stuck" on an m-dimensional manifold in $\mathcal{Q}$ or if, through a clever sequence of inputs, we can span the entire space.

For the finite-dimensional system (4), this question is answered by computing the Lie algebra generated by $g_1, \dots, g_m$. Recall that the configuration reached from $q_0 \in \mathcal{Q}$ by applying the sequence of inputs $u_i = 1$, $u_j = 1$, $u_i = -1$, and $u_j = -1$, all for some small time t, is given approximately by the Taylor series

$$q(4t) = q_0 + t^2 \left(\frac{\partial g_j}{\partial q} g_i - \frac{\partial g_i}{\partial q} g_j \right) + O(t^3),$$

where we recognize the t^2 term as the Lie bracket

$$[g_i, g_j] = \frac{\partial g_j}{\partial q} g_i - \frac{\partial g_i}{\partial q} g_j.$$

This bracket is a new control vector field that can be followed approximately, although at a slower rate than either g_i or g_j. The Lie algebra is the linear span of all such vector fields, generated by repeated bracketing. If the Lie algebra has rank n, then a finite-dimensional drift-free control system of the form (4) is controllable [8]. As an example, the Lie algebra of (1) is given by the span of g_1, g_2, and

$$[g_1, g_2] = \begin{bmatrix} \sin q_3 \\ -\cos q_3 \\ 0 \end{bmatrix},$$

where we note that $[g_1, g_2]$ is the previously forbidden control vector field that is normal to the direction of rolling. This Lie algebra has rank 3 everywhere, so (1) is controllable.

For infinite-dimensional systems like (3), we cannot directly apply the Lie algebra rank condition because the configuration space is no longer compact. Instead, our approach will be to explicitly approximate any desired control vector field, in this case some smooth function $f(\epsilon) \in T_q\mathcal{Q}$ of $\epsilon \in [1-\delta, 1+\delta]$, by a polynomial approximation in powers of ϵ generated using a sequence of higher-order Lie bracket motions. This approach is inspired by the one taken in [1]–[7].

B. Finding a Controllable Subsystem

Unfortunately, the ensemble control system (3), as we have expressed it so far, is not controllable. This result was suggested by [3] and we will now prove it by construction. In particular, notice that for any u_1 and u_2 we have

$$\dot{q}_3(t, \epsilon) = \epsilon u_2(t).$$

As a consequence, if we define an auxiliary state $\gamma(t) \in S^1$ such that $\gamma(0) = 0$ and

$$\dot{\gamma}(t) = u_2(t),$$

then it is clear that

$$q_3(t, \epsilon) = q_3(0, \epsilon) + \epsilon\gamma(t)$$

for all $\epsilon \in [1-\delta, 1+\delta]$. In other words, the change in heading of each robot in the ensemble after the application of any input is a linear function of ϵ. This result implies that if

$$\frac{q_3(T, \epsilon) - q_3(0, \epsilon)}{\epsilon} \neq \text{constant},$$

as will certainly hold in general, then it is not possible to achieve any arbitrary $q_{\text{goal}} \in \mathcal{Q}$. Hence, the system (3) is not controllable.

However, this result also suggests the construction of a subsystem that, as we will show in the following section, is controllable. We write the configuration of this subsystem as

$$p(t, \epsilon) = (x(t, \epsilon), y(t, \epsilon), \gamma(t)),$$

where $\gamma(t)$ is the auxiliary state we have just introduced above. We will denote the configuration space by $\mathcal{P}$. We have just shown that the evolution of this subsystem is governed by the alternate kinematic model

$$\dot{p}(t, \epsilon) = \epsilon h_1 (p(t, \epsilon), \epsilon) \, u_1(t) + h_2 (p(t, \epsilon), \epsilon) \, u_2(t), \tag{5}$$

where

$$\begin{aligned} h_1 (p(t, \epsilon), \epsilon) &= \begin{bmatrix} \cos (q_3(0, \epsilon) + \epsilon p_3(t, \epsilon)) \\ \sin (q_3(0, \epsilon) + \epsilon p_3(t, \epsilon)) \\ 0 \end{bmatrix} \\ h_2 (p(t, \epsilon), \epsilon) &= \begin{bmatrix} 0 \\ 0 \\ 1 \end{bmatrix} \end{aligned} \tag{6}$$

and $q_3(0, \epsilon)$ is the initial heading given by q_{start}, as before. For convenience, we will abbreviate

$$\begin{aligned} c_3(t, \epsilon) &= \cos (q_3(0, \epsilon) + \epsilon p_3(t, \epsilon)) \\ s_3(t, \epsilon) &= \sin (q_3(0, \epsilon) + \epsilon p_3(t, \epsilon)) \end{aligned}$$

so that

$$h_1 (p(t, \epsilon), \epsilon) = \begin{bmatrix} c_3(t, \epsilon) \\ s_3(t, \epsilon) \\ 0 \end{bmatrix}. \tag{7}$$

Since there is no longer any functional dependence of $p_3(t, \epsilon)$ on ϵ, it is clear that we have removed the feature of (3) that allowed us to conclude a lack of controllability. We will see that the resulting subsystem (5) is, in fact, controllable.

C. Controllability Via Polynomial Approximation

We will now show that the infinite-dimensional subsystem (5) with control vector fields defined by (6)-(7) is controllable.

Taking Lie brackets, we have

$$\begin{aligned}
[\epsilon h_1, h_2] &= \epsilon\left(\frac{\partial h_2}{\partial p}h_1 - \frac{\partial h_1}{\partial p}h_2\right) \\
&= 0 - \epsilon \begin{bmatrix} 0 & 0 & -\epsilon s_3 \\ 0 & 0 & \epsilon c_3 \\ 0 & 0 & 0 \end{bmatrix} \begin{bmatrix} 0 \\ 0 \\ 1 \end{bmatrix} \\
&= \epsilon^2 \begin{bmatrix} s_3 \\ -c_3 \\ 0 \end{bmatrix}
\end{aligned}$$

and

$$\begin{aligned}
[[\epsilon h_1, h_2], h_2] &= 0 - \epsilon^2 \begin{bmatrix} 0 & 0 & \epsilon c_3 \\ 0 & 0 & \epsilon s_3 \\ 0 & 0 & 0 \end{bmatrix} \begin{bmatrix} 0 \\ 0 \\ 1 \end{bmatrix} \\
&= -\epsilon^3 \begin{bmatrix} c_3 \\ s_3 \\ 0 \end{bmatrix} \\
&= -\epsilon^3 h_1.
\end{aligned}$$

Let us define

$$h_3 = \begin{bmatrix} -s_3 \\ c_3 \\ 0 \end{bmatrix},$$

so that $[\epsilon h_1, h_2] = -\epsilon^2 h_3$. All possible control vector fields that can be generated by repeating this process to an arbitrary level of bracketing k can be written in the form

$$f(\epsilon) = ch_2 + \sum_{i=0}^{k}\left(a_i \epsilon^{2i+1} h_1 + b_i \epsilon^{2i+2} h_3\right) \tag{8}$$

for some freely chosen set of coefficients a_i, b_i, and c. Just as for finite-dimensional systems, a Taylor series analysis tells us that each of these control vector fields can be followed approximately by some sequence of inputs u_1 and u_2.

For a given configuration $p(t,\epsilon)$ and a given value of ϵ, the vector fields h_1, h_2, and h_3 are orthonormal, so that

$$h_1^T h_1 = h_2^T h_2 = h_3^T h_3 = 1$$

and

$$h_1^T h_2 = h_2^T h_3 = h_3^T h_1 = 0.$$

As a consequence, we may write (8) as a set of three scalar equations

$$\begin{aligned}
h_1^T f(\epsilon) &= \sum_{i=0}^{k} a_i \epsilon^{2i+1} \\
h_2^T f(\epsilon) &= c \\
h_3^T f(\epsilon) &= \sum_{i=0}^{k} b_i \epsilon^{2i+2}.
\end{aligned}$$

We have already established that p_3 is constant in ϵ, so that by necessity $h_2^T f(\epsilon)$ is also constant in ϵ. Furthermore, the Stone-Weierstrass Theorem [31] tells us that given $\eta > 0$ and a continuous real function

$$\nu(\epsilon)\colon [1-\delta, 1+\delta] \to \mathbb{R},$$

there exists a polynomial function $\rho(\epsilon)$ such that

$$|\rho(\epsilon) - \nu(\epsilon)| < \eta$$

for all $\epsilon \in [1-\delta, 1+\delta]$. As a corollary, as long as $\delta < 1$, it is possible to choose a_i and b_i so that

$$\begin{aligned}
h_1^T f(\epsilon) &\approx \sum_{i=0}^{k} a_i \epsilon^{2i+1} \\
h_3^T f(\epsilon) &\approx \sum_{i=0}^{k} b_i \epsilon^{2i+2},
\end{aligned}$$

with error vanishing in k. This result implies that we can get arbitrarily close to any desired control vector field $f(\epsilon) \in T_p\mathcal{P}$. We immediately conclude that the subsystem (5) is controllable. In other words, it is always possible to find inputs $u_1(t)$ and $u_2(t)$ that steer an uncountably infinite number of unicycles to any arbitrary position $x(T,\epsilon)$ and $y(T,\epsilon)$, even if the heading $\theta(T,\epsilon)$ must remain a linear function of ϵ.

IV. Motion Planning Algorithm

In the previous section, we showed that the subsystem (5) is controllable (i.e., that the ensemble control system (3) is controllable with respect to position). Based on this result, we will now derive a motion planning algorithm that steers the system (5) from start to goal. We will assume that "start" and "goal" are given as continuous functions $p_{\text{start}}(\epsilon)$ and $p_{\text{goal}}(\epsilon)$ of the parameter ϵ. For our application of interest—where (5) captures the range of possible outcomes for a single vehicle—it will always be the case that these functions are constant and have the form

$$\begin{aligned}
p_{\text{start}}(\epsilon) &= (x_{\text{start}}, y_{\text{start}}, 0) \\
p_{\text{goal}}(\epsilon) &= (x_{\text{goal}}, y_{\text{goal}}, 0).
\end{aligned}$$

However, our motion planning algorithm does not depend on this structure, so we ignore it for now. Our general strategy will be to use piecewise-constant inputs of a particular form, as suggested by [32] and reviewed in [8], [29], [33].

Before proceeding, notice that the vector field h_1 in (6) may be expressed

$$h_1(p(t,\epsilon),\epsilon) = R(\epsilon)\begin{bmatrix} \cos \epsilon p_3 \\ \sin \epsilon p_3 \\ 0 \end{bmatrix}$$

where

$$R(\epsilon) = \begin{bmatrix} \cos q_3(0,\epsilon) & -\sin q_3(0,\epsilon) & 0 \\ \sin q_3(0,\epsilon) & \cos q_3(0,\epsilon) & 0 \\ 0 & 0 & 1 \end{bmatrix},$$

so if we apply the transformation

$$\begin{aligned}
p'_{\text{start}}(\epsilon) &= 0 \\
p'_{\text{goal}}(\epsilon) &= R(\epsilon)^T\left(p_{\text{goal}}(\epsilon) - p_{\text{start}}(\epsilon)\right),
\end{aligned}$$

then without loss of generality it is always possible to assume $q_3(0,\epsilon) = 0$ for all ϵ. For convenience here, we will also assume that inputs are constrained by $|u_1| \le 1$ and

$|u_2| \leq 1$ but that there is no minimum turning radius, corresponding to the differential-drive vehicle that we will consider in Section V. This assumption is easily relaxed.

A. Motion Primitives with Piecewise-Constant Inputs

Now, for a non-negative integer i, a positive angle ϕ, and freely chosen $a'_i, b'_i \in \mathbb{R}$, consider a motion primitive of the following form:

$$(u_1, u_2) = \begin{cases} (0,1) & 0 \leq t < i\phi \\ (1,0) & i\phi \leq t < i\phi + a'_i \\ (0,-1) & i\phi + a'_i \leq t < 3i\phi + a'_i \\ (1,0) & 3i\phi + a'_i \leq t < 3i\phi + a'_i + b'_i \\ (0,1) & 3i\phi + a'_i + b'_i \leq t \leq 4i\phi + a'_i + b'_i. \end{cases}$$

This primitive specifies the following sequence of actions:

1) turn left in place an angle $i\phi$
2) drive straight a distance a'_i
3) turn right in place an angle $-2i\phi$
4) drive straight a distance b'_i
5) turn left in place an angle $i\phi$.

We can think of this sequence as a simple piecewise-constant approximation to a sinusoidal input of magnitude $i\phi$. It is easy to show by direct calculation that the result, if applied to the system (5) and assuming that $p_3(0, \epsilon) = 0$, is to achieve

$$\Delta p(\epsilon) = \begin{bmatrix} (a'_i + b'_i)\epsilon \cos \epsilon i\phi \\ (a'_i - b'_i)\epsilon \sin \epsilon i\phi \\ 0 \end{bmatrix}.$$

With the input transformation

$$a'_i = \frac{a_i + b_i}{2}$$
$$b'_i = \frac{a_i - b_i}{2}$$

for freely chosen $a_i, b_i \in \mathbb{R}$, we can write this expression in the decoupled form

$$\Delta p(\epsilon) = \begin{bmatrix} a_i \epsilon \cos \epsilon i\phi \\ b_i \epsilon \sin \epsilon i\phi \\ 0 \end{bmatrix}.$$

Because primitives of this form leave p_3 invariant, we are free to concatenate them. In particular, the result after applying $k+1$ primitives with $i = 0, \ldots, k$, followed by a final turn in place through an angle c, is to achieve

$$\Delta p(\epsilon) = \begin{bmatrix} \sum_{i=0}^{k} a_i \epsilon \cos \epsilon i\phi \\ \sum_{i=0}^{k} b_i \epsilon \sin \epsilon i\phi \\ c \end{bmatrix}.$$

Just as in Section III, our problem has been reduced to function approximation. Denote the first two components of p'_{goal} by $x(\epsilon)$ and $y(\epsilon)$, respectively. We have assumed that both $x(\epsilon)$ and $y(\epsilon)$ are continuous, so we can select coefficients a_i and b_i making the following series converge as $k \to \infty$:

$$x(\epsilon) \approx \sum_{i=0}^{k} a_i \epsilon \cos \epsilon i\phi \tag{9}$$

$$y(\epsilon) \approx \sum_{i=0}^{k} b_i \epsilon \sin \epsilon i\phi. \tag{10}$$

B. Numerical Computation

To actually compute a_i and b_i in (9)-(10), we solve a linear programming problem. It is clear that each set of coefficients may be computed independently, and we apply the same procedure in both cases, so for ease of exposition we will focus on a_i. First, we sample N values $\epsilon_1, \ldots, \epsilon_N$ equally spaced in $[1-\delta, 1+\delta]$. Then, we construct the matrix

$$W = \begin{bmatrix} \epsilon_1 & \epsilon_1 \cos \epsilon_1 1\phi & \ldots & \epsilon_1 \cos \epsilon_1 k\phi \\ \vdots & \vdots & & \vdots \\ \epsilon_N & \epsilon_N \cos \epsilon_N 1\phi & \ldots & \epsilon_N \cos \epsilon_N k\phi \end{bmatrix}$$

and the vector

$$v = \begin{bmatrix} x(\epsilon_1) \\ \vdots \\ x(\epsilon_N) \end{bmatrix},$$

so that (9) may be written in discrete form as

$$Wa \approx v. \tag{11}$$

To minimize the resulting path length, we want to find the vector of coefficients a minimizing $\|a\|_\infty$ subject to the constraint that $\|v - Wa\|_\infty \leq e$ for a given error tolerance $e > 0$. Introducing an auxiliary decision variable s, we can write this linear program in the standard form

$$\begin{aligned} \text{minimize} \quad & s \\ \text{subject to} \quad & -\mathbf{1}e \leq v - Wa \leq \mathbf{1}e \\ & -\mathbf{1}s \leq a \leq \mathbf{1}s. \end{aligned}$$

As we discussed above, some k always exists making this linear program feasible for any $e > 0$. By increasing the number of sampled points N, we can also guarantee by continuity of $x(\epsilon)$ that (11) is a good approximation of (9).

C. Results in Simulation

Figure 1 shows the results of applying our motion planning algorithm to an ensemble control system of the form (5) for which $\delta = 0.2$ and $p_{\text{goal}}(\epsilon) = (1, 0, 0)$. In this example, we specified the error tolerance $e = 10^{-4}$ and used $k = 6$ motion primitives. To generate a clean plot, we chose $N = 100$, but the results are nearly identical even with coarse sampling of ϵ, for example when $N = 10$. We claimed in the presentation of our planning algorithm that the minimum achievable error decreases with the number of motion primitives k. Figure 2 provides empirical evidence supporting this claim for the same example as in Fig. 1.

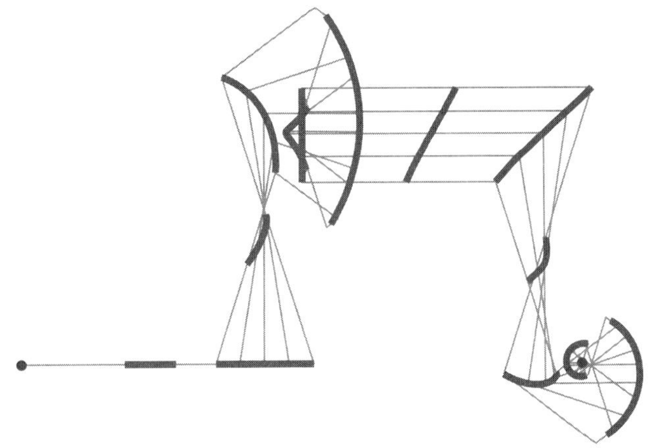

Fig. 1. An ensemble with $\epsilon \in [0.8, 1.2]$ moving a unit distance in the x direction, achieving an error tolerance of $e = 10^{-4}$. Thin lines show the path followed by robots with particular values of ϵ. Thick lines show the entire ensemble at instants of time.

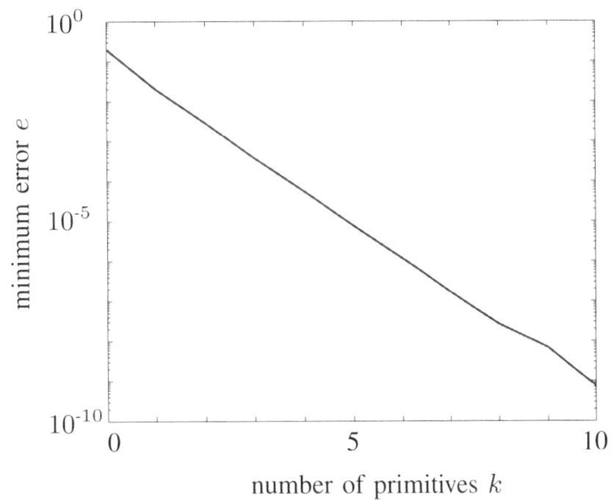

Fig. 2. For the example in Fig. 1, the minimum error e that can be achieved decreases with the number of motion primitives k.

V. Hardware Experiments

In this section we will apply our motion planning algorithm to a differential-drive robot with unknown but bounded wheel radius. First, we will describe the robot we used in our experiments. Then, we will show that (2) is an appropriate model of this robot. Finally, we will show the results of hardware experiments.

A. Experimental Setup

Figure 3 shows the robot we used in our experiments. It is a differential-drive robot with a caster wheel in front for stability. It moves on a flat tile floor without obstacles and uses only dead-reckoning for navigation. In particular, the robot runs a feedback control loop to read the wheel encoders, update a dead-reckoning position estimate, and regulate the speed of each motor. Although we use no other sensors for feedback control, global position data is available from an off-board vision system for later analysis. This vision system records pose information at 27Hz with a position accuracy of 2cm and an orientation accuracy of $1°$.

Fig. 3. The differential-drive robot used for experimental validation.

Before conducting our experiments, we applied a standard calibration procedure to find the effective wheelbase and wheel radius in order to reduce systematic dead-reckoning error [34]. The calibration was done with wheels of diameter 12.7cm. However, these wheels are interchangeable—in our experiments, we used four sets that varied between 10.16-15.24cm in diameter, as shown in Fig. 4. We did not recalibrate for these other wheels, and assumed that the wheel diameter was unknown but bounded in the set $[10.2, 15.2]$, or in other words the set $[0.8, 1.2]$ relative to the nominal diameter 12.7cm.

B. Application of the Model to a Differential-Drive Robot

We will show that

$$\dot{q}(t) = \epsilon\left(g_1(q(t))u_1(t) + g_2(q(t))u_2(t)\right)$$

is a valid kinematic model of our robot, where

$$g_1(q) = \begin{bmatrix}\cos q_3 \\ \sin q_3 \\ 0\end{bmatrix} \qquad \text{and} \qquad g_2(q) = \begin{bmatrix}0 \\ 0 \\ 1\end{bmatrix}.$$

It suffices to show that the forward speed v and turning rate ω of a differential-drive robot are given by $v = \epsilon u_1$ and $\omega = \epsilon u_2$, respectively, for control inputs $u_1, u_2 \in \mathbb{R}$. Recall that for wheel radius r and wheel separation b, the forward speed and turning rate of a differential-drive robot are given by

$$v = \frac{r(\omega_R + \omega_L)}{2} \qquad \text{and} \qquad \omega = \frac{r(\omega_R - \omega_L)}{b},$$

where ω_R and ω_L are the angular velocities of the right and left wheels, respectively. Assume that the wheel radius, a positive constant, is unknown but bounded according to $r \in [r_{\text{min}}, r_{\text{max}}]$. If we define

$$\bar{r} = \frac{r_{\text{max}} + r_{\text{min}}}{2} \qquad \text{and} \qquad \delta = \frac{r_{\text{max}} - r_{\text{min}}}{2\bar{r}}$$

then we can write $r = \epsilon\bar{r}$ for some $\epsilon \in [1 - \delta, 1 + \delta]$, so that

$$v = \epsilon\left(\frac{\bar{r}(\omega_R + \omega_L)}{2}\right) \qquad \text{and} \qquad \omega = \epsilon\left(\frac{\bar{r}(\omega_R - \omega_L)}{b}\right).$$

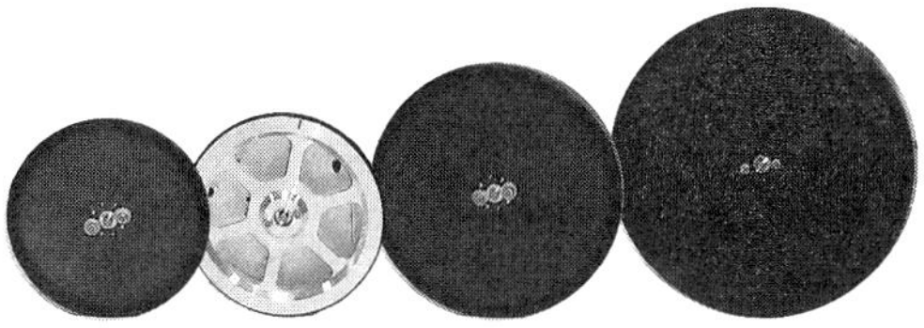

Fig. 4. We used four wheel sizes for experimental validation. These wheels are 10.16, 10.48, 12.7 and 15.24 cm in diameter.

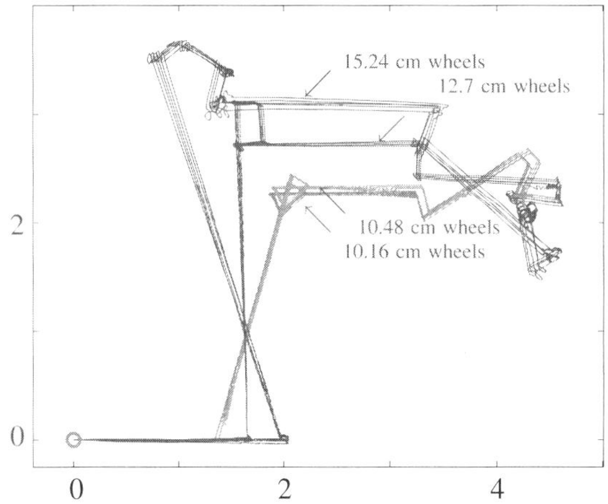

Fig. 5. Ground truth data gathered from camera system, starting at $(0, 0)$ with goal $(4.25, 2.25)$. Five runs for each wheel set are shown. Loops at the corners are artifacts from the camera system. Plots correspond to 10.16, 10.48, 12.7 and 15.24 cm wheels. Units are in meters.

This expression simplifies if we select wheel angular velocities

$$\omega_R = \frac{2u_1 + bu_2}{2\bar{r}} \quad \text{and} \quad \omega_L = \frac{2u_1 - bu_2}{2\bar{r}}$$

for any given $u_1, u_2 \in \mathbb{R}$, so that

$$v = \epsilon u_1 \quad \text{and} \quad \omega = \epsilon u_2,$$

and we have our result.

C. Experimental Results

Figure 5 shows the results of our experiments, which successfully validated our approach. We selected a goal location at $x = 4.25, y = 2.25$ m with the robot starting at the origin, and an error tolerance of 2cm. We applied the algorithm described in Section IV to plan a single sequence of inputs—i.e., single path—that will take the robot from start to goal regardless of its wheel size. Five runs were recorded for each wheel size. Fig. 5 shows the resulting trajectories, all of which reached a small neighborhood of the goal position.

The effects of dead-reckoning drift are obvious in Fig. 6. This drift is due to wheel slip, gear backlash, surface irregularities, wheel flex and other unmodeled disturbances. The vision system adds additional error. Ground truth position information was calculated from markings on the top of the robot. These markings were level and centered over the wheelbase for the 10.48 cm wheels, but tilted by 10° for the largest wheels. These effects are most noticeable when the robot turns in place.

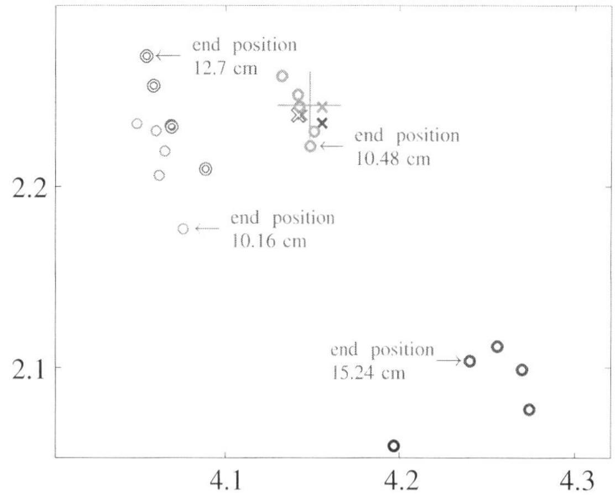

Fig. 6. Ending position for each run. Large '+' for goal position, 'x' for expected ending position under zero odometry drift, 'o' for actual ending positions. The four symbol types correspond to 10.16, 10.48, 12.7 and 15.24 cm wheels. Units are in meters.

TABLE I
IN-GROUP ERROR MEASUREMENTS

wheel diam(cm)	distance mean(m)	distance var (m^2)	θ mean(rad)	θ var(rad)^2
10.16	0.10	3.2e-5	-0.002	1.9e-6
10.48	0.02	4.1e-5	-0.002	6.1e-7
12.70	0.08	1.2e-4	-0.008	2.5e-5
15.24	0.19	2.2e-4	-0.017	1.9e-4

Table I shows the in-group mean and variance for the different wheel sets. The smallest variance is seen for the 10.48 wheels because the system was calibrated for these wheels. The 10.48 cm wheels are aluminum with rubber o-rings stretched over the rim, while the other wheels are 0.64 cm thick ABS plastic with a molded rubber traction ring covering the rim. The edge of the plastic wheels has a rectangular cross section, making the effective wheel base slightly larger than for the aluminum wheels. This causes larger errors on the ABS wheels. The means and variances tend to increase with the wheel diameter because the wheel diameter linearly scales all disturbance perturbations. The group mean error for distance is 0.098 m, five times larger than the specified terminal region, and the group variance is 0.004 m^2. The group mean error for heading is -0.007 rad and the group variance is 0.0001 rad^2.

VI. CONCLUDING REMARKS

In this paper we applied the framework of ensemble control theory to derive a motion planning algorithm that steers a nonholonomic unicycle from start to goal despite uncertainty that scales both the forward speed and the turning rate by an unknown but bounded constant. We validated our approach using a differential-drive robot with unknown but bounded wheel radius and showed the results with hardware experiments.

Ensemble control theory is a recent development, and we feel that its application to robotics has raised a number of interesting questions for future work. For example, our algorithm constructs feasible inputs, but what are the optimal ones? We have shown controllability, but what about small-time local controllability that would allow us to deal in a systematic way with obstacles? (The fact that Fig. 1 simply scales with distance traveled indicates that small-time local controllability likely holds, but we require a proof.)

We would also like to address what may seem like an obvious flaw in our approach, particularly in the context of our chosen hardware demonstration. Within the scope of this paper, the plans we construct are paths. By ignoring the error that would result if a robot actually followed these paths, we are implicitly assuming the existence of a feedback control policy. Such a policy would require sensors, and these sensors might also be used to simply identify the unknown parameter ϵ, thus obviating the problem we consider. This objection is certainly reasonable, but there are two points to keep in mind. First, sensor data may be sufficiently noisy or come at a sufficiently low rate to make online system identification impractical while still allowing periodic course corrections (perhaps more intriguing, the feedback control policy might be sufficient to drive system uncertainty back to a 1-D manifold, but not to resolve position on this manifold). Second, although we focus here on planning for one robot with unknown ϵ, some applications—in particular, those that involve micro/nano-scale robotic systems—require planning for many robots, each with slightly different ϵ. In this latter case, online system identification is essential but does not remove the problem we consider. Regardless, this topic is an opportunity for future inquiry.

VII. Acknowledgements

The authors wish to thank Chad Burns, Dušan Stipanović and Dan Block for the hardware demonstration facilities and Seth Hutchinson for helpful discussion. This material is based upon work supported by the National Science Foundation under Grant Nos. 0931871 and 0956362.

References

[1] R. W. Brockett and N. Khaneja, "On the control of quantum ensembles," in *System Theory: Modeling, Analysis and Control*, T. Djaferis and I. Schick, Eds. Kluwer Academic Publishers, 1999.

[2] N. Khaneja, "Geometric control in classical and quantum systems," Ph.D. dissertation, Harvard University, 2000.

[3] J.-S. Li, "Control of inhomogeneous ensembles," Ph.D. dissertation, Harvard University, May 2006.

[4] J.-S. Li and N. Khaneja, "Control of inhomogeneous quantum ensembles," *Physical Review A (Atomic, Molecular, and Optical Physics)*, vol. 73, no. 3, p. 030302, 2006.

[5] ——, "Ensemble controllability of the bloch equations," in *IEEE Conf. Dec. Cont.*, San Diego, CA, Dec. 2006, pp. 2483–2487.

[6] ——, "Ensemble control of linear systems," in *IEEE Conf. Dec. Cont.*, New Orleans, LA, USA, Dec. 2007, pp. 3768–3773.

[7] ——, "Ensemble control of bloch equations," *IEEE Trans. Autom. Control*, vol. 54, no. 3, pp. 528–536, Mar. 2009.

[8] R. Murray, Z. X. Li, and S. Sastry, *A Mathematical Introduction to Robotic Manipulation*. Boca Raton, FL: CRC Press, 1994.

[9] S. V. D. Meer, "Stochastic cooling and the accumulation of antiprotons," in *Nobel Lectures in Physics*, T. Frängsmyr, Ed. Singapore: World Scientific, 1993.

[10] N. Michael and V. Kumar, "Controlling shapes of ensembles of robots of finite size with nonholonomic constraints," in *Robotics: Science and Systems*, Zurich, Switzerland, 2008.

[11] P. Rouchon, "Motion planning, equivalence, infinite dimensional systems," *Int. J. Appl. Math. Comput. Sci.*, vol. 11, no. 1, pp. 165–188, 2001.

[12] M. T. Mason, *Mechanics of Robotic Manipulation*. MIT Press, 2001.

[13] S. Thrun, W. Burgard, and D. Fox, *Probabilistic Robotics*. The MIT Press, 2005.

[14] S. M. LaValle, *Planning algorithms*. New York, NY: Cambridge University Press, 2006.

[15] M. Erdmann, "Using backprojections for fine motion planning with uncertainty," *International Journal of Robotics Research*, vol. 5, no. 1, pp. 19–45, 1986.

[16] R. Alterovitz, T. Siméon, and K. Goldberg, "The stochastic motion roadmap: A sampling framework for planning with markov motion uncertainty," in *Robotics: Science and Systems III*, 2008, pp. 233–241.

[17] M. A. Erdmann and M. T. Mason, "An exploration of sensorless manipulation," *IEEE J. Robot. Autom.*, vol. 4, no. 4, pp. 369–379, Aug. 1988.

[18] K. Y. Goldberg, "Orienting polygonal parts without sensors," *Algorithmica*, vol. 10, no. 2, pp. 201–225, 1993. [Online]. Available: http://dx.doi.org/10.1007/BF01891840

[19] K. M. Lynch, M. Northrop, and P. Pan, "Stable limit sets in a dynamic parts feeder," *IEEE Trans. Robot. Autom.*, vol. 18, no. 4, pp. 608–615, Aug. 2002.

[20] T. D. Murphey, J. Bernheisel, D. Choi, and K. M. Lynch, "An example of parts handling and self-assembly using stable limit sets," in *Int. Conf. Int. Rob. Sys.*, Aug. 2005, pp. 1624–1629.

[21] O. C. Goemans, K. Goldberg, and A. F. van der Stappen, "Blades: a new class of geometric primitives for feeding 3d parts on vibratory tracks," in *Int. Conf. Rob. Aut.*, May 2006, pp. 1730–1736.

[22] T. H. Vose, P. Umbanhowar, and K. M. Lynch, "Friction-induced velocity fields for point parts sliding on a rigid oscillated plate," in *Robotics: Science and Systems*, Zurich, Switzerland, June 2008.

[23] S. Akella, W. H. Huang, K. M. Lynch, and M. T. Mason, "Parts feeding on a conveyor with a one joint robot," *Algorithmica*, vol. 26, no. 3, pp. 313–344, 2000.

[24] A. van der Stappen, R.-P. Berretty, K. Goldberg, and M. Overmars, "Geometry and part feeding," *Sensor Based Intelligent Robots*, pp. 259–281, 2002.

[25] K. Goldberg, B. V. Mirtich, Y. Zhuang, J. Craig, B. R. Carlisle, and J. Canny, "Part pose statistics: estimators and experiments," *IEEE Trans. Robot. Autom.*, vol. 15, no. 5, pp. 849–857, Oct. 1999.

[26] M. Moll and M. Erdmann, "Manipulation of pose distributions," *The International Journal of Robotics Research*, vol. 21, no. 3, pp. 277–292, 2002.

[27] S. Akella and M. T. Mason, "Using partial sensor information to orient parts," *The International Journal of Robotics Research*, vol. 18, no. 10, pp. 963–997, 1999. [Online]. Available: http://ijr.sagepub.com/cgi/content/abstract/18/10/963

[28] T. D. Murphey and J. W. Burdick, "Feedback control methods for distributed manipulation systems that involve mechanical contacts," *The International Journal of Robotics Research*, vol. 23, no. 7-8, pp. 763–781, 2004.

[29] J.-P. Laumond, Ed., *Robot Motion Planning and Control*, ser. Lecture Notes in Control and Information Sciences. Springer Berlin / Heidelberg, 1998, vol. 229/1998.

[30] H. Choset, K. Lynch, S. Hutchinson, G. Kanto, W. Burgard, L. Kavraki, and S. Thrun, *Principles of Robot Motion: Theory, Algorithms, and Implementations*. MIT Press, 2005.

[31] H. L. Royden, *Real Analysis*, 3rd ed. Prentice Hall, 1988.

[32] G. Lafferriere and H. Sussmann, "A differential geometric approach to motion planning," in *Nonholonomic Motion Planning*, Z. Li and J. Canny, Eds. Kluwer, 1993, pp. 235–270.

[33] S. Sastry, *Nonlinear Systems: Analysis, Stability, and Control*. Springer, 1999.

[34] J. Borenstein and L. Feng, "Umbmark: a benchmark test for measuring odometry errors in mobile robots," in *SPIE*, W. J. Wolfe and C. H. Kenyon, Eds., vol. 2591, no. 1, 1995, pp. 113–124.

Remotely Powered Propulsion of Helical Nanobelts

Gilgueng Hwang*, Sinan Haliyo**, and Stéphane Régnier**
*Laboratory for Photonics and Nanostructures CNRS, route de Nozay 91460 Marcoussis, France
**Insitut des Systèmes Intelligents et de Robotique, Université Pierre et Marie Curie, 4 Place Jussieu, Paris, France
e-mail: gilgueng.hwang@lpn.cnrs.fr

Abstract—**Nanorobots capable of self-propulsion in low Reynolds number fluids would have a big technological impact in many fields. Few known mechanisms are able to propel such devices. Here we demonstrate that helical nanobelts can swim in liquid when actuated by an electric field generated by electro-osmotic force. Moreover, we show these devices achieve speeds at or slightly above those demonstrated by natural bacteria. The helical nanobelts were designed and fabricated with a head and tail to mimic natural micro-organisms such as bacteria and their flagellae. The electro-osmotic propulsion of helical nanobelts might be used as biomedical carriers, wireless manipulators, and as local probes for rheological measurements.**

I. Introduction

In this paper, we describe the design of helical nanobelt (HNB) based nanorobots and demonstrate their electro-osmotic propulsion for in-vivo biomedical applications. Electro-osmotic field propelled HNBs have many potential applications, as biomedical nanorobots targeting and treating cancer cells, or for neural cell probing.

Recently, in-vivo biomedical nanorobots have drawn much interest especially in discovering new and efficient wireless power transfer methods and in controlled locomotion mechanisms. There are many approaches to mimic nature's bacterial swimming behaviors for the locomotion of artificial nanostructures. The swimming of real bacteria is mainly divided into cork-screw type rotating propulsion and the flapping flagella [1]. 30 μm scale bacteria propulsion was demonstrated by an attached ferromagnetic metal pad under an external rotating magnetic field [2-4]. The first rotating magnetic field was used to drive the flagella like tail in [5,6]. Artificial bacteria flagella-like propulsion of helical nanobelts was demonstrated using rotating external magnetic field [2-4]. Further size reduction and applying the same principle to drive artificial bacteria was achieved in [7].

Concerning the driving power source, there are several other limitations of using external field. External field gradient rapidly decreases with the distance from the source. Therefore, if we consider such limitations, helical propellers should be efficient at low Reynolds numbers [8].

Although the helical morphology is advantageous to enable their propulsion at low Reynolds numbers by reducing the viscous drag, the swimming performance was still much lower than the natural bacteria. To improve the swimming performance, increasing magnetic field intensity could be considered.

However, the major challenges toward their potential in vivo biomedical applications should be noted as the lack of closed loop motion control either by teleoperation or autonomous navigation. This imposes the necessity of localization based on their position tracking. Widely used non-invasive biomedical sensors or imaging devices include radiography, computed tomography (CT), ultrasonography and magnetic resonance imaging (MRI). Considering the realtime navigation and high spatial resolution, MRI has been one of the most promising device among others. Miniaturized objects like 1.5 mm diameter ferromagnetic beads were tracked and navigated under realtime MRI [9]. For the tracking and navigation by MRI of smaller objects at low Reynolds numbers, the addition of the propulsion dedicated field gradient generator in the MRI bore for including propulsion force is necessary. However, as-mentioned gradient coils based propulsion would be largely limited when it has to be used with an external biomedical observation device such as MRI, because of the conflict between the imaging and propelling magnetic fields. Therefore it is inevitable to utilize non-magnetic force propulsion.

As non-magnetic approaches, the flapping or undulating motion of flagella was achieved in meso-scale using a commercial piezoelectric bimorph [10]. Microrobots that harness natural bacteria have also been demonstrated [11-13]. However it still remains unclear which propulsion method is optimal for small scale robotics applications. There are very few works on electric-field based wireless micro/nanorobots. For example, diodes were actuated using electro-osmotic pumping [14, 15]. It showed that electro-osmotic field can be a good candidate for actuating micro/nano robots. However the work was only demonstrated in macro scale. Thus scaling down is

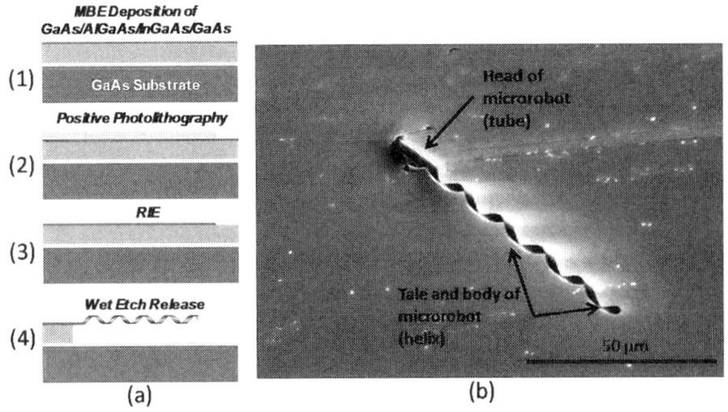

Fig. 1. Basic fabrication process sequence of HNBs (a)(1-4): initial planar bilayer, patterned through conventional microfabrication techniques folds itself into a 3-D nanostructure during wet etch release, (b): created HNB with a tail and a head (SEM photo).

necessary. Furthermore HNBs were not attempted to propel by non-magnetic force.

Therefore in this paper we show a new efficient locomotion method of HNBs by electro-osmotic force. As different types of HNBs showed different swimming performances, we optimized the design of HNBs through the choices of surface material and geometry. The fabrication processes of different types of HNBs are described in next section. The electro-osmotic propulsion experimental setups and the schematics are shown in section 3 and section 4 includes the experimental results of electro-osmotic propulsion of HNBs.

II. Fabrication of 3-D Helical Nanobelts

A. *Why 3-D Helical Nanobelts?*

There are many different types of bacterial locomotion, such as rotating and undulating motions. However, their major features, especially the surface condition, geometry, and locomotion, have been evolved to be able to swim efficiently at low Reynolds numbers. Therefore they could survive within extreme environments, such as blood vessels inside the human body. For example, *E. coli* bacteria consist of a highly elastic body formed of protein. Therefore they can self-transform their morphology to increase their swimming velocity in a viscous liquid environment. We consider that such a mobility of natural bacteria is mainly attributed to the nonlinear mechanics of these natural microorganisms.

To mimic such mechanics of natural bacteria, the ultra-flexible nanostructures with large non-constant mechanics are desirable. HNBs are one of the most flexible nanostructures that can be fabricated using standard microfabrication controlled from the geometry design.

Therefore the HNBs used in the paper are well designed for electro-osmotic propulsion by excelling the swimming performance of the natural bacteria. They can also be mass produced by standard microfabrication. Because the process is also controlled, the HNBs with necessary features can be fabricated.

B. *Fabrication Process*

Figure 1 illustrates the fabrication process. The initial layers were grown on semi-insulating GaAs (Freiberger Compound Materials) using a molecular beam epitaxy system (VEECO, Gen III MBE) equipped with a valved cracker for As and solid sources for Ga and In (Figure 1a). For n-type doping, a solid Si source was used. Substrate temperature was measured by a pyrometer. After a thin GaAs buffer layer, the 500 nm thick sacrificial AlGaAs layer is deposited. The layer contains 20% Ga in order to prevent oxidation. The sacrificial layer in previous designs was made of AlAs. The oxidation of these layers after exposure damaged the InGaAs/GaAs bilayer. On top of the sacrificial layer the InGaAs/GaAs layer is deposited which later self-forms into the nanosprings. An In content of 15% in the InGaAs layer was determined by X-ray diffraction (XRD) measurements. The thickness of this layer must be smaller than the critical thickness to maintain elastic strain. The layer properties along with other specifications of the structures are summarized in Table 1.During the deposition of the InGaAs/GaAs bilayer on this wafer, we attempted to get a slightly lower doping concentration than on the wafer that was used for HNB 2, 3, and 4. From the results it seems that the doping concentration was too high which resulted in a partial self-compensation and, therefore, in a decrease in the effective doping concentration. The doping concentrations of the structures are 4.4×10^{18} cm^{-3} and 8.5×10^{18} cm^{-3} (Table 1).

The initial pattern can be created through photolithography. Reversible photoresist AZ5214 was used as a resist. After the development of the resist, ion etching (RIE) with a $SiCl_4$ gas was used to transfer the pattern to the InGaAs/GaAs bilayers [16]. For HNB 4, the thickness of photoresist was reduced around 50 nm with O_2 plasmas. For metallic HNB 3, the Cr/Ni layers on the surface and on the heads to which the structures are fixed at the end were created before RIE through a lift-off process with negative photoresist AZ5214.

The Cr layer is 10 nm thick and serves as an adhesion layer. The 10 nm thick Ni layer is used for metal HNB 3.

TABLE I
Specifications of Helical Nanobelts Used in Experiments and Simulations

	HNB 1 low	HNB 2 high	HNB 3 Cr/Ni	HNB 4 Dielectric
Thickness $In_{0.15}Ga_{0.85}As/GaAs$	11.6 / 15.6 nm	8 ~ 10 /15.6 nm	11.6 / 15.6 nm	
Thickness Cr/Ni	NA		10 / 10 nm	Photoresist ~ 50
Diameter (tail=head)	2.1 µm			
Pitch & width (tail)	14 µm & 2.5 µm			
Number of turns (tail)	4.5 turns			
Length (head)	12 µm			
Length (total)	74 µm			
InGaAs/GaAs Doping (N_D)	$4.4 \cdot 10^{18}$ cm^{-3}	$8.5 \cdot 10^{18}$ cm^{-3}		

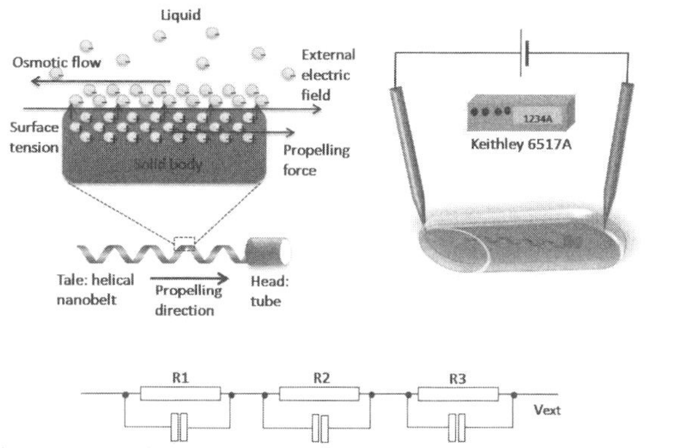

Fig. 2. Schematic diagrams of the experiments for HNB swimming. A DC bias was applied between two electrodes located through axial direction of HNB body. Thus the propulsion direction can be controlled and predicted by the configuration of electrodes. For controlling the direction of electric field, micromanipulators with tungsten probes were utilized to form electrodes.

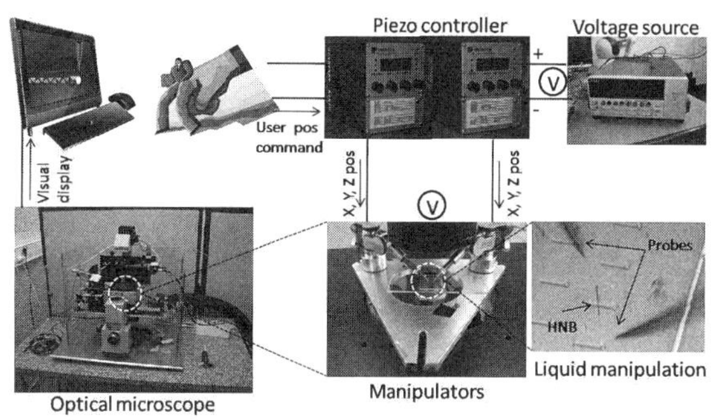

Fig. 3. System setup for electro-osmotic propulsion experiments: Optical microscope with a camera was used to give user visual feedback and thus proper manipulation in liquid. User operates the piezo controller to manipulate tungsten probes, which were utilized to manipulate HNBs and to apply electric field generated by an external voltage source.

Finally, a 2% HF aqueous solution at 4°C was used to selectively etch the AlGaAs sacrificial layer under the InGaAs/GaAs heterostructures for the self-forming of the nanostructures (Figure 1a). During this wet etch, the patterned bilayer curled up along a <100> direction releasing the internal strain and forming 3-D structures. The direction of the scrolling is determined by the anisotropy in stiffness of the InGaAs/GaAs bilayer. After the wet etch release, the chips were rinsed in deionized water and subsequently in isopropyl alcohol [17]. Samples are also conserved in isopropyl alcohol in order to prevent oxidation by water of heads to which the structures are fixed.

As a result, four different types of HNBs (metal, high, low doping and dielectric HNBs) with similar geometry features were created based on the described process (Figure 1b). The detailed specifications of each HNB are summarized in the Table 1.

III. Electro-Osmotic Force Propulsion

In this section, the basic principle and experimental setup to generate electro-osmosis effect is described.

A. Principle of HNB Electro-osmotic Propulsion

The propulsion of HNBs involves two conversions: transforming energy from an external source to a force and transforming this force into a motion. The proposed method to supply power remotely to nanorobots is the electrokinetic process. Its main advantage for in-vivo applications is its compatibility with medical Magnetic Resonance Imaging.

In general, the surface charge that develops at the solid-liquid interfaces plays an important role in a number of electrokinetic processes. Due to the presence of a thin interfacial layer named the Stern layer, it results in a non-zero electric potential at the liquid interface, also known as the zeta potential ζ. This potential is screened by mobile counter-ions diffusing in the liquid in a layer generally named Electrical Diffuse Layer (EDL), which has a typical thickness of a few tens of nanometers depending on the solution ionic strength. Under an electric potential difference, the resulting electric field puts in motion the excess counter-ions in the EDL. This moving layer drags the whole liquid inside the fluidic medium, resulting in "plug-like" flows also referred to as Electro-Osmotic Flows (EOF). In addition, one should also consider the surface charge developing at the HNB surface. In the case of a nanorobot floating in liquid, its own Stern layer generates a flow which applies a hydrodynamic pressure on the surface of the robot, propelling it in the opposite direction.

Figure 2 depicts the electro-osmotic propulsion mechanism through the interface between HNB`s thin membrane surface and liquid medium solution. Since the propulsion force occurs through the whole surface, the large surface to volume ratio of HNBs is very advantageous.

We proved that the HNB motility results from a local electro-osmotic flux powered by the external field. The specific direction of HNB`s propulsion along the cathode and anode probes indicates that a DC field along the HNB is responsible for this propulsion. The equivalent circuit can predict that the electro-osmotic propulsion force is affected by the difference of electrical resistivity between the medium solution and the HNB (Figure 2). The electric voltage applied to the HNB by the external field can be estimated from a serially-connected resistors model describing the ionic conductance through the liquid medium and capacitors for the ionic layers. In a DC electric field, the resistance is the leading contribution. The resulting DC voltage of magnitude V_d induced in the HNB is modeled as

$$V_d = \frac{R_2}{R_1 + R_2 + R_3} V_{ext} \tag{1}$$

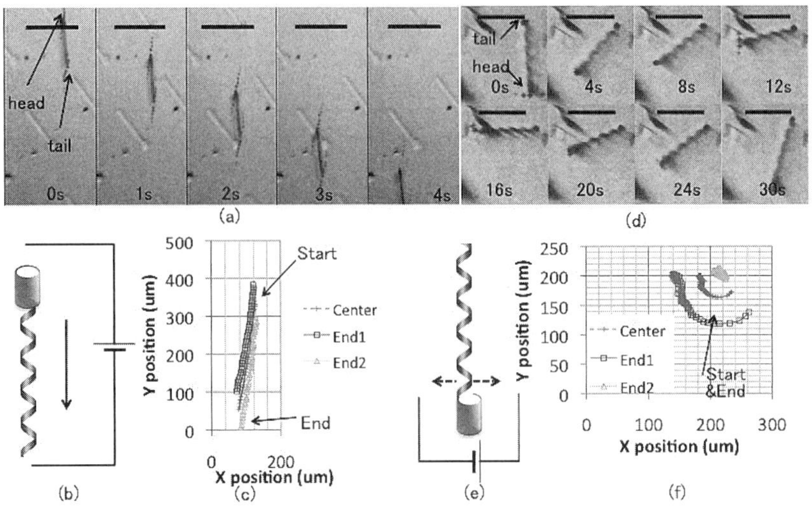

Fig. 4. Different swimming propulsions of HNBs are achieved by modifying electrode's configuration. Series of snap shots during backward (a) and rotating, (d) propulsions were taken during the experiments (b and e). The trajectories of each swimming experiment are displayed in (c) and (f). The scale bars in (a) and (d) are 80 μm and 60 μm.

Where V_{ext} is the DC input voltage applied to the probes in the petri-dish and R_1 (resistance between left electrode and HNB), R_2 (resistance of HNB), R_3 (resistance between HNB and right electrode) are depicted in Fig. 2. Assuming that the resistance of the liquid is linearly proportional to the distance between the probes, equation (1) can be simplified as $V_d = E_{ext}l_d$, where l_d is the length of the HNB body and E_{ext} is the external DC field. The external DC field between the electrodes makes electro-osmotic fluid flow along the HNB body.

We consider a liquid electrolyte, consisting of positive and negative particles in liquid solution. The interface between the electrolyte and the container wall generally forms a double layer in equilibrium, where a nonzero surface charge is screened by a very thin diffuse layer of excess ionic charge of width λ, the Debye screening length (typically 1-100nm). The double layer is effectively a capacitor skin at the interface, which has a small voltage ζ across called the zeta potential. We describe the effect of the HNB and also the environment under electric field more clearly by further detailed model which decomposes the part of intrinsic electrophoretic force directly generated to HNB and the electro-osmotic flow from the surface interaction. Now we can consider a tangential electric field E_{ii} applied in parallel to a flat surface. The electric field acts on ions in the diffuse part of the double layer, which drag the fluid to produce an effective slip velocity outside the double layer by the Helmholtz-Smoluchowski formula (Eq. 2) [18],

$$u_{11} = \left(\frac{2\varepsilon_0\varepsilon_r\zeta_{HNB}H}{3\eta} - \frac{\varepsilon_0\varepsilon_r\zeta}{\eta} \right)\vec{E} \qquad (2)$$

where $\varepsilon_0\varepsilon_r$ and η are the electrolyte permittivity and fluid viscosity, which are constants (environment parameters). Increasing the speed of the HNB requires to increase ζ_{HNB} and H which are the zeta potential around the surface of the HNB and the Henri function, which depends on the geometry and hydrodynamic properties, and the external field (E). The Zeta potential ζ on the borders of the surface is inversely proportional to the speed, thus it can be minimized and controlled by applying experiments under closed microfluidic channel.

An important application of electro-osmotic flow is capillary electro-osmosis, where an electric field is applied down a capillary tube to generate an uniform plug flow (Eq. 2), driven by the slip at the surface. Typical flow speeds of 100 μm/s (6 mm/min) are produced by a field of 100 Volt/cm. This value will be used to compare the swimming performance of HNBs in the next section. It should be noted here that the electro-osmotic propulsion does not depend on the size of the micro-robots. Thus microscopic HNB could move about as fast as a macrosized one.

For the experiments in this paper, we used isopropyl alcohol as medium solution to minimize the possible aging effect or oxidation of HNBs by water. However, the relation between swimming performance and medium solution's pH and/or resistivity is not explored here.

B. Electro-osmotic Propulsion Experimental Setup

To cut and release the fixed end of HNBs from chips and generate electric field through the liquid medium and HNBs, at least two probes with high precision positioning accuracy are required. Every experiment is monitored and recorded for the later analysis of the swimming performances of HNBs. Therefore multiple nanorobotic manipulators are installed under the optical microscope. The complete experimental setup is shown in Figure 3. Two nanorobotic manipulators (Kleindiek, MM3A) are installed under optical microscope; each has three degrees of freedom and respectively 5 nm, 3.5 nm, and 0.25 nm resolutions in X, Y, and Z directions. For the application of the electrical field and high resistance measurement, an electrometer with a DC power supply (Keithley 6517A) was used. Two probes attached to each side of manipulators are positioned in less than 1mm distance. To avoid the optical reflection from the meniscus between probes and liquid medium interface, longer probes (~5mm) with sharp tips were used.

IV. Electro-Osmotic Propulsion Experiment

A. Propulsion Experiment by Electro-osmotic Pump

Figure 4a describes the series of photos taken from the recorded video and the tracked HNB`s trajectories in forward propulsion. An external DC field of 118 V/mm was applied for the forward propulsion and results in backward motion in negative potential. The velocity achieved with this field and configuration is 126 µm/s, which exceeds the propulsion velocity of *E. coli* bacteria with a rotating frequency of 100 Hz [19,20]. Since the size of HNBs and natural bacteria (e.g. *E. coli* bacteria) are different (HNB is 10 times larger than *E. coli*), relative velocities were compared using units of body-length.

Through the experiments at an electric field of 239 V/mm, a velocity of 1,785 µm/s is achieved with HNBs, which is 24 body-lengths per second. It should be noted that *E. coli* bacteria moves at a speed around 10 body-lengths per second [19].

B. Controlled Swimming Experiments of HNBs

Figure 4 demonstrates the forward and rotating propulsion of HNBs. Experiments are depicted in Fig. 4b and e. The probes attached to manipulators can change the configuration to generate the desired electric field gradient. Please note that the rotation was achieved using a HNB fixed in one extremity while the other is free. In this case, the electrode configuration was fixed to make the HNB rotate. The head was fixed with van der waals interaction with the surface. Therefore this surface interaction can be reversibly reduced to release HNBs from substrate by applying a small amount of mechanical force using a manipulator. Furthermore, the forward and backward swimming direction can be converted without the complicated tumbling motion of *E. coli* bacteria [20]. The swimming propulsions of HNBs are achieved by modifying the electrodes` configuration. A series of snap shots during forward (Figure 4a) and rotating (Figure 4d) propulsions were taken during the propulsion experiments. The trajectories of each swimming experiment are displayed in Fig. 4c and 4f. For trajectories, a video analysis software and a custom-made C++ program were utilized. Rotation of HNBs was achieved by their aligning to the applied field direction.

We have confirmed that HNBs swimming directions could be easily controlled by differentiating the field gradient. For the steering control, the field gradient needs to be aligned to the rotational axes of the HNBs. The high surface-to-volume ratio of HNBs increases the swimming efficiency by increasing the charged area in a given volume and mass.

The maneuverability of the currently demonstrated electro-osmotic HNBs is still limited to the basic motions compared to the magnetic field based approaches. However it can be overcome by making addressable swimming using an array of embedded electrodes.

In the paper, one side of the HNB was fixed to the substrate and the other side was free to demonstrate the rotation. Of course the rotation of a non-tethered HNB is also possible. Recently, shark-like motion and rotation motion have also been demonstrated. By controlling the field gradient in more detail, the maneuverability of a HNB can be further improved.

C. Force Estimation

We estimate the force generated by the HNB`s swimming from the drag it overcomes, considering its shape approximately as an ellipsoid [7].

$$F = \frac{4\eta\pi a v}{\left[\ln\left(\frac{2a}{b}\right) - \frac{1}{2}\right]} \quad (3)$$

where a and b are the dimensions along the longitudinal and lateral axes respectively, η is the viscosity of the liquid medium, and v is speed of the HNB. For a 74 µm long HNB with a diameter of 2.1µm moving at 1.8 mm/s in isopropyl alcohol medium, the generated force of 1.3 nN (Eq. 3) is an order of magnitude higher than the conventional optical trap force which is less than 200 pN. Furthermore, HNBs can apply a large pressure, approximately 91 N/m^2, which can be advantageous used for local manipulation, such as on biological membranes [18,19].

D. Swimming Performances of Different HNBs

We analyzed the swimming performances of different HNBs as a function of the parameters controlling the electro-osmotic propulsion force. The maximum velocity of an HNB in isopropyl alcohol medium as a function of electric field intensity is characterized in Fig. 5. A 74 µm long HNB is used for the tests. The intensity of the field is increased from 24.99 to 239.66 V/mm. The maximum velocity was not linearly proportional to the applied field intensity, especially at higher field intensity above 87.75 V/mm. This can be explained by the fact that HNBs swimming in isopropyl medium have relatively higher Reynolds numbers in this region compared to the water. Reynolds number can be calculated by the equation [1].

(4)

Where L is the length of HNB, v is the velocity, ρ is the density of the liquid and μ is the viscosity of the liquid. For

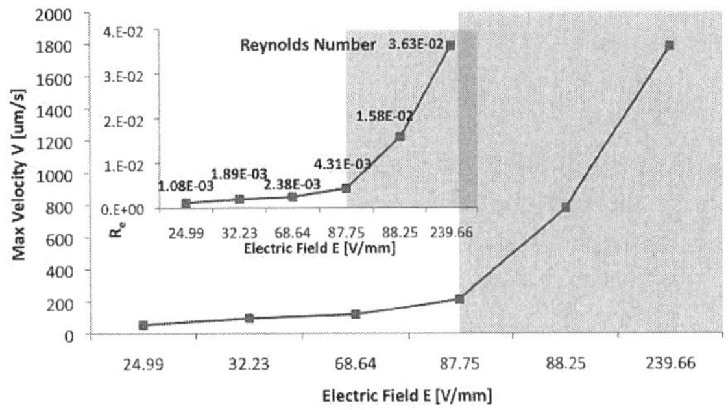

Fig. 5. Dependence of HNB velocity on applied electric field intensity. The plot was obtained during the swimming experiments of a 74 μm long HNB in isopropyl alcohol medium. The nonlinearity of the swimming performance is demonstrated at the region with higher electric field intensity than 87.75 V/mm. The inset figure shows the Reynolds numbers in different velocities *vs.* applied electric field.

isopropyl alcohol, the kinematic viscosity (μ/ρ) is approximately 3.6×10^{-2} cm^2/s thus the Reynolds number is proportional to the length and velocity of the HNB. In natural bacterial propulsion with 1 μm length and a velocity of 30 μm/s, the Reynolds number is 3×10^{-5} in water [1]. For example, 74 μm long HNBs swimming at 1,785 μm/s in isopropyl alcohol medium have a Reynolds number of 3.6×10^{-2}, which is much higher than bacterial propulsion. It involves more inertial effect during the propulsion and explains the non-linearity of the velocity and electric field. In our experiments, the swimming velocity of HNBs is between 4.4 μm/s and 1,785 μm/s. The Reynolds numbers are thus between 3.3×10^{-4} and 3.6×10^{-2}. Fig. 5 and the inset plot depict the Reynolds numbers of HNBs with different velocities. It is apparent that the linear relation between the field intensity and the velocity is achieved with a velocity less than 87.75 V/mm and Reynolds numbers around 4.3×10^{-3}. All other previous bacteria-like propulsions were demonstrated within this velocity around 200 μm/s [3,4,7]. This number could be at the marginal limit representing natural bacterial propulsion. The unusually fast swimming propulsion demonstrated in this work is due to the generated inertial force to overcome the viscous drag by reaching at a higher Reynolds numbers. The inertial effect of a rigid-body swimmer at low Reynolds number is tiny so ignorable. According to the speculation based on empirical experiments, however the extremely flexible HNBs (the stiffness of HNBs can be around 0.001 N/m and 0.00001 N/m in longitudinal and bending respectively) can reduce the viscous drag by oscillating the HNB. Therefore they can undergo the inertial effect. In fact, the HNB continued to swim for a while even though the field was cut off when it is accelerated to a velocity faster than the critical velocity. This deformation can generate the surface charges of HNBs to enhance local electro-osmotic force. The HNBs showed the giant piezoresistivity partly due to the piezoelectric effect [17]. Therefore the surface charge added the induced propulsion force thus contributed to the nonlinear curve of the maximum velocity. It can explain the continued swimming phenomenon even after external field is cut off.

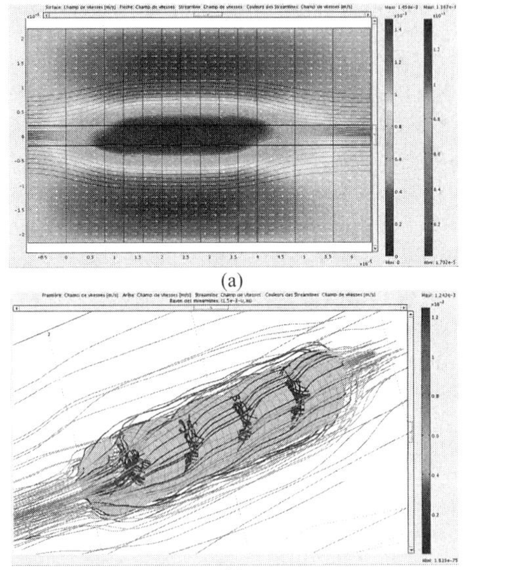

(a)

(b)

Fig. 6. Hydrodynamic simulations using finite element modeling: (a) 2-D stream around HNB, (b) coarse 3-D model around HNB.

Furthermore, the non-linearity and fluctuations of the curve in Figure 5 are also attributed to intermolecular interactions of the HNB with the substrate. Especially the heavier tube type heads of HNBs were apt to submerge more than the body and tail, which makes the initial energy to break this interaction difficult to predict.

To explain the unusually fast swimming HNB, finite element simulations in liquid medium were performed. Figure 6a shows the cross-sectional view of the HNB moving with a velocity of 1 mm/s. Although the inside of HNB is a bit undulated, a very regular laminar flow was estimated around it. As was expected, the flow speed was slower inside of HNB and was faster outside. This laminar flow is due to the helical morphology with ultra-thin layers of HNB, even though the flow speed was fast for the given dimensions. The Reynolds number was around 10^{-2}, which acts like a very viscous liquid. To further estimate the undulation inside the HNB, a 3-D model of a HNB was created from the assembled hexahedral blocks (Figure 6b). This coarse model reduces the calculation time of the finer model. The same phenomena with a very slower laminar flow inside and a faster one outside HNB were estimated. In this case, the flow inside HNB was winded slowly. This reveals that the fast swimming HNB can generate inner torque thus rotates the structure to balance the energy. The rotation energy contributes to keep the flow around HNB very laminar while it swims very fast. Therefore, it is predicted that swimming performance would highly depend on the surface geometry, especially inside the HNB.

To further understand the repeatable swimming behaviors due to an electro-osmotic force and to find the

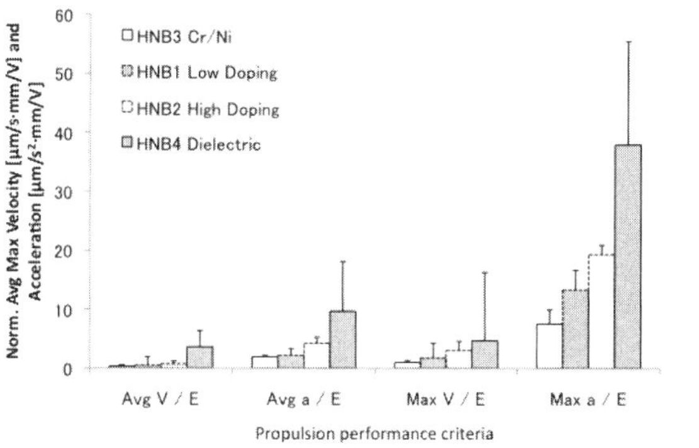

Fig. 7. Dependence of normalized velocity and acceleration of propulsion on 4 types of HNBs. High and low doped semiconductor HNB showed higher swimming performance than metal (Cr/Ni) coated HNB and best performances are achieved with type 4 (Dielectric).

optimized design parameters, especially the surface profile, four different types of HNB, performance was investigated were evaluated in their performances. High and low doping HNB, a metallic (Cr/Ni) one and a dielectric counterpart were analyzed. Figure 6 shows the average and maximum propulsion velocity and acceleration normalized by the applied electric field. HNB performance was investigated in several steps (6 - 11 steps) ranging from 24 to 401 V/m. It was difficult to compare four HNB in identical single field intensity because the experimental configuration of probes and HNB are difficult to maintain in-between each experiment. Therefore, a statistical average velocity was chosen to describe the general behavior of different HNB. It should also be noted that experiments on different types of HNB were conducted in different days which might slightly change electrical resistivity of the liquid medium. Even though the exact resistance values of the utilized HNBs were not verified because of the difficulty in collecting them after experiments, the intrinsic resistance of HNB can be predicted as $R_{HNB4} > R_{HNB1} > R_{HNB2} > R_{HNB3}$. Hence, the average and maximum velocities and accelerations depend on the resistance of HNB as was expected from the equation (3). Highly doped HNB2 showed better swimming performance than low doped HNB1. This can be explained by the fact that the InGaAs layer thickness (estimated as 8 ~ 10 nm) of HNB2 is slightly thinner than the one of HNB1 (11 nm). It results in better performance in HNB2 than HNB1 from the reduced gravity effect and hydrodynamic friction regardless of the higher conductivity. It reveals that the doping concentration does not play much role in the swimming performance compared to the geometry or volume effect. It should also be noted that the HNB3 and HNB4 are slightly thicker than HNB1 and HNB2 because of additional metallic layers (Cr/Ni 10/10nm, AZ5214 photoresist around 50 nm, Table 1). However, the HNB3 had lower performance while HNB4 was better than the others. In the case of HNB3, this can be explained by the increase in hydrodynamic friction during swimming which

TABLE II
PROPULSION PERFORMANCE COMPARISON OF HELICAL STRUCTURES

	Magnetic 1 [3,4]	Magnetic 2 [7]	**Electro-osmotic**	E. Coli [16,17,20]
Length / Diameter [µm]	50 / 2.8	2 / 0.2	**74 / 2.1**	2 / 0.5
Max. Vel. [µm/s]	18	40	**1785**	20
Body-length / second	0.4 ×	20 ×	**24 ×**	10 ×
Force [pN]	3.0	~ 1.0	**1300**	0.6

slow down HNB. Furthermore, in case of metallic HNB3, the gravitational force can also increase the surface interaction force by dragging it down to the substrate. The detailed calculation and comparison of buoyancy force and gravitational force are not described here, however the resulting nonfluidic forces (as the difference between the gravitational force and the buoyancy force) on HNB3 is 0.94 pN, which is two times higher than on HNB1 (0.43 pN). These negative effects on HNB3 reduce its swimming performance compared to the others (Figure 7). The result of HNB4 is contrary to the HNB3 while additional layer was deposited onto the surface. It can be explained by the fact that the nonfluidic force of HNB4 was decreased by the increased buoyancy force from the photoresist, which has much lower density than both gold and nickel.

The result clearly shows the high dependency on the surface charge and thus zeta potential in dielectric surface coating to HNB compared to conductive or semiconductor materials (Figure 7).

Finally the swimming performance of electro-osmotic propulsion HNBs was compared with other helical nanostructures (Table II). Compared to two rotating magnetic propulsions of helical nanostructures, the demonstrated electro-osmotic propulsion of HNBs showed much higher swimming performance in terms of maximum swimming velocity and manipulation force. Furthermore it even excels the one of natural E. coli bacterial propulsion. The demonstrated velocity (1785 µm/s) can reach 90 times faster than others. Concerning to the manipulation force reaching at around 1.3 nN, it is 1,000 times higher than others. It is even much higher than widely used optical tweezers (~ 200 pN). The swimming efficiency can further be improved by further size reduction and/or surface coating with dielectric materials.

The absolute voltage applied in the work ranges from a few volts to a few tens of volts and it limits their biomedical applications in this form. However it should be noted here that the electro-osmotic swimming propulsion was demonstrated in the open channel more precisely in the petri-dish. Therefore all experiments in this type of open channel assume certain amount of offset voltage compared to the experiments in a tiny closed channel.

In the case of a closed channel, like a human blood vessel, the propulsion force can further be enhanced by the reduced gravitational force thus increasing buoyancy force. Furthermore the surface interaction from the petri-dish can be minimized as well in case of the closed channel experiments. We are currently working on HNB swimming inside a microfluidic channel for controlling the swimming environment.

For in-vivo biomedical applications of the electro-osmotic HNB swimmers, non-invasive and external electric field generation is inevitable. As one of solutions, patch-type external embedded electrodes can be considered to control the swimming motions of HNBs remotely.

V. Conclusion

The demonstrated direct pulling of HNBs by electro-osmotic pumping was proven to be an efficient energy conversion mechanism for microscopic artificial swimming objects and overcomes easily the viscous drag and gravitational forces in an aqueous environment at low Reynolds numbers. Swimming velocities as fast as 1.8 mm/s (24 times the body length per second) and pushing forces as high as 1.3 nN are achieved by HNBs propelled by an electro-osmotic force under an external electric field. These performances are higher than other state-of-the-art microscale artificial swimmers. Please note that the velocity of 24 body lengths per second is even faster than natural bacterial propulsion. An inertia effect was observed above a critical velocity and was speculated as the reason why the electro-osmotic HNB swimmers could outrun natural bacteria. Hydrodynamic simulation of this unusually fast swimmer revealed a very laminar flow around the HNB with a torque inside them. Although helical morphology with a thin-film is beneficial to such a fast swimming at low Reynolds numbers, the swimming performance appears to depend highly on the surface conditions. Among the four different types, dielectric polymer coated HNB showed the best swimming performance compared to others with metallic or semiconductor surfaces. These HNB can be used as wireless liquid manipulators and assemblers for biomedical or MEMS/NEMS applications or remote physical or chemical detection by functionalizing with proper read-outs. For these complex tasks, it would be necessary to improve the control schemes and methods, implemented through finer hydrodynamic finite element model simulations.

References

[1] E. M. Purcell, "Life at Low Reynolds Number," in *American Journal of Physics,* vol. 45, 1977, pp. 3–11.

[2] D. J. Bell, S. Leutenegger, K. M. Hammar, L. X. Dong, B. J. Nelson D. X. Dong, "Flagella-like Propulsion for Microrobots Using a Magnetic Nanocoil and Rotating Electromagnetic Field." in *IEEE International Conference on Robotics and Autumation*, 2007, pp. 1128–1133.

[3] L. Zhang, J. J. Abbott, L. Dong, B. E. Kratochvil, D. Bell, B. J. Nelson, "Artificial Bacterial Flagella: Fabrication and Magnetic Control" in *Applied Physics Letters*, vol. 94, no. 6, 2009, pp. 064107-3.

[4] L. Zhang, J. J. Abbott, L. Dong, K. E. Peyer, B. E. Kratochvil, H. Zhang, C. Bergeles, B. J. Nelson, "Characterizing the Swimming Properties of Artificial Bacterial Flagella" in *Nano Letters*, vol. 9, no. 10, 2009, pp. 3663–3667.

[5] T. Honda, K. I. Arai, K. Ishiyama, "Micro swimming mechanisms propelled by external magnetic fields," in *IEEE Transactions on Magnetics*, vol. 32, no. 5, 1996, pp. 5085–5087.

[6] K. Ishiyama, K. I. Arai, M. Sendoh, A. Yamazaki, "Spiral-type micro-machine for medical applications," in *Journal of Micromechatronics*, vol. 2, no. 1, 2003, pp. 77-86.

[7] A. Ghosh, P. Fischer, "Controlled Propulsion of Artificial Magnetic Nanostructured Propellers," in *Nano Letters*, vol. 9, no. 6, 2009, pp. 2243-2245.

[8] J. J. Abbott, K. E. Peyer, M. C. Lagomarsino, L. Zhang, L. Dong, I. K. Kaliakatsos, B. J. Nelson, "How Should Microrobots Swim?," in *International Journal of Robotics Research*, doi: 10.1177/0278364909341685.

[9] S. Martel, J. Mathieu, O. Felfoul, A. Chanu, E. Aboussouan, S. Tamaz, P. Pouponneau, L. Yahia, G. Beaudoin, G. Soulez, M. Mankiewicz, "Automatic navigation of an untethered device in the artery of a living animal using a conventional clinical magnetic resonance imaging system," in *Applied Physics Letters*, vol. 90, 2007, pp. 114105.

[10] G. Kosa, P. Jakab, N. Hata, F. Jolesz, Z. Neubach, M. Shoham, M. Zaaroor, G. Szekely, "Flagella Swimming for Medical Micro Robots: Theory, Experiments and Application," in *IEEE International Conference on Biomedical Robotics and Biomechatronics*, 2008, pp. 258–263.

[11] S. Martel, C. Tremblay, S. Ngakeng, G. Langlois, "Controlled manipulation and actuation of micro-objects with magnetotactic bacteria," in *Applied Physics Letters*, vol. 89, 2006, pp. 233904.

[12] B. Behkam, M. Sitti, "Bacterial flagella-based propulsion and on/off motion control of microscale objects," in *Applied Physics Letters*, vol. 90, no. 2, 2007, pp 1-3.

[13] E. Steager, C. B. Kim, C. Naik, J. Patel, S. Bith, L. Reber, M. J. Kim, "Control of microfabricated structures powered by flagellated bacteria using phototaxis," in *Applied Physics Letters*, vol. 90, no. 26, 2007, pp. 263901.

[14] S. T. Chang, V. N. Paunov, D. N. Petsev, O. D. Velev, "Remotely Powered Self-propelling Particles and Micropumps based on Miniature Diodes," in *Nature Materials*, vol. 6, no. 3, 2007, pp. 235-240.

[15] S. T. Chang, E. Beaumont, D. N. Petsev, O. D. Velev, "Remotely powered distributed microfluidic pumps and mixers based on miniature diodes," in *Lab on a Chip*, vol. 8, 2008, pp. 117-124.

[16] S. Varoutsis, S. Laurent, I. Sagnes, A. Lemaitre, L. Ferlazzo, C. Mériadec, G. Patriarche, I. Robert-Philip, I. Abram, "Reactive-ion etching of high-Q and submicron-diameter GaAs/AlAs micropillar cavities," in *Journal of Vacuum Science and Technology B*, vol. 23, no. 6, 2005, pp. 2499-2503.

[17] G. Hwang, D. J. Bell, L. Dong, H. Hashimoto, S. Schon, B. J. Nelson, "Piezoresistive InGaAs/GaAs Nanosprings with Metal Connectors," in *Nano Letters*, vol. 9, no. 2, 2009, pp. 554–561.

[18] R. J. Hunter, "*Zeta potential in colloid science principles and applications*," Academic Press, London a.o., 1981.

[19] H. C. Berg, D. A. Brown, "Chemotaxis in Escherichia coli analysed by Three-dimensional Tracking," in *Nature*, vol. 239, no. 5374, 1972, pp. 500–504.

[20] H. C. Berg, "*E. coli in Motion*," Springer, New York, 2004.

[21] W. C. Little, M. L. Smith, U. Ebneter, V. Vogel, "Assay to mechanically tune and optically probe fibrillar fibronectin conformations from fully relaxed to breakage," in *Matrix Biology*, vol. 27, 2008, pp. 451-461.

[22] S. J. Park, M. B. Goodman, B. L. Pruitt, "Analysis of nematode mechanics by piezoresistive displacement clamp," in *Proceedings of the National Academy of Science of the United States of America (PNAS)*, vol. 104, 2007, pp. 17376-17381.

[23] S. Chattopadhyay, R. Moldovan, C. Yeung, X. L. Wu, "Swimming efficiency of bacterium Escherichia coli," in *Proceedings of the National Academy of Science of the United States of America (PNAS)*, vol. 103, 2006, pp. 13712-13717.

A Molecular Algorithm for Path Self-Assembly in 3 Dimensions

Rebecca Schulman
Department of Physics
University of California, Berkeley
Berkeley, CA, USA
Email: rschulman@berkeley.edu

Bernard Yurke
Materials Science and Engineering Department &
Electrical Engineering Department
Boise State University
Boise, ID, USA
Email: bernardyurke@boisestate.edu

***Abstract*— A molecular algorithm is a set of molecular interactions that carry out a particular task. We describe a molecular algorithm for self-assembling a path between two stationary points when the locations of these points are not known in advance. While efficient path finding algorithms for electronic robots exist, molecules lack the centralized memory or computing power to implement them. The algorithm takes advantage of the inherent physics at the molecular scale, and unlike other biomimetic algorithms for path finding, is designed to work in an unstructured environment and does not require complex molecular components. Designed molecules self-assemble a DNA nanotube starting from a "seed" molecule attached to the start point. During growth, the DNA nanotube's end diffuses through space, and this diffusion is harnessed as a search process: when the DNA nanotube's end contacts the destination, it attaches to it and stops growing, forming a stable path.**

We use simulations and analysis to predict that paths of up 10 microns are formed with more than 99% probability if the destination is larger than 500 nanometers in diameter, making it practical. However, the probability of the DNA nanotube missing the destination is highly dependent on destination distance and size. It increases as $\exp\left[-1/\Omega\left(r^3\right)\right]$, where r is the Euclidean distance from the start point to the destination, and scales approximately as $\exp\left[-\Omega(d)\right]$, where d is the diameter of the destination. A path finding algorithm that could work over longer distances or with small destinations will likely require new molecular components; we describe how new components could be used to solve the molecular path finding problem using a divide and conquer approach.

I. Introduction

Bottom-up construction is an increasingly promising way to inexpensively create materials with designed nanoscale features [46, 21]. In a bottom-up construction process, designed molecules autonomously assemble the desired device. Molecules, unlike electronic devices, have almost no capacity for computation or information storage. Instead, a series of stochastic molecular attachments and detachments must collectively converge on the desired structure. These collective actions, i.e. molecular algorithms, are surprisingly powerful: molecules can perform universal construction [43] and universal computation [32, 11]. Molecular algorithms are also capable of error-correction and robust behavior [10]. While in practice most are still fragile and scaling is difficult [40, 2, 13], simple algorithms have been demonstrated experimentally. For example, designed molecules can self-assemble a Sierpinski gasket [36], and 2- and 3-dimensional structures with arbitrary features [34, 13], and can perform tasks such as walking [5, 30, 23] and cargo transfer [16].

One way to make a molecular algorithm robust is to make it adaptive. Scaling is more feasible if individual construction algorithms (i.e. modules) can respond to errors made in the execution of a previous construction process. Environmentally adaptive molecular algorithms would be useful in unstructured environments such as cell cultures where the geometry may be uncertain and algorithms that can deduce and respond to these geometries would be required. Biological algorithms for assembly are environmentally adaptive—cells use chemical and mechanical signals to determine when and where to grow and move. However, devising ways for synthetic molecules to collectively sense the local environment and act on the results of the information received can be difficult.

In this paper we describe a molecular algorithm for path finding and characterize its performance. In the path finding problem, the inputs are fixed start and destination points of unknown location, and the desired output is a physical link between them. An algorithm to solve the path finding problem is inherently environmentally adaptive: the location of the path that is formed depends on the locations of the start and destination in the environment. Path finding is also a basic primitive for bottom-up construction and could be used to assemble wires such that they connect to both contacts when the location of the contacts is uncertain, or to immobilize an object by creating a physical tether between the object and a surface.

Biology provides an example bottom-up path finding algorithm, the separation of chromosomes during somatic animal cell division [18] that occurs through the assembly of the *mitotic spindle*. After a cell's chromosomes are duplicated, they remain grouped at the *spindle equator*. They then must be separated such that the mother and daughter cells each receive one set. A pair of *kinetochores* form on each pair of chromosomes and *centromeres* arise on either side of the chromosomes. Microtubules, nanotubes assembled from protein monomers, grow from the centromeres. If a growing microtubule encounters a kinetochore, it stays put, forming a permanent link. The kinetochores are oriented on pairs of chromosomes such that the two attaching microtubules come from different centromeres. The centromeres pull on the

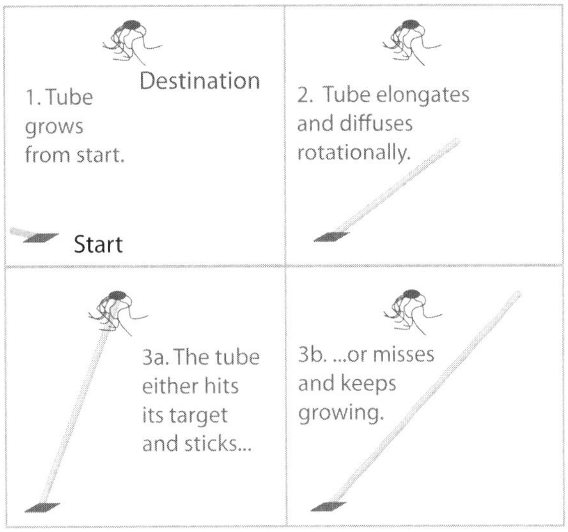

Fig. 1. **The path finding algorithm.** A stationary start marker attached to a flat surface initiates DNA nanotube growth. A stationary destination marker can bind the end of a growing DNA nanotube. As the DNA nanotube grows (1–2), its end diffuses rotationally, effectively searching for the destination. The DNA nanotube either encounters and sticks to the destination (3a) or gets too long to hit the destination (3b) before finding it.

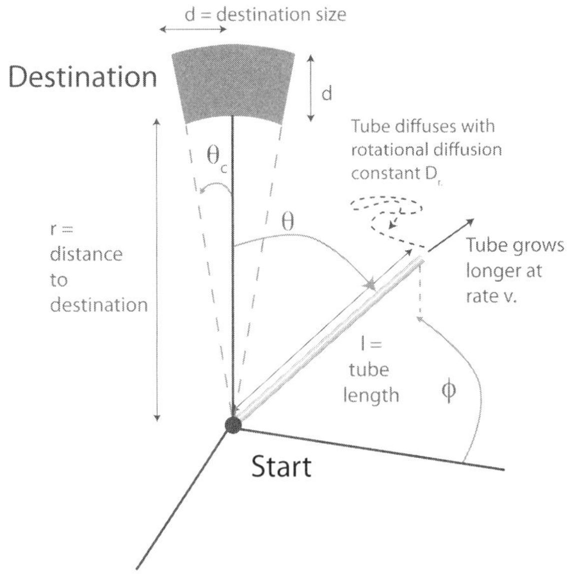

Fig. 2. **Schematic for the molecular path finding problem.** A start point (black dot) can produce a single growing DNA nanotube (gray cylinder). If the DNA nanotube end enters the destination region ($r < l < r + d, \theta < \theta_c$), it sticks. The location of the DNA nanotube's end is given using spherical coordinates.

attached microtubules, separating the chromosomes into two complete sets.

The example of the mitotic spindle suggests that polymer search and capture could be used to form synthetic links between start and destination locations. The mitotic spindle is complex and designed specifically for separating chromosomes, so we propose a simpler molecular algorithm in the same spirit.

The molecules in the path finding algorithm are DNA monomer components that can assemble to form rigid DNA nanotubes (analogous to microtubules), a seed molecule which attaches to the start point (analogous to the centromere), and a capture molecule which attaches to the destination point that growing DNA nanotubes stick to (analogous to the kinetochore). DNA nanotube growth is initiated by a seed [2] at the start point. If the DNA nanotube's end encounters the destination, it sticks and the DNA nanotube forms a path between the start and destination (Figure 1).

Here we seek to understand how this algorithm's performance scales with the size of the destination and the distance between the start and the destination. To do so, we analyze the physics of DNA nanotube diffusion and derive an analytical approximation for the probability that the DNA nanotube will hit the destination. We then determine whether the algorithm will perform well for destination sizes and distances of interest. We use existing estimates for the physical parameters that govern the diffusion and growth rates of DNA nanotubes to numerically compute the probability of forming a path for typical destination sizes and distances from the start point.

In Section II we describe prior work on path finding. In Section III we more formally describe the problem and review the physics of DNA nanotube growth and diffusion that will be required to analyze the algorithm's performance. We use this material in Section IV to give expressions for how the algorithm's performance scales with destination distance and size. We show that while path finding via diffusion is very effective for relatively short distances and large destinations, the probability of successful path finding drops exponentially with increasing destination distance and decreasing destination size. In Section V we therefore discuss some possible ways to create algorithms that will perform better when the destination is small and/or far from the start point.

II. Background

The problem of directing an autonomous agent to find a path through an environment is a classic problem in robotics. Any mobile robot in an uncertain environment must solve some instance of this problem [19]. However, algorithms designed for a single autonomous agent with a large amount of memory and the capacity for efficient, large-scale computation (e.g. [3, 20]) cannot generally be applied to molecular systems.

Path finding has also been studied in biomimetic systems, where large numbers of very simple components with little capacity for computation and memory storage work together to accomplish a goal. In amorphous computing, a collection of fixed, unreliable agents are arrayed randomly in space [1]. Each agent is capable of limited computation and communication with neighboring agents. Paths between specified objects can be discovered in this system using a gradient of morphogens: a destination emits a signal which is passed to neighbors and a path forms in response to these signals [28]. The spatial computing framework, similar to the amorphous computing architecture, provides a language in which path finding can also be solved using gradients [6]. While gradient-

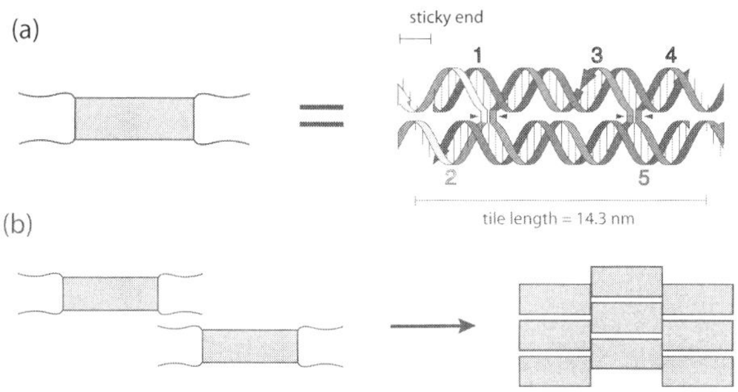

Fig. 3. **DAE-E DNA tiles are the building blocks of DAE-E DNA nanotubes.** (a) DAE-E DNA tiles consist of 5 short DNA strands that fold into the structure shown on the right because of a preference for Watson-Crick complementarity. (b) Complement sticky end regions hybridize, and the sequences of tiles are designed to hybridize (attach) as shown. Repeated hybridization produces a tile lattice. The particular geometry of DAE-E DNA tiles leads to a slight curved lattice, which is stable as a closed tube [35].

following is efficient and finds paths with high probability, it is still too complex for molecular components to execute.

It is possible to design molecular construction elements with prescribed shape [41, 34] and prescribed affinity for other elements [48, 25] using synthetic DNA. These construction elements stochastically attach to and detach from each other and move through solution via Brownian motion. DNA-based algorithms for computation and construction exploit the specific affinities of molecules and the fact that when two elements are attached, their affinity for attaching and detaching to certain types other elements can change, which serves as a basic form of state [32, 47, 24].

One widely studied model of molecular construction is the DNA tile self-assembly model [47, 37]. In this model, DNA components called tiles are equipped with single-stranded sticky ends, and tiles bind to other tiles by these sticky ends, forming lattices (Figure 3). Because tiles can bind only to locations where their sticky ends are complementary to those available, the attachment of a tile to a location where it attaches to two tiles at the same time can be viewed as an information transfer step [47].

The DNA tile assembly model can be used to solve the path finding problem on a 2-dimensional surface, even when obstacles are present [7]. In this method, the start and destination are molecules that interact with tiles. The tiles are designed so that they assemble all possible paths from the start to the destination. The possible paths are attached in a single lattice that eventually integrates the destination molecule. Every tile in the lattice that is part of a complete path is strongly attached to other tiles on the path, while tiles not part of a path are more weakly attached. When the temperature is increased, all attachments are weakened and tiles that are not part of a complete path melt away from the lattice. While this path finding algorithm finds optimal paths using molecular components, the algorithm is not robust to noise in molecular information processing [2] and would be very difficult to implement because it requires a highly structured environment at the molecular scale. The algorithm we describe in this paper is robust, implementable, and also forms 3-dimensional paths in addition to 2-dimensional ones.

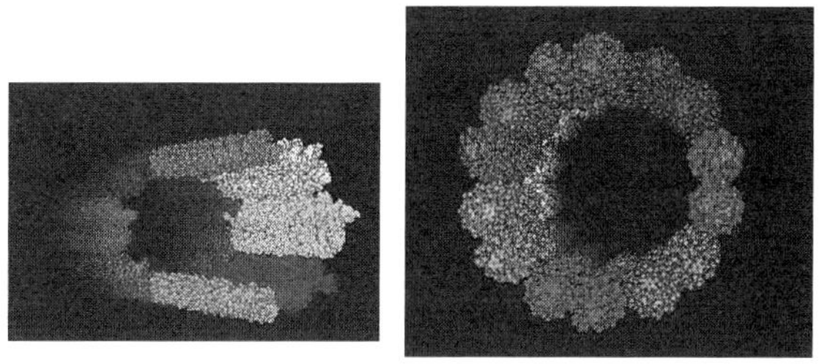

Fig. 4. **3-dimensional models of a 7-tile wide DAE-E DNA nanotube.** Each tile is drawn in a different shade. Diagrams were prepared using NAMOT [8] and PyMol [31] software.

III. Preliminaries

A. Problem Definition

Formally, the goal of a path-finding problem is to connect a start location s with a destination region $\mathcal{D}$ by self-assembling a path between them. s is located at the origin and other points are referred to in spherical coordinates, i.e. (r, θ, ϕ).

We will first consider an *idealized* version of the path finding problem. In the idealized path finding problem, a DNA nanotube grows from s and can diffuse in any angle in three dimensions. The *destination object*, $\mathcal{D}$ of destination size d is the set of points $\mathcal{D} = \{(p, \theta, \phi)$ such that $r < p < (r+d), \theta < \theta_c\}$ where $\theta_c = \arctan \frac{d}{r+d}$.

In practice, the start and destination objects are attached to large flat surfaces (Figure 1), and DNA nanotubes cannot grow or diffuse through these surfaces. In the *implementation* version of the path finding problem, a DNA nanotube grows from s and can diffuse only in the region $\theta \leq \pi/2$.

B. DNA Nanotube Growth

In this section we describe the construction of DNA nanotubes and how DNA nanotube architecture and the physical conditions for growth determine DNA nanotube growth rates.

We consider the growth of DAE-E DNA tile nanotubes. DAE-E DNA tile nanotubes consist of DAE-E DNA tiles (Figure 3) that assemble into a structure consisting of a radial array of parallel DNA helices (Figure 4). A DAE-E DNA tile consists of 5 short, synthetic DNA strands and is 14.3 nanometers long and about 4 nanometers wide. DAE-E DNA tiles attach via the hybridization of the single strand sticky ends on their four edges. A DAE-E DNA tile nanotube with n helices has $n/2$ tiles in each layer and in practice, DAE-E DNA nanotubes with 8–24 helices are known to be stable [35]. In this paper, DAE-E DNA tiles and DAE-E DNA tile nanotubes will be henceforth referred to simply as *DNA tiles* and *DNA nanotubes*.

Experiments with DNA tiles [36, 38, 29] support a model where tile attachment and detachment rates are first order chemical reactions [48]. The rate k_{on} at which a tile attaches at such a site is given by

$$k_{on} = k_f[t] \tag{1}$$

where k_f is a diffusion controlled forward rate [45] and $[t]$ is the concentration of the tiles in the environment. This rate means that tiles arrive at an attachment site at exponentially distributed times, where the mean is k_{on}. A tile attached to a nanotube by n sticky end bonds detaches from the DNA nanotube at rate $k_{off,n}$:

$$k_{off,n} = k_f e^{-\frac{\Delta G_n}{RT}} \qquad (2)$$

where ΔG_n is the Gibbs free energy of attachment by n sticky end bonds, R is the universal gas constant, and T is absolute temperature. Sticky end binding is cooperative, so that ΔG_n increases with n [33, 9, 29].

When $k_{off,2} < k_{on} \ll k_{off,1}$, i.e. when the attachment of a tile to a DNA nanotube is favorable only if the tile attaches by at least 2 bonds, DNA nanotubes grow quickly from a seed structure where tiles can attach by 2 bonds, but slowly in free solution, where attachments by just 1 bond are required to initiate the growth of a new DNA nanotube [14, 38, 39]. Under these conditions, DNA nanotubes grow by the attachment of tiles at the DNA nanotube's distal end, and most attaching tiles form two sticky end bonds.

We used a simple stochastic kinetic simulation of DNA nanotube growth [15, 48] to estimate nanotube growth rates. In the simulation, a tile could attach to a DNA nanotube in a location where it formed two bonds and detach from a DNA nanotube if it was attached by no more than two bonds. We used experimentally measured values of k_f, ΔH and ΔS in the simulation. k_f for a similar type of DNA tiles has been measured and is approximately 3,000,000 per Molar per second, where Molar is moles per liter [17]. $\Delta G = \Delta H - T\Delta S$ has been measured for similar DNA tiles, where T is absolute temperature. $\Delta H = -102.4$ kilocalories / mol and $\Delta S = -0.300$ kilocalories / mol [38]. Our estimate for the growth rate of a 12-tile-wide DNA nanotube in tile rows per unit time was the average time it took to grow a row in the simulated growth of about 2000 tiles. For experiments that take place at around 305 Kelvin and 40 nM tile concentration, we found that a 6-tile wide DNA nanotube grows 14.3 nm (the length of 1 tile row) about every 94.2 ± 1.6 seconds (1.6 seconds is 2 standard deviations of the predicted time per row in 100 iterations of this simulation). We used this value for the growth rate v in the next section. The fluctuations in length due to the stochastic nature of growth should be on order the square root of the DNA nanotube length (in tile rows) and were ignored in our analysis. The growth rate can be made faster or slower by increasing or decreasing the tile concentration.

C. Polymer Dynamics

Previous studies indicate that DAE-E DNA nanotube stiffness and thus diffusion rates are consistent with existing polymer physics models. This section describes the basic components of these models that are used in the analysis of the path finding algorithm in the next section.

Because a polymer's repeating units are physically connected, the directions of the units' Brownian motion are correlated, and the degree of correlation in this motion is the polymer's stiffness. The basic unit of polymer stiffness is its *persistence length*, formally the length scale over which correlations in the direction of the polymer's tangent vector are lost. The persistence length of a polymer is intrinsic to its repeating structure, i.e. there is a persistence length for double-stranded DNA, which is p_{helix} = 50 nanometers [42]. Assuming that the Young's modulus of a DNA helix and a DNA nanotube are the same, the persistence length of a DNA nanotubes is predicted to be [35]:

$$\frac{p_{tube}}{p_{helix}} = 2N\left[1 + 2\left(\frac{R}{r}\right)^2\right] \qquad (3)$$

where p_{tube} is the persistence length of the nanotube, N is the number of helices in the nanotube, R is the nanotube radius and $r \approx 1$ nanometer is the radius of a DNA double helix. This equation predicts that 6-tile-wide to 12-tile-wide nanotubes should have a persistence lengths from 10 to 80 μm respectively. These values are qualitatively consistent with experimentally measured values of nanotube persistence length [35, 14]. The width of the DNA nanotubes used in the path finding algorithm can be controlled by designing the seed for the DNA nanotube's to grow DNA nanotubes of the desired width [2]. It seems feasible to make DNA nanotubes with a persistence length of up to 12 tiles around, or ≈ 80 μm.

When a polymer is much shorter than its persistence length, it is classified as *rigid* and behaves much like a straight rod. Since in our path finding algorithm one end of the growing nanotube is attached to the start point, a rigid nanotube diffuses by rotation only. In this case, if a DNA nanotube has length l, the end of the DNA nanotube is effectively diffusing around the surface of a sphere with radius l. The diffusion speed is characterized by a rotational diffusion constant, which we will call D_{rot}. If a nanotube has length l and the location of a nanotube's end point at time t is $u(t)$, then $\langle (u(t) - u(0))^2 \rangle = 4D_{rot}l^2t$ [12]. Here we consider tube lengths up to 25 μm, which is much less than the 80 μm persistence length of some DNA nanotubes, so we assume that the diffusing tube is rigid.

The rotational diffusion constant of an isolated polymer in a fluid appears to be dependent mostly on the polymer's size rather than its molecular type [50], and enough information about rotational diffusion constant exists to allow us to predict the changes that DNA nanotubes will find their destination in this regime.

IV. Path Finding Efficiency

A. Analysis of Path Finding Efficiency in an Idealized Geometry

We are interested in the question of whether a DNA nanotube finds its destination via diffusion as it grows. When a DNA nanotube length's l is less than the length to the edge of the destination r, the probability of the DNA nanotube's end hitting the destination is 0 (Figure 2). Similarly, when the DNA nanotube's end is longer than $r + d$, where d is the destination size, the probability of the DNA nanotube's end hitting the destination is also 0. We are therefore interested in the chance that the DNA nanotube finds the destination when $r \le l \le r + d$. The time during which the DNA nanotube's length is in this regime is $t = d/v$, where v is the growth speed of the DNA nanotube.

If d_{tube} is the DNA nanotube's cross-sectional diameter, T is absolute temperature, k_b is Boltzmann's constant and η_s the is viscosity of the solution the DNA nanotube is grown in, the rotational diffusion constant (in distance squared per time) is approximately [50]:

$$D_{rot} = \frac{3k_bT}{\pi\eta_s l}(\delta - \zeta) \tag{4}$$

$$\delta = \log\left(\frac{2l}{d_{tube}}\right) \tag{5}$$

$$\zeta = 1.45 - 7.5\left(\tfrac{1}{\delta} - 0.27\right)^2 \tag{6}$$

where δ and ζ are hydrodynamic correction factors. When $l < r$, the probability density that the DNA nanotube is oriented at angles (θ, ϕ) is uniform, i.e. $\rho_l(\theta, \phi) = \frac{1}{4\pi l^2}$. If $\theta < \theta_c = \arctan\frac{d}{r+d}$ when the DNA nanotube first grows to a length $l \ge r$, the DNA nanotube immediately attaches to the destination and has succeeded in building a path. The probability that the DNA nanotube hits immediately, P_0 is

$$P_0(r, \theta_c) = \int_0^{2\pi}\int_0^{\theta_c} r^2 \sin(\theta)\, \rho_r(\theta, \phi) \tag{7}$$

$$= \tfrac{1}{2}\left(1 - \cos(\theta_c)\right) \tag{8}$$

If upon reaching length r the DNA nanotube does not hit the destination immediately, the probability density of its orientation is uniform over all points $\theta > \theta_c$. We can use the diffusion equation over the probability distribution that a DNA nanotube is located at some angle (i.e. the Smoluchowski equation) to determine the fraction of DNA nanotubes that find the destination by diffusing into it, P_{diff}.

The diffusion equation in spherical coordinates is given by

$$\frac{\partial\rho}{\partial t} = \frac{D}{r^2}\left[\frac{\partial}{\partial r}\left(r^2\frac{\partial\rho}{\partial r}\right) + \frac{1}{\sin^2\theta}\frac{\partial^2\rho}{\partial\phi^2} + \frac{1}{\sin\theta}\frac{\partial}{\partial\theta}\left(\sin\theta\frac{\partial\rho}{\partial\theta}\right)\right] \tag{9}$$

Because we ignore the small fluctuations in DNA nanotube length, there is no diffusion in the $\hat{\mathbf{r}}$ direction, and because the destination is radially symmetric with respect to $\hat{\phi}$, ρ is independent of ϕ. The diffusion equation therefore simplifies to:

$$\frac{\partial\rho}{\partial t} = \frac{D_{rot}(r)}{r^2\sin(\theta)}\frac{\partial}{\partial\theta}\left(\sin(\theta)\frac{\partial\rho}{\partial\theta}\right) \tag{10}$$

with boundary conditions

$$\rho = 0 \text{ at } \theta = \theta_c \tag{11}$$

$$\frac{\partial\rho}{\partial\theta} = 0 \text{ at } \theta = \pi \tag{12}$$

The boundary conditions stipulate that if a DNA nanotube hits the destination it is absorbed, and if it hits the pole opposite the destination it continues to diffuse.

In the case where $r \gg d$, i.e. where the destination is small and far away, the diffusion constant does not change significantly over the time that the DNA nanotube could hit it. In this case, D_{rot} is approximately constant for $r \le l \le r + d$, and the time it takes before the DNA nanotube first diffuses into the destination is approximately exponentially distributed [44] with mean [22]:

$$\langle t_{hit}\rangle \approx 2\frac{r^2}{D_{rot}}\left[\frac{\log\frac{2}{1-\cos\theta_c}}{1+\cos\theta_c} - 1\right] \tag{13}$$

The length of time during which a DNA nanotube can hit the destination is $t_{max} = d/v$. The probability that the DNA nanotube diffuses into the destination, P_{diff}, is therefore

$$P_{diff} \approx 1 - \exp\left[\frac{t_{max}}{\langle t_{hit}\rangle}\right], \tag{14}$$

so the total probability that the DNA nanotube successfully finds the destination is

$$P_{hit} \approx P_0 + (1 - P_0)\, P_{diff}. \tag{15}$$

Combining Equations 14 and 13 gives

$$P_{diff} \approx 1 - \exp\left[\frac{-d/v}{2\frac{r^2}{\frac{3k_bT}{\pi\eta_s l}(\delta-\zeta)}\left[\frac{\log\frac{2}{1-\cos\theta_c}}{1+\cos\theta_c} - 1\right]}\right] \tag{16}$$

Some algebra simplifies this equation to

$$P_{diff} \approx 1 - \exp\left[\frac{-dC(\delta-\zeta)}{vr^2 l\left[\frac{\log\frac{2}{1-\cos\theta_c}}{1+\cos\theta_c} - 1\right]}\right] \tag{17}$$

where $C = \frac{3k_bT}{2\pi\eta_s}$, a constant that does change with either destination distance or size.

Figures 5 and 6 show P_{hit} for the rotational diffusion constant in Equation 4 and the DNA nanotube growth rate given in Section III. The viscosity of the solution was assumed to be the viscosity of water at 305K, 7.98×10^{-4} Pa·s. Probabilities were computed via numerical integration of Equation 10 and its approximate analytical solution given in Equation 17. The results of numerical integration and of the analytical approximation are similar and both methods suggest that for short distances (< 10 μm), virtually all DNA nanotubes should find a destination with a diameter of 500 nanometers. The probability of finding the destination drops sharply as both r increases and d decreases, however.

To quantitatively determine how the probability of finding the destination scales with increasing r and decreasing d, note that $\left(\frac{\log\frac{2}{1-\cos\theta_c}}{1+\cos\theta_c} - 1\right)$ increases as θ_c decreases. P_{diff} therefore decreases with decreasing θ_c. In contrast, the factor $\delta - \zeta$ increases with $\log r$.

Thus, as r increases for constant d, $1 - P_{diff}$ increases as $\exp\left[-1/\Omega(r^3)\right]$. That is, as the destination becomes further away, the probability of a DNA nanotube hitting the destination before growing too long to do so drops exponentially as the cube of the distance! As d decreases for a constant destination distance r, $1 - P_{diff}$ scales as approximately $\exp\left[-\Omega(d)\right]$. Since $P_0 \ll P_{diff}$ for all but the closest, largest destinations, $1 - P_{hit}$ scales approximately as $1 - P_{diff}$.

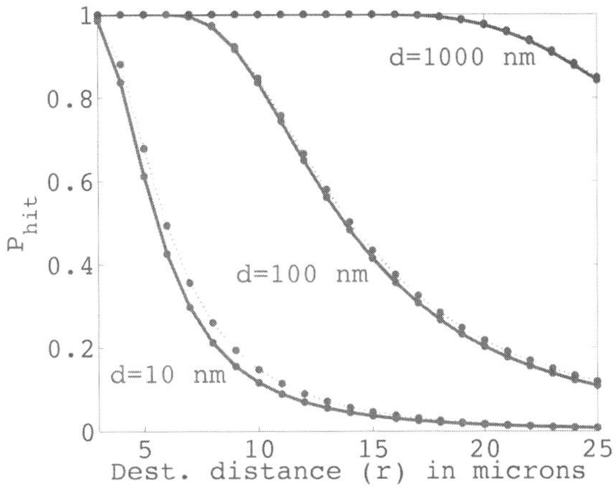

Fig. 5. **Predicted performance of the path finding algorithm as the destination distance r varies for destination sizes $d = 10$ nm, $d = 100$ nm and $d = 1000$ nm. Solid lines show the approximate analytical solution to Equation 17. Dashed lines are the numerical integration of Equation 10.**

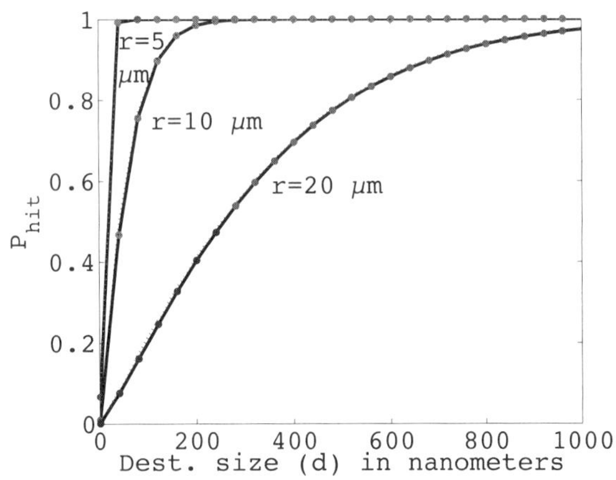

Fig. 6. **Predicted performance of the path finding algorithm as the destination size d varies, for destination distances $r = 5$ μm, $r = 10$ μm and $r = 20$ μm. Solid lines show the approximate analytical solution to Equation 17. Dashed lines are the numerical integration of Equation 10, but are too close to the analytical solution to be distinguished.**

While the chances of hitting the destination decrease exponentially with increasing destination distance or decreasing destination size, decreasing the speed of DNA nanotube growth can drastically improve the performance. $\log(1 - P_{diff})$ changes as $\Theta(1/v)$. That is, halving the speed of growth roughly squares the probability of missing the destination.

B. Simulation of Path Finding in an Implementation Geometry

In a path finding implementation, the start and destination markers are attached to solid objects that prevent a growing DNA nanotube from moving freely in every direction. Additionally, the geometry of the destination marker is unlikely to have a shape, as shown in Figure 2, where for each length l in the range $r < l < r + d$, the extent of the destination structure is described by the same θ_c. These features of

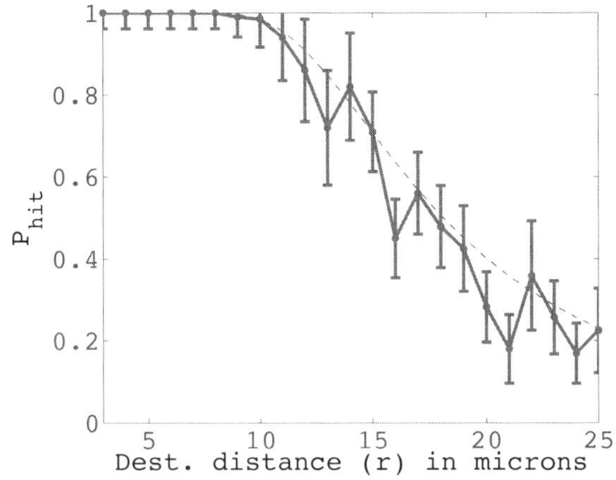

Fig. 7. **Simulated probability of hitting a target in the implementation geometry.** The thick line shows the simulated probability of hitting a destination, a half sphere of radius 100 nm located on the same flat surface as the start point. Error bars are two standard deviations. The dashed line is the approximate analytical values of P_{hit} for the idealized geometry with destination size 200 nm.

an implementation can change the rate and pattern of DNA nanotube end diffusion, and therefore path finding efficiency.

We developed a stochastic simulation of DNA nanotube diffusion to predict path finding efficiency in geometries containing irregular environment and destination marker geometries. We tested this simulation on a particularly simple *implementation geometry* where both the start and markers are attached to a solid surface and where the DNA nanotube is otherwise free to diffuse.

The destination in the implementation geometry is a half sphere with radius 100 nm centered on the surface. The seed size is 100 nm and DNA nanotubes were 12 tiles wide. DNA nanotubes can diffuse in any direction that did not collide with the surface to which the destination and surface markers were attached. The temperature, tile concentration, and sample viscosity in the simulation were the same as those used in Section IV-A.

The stochastic kinetic attachment and detachment of tiles governed DNA nanotube length as described in Section III-B. Diffusion was modelled via a standard algorithm for discretized Brownian dynamics on a sphere [12] where diffusive motion vectors were generated by moving the DNA nanotube a randomly angle on the plane tangent to the sphere and projecting the final point back onto the surface of the sphere. We used a time step of 10 μs, or an average DNA nanotube movement of 11-12 nm per time step for $l = 3$ microns. Approximately 100 DNA nanotubes were simulated per destination distance.

The results of the simulations are given in Figure 7. P_{hit} in the implementation geometry scales similarly to P_{hit} in the idealized geometry, but the absolute probability is slightly less in the simulated implementation geometry than in the idealized geometry. Both of these results are to be expected. Unless an obstacle prevents the DNA nanotube from reaching the destination (in which case the probability of forming a path is 0), the efficiency is related to the ratio of the volume that the DNA nanotube's end can diffuse through and the volume covered by the destination. This ratio scales at the same rate for both the idealized geometry and the implementation geometry. In the implementation geometry, the probability of hitting the destination, a half sphere, changes significantly for $r - d \leq l \leq r + d$, whereas in the idealized geometry, the probability of the tube hitting the target does not change over this length interval. Because the overall volume of the destination in the implementation geometry is smaller than the volume of the destination in the idealized the geometry, the probability of hitting is slightly less in the implementation geometry.

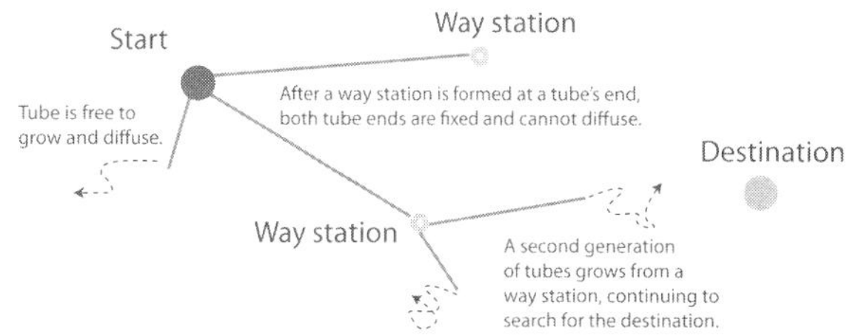

Fig. 8. **A divide and conquer approach to molecular path finding.** Instead of a single DNA nanotube growing and looking for the destination by diffusion, an initial set of DNA nanotubes could each stop growing and create fixed "way stations" from which a second set of DNA nanotubes could grow. The arrangement of way stations would be distributed evenly in space if each way station emitted a gradient preventing other way stations from forming nearby. This process could happen recursively, allowing DNA nanotubes to locate faraway destinations efficiently.

V. Conclusions and Future Directions

We've described an algorithm for assembling a molecular path between two points and used analysis and simulations to determine how the algorithm's probability of success scales with the size of the destination and its distance from the start point. The probability that the self-assembly process will complete when the necessary path is short (less than 10 microns) and the destination is large (more than 500 nanometers in radius) are high, meaning that the algorithm is practical for many situations.

However, we've also shown that the performance of the algorithm falls off superexponentially as the distance to the destination increases and at least exponentially as the destination size decreases. These performance drop-offs are not optimal: a linear search for non-growing tubes would complete successfully with probability 1 in a time that decreased quadratically as the destination distance increased and as the destination size decreased. It is therefore worth considering mechanisms that would improve the performance of the search process. It may be difficult to do so with the existing capacities of designed molecules, but an evaluation of the kinds of information processing that would provide a qualitative improvement in path finding performance could be used to guide the development of new molecular components.

The current algorithm's exponential performance decreases are caused by the random search for the destination and the limited time window during which a DNA nanotube can find and attach to a destination. Speeding up or changing the diffusion process would therefore simply change the exponent of performance falloff by a constant factor. But if a DNA nanotube could specifically grow *toward* the destination, performance might be improved. This would be possible if the destination could somehow communicate information to the growing DNA nanotube about its location in advance of contacting it. This type of communication occurs during growth of the *Xenopus laevis* mitotic spindle, where a gradient of small molecules forms around the destination for microtubule attachment [4]. One might be tempted to use such a gradient to increase the DNA nanotube's growth rate as it neared the destination, but such a mechanism might actually decrease the probability of finding the destination, because as the DNA nanotube approached the destination it would grow faster and faster, decreasing the time available for searching for the destination.

In *Xenopus laevis*, the gradient directs microtubule growth via a more subtle mechanism. Microtubules exhibit *dynamic instability*—i.e. they grow with a constant rate for a period of time, and then occasionally undergo *catastrophes*, periods of abrupt shrinking [27]. By allowing length decreases as well as increases, dynamic instability would improve the chances that a DNA nanotube would find the destination eventually, by giving the DNA nanotube more than one chance to explore the area that is the right distance from the start point. Dynamically unstable growth combined with a gradient of molecules emanating from the destination that decreased the frequency of catastrophe might create a qualitative improvement in the probability of finding the destination: a DNA nanotube would remain short, diffusing quickly, until it found the generally correct direction to hunt for the destination where the concentration of molecules that reduced the probability of catastrophes was highest [49]. So long as the DNA nanotube continued to diffuse in an area where the concentration of the catastrophe-reducing molecule was high, it would continue to grow. Further, if the DNA nanotube missed the destination, it would undergo catastrophes that would shorten the DNA nanotube until it became approximately a length that could reach the destination.

Another possibility would be to consider parallel searching. In one such approach, the DNA nanotubes could occasionally attach to a nearby surface and stop growing, creating a "way station" from which several new DNA nanotubes could begin growing. If way stations emitted gradient signals discouraging the creation of other way stations nearby, DNA nanotubes might distribute such stations evenly in space (Figure 8). A way station would have the effect of speeding up the search because DNA nanotubes growing from the way stations would be shorter than DNA nanotubes that had grown all the way from the start point and therefore would diffuse more quickly. However, this algorithm's application would be limited to environments where surfaces or other stationary structures were available for way stations to attach to. Implementing a clean-up process to remove way stations and DNA nanotubes that are not part of a successful path would also be necessary.

REFERENCES

[1] Abelson, H., Allen, D., Coore, D., Hanson, C., Homsy, G., Knight, T. F., Nagpal, R., Rauch, E., Sussman, G. J., Weiss, R. "Amorphous Computing," *Communiciations of the ACM*, vol. 43, iss. 5, pp 74–82, 2000.

[2] Barish, R., Schulman, R., Rothemund, P. W. K., Winfree E. "An Information-Bearing Seed for Nucleating Algorithmic Self-Assembly," *Proceedings of the National Academy of Sciences USA*, vol. 106, no. 15, pp 6054–6059, 2009.

[3] Barraquand, J., Langlois, B., Latombe, J. "Numerical Potential Field Techniques for Robot Path Planning," *IEEE Transactions on Systems, Man and Cybernetics*, vol. 22, no. 2, pp 224–241, 1992.

[4] Bastiaens, P., Caudron, M., Niethammer, P., Karsenti, E. "Gradients in the self-organization of the mitotic spindle," *Trends in Cell Biology*, vol. 16, no. 3, pp. 125–134, 2006.

[5] Bath, J., Green, S. J., Turberfield, A. J. "A free-running DNA motor powered by a nicking enzyme," *Angewandte Chemie International Edition*, vol. 44, pp. 4358-4361, 2005.

[6] Beal, J., Bachrach, J., "Infrastructure for engineered emergence on sensor/actuator networks," *IEEE Intelligent Systems*, vol 21, iss. 2, pp 10–19, Mar. 2006.

[7] Brun, Y. and Reichus, D. "Path finding in the tile assembly model," *Theoretical Computer Science*, vol. 410, pp 1461–1472, 2009.

[8] Carter, II, E. S., Tung, C.-S. "NAMOT2 – a redesigned nucleic acide modelling tool: construction of non-canonical DNA structures," *Bioinformatics*, vol. 12, no. 1, pp 25–30, 1996.

[9] Chen, H.-L., Schulman, R., Goel, A., Winfree, E. "Preventing Facet Nucleation During Algorithmic Self-Assembly," *Nano Letters*, vol. 7, no. 9, pp. 2912–2919, 2007.

[10] Chen, H.-L., Goel, A. "Error Free Self-Assembly using Error Prone Tiles," *DNA Computing 10*, Milan, It., pp 62–75, Jun. 2004.

[11] Cook, M., Soloveichik, D., Winfree, E., Bruck, J. "Programmability of Chemical Reaction Networks," *Algorithmic Bioprocesses*, Berlin Heidelberg: Springer, 2009, pp 543–584.

[12] Doi, M. and Edwards, S. F., *The Theory of Polymer Dynamics*. Oxford, England: Oxford University Press, 1986.

[13] Douglas, S. M., Dietz, H., Liedl, T., Högberg, B., Graf, F., Shih, W. M. "Self-assembly of DNA into nanoscale three-dimensional shapes," *Nature*, vol. 459, pp 414–418, 2009.

[14] Ekani-Nkodo, A., Kumar, A., Fygenson, D. K. "Joining and Scission in the Self-Assembly of Nanotubes from DNA Tiles," *Physical Review Letters*, vol. 93, pp 268301, 2004.

[15] Gillepsie, D. T., "Exact stochastic simulation of coupled chemical reactions," *Journal of Physical Chemistry*, vol. 81, no. 25, pp. 2340–2361, 1977.

[16] Gu, H., Chao, J., Xiao, S.-J., Seeman, N. C. "A proximity-based programmable DNA nanoscale assembly line," *Nature*, vol. 465, pp. 202-205, 2010.

[17] Hariadi, R., Yurke, B., Winfree E. Mauscript in preparation.

[18] Karsenti, E.,Vernos, I. "The Mitotic Spindle: A Self-Made Machine," *Science*, vol. 294, pp 543–547, 2001.

[19] Latombe, J. *Robot Motion Planning*, Dordrecht, Neth.: Kluwer Academic, 1990.

[20] LaValle, S. M. *Planning Algorithms*, Cambridge, UK: Cambridge University Press, 2006.

[21] Lehn, J.-M., "Supramolecular Chemistry," *Science*, vol. 260, iss. 5115, pp. 1762–1764, 1993.

[22] Linderman, J. J., Lauffenberger, D. A. "Analysis of Intracellular Receptor/Ligand Sorting: Calculation of Mean Surface and Bulk Diffusion Times within a Sphere," *Biophysical Journal*, vol. 50, pp 295–305, 1986.

[23] Lund, K., Manzo, A. J., Dabby, N., Michelotti, N., Johnson-Buck, A., Nangreave, J., Taylor, S., Pei, R., Stojanovic, M., Walter, N. G., Winfree, E., Yan, H. "Molecular robots guided by prescriptive landscapes," *Nature*, vol. 465, pp. 206–210, 2010.

[24] Magnasco, M. O. "Chemical Kinetics is Turing Universal," *Physical Review Letters*, vol. 78, no. 6, pp. 1190–1193, 1997.

[25] Mao. C., LaBean, T. H., Reif, J. H., Seeman, N. C. "Logical Computation using algorithmic self-assembly of DNA triple-crossover molecules," *Nature*, vol. 407, no. 6803, pp. 493–496, 2000.

[26] Markov, I. V. *Crystal Growth for Beginners*, Singapore: World Scientific, 2003.

[27] Mitchison, T., Kirschner, M. "Dynamic instabiity of microtubule growth," *Nature*, vol. 312, pp. 237–242, 1984.

[28] Nagpal, R. "Programmable Self-Assembly Using Biologically-Inspired Multiagent Control." *Conference on Autonomous Agents and Multiagent Systems (AAMAS)*, Bologna, It., July 2002.

[29] Nangreave, J., Yan, H., Liu, Y. "Studies of Thermal Stability of Multivalent DNA Hybridization ina Nanostructured System," *Biophysical Journal*, vol. 97, iss. 2, pp 563–571, 2009.

[30] Omabegho, T., Sha, R., Seeman, N. C. "A Bipedal DNA Brownian motor with coordinated legs," *Science* vol. 324, pp. 67-71, 2009.

[31] The PyMOL Molecular Graphics System, Version 1.3, Schrdinger, LLC.

[32] Rothemund, P. W. K. "A DNA and Restriction Enzyme Implementation of Turing Machines," *DNA Based Computers*, Princeton, NJ, pp. 75–119, Apr. 1995.

[33] Rothemund, P. W. K. "Theory and Experiments in Algorithmic Self-Assembly," PhD thesis. University of Southern California, Los Angeles, 2001.

[34] Rothemund, P. W. K. "Folding DNA to create nanoscale shapes and patterns," *Nature*, vol. 440, pp. 297–302, 2006.

[35] Rothemund, P. W. K., Ekani-Nkodo, A., Papadakis, N., Kumar, A., Fygenson, D. K., Winfree, E. "Design and Characterization of Programmable DNA Nanotubes," *Journal of the Americal Chemical Society*, vol. 126, no. 50, pp 16344–16353, 2004.

[36] Rothemund, P. W. K., Papadakis, N., Winfree, E. "Algorithmic Self-Assembly of DNA Sierpinski Triangles," *PLoS Biology*, vol. 2, iss. 12, pp 424–436, 2004.

[37] Rothemund, P. W. K., Winfree, E. "The Program-Size Complexity of Self-Assembled Squares," *Symposium on Theory of Computing (STOC)*, Portland, Or., pp 459–468, May 2000.

[38] Schulman, R., Winfree, E. "Synthesis of crystals with a programmable kinetic barrier to nucleation," *Proceedings of the National Academy of Sciences USA*, vol. 104, no. 39, pp 15236–15241, 2007.

[39] Schulman, R., Winfree, E. "Programmable Control of Nucleation for Algorithmic Self-Assembly," *SIAM Journal on Computing*, vol. 39, iss. 4, pp 1581–1616, 2009.

[40] Seelig, G., Soloveichik, D., Zhang, D. Y., Winfree, E. "Enzyme-Free Nucleic Acid Logic Circuits," *Science*, vol. 314, pp. 1585–1588, 2006.

[41] Seeman, N. C. "An Overview of Structural DNA Nanotechnology," *Molecular Biotechnology*, vol. 37, no. 3, pp 246–257, 2007.

[42] Smith, S. B., Finzi, L., Bustamante, C. "Direct mechanical measurements of the elasticity of single DNA molecules by using magnetic beads," *Science*, vol. 258, pp. 11221126, 1992.

[43] Soloveichik, D., Winfree, E., "Complexity of self-assembled shapes," *SIAM Journal on Computing*, vol. 36, no. 6, pp 1544–1569, 2007.

[44] Szabo A., Schulten, K., Schulten, Z. "First passage time approach to diffusion-controlled reactions," *Journal of Chemical Physics*, vol. 72, pp. 4350–4357, 1980.

[45] Wetmur, J. G., Fresco, J. "DNA Probes: Applications of the Principles of Nucleic Acid Hybridization," *Critical Reviews in Biochemistry and Molecular Biology*, vol. 26, no. 3–4, pp 227–259, 1991.

[46] Whitesides, G. M., Mathias, J. P., Seto, C. "Molecular self-assembly and nanochemistry: a chemical strategy for the synthesis of nanostructures," *Science*, vol. 254, iss. 5036, pp 1312–1319, 1991.

[47] Winfree, E. "On the Computational Power of DNA Annealing and Ligation," *DNA Based Computers*, Princeton, NJ, pp. 187–198, Apr. 1995.

[48] Winfree, E. "Simulations of Computing by Self-Assembly," California Institute of Technology Technical Report CS-TR:1998.22, 1998.

[49] Wollman, R., Cytrynbaum, E. N. Jones, J. T., Meyer, T. Scholey, J. M., Mogilner, A., "Efficient chromosome capture requires a bias in the "Search-and-Capture" process during mitotic spindle assembly", *Current Biology*, vol. 15, pp. 828–832, 2005.

[50] Zero, K. M., Pecora, R. "Rotational and Translational Diffusion in Semidilute Solutions of Rigid-Rod Macromolecules," *Macromolecules*, vol. 15, pp 87–93, 1982.